Bionanotechnology

Global Prospects

II

Bionanotechnology II

Global Prospects

II

Edited by
David E. Reisner

CRC Press
Taylor & Francis Group
Boca Raton London New York

CRC Press is an imprint of the
Taylor & Francis Group, an **informa** business

CRC Press
Taylor & Francis Group
6000 Broken Sound Parkway NW, Suite 300
Boca Raton, FL 33487-2742

First issued in paperback 2017

© 2012 by Taylor & Francis Group, LLC
CRC Press is an imprint of Taylor & Francis Group, an Informa business

No claim to original U.S. Government works

ISBN-13: 978-1-4398-0463-6 (hbk)
ISBN-13: 978-1-138-07678-5 (pbk)

Library of Congress Cataloging-in-Publication Data

Bionanotechnology II : global prospects / editor, David E. Reisner.
    p. ; cm.
  Bionanotechnology 2
  Bionanotechnology two
  Includes bibliographical references and index.
  ISBN 978-1-4398-0463-6 (hardcover : alk. paper)
  I. Reisner, David Evans. II. Title: Bionanotechnology 2. III. Title:
Bionanotechnology two.
  [DNLM: 1. Nanotechnology. 2. Biotechnology. QT 36.5]

  LC classification not assigned
  660.6--dc23   2011030273

Visit the Taylor & Francis Web site at
http://www.taylorandfrancis.com

and the CRC Press Web site at
http://www.crcpress.com

Contents

Preface

Much has happened in the world in the past 36 months since the publication of my previous work, *Bionanotechnology: Global Prospects*. Let's take a look. Within a few weeks of publication, the 2008 global financial crisis bottomed out with the collapse of three of the biggest financial institutions over the so-called Lehman-AIG-Merrill weekend, a bitter testimonial to technology run amok in the form of "financial engineering" allowed to flounder in an era of deregulation. Complex financial derivative instruments such as the credit default obligation (CDO) and the credit default swap (CDS) are all captured in the 2011 Academy Award-winning documentary *Inside Job*. It is ironic and relevant to this book that the financial services sector in the United States continues to drain the best and brightest brains from Ivy League schools, gifted thinkers that otherwise could populate science and engineering jobs in the United States.

More benign (usually) social networking technologies now pervade our lives in ways unimaginable in 2008, even when Tim Berners-Lee declares in *Wired* magazine last September that "the Web is dead," referring to fragmentation caused by the exponential rise in "apps." Although Twitter reached a tipping point at the South by Southwest Festival (SXSW) in 2007, it has now reached upwards of 100 million tweets per day, dwarfing its mainstream impact of just a few years ago. Are we too connected? Maybe, but immediate Twitter pictures from the 2010 Haitian earthquake alerted the world to the extent of the tragedy, as did tweets from protesters in the Middle East that went viral across the region. A debate rages between the Malcolm Gladwell/Evgeny Morozov and Clay Shirky/Jared Cohen camps as to the true impact of Twitter, but the subsequent overthrow of authoritarian leaders in Tunisia and Egypt leaves little doubt of the power of social media.

For its part, WikiLeaks has grown to become a de facto purveyor of its own rogue Freedom of Information Act (FOIA) rules for whomever, empowering every human being with an Internet connection. Indeed, a Tunisian version (Tunileaks), played a crucial role in the Arab Spring. Futhermore, a cottage industry is blooming in the area of censorship circumvention software/hardware in response to Government shutdown of the Internet. Mobile Ad-hoc Networks or Wireless Mesh Networks (WMNs) can function phone-to-phone without cell tower infrastructures and are enabling "Internet-in-a suitcase" dreams to become reality.

And empowerment is a key to understanding the nature of technology development and deployment in the world today. The Internet is the great equalizer, enabling all countries to produce world-class intellectual property (IP). This is particularly evident in the oil-rich countries, where crystal balls are revealing peak oil milestones as soon to arrive or already passed. These countries understand that in the future, they will need to build economies on IP, not oil. In the preface to my first edition, I emphasized the "flat world" of Tom Friedman. Amy Chua's *World on Fire* challenged that notion, examining the disproportionate economic and political influence of "market-dominant minorities" and how less affluent majorities harbor resentment against them, a problem exacerbated by flat-world dynamics. As testimonial, we see such seemingly benign social networks as Facebook capable of fomenting popular revolts.

The hottest venture capital investment opportunities are now energy related and cleantech (nanotechnology thinly disguised). Bionanotechnology is also positioning itself as a front-runner. Its implications in more efficient health care delivery are compelling. Bionanotech-based products are already in the market place, including drug-releasing nanocoated vascular stents and numerous FDA-approved drug delivery nanotherapeutics. Various other bionanotech-based products are undergoing clinical trials. Examples range from perfluorocarbon nanoparticles (for angiogenesis in atherosclerosis diagnosis) to dendrimer-based topical gel (for HIV and HPV) to nanoliposomes (for various cancers).

This follow-up volume to *Bionanotechnology: Global Prospects* echoes the format of my previous effort. It is in no way meant to be comprehensive; in fact, it is a random walk of sorts, a result of having engaged authors in my professional travels. As in the first volume, I make no claims as to editing of any technical contents—the field is just too expansive and interdisciplinary. I am relying on reputation and credibility of authorship. I have sought out chapters from the traditional geographic "hotspots" of nanotechnology, though this distinction is somewhat fading. However, I have also actively pursued authors in parts of the world that seemed to be under the radar, for example, Thailand, Turkey, and Jordan. In fact, nearly a dozen countries are represented and include cities as far ranging as Ankara, Tehran, Randburg, Cairo, Singapore, Irbid, Chiang Mai, Bangkok, Exeter, and Buenos Aires.

Chapter title pages contain only the principal affiliation, with a few exceptions, in the event the work was evenly distributed across separate institutions. Author's given names are spelled out for the book format. This is purely a formatting consideration. I hope I have not offended any of my authors. Detailed affiliations are to be found in the list of contributors. By the same token, the publisher has kindly provided a color insert section. Note the color images also appear as grayscale images in the text. I made efforts to pick those color images that appeared to benefit most from a color display.

This collection includes a strong showing from the Middle East. I believe the region is witnessing the onset of a paradigm shift, in which IP starts to take hold in lands that are now known primarily for oil. Bahrain and Dubai are heavily focused on becoming world-class financial service centers. All the oil-rich countries realize that they need to search for the next big thing and create the infrastructure to do so. A good example is the new international university KAUST in the Kingdom of Saudi Arabia, focusing heavily on research. I am especially pleased to highlight in the opening chapter and on the book cover a collaboration between the Hebrew University in Jerusalem and Al Quds University in East Jerusalem, to form one-dimensional nanostructures. This work has been funded by the Israeli-Palestinian Science Organization (IPSO) and the Deutsche Forschungsgemeinschaft.

Topic areas span a wide range of subject areas that fall under the bionanotechnology banner, either squarely or tangentially. In the final accounting, this volume has a strong emphasis on nanomedicine, both drug delivery and tissue engineering. Other areas include water treatment, energy conversion (solar), nanoelectronics, photonics, and agriculture. It is my sincere hope that this book inspires readers to get involved in the burgeoning area of bionanotechnology, as much good innovation occurs at the crossroads of disparate disciplines. This book can only be described as a labor of love, what Shirky would categorize as a classic case of "intrinsic motivation."

David E. Reisner
The Nano Group, Inc.
Manchester, Connecticut

All the, small things
True care, truth brings
I'll take, one lift
Your ride, best trip

—"All the Small Things," lyrics by Blink-182

The Editor

David E. Reisner, Ph.D., is a well known early pioneer and entrepreneur in the burgeoning field of nanotechnology, who, in 1996, cofounded two nanotech companies in Connecticut, Inframat® and US Nanocorp®. Since founding, for nearly 15 years, he was CEO of both companies, which were recognized in Y2002 - Y2005 for their fast revenue growth as Deloitte & Touche *Connecticut Technology Fast50 Award* recipients. In 2004, The Nano Group, Inc. was formed as a parent holding company for investment. Reisner and cofounders were featured in *Forbes* magazine in 2004.

Reisner has over 175 publications and is an inventor on 10 issued patents. He is the editor for the Bionanotechnology section of the third edition of *The Biomedical Engineering Handbook* and is editor of the first edition of *Bionanotechnology: Global Prospects*. He has written articles on the business of nanotechnology in *Nanotechnology Law & Business* as well as in the Chinese publication *Science & Culture Review*.

Reisner served a three-year term as a Technology Pioneer for the World Economic Forum and was a panelist at the 2004 Annual Meeting in Davos. He is on the board of the Connecticut Venture Group and was chairman of the board of the Connecticut Technology Council from 2005-2009. Reisner was a NASA *NanoTech Briefs* Nano50 awardee in 2006. For his efforts in the field of medical implantable devices, Reisner won the 1st annual BEACON award for Medical Technology in 2004. He is a member of the Connecticut Academy of Science and Engineering.

Reisner is a 1978 University Honors graduate from Wesleyan University and received his Ph.D. at MIT in 1983 in the field of chemical physics. An avid hiker, he summitted Kilimanjaro in 1973. Reisner was recognized for his historic preservation efforts in 1994 when he received the Volunteer Recognition Award from the Connecticut Historical Commission and the Connecticut Trust for Historic Preservation. He is known nationally for his expertise in vintage Corvette restoration and documentation. Reisner did volunteer work in Haiti, soon after the 2010 earthquake, in Jérémie.

Contributors

Nasser Aghdami
Department of Regenerative Medicine
Royan Institute for Stem Cell Biology and
    Technology, ACECR
Tehran, Iran

Hassan M. E. Azzazy
Department of Chemistry
Yousef Jameel Science & Technology
    Research Center
The American University In Cairo
New Cairo, Egypt

Hojae Bae
Harvard-MIT Division of Health Sciences
    and Technology
Department of Medicine, Brigham and
    Women's Hospital
Harvard Medical School
Cambridge, Massachusetts, USA

Bernardo Barbiellini
Department of Physics
Northeastern University
Boston, Massachusetts, USA

Randall W. Barton
NanoViricides, Inc.
West Haven, Connecticut, USA

Karen Bellman
Department of Mechanical and Aerospace
    Engineering
The Ohio State University
Columbus, Ohio, USA

Arda Buyuksungur
BIOMATEN and Department of
    Biotechnology
Middle East Technical University
Ankara, Turkey

Shelton D. Caruthers
Center for Translational Research in
    Advanced Imaging and Nanomedicine
Washington University School of
    Medicine
St. Louis, Missouri, USA

Aaron Chen
Department of Chemical Engineering and
    Material Science
University of California, Irvine
Irvine, California, USA

Fanqing Frank Chen
Life Sciences Division
Lawrence Berkeley National Laboratory
Berkeley, California, USA

Heather C. Chiamori
Berkeley Sensor and Actuator Center
    (BSAC)
Department of Mechanical Engineering
University of California, Berkeley
Berkeley, California, USA

Michael Chin
Department of Earth & Environmental
    Engineering
Langmuir Center for Colloids and
    Interfaces
Columbia University
New York, New York, USA

Andrew Y. Choo
Inovio Pharmaceuticals
Blue Bell, Pennsylvania, USA

Birsen Demirbag
BIOMATEN and Department of
    Biotechnology
Middle East Technical University
Ankara, Turkey

Utkan Demirci
Bio-Acoustic-MEMS in Medicine (BAMM)
    Laboratory
Center for Biomedical Engineering
Department of Medicine
Brigham and Women's Hospital
Harvard Medical School
Boston, Massachusetts, USA
Harvard-MIT Health Sciences and
    Technology
Cambridge, Massachusetts, USA

Or Dgany
The Robert H. Smith Institute of Plant
    Sciences and Genetics in Agriculture
Faculty of Agricultural, Food and
    Environmental Quality Sciences
The Hebrew University of Jerusalem
Rehovot, Israel

Anil R. Diwan
NanoViricides, Inc.
West Haven, Connecticut, USA

Gozde Eke
BIOMATEN and Department of Micro
    and Nanotechnology
Middle East Technical University
Ankara, Turkey

Khaled N. Elshuraydeh
Higher Council for Science and
    Technology
Amman, Jordan

Tugba Endoğan
Department of Polymer Science and
    Technology
Middle East Technical University
Ankara, Turkey

Si-Shen Feng
Division of Bioengineering
National University of Singapore
Singapore, Singapore

Barry S. Flinn
Institute for Advanced Learning and
    Research
Danville, Virginia, USA

Mariekie Gericke
Biotechnology Division
Mintek
Randburg, South Africa

Nesrin Hasirci
Department of Chemistry
BIOMATEN and Middle East Technical
    University
Ankara, Turkey

Vasif Hasirci
BIOMATEN and Department of Biological
    Sciences
Middle East Technical University
Ankara, Turkey

Mona Hassuneh
University of Jordan
Amman, Jordan

Richard Helferich
MetaMateria Technologies LLC
Columbus, Ohio, USA

Arnon Heyman
The Robert H. Smith Institute of Plant
    Sciences and Genetics in Agriculture
Faculty of Agricultural, Food and
    Environmental Quality Sciences
The Hebrew University of Jerusalem
Rehovot, Israel

Linh Hoang
University of California, Santa Cruz
Santa Cruz, California, USA

Alexander Ip
Bio-Acoustic-MEMS in Medicine (BAMM)
    Laboratory
Center for Biomedical Engineering
Department of Medicine
Brigham and Women's Hospital
Harvard Medical School
Boston, Massachusetts, USA

Yun-Ho Jang
Harvard-MIT Division of Health Sciences
   and Technology
Department of Medicine, Brigham and
   Women's Hospital
Harvard Medical School
Cambridge, Massachusetts, USA

Hirokazu Kaji
Harvard-MIT Division of Health Sciences
   and Technology
Department of Medicine, Brigham and
   Women's Hospital
Harvard Medical School
Cambridge, Massachusetts,
   USA
Department of Bioengineering and
   Robotics, Graduate School of
   Engineering
Tohoku University
Sendai, Japan

Sinem Kardesler
BIOMATEN and Department of
   Biotechnology
Middle East Technical University
Ankara, Turkey

Hasan Onur Keles
Bio-Acoustic-MEMS in Medicine (BAMM)
   Laboratory
Center for Biomedical
   Engineering
Department of Medicine
Brigham and Women's
   Hospital
Harvard Medical School
Boston, Massachusetts, USA

Halime Kenar
Department of Biotechnology
Middle East Technical
   University
Ankara, Turkey

Ali Khademhosseini
Harvard-MIT Division of Health Sciences
   and Technology
Wyss Institute for Biologically Inspired
   Engineering
Department of Medicine, Brigham and
   Women's Hospital
Harvard Medical School
Cambridge, Massachusetts, USA
WPI-Advanced Institute for Materials
   Research
Tohoku University
Sendai, Japan

Amir S. Khan
Inovio Pharmaceuticals
Blue Bell, Pennsylvania, USA

Michelle Khine
Department of Biomedical Engineering
Department of Chemical Engineering and
   Materials Science
University of California, Irvine
Irvine, California, USA

J. Joseph Kim
Inovio Pharmaceuticals
Blue Bell, Pennsylvania, USA

Aysu Kucukturhan
BIOMATEN and Department of
   Biomedical Engineering
Middle East Technical University
Ankara, Turkey

Gregory M. Lanza
Center for Translational Research in
   Advanced Imaging and Nanomedicine
Washington University School of Medicine
St. Louis, Missouri, USA

Urszula Tylus Latosiewicz
Department of Chemistry and Chemical
   Biology
Northeastern University
Boston, Massachusetts, USA

Ang Li
Singapore-MIT Alliance for Science &
    Technology
Singapore, Singapore

Chwee Teck Lim
Mechanobiology Institute
Division of Bioengineering & Department
    of Mechanical Engineering
National University of Singapore
Singapore, Singapore

Liwei Lin
Berkeley Sensor and Actuator Center (BSAC)
Department of Mechanical Engineering
University of California, Berkeley
Berkeley, California, USA

Wentai Liu
University of California, Santa Cruz
Santa Cruz, California, USA

Hanan I. Malkawi
Yarmouk University
Irbid, Jordan

Izhar Medalsy
Institute of Chemistry and Center for
    Nanoscience and Nanotechnology
The Hebrew University of Jerusalem
Jerusalem, Israel

Maria Jose Morilla
Programa de Nanomedicinas
Universidad Nacional de Quilmes
Buenos Aires, Argentina

Albana Ndreu
Department of Biotechnology
Middle East Technical University
Ankara, Turkey

Artphop Neamnark
The Petroleum and Petrochemical College
Chulalongkorn University
Bangkok, Thailand

Dina Nemr
Department of Chemistry
    The American University in Cairo
New Cairo, Egypt

Diep Nguyen
Department of Biomedical Engineering
University of California, Irvine
Irvine, California, USA

Joseph Noyes
School of Physics
University of Exeter
Exeter, United Kingdom

Suriya Ounnunkad
Department of Chemistry
Materials Science Research Center
Chiang Mai University
Chiang Mai, Thailand

Sukon Phanichphant
Department of Chemistry
Materials Science Research Center
Chiang Mai University
Chiang Mai, Thailand

Danny Porath
Institute of Chemistry and Center for
    Nanoscience and Nanotechnology
The Hebrew University of Jerusalem
Jerusalem, Israel

Shaurya Prakash
Department of Mechanical and Aerospace
    Engineering
The Ohio State University
Columbus, Ohio, USA

Heni Rachmawati
Bandung Institute of Technology
Bandung, Indonesia

Venkatesan Renugopalakrishnan
 Children's Hospital
 Harvard Medical School
 Boston, Massachusetts, USA
 Department of Chemistry and Chemical
    Biology
 Northeastern University
 Boston, Massachusetts, USA

Eder Lilia Romero
Programa de Nanomedicinas
Universidad Nacional de Quilmes
Buenos Aires, Argentina

Bruce Russell
Singapore Immunology Network, Biopolis
Agency for Science Technology and
  Research
Singapore, Singapore

Pakakrong Sansanoh
The Petroleum and Petrochemical
  College (PPC)
Center for Petroleum, Petrochemical, and
  Advanced Materials (CPPAM)
Chulalongkorn University, Pathumwan
Bangkok, Thailand

Niranjan Y. Sardesai
Inovio Pharmaceuticals
Blue Bell, Pennsylvania, USA

J. Richard Schorr
MetaMateria Technologies LLC
Columbus, Ohio, USA

Suvankar Sengupta
MetaMateria Technologies LLC
Columbus, Ohio, USA

Mark A. Shannon
Department of Mechanical Science and
  Engineering
University of Illinois, Urbana-Champaign
Urbana, Illinois, USA

Himanshu Sharma
Department of Chemical Engineering and
  Materials Science
University of California, Irvine
Irvine, California, USA

Oded Shoseyov
The Robert H. Smith Institute of Plant
  Sciences and Genetics in Agriculture
Faculty of Agricultural, Food and
  Environmental Quality Sciences
The Hebrew University of Jerusalem
Rehovot, Israel

Ponisseril Somasundaran
Department of Earth & Environmental
  Engineering
Langmuir Center for Colloids and Interfaces
Columbia University
New York, New York, USA

Ndabenhle M. Sosibo
Nanotechnology Innovation Centre
Advanced Materials Division, Mintek
200 Malibongwe Drive
Randburg, South Africa

Mukhles Sowwan
Nanotechnology Research Laboratory
Al Quds University
East Jerusalem, Israel

Bingfeng Sun
Division of Bioengineering
National University of Singapore
Singapore, Singapore

Pitt Supaphol
The Petroleum and Petrochemical
  College (PPC)
Center for Petroleum, Petrochemical, and
  Advanced Materials (CPPAM)
Chulalongkorn University, Pathumwan
Bangkok, Thailand

Orawan Suwantang
School of Science
Mae Fah Luang University
Tasod, Muang
Chiang Rai, Thailand

Jayant G. Tatake
NanoViricides, Inc.
West Haven, Connecticut, USA

Robert T. Tshikhudo
Nanotechnology Innovation Centre
Advanced Materials Division, Mintek
Randburg, South Africa

Harry L. Tuller
Department of Materials Science and
  Engineering
Massachusetts Institute of Technology
Cambridge, Massachusetts, USA

Sarah Ranjbar Vaziri
Department of Stem Cells and
    Developmental Biology
Royan Institute for Stem Cell Biology and
    Technology, ACECR
Tehran, Iran

Peter Vukusic
School of Physics
University of Exeter
Exeter, United Kingdom

ShuQi Wang
Bio-Acoustic-MEMS in Medicine (BAMM)
    Laboratory
Center for Biomedical Engineering
Department of Medicine
Brigham and Women's Hospital
Harvard Medical School
Boston, Massachusetts, USA

Samuel A. Wickline
Center for Translational Research in
    Advanced Imaging and Nanomedicine
Washington University School of
    Medicine
St. Louis, Missouri, USA

Stephen R. Wilson
Transformation Nanotechnologies
Danville, Virginia, USA

Feng Xu
Bio-Acoustic-MEMS in Medicine (BAMM)
    Laboratory
Center for Biomedical Engineering
Department of Medicine
Brigham and Women's Hospital
Harvard Medical School
Boston, Massachusetts, USA

Zhi Yang
University of California, Santa Cruz
Santa Cruz, California, USA

Deniz Yucel
Department of Biotechnology
Middle East Technical University
Ankara, Turkey

Soroush M. Mirzaei Zarandi
Department of Biomedical Engineering
University of California, Irvine
Irvine, California, USA

Wenwei Zheng
Department of Nutritional Science and
    Toxicology
University of California, Berkeley
Berkeley, California, USA
Life Sciences Division
Lawrence Berkeley National Laboratory
Berkeley, California, USA

1

SP1 Protein-Gold Nanoparticle Hybrids as Building Blocks for Nanofabrication of One-Dimensional Systems

Izhar Medalsy,[1] **Or Dgany,**[2] **Arnon Heyman,**[2] **Oded Shoseyov,**[2] **Mukhles Sowwan**[3]**, and Danny Porath**[1]

[1]*Institute of Chemistry and Center for Nanoscience and Nanotechnology, The Hebrew University of Jerusalem, Jerusalem, Israel*

[2]*The Robert H. Smith Institute of Plant Sciences and Genetics in Agriculture, Faculty of Agriculture Food, and Environment, The Hebrew University of Jerusalem, Rehovot, Israel*

[3]*Nanotechnology Research Laboratory, Al Quds University, East Jerusalem, Israel*

CONTENTS

1.1 Introduction

One of the main challenges within the realm of nanoscience and nanotechnology is to fabricate wires where the internal structure along the wires can be well controlled and manipulated.[1–5] These structures hold the potential of miniaturizing current devices by embedding the computational unit within the wire.[6] Examples of other one-dimensional (1D) structures are DNA molecules and 1D crystals like nanotubes made of carbon or silicon.[7–9] These structures have several drawbacks; DNA was proven to conduct only over short segments,[10,11] while the 1D crystals usually have homogeneous structure and therefore lack recognition and structural manipulation capabilities.[12] Self-assembly of wires made of suitable building blocks that enable structuring and well-defined formation can be an attractive alternative route to overcome these limitations. The building blocks can be used to connect short DNA segments or form by themselves conductive wires. Candidates to serve as building blocks must demonstrate durability and a well-controlled structure.

Proteins can serve as useful building blocks for nanoelectronic implementations, among other reasons because of the ability to modify their structure using genetic engineering.[13] Such manipulations enable optimization of the desired building block properties, such as self-assembly and selective attachment of various nanoparticles.[14,15] One of the drawbacks

1

of proteins, however, is their sensitivity to different environmental conditions: temperature, pH, proteases, etc. A protein that is both durable and structurally diverse is of great interest. SP1 was reported to form highly ordered 2D arrays, selectively attach to different surfaces, and serve as a logic and memory unit based on its capability to attach a silicon nanoparticle (NP) to its pore.[13,16,17] Here we show the use of the SP1 proteins as building blocks for future nanowires, based on their extreme stability and on their ability to connect gold nanoparticle (GNP) to their inner pores, followed by forming junctions between DNA strands or by self-assembly of nanotubes.[18]

1.2 The SP1-GNP Hybrid

SP1 is a circular homo-oligomeric dodecameric protein, isolated from aspen (*Populus tremula*) plants.[19,20] One of SP1's great advantages is its extreme stability. SP1 is a boiling stable protein with a melting temperature of 109°C that can withstand a broad range of extreme conditions, such as a wide pH range, protease, detergents, chaotropic agents, and organic solvents. Its crystal structure reveals a ring, 11 nm in diameter, with an inner pore of 2–3 nm and width of 4–5 nm.[21] The N terminil region of the SP1 dodecamer faces the inner pore of the ring and is tolerant to amino acid deletions and additions.[22] Adding six-histidine amino acid groups to each one of the 12 SP1 subunits' N termini led to the formation of a mutant capable of attaching a Ni-NTA-covered gold nanoparticle (GNP) to its inner pore. The interaction of Ni-NTA and histidines is extremely strong, about 10^{-7} to 10^{-13} M. This mutant, called 6His-SP1, shows the same high durability and boiling resistance as the nonmutated SP1.[18] Commercially available 1.8 nm GNPs covered by Ni-NTA ligands (Nanoprobes, Yaphank NY USA) were attached to the N termini of the 6His-SP1. Incubation of this variant with Ni-NTA GNPs in the presence of 3 M GuHCl (presumably enabling exposure of the histidine tags by loosening the protein structure[16]) targeted the GNPs to the center of the dodecamer rings (Figure 1.1). The diameter of the Ni-NTA GNP (1.8 nm) fits well into the diameter of the dodecamer inner pore (~3 nm).

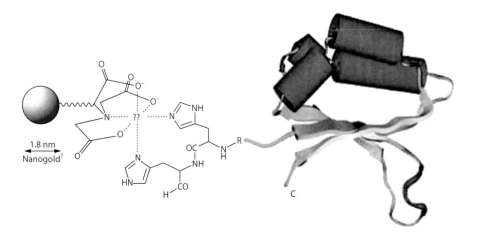

FIGURE 1.1
6His-SP1-GNP hybrids. The 6His amino acid addition to the end of each monomer N termini has high affinity to the Ni-NTA groups covering the 1.8 GNP. Addition of GuHCl causes the exposure of the N termini to the Ni^{2+} groups, enabling the formation of the hybrid.

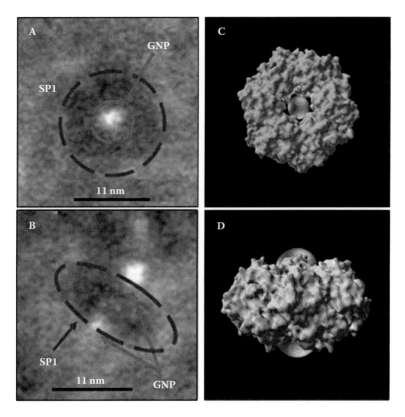

FIGURE 1.2
6His-SP1-GNP hybrids. (A,B) HAADF-STEM (vanadium staining) images of the 6His-SP1-GNP hybrid in top and side views, respectively. (C,D) Computer simulations of the 6His-SP1-GNP hybrid top and side views, respectively. (From Medalsy et al. 2008.)

The formation of SP1-GNP hybrids was validated using electron microscopy and atomic force microscopy (AFM). High-angle annular dark-field scanning transmission electron microscopy (HAADF-STEM) revealed the existence of the GNP in the SP1 inner pore. Figure 1.2A shows the GNP (bright white) centered within the darker round protein ring. The side view (Figure 1.2B) shows GNPs attached to both faces of the dodecameric ring. This hybrid formation is consistent with the crystal structure, as there are six N termini on each face of the 6His-SP1, pointing in opposite directions. Figure 1.2C and D shows illustrations of 6His-SP1 attached to two GNPs on both sides in top and side views, respectively.

Electrostatic force microscopy (EFM) is capable of distincting between a polarizable object, for instance, a GNP, and an isolating object like the SP1 protein. By scanning over the investigated object in EFM mode, the biased tip induces an electric field component normal to the surface. The charge carriers in a polarizable object are polarized according to the field direction, and therefore, for both positive and negative tip voltages, there will be an electrostatic attraction between the tip and the object. An insulating object will not respond to the electric field, and thus will not produce any electrostatic signal.[23,24]

EFM measurements performed on 6His-SP1 and on 6His-SP1-GNP hybrids (Figure 1.3) revealed a clear difference between the two species. The EFM scan was repeated for −5 V (Figure 1.3B), 0 V (Figure 1.3C), and +5 V (Figure 1.3D). The phase shift reaction above

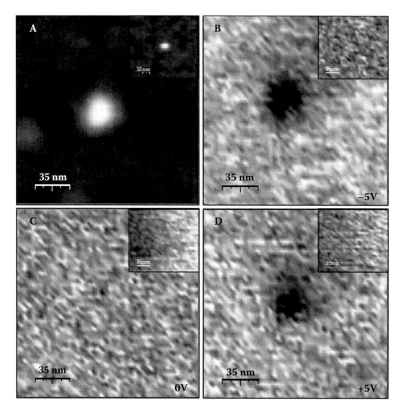

FIGURE 1.3
(A) Topographic AFM image of a single 6His-SP1-GNP hybrid, and 6His-SP1 with no GNP attached to it (inset). (B–D) EFM measurements of a 6His-SP1-GNP hybrid: the phase shift images were measured 50 nm above the set point height at negative (–5 V) (B), zero (C), and positive (+5 V) (D) tip bias voltages. The insets show similar measurements done on 6His-SP1 with no GNP attached to it. No electrostatic interaction signal is observed. (From Medalsy et al. 2008.)

the 6His-SP1-GNP hybrid location, for both positive and negative bias voltages, is a clear evidence for the existence of a polarizable object, metallic for this signal strength, within the 6His-SP1 protein. The lack of electrostatic signal from the 6His-SP1 (insets in Figure 1.3B–D) ensures that the origin of the signal is the GNP bound to the 6His-SP1 and not the protein itself. The combined transmission electron microscopy (TEM) and EFM information confirms the attachment of the GNP to the 6His-SP1 protein and the formation of the hybrid structure.

1.3 SP1-GNP-DNA Tri-Block Conjugates

One-dimensional wires made of SP1-GNP hybrids anchored at various positions along a dsDNA were formed using a combination of thiolated ssDNA that partially overlaps with other thiolated ssDNA and 6His-SP1-GNP hybrids. In these wires the thiol groups are used as anchoring groups to the GNPs in the hybrids. The partial overlap between the

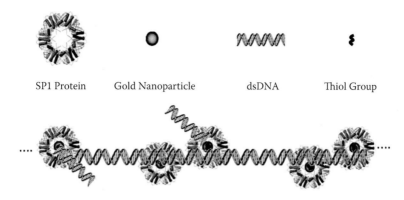

FIGURE 1.4
Schematic of a 1D nanostructure of a 6His-SP1-GNP-DNA tri-block conjugate. The basic components are the 6His-SP1 protein, GNPs, dsDNA molecules, and thiol groups placed at different places on the DNA structure. Through thiol gold connection the system can create complex structures.

complementary ssDNA determines the position of the thiol groups along the dsDNA template. These positions can be varied and controlled by the specific choice of the sequence length and overlap.

The ssDNA sequences used to form the dsDNA wires in this experiment were 80-mers long. The various sequences had an overlap with partially complementary sequences with thiols at the 5′ ends.[25] The sequences were designed to form long DNA templates with thiol groups repeated periodically every 22, 54, and 80 bp and their multiplications (Figure 1.4).

The dsDNA templates, at equimolar concentration (100 nM) of sense and partially complementary antisense oligomers, were mixed and annealed. The mixture was heated to 95°C for 4 minutes, and then incubated at 70°C for 10 minutes. Finally, the temperature was slowly decreased to 4°C (0.5°C/minute), and kept at 4°C for 10 minutes. To ensure annealing, the samples were analyzed by agarose gel electrophoresis. Only after annealing could the DNA band be seen in ethidium bromide-stained gel, confirming the formation of the laddered dsDNA wires. Binding the hybrids to the template dsDNA nanowires was done by mixing the 6His-SP1-GNP solution with the dsDNA solution and incubating at 37°C for 3 hours at different concentrations.

Figure 1.5A shows an AFM topography image of the formed 1D nanostructures. The 6His-SP1-GNP hybrids are anchored to the template DNA through the thiol groups in predefined positions. Repeating the same procedure of incubation with nonthiolated DNA did not lead to formation of the tri-block conjugates. Figure 1.5B and C show wires and junctions based on the tri-block conjugates. These junctions are composed of several DNA segments connected together by the 6His-SP1-GNP hybrid, seen as a bright spot in the topographic image.

To verify the existence of the GNP in the 6His-SP1 pore, EFM was performed on the tri-block conjugate. Figure 1.5D shows an EFM image of the nanostructure shown in Figure 1.5B. One can see a polarization signal at the positions of 6His-SP1-GNP. This proves that a GNP is attached to the central cavity of the protein, which is likely bound to the DNA through the thiol group serving as the connecting unit between two DNA strands.

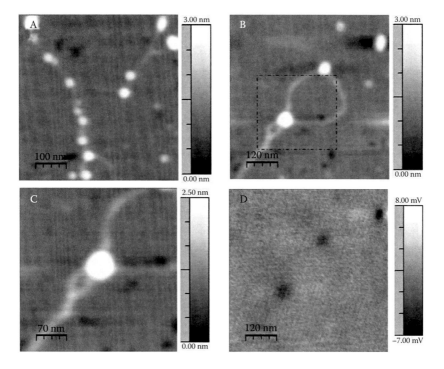

FIGURE 1.5
6His-SP1-GNP-DNA tri-block conjugates. (A) AFM topography of a 1D structure based on the tri-block conju-gate. The 6His-SP1-GNP hybrids appear as bright spots on the DNA strand. (B,C) DNA junction based on the tri-block conjugates. The enlarged image in Figure 1.5C shows multiple DNA molecules anchored to the same hybrid (dashed square in B). (D) EFM image of the junction in Figure 1.4B. One can see a polarization signal at the positions of 6His-SP1-GNP. The SP1 protein is electrically silent and exhibits no polarization signal.

1.4 SP1-GNP Tubes

Another 1D structure formed using the SP1-GNP hybrids is a tubular chain composed of alternating stacked units of protein and GNP. Such structures spontaneously form upon incubation of 6His-SP1-GNP hybrids in buffer solution. The inter-GNP separations within the protein-GNP chains made of this protein variant are irregular and depend on the rela-tive position of the GNP with regard to the protein, mainly due to the flexible long 6His tags (at the end of each of the 12 N termini), such that the GNPs are not in direct electrical contact. Figure 1.6A shows a scheme of a 6His-SP1 protein-GNP chain. A TEM image of several 6His-SP1-GNP chains can be seen in Figure 1.6B: the GNPs appear as dark spots between the 6His-SP1 rings (that are "standing" in a side view). A HAADF-TEM image of one of the protein-GNP chains is shown in Figure 1.6C. The GNPs appear as bright spots, revealing a distance of ~4 nm.

In order to improve the packing of the 6His-SP1s and reduce the inter-GNP separation (and 6His-SP1 separations), another SP1 derivative was engineered by adding a six-histi-dine tag to a truncated N terminal SP1 (five N terminal amino acid residues were deleted

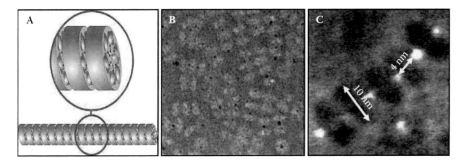

FIGURE 1.6
6His-SP1-GNP hybrids form protein-GNP chains with different GNP separations. (A) Scheme representing the organization of protein-GNP chains composed of 6His-SP1-GNP hybrids. (B) TEM image (phosphotungstic acid staining) of 6His-SP1-GNP chains (marked by solid lines). (C) HAADF-STEM image of 6His-SP1-GNP chains with 4 nm GNP separation (vanadium staining). From Medalsy et al. 2008.

from the protein, which is termed ΔSP1), thereby forming the variant 6HΔSP1. This variant of the SP1 protein should have an inner pore that may enable a deeper protrusion of the GNP into the protein pore, as 30 amino acids were deleted on each side of the SP1 ring. As expected, protein-GNP chains composed of 6HΔSP1-GNP hybrids have shorter inter-GNP separations and higher persistence length in comparison with the 6His-SP1-GNP chains (Figure 1.7A–C).

Imaging of the 6His-SP1-GNP chains deposited on a freshly cleaved mica surface covered with a ~5 nm thick layer of evaporated amorphous carbon was performed using AFM, revealing the dimensions of the tubular chains. The carbon layer was obtained using a standard carbon evaporator. The carbon layer was added to passivate the negatively charged mica surface. Figure 1.8A reveals the expected height of the tubes, 9–10 nm. The

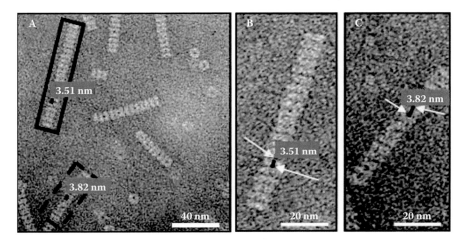

FIGURE 1.7
(A) TEM image (Nanovan staining) of 6HΔSP1-GNP chains with 3.5 and 3.8 nm GNP separation. (B and C) Enlargement of the bold boxes from A (the arrows in B and C point to the adjacent GNPs). From Medalsy et al. 2008.

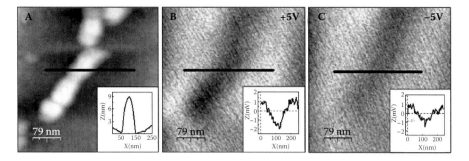

FIGURE 1.8
(See color insert.) (A) Topographic AFM image of 6His-SP1-GNP chain formation on mica (insets shows a cross-section on the chain). (B and C) EFM at positive and negative bias voltage of 6His-SP1-GNP chain on mica, showing clear polarization of the embedded GNPs (no signal is observed at 0 V). The insets show the signal strength. From Medalsy et al. 2008.

existence of the GNP in the 6His-SP1-GNP chain was confirmed by EFM measurements in both positive and negative bias voltages, as shown in Figure 1.8B and C, respectively. From the dark signal obtained for both positive and negative bias voltage we can conclude that polarizable objects, i.e., the GNPs, are positioned within the tube.

Measurements of the physical strength of the 6His-SP1-GNP chain were performed on a highly oriented polarized graphite (HOPG) surface using the AFM tip in contact mode. After deposition of 6His-SP1-GNP solution on HOPG short tubes were observed (Figure 1.9A).

As can be seen in Figure 1.9, the formation of tubes on HOPG is evident. The height and structure of the tubes are 9 nm, as expected (Figure 1.9B). By using contact mode AFM on a single line, we were able to exert gentle lateral sheer forces in the direction of movement of the cantilever on the side of the tube by the AFM tip (arrow in Figure 1.9B). Because of the scratch, the tube was moved but not broken, as seen in Figure 1.9C. The resistivity of the tube to sheer forces shows that strong attachment exists between the chain elements.

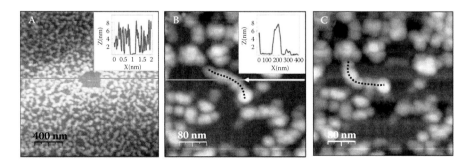

FIGURE 1.9
(See color insert.) 6His-SP1-GNP tube structures on HOPG. (A) Large-area topographic imaging shows multiple tubes. (B) A single tube with a height of ~9 nm (dashed line). (C) With contact mode AFM, a single line was scratched, thus applying lateral sheer force on the tube (the arrow in B). The tube was moved and bent but not torn, indicating the physical strength of the structure.

1.5 Summary

One-dimensional systems and wires are of great interest to various fields in the nanoscience and nanotechnology communities, especially for nanoelectronics. In this report, we demonstrate the possibility of using protein-NP hybrids as building blocks for various 1D systems that may later be controlled in several parameters: length (number of consecutive units), interunit distance, conductivity, selective attachment to surfaces, formation methods, and more. Using the SP1-GNP hybrids in a preprogrammed way will enable us to gain good control over these parameters. Moreover, the combination of a conducting core in a modifiable isolating scaffold gives these systems additional advantages over other exposed wires with uniform structure and chemistry. We demonstrated spontaneous formation of nanowires made from the SP1-GNP hybrid building blocks. We also demonstrated wires that are formed by combining the SP1 protein and GNP hybrids with DNA. Further development of these structures, e.g., by reducing the interunit distance between the GNPs, will produce a highly controlled conductive wire. By controlling the distance, the composition, and the size of the inner particle, we have a new way of investigating 1D conductance mechanisms at hand. Combined with nonlinear conductance behavior, this family of conducting wires can serve as future nanodevices.

Acknowledgments

We thank Dr. Igor Brodsky for helpful discussions and technical assistance. We acknowledge financial support from the DFG Grant "Single Molecule Based Ultra High Density Memory" CU 44/3-2 and the French Ministry of External Affairs, Israeli-Palestinian Science Organization, the Friends of IPSO, USA (with funds donated by the Meyer Foundation), and the James Franck Program.

References

1. Bachtold, A., et al. Logic circuits with carbon nanotube transistors. *Science*, 2001, 294(5545): 1317–1320.
2. Lieber, C.M. The incredible shrinking circuit—Researchers have built nanotransistors and nanowires. Now they just need to find a way to put them all together. *Scientific American*, 2001, 285(3): 58–64.
3. McMillan, R.A., et al. Ordered nanoparticle arrays formed on engineered chaperonin protein templates. *Nature Materials*, 2002, 1(4): 247–252.
4. Tucker, J.R. Complementary digital logic based on the coulomb blockade. *Journal of Applied Physics*, 1992, 72(9): 4399–4413.
5. Welser, J.J., et al. Room temperature operation of a quantum-dot flash memory. *IEEE Electron Device Letters*, 1997, 18(6): 278–280.
6. Chen, Z.H., et al. An integrated logic circuit assembled on a single carbon nanotube. *Science*, 2006, 311(5768): 1735–1735.

7. Rutherglen, C., D. Jain, and P. Burke. Nanotube electronics for radiofrequency applications. *Nat Nano*, 2009, 4(12): 811–819.

8. Huang, Y., et al. Directed assembly of one-dimensional nanostructures into functional networks. *Science*, 2001, 291(5504): 630–633.

9. Martel, R., et al. Single- and multi-wall carbon nanotube field-effect transistors. *Applied Physics Letters*, 1998, 73(17): 2447–2449.

10. Porath, D., et al. Direct measurement of electrical transport through DNA molecules. *Nature*, 2000, 403(6770): 635–638.

11. Cohen, H., et al. Direct measurement of electrical transport through single DNA molecules of complex sequence. *Proceedings of the National Academy of Sciences of the United States of America*, 2005, 102(33): 11589–11593.

12. Maune, H.T., et al. Self-assembly of carbon nanotubes into two-dimensional geometries using DNA origami templates. *Nat Nano*, 2010, 5(1): 61–66.

13. Medalsy, I., et al. Logic implementations using a single nanoparticle-protein hybrid. *Nat Nano*, 2010, 5, 451–457.

14. Mann, S. Self-assembly and transformation of hybrid nano-objects and nanostructures under equilibrium and non-equilibrium conditions. *Nature Materials*, 2009, 8(10): 781–792.

15. Whyburn, G.P., Y.J. Li, and Y. Huang. Protein and protein assembly based material structures. *Journal of Materials Chemistry*, 2008, 18(32): 3755–3762.

16. Heyman, A., et al. Protein scaffold engineering towards tunable surface attachment. *Angewandte Chemie International Edition*, 2009, 48(49): 9290–9294.

17. Heyman, A., et al. Float and compress: Honeycomb-like array of a highly stable protein scaffold. *Langmuir*, 2009, 25(9): 5226–5229.

18. Medalsy, I., et al. SP1 protein-based nanostructures and arrays. *Nano Letters*, 2008, 8(2): 473–477.

19. Wang, W.X., et al. Crystallization and preliminary x-ray crystallographic analysis of SP1, a novel chaperone-like protein. *Acta Crystallographica Section D—Biological Crystallography*, 2003, 59: 512–514.

20. Wang, W.X., et al. Characterization of SP1, a stress-responsive, boiling-soluble, homo-oligomeric protein from aspen. *Plant Physiology*, 2002, 130(2): 865–875.

21. Dgany, O., et al. The structural basis of the thermostability of SP1, a novel plant (*Populus tremula*) boiling stable protein. *Journal of Biological Chemistry*, 2004, 279(49): 51516–51523.

22. Wang, W.X., et al. Aspen SP1, an exceptional thermal, protease and detergent-resistant self-assembled nano-particle. *Biotechnology and Bioengineering*, 2006, 95(1): 161–168.

23. Marchi, F., et al. Characterisation of trapped electric charge carriers behaviour at nanometer scale by electrostatic force microscopy. *Journal of Electrostatics*, 2008, 66(9–10): 538–547.

24. Qiu, X.H., et al. Electrostatic characteristics of nanostructures investigated using electric force microscopy. *Journal of Solid State Chemistry*, 2008, 181(7): 1670–1677.

25. The sequence is 5-thiolTAACAGGATTAGCAGAGCGAGGAATCATACGTACTCAACTGCTG-GGAGCGAGACGATTAGGACAATAACTTGGTATGCT3′,5′-thiolCCTCGCTCTGCT, purchased from BioMers Ltd., Germany.

2

Bionanotechnology Strategies for Cell Detection

Nasser Aghdami and Sarah Ranjbar Vaziri

Department of Regenerative Medicine for Stem Cell Biology and Technology, Royan Institute, ACECR, Tehran, Iran

CONTENTS

2.1 Introduction

Every time we start to read about new technologies or new sciences, we can find something relevant from ancient mythologies of different civilizations. It seems that human beings live in a closed historical cycle; at least our thinking or ideas remain unchangeable in history, and in this long story, science just helps us to bring ideas and imagination to practical realities and applications. A hero who never dies, and after any injury his body regenerates the lesion, is a classic example to think about the regenerative potential of stem cells. In ancient Iranian mythology, Esfandiyar was such a heroic character; he would recover from his wounds after every encounter with his enemies [Ferdowsi, Shahnameh, 977 A.D]. Nowadays we live in a faster world, where there are more accidents that lead to loss of organs or functions, and anyone can aspire to have this power to regenerate lost organs or acquire normal function again. Perhaps it is here that modern stem cell technology follows the ancient mythology, and in few years we may see yet another miracle of the application of scientific knowledge.

The increase in average life expectancy leads to an aging society where human beings are particularly susceptible to progressive cell destruction and irreversible loss of tissue function, and ultimately degenerative diseases. Regenerative medicine is expected to play a powerful role in this scenario in the future. It promises to develop effective therapies to repair the underlying pathobiology and restore the native cellular architecture and organ function through the use of natural (e.g., living cells) or synthetic materials. The term *regenerative medicine*, coined by William A. Haseltine in 2000, subsumes pharmacotherapy, tissue engineering, gene therapy, and cellular therapies, including stem cells.

Cells have a central role in tissue regeneration, and understanding the principles of cell biology contributes to the decoding of tissue homeostasis and repair mechanisms.[1] Somatic, adult, or embryonic stem cells are the main candidates in this scenario. In addition, versions of embryonic-like cells, called induced pluripotent stem cells, have been recently used for this purpose. These cells are reprogrammed from adult cells,[2] and can be obtained very easily. Stem cells provide the active ingredient of regenerative therapy, as they can continually self-regenerate and, under the right conditions, differentiate into specialized cell types to form new tissues.[3]

Specialized applications of therapeutic repair are being explored with the use of stem cell-based platforms and their cell progeny derivatives for cell therapy of degenerative disorders.[1] However, regeneration of organs from stem cells is a complex continuous process, and the ability of transplanted cells to migrate from the transplantation site to the lesioned area and to survive, differentiate, and replace lost cells or produce growth factors and cytokines is crucial for long-lasting tissue regeneration.[4] Stem cells are placed in tissue-specific microenvironments that nurture cells and maintain tissue homeostasis. This microenvironment, called a niche, is essential to maintain a balance between stem cell quiescence and activity.[5–7] Understanding the biological signals that result in stem cell homing and regeneration of target tissue is a key factor in successful stem cell therapy. Constant observation of tissue regeneration, at the single-cell level, would markedly improve our knowledge about the underlying cellular mechanisms and is essential for preclinical studies in rodents and potentially in humans.[8]

Conventionally, the study of stem cell engraftment was mainly confined to invasive and irreversible histological and molecular biology techniques,[9] which truly represent the problem of "finding a needle in a haystack." Invasive biopsies, needed to pursue histological techniques, can cause tissue damage and are also ethically questionable, as these can result in iatrogenic complications. Additionally, the accuracy of conventional histology and molecular biology as the predominant ex vivo assessment techniques depends on adequate sampling. For example, if the transplanted cells are engrafted in a large organ, such as the brain, the biopsy might sample a part of the organ where the cells were not engrafted and the result would be a false negative. Even if this approach is justified based on the use of a genetic marker, such as bromodexyuridine (BrdU), repeated biopsy to follow the cell is technically almost impossible. Therefore, histological studies based on biopsy have considerable limitations for cell tracking, and it is impossible to use these techniques for cell monitoring in stem cell therapy settings.

Another major limitation of conventional histological and molecular biology assessments is that they are very poor in live cellular detection, as they lack application in living organisms, and thus cannot provide deep tissue information for stem cell homing. Specific imaging techniques therefore are a pressing need in biomedical sciences to enhance the in vivo study of cells that is crucial for the development of cellular medicine.[8] Developments of new imaging technologies have long played a pivotal role in the progress of biomedical sciences, as shown by the number of Nobel Prizes awarded for this progress: for phase-contrast microscopy (1953), computer-assisted tomography (1979), electron microscopy (1986), scanning tunneling microscopy (1986), nuclear magnetic resonance spectroscopy (2002), magnetic resonance imaging (MRI) (2003), and gene-targeted imaging with green fluorescent protein (GFP) (2008).

A stem cell has the potential to differentiate into other cell types and proliferate, survive, and self-renew over a long period of time. Due to the lack of a specific culture system that allows stem cell growth for the long term, in vitro imaging is not sufficient to provide reliable information. Furthermore, most stem cell characteristics depend on their cell-cell contact and the three-dimensional (3D) structure of their niche. Although there are some techniques for in vivo cell imaging, such as genetic markers, stem cell detection requires some specific considerations. Stem cells are rare in tissues,[10,11] so unlike observing a number of homogeneous cells (such as in T cell imaging),[12] a strong signal of higher resolution is likely to be required for the detection and distribution of stem cells within tissues and organs, especially at the single-cell level. In addition, study of different stages of cell maturations and the intrinsic characters of stem cells can be carried out only by using a combination of molecular markers,[13] and therefore an imaging modality that allows multi-color labeling and detection is crucial. High costs for generating enough data to carry out a reliable statistical analysis that is extremely time-consuming are another requirement.

Of course, in this scientific court, these life-saving ideas and imaginations are innocent, and our own creativity is main accused. For example, when Anton van Leeuwenhoek discovered sperm in 1678 with his microscope. He originally believed them to be parasitic animals living within the semen (hence the term *spermatozoa*, meaning "sperm animals"), but he later came to believe that each sperm contained a preformed embryo. Nicolas Hartsoeker, the other codiscoverer of sperm, drew a picture of *what he hoped to find*: a preformed human ("homunculus") within the human sperm.[14] This story indicates that researchers tend to see what they expect to find. The brain can filter and interpret information in a way that matches with its own inclinations. This means that the right data can result in inappropriate or distorted conclusions, and vice versa. Thus, it is almost impossible to understand the complex continuous processes involved in organ regeneration through stem cells if the analyses rely on individual snapshots of cell populations. Constant, accurate, and specific scientific observation with high-resolution sophisticated techniques is indispensable.

Determining the fate of a particular cell population within a heterogeneous environment is the common goal of all cell tracking assays. Specific labels, known as contrast agents, are necessarily used to highlight and distinguish a particular cell among other cells within tissue. Along with the development of new imaging modalities, use of better contrast agents and tracers has led to enhanced sensitivity ("more bang for the buck"), increased cellular uptake, and has enabled the modulation of energy (light, sound, and electron beams). Although in past decades major progress was seen in the development of tracking agents, nanoparticles form an important class of materials with unique features that make them highly suitable for biomedical imaging. They can dramatically amplify signal changes when used as contrast agents in magnetic resonance imaging, exhibit surface properties that improve cell entrance and selective uptake by specific cells (e.g., macrophages), and possess unique chemical properties that make them visible (e.g., quantum dots in fluorescent imaging). Nanoparticles are now widely used with many commercially available preparations. Several of these are being used clinically and have undergone FDA approval. For many nanoparticles, off-label uses are popular within the wider biomedical community.

In this chapter, we will first discuss current cell tracking agents, and then describe in detail the unique role of nanoparticles in biomedical imaging, the physicochemical structure and applications of quantum dots and superparamagnetic particles, and their applications in biomedical imaging research.

2.2 Cell Tracking Agents

Biomedical imaging involves the techniques and process used to create images of biological material (or parts and function) for research or clinical purposes. Although usually light, photonic waves are used to make cells or tissue visible, in biomedical imaging this process can be done by other kinds of waves, such as x-ray or magnetic field, and all of them are a type of energy. Thus, imaging needs an energy source, detectable waves, and detectors. Contrast or staining agents are substances used to enhance the visibility of target structures by increasing the energy reflection from the targets. Suitable agents for labeling the cellular property of interest are crucial, and a lack of these agents often limits study results. Long-term biological compatibility, safety, brightness, stability, and the possibility to distinguish several markers simultaneously are the major requirements for these agents.

It is relatively easy to create a wish list of the properties describing an ideal cell staining and tracking agent. However, there is no single agent that allows a comprehensive approach to stem cell imaging, as stem cell development involves a heterogeneous and interactive cast of players that evolves over time. Thus, it is logical to have a spectrum of agents from low to high resolution to label a wide variety of cell types with sensitive and quantitative multiparameter properties for monitoring how a selected subpopulation of cells would subsequently expand or evolve. At one end of this spectrum, positron emission tomography, magnetic resonance imaging (MRI), and optical coherence tomography provide real-time readouts from animal or human subjects, with resolutions of approximately ~1, ~100, and ~10 μm, respectively. At the other end, electron microscopy provides near-molecular-level spatial resolution; however, the cell must be fixed, which is invasive and prevents dynamic imaging. Fluorescence modalities provide a range of spatial and temporal resolution between these two resolution extremes, but it is not enough to have suitable resolution for in vivo imaging. Nanoparticle-based contrast agents extended this range to have appreciable resolution for cell imaging. However, to precisely depict the role of nanoparticles in biological imaging, properties of commonly used imaging modalities should first be clarified. As the first step, the advantages and disadvantages of common conventional contrast agents will be discussed, and then the quantum dots and superparamagnetic particles used as nanoparticles will be described in detail.

The most important imaging modalities for live cell tracking at single-cell resolution use fluorescence or MRI. Radio-nucleated magnetic resonance and fluorescent probes are the main contrast agents used in medical imaging.[15] Radiolabels, as the oldest contrast agents, such as 51Cr, 125I, or 111In, are sensitive detectors for very small numbers of cells, and labeling procedures are relatively simple. These labels have traditionally been used for tagging targets used in cell-mediated cytotoxicity assays, in vivo biodistribution studies, and whole body radioimaging.[16,17] The major limitations for radiolabeled cell tracking include radiotoxicity, short-term half-life, probe leakage (due to weak intracellular protein binding), significant regulatory compliance, and waste disposal costs.[15] There is a growing trend toward the use of magnetic, bioluminescent, or fluorescent probes for most of the cell tracking applications, except those in which high-sensitivity whole body distribution data are required. We will first focus on fluorescent and magnetic probes in the next sections.

2.3 Fluorescence

Many biological and natural materials give off light of a particular wavelength when exposed to light of another wavelength, a property that is called *luminescence*. If the light is emitted rapidly after illumination (within 1 millionth of a second), the luminescence is *fluorescence*. If the light emission takes longer, the luminescence is *phosphorescence*. Fluorescent materials always emit light at a longer wavelength than the wavelength of the exciting light. The difference between the wavelength of the exciting light and the emitted light is referred to as the Stoke's shift, and is based on Stoke's law. The range of wavelengths that excite fluorescence is known as the absorption spectrum. The emitted light, which also covers a range of wavelengths, is known as the emission spectrum. Since the Stoke's shift for most materials is not that great, there is generally some overlap between the excitation and the emission spectra.

Many biological materials, in particular, many vitamins, some hormones, and a variety of biological enzymes, are naturally fluorescent, so they can fluoresce strongly enough as to interfere with specific fluorescence labeling studies, called background fluorescence, or autoflourescence. Narrow bandpass filters that allow only a limited range of wavelengths of light are used for both exciting light sources and emitted light paths to solve this problem. The rationale for this is that by restricting the exciting light wavelengths, some of the autofluorescence can be reduced. In addition, by restricting the emitted light, less of the autofluorescence light interferes with the desired specific fluorescence. Excitation filters are chosen to match the excitation maximum of the fluorescence label being used, and the emission filters are matched for maximum emission. Regardless of how carefully one matches the excitation and emission filters to a given label, there will still be some background autofluorescence, and this will have the effect of reducing the perceived contrast between the "real" or actual label fluorescence and the specimen background.

On continued stimulation most fluorescent materials fade, a phenomenon known as photobleaching. Although different fluorescent materials fade at different rates and some specific preparation methods can reduce the rate of fading, all the fluorescent materials eventually fade, and this effect is irreversible. Therefore, the specimens should be illuminated only while aligning and focusing, and during actual image collection. At all other times, the excitation light source should be turned off.

There is growing interest in the field of fluorescent imaging for establishing the efficacy of stem cells for therapeutic applications. One of the major methods to label cells is the use of a reporter gene system, which is based on using cells that were genetically engineered ex vivo. Homologs of the green fluorescent protein from *Aequorea victoria* (avGFP) are widely used in biotechnology and as stable genetic tags for very long-term cell tracking.[18,19] It continues to be used to enable detection of labeled tumor cells in intact organs or whole animals at increasingly high sensitivity by employing improved detection methods and color-coded host tumor models. However, genetic tagging is time-consuming, complex, and frequently less efficient and stable than alternative methods, particularly when trying to tag primary cells. Bioluminescent tags, when used as genetic probes, simplify in vivo detection; however, their signal is oxygen limited, making it difficult to determine whether site-to-site intensity differences are due to varying numbers of cells or varying luminescence efficacy. Most of the genetic tags do not allow a distinction between parental cells and progeny, and so it is not possible to determine the extent of clonal expansion, and it can be difficult to quantify the fraction of administered cells reaching their targets.

Carboxyfluorescein succinimidyl ester (CFSE), which was originally developed for lymphocyte labeling and to track their migration within animals for many months, is a fluorescent cell staining dye.[20] Subsequent studies revealed that CFSE can react rapidly with membrane and intracellular proteins under physiologic conditions, and can be used to monitor cell proliferation, both in vitro and in vivo, due to the progressive halving of CFSE fluorescence within daughter cells following each cell division.[21] Despite these advantages, since CFSE randomly labels various cell proteins, it can alter the cell function due to reaction with specific functional proteins.[15] pH dependence of the carboxyfluorescein fluorescence intensity is another problem with this probe, which can complicate analysis of intensity distributions, as alterations in subcellular distribution pH may lead to changes in intensity that may be misinterpreted as cell division. Additionally, cell death causes the same misinterpretation due to loss of intensity.

Lipophilic fluorescent dyes, such as membrane dyes, label the lipid regions of cell membranes, rather than proteins, and cause less alteration in cell functions. PKH26, as a lipophilic dye, because of its excellent membrane retention and biochemical stability, has been used as the probe of choice for in vivo tracking of ex vivo-labeled stem cells and tumor- or antigen-specific lymphocytes. Like CFSE, lipophilic fluorescence intensities are halved with each successive generation in dividing cells. However, it remains stable for weeks to months in nondividing cells. In addition to different signal strengths due to dye redistribution after phagocytosis, these dyes have other limitations. As the membrane dyes bind noncovalently, exposure of labeled cells to solvents that extract lipids may alter the dyes' intensity, and this should be considered when using standard tissue fixation and processing methods. A related limitation of lipophilic membrane dyes is that obtaining reproducible results requires greater precision of dilution, because increasing dye concentration or decreasing cell concentration (or cell size) results in increasing numbers of dye molecules incorporated per cell.

Despite the widespread applications of fluorescence and bioluminescence in bioimaging, many critical parameters, such as tissue-to-detector geometry, autofluorescence, tissue optical properties, absorption, and scattering, remain unaccounted for in the current state of data analyses. Furthermore, organic fluorophores suffer from fast photobleaching and broad, overlapping emission wavelengths, and therefore are limited in applications involving long-term imaging and multicolor detection. Advances in synthesis and biofunctionalization of colloidal semiconductor nanocrystals, which are called quantum dots (QDs), during the past decade have generated increasingly widespread interest among investigators in the fields of biology and medicine. Quantum dots are robust fluorescence emitters with size-dependent emission wavelengths. Their extreme brightness and resistance to photobleaching enable the use of very low laser intensities over extended time periods, making them especially useful for cell tracking.

2.4 Quantum Dots

After the first publications in 1998 by Bruchez et al.[22] and Chan and Nie[23] on the possibility of using QDs as competent flourescent probes in biological imaging and related applications, QD-based research has progressively improved. In less than a decade, QDs have prevailed over many of the intrinsic limitations encountered by conventional organic dyes and genetically encoded fluorescent proteins and have become powerful tools in many fields, such as molecular biology, cell biology, molecular imaging, and medical diagnostics.[24]

Quantum dots are luminescent colloidal semiconductor nanocrystals typically in the size range of 2–10 nm comprising 200–10,000 atoms,[25] in which the electron motion is confined by potential barriers in all three dimensions.[26] The general structure of QDs is composed of an inorganic core that frequently consists of group II–VI (e.g., CdSc, CdTe, CdS, and ZnSe) or III–V (e.g., InP and InAs) atoms of the periodic table.[27] The core is overcoated by a shell of another semiconductor with a larger spectral band gap, mostly ZnS,[28] and a subsequent aqueous organic coating is utilized, to which the biomolecules are conjugated,[29] as shown in Figure 2.1.

QDs exhibit several unique optical and electrical properties, including broad excitation and size-tunable narrow emission spectrum, long luminescence lifetime, exceptional photochemical stability, high brightness, and large two-photon action cross sections.[22,23,30,31] These remarkable properties make QDs fluorophores of choice in applications requiring long-term, multitarget, and high-sensitivity detection, such as dynamic live cell imaging and in vivo whole organism imaging.[28,32]

QDs gained their name from the quantum confinement effect, which only occurs in structures with physical dimensions smaller than the Bohr exciton radius (electron-hole spacing in excited state).[33,34] A direct consequence of this phenomenon is that the energy levels of the exciton (electron-hole pair) will become quantized, and therefore the size of

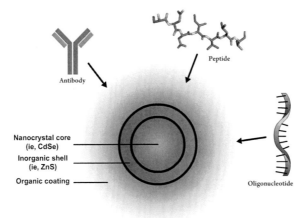

FIGURE 2.1
The general structure of a QD. The nanocrystal core is passivated by a semiconductor shell, mostly ZnS. An organic coating is used that enables conjugation of different functional groups (antibody, peptide, oligonucleotide) for desired bioactivity.

the core determines the energy of the emitted photon,[28] which can be continuously tuned from the ultraviolet[35] to near infrared,[36] and even to the mid-infrared.[37] The smaller the diameter of a semiconductor QD, the larger the band gap becomes, and thus the fluorescence of the shorter wavelength is achieved.[28] A photon with greater energy than that of the semiconductor band gap is required to generate an exciton that has a lifetime of about tens of nanoseconds in QDs.[38] This long luminescence duration along with broad Stoke's shift can be as large as 300–400 nm, depending on the wavelength of the excitation light,[39] and enables time-gated imaging, which improves the sensitivity of fluorescence detection in live cell imaging by reducing background autofluorescence.[40] QDs can be excited efficiently at any wavelength shorter than their emission peak,[41] which results in a broadband absorption spectrum extending from the ultraviolet to a cutoff visible spectrum wavelength.[42] However, it has an increased probability toward shorter wavelengths. This allows simultaneous excitation of many spectrally distinct QD emitters with a single excitation source, in contrast to the organic fluorophores for which a certain wavelength for every different fluorophore is required.[28] When the exited electron turns back to its lower energy state, a photon with a narrow symmetric energy band (typically 20–40 nm full width at half maximum (FWHM)) is emitted upon radiative recombination.[43] Therefore, QDs with different emissions can be used for simultaneous visualization of multiple cellular targets or different cell populations without any spectra overlap, which is of particular interest for both in vitro and in vivo multicolor imaging, besides their ability for polychromatic extracellular membrane labeling with up to eight colors in flow cytometric analysis.[26,44–46] By using different QD conjugates and a single excitation source, various cellular components, including microtubules, mitochondria, and nucleosomes, have been detected in a HeLa cell.[47] Negatively charged dihydroxylipoic acid (DHLA)-capped QDs with different emission wavelengths have also been utilized for real-time tracking of metastatic cells in a living mouse. By the use of fluorescence emission scanning microscopy, five different QD-tagged populations of melanoma cells have been simultaneously identified as they extravasated into lung tissue.[48]

The high photobleaching threshold and the strong resistance to chemical and metabolic degradation[49,50] in QDs permit continuous monitoring and imaging over an extended period of time.[51] In the study designed by Wu et al., nuclear antigens and microtubules were simultaneously stained with Alexa Fluor488 (green) and QD630 (red) in the same cell. Regardless of which target was labeled with Alexa dye, the fluorescent signal faded quickly and became undetectable within 2 min, in contrast to QD630, which was highly photostable for the entire 3 min illumination period.[45] It is believed that the QD photobleaching process is related to a prolonged photochemical destruction of the core components. This hypothesis is supported by the observation of a blue shift in the QD emission spectra upon continuous UV illumination. Henglein and co-workers speculated that optical excitation is responsible for initiating the CdS decomposition process by the formation of S or SH radicals. These radicals are able to form an SO_2 complex via an air oxidation reaction that ultimately leads to slow nanoparticle degradation.[52] An appropriate capping with a thick semiconductor layer is applied to improve the photostability of QDs by protecting the core surface atoms from oxidation.[53,54] Moreover, in contrast to organic fluorophores, QDs are less susceptible to metabolic degradation due to their inorganic nature.[55,56] It was shown that QDs remained intact in live cells and organisms for several weeks to months, with no signs of toxicity or QD breakdown.[32,50]

The superior brightness of QDs arises from a combination of a high quantum yield (the ratio of emitted to absorbed photons) and a large extinction coefficient, so that they are able to re-emit much of the absorbed light.[57,58] The quantum yield, which could be up to 85% in high-quality QDs, is not affected by the bioconjugation process, in contrast to the organic fluorophores, whose yield is dramatically reduced during functionalization with biomolecules.[59] Bawendi and co-workers[60] estimated that the molar extinction coefficients of CdSe QD are about 10^5–$10^6 M^{-1}$ cm^{-1}, depending on the size and excitation wavelength of the particle. Since this value is about 10–100 times larger than that of organic dyes,[60,61] QD absorption rates will be ~10–50 times faster at the same excitation photon flux (number of incident photons per unit area). This superiority makes them brighter probes under photon-limited in vivo conditions, where light is extremely diminished by scattering and absorption.[39] The large two-photon absorption cross section of QDs, which is two to three orders of magnitude greater than traditional organic fluorophores, permits more efficient probing of thick specimens by the use of multiphoton excitation microscopy.[31,48] With this technique, fluorescence signals of CdSe-ZnS QDs have been detected hundreds of micrometers deep through the intact skin of living mice.[31] In a report by Stroh et al.,[62] tumor vessels have been differentiated from both the perivascular cells and the extracellular matrix by the use of micelle-encapsulated QDs and multiphoton microscopy in GFP-expressing transgenic mice. In comparison with the traditional high molecular weight dextran, a much sharper boundary was visualized between intra- and extravascular spaces. In addition, the relatively broad excitation spectrum of QDs permitted the simultaneous monitoring of demarcated vessels and GFP-expressing perivascular cells.

Finally, the sufficient electron density of the QD's core enables transmission electron microscopy imaging. This ability, coupled with fluorescent imaging of the same sample, can provide information on both temporal dynamics and high spatial resolution of the precise localization of nanoparticles within cells and tissues.[63]

2.5 Synthesis, Solubilization, and Bioconjugation

There has been progressive research based on quantum dots in the field of nanobiotechnology, which accounts for the considerable improvements in their synthesis and surface chemistry since 1982.[64,65] Typical synthesis strategies of semiconductor QDs involve the growth of nanocrystals using molecular precursors in either aqueous or organic solutions.[54] Before 1993, QDs were mainly prepared in aqueous solutions containing stabilizing agents (e.g., thioglycerol or polyphosphate). This procedure generates nanocrystals with low quantum yields and large size variation, resulting in broad emission spectra (~50 nm FWHM).[41] In 1993, Bawendi's group[60] developed a novel approach for synthesis of highly luminescent monodispersed QDs by using a high-temperature organometallic process. In contrast to the former synthesis methods, the nanocrystals had nearly perfect crystalline structures and fewer size variations. However, the fluorescence quantum yields were still relatively low. The deposition of a surface-capping layer such as ZnS was found to dramatically improve their fluorescence quantum yields.[61,66,67] Alternative precursor materials such as CdO could also be utilized in the synthesis of CdS, CdSe, and CdTe nanocrystals, which resulted in the creation of QDs with a relatively narrow FWHM (45–65 nm) and acceptable quantum efficiency without the need for an inorganic capping layer surface coating.[68]

Generally, the QDs synthesized in organic solvents have hydrophobic surface ligands, which should be rendered water dispersible for their utilization in biological applications.[69] This could be achieved by two general strategies: ligand exchange and coating with an amphiphilic polymer. Ligand exchange involves the substitution of the coordinating hydrophobic surfactants on the nanocrystal surface with heterobifunctional ligands, each presenting a surface anchoring moiety to interact with the inorganic QD surface (e.g., thiol) and an opposing hydrophilic end group (e.g., hydroxyl, carboxyl) to achieve water compatibility.[23,70–73] Although the generated QDs are useful for biological assays, ligand exchange results in reduced fluorescent efficiency and a tendency to aggregate and precipitate in biological buffers. In order to alleviate these problems, the hydrophobic QDs are coated with amphiphilic polymers.[45,74,75] Due to the protective hydrophobic bilayer encapsulating each QD, the resulting nanoparticles are water soluble and highly stable for long periods of time. Typically, the QDs coated with amphiphiles have a larger hydrodynamic size than those with a monolayer of ligand,[25] which could be a drawback when constructing Förster resonance energy transfer (FRET) pairs, or for in vivo applications in humans.[76]

QD synthesis is generally followed by a subsequent step to functionalize the nanoparticles. For this purpose, certain biomolecules, such as oligonucleotides,[74,77,78] proteins,[23,45,50] peptides,[79,80] or small molecules,[22,41,81] are bound to the surface of nanomaterials either covalently or noncovalently. The relatively large surface area of QDs allows simultaneous attachment of various functional molecules, including targeted moieties, therapeutic agents, and imaging probes, providing new potential for many clinical therapies and diagnostics. In most cases, the biological function of the conjugated molecules, such as the ability of molecular recognition, is preserved upon conjugation to colloidal QDs. The oligonucleotides conjugated to QDs could successfully hybridize to their complementary sequences.[74,77,78] Several peptides,[79,80] proteins,[23,82] and antibodies[45,50,80] have also been specifically targeted by certain bioconjugated QDs.[83]

There are several determining factors that should be controlled during nanomaterial bioconjugation, including the ratio of biomolecules per QD surface, their orientation, the interval distance between biomolecule and QD surface, and the strength of the biomolecules' attachment.[44] So far, none of the current methods, such as bifunctional linkage,[23] silanization,[22] incorporation in micro- or nanobeads,[84] and hydrophobic or electrostatic interactions,[48,70] have fulfilled all of these criteria together, so that any single or specific conjugation strategy has its own particular advantages and disadvantages.

2.6 Extracellular and Intracellular Targeting

Typically, QD surface functionalization is carried out to achieve extracellular or intracellular targeting; however, there are some intracellular delivering methods, such as microinjection, electroporation, and nonspecific endocytosis, that do not require any bioconjugation.[28,50,85]

QDs conjugated with particular targeting molecules, such as proteins, antibodies, and ligands, are utilized for labeling different biomarkers that are only expressed on the surface of interested cells.[30,86,87] One classical example in which specific antigens on cellular membrane were effectively targeted with conjugated QDs is the localization of a QD-secondary antibody conjugated to Her2, an overexpressed growth factor receptor on the surface of human breast cancerous cells, by Bruchez and co-workers.[45,76] In another study, serotonin-conjugated QDs were targeted to neurotransmitter receptors expressed on human epithelial kidney cells (HEK) cell membrane.[81] Lymphocytes surface antigens, including CD5, CD19, and CD45, have also been labeled by antibody-conjugated QDs.[88]

The intense brightness of QDs, as well as their exceptional photostability, enables real-time tracking of a single particle over an extended period of time, which is a potent approach to study the molecular behavior underlying biological events such as cell membrane organization, lateral diffusion of molecules, and molecular interactions.[89–91] Through the use of streptavidin-functionalized CdSe-ZnS QDs, individual glycine receptors have been monitored in the synaptic cleft of live spinal neurons for durations longer than 20 min.[63] Antibody-conjugated QDs have been used for the visualization of integrin dynamics during osteogenic differentiation of human bone marrow-derived progenitor cells (BMPCs).[92] Lidke et al. used the epidermal growth factor (EGF)-conjugated QD to directly image the internalization of EGF receptors and the subsequent signaling pathways.[93]

Two distinct methods are followed in intracellular labeling: first, whole cell labeling, which is particularly of interest when a specific cell population is aimed to be detected or tracked, as in cell lineage studies,[28] and second, when particular subcellular compartments inside a living cell are desired to be targeted and tracked, which represents a powerful tool in molecular cell biology.[94] In contrast to cell membrane targeting, the intracellular delivery of QDs in living cells is confronted with some difficulties. A major problem is the lack of an efficient method to transport QDs across the plasma membrane barrier without disturbing the cell membrane and, in the case of intracellular specific targeting, prevention of entrapment in the endocytic pathway or aggregation in the cytoplasm.

Intracellular delivery of nanoparticles in living cells can be achieved by both physical and biochemical strategies (Figure 2.2), except nonspecific endocytosis (Figure 2.2A), which depends on the electrostatic interaction among nanoparticles and cellular membrane, as well as the innate capacity of most cells to uptake nanoparticles from extracellular space via endocytosis in a size- and charge-dependent manner.[50,95] The main advantage of this passive delivery is its simplicity and the ability to deliver QDs to a large number of cells simultaneously. However, endosomal escaping and the potential for cytotoxicity

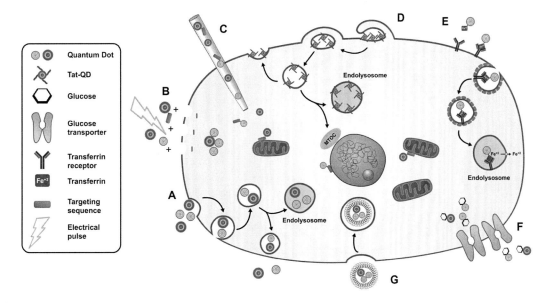

FIGURE 2.2
Intracellular delivery of QDs by using different strategies. A) non-specific endocytosis, B) electroporation, C) microinjection, D) peptide-mediated uptake, E) protein-mediated delivery, F) small molecules, and G) polymer-mediated delivery.

will remain an impediment to its usage as an efficient technique.[27] Physical methods that involve direct manipulation of the cells encompass microinjection and electroporation. High-voltage electric pulses are utilized in electroporation (Figure 2.2B) to temporarily increase the permeability of the plasma membrane.[76,96] This method, similar to nonspecific endocytosis, is suited for delivering QDs to a large population of cells. The direct delivery of nanoparticles into the cytoplasm without requiring subsequent endosomal escape is also advantageous. However, intracellular aggregation of the electroporated particles and the high rate of cellular death due to the applied electrical shock are major problems with this technique.[94] By using a fine-tipped glass micropipette, it is possible to directly microinject (Figure 2.2C) the QDs into the cytoplasm of individual cells in a monodispersed form. In spite of low delivery efficiency, this method results in a lower rate of cellular death[74] than electroporation. The chemically mediated delivery system facilitates plasma membrane translocation by using different biomolecules, including cationic peptides (Figure 2.2D) (e.g., HIV-derived TAT peptide),[97,98] various proteins (Figure 2.2E) (e.g., transferrin),[99,100] small molecules (Figure 2.2F), such as nutrients or cofactors,[101] or overcoating with amphiphilic polymers (Figure 2.2G) (e.g., phospholipid micelles).[102] For high delivery efficiency, short incubation times, minimal toxicity, a relatively small size of the conjugated particles, and an ability for noncovalent assembly (i.e., biotin-avidin), peptide-mediated delivery appears to be the most efficient approach for delivering nanoparticles into the living cells; however, this method, like others, has its own limitations.[103] One of the main issues in this strategy is nanoparticles' endosomal entrapment,[23,50,104] which is not an impediment in applications, such as whole cellular tracking and cytometery.[94] However, targeting and imaging of particular subcellular components such as cytoplasmic proteins, organelles, and complex biomolecular processes in living cells[55] requires access of the bioconjugated QDs to the cytoplasm or nuclear region. Using microinjection of QDs conjugated to organelle-targeting peptides, Derfus et al. demonstrated the ability of delivery and specific targeting of either cellular mitochondria or nuclei in fibroblasts.[94] Correspondingly, electroporated peptide-QD conjugates have been used for targeting and imaging of the nuclei in living cells.[105] Antibody-conjugated QDs encapsulated within a biodegradable nanocomposite have been designed by Kim et al.[106] for facilitating multiplexed labeling of subcellular structures inside living cells. This bioresponsive delivery system, which undergoes endolysosomal-to-cytosolic translocation, enables controllable release of the functionalized QDs through hydrolysis of the polymeric capsule within the cytoplasmic compartment. Noninvasive monitoring of multiple organelles, structures, and intracellular processes by using QDs can open up the probability of studying different molecular interactions and complex biological processes inside living cells.

2.7 In Vivo Application of Quantum Dots

The superior photophysical aspects of QDs, together with the significant improvement in their synthesis, coating, and conjugation techniques, have made QDs the fluorophores of choice in wide-ranging cellular and molecular imaging applications,[55] based on their use as fluorescent labels or contrast agents.[107,108] The ability of real-time tracking of single-cell migration and differentiation is assumed to be important in a number of research areas, such as embryogenesis, cancer metastasis, and stem cell therapeutics.[59] In addition,

QD-based contrast agents can be utilized in many biomedical applications, such as tumor targeting and imaging,[75] lymph node mapping,[109] and vascular imaging.[79]

2.8 Cellular Fate Tracking in Developmental Biology

Developmental biology is among the rapidly growing and interesting fields in biology that study the initiation and maturation of living organisms. It can integrate different levels of biology, such as molecular biology, physiology, cell biology, genetics, etc., so that the study of development is indispensable for a better understanding of different fields in biology. Development is a progressive process in which a single cell is able to generate a multicellular organism with different cell types. Therefore, producing cellular diversity within each generation is one of the main goals followed during development, which leads to several fundamental questions, such as: How can a single cell differentiate into hundreds of different cell types? Which cell in an early-stage embryo will finally develop into which part of the final creature?

Because of QDs exceptional photostability, symmetrical diffusion among daughter cells during cell divisions, and remaining within labeled cells, these nanoparticles can be effectively exploited for labeling early-stage embryonic cells in order to perform in vivo fate mapping and monitoring of migration and differentiation of individual cells, which is still unachievable with currently available tracing agents. In an elegant set of experiments designed by Dubertret et al., micelle-encapsulated CdSe-ZnS QDs were microinjected into individual cells in early-stage *Xenopus* embryos. In particular, the QD micelles were equally transferred to daughter cells, and they were highly stable and compatible with normal embryonic development, resulting in real-time monitoring of cell lineage and differentiation up to the tail bud stage.[74] The embryonic development of zebra fish as the vertebrate model organism has also been studied by Rieger's group[110] due to its optical clarity, external development, and ease of manipulation. Red light-emitting QDs have been coinjected with mRNA-encoding GFP, an established lineage tracer in zebra fish embryos, into individual blastomeres at the 32- and 64-cell stage. The results demonstrated that QDs can reliably mark a large subfraction of the descendants, besides the injected embryos, to up to 10^8 QDs/cell, did not show any malformations or developmental problems during embryogenesis. In another study, QDs coated with commercial phospholipids were electroporated into mouse embryonic neural stem and progenitor cells (NSPCs). Consequently, QD labeling had no adverse effects on the normal developmental ability of such cells either in vitro or in utero following implantation of the labeled embryos. Using an in utero ultrasound-guided electroporation system, Slotkin et al. revealed that the labeled NSPCs were able to survive, migrate, and differentiate into nervous system lineages.[111] These studies demonstrated that QDs are favorably stable in physiological conditions, and are biocompatible with the highly controlled and sensitive process of embryonic development,[47] which makes them promising intravital lineage tracer agents for analyzing many aspects of embryogenesis within the living and developing organisms.[110]

2.9 Cell Tracking in Stem Cell Therapy

Quantum dot imaging holds great promise as a competent agent for tracking stem cell fate with single-cell resolution while allowing quantification of exact cell numbers at any anatomic location, which is particularly of interest in cell-based therapies. Accordingly, in a study performed by Rosen et al.,[112] human Mesenchymal stem cells (MSCs) have been loaded by red emission QDs for identifying the spatial distribution of cells following their transplantation into a canine ventricle on an extracellular matrix scaffold. Despite the high levels of autofluorescence in the cardiac milieu, individual QD signals have been clearly detected in histologic sections for up to 8 weeks. Besides, some of the cells had developed an endothelial phenotype, which represented the in vivo compatibility of QDs with stem cell differentiation potential. In another set of experiments, QD-hMSCs committed to a cardiac lineage in vitro have been delivered into the canine heart on a patch using the ventricular defect model. It was shown that QD-containing cardiogenic cells differentiated into mature myocytes with normal sarcomere spacing.[113,114] The biodistribution of intravenously administered QD-labeled MSCs into a SCID nude mice also has been tracked as they localized into the liver, lung, and spleen.[115]

The inherent long-term fluorescence intensity in QDs permits both in vitro and in vivo monitoring of stem cell survival, localization, and differentiation,[111,115–118] which is critically needed in the field of regenerative medicine. Arginine-glycine aspartic acid (RGD)-conjugated QDs have been used for labeling of hMSCs during self-replication and multilineage differentiation into adipogenic, osteogenic, and chondrogenic cells. The study demonstrated that during the test period of 22 days, nanoparticles did not adversely affect cellular viability and proliferation. Additionally, QD-labeled cells had similar differentiation potential, as evidenced by the expression of matrix biosynthesis markers, compared to unlabeled cells of the same subpopulation.[116] Similarly, in a recent experiment by Yukawa et al., adipose tissue-derived stem cells (ASCs) labeled with negatively charged QDs emitting red light of 655 nm wavelength also retained their multilineage potential.[119] In an attempt to detect cell fusion events in vitro, Murasawa et al.[120] designed a coculture system that consisted of both labeled human endothelial progenitor cells (EPCs) and a rat cardiac myoblast cell line (H9C2). Through the use of fluorescent microscopy for the visualization of the QD signals, the frequency of cell fusion was evaluated to be less than 1%, demonstrating the fact that transdifferentiation, not cell fusion, is dominant for EPCs committed to myocardial lineage cells.

The high sensitivity of QDs, along with virtually an unlimited number of well-separated colors, all excitable by a single light source, permits real-time multiplex tracking of particular stem cell subpopulations,[121] which is certainly advantageous because of the rarity of these cells and the requirement of multiple markers for their correct identification in most cases.[122] In order to confirm the QDs' ability for in vivo multicolor imaging, murine embryonic stem cells have been efficiently labeled with six different QDs by using a peptide-mediated strategy.[117] The labeled cells were then injected into the various locations on the back of immunodeficient mice. With one single-excitation wavelength, six colors, ranging from visible to near infrared (NIR), have been simultaneously detected in living mice. However, only the QDs that had 800 nm emitting fluorescence could be imaged up to 14 days, due to low tissue absorption, scattering, and autofluorescence, as well as QD800 high-photon penetration depth,[24,123,124] revealing that near-infrared QDs are more feasible probes for cell tracking within deep tissues.[109,125] Despite the ability of these fluorophores for deep tissue imaging in animal models, the NIR region is limited to only 4 to 10 cm of tissue,[121,126] so that their clinical use is only available for near-surface applications such as intraoperative imaging.[109,127] Through use of multiphoton intervital microscopy, bone marrow precursor cells labeled with QD590-Tat conjugates (orange) have been tracked as they navigated to the tumor vessels highlighted with QD470 micelles (blue).[128] Multicolor tracking of living stem cells using different-sized QDs could be limited due to the nanoparticle escape in which small-sized QDs have been shown to rapidly disappear from the cells.[121] Asymmetric division,[129] which results in nonequivalent distribution of endosomes between daughter cells and leads to the appearance of nonfluorescent cells,[98] and the possibility of uptake of the tracing agents by unlabeled cells after cellular death or leakage through gap junction channels, resulting in false positive data,[127] are among the important impediment factors in efficient stem cell tracking with fluorescent nanoparticles. In order to evaluate possible particle transfer from loaded to unloaded cells, QD-hMSCs were cocultured with hMSCs transfected to express GFP. The results indicated no evidence of QD internalization by GFP-expressing cells, which was consistent with the known diameters of QDs (~10b nm) and gap junction channels (~1 nm). Moreover, incubation of canine myocytes with the mechanically disrupted QD-hMSC lysate, in order to simulate the in vivo condition of dying cells, demonstrated that QD clusters were not internalized by cells for up to 24 h.[112]

2.10 Using Quantum Dots as Optical Contrast Agents

QDs can also be utilized as promising contrast agents for discerning the anatomical structures of interest from surrounding tissues, and thus improving the signal-to-noise ratio, which is highly desired in diagnostic imaging. Typically, there are three different ways for delivering QD-based contrast agents to the target organ, including local injection and nontargeted and targeted systemic administration.[83] QDs, similar to other nanoparticles, are nonspecifically taken up by the reticuloendothelial system (RES), which is mainly located in the spleen, liver, and lymph nodes. Consequently, nontargeted systemically administered nanoparticles will eventually be accumulated in these organs. For most of the uses in vivo, reducing nonspecific QD uptake and prolonging their circulation lifetime is essential. Exceptions are settings that require defining some tumor types or detection of sentinel lymph nodes. In these situations, rapid deposition in RES is potentially useful.[32] It was demonstrated by Ballou et al. that the half-life of QDs in the bloodstream of mice is markedly increased by the use of long-chain polyethylene glycol (PEG) surfaces,[32] an effect that also has been documented for other types of nanoparticles and small molecules. This is attributed, in part, to the reduced nonspecific adsorption and decreased antigenicity of the nanoparticles.[130] The hydrodynamic diameter has also been considered an effective parameter for particle distribution kinetics and their clearance from the organism,[27] in which larger particles are cleared much faster from the circulation than smaller particles, which are more likely to be taken up by other cells within the body, including targeted cells.[131]

2.11 Tumor Targeting and Imaging

One of the important potential uses of QDs is for the specific and effective detection of particular cells via recognition of their unique markers. These properties have opened new possibilities for tumor targeting and imaging, as some specific markers are particularly expressed by cancerous cells. For this purpose Gao et al.[75] used ABC triblock copolymer-coated QDs conjugated to monoclonal antibodies directed against human prostate-specific membrane antigen (PSMA) to target and image carcinoma cells transplanted in nude mice. Following intravenous injection, antibody-conjugated QDs could recognize and bind to the extracellular domain of PSMA at the tumor site, which indicates that active targeting of specific cells and tissues leads to more effective particle delivery. However, it is just possible to capture the specific processes involved in their postinjection delivery via in vivo real-time single-particle tracking. This is unachievable by conventional modalities of in vivo imaging, such as computed tomography, magnetic resonance imaging, positron emission tomography, and organic fluorescence or luminescence imaging, due to their insufficient resolution at the single-particle level in vivo.[132] In order to evaluate the ability of QDs for long-time trajectory tracking of single particles, CdSe nanoparticles conjugated to monoclonal anti-HER2 antibody were injected into the tail vein of mice with HER2-overexpressing breast cancer. With a dorsal skinfold chamber model and a high-resolution intravital imaging system, six stages of QD-antibody delivery processes from a capillary vessel to cancerous cells were identified as follows: circulation within a blood vessel, extravasation, reaching the extracellular region, binding to human epidermal growth factor receptor 2 (HER2) on the cell membrane, permeation from the cell membrane to the perinuclear region by endocytosis, and finally, in the perinuclear region. These results provide valuable quantitative information about active targeting with conjugation of tumor-specific antibodies that, in turn, allow rational improvements in particle designing to increase the therapeutic index of the tumor-targeting nanocarriers.[133]

2.12 Lymph Node and Vascular Mapping

Fluorescent imaging of vasculature, lymphatic organs, and lymph node draining in both normal and cancerous tissues is achievable by injecting QD nanoparticles at proper sites of living animals.[134] Lymphatic mapping and sentinel lymph node (SLN) identification are among the most progressive advances that have revolutionized surgical oncology in recent years.[135] Imaging of lymph nodes in the vicinity of cancerous tissue is very critical for monitoring the presence of any metastatic cells during surgical resection.[57] The high-background signals, unpredictable drainage patterns, and inability for real-time imaging of lymphatic tracers are the limiting factors found in currently available SLN mapping techniques.[136] Lymph node imaging in animal models by using QDs was first reported by Kim and co-workers.[109] They intradermally injected NIR type II QDs with a CdTe/CdSe core-shell system in mice and living pigs and demonstrated real-time imaging of SLNs at 1 cm tissue depth. The spectroscopic properties of NIR type II QDs, offering virtually background-free imaging, and the optimal hydrodynamic diameter (~10 nm) of particles, ensuring their retention in SLNs, were shown to be two advantageous factors during this study.[24] The versatility of this method has been further proved by successful real-time intraoperative SLN mapping in the gastrointestinal tract[136] and pulmonary system[137] of adult pigs. These results indicate the probability of NIR imaging with QDs for rapid and precise location, delineation, and removal of SLNs, or even other types of cancerous lesions in humans.[109,138]

The ability of multicolor labeling and imaging with QDs can be used in tumor diagnosis, considering that SLNs could be labeled with nonspecific QDs, while appropriate conjugated QDs could be utilized for image-guided localization of metastatic cells within lymph nodes.[139] For studying the anatomy and flow within the lymphatic system, five different QDs with distinct visible-range emissions have been used for in vivo real-time multicolor lymphatic imaging in the neck and axiliary region of mice. This method enabled simultaneous and separate monitoring of multiple lymphatic basins in real time, which was possible for up to at least 7 days.[140]

QDs have also been used as potent fluorescent angiography contrast agents to passively image the vascular systems of different animal models. Through the use of multiphoton microscopy, capillaries as deep as several hundred micrometers have been imaged through the intact skin of a living mouse injected with green light-emitting QDs.[31] The coronary vasculature has also been visualized by Lim et al. by intravenous injection of NIR-(CdTe) CdSe QDs into an adult rat.[38] In another study, red QDs were used for monitoring vasculature development in transparent zebra fish embryos. As a result, QDs enabled obtaining comparable details of the embryonic pattern of vasculogenesis, ranging from early differentiation to the late larval stages, as the QDs marked the newly formed vessels.[110]

In order to generate contrast at specified locations, QDs should be effectively, specifically, and reliably directed to intended sites without any alteration. In a pioneering study, demonstrating that QDs can be targeted in vivo with a high level of specificity, Akerman and co-workers[79] conjugated CdSe-ZnS nanoparticles with tissue-specific peptides. These conjugated QDs were then injected to breast carcinoma xenograft-bearing nude mice for targeting lung vasculature, tumor blood vessels, or tumor lymphatic vessels. Ex vivo histological analysis revealed that QDs specifically homed into the tumor vasculature and targed organs, but not to the surrounding tissues, guided by the surface peptide molecules. NIRQD-RGD[141] have also been utilized for imaging of tumor vessels by targeting integrin $a_v b_3$, a cell adhesion molecule, which is overexpressed on the surface of invasive cancerous cells and most tumor vasculature, but not in quiescent endothelium and normal tissues.[142,143] NIR vasculature imaging has been successfully achieved by intravenous administration of QD705-RGD in athymic nude mice having subcutaneous human glioblastoma tumors. Integrin-targeted NIR fluorescence imaging has introduced new perspectives in cancer diagnosis, as well as image-guided surgery and therapy.[141]

Despite all advantages of using NIR-QDs for deep tissue imaging, whole animal optical imaging is still limited due to low excitation efficiency and significant background autofluorescence. Short wavelengths that are required for the efficient excitation of any QDs suffer from the inability of deep tissue penetration, so that the amount of incident excitation light that reaches the fluorescent probe is reduced along with the endogenous fluorophore excitation. Bioluminescence resonance energy transfer (BRET), which converts chemical energy into photon emission, has been used to circumvent the issues associated with an external excitation source, by creating self-illuminating QDs. In the study performed by Rao and co-workers, multiple molecules of Luc8, a variant of *Rotylenchulus reniformis* luciferase, were covalently conjugated to a single QD. Oxidation of coelenterazine, the substrate of luciferase, via its binding results in blue light emission, which could subsequently excite the conjugated nanoparticle due to the complete overlap between Luc8 emission and QD absorption spectra. In vivo imaging of administered bioluminescent QD conjugates emitting long wavelengths demonstrated the feasibility of multiplex molecular imaging in small animals with the highly enhanced signal-to-background ratio, and no background in nonsuperficial locations.[59,144,145] One of the important issues that should be addressed before the widespread utilization of bioluminescent QD technology for in vivo imaging is targeting specificity, considering the fact that a majority of QD surface areas are covered by luciferase molecules, and so there should be plenty of space for targeting biomolecules.[24]

2.13 Bimodal Molecular Imaging Probes

Dual- or multimodality molecular imaging is the synergistic combination of two or more detection techniques in order to compensate for the inadequacy of each imaging modality alone. A major advantage is that it can provide complementary information with more accuracy and reliability.[146] For instance, the high sensitivity of optical fluorescence imaging permits the obtaining of detailed information at subcellular levels, although it has limited deep tissue penetration, and lack of quantification, anatomic resolution, and spatial information. MRI, on the other hand, is more suited for following the in vivo distribution of molecules and providing anatomical reference points. However, it has impartially low contrast agent sensitivity.[26] Accordingly, there is considerable interest in developing dual-modality contrast agents for combined optical and MR imaging, due to their exceptional tissue contrast and spatial resolution, as well as their probable potential in clinical utility.[147] As dual-mode tumor-targeted agents, MRI could be employed for lesion localization, while optical methods could guide tumor border identification during subsequent surgical resection. Another application would be targeting atherosclerotic plaques in which MRI could be utilized for in vivo plaque identification, whereas optical emission would be used to guide endoscopic interventions at identified locations. Such bimodal imaging techniques hold great promise for improving diagnosis and therapy of several diseases.[148]

A series of core-shell CdSe-Zn1-xMnxS nanoparticles with high quantum yield and relaxivity have been synthesized for use in dual-mode optical and MRI techniques. The amount of Mn^{2+} that can be doped into the shell varies with the thickness of the shell or the amount of Mn^{2+} introduced to the reaction. The quantum yield and Mn^{2+} concentration in these nanoparticles were adequate to produce contrast for both modalities at a relatively low concentration.[148] Bifunctional silica-coated magnetic QDs consisting of Fe_2O_3 and CdSe have been synthesized with superparamagnetism and an efficient quantum yield over a broad range of colors without the need for ZnS capping. Different cellular membranes have been targeted by these synthesized nanocomposites for biolabeling and imaging of living cells.[149] MRI detectable and targeted QDs have been developed by paramagnetically coated CdSe-ZnS and conjugated with cyclic RGD peptides to target cultured human endothelial cells. The bimodal character, high relaxivity, and specificity of these nanoparticles make them desired contrast agents for the detection of tumor angiogenesis.[150] For exploring the adequacy of dual-modality nanoparticles to be used for in vivo medical imaging, paramagnetic gadolinium chelates have been attached to polymer-coated QDs. Preliminary cellular and in vivo animal studies demonstrated the promising potential of these heterostructures for in vivo tumor imaging in animal models.[151] Bimodal AnxA5-QD-Gd-wedge nanoparticles have been used to target phosphatidylserine on the surface of apoptotic cells in order to assess anticancer therapy efficacy.[152] In vitro experiments demonstrate the competency of these nanoparticles for both anatomic and subcellular imaging of structures in the vessel wall by the use of MRI and photon laser scanning microscopy.[153] Polymer-coated Fe_2O_3 cores overcoated by a CdSe-ZnS QD shell and functionalized with tumor-specific antibodies have been utilized for magnetic separation of breast cancer cells that were detected by fluorescence imaging.[154] Heterodimer nanoparticles composed of CdS-FePt have been synthesized by taking advantage of lattice mismatching and selective annealing at relatively low temperatures.[155] Spherical-shaped nanocomposites with an iron oxide Fe_3O_4 magnetic core and CdSe-ZnS quantum dot shell have been synthesized by Zhang et al.[170] that have good magnetic response, photostability, chemical activity, and water miscibility. The multifunctional nanoparticles are well suited for wide-ranging applications, such as drug and gene delivery, bioseparation, and biological detection. The area of multimodality molecular imaging is moving from a research curiosity to one having preclinical and clinical applications.

2.14 Cytotoxicity

With the wide-ranging use of QDs for bioimaging and in vivo animal imaging there have been increasing concerns over their short- and long-term toxicity,[156] which could be one of the important restricting factors for their use as potential bioprobes in future clinical applications.[27] For the purpose of introducing this technology in a clinical setting, numerous in vitro and in vivo animal-based assays have been focused on preclinical testing. However, presently, it is difficult to make a firm conclusion about the toxic potential of these nanoparticles, due to the contradictory results found in current literature,[157] which could be attributed to the lack of any standardized synthesis protocol, surface coating, solubilization ligands, dose, exposure time, and finally, the studied cell system.[158] Several factors are generally involved in QD cytotoxicity, as with other nanoparticles, that are associated with inherent particles' physiochemistry as well as the external environmental conditions. The toxicity induced by physiochemical properties of a particular QD originates from its overall size, electrical charge, concentration, surface coating, and oxidative, photolytic, and mechanical stability, all of which should be considered individually for toxicological assessment.[24]

There are several studies that confirmed the relative dependency between particle size and the resulting toxicity at both the intracellular and animal levels. The initial study presented by Lovric et al. demonstrated that 2.2 nm CdTe QDs, compared to the 5.2 nm ones, led to higher cellular death, which could be related to the different subcellular distribution, as larger nanoparticles have been found distributed inside the cytoplasm, with none entering the nucleus, while smaller particles have been mainly localized within the nuclear compartments. The interaction between QDs and nuclear DNA or proteins has been found to induce different types of genotoxicity.[157,159] In another study by Kirchner and co-workers different cell lines, including NRK fibroblasts, MDA-MB-435S breast cancer cells, CHO cells, and RBL cells, were exposed to the same concentration of CdSe-ZnS QDs. It was suggested that the higher surface-to-volume ratio in smaller particles is responsible for their greater cytotoxic effects.[160]

The surface charge is also a key factor for particle dispersion and the behavior of particles in living organisms. For evaluating the toxicity related to surface charge in hepatocytes, Hoshino et al.[161] coated CdSe-ZnS QDs with mercaptoundecanoic acid (MUA) (QD-COOH), cysteamine (QD-NH2), and thioglycerol (QD-OH), as well as their combinations.[162] The results revealed that the toxicity depends on the type of ligand used for surface modification, but not on the core material; the highly negative charge in MUA-coated QDs induced DNA damage after 2 h, while thioglycerol with a lower negative charge had the least genotoxicity and therefore cellular damage. In another study, neutral polyethylene glycol (PEG)-coated QDs were found to be less toxic in human epidermal keratinocytes (HEKs),in an established in vitro model for epidermal toxicity, when compared to cationic PEG-amine or anionic carboxylic acid-coated particles.[163]

QD concentration is another parameter that requires more consideration. In a study by Hoshino et al.[164] EL-4 cells were incubated with different concentrations of CdSe-ZnS-SSA

QDs. It was shown that the cellular viability decreased at higher QD concentrations, in which most of the cells died after 6 h. A high rate of cellular death has also been observed in adipose tissue-derived stem cells transfected with more than 2.0 nM QD655.[119] Injection of 5×10^9 QDs/cell into xenopus blastomeres resulted in various abnormalities in embryos, such as changes in cell size, cell movement, axis elongation, and posterior truncations, while embryos injected with 2×10^9 QDs/cell were statically similar to uninjected embryos.[74] High levels of malformations and embryonic death have also been found in zebra fish embryos injected with excessive QDs.[110] Live embryos are ideal models for investigating toxic effects due to the extreme sensitivity to even slight perturbations, which generally results in phenotype abnormalities.[83] Dose-dependent cytotoxic effects suggest that for continuous in vivo monitoring of transplanted cells, QD exposure should be limited to minimally low concentrations.[165]

The consensus of many studies for in vitro and in vivo toxicity is that encapsulation of QDs within a stable shell has been shown to predominantly reduce toxicity by providing a physical barrier against degradation or oxidation of the nanoparticles. In human breast cancer MCF-7 cells, bare CdTe results in cellular alteration, while coating with different capping materials, such as mercaptopropionic acid, cysteamine, and n-acetylcysteine, decreased the relative toxicity for the time periods studied.[166,167] However, the toxicity of the capping materials should also be considered, since several groups have found increased cytotoxicity associated with capping materials such as mercaptoacetic acid (MAA).[168]

The stability of QDs under certain conditions is anticipated as the most important aspect of their cell-induced toxicity. Despite many modifications for increasing nanoparticle stability, several studies suggested that under oxidative and photolytic conditions, QDs are labile and thus are subjected to degradation.[27] In addition, the long nanoparticle persistence inside living animals, which results in their subsequent degradation over time, also requires more consideration for improving their pharmacokinetics and accelerating their biotransformation and removal. The large hydrodynamic size of QDs, small pore sizes in the mammalian vasculature, and the glomerular barrier are the limiting factors in the nanoparticles' body clearance. By choosing appropriate surface coatings and optimal hydrodynamic diameters,[169,170] we can effectively clear most of the QDs from the organism.[76]

At least four possible mechanisms have been assumed to be responsible for the toxicity induced by QDs in living cells: (1) liberation of toxic components of the core,[160,171,172] (2) generation of free radicals, particularly reactive oxygen species (ROS),[159,162,173] (3) interaction of QDs with intracellular components,[27] and (4) nanoparticle aggregation and precipitation on the cellular surface.[160] The exact recognition of the specific mechanisms involved in QD-induced cellular toxicity is highly crucial before any clinical translation.

Regarding the high cytotoxic potential of divalent cations, their liberation from the nanoparticle's core following photodegradation, air oxidation, or intracellular decomposition is of serious concern. In order to evaluate the toxicity induced by free Cd^{2+}, rat primary hepatocytes used as a liver model were cultured in bare CdSe QD solution and subsequently exposed to ultraviolet light (UV) or air. It was suggested that the nanoparticle surface oxidation, either air induced or UV catalyzed, was responsible for free cadmium level elevation,[61,174] and a direct correlation existed between Cd^{2+} concentration and the extent of hepato-toxicity. However, it was only assumed that the underlying mechanism of cytotoxicity was the inactivation of essential mitochondrial proteins through interaction between free cadmium and the sulfhydryl group[175]; thiol group inactivation then resulted in oxidative stress and mitochondrial dysfunction. Metallothionein, a protein found in the cytoplasm of hepatocytes, detoxifies cadmium by sequestering it into an inert complex.

Yet, small amounts of this protein, normally present in animals, are not sufficient when cells are exposed to high levels of Cd^{2+} ions.[171]

Protons in low-pH environments, such as lysosomes and peroxisomes, on the other hand, can lead to detachment of coordination groups from the QD surface, releasing cationic ions, resulting in subsequent cellular toxicity.[176,177] In addition, it is known that cellular toxicity of cadmium is, in part, related to alteration of intracellular calcium homeostasis, which can competitively reduce extracellular calcium influx or increase intracellular calcium levels by inhibiting calcium-dependent ATPase.[178,179]

Free radical formation is considered an alternative mechanism involved in QD cytotoxicity.[162,173] During the excited state, QDs are able to transfer electrons or energy to nearby oxygen molecules, thus generating a reactive oxygen species (ROS), such as hydroxyl radical, super oxide, or singlet oxygen, depending on their composition.[180] ROS elevation could induce oxidative stress by releasing the endogenous antioxidants, and thus their subsequent depletion.[181] Oxidative stress in turn results in irreversible cellular damage or death by peroxidizing lipids, altering proteins, disrupting DNA, interfering with signaling functions, and modulating gene transcription.[182,183] However, sensitivity of cells to ROS may depend on the cell type, the level and duration of oxidant production, the species of ROS generated, and the specific site of ROS production. Furthermore, oxidative stress in living organisms may have a role in the induction or enhancement of inflammation through upregulation of redox-sensitive transcription factors and the related kinases.[184,185]

Considering the nanoscale size of QDs, which is smaller than cellular organelles, they are able to penetrate into the cells and interact with intracellular components, resulting in their structural alterations and dysfunction. QD aggregates were found localized around the mitochondria, and after 72 h morphological effects included swollen mitochondria and enlarged Golgi cisterns.[157] Kirchner et al. have proposed that in addition to the release of Cd^{2+} ions from semiconductor QDs, their precipitation behavior also contributes to cytotoxicity after studying the effect of QDs on MDA-MB-435S breast cancer cells.[160]

2.15 Magnetic Resonance Imaging

Magnetic resonance imaging uses hydrogen nuclei signals for image generation. Recall from the periodic table that a hydrogen atom consists of a nucleus containing a single proton with positive charge and a single electron with negative charge. The proton rotates about its axis like a spinning top, and so has an angular momentum that strives to retain the spatial orientation of its rotation axis, and magnetic moment as a rotating mass with an electrical charge. The orientation of the proton rotation axis can be identified by magnetization vector B and can be measured by signal generation in a receiver coil.

Electromagnetic waves and an external magnetic field can alter the orientation of the hydrogen nuclei rotational axis to align the magnetic moments with the direction of the external magnetic field, B0, a process that is called precession. However, in the quantum field proton alignment changes are due to an external force and have some limitation. Like electrons, the protons can only spin in specific orbits, so they cannot align exactly in external field direction, and there is an angle width between the nuclei magnetic momentum and the external magnetic field. Therefore, after this enforcement, protons align in different phases or energy levels: one low energy and one high energy, which are separated by a very small splitting energy. Thus, phase refers to the position of a magnetic moment on its circular precessional path and is expressed as an angle. However, the sums of the magnetic vectors are in longitudinal magnetization (z axis). Radiofrequency (RF) pulse synchronizes the phases and rotates all the longitudinal magnetization into the transverse plane. To make it easy, imagine that the external magnetic field aligns the protons in the z axis (longitudinal magnetization) and RF puts them in the xy plane (transverse magnetization).

After excitation, the xy plane of the magnetic moments gradually realigns with the z axis of the external magnetic field B0. This process is known as longitudinal relaxation recovery and occurs exponentially with a time constant T1. T1 values depend on the strength of the external magnetic field and the internal motion of the molecules (Brownian motion). Biological tissues have T1 values of half a second to several seconds at 1.5 T.

Immediately after the RF pulse, the spins are in the same phase. This state is called phase coherence. This phase coherence is gradually lost as some spins are in a high level while others fall to a low level. So the transverse magnetization, which comes from the vector sums, becomes smaller and smaller and finally disappears because spins lose coherence (dephasing). Dephasing, because of energy transfer between spins, occurs with the time constant T2 and is more or less independent of the strength of the external magnetic field, B0. T1 is thus associated with the enthalpy of the spin system (the number of nuclei with parallel vs. antiparallel spin), while T2 is associated with its entropy (the number of nuclei in phase). However, intrinsic inhomogeneties exist due to the magnetic field generator and the person who is being imaged and contribute to dephasing even faster than T2. This second type of decay occurs with the time constant T2*. The T2* effect occurs at tissue borders, particularly at the air-tissue interface, or is induced by the local magnetic field (e.g., iron particles). Depending on the tissue and the external magnetic field, T1, T2, and T2* values are different, but T1 and T2 relaxation are completely independent of each other, although they occur more or less simultaneously. MR signal decay due to T2 relaxation occurs within the first 100 to 300 ms, which is long before there has been T1 relaxation at 0.5 to 5 s.

MRI is a well-established clinical tool for noninvasive imaging of anatomical changes of the whole body. It offers several advantages over other imaging techniques, including greater speed, more direct anatomical correlation, and lower cost.[186] Although not able to match the resolution of photon base imaging, a number of strategies have been employed to improve its resolution and sensitivity by using contrast agents or increasing the magnitude of induced nuclear magnetism. Due to limited resolution, the presence of a cluster of cells ranging from 1×10^5 to 5×10^5 is required to be imaged by MRI techniques.[187,188]

In addition to proton density, which determines the maximum signal obtained from a given tissue, two other intrinsic features of a biological tissue contribute to its signal intensity or brightness on MR images: T1 and T2 times, which alters with contrast agents. Proton density and T1 and T2 times may vary widely from ones, tissue to the next one therefore, depending on which of these parameters is focused, the results of MRI are different in a tissue. As explained later, the T1 time of a tissue is the time that the excited spin takes to recover and is available for the next excitation, and T2 is the time referred to how quickly an MR signal fades after excitation. Images with contrasts that are mainly determined by T1 or T2 are called T1-weighted images (T1w) and T2-weighted images (T2w), respectively.

Manipulating the chemical environment around the protons can alter the signals. MR contrast agents have been developed as a way to manipulate the inside an organism, and they can significantly improve the MRI resolution in cellular imaging. Most cells, including stem cells, do not have endogenous magnetic resonance contrast, and for visualizing, they need to be labeled by either endogenous or exogenous contrast agents. In the external magnetic field, contrast agents generate an induced magnetic field, which perturbs the magnetic relaxation processes of the protons of the hydrogen nucleus surrounding the magnetic contrast agents. In fact, using the contrast agent shortens the relaxation time and results in a consequent darkening of the MR images (Figure 2.2). Contrast strength is quantitatively represented as the relaxivity: r1 for the longitudinal rate (1/T1) or r2 for the transverse rate (1/T2). T1 agents usually have r2/r1 ratios of 1–2, whereas the value for T2 agents is as high as 10 or more.[189]

The small but positive magnetic natures of paramagnetic materials make them ready to change their alignment in the external magnetic field to act as atomic magnetic dipoles. Their interaction with water protons through dipole-dipole interaction shortens the water MR relaxation times, which results in an increase of relaxation rates (1/T1 and 1/T2). Paramagnetic materials such as Mn^{2+} or Gd^{3+} commonly act like positive (bright) contrast agents. They reduce both T1 and T2 times, but since T1 is usually 1–2 orders of magnitude greater than T2, signal increase due to T1 shortening is more significant on T1-weighted images.

Superparamagnetic particles as negative (dark) agents are a different class of MRI contrast agents that are extremely strong in enhancing proton relaxation. Superparamagnetism is primarily associated with very small crystalline iron oxides (1–14 nm).[190] The single-crystal nature of superparamagnetic iron oxides (SPIOs) leads them to form a single magnetic domain after applying the external magnetic field. They shorten T2 and T2* times more than T1 and generate a negative (dark) effect on MRI.[191,192]

Gd chelates (Gd-DTPA, Gd-DOTA, and Gd-DO3A) and Mn chloride are paramagnetic contrast agents used in experimental and clinical studies. Gadopentate dimeglumine was the first safe MRI contrast agent approved by the U.S. Food and Drug Administration, and is widely used to enhance pathology in the brain, heart, liver, tumors, and other tissues imaged on MRI. Intravenous administration of Gd-DTPA, Gd-DOTA, or Gd-DO3A has relatively minimal side effects, and more than 99% of the dose is rapidly cleared and excreted along the renal pathway.[193] Crich and colleagues used Gd-HPDO3A for tracking endothelial progenitor cells (EPCs). They injected 10^6 labeled cells under the mouse kidney capsule and studied the cell locations on T1-weighted MRI.[194] They estimated that the number of Gd chelates needed to visualize cells is approximately 10 to 100 µM per cell, which could have quite a toxic effect. However, it seems that Gd chelates are not safe for cell labeling, because of the unknown clearance mechanism, long-term retention of GD chelates inside the cell, and production of Gd^{3+} ions through rapid dechelation due to low pH in the lysosomes and endosomes of cells.[195] Recently, Gd fullerenol has been used for tracking mesenchymal stem cells at 7T MRI following direct injection of 10^6 cells into the rat thigh.[196] Anderson et al. showed that Gd fullerenol initially decreased the mesenchymal stem cell's ability to proliferate by altering the mitochondrial function.

Mn chloride is another paramagnetic contrast agent that is employed for in vivo imaging. It is taken up by cells through calcium channels in the cell membrane.[197,198] Lymphocytes can be successfully labeled with Mn chloride and have been studied on T1-weighted MRI. However, they lack sufficient contrast enhancement to be detected as labeled cells in vivo.[199] The major limitation of using Mn as an MRI contrast agent is its significant cardiotoxicity.[191]

Uniform and high magnetic moment, high stability under physiological conditions (e.g., high salt concentration and pH changes), ability to escape from the reticuloendothelial system, low toxicity, high biocompatibility, and linking ability to biologically active material (e.g., nucleic acid, proteins) are principal properties of suitable MR contrast agents. Magnetic nanoparticles consist of a core of superparamagnetic iron oxide and polymer coating, to have high magnetic moment and functionality to be linked to biologically active species. Moreover, their provision of the maximum change in signal (albeit hypointensity) per unit of metal, in particular on T2*-weighted images, their easy detection by light and electron microscopy, their changing of magnetic properties according to their size, and their biocompatibility due to their iron-made structure make SPIO nanoparticles a preferred contrast agent in MRI. Ferumoxides (Feridex®, Bayer Healthcare) were approved by the U.S. Food and Drug Administration in 1996 for clinical cell tracking.

2.16 Superparamagnetic Nanoparticles

Superparamagnetic (SPIO) nanoparticles are being used as MRI contrast agents. They can dephase neighboring protons in a region equivalent to many times the size of metal itself, in particular on T2*-weighted images, which is caused by a crystal ordering (spinels). SPIOs consist of two components, a superparamagnetic magnetite (Fe_3O_4) or maghemite (γFe_2O_3) core, which is biocompatible and recycled in normal iron metabolism pathways, and a hydrophilic coating, which allows the chemical linkage of functional groups and ligands. SPIO crystals, which are smaller than a magnetic domain (approximately 30 nm), have random orientation in the absence of an external magnetic field. However, external magnetic fields can reorient the crystal magnetic dipoles to induce a net magnetic field, which is larger than the sum total of individual unpaired magnetic dipoles.

Jun et al. demonstrated that magnetic moments of nanoparticles are size dependent.[200] By controlling the synthesis conditions, they prepared magnetic Fe_3O_4 nanoparticles in different sizes through thermal decomposition of Fe(acac)3 (acac = acetylacetonate) in a hot organic solvent. The T2-weighted spin-echo MR images at 1.5 T changed from white to black via gray colors as the dimension of magnetic nanoparticles increased from 4 to 12 nm. Such an effect makes them suitable to identify a variety of biological events. Nanoparticles larger than 30 nm are taken up by phagocytes (dark cells), so the tumor cells that cannot phagocyte these particles (white cells) can be distinguished from other cells by their white color.[201,202] Smaller magnetic nanoparticles can be conjugated with target-specific biomolecules to detect target cells via interaction between conjugated particles and expressed markers on target cells.[203]

Numerous chemical methods have been used for magnetic nanoparticle synthesis. The solution method is the simplest and most efficient technique to produce magnetic nanoparticles with a monodisperse population of suitable size. Iron oxides are prepared by the injection of precursors (e.g., ferrous and ferric salts) into an aqueous medium containing surfactants at high temperature, followed by aging and a size control process. Use of neutral organometallic precursors in coordinating alkyl solutions with a high boiling point in the hot-injection solvothermal process represents a promising method, compared with conventional solution techniques in which ionic precursors in water or another polar solvent are used.[204]

Although the conventional synthesis techniques control the mean size of nanoparticles by pH adjustment, ionic strength, temperature, or the Fe(II)/Fe(III) ratio, their novel electrical, chemical, magnetic, and optical properties need extensive efforts to develop uniform nanometer-sized particles for scientific and biological studies. Recently, a microemulsion synthesis method has been described to provide water-soluble nanoparticles with precise control of size disparity.[205] In this technique, magnetic nanoparticles with a diameter of 4 nm were prepared by controlled hydrolysis of an aqueous solution of ferrous and ferric chloride with ammonium hydroxide in a microemulsion formed by sodium bis(2-ethyl-hexyl) sulfosuccinate as the surfactant and heptanes as the continuous phase.

In general, iron oxide nanoparticles require coating to prevent aggregation and to make the nanoparticles soluble in aqueous solution or amenable to further biochemical manipulation. A very high density of coating is often needed to effectively stabilize iron oxide nanoparticles, and many polymeric coating materials, such as dextran, carboxymethylated dextran, carboxydextran, starch, polyethylene glycol (PEG), arabinogalactan, organic siloxane, and sulfonated styrene-divinylbenzene, have been used.[206] The type of specific coating depends on the end application and is chosen depending upon its application.

Dextran is the most commonly used material for iron oxide nanoparticle coating. It represents high colloidal stability under harsh physiological conditions, and hence it is a suitable platform for further biochemical manipulation. Dextran-coated iron oxide nanoparticles are prepared by coprecipitation from aqueous solution containing a ferrous salt and dextran by adding an alkaline solution.[207] Feridex (Endorem) (5–6 nm core and 80–150 nm total size range), Resovist® (Schering) (4.2 nm core and 62 nm total size), the ultra-small superparamagnetic iron oxide (USPIO) Combidex® (Sinerem) (4–6 nm core and 20–40 total size range), monocrystalline iron oxide nanoparticle (MION), and its derivative, cross-linked iron oxide (CLIO) (2.8 nm core and 10–30 total size range) are currently available dextran-coated iron oxide nanoparticles.[192] Long blood half-life and relatively small size (Table 2.1) make the dextran-coated SPIO highly valuable for in vivo cellular and molecular MRI.

Apoferritin protein coating is another approach to prevent destabilization and agglomeration of iron oxide nanoparticles. Ferritin, a well-known iron storage protein, is composed of a hydrated iron oxide core and a polypeptide apoferritin shell. In the magnetoferritin

TABLE 2.1

Available Coated Iron Oxide Contrast Agents

Agent	Iron Oxide Core Size (nm)	Total Size (nm)	Coating Material	Magnetization at 1.5 T (emu g^{-1})	Half-Life in Blood
Ferumoxides (AM125)	5	120–150	Dextran	94	~2 h
Ferumoxtran-10 (AM1227)	6	15–30	Dextran	95	>24 h
Ferucarbotran A (SHU555A)	4	65	Carboxydextran	NA	3 min
Ferumoxytol (C7228)	~7	35	Carboxymethyl dextran	NA	10–14 h
Lumirem (AMI121)	10	300	Silica	NA	<5 min
Feruglose (NC100150)	6	15–20	Methylcellulose	86	6 h
VSOP (C184)	4	~9	Citrate	NA	30–60 min
Magnetoferritin	~6		Ferritin	NA	~10 min
Magnetodendrimer	7–8	20–30	Carboxy-terminated dendrimer	~94	
Magnetoliposome	~16	~40	Liposome	NA	
MION, CLIO	~3	10–30	Dextran	68	>10 h

nanoparticle, the native hydrated iron oxide is removed by acidic dialysis, and then apo-ferritin is reconstituted by Fe(II) under slow oxidative conditions at 600°C and pH 8.5.[208] Magnetoferritin particles are cleared rapidly from the blood by the liver, spleen, and lymph nodes.[209]

Oxidation of Fe(II) at slightly elevated pH and temperature forms a highly soluble nano-composite of iron oxides and dendrimers and was described by Bulte et al. as a class of magnetic contrast agent designed to achieve a high degree of nonspecific intracellular magnetic labeling for any mammalian cell, including human stem cells.[210] These aggregates have an overall size of 20–30 nm, and labeled oligodendroglial progenitors can be readily detected in vivo as long as 6 weeks after transplantation. Similarly, liposomes, the drug and gene delivery agents, were used as a coating material to solubilize iron oxide nanoparticles by Bulte and co-workers.[211] Liposomes have the same bilayer structure as cell membrane and are made up of surfactant molecules with a hydrophilic head and a hydrophobic tail that can be filled with iron oxide. The magnetoliposome core size and total size are 16 and 40 nm, respectively, and were used for successfully tracking bone marrow stem cells.[211]

In contrast to conventional iron oxide nanoparticles that are taken up by the reticulum endoplasmic system (RES) rapidly, particles that are made via a nonhydrolytic, high-temperature growth method are able to escape from RES, and possess an active functional site to link with biological active molecules.[192] However, as nonhydrolytically synthesized molecules, they are soluble only in organic media and need surface modification, including a bifunctional ligand,[200,212] micellular,[213,214] polymer,[215] and siloxane linking procedure,[216,217] to be water-soluble nanoparticles. Nitin and co-workers developed functionalized superparamagnetic iron oxide nanoparticles with a polyethylene glycol (PEG)-modified coating, leading to high water solubility and stability, and allowing for bioconjugation of various moieties, including fluorescent dyes and the Tat peptide.[214] The PEGylated nanoparticles were delivered into living cells for further study by using MRI.

Polymeric phosphine oxide binds tightly to iron oxide nanoparticle surfaces through multidendate bondings. Kim and co-workers brought a variety of nanoparticles into water using easily synthesized polymers consisting of phosphine oxide and PEG.[218] Since high iron concentrations intensify high T1-weighted and low T2-weighted signals, in spite of conventional iron oxide nanoparticles, the phosphine-stabilized γFe_2O_3 particles are potent T1 and T2 relaxation agents. Similarly, 2,3-dimercaptosuccinic acid (DMSA) ligands have been used to produce water-soluble iron oxide (WSIO) nanoparticles, which are able to bind to target-specific biomolecules through the remaining free thiol group of the ligand.[200] These nanoparticles, upon conjugation with Herceptin, known as a cancer-targeting antibody, have been used for in vitro and in vivo cancer cell imaging.[200,212]

The particle entrance into a cell or a particle's link with other biological molecules is essential for cell labeling. Superparamagnetic nanoparticles' internalization as nanoparticles depends on specific transmembrane mechanisms, which we have discussed before for quantum dots. Phagocytosis, receptor-mediated endocytosis, and pinocytosis are the main mechanisms for supernanoparticle cell entrance. However, the coating design can control the internalization rate and provide the optimum timing for MRI contrast agents.[219] Specific surface receptors are used to facilitate labeling of the cells without phagocyte capacity. Monoclonal antibodies for specific proteins or receptors, nanoparticle surface modifiers such as magnetodendrimers, and surface charge modifiers such as anionic maghemite nanoparticles are the most used protocols for receptor-mediated labeling.

Although transfection agents, which are highly charged macromolecules to introduce DNA into cells, are toxic to cells, poly-L-ornithine,[220] poly-L-lysine,[221] and HIV-1 Tat peptide[222] have been successfully used to enhance the cell labeling capacity of SPIO nanoparticles. Arbab and co-workers have shown long-term (28-day) viability for mesenchymal and hematopoietic stem cells that were transfected with Feridex as the SPIO and protamine sulfate, a cationic transfection agent.[223] Zhang et al. showed that the transfection agent Lipofectamine™ 2000 with SPIO Endorm® could label pig skeletal myoblasts nearly 100%, whereas the mean individual cell labeling without transfection was 70%.[224]

2.17 SPIO Applications in Cell Tracking

Noninvasive, high-resolution MRI is poised to become increasingly used in the characterization and evaluation of microstructure and function in tissue engineering and regenerative medicine. The spatial MRI resolution, near-cellular resolution (30 mm), and use of iron oxide contrast agents have expanded MRI's ability to image specific cells at the single-cell level. This ability in single-cell imaging is proposed for visualization of cell seeding in biological or synthetic 3D scaffolds and has great potential for screening different cell types in a 3D scale, whereas light microscopy can only permit 2D or limited 3D imaging. Contrast agents are another strategy to enhance resolution. SPIO nanoparticles as contrast agents are special for cell labeling, as they can be loaded on cells within 1 h.[225] Their intracellular concentration (5–10 µmol/L) is sufficient to generate detectable signals on T2-weighted images.[226]

Poirier-Quinot et al. showed that high-resolution MRI on a clinical 1.5 T device is a reliable technique to assess magnetically labeled mesenchymal cell seeding on 3D porous scaffolds before and after in vivo implantation.[227] They detected mesenchymal stem cells with different iron loads (5, 12, and 31 pg of iron per cell) after seeding on scaffolds, using gradient-echo MR images according to phase distortions and areas of intensely low signal. In vivo monitoring of the implanted scaffolds, tracking of the seeding cells, and the behavior of tissue surrounding the implant are other features of MRI. Ko et al. used MRI to monitor the cellular behavior around the implant in gel/sponge scaffolds that were seeded with magnetically labeled mesenchymal cells and correlated their results to histological data.[228] Moreover, MRI can be used to monitor the cell proliferation in other 3D culturing equipment, such as hollow fibers, where visualizing the cells is the main limitation.[229]

Endothelial progenitor cells can migrate to the ischemic site under angiogenic site tropism, and studying the migration mechanism in in vitro models can offer a promising therapeutic alternative to the angiogenic growth factors. High-resolution MRI has been used successfully to follow the formation of vascular type networks in Matrigel.[230] The main limitation of MRI in single-cell detection is that the current high-resolution systems, such as high-temperature superconducting cryo-detection coils, permit high-resolution imaging in a relatively small volume, close to the coil surface (<15 mm) and cannot be used in a bigger area.

Numerous cell types have been investigated using superparamagnetic nanoparticles in experimental disease models, including central nervous system diseases. Endorem-labeled human mesenchymal stem cells were grafted either intracerebrally or intravenously in experimental models of stroke.[231] Using a 4.7 T Bruker spectrometer, 1 week after cell transplantation, a hypointense signal was found in the lesion. The signals were intensified during the second and third weeks after grafting, regardless of the route of administration, and their intensity corresponded to the Prussian blue staining, anti-BrdU staining, or GFP labeling. The myelin-deficient Long Evans Shaker rats can be recovered by oligodendrocyte progenitor cell transplantation. Bulte et al. have shown the in vivo migration and remyelination capacity of SPIO-labeled oligodendrocyte progenitor cells in the spinal cord[232] and brain.[210] MRI has also been used to monitor olfactory ensheathing and Schwann cells in spinal cord injury[233] and in demyelinated spinal cord lesions[234] by direct injection.

Neural precursor cells showed beneficial effects in experimental autoimmune encephalomyelitis (EAE), an experimental model relevant to human multiple sclerosis.[235] The migration of transplanted PLL-Feridex-labeled neural precursor cells, which were demonstrated by serial in vivo MRI, can ameliorate the brain inflammation in a mouse model of EAE. The extent of migration correlated well with the clinical severity of the disease. Stem cells including bone marrow-derived endothelial precursor cells can be followed as delivery vehicles by MRI. Transplantation of SPIO-labeled cells in glioma-bearing severely combined immunodeficient mice induced hypointense regions within the tumor area.[236] Histological studies demonstrated expression of CD31 and von Willebrand factors on labeled cells that had infiltrated the tumor mass.

Scar formation due to ischemia-induced death of cardiac myocytes can be recovered by infusion or injection of stem or progenitor cells. Furthermore, endothelial progenitor cells have been shown to augment blood flow. Despite promising results from primary clinical trials, many practical issues and mechanisms remain highly controversial. Visualization and quantification of injected cells within the injured myocardial tissue obscure understanding of mechanisms relating to stem cell transplantation. Kraitchman et al. studied the MRI potential of tracking magnetically labeled bone marrow-derived mesenchymal stem cells in a swine model of myocardial infarction. Intramyocardial injection of 2.8 to 16×10^7 cells was tracked successfully using a 1.5 T MR scanner for up to 8 weeks after cell transplantation.[237] An extremely high heart rate and small tissue are major challenges in heart MRI in small animals. However, Cahill et al. have used ferumoxide:polycation complexes for the monitoring of therapeutic muscle-derived stem cell transplants in murine dystrophic muscle.[238] Endosomal accumulation of superparamagnetic iron oxide resulted in changes in the MR contrast T2 and T2* at the left anterior wall of the animals receiving labeled transplants in correlation with histological images obtained by conventional microscopy. Furthermore, magnetically labeled adipose stem cells,[239] mesenchymal stem cells,[240] human amniotic fluid stem cells,[241] and embryonic stem cells have been tracked successfully in heart disease models in small animals by MR techniques.

The ability to track hematopoietic stem cells' migration would provide new insight into the mechanisms of cellular engraftment after hematopoietic stem cell transplantation. Akbari et al. have used two commercially available FDA-approved SPIO nanoparticles, ferumoxide and protamine sulfate, for in vivo homing of human CD34+ cells in irradiated mice. They found that the intravenous injection of 0.5–1.5×10^7 labeled CD34+ cells can decrease MR signal intensity in bone marrow at 24 and 48 h after injection.[242] The same results have been reported by Daldrup-Link et al. for cord blood-derived hematopoietic stem cells using ferumoxides.[243]

Unlike quantum dots, iron is naturally present in the human body in the amount of almost 4 g in an average adult. Ferumoxides, the only pharmaceutical-grade MRI contrast agents, are completely biodegraded in approximately 2–4 weeks. The iron core of ferumoxide is recycled into the normal blood poll, including hemoglobin.[244] Thus, the approved contrast agents can be used in clinical cellular therapy trials. However, the translation from experimental studies to clinical applications requires much better preparation and perseverance to be evaluable in phase I clinical trials.

Initially SPIO nanoparticles Feridex-USA (Endorem-Europe) and Combidex-USA (Sinerem-Europe), which are now FDA approved, were used for cell detection in the reticuloendothelilal system, including liver[245] and lymph nodes.[246] However, in April 2004, de Vries and co-workers for the first time used clinical-grade SPIO (Endorm) for tracking the autologous dendritic cells of a cancer vaccine in melanoma patients.[247] Using MRI at 3 and 7 T, they showed that tracking of magnetically labeled dendritic cells is feasible in humans for detecting very low numbers of cells in conjunction with detailed patterns of inter- and intranodal cell migration. In the same year, Zhu and co-workers in Huashan Hospital, Shanghai, China, used Feridex and Effectene (a nonclinical-grade lipofection reagent) for tracking autologous neural stem cells in a 34-year-old man with brain trauma.[248] They used a gradient-echo MRI at 3 T and showed that neural stem cells that were implanted stereotactically around the region of brain damage moved from the injection site to the border zone of the lesion. The signal intensity disappeared after 7 weeks, possibly due to cell proliferation. In another study, Toso et al. labeled human cadaveric islet cells with ferucarbotran (a supermagnetic iron oxide agent) and transplanted them intraportally into four patients with type 1 diabetes.[249] The magnetically labeled islet cells induced hypointense spots within the liver and continued to do so for 6 months after transplantation. The study confirmed that SPIO nanoparticles are safe as cell labeling agents and have no harmful effects on islet cell function.

2.18 Limitations

The false positive effect is the main limitation for SPIO nanoparticles as contrast agents in cellular monitoring after transplantation because iron aggregation after bleeding in the injection site, or even in the target organ, is susceptible to artifact production in SPIO-induced magnetic resonance images. Contrast transfer from the dead cells to nontarget cells, such as macrophages, is another source of the false positive effect during imaging acquisition. Although SPIO nanoparticles did not affect the viability of magnetically labeled cells, Bulte et al. showed that Feridex loading can block mesenchymal differentiation to chondrocytes.[250] Even though other studies do not prove these data, the clinical safety of SPIO nanoparticles when they are used for stem cell research remains controversial.

Finally, there are some limitations related to MRI itself. Using MRI in patients with pacemakers or implantable cardioverters/defibrillators is not possible. Patients with these devices may need stem cell therapy more than others, and in this case we have to wait for more advances in developing safer MRI technology or better implantable devices.

References

1. Nelson, T.J., A. Behfar, and A. Terzic. Strategies for therapeutic repair: The "R" regenerative medicine paradigm. *Clin Transl Sci*, 2008, 1(2): 168–71.
2. Yu, J., M.A. Vodyanik, K. Smuga-Otto, J. Antosiewicz-Bourget, J.L. Frane, S. Tian, J. Nie, G.A. Jonsdottir, V. Ruotti, R. Stewart, I.I. Slukvin, and J.A. Thomson. Induced pluripotent stem cell lines derived from human somatic cells. *Science*, 2007, 318(5858): 1917–20.

3. Daley, G.Q., and D.T. Scadden. Prospects for stem cell-based therapy. *Cell*, 2008, 132(4): 544–48.

4. Sykova, E., P. Jendelova, and V. Herynek. MR tracking of stem cells in living recipients. *Methods Mol Biol*, 2009, 549: 197–215.

5. Yin, T., and L. Li. The stem cell niches in bone. *J Clin Invest*, 2006, 116(5): 1195–201.

6. Anversa, P., J. Kajstura, A. Leri, and R. Bolli. Life and death of cardiac stem cells: a paradigm shift in cardiac biology. *Circulation*, 2006, 113(11): 1451–63.

7. Moore, K.A., and I.R. Lemischka. Stem cells and their niches. *Science*, 2006, 311(5769): 1880–85.

8. Schroeder, T. Imaging stem-cell-driven regeneration in mammals. *Nature*, 2008, 453(7193): 345–51.

9. Brazelton, T.R., and H.M. Blau. Optimizing techniques for tracking transplanted stem cells *in vivo*. *Stem Cells*, 2005, 23(9): 1251–65.

10. Barker, N., J.H. van Es, J. Kuipers, P. Kujala, M. van den Born, M. Cozijnsen, A. Haegebarth, J. Korving, H. Begthel, P.J. Peters, and H. Clevers. Identification of stem cells in small intestine and colon by marker gene Lgr5. *Nature*, 2007, 449(7165): 1003–7.

11. Tumbar, T., G. Guasch, V. Greco, C. Blanpain, W.E. Lowry, M. Rendl, and E. Fuchs. Defining the epithelial stem cell niche in skin. *Science*, 2004, 303(5656): 359–63.

12. Sumen, C., T.R. Mempel, I.B. Mazo, and U.H. von Andrian. Intravital microscopy: visualizing immunity in context. *Immunity*, 2004, 21(3): 315–29.

13. Kiel, M.J., O.H. Yilmaz, T. Iwashita, O.H. Yilmaz, C. Terhorst, and S.J. Morrison. SLAM family receptors distinguish hematopoietic stem and progenitor cells and reveal endothelial niches for stem cells. *Cell*, 2005, 121(7): 1109–21.

14. Hill, K.A. Hartsoeker's homonculus: a corrective note. *J Hist Behav Sci*, 1985, 21(2): 178–79.

15. Wallace, P.K., and K.A. Muirhead. Cell tracking 2007: a proliferation of probes and applications. *Immunol Invest*, 2007, 36(5–6): 527–61.

16. Fisher, B., B.S. Packard, E.J. Read, J.A. Carrasquillo, C.S. Carter, S.L. Topalian, J.C. Yang, P. Yolles, S.M. Larson, and S.A. Rosenberg. Tumor localization of adoptively transferred indium-111 labeled tumor infiltrating lymphocytes in patients with metastatic melanoma. *J Clin Oncol*, 1989, 7(2): 250–61.

17. Morse, M.A., R.E. Coleman, G. Akabani, N. Niehaus, D. Coleman, and H.K. Lyerly. Migration of human dendritic cells after injection in patients with metastatic malignancies. *Cancer Res*, 1999, 59(1): 56–58.

18. Chapman, S., K.J. Oparka, and A.G. Roberts. New tools for *in vivo* fluorescence tagging. *Curr Opin Plant Biol*, 2005, 8(6): 565–73.

19. Chudakov, D.M., S. Lukyanov, and K.A. Lukyanov. Fluorescent proteins as a toolkit for *in vivo* imaging. *Trends Biotechnol*, 2005, 23(12): 605–13.

20. Parish, C.R. Fluorescent dyes for lymphocyte migration and proliferation studies. *Immunol Cell Biol*, 1999, 77(6): 499–508.

21. Lyons, A.B., and C.R. Parish. Determination of lymphocyte division by flow cytometry. *J Immunol Methods*, 1994, 171(1): 131–37.

22. Bruchez, M., Jr., M. Moronne, P. Gin, S. Weiss, and A.P. Alivisatos. Semiconductor nanocrystals as fluorescent biological labels. *Science*, 1998, 281(5385): 2013–16.

23. Chan, W.C., and S. Nie. Quantum dot bioconjugates for ultrasensitive nonisotopic detection. *Science*, 1998, 281(5385): 2016–18.

24. Cai, W., A.R. Hsu, Z. Li, and X. Chen. Are quantum dots ready for *in vivo* imaging in human subjects? *Nanoscale Res Lett*, 2007, 2(6): 265–81.

25. Smith, A.M., G. Ruan, M.N. Rhyner, and S. Nie. Engineering luminescent quantum dots for *in vivo* molecular and cellular imaging. *Ann Biomed Eng*, 2006, 34(1): 3–14.

26. Alivisatos, P. The use of nanocrystals in biological detection. *Nat Biotechnol*, 2004, 22(1): 47–52.

27. Hardman, R. A toxicologic review of quantum dots: toxicity depends on physicochemical and environmental factors. *Environ Health Perspect*, 2006, 114(2): 165–72.

28. Michalet, X., F.F. Pinaud, L.A. Bentolila, J.M. Tsay, S. Doose, J.J. Li, G. Sundaresan, A.M. Wu, S.S. Gambhir, and S. Weiss. Quantum dots for live cells, *in vivo* imaging, and diagnostics. *Science*, 2005, 307(5709): 538–44.

29. Iga, A.M., J.H. Robertson, M.C. Winslet, and A.M. Seifalian. Clinical potential of quantum dots. *J Biomed Biotechnol*, (http://downloads.hindawi.com/journals/jbb/2007/076087.pdf).

30. Medintz, I.L., H.T. Uyeda, E.R. Goldman, and H. Mattoussi. Quantum dot bioconjugates for imaging, labelling and sensing. *Nat Mater*, 2005, 4(6): 435–46.

31. Larson, D.R., W.R. Zipfel, R.M. Williams, S.W. Clark, M.P. Bruchez, F.W. Wise, and W.W. Webb. Water-soluble quantum dots for multiphoton fluorescence imaging *in vivo*. *Science*, 2003, 300(5624): 1434–36.

32. Ballou, B., B.C. Lagerholm, L.A. Ernst, M.P. Bruchez, and A.S. Waggoner. Noninvasive imaging of quantum dots in mice. *Bioconjug Chem*, 2004, 15(1): 79–86.

33. Alivisatos, A.P. Semiconductor clusters, nanocrystals, and quantum dots. *Science*, 1996, 271 (5251): 933–37.

34. Nirmal, M., and L. Brus. Luminescence photophysics in semiconductor nanocrystals. *Acc Chem Res*, 1999, 32(5): 407–14.

35. Zhong, X., Y. Feng, W. Knoll, and M. Han. Alloyed Zn(x)Cd(1-x)S nanocrystals with highly narrow luminescence spectral width. *J Am Chem Soc*, 2003, 125(44): 13559–63.

36. Kim, S., B. Fisher, H. Eisler, and M. Bawendi. Type-II quantum dots: CdTe/CdSe(core/shell) and CdSe/ZnTe(core/shell) heterostructures. *J Am Chem Soc*, 2003, 125(38): 11466–67.

37. Pietryga, J.M., R.D. Schaller, D. Werder, M.H. Stewart, V.I. Klimov, and J.A. Hollingsworth. Pushing the band gap envelope: mid-infrared emitting colloidal PbSe quantum dots. *J Am Chem Soc*, 2004, 126(38): 11752–53.

38. Lim, Y.T., S. Kim, A. Nakayama, N.E. Stott, M.G. Bawendi, and J.V. Frangioni. Selection of quantum dot wavelengths for biomedical assays and imaging. *Mol Imaging*, 2003, 2(1): 50–64.

39. Gao, X., L. Yang, J.A. Petros, F.F. Marshall, J.W. Simons, and S. Nie. *In vivo* molecular and cellular imaging with quantum dots. *Curr Opin Biotechnol*, 2005, 16(1): 63–72.

40. Watson, A., X. Wu, and M. Bruchez. Lighting up cells with quantum dots. *Biotechniques*, 2003, 34(2): 296–300, 302–3.

41. Chan, W.C., D.J. Maxwell, X. Gao, R.E. Bailey, M. Han, and S. Nie. Luminescent quantum dots for multiplexed biological detection and imaging. *Curr Opin Biotechnol*, 2002, 13(1): 40–46.

42. Fortina, P., L.J. Kricka, S. Surrey, and P. Grodzinski. Nanobiotechnology: the promise and reality of new approaches to molecular recognition. *Trends Biotechnol*, 2005, 23(4): 168–73.

43. Bruchez, M.P. Turning all the lights on: quantum dots in cellular assays. *Curr Opin Chem Biol*, 2005, 9(5): 533–37.

44. Medintz, I.L., A.R. Clapp, F.M. Brunel, T. Tiefenbrunn, H.T. Uyeda, E.L. Chang, J.R. Deschamps, P.E. Dawson, and H. Mattoussi. Proteolytic activity monitored by fluorescence resonance energy transfer through quantum-dot-peptide conjugates. *Nat Mater*, 2006, 5(7): 581–89.

45. Wu, X., H. Liu, J. Liu, K.N. Haley, J.A. Treadway, J.P. Larson, N. Ge, F. Peale, and M.P. Bruchez. Immunofluorescent labeling of cancer marker Her2 and other cellular targets with semiconductor quantum dots. *Nat Biotechnol*, 2003, 21(1): 41–46.

46. Chattopadhyay, P.K., D.A. Price, T.F. Harper, M.R. Betts, and J. Yu. Quantum dot semiconductor nanocrystals for immunophenotyping by polychromatic flow cytometry. *Nat Med*, 2006, 12(8): 972–77.

47. Wu, X., and M.P. Bruchez. Labeling cellular targets with semiconductor quantum dot conjugates. *Methods Cell Biol*, 2004, 75: 171–83.

48. Voura, E.B., J.K. Jaiswal, H. Mattoussi, and S.M. Simon. Tracking metastatic tumor cell extravasation with quantum dot nanocrystals and fluorescence emission-scanning microscopy. *Nat Med*, 2004, 10(9): 993–98.

49. Resch-Genger, U., M. Grabolle, S. Cavaliere-Jaricot, R. Nitschke, and T. Nann. Quantum dots versus organic dyes as fluorescent labels. *Nat Methods*, 2008, 5(9): 763–75.

50. Jaiswal, J.K., H. Mattoussi, J.M. Mauro, and S.M. Simon. Long-term multiple color imaging of live cells using quantum dot bioconjugates. *Nat Biotechnol*, 2003, 21(1): 47–51.

51. Santra, S., and D. Dutta. *Quantum dots for cancer imaging in nanoparticles in biomedical imaging.* Springer, New York, 2008, pp. 463–85.

52. Henglein, A. Small-particle research: physicochemical properties of extremely small colloidal metal and semiconductor particles. *Chem Rev*, 1989, 89(8): 1861–73.

53. Sukhanova, A., J. Devy, L. Venteo, H. Kaplan, M. Artemyev, V. Oleinikov, D. Klinov, M. Pluot, J.H. Cohen, and I. Nabiev. Biocompatible fluorescent nanocrystals for immunolabeling of membrane proteins and cells. *Anal Biochem*, 2004, 324(1): 60–67.

54. Singhal, A., H. Fischer, J. Wang and W. Chan. Biomedical applications of semiconductor quantum dots. In *Bioelectric Engineering.* Springer, New York, 2004, 37–50.

55. Jaiswal, J.K., and S.M. Simon. Potentials and pitfalls of fluorescent quantum dots for biological imaging. *Trends Cell Biol*, 2004, 14(9): 497–504.

56. Wang, H., H. Zhang, X. Li, J. Wang, and Z. Huang. Solubilization and bioconjugation of QDs and their application in cell imaging. *J Biomed Mater Rest A*, 2007: 833–41.

57. Bentolila, L.A., Y. Ebenstein, and S. Weiss. Quantum dots for *in vivo* small-animal imaging. *J Nucl Med*, 2009, 50(4): 493–96.

58. Sukhanova, A., and I. Nabiev. Fluorescent nanocrystal quantum dots as medical diagnostic tools. *Expert Opin Med Diagn*, 2008, 2: 429–47.

59. Xing, Y., and J. Rao. Quantum dot bioconjugates for *in vitro* diagnostics and *in vivo* imaging. *Cancer Biomark*, 2008, 4(6): 307–19.

60. Murray, C., D. Norris, and M. Bawendi. Synthesis and characterization of nearly monodisperse CdE (E=S, Se, Te) semiconductor nanocrystallites. *J Am Chem Soc*, 1993, 115: 8706–15.

61. Dabbousi, B.O., J. Rodriguez-Viejo, F.V. Mikulec, and J.R. Heine. (CdSe)ZnS core-shell quantum dots: synthesis and characterization of a size series of highly luminescent nanocrystallites. *J Phys Chem B*, 1997, 101: 9463–75.

62. Stroh, M., J.P. Zimmer, D.G. Duta, T.S. Levchenenko, and K.S. Cohen. Quantum dots spectrally distinguish multiple species within the tumor milieu *in vivo. Nat Med*, 2005, 11(6): 678–82.

63. Dahan, M., S. Levi, C. Luccardini, P. Rostaing, B. Riveau, and A. Triller. Diffusion dynamics of glycine receptors revealed by single-quantum dot tracking. *Science*, 2003, 302(5644): 442–55.

64. Ekimov, A., and A. Onushchenko. Quantum size effect in the optical-spectra of semiconductor micro-crystals. *Sov Phys Semicond*, 1982, 16: 775–78.

65. Efros, A.L., and A.L. Efros. Interband absorption of light in a semiconductor sphere. *Sov Phys Semicond*, 1982, 16: 772–75.

66. Peng, X., M.C. Schlamp, A.V. Kadavanich, and A.P. Alivisatos. Epitaxial growth of highly luminescent CdSe/CdS core/shell nanocrystals with photostability and electronic accessibility. *J Am Chem Soc*, 1997, 119(30): 7019–29.

67. Hines, M.P., and P. Guyot-Sionnest. Synthesis and characterization of strongly luminescing ZnS-capped CdSe nanocrystals. *J Phys Chem*, 1996, 100(2): 468–471.

68. Peng, Z.A., and X. Peng. Formation of high-quality CdTe, CdSe, and CdS nanocrystals using CdO as precursor. *J Am Chem Soc*, 2001, 123(1): 183–84.

69. Yua, W.W., E. Changb, R. Drezekb, and V.L. Colvin. Water-soluble quantum dots for biomedical applications. *Biochem Biophys Res Commun* 2006, 348: 781–86.

70. Mattoussi, H., J.M. Mauro, E.R. Goldman, G.P. Anderson, V.C. Sundar, and F.V. Mikulec. Self-assembly of CdSe–ZnS quantum dot bioconjugates using an engineered recombinant protein. *J Am Chem Soc*, 2000, 122(49): 12142–50.

71. Mitchell, G.P., C.A. Mirkin, and R.L. Letsinger. Programmed assembly of DNA functionalized quantum dots. *J Am Chem Soc*, 1999, 121(35): 8122–23.

72. Goldman, E.R., E.D. Balighian, and H. Mattoussi. Avidin: a natural bridge for quantum dot-antibody conjugates. *J Am Chem Soc*, 2002, 124(22): 6378–82.

73. Uyeda, H.T., I.L. Medintz, J.K. Jaiswal, S.M. Simon, and H. Mattoussi. Synthesis of compact multidentate ligands to prepare stable hydrophilic quantum dot fluorophores. *J Am Chem Soc*, 2005, 127(11): 3870–78.

74. Dubertret, B., P. Skourides, D.J. Norris, V. Noireaux, A.H. Brivanlou, and A. Libchaber. *In vivo* imaging of quantum dots encapsulated in phospholipid micelles. *Science*, 2002, 298(5599): 1759–62.

75. Gao, X., Y. Cui, R.M. Levenson, L.W. Chung, and S. Nie. *In vivo* cancer targeting and imaging with semiconductor quantum dots. *Nat Biotechnol*, 2004, 22(8): 969–76.

76. Biju, V., T. Itoh, A. Anas, A. Sujith, and M. Ishikawa. Semiconductor quantum dots and metal nanoparticles: syntheses, optical properties, and biological applications. *Anal Bioanal Chem*, 2008, 391(7): 2469–95.

77. Gerion, D., W.J. Parak, S.C. Williams, D. Zanchet, C.M. Micheel, and A.P. Alivisatos. Sorting fluorescent nanocrystals with DNA. *J Am Chem Soc*, 2002, 124(24): 7070–74.

78. Pathak, S., S.K. Choi, N. Arnheim, and M.E. Thompson. Hydroxylated quantum dots as luminescent probes for *in situ* hybridization. *J Am Chem Soc*, 2001, 123(17): 4103–4.

79. Akerman, M.E., W.C. Chan, P. Laakkonen, S.N. Bhatia, and E. Ruoslahti. Nanocrystal targeting *in vivo*. *Proc Natl Acad Sci USA*, 2002, 99(20): 12617–21.

80. Winter, J.O., T.Y. Liu, B.A. Korgel, and C.E. Schmidt. Recognition molecule directed interfacing between semiconductor quantum dots and nerve cells. *Adv Mater*, 2001, 13: 1673–77.

81. Rosenthal, S.J., I. Tomlinson, E.M. Adkins, S. Schroeter, S. Adams, L. Swafford, J. McBride, Y. Wang, L.J. DeFelice, and R.D. Blakely. Targeting cell surface receptors with ligand-conjugated nanocrystals. *J Am Chem Soc*, 2002, 124(17): 4586–94.

82. Kloepfer, J.A., R.E. Mielke, M.S. Wong, K.H. Nealson, G. Stucky, and J.L. Nadeau. Quantum dots as strain- and metabolism-specific microbiological labels. *Appl Environ Microbiol*, 2003, 69(7): 4205–13.

83. Parak, W.J., T. Pellegrino, and C. Plank. Labelling of cells with quantum dots. *Nanotechnology*, 2005, 16: R9–25.

84. Han, M., X. Gao, J.Z. Su, and S. Nie. Quantum-dot-tagged microbeads for multiplexed optical coding of biomolecules. *Nat Biotechnol*, 2001, 19(7): 631–35.

85. Mattheakis, L.C., J.M. Dias, Y.J. Choi, J. Gong, M.P. Bruchez, J. Liu, and E. Wang. Optical coding of mammalian cells using semiconductor quantum dots. *Anal Biochem*, 2004, 327(2): 200–8.

86. Miyawaki, A., A. Sawano, and T. Kogure. Lighting up cells: labelling proteins with fluorophores. *Nat Cell Biol*, 2003 (Suppl): S1–7.

87. Xing, Y., Q. Chaudry, C. Shen, K.Y. Kong, H.E. Zhau, L.W. Chung, J.A. Petros, R.M. O'Regan, M.V. Yezhelyev, J.W. Simons, M.D. Wang, and S. Nie. Bioconjugated quantum dots for multiplexed and quantitative immunohistochemistry. *Nat Protoc*, 2007, 2(5): 1152–65.

88. Zheng, J., A.A. Ghazani, Q. Song, and S. Mardyani. Cellular imaging and surface marker labeling of hematopoietic cells using quantum dot bioconjugates. *Lab Hematol*, 2006, 12: 94–98.

89. Dahan, M. From analog to digital: exploring cell dynamics with single quantum dots. *Histochem Cell Biol*, 2006, 125(5): 451–56.

90. Saxton, M.J., and K. Jacobson. Single-particle tracking: applications to membrane dynamics. *Annu Rev Biophys Biomol Struct*, 1997, 26: 373–99.

91. Bannai, H., S. Levi, C. Schweizer, M. Dahan, and A. Triller. Imaging the lateral diffusion of membrane molecules with quantum dots. *Nat Protoc*, 2006, 1(6): 2628–34.

92. Chen, H., I. Titushkin, M. Stroscio, and M. Cho. Altered membrane dynamics of quantum dot-conjugated integrins during osteogenic differentiation of human bone marrow derived progenitor cells. *Biophys J*, 2007, 92: 1399–408.

93. Lidke, D., P. Nagy, R. Heintzmann, D. Arndt-Jovin, J. Post, H. Grecco, E. Jares-Erijman, and T. Jovin. Quantum dot ligands provide new insights into receptor–mediated signal transduction. *Nat Biotechnol*, 2004, 22: 198–203.

94. Derfus, A.M., W.C.W. Chan, and S.N. Bhatia. Intracellular delivery of quantum dots for live cell labeling and organelle tracking. *Adv Mater*, 2004, 16(12): 961–66.

95. Nabiev, I., S. Mitchell, A. Davies, Y. Williams, D. Kelleher, and R. Moore. Nonfunctionalized nanocrystals can exploit a cell's active transport machinery delivering them to specific nuclear and cytoplasmic compartments. *Nano Lett*, 2007, 7(11): 3452–61.

96. Dower, W.J., J.F. Miller, and C.W. Ragsdale. High efficiency transformation of *E. coli* by high voltage electroporation. *Nucleic Acids Res*, 1988, 16(13): 6127–45.

97. Gao, X., W.C. Chan, and S. Nie. Quantum-dot nanocrystals for ultrasensitive biological labeling and multicolor optical encoding. *J Biomed Opt*, 2002, 7(4): 532–37.

98. Lagerholm, B.C., M. Wang, L.A. Ernst, D.H. Ly, and H. Liu. Multicolor coding of cells with cationic peptide coated quantum dots. *Nano Lett*, 2004, 4(10): 2019–22.

99. Zheng, J., A.A. Ghazani, Q. Song, S. Mardyani, W.C. Chan, and C. Wang. Cellular imaging and surface marker labeling of hematopoietic cells using quantum dot bioconjugates. *Lab Hematol*, 2006, 12(2): 94–98.

100. Pan, Y.L., J.Y. Cai, L. Qin, and H. Wang. Atomic force microscopy-based cell nanostructure for ligand-conjugated quantum dot endocytosis. *Acta Biochim Biophys Sin (Shanghai)*, 2006, 38(9): 646–52.

101. de Farias, P.M.A., B.S. Santos, F.D. Menezes, J.A.G. Brasil, and R.M.A. Ferreira. Highly fluorescent semiconductor core–shell CdTe–CdS nanocrystals for monitoring living yeast cells activity. *Appl Phys A*, 2007, 89: 957–61.

102. Schroeder, J.E., I. Shweky, H. Shmeeda, U. Banin, and A. Gabizon. Folate-mediated tumor cell uptake of quantum dots entrapped in lipid nanoparticles. *J Control Release*, 2007, 124(1–2): 28–34.

103. Delehanty, J.B., I.L. Medintz, T. Pons, F.M. Brunel, P.E. Dawson, and H. Mattoussi. Self-assembled quantum dot-peptide bioconjugates for selective intracellular delivery. *Bioconjug Chem*, 2006, 17(4): 920–27.

104. Parak, W.J., R. Boudreau, M. Le Gros, D. Gerion, D. Zanchet, and C.M. Micheel. Cell motility and metastatic potential studies based on quantum dot imaging of phagokinetic tracks. *Adv Mater*, 2002, 14: 882–85.

105. Chen, F., and D. Gerion. Fluorescent Cdse/Zns nanocrystal-peptide conjugates for long-term, nontoxic imaging and nuclear targeting in living cells. *Nano Lett*, 2004, 4: 1827–32.

106. Kim, B.Y., W. Jiang, J. Oreopoulos, C.M. Yip, J.T. Rutka, and W.C. Chan. Biodegradable quantum dot nanocomposites enable live cell labeling and imaging of cytoplasmic targets. *Nano Lett*, 2008, 8(11): 3887–92.

107. Gerion, D., J. Herberg, R. Bok, E. Gjersing, and E. Ramon. Paramagnetic silica-coated nanocrystals as an advanced MRI contrast agent. *J Phys Chem C*, 2007, 111: 12542–51.

108. Mulder, W.J., A.W. Griffioen, G.J. Strijkers, D.P. Cormode, K. Nicolay, and Z.A. Fayad. Magnetic and fluorescent nanoparticles for multimodality imaging. *Nanomed*, 2007, 2(3): 307–24.

109. Kim, S., Y.T. Lim, E.G. Soltesz, A.M. De Grand, J. Lee, A. Nakayama, J.A. Parker, T. Mihaljevic, R.G. Laurence, D.M. Dor, L.H. Cohn, M.G. Bawendi, and J.V. Frangioni. Near-infrared fluorescent type II quantum dots for sentinel lymph node mapping. *Nat Biotechnol*, 2004, 22(1): 93–97.

110. Rieger, S., R.P. Kulkarni, D. Darcy, S.E. Fraser, and R.W. Koster. Quantum dots are powerful multipurpose vital labeling agents in zebrafish embryos. *Dev Dyn*, 2005, 234(3): 670–81.

111. Slotkin, J.R., L. Chakrabarti, H.N. Dai, R.S. Carney, T. Hirata, B.S. Bregman, G.I. Gallicano, J.G. Corbin, and T.F. Haydar. *In vivo* quantum dot labeling of mammalian stem and progenitor cells. *Dev Dyn*, 2007, 236(12): 3393–401.

112. Rosen, A.B., D.J. Kelly, A.J. Schuldt, J. Lu, I.A. Potapova, S.V. Doronin, K.J. Robichaud, R.B. Robinson, M.R. Rosen, P.R. Brink, G.R. Gaudette, and I.S. Cohen. Finding fluorescent needles in the cardiac haystack: tracking human mesenchymal stem cells labeled with quantum dots for quantitative *in vivo* three-dimensional fluorescence analysis. *Stem Cells*, 2007, 25(8): 2128–38.

113. Potapova, I., S. Doronin, and D. Kelly. Functional regeneration of the canine ventricle using adult human mesenchymal stem cells committed *in vitro* to a cardiac lineage. *Circulation* 2006, 99: E19.

114. Rosen, A., D. Kelly, and P. Brink. Finding fluorescent needles in the cardiac haystack: reconstructing the 3-dimensional distribution of quantum dot-loaded human mesenchymal stem cells injected into the rat ventricle *in vivo*. *Circulation*, 2006, 99: E34.

115. Lei, Y., H. Tang, L. Yao, R. Yu, M. Feng, and B. Zou. Applications of mesenchymal stem cells labeled with Tat peptide conjugated quantum dots to cell tracking in mouse body. *Bioconjug Chem*, 2008, 19(2): 421–27.

116. Shah, B.S., P.A. Clark, E.K. Moioli, M.A. Stroscio, and J.J. Mao. Labeling of mesenchymal stem cells by bioconjugated quantum dots. *Nano Lett*, 2007, 7(10): 3071–79.
117. Lin, S., X. Xie, M.R. Patel, Y.H. Yang, Z. Li, F. Cao, O. Gheysens, Y. Zhang, S.S. Gambhir, J.H. Rao, and J.C. Wu. Quantum dot imaging for embryonic stem cells. *BMC Biotechnol*, 2007, 7: 67.
118. Ferreira, L., J.M. Karp, L. Nobre, and R. Langer. New opportunities: the use of nanotechnologies to manipulate and track stem cells. *Cell Stem Cell*, 2008, 3(2): 136–46.
119. Yukawa, H., S. Mizufune, C. Mamori, and Y. Kagami. Quantum dots for labeling adipose tissue-derived stem cells. *Cell Transpl*, 2009, 18: 591–599.
120. Murasawa, S., A. Kawamoto, M. Horii, S. Nakamori, and T. Asahara. Niche-dependent translineage commitment of endothelial progenitor cells, not cell fusion in general, into myocardial lineage cells. *Arterioscler Thromb Vasc Biol*, 2005, 25: 1388–94.
121. Seleverstov, O., O. Zabirnyk, M. Zscharnack, L. Bulavina, and M. Nowicki. Quantum dots for human mesenchymal stem cells labeling. A size-dependent autophagy activation. *Nano Lett*, 2006, 6: 2826–32.
122. Tholouli, E., E. Sweeney, E. Barrow, and V. Clay. Quantum dots light up pathology. *J Pathol* 2008, 216: 275–85.
123. Frangioni, J.V. *In vivo* near-infrared fluorescence imaging. *Curr Opin Chem Biol*, 2003, 7(5): 626–34.
124. Bremer, C., C.H. Tung, and R. Weissleder. *In vivo* molecular target assessment of matrix metalloproteinase inhibition. *Nat Med*, 2001, 7(6): 743–48.
125. Bailey, R.E., J.B. Strausburg, and S. Nie. A new class of far-red and near-infrared biological labels based on alloyed semiconductor quantum dots. *J Nanosci Nanotechnol*, 2004, 4(6): 569–74.
126. Sevick-Muraca, E., J. Houston, and M. Gurfinkel. Fluorescence-enhanced, near infrared diagnostic imaging with contrast agents. *Curr Opin Chem Biol*, 2002, 6: 642–50.
127. Frangioni, J., and R. Hajjar. *In vivo* tracking of stem cells for clinical trials in cardiovascular disease. *Circulation*, 2004, 110: 3378–83.
128. Stroh, M., J. Zimmer, D. Duda, T.S. Levchenko, K. Cohen, E. Brown, and D. Scadden. Quantum dots spectrally distinguish multiple species within the tumor milieu *in vivo*. *Nat Med*, 2005, 11(6): 678–82.
129. Morrison, S.J., and J. Kimble. Asymmetric and symmetric stem-cell divisions in development and cancer. *Nature*, 2006, 441: 1068–74.
130. Roberts, M.J., M.D. Bentley, and J.M. Harris. Chemistry for peptide and protein PEGylation. *Adv Drug Deliv Rev*, 2002, 54(4): 459–76.
131. Owens, D.E., 3rd, and N.A. Peppas. Opsonization, biodistribution, and pharmacokinetics of polymeric nanoparticles. *Int J Pharm*, 2006, 307(1): 93–102.
132. Lyons, S.K. Advances in imaging mouse tumour models *in vivo*. *J Pathol*, 2005, 205(2): 194–205.
133. Tada, H., H. Higuchi, T.M. Wanatabe, and N. Ohuchi. *In vivo* real-time tracking of single quantum dots conjugated with monoclonal anti-HER2 antibody in tumors of mice. *Cancer Res*, 2007, 67(3): 1138–44.
134. Ballou, B., L.A. Ernst, S. Andreko, J.A. Fitzpatrick, B.C. Lagerholm, A.S. Waggoner, and M.P. Bruchez. Imaging vasculature and lymphatic flow in mice using quantum dots. *Methods Mol Biol*, 2009, 574: 63–74.
135. Torne, A., and L.M. Puig-Tintore. The use of sentinel lymph nodes in gynaecological malignancies. *Curr Opin Obstet Gynecol*, 2004, 16: 57–64.
136. Soltesz, E., S. Kim, S. Kim, R. Laurence, and A. De Grand. Sentinel lymph node mapping of the gastrointestinal tract by using invisible light. *Ann Surg Oncol*, 2006, 13(3): 386–96.
137. Soltesz, E.G., S. Kim, R.G. Laurence, A.M. DeGrand, C.P. Parungo, D.M. Dor, L.H. Cohn, M.G. Bawendi, J.V. Frangioni, and T. Mihaljevic. Intraoperative sentinel lymph node mapping of the lung using near-infrared fluorescent quantum dots. *Ann Thorac Surg*, 2005, 79(1): 269–77; discussion, 269–77.
138. Frangioni, J.V., S.W. Kim, S. Ohnishi, S. Kim, and M.G. Bawendi. Sentinel lymph node mapping with type-II quantum dots. *Methods Mol Biol*, 2007, 374: 147–59.

139. Ballou, B., L.A. Ernst, S. Andreko, T. Harper, J.A. Fitzpatrick, A.S. Waggoner, and M.P. Bruchez. Sentinel lymph node imaging using quantum dots in mouse tumor models. *Bioconjug Chem*, 2007, 18(2): 389–96.

140. Kosaka, N., M. Ogawa, N. Sato, P.L. Choyke, and H. Kobayashi. *In vivo* real-time, multicolor, quantum dot lymphatic imaging. *J Invest Dermatol*, 2009, 129(12): 2818–22.

141. Cai, W., D.W. Shin, K. Chen, O. Gheysens, Q. Cao, S.X. Wang, S.S. Gambhir, and X. Chen. Peptide-labeled near-infrared quantum dots for imaging tumor vasculature in living subjects. *Nano Lett*, 2006, 6(4): 669–76.

142. Hood, J.D., and D.A. Cheresh. Role of integrins in cell invasion and migration. *Nat Rev Cancer*, 2002, 2(2): 91–100.

143. Hynes, R.O. Integrins: bidirectional, allosteric signaling machines. *Cell*, 2002, 110(6): 673–87.

144. Gao, J., and B. Xu. Applications of nanomaterials inside cells. *Nano Today* 2009, 4: 37–51.

145. So, M.K., C. Xu, A.M. Loening, S.S. Gambhir, and J. Rao. Self-illuminating quantum dot conjugates for *in vivo* imaging. *Nat Biotechnol*, 2006, 24(3): 339–43.

146. Jennings, L.E., and N.J. Long. 'Two is better than one'—probes for dual-modality molecular imaging. *Chem Commun (Camb)*, 2009(24): 3511–24.

147. Rhyner, M.N., A.M. Smith, X. Gao, H. Mao, L. Yang, and S. Nie. Quantum dots and multifunctional nanoparticles: new contrast agents for tumor imaging. *Nanomed*, 2006, 1(2): 209–17.

148. Wang, S., B.R. Jarrett, S.M. Kauzlarich, and A.Y. Louie. Core/shell quantum dots with high relaxivity and photoluminescence for multimodality imaging. *J Am Chem Soc*, 2007, 129(13): 3848–56.

149. Selvan, S.T., P.K. Patra, C.Y. Ang, and J.Y. Ying. Synthesis of silica-coated semiconductor and magnetic quantum dots and their use in the imaging of live cells. *Angew Chem Int Ed Engl*, 2007, 46(14): 2448–52.

150. Mulder, W.J., R. Koole, R.J. Brandwijk, G. Storm, P.T. Chin, G.J. Strijkers, C. de Mello Donega, K. Nicolay, and A.W. Griffioen. Quantum dots with a paramagnetic coating as a bimodal molecular imaging probe. *Nano Lett*, 2006, 6(1): 1–6.

151. Rhyner, M.N., A.M. Smith, X. Gao, H. Mao, L. Yang, and S. Nie. Quantum dots and targeted nanoparticle probes for *in vivo* tumor imaging. In *Nanoparticles in biomedical imaging*, ed. J.W.M. Bulte. Springer, New York, 2008, pp. 413–27.

152. Haas, R.L., D. de Jong, R.A. Valdes Olmos, C.A. Hoefnagel, I. van den Heuvel, S.F. Zerp, H. Bartelink, and M. Verheij. *In vivo* imaging of radiation-induced apoptosis in follicular lymphoma patients. *Int J Radiat Oncol Biol Phys*, 2004, 59(3): 782–87.

153. Prinzen, L., R.J. Miserus, A. Dirksen, T.M. Hackeng, N. Deckers, N.J. Bitsch, R.T. Megens, K. Douma, J.W. Heemskerk, M.E. Kooi, P.M. Frederik, D.W. Slaaf, M.A. van Zandvoort, and C.P. Reutelingsperger. Optical and magnetic resonance imaging of cell death and platelet activation using annexin a5-functionalized quantum dots. *Nano Lett*, 2007, 7(1): 93–100.

154. Wang, D., J. He, N. Rosenzweig, and Z. Rosenzweig. Superparamagnetic Fe2O3 Beads–CdSe/ZnS quantum dots core–shell nanocomposite particles for cell separation. *Nano Lett*, 2004, 4(3): 409–13.

155. Gu, H., R. Zheng, X. Zhang, and B. Xu. Facile one-pot synthesis of bifunctional heterodimers of nanoparticles: a conjugate of quantum dot and magnetic nanoparticles. *J Am Chem Soc*, 2004, 126(18): 5664–65.

156. Chang, E., N. Thekkek, W.W. Yu, V.L. Colvin, and R. Drezek. Evaluation of quantum dot cytotoxicity based on intracellular uptake. *Small*, 2006, 2(12): 1412–17.

157. Lewinski, N., V. Colvin, and R. Drezek. Cytotoxicity of nanoparticles. *Small*, 2008, 4(1): 26–49.

158. Maysinger, D., J. Lovric, A. Eisenberg, and R. Savic. Fate of micelles and quantum dots in cells. *Eur J Pharm Biopharm*, 2007, 65(3): 270–81.

159. Lovric, J., S.J. Cho, F.M. Winnik, and D. Maysinger. Unmodified cadmium telluride quantum dots induce reactive oxygen species formation leading to multiple organelle damage and cell death. *Chem Biol*, 2005, 12(11): 1227–34.

160. Kirchner, C., T. Liedl, S. Kudera, T. Pellegrino, A. Munoz Javier, H.E. Gaub, S. Stolzle, N. Fertig, and W.J. Parak. Cytotoxicity of colloidal CdSe and CdSe/ZnS nanoparticles. *Nano Lett*, 2005, 5(2): 331–8.

161. Hoshino, A., K. Fujioka, T. Oku, M. Suga, Y. Sasaki, T. Ohta, and M. Yasuhara. Physicochemical properties and cellular toxicity of nanocrystal quantum dots depend on their surface modification. *Nano Lett*, 2004, 4(11): 2163–69.

162. Ipe, B.I., M. Lehnig, and C.M. Niemeyer. On the generation of free radical species from quantum dots. *Small*, 2005, 1(7): 706–9.

163. Ryman-Rasmussen, J.P., J.E. Riviere, and N.A. Monteiro-Riviere. Surface coatings determine cytotoxicity and irritation potential of quantum dot nanoparticles in epidermal keratinocytes. *J Invest Dermatol*, 2007, 127: 143–53.

164. Hoshino, A., K. Hanaki, K. Suzuki, and K. Yamamoto. Applications of T-lymphoma labeled with fluorescent quantum dots to cell tracing markers in mouse body. *Biochem Biophys Res Commun*, 2004, 314: 46–53.

165. Muller-Borer, B.J., M.C. Collins, P.R. Gunst, W.E. Cascio, and A.P. Kypson. Quantum dot labeling of mesenchymal stem cells. *J Nanobiotechnol*, 2007, 5: 9.

166. Bakalova, R., Z. Zhelev, R. Jose, T. Nagase, H. Ohba, M. Ishikawa, and Y. Baba. Role of free cadmium and selenium ions in the potential mechanism for the enhancement of photoluminescence of CdSe quantum dots under ultraviolet irradiation. *J Nanosci Nanotechnol*, 2005, 5(6): 887–94.

167. Rzigalinski, B.A., and J.S. Strobl. Cadmium-containing nanoparticles: perspectives on pharmacology and toxicology of quantum dots. *Toxicol Appl Pharmacol*, 2009, 238(3): 280–88.

168. Smith, A.M., H. Duan, A.M. Mohs, and S. Nie. Bioconjugated quantum dots for *in vivo* molecular and cellular imaging. *Adv Drug Deliv Rev*, 2008, 60(11): 1226–40.

169. Ballou, B. Quantum dot surfaces for use *in vivo* and *in vitro*. *Curr Top Dev Biol*, 2005, 70: 103–20.

170. Zhang, T., J.L. Stilwell, D. Gerion, L. Ding, O. Elboudwarej, P.A. Cooke, J.W. Gray, A.P. Alivisatos, and F.F. Chen. Cellular effect of high doses of silica-coated quantum dot profiled with high throughput gene expression analysis and high content cellomics measurements. *Nano Lett*, 2006, 6(4): 800–8.

171. Derfus, A.M., W.C.W. Chan, and S.N. Bhatia. Probing the cytotoxicity of semiconductor quantum dots. *Nano Lett*, 2004, 4(1): 11–18.

172. Mancini, M.C., B.A. Kairdolf, A.M. Smith, and S. Nie. Oxidative quenching and degradation of polymer-encapsulated quantum dots: new insights into the long-term fate and toxicity of nanocrystals *in vivo*. *J Am Chem Soc*, 2008, 130(33): 10836–37.

173. Green, M., and E. Howman. Semiconductor quantum dots and free radical induced DNA nicking. *Chem Commun (Camb)*, 2005(1): 121–23.

174. Aldana, J., Y.A. Wang, and X. Peng. Photochemical instability of CdSe nanocrystals coated by hydrophilic thiols. *J Am Chem Soc*, 2001, 123(36): 8844–50.

175. Rikans, L.E., and T. Yamano. *J Biochem Mol Toxicol*, 2000, 14: 110–17.

176. Aldana, J., N. Lavelle, Y. Wang, and X. Peng. Size-dependent dissociation pH of thiolate ligands from cadmium chalcogenide nanocrystals. *J Am Chem Soc*, 2005, 127(8): 2496–504.

177. Tsay, J.M., and X. Michalet. New light on quantum dot cytotoxicity. *Chem Biol*, 2005, 12(11): 1159–61.

178. Li, M., T. Kondo, Q.L. Zhao, F.J. Li, K. Tanabe, Y. Arai, Z.C. Zhou, and M. Kasuya. Apoptosis induced by cadmium in human lymphoma U937 cells through Ca2+-calpain and caspase-mitochondria-dependent pathways. *J Biol Chem*, 2000, 275(50): 39702–9.

179. Lemarie, A., D. Lagadic-Gossmann, C. Morzadec, N. Allain, O. Fardel, and L. Vernhet. Cadmium induces caspase-independent apoptosis in liver Hep3B cells: role for calcium in signaling oxidative stress-related impairment of mitochondria and relocation of endonuclease G and apoptosis-inducing factor. *Free Radic Biol Med*, 2004, 36(12): 1517–31.

180. Samia, A.C., X. Chen, and C. Burda. Semiconductor quantum dots for photodynamic therapy. *J Am Chem Soc*, 2003, 125(51): 15736–37.

181. Osseni, R.A., C. Debbasch, M.-O. Christen, and P. Rat. Tacrine-induced reactive oxygen species in a human liver cell line: the role of anethole dithiolethione as a scavenger. *Toxicol In Vitro*, 1999, 13: 683–88.

182. Brown, D.M., K. Donaldson, P.J. Borm, R.P. Schins, M. Dehnhardt, P. Gilmour, L.A. Jimenez, and V. Stone. Calcium and ROS-mediated activation of transcription factors and TNF-alpha cytokine gene expression in macrophages exposed to ultrafine particles. *Am J Physiol Lung Cell Mol Physiol*, 2004, 286(2): L344–53.

183. Cho, S.J., D. Maysinger, M. Jain, B. Roder, S. Hackbarth, and F.M. Winnik. Long-term exposure to CdTe quantum dots causes functional impairments in live cells. *Langmuir*, 2007, 23(4): 1974–80.

184. Rahman, I., S.K. Biswas, L.A. Jimenez, M. Torres, and H.J. Forman. Glutathione, stress responses, and redox signaling in lung inflammation. *Antioxid Redox Signal*, 2005, 7(1–2): 42–59.

185. Aillon, K.L., Y. Xie, N. El-Gendy, C.J. Berkland, and M.L. Forrest. Effects of nanomaterial physicochemical properties on *in vivo* toxicity. *Adv Drug Deliv Rev*, 2009, 61(6): 457–66.

186. Stroh, A., C. Faber, T. Neuberger, P. Lorenz, K. Sieland, P.M. Jakob, A. Webb, H. Pilgrimm, R. Schober, E.E. Pohl, and C. Zimmer. *In vivo* detection limits of magnetically labeled embryonic stem cells in the rat brain using high-field (17.6 T) magnetic resonance imaging. *Neuroimage*, 2005, 24(3): 635–45.

187. Matuszewski, L., T. Persigehl, A. Wall, W. Schwindt, B. Tombach, M. Fobker, C. Poremba, W. Ebert, W. Heindel, and C. Bremer. Cell tagging with clinically approved iron oxides: feasibility and effect of lipofection, particle size, and surface coating on labeling efficiency. *Radiology*, 2005, 235(1): 155–61.

188. Daldrup-Link, H.E., M. Rudelius, R.A. Oostendorp, M. Settles, G. Piontek, S. Metz, H. Rosenbrock, U. Keller, U. Heinzmann, E.J. Rummeny, J. Schlegel, and T.M. Link. Targeting of hematopoietic progenitor cells with MR contrast agents. *Radiology*, 2003, 228(3): 760–67.

189. Caravan, P., J.J. Ellison, T.J. McMurry, and R.B. Lauffer. Gadolinium(III) chelates as MRI contrast agents: structure, dynamics, and applications. *Chem Rev*, 1999, 99(9): 2293–352.

190. Thorek, D.L., A.K. Chen, J. Czupryna, and A. Tsourkas. Superparamagnetic iron oxide nanoparticle probes for molecular imaging. *Ann Biomed Eng*, 2006, 34(1): 23–38.

191. Arbab, A.S., W. Liu, and J.A. Frank. Cellular magnetic resonance imaging: current status and future prospects. *Expert Rev Med Devices*, 2006, 3(4): 427–39.

192. Bulte, J.W., and D.L. Kraitchman. Iron oxide MR contrast agents for molecular and cellular imaging. *NMR Biomed*, 2004, 17(7): 484–99.

193. Bulte, J.W. *In vivo* MRI cell tracking: clinical studies. *AJR Am J Roentgenol*, 2009, 193(2): 314–25.

194. Crich, S.G., L. Biancone, V. Cantaluppi, D. Duo, G. Esposito, S. Russo, G. Camussi, and S. Aime. Improved route for the visualization of stem cells labeled with a Gd-/Eu-chelate as dual (MRI and fluorescence) agent. *Magn Reson Med*, 2004, 51(5): 938–44.

195. Thomsen, H.S. Nephrogenic systemic fibrosis: a serious late adverse reaction to gadodiamide. *Eur Radiol*, 2006, 16(12): 2619–21.

196. Anderson, S.A., K.K. Lee, and J.A. Frank. Gadolinium-fullerenol as a paramagnetic contrast agent for cellular imaging. *Invest Radiol*, 2006, 41(3): 332–38.

197. Wolf, G.L., K.R. Burnett, E.J. Goldstein, and P.M. Joseph. Contrast agents for magnetic resonance imaging. *Magn Reson Annu*, 1985: 231–66.

198. Mendonca-Dias, M.H., E. Gaggelli, and P.C. Lauterbur. Paramagnetic contrast agents in nuclear magnetic resonance medical imaging. *Semin Nucl Med*, 1983, 13(4): 364–76.

199. Aoki, I., Y. Takahashi, K.H. Chuang, A.C. Silva, T. Igarashi, C. Tanaka, R.W. Childs, and A.P. Koretsky. Cell labeling for magnetic resonance imaging with the T1 agent manganese chloride. *NMR Biomed*, 2006, 19(1): 50–9.

200. Jun, Y.W., Y.M. Huh, J.S. Choi, J.H. Lee, H.T. Song, S. Kim, S. Yoon, K.S. Kim, J.S. Shin, J.S. Suh, and J. Cheon. Nanoscale size effect of magnetic nanocrystals and their utilization for cancer diagnosis via magnetic resonance imaging. *J Am Chem Soc*, 2005, 127(16): 5732–33.

201. McLachlan, S.J., M.R. Morris, M.A. Lucas, R.A. Fisco, M.N. Eakins, D.R. Fowler, R.B. Scheetz, and A.Y. Olukotun. Phase I clinical evaluation of a new iron oxide MR contrast agent. *J Magn Reson Imaging*, 1994, 4(3): 301–7.

202. Weissleder, R. Liver MR imaging with iron oxides: toward consensus and clinical practice. *Radiology*, 1994, 193(3): 593–95.

203. Kresse, M., S. Wagner, D. Pfefferer, R. Lawaczeck, V. Elste, and W. Semmler. Targeting of ultra-small superparamagnetic iron oxide (USPIO) particles to tumor cells *in vivo* by using transferrin receptor pathways. *Magn Reson Med*, 1998, 40(2): 236–42.

204. Laurent, S., D. Forge, M. Port, A. Roch, C. Robic, L. Vander Elst, and R.N. Muller. Magnetic iron oxide nanoparticles: synthesis, stabilization, vectorization, physicochemical characterizations, and biological applications. *Chem Rev*, 2008, 108(6): 2064–110.

205. Carrero, P., J.L. Burguera, M. Burguera, and C. Rivas. A time-based injector applied to the flow injection spectrophotometric determination of boron in plant materials and soils. *Talanta*, 1993, 40(12): 1967–74.

206. Gupta, A.K., and M. Gupta. Synthesis and surface engineering of iron oxide nanoparticles for biomedical applications. *Biomaterials*, 2005, 26(18): 3995–4021.

207. Sjogren, C.E., C. Johansson, A. Naevestad, P.C. Sontum, K. Briley-Saebo, and A.K. Fahlvik. Crystal size and properties of superparamagnetic iron oxide (SPIO) particles. *Magn Reson Imaging*, 1997, 15(1): 55–67.

208. Meldrum, F.C., B.R. Heywood, and S. Mann. Magnetoferritin: *in vitro* synthesis of a novel magnetic protein. *Science*, 1992, 257(5069): 522–23.

209. Bulte, J.W., T. Douglas, S. Mann, J. Vymazal, P.G. Laughlin, and J.A. Frank. Initial assessment of magnetoferritin biokinetics and proton relaxation enhancement in rats. *Acad Radiol*, 1995, 2(10): 871–78.

210. Bulte, J.W., T. Douglas, B. Witwer, S.C. Zhang, E. Strable, B.K. Lewis, H. Zywicke, B. Miller, P. van Gelderen, B.M. Moskowitz, I.D. Duncan, and J.A. Frank. Magnetodendrimers allow endosomal magnetic labeling and *in vivo* tracking of stem cells. *Nat Biotechnol*, 2001, 19(12): 1141–17.

211. Bulte, J.W., M. de Cuyper, D. Despres, and J.A. Frank. Short- vs. long-circulating magnetoliposomes as bone marrow-seeking MR contrast agents. *J Magn Reson Imaging*, 1999, 9(2): 329–35.

212. Huh, Y.M., Y.W. Jun, H.T. Song, S. Kim, J.S. Choi, J.H. Lee, S. Yoon, K.S. Kim, J.S. Shin, J.S. Suh, and J. Cheon. *In vivo* magnetic resonance detection of cancer by using multifunctional magnetic nanocrystals. *J Am Chem Soc*, 2005, 127(35): 12387–91.

213. Pileni, M.P. The role of soft colloidal templates in controlling the size and shape of inorganic nanocrystals. *Nat Mater*, 2003, 2(3): 145–50.

214. Nitin, N., L.E. LaConte, O. Zurkiya, X. Hu, and G. Bao. Functionalization and peptide-based delivery of magnetic nanoparticles as an intracellular MRI contrast agent. *J Biol Inorg Chem*, 2004, 9(6): 706–12.

215. Konwarh, R., N. Karak, S.K. Rai, and A.K. Mukherjee. Polymer-assisted iron oxide magnetic nanoparticle immobilized keratinase. *Nanotechnology*, 2009, 20(22): 225107.

216. Kim, K.S., and J.K. Park. Magnetic force-based multiplexed immunoassay using superparamagnetic nanoparticles in microfluidic channel. *Lab Chip*, 2005, 5(6): 657–64.

217. Kohler, N., G.E. Fryxell, and M. Zhang. A bifunctional poly(ethylene glycol) silane immobilized on metallic oxide-based nanoparticles for conjugation with cell targeting agents. *J Am Chem Soc*, 2004, 126(23): 7206–11.

218. Kim, S.W., S. Kim, J.B. Tracy, A. Jasanoff, and M.G. Bawendi. Phosphine oxide polymer for water-soluble nanoparticles. *J Am Chem Soc*, 2005, 127(13): 4556–57.

219. Rogers, W.J., C.H. Meyer, and C.M. Kramer. Technology insight: *in vivo* cell tracking by use of MRI. *Nat Clin Pract Cardiovasc Med*, 2006, 3(10): 554–62.

220. Arbab, A.S., G.T. Yocum, L.B. Wilson, A. Parwana, E.K. Jordan, H. Kalish, and J.A. Frank. Comparison of transfection agents in forming complexes with ferumoxides, cell labeling efficiency, and cellular viability. *Mol Imaging*, 2004, 3(1): 24–32.

221. Arbab, A.S., L.A. Bashaw, B.R. Miller, E.K. Jordan, B.K. Lewis, H. Kalish, and J.A. Frank. Characterization of biophysical and metabolic properties of cells labeled with superparamagnetic iron oxide nanoparticles and transfection agent for cellular MR imaging. *Radiology*, 2003, 229(3): 838–46.

222. Zhao, M., M.F. Kircher, L. Josephson, and R. Weissleder. Differential conjugation of Tat peptide to superparamagnetic nanoparticles and its effect on cellular uptake. *Bioconjug Chem*, 2002, 13(4): 840–44.

223. Arbab, A.S., G.T. Yocum, H. Kalish, E.K. Jordan, S.A. Anderson, A.Y. Khakoo, E.J. Read, and J.A. Frank. Efficient magnetic cell labeling with protamine sulfate complexed to ferumoxides for cellular MRI. *Blood*, 2004, 104(4): 1217–23.

224. Zhang, Z., E.J. van den Bos, P.A. Wielopolski, M. de Jong-Popijus, D.J. Duncker, and G.P. Krestin. High-resolution magnetic resonance imaging of iron-labeled myoblasts using a standard 1.5-T clinical scanner. *Magma*, 2004, 17(3–6): 201–9.

225. Josephson, L., C.H. Tung, A. Moore, and R. Weissleder. High-efficiency intracellular magnetic labeling with novel superparamagnetic-Tat peptide conjugates. *Bioconjug Chem*, 1999, 10(2): 186–91.

226. Hoshino, K., H.Q. Ly, J.V. Frangioni, and R.J. Hajjar. *In vivo* tracking in cardiac stem cell-based therapy. *Prog Cardiovasc Dis*, 2007, 49(6): 414–20.

227. Poirier-Quinot, M., G. Frasca, C. Wilhelm, N. Luciani, J.C. Ginefri, L. Darrasse, D. Letourneur, C. Le Visage, and F. Gazeau. High resolution 1.5T magnetic resonance imaging for tissue engineering constructs: a noninvasive tool to assess 3D scaffold architecture and cell seeding. *Tissue Eng Part C Methods*, 2009.

228. Ko, I.K., H.T. Song, E.J. Cho, E.S. Lee, Y.M. Huh, and J.S. Suh. *In vivo* MR imaging of tissue-engineered human mesenchymal stem cells transplanted to mouse: a preliminary study. *Ann Biomed Eng*, 2007, 35(1): 101–8, 2010, 16(2): 185–200.

229. Bartusik, D., B. Tomanek, E. Lattova, H. Perreault, J. Tuszynski, and G. Fallone. The efficacy of new colchicine derivatives and viability of the T-Lymphoblastoid cells in three-dimensional culture using 19F MRI and HPLC-UV ex vivo. *Bioorg Chem*, 2009, 37(6): 193–201.

230. Wilhelm, C., L. Bal, P. Smirnov, I. Galy-Fauroux, O. Clement, F. Gazeau, and J. Emmerich. Magnetic control of vascular network formation with magnetically labeled endothelial progenitor cells. *Biomaterials*, 2007, 28(26): 3797–806.

231. Jendelova, P., V. Herynek, J. DeCroos, K. Glogarova, B. Andersson, M. Hajek, and E. Sykova. Imaging the fate of implanted bone marrow stromal cells labeled with superparamagnetic nanoparticles. *Magn Reson Med*, 2003, 50(4): 767–76.

232. Bulte, J.W., S. Zhang, P. van Gelderen, V. Herynek, E.K. Jordan, I.D. Duncan, and J.A. Frank. Neurotransplantation of magnetically labeled oligodendrocyte progenitors: magnetic resonance tracking of cell migration and myelination. *Proc Natl Acad Sci USA*, 1999, 96(26): 15256–61.

233. Lee, I.H., J.W. Bulte, P. Schweinhardt, T. Douglas, A. Trifunovski, C. Hofstetter, L. Olson, and C. Spenger. *In vivo* magnetic resonance tracking of olfactory ensheathing glia grafted into the rat spinal cord. *Exp Neurol*, 2004, 187(2): 509–16.

234. Dunning, M.D., A. Lakatos, L. Loizou, M. Kettunen, C. Ffrench-Constant, K.M. Brindle, and R.J. Franklin. Superparamagnetic iron oxide-labeled Schwann cells and olfactory ensheathing cells can be traced *in vivo* by magnetic resonance imaging and retain functional properties after transplantation into the CNS. *J Neurosci*, 2004, 24(44): 9799–810.

235. Ben-Hur, T., R.B. van Heeswijk, O. Einstein, M. Aharonowiz, R. Xue, E.E. Frost, S. Mori, B.E. Reubinoff, and J.W. Bulte. Serial *in vivo* MR tracking of magnetically labeled neural spheres transplanted in chronic EAE mice. *Magn Reson Med*, 2007, 57(1): 164–71.

236. Anderson, S.A., J. Glod, A.S. Arbab, M. Noel, P. Ashari, H.A. Fine, and J.A. Frank. Noninvasive MR imaging of magnetically labeled stem cells to directly identify neovasculature in a glioma model. *Blood*, 2005, 105(1): 420–25.

237. Kraitchman, D.L., A.W. Heldman, E. Atalar, L.C. Amado, B.J. Martin, M.F. Pittenger, J.M. Hare, and J.W. Bulte. *In vivo* magnetic resonance imaging of mesenchymal stem cells in myocardial infarction. *Circulation*, 2003, 107(18): 2290–93.

238. Cahill, K.S., G. Gaidosh, J. Huard, X. Silver, B.J. Byrne, and G.A. Walter. Noninvasive monitoring and tracking of muscle stem cell transplants. *Transplantation*, 2004, 78(11): 1626–33.

239. Liu, Z.Y., Y. Wang, G.Y. Wang, X.H. Li, Y. Li, and C.H. Liang. [*In vivo* magnetic resonance imaging tracking of transplanted adipose-derived stem cells labeled with superparamagnetic iron oxide in rat hearts]. *Zhongguo Yi Xue Ke Xue Yuan Xue Bao*, 2009, 31(2): 187–91.

240. Kim, Y.J., Y.M. Huh, K.O. Choe, B.W. Choi, E.J. Choi, Y. Jang, J.M. Lee, and J.S. Suh. *In vivo* magnetic resonance imaging of injected mesenchymal stem cells in rat myocardial infarction; simultaneous cell tracking and left ventricular function measurement. *Int J Cardiovasc Imaging*, 2009, 25 (Suppl 1): 99–109.

241. Delo, D.M., J. Olson, P.M. Baptista, R.B. D'Agostino, Jr., A. Atala, J.M. Zhu, and S. Soker. Noninvasive longitudinal tracking of human amniotic fluid stem cells in the mouse heart. *Stem Cells Dev*, 2008, 17(6): 1185–94.

242. Akbari, A.A., H. Mozdarani, S. Akhlaghpoor, A.A. Pourfatollah, and M. Soleimani. Evaluation of the homing of human CD34+ cells in mouse bone marrow using clinical MR imaging. *Pak J Biol Sci*, 2007, 10(6): 833–42.

243. Daldrup-Link, H.E., M. Rudelius, G. Piontek, S. Metz, R. Brauer, G. Debus, C. Corot, J. Schlegel, T.M. Link, C. Peschel, E.J. Rummeny, and R.A. Oostendorp. Migration of iron oxide-labeled human hematopoietic progenitor cells in a mouse model: *in vivo* monitoring with 1.5-T MR imaging equipment. *Radiology*, 2005, 234(1): 197–205.

244. Yocum, G.T., L.B. Wilson, P. Ashari, E.K. Jordan, J.A. Frank, and A.S. Arbab. Effect of human stem cells labeled with ferumoxides-poly-L-lysine on hematologic and biochemical measurements in rats. *Radiology*, 2005, 235(2): 547–52.

245. Stark, D.D., R. Weissleder, G. Elizondo, P.F. Hahn, S. Saini, L.E. Todd, J. Wittenberg, and J.T. Ferrucci. Superparamagnetic iron oxide: clinical application as a contrast agent for MR imaging of the liver. *Radiology*, 1988, 168(2): 297–301.

246. Weissleder, R., G. Elizondo, J. Wittenberg, A.S. Lee, L. Josephson, and T.J. Brady. Ultrasmall superparamagnetic iron oxide: an intravenous contrast agent for assessing lymph nodes with MR imaging. *Radiology*, 1990, 175(2): 494–98.

247. de Vries, I.J., W.J. Lesterhuis, J.O. Barentsz, P. Verdijk, J.H. van Krieken, O.C. Boerman, W.J. Oyen, J.J. Bonenkamp, J.B. Boezeman, G.J. Adema, J.W. Bulte, T.W. Scheenen, C.J. Punt, A. Heerschap, and C.G. Figdor. Magnetic resonance tracking of dendritic cells in melanoma patients for monitoring of cellular therapy. *Nat Biotechnol*, 2005, 23(11): 1407–13.

248. Zhu, J., L. Zhou, and F. XingWu. Tracking neural stem cells in patients with brain trauma. *N Engl J Med*, 2006, 355(22): 2376–78.

249. Toso, C., J.P. Vallee, P. Morel, F. Ris, S. Demuylder-Mischler, M. Lepetit-Coiffe, N. Marangon, F. Saudek, A.M. James Shapiro, D. Bosco, and T. Berney. Clinical magnetic resonance imaging of pancreatic islet grafts after iron nanoparticle labeling. *Am J Transplant*, 2008, 8(3): 701–6.

250. Kostura, L., D.L. Kraitchman, A.M. Mackay, M.F. Pittenger, J.W. Bulte, Feridex labeling of megenchymal stem cells inhibits chongrogenesis but not adipogenesis or osteogenesis. *NMR Biomed*, 2004, 17(7): 513–517.

3

Gold Nanoparticles in Biomedicine

Dina Nemr and Hassan M.E. Azzazy

Department of Chemistry, the American University in Cairo, New Cairo, Egypt

CONTENTS

3.1 Introduction

Gold nanoparticles (AuNPs) were first recognized in the form of colloidal gold. Dating back to the fourth and fifth centuries B.C., colloidal gold was used in glass and ceramic staining, as exemplified by the Lycurgus Cup, the color of which is ruby red upon transmission of light and green upon reflectance of light.[1] Drinking colloidal gold was thought to have healing powers and was recorded as such in several documents historically. It was Michael Faraday, in 1857, who is said to have first reported the synthesis of red colloidal gold by the reduction of chloroaurate ($AuCl_4^-$) and explored its optical properties.[1]

Indeed, it is the unique optical properties of AuNPs that make them such attractive tools in biomedicine, in addition to their inertness, biocompatibility, and high surface-to-volume ratio. At the nanoscale, AuNPs exhibit properties that are very different from gold in its bulk form, and surface properties become more pronounced. In this chapter, we discuss the properties of AuNPs, their synthesis by chemical and biological methods, and their applications in biomedicine. Assessments of cytotoxicity and biodistribution, as well as future prospects and challenges of AuNPs in biomedical applications, are also discussed herein.

3.2 Gold Nanostructures

AuNPs can be made in a variety of shapes that have an effect on their optical features (Figure 3.1). The oldest and most popular gold nanoparticles are the gold nanospheres, which range in size from 0.8 to 250 nm.[2] Also spherical in shape are the gold nanoshells, which are composed of a dielectric core enclosed in a thin gold shell, where the core is usually made of silica or gold sulfide.[3] The dielectric core helps shift the plasmon resonance into the NIR region: an optical property which is further discussed in the next

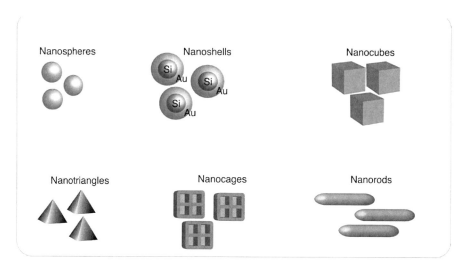

FIGURE 3.1
Most commonly used gold nanostructures in biomedical applications.

section.[3] Nanoshells largely in use range in size from 80 to 150 nm.[4,5] Smaller nanoshells ranging from 20 to 60 nm with a core of iron oxide nanocrystals have also been explored.[6] An advantage of using iron oxide is its magnetic properties, which can be harnessed for example in directing the magnetic AuNPs to a target site using an external magnet.

Though the spherical gold nanostructures have enjoyed wide popularity and applications, they do have their limitations with regard to their physical properties, such as their limited tunability into the near-infrared (NIR) region.[7] This urged scientists to develop new and innovative methods by which anisotropic nanoparticles can be obtained,[8,9] paving the way for a multitude of different shapes and sizes of AuNPs, as well as applications. Now AuNPs can be synthesized with geometries such as nanorods,[7,10] nanocages,[11,12] triangular nanoprisms,[7,13–15] nanocubes,[7,15–17] and ellipsoidal,[18] bipyramidal,[17] and branched[7,17] AuNPs.

Gold nanorods, as opposed to the spherical shapes, have two dimensions: length and diameter, also referred to as short and long axes, with transverse and longitudinal bands.[19] Gold nanocages, on the other hand, are hollow nanocubes, like boxes, with holes in their walls.[20]

3.3 Properties

3.3.1 Optical Properties (Radiative)

3.3.1.1 Surface Plasmon Resonance

Metals have conductive electrons, which, upon undergoing collective excitation, are referred to as plasmons.[21] The free conduction band electrons on the surface of AuNPs interact with the electromagnetic component of light and oscillate in a collective dipolar manner, known as surface localized plasmon.[21] When the surface plasmon frequency is close to that of the excitation of light, it is referred to as surface plasmon resonance (SPR), which lies within visible frequencies.[22]

Controlling the size and shape of AuNPs alters their properties by controlling the motion and degree of confinement of their electrons.[19] This is why AuNPs exhibit properties that are very different from gold in its bulk form, and surface properties are more pronounced and dominant.[19,22] As a result, the SPR enhances the radiative properties of AuNPs, such as light scattering and absorption.[19]

The size, shape, and composition of the gold nanostructures affect their SPR properties. The SPR of gold nanospheres lies within the visible region, and increasing their size causes a limited shift into the NIR.[19] In nanoshells, changing the shell thickness-to-core ratio causes a greater shift in SPR from the visible region to as far as the mid-infrared region.[23,24] Gold nanorods, on the other hand, have two axes, upon which oscillations occur in two directions. The short axis has a transverse SPR band similar to that of a nanosphere, and therefore exhibits similar absorption and scattering properties in the visible region, while the longitudinal axis exhibits a much stronger absorption band.[19] The nanorods can be tuned to absorb 650–1,500 nm,[21] not necessarily by changing the diameter or short axis,[18] but rather by changing the aspect ratio: length/diameter.[21] The higher the aspect ratio, the more red shifted the longitudinal axis becomes.[19] In nanospheres, nanoshells, and nanorods at a fixed aspect ratio, absorption increases in smaller sizes and scattering increases in larger sizes. Thus, a nanoparticle of 20 nm mainly absorbs incident light with little or no scattering, whereas for a nanoparticle of 40 nm, scattering begins to emerge, and its contribution to total light extinction (sum of absorption and scattering) increases as the particle size increases.[25] This allows the utilization of AuNPs with different scattering/absorption properties in different applications. For example, large nanoparticles could be used in reflectance-based imaging due to their higher scattering abilities, while small nanoparticles could be used in absorption-based imaging due to their higher absorption abilities, or in photothermal therapy, where light energy is converted into heat.[25]

The absorption and scattering wavelengths are determined by the size and shape of AuNPs, which account for their different colors. Nanospheres of 20 nm have SPR wavelengths in the 520 nm visible region, appearing as red,[26] while nanospheres larger than 100 nm have a grimy yellowish color.[27] Nanospheres generally exhibit a yellow, orange, red, pink, or purple color; nanorods: green, blue, magenta[28]; triangular nanoprisms: yellow, orange, pink, or green.[29] These color ranges depend on particle size and aspect ratio, as well as the refractive index of the solvent they are in.[26]

3.3.1.2 Plasmon-Plasmon Interactions

Another property of AuNPs is the plasmon-plasmon interaction. Spherical AuNPs are synthesized in the form of a colloidal solution that has an intense red color. If the AuNPs are brought closer to one another, electromagnetic coupling occurs and the AuNPs aggregate.[30] These interactions between adjacent gold nanoparticles lead to a decrease in the SPR energy absorption band. This increases the absorbance of the red wavelength by the aggregated AuNPs[30] and changes the colloid color from red to blue: a phenomenon known as the red shift.[24,31,32] Some changes in conditions may induce the aggregation of AuNPs, such as the addition of NaCl to citrate capped or negatively charged AuNPs. Citrate-capped colloidal AuNPs are negatively charged and remain in suspension due to electrostatic repulsions. The addition of NaCl neutralizes these negative charges, and the AuNPs become destabilized and aggregate, the color of their solution turning from red to blue. This property is especially useful in colorimetric detection methods in diagnostic applications.

3.3.2 Physical Properties (Nonradiative)

When AuNPs absorb light, the electron oscillations lead to electron-electron collisions, which convert light energy into heat.[33] This is followed by a transfer of the energy through electron-phonon interactions, causing a further rise in temperature. The fate of the heat generated is either to be transferred to the surrounding environment, or melt or ablate the nanoparticles.

3.4 Synthesis of AuNPs

3.4.1 Chemical Synthesis

3.4.1.1 Colloidal AuNPs

Colloidal AuNPs can be synthesized chemically by the reduction of tetrachloroauric acid ($HAuCl_4$) using a reducing agent.[21] Citrate is commonly used as both a reducing and capping agent when added to $HAuCl_4$ in water,[23] producing AuNPs with an average size of 10–60 nm.[34] The size can be controlled by changing the gold-to-citrate concentration ratio.

3.4.1.2 Thiolated AuNPs

The Brust–Schiffrin method produces thiolated AuNPs with a size range from 1 to 6 nm (depending on the gold/thiol ratio) by reacting the $AuCl_4^-$ salt with sodium borohydride in the presence of a thiol ligand. Brust et al.[35] used a two-phase system, where tetraoctylammonium bromide is used to transfer $AuCl_4^-$ from aqueous solution to toluene, followed by reduction using sodium borohydride in the presence of dodecanethiol. This yielded 1–3 nm thiol-stabilized AuNPs. In a later experiment, Brust et al.[36] further developed the method to a single-phase reduction of $AuCl_4^-$ by sodium borohydride to form 2 nm gold nanoclusters stabilized by mercaptophenol. The Brust–Schiffrin method has thus paved the way for the simultaneous synthesis and functionalization of AuNPs with thiol groups. The thiol groups impart higher stability to AuNPs and can be used to conjugate AuNPs to other functional groups.

3.4.1.3 Gold Nanoshells

For gold nanoshells, a dielectric core, such as silica, is used in a four-step process developed by Oldenburg et al.[37] to produce a silica core enclosed within a thin gold shell. The silica nanoparticles are synthesized using the Stober method,[38] where alkyl silicates are hydrolyzed followed by the condensation of silicic acid in alcohol in the presence of an ammonia catalyst. The silica nanoparticles then undergo amination by means of surface adsorption of organosilane (3-aminopropyltriethoxysilane), with amine terminals extending from the surface of the silica nanoparticles. A gold colloid solution is prepared by reduction of $HAuCl_4$ by tetrakis(hydroxymethyl)-phosphonium chloride (THPC), producing 1–2 nm sized spherical AuNPs.[39] Addition of the colloidal gold to the silica nanoparticle solution leads to their attachment via covalent bonding. The Au-silica particles are then added as seeds to a solution of aged $HAuCl_4$ and potassium carbonate, and reduced by sodium borohydride or citrate, to give the final uniform thin gold shell with a silica core. The aspect ratio (core/shell ratio) is controlled by varying the size of the silica core or the initial colloidal AuNPs. Nanoshells are typically synthesized with total diameters between 80 and 150 nm,[4,5] though smaller nanoshells with diameters between 10 and 60 nm have also been produced.[40] Alternatively, the silica core can be replaced with an iron oxide core to give sizes between 45 and 60 nm.[41]

3.4.1.4 Gold Nanorods

The seed-mediated two-step method is most popular for the synthesis of gold nanorods.[42] In the first step, sodium borohydride is added to a solution of cetyl trimethylammonium bromide (CTAB) and tetrachloroauric acid ($HAuCl_4$) to form the initial gold seeds. In the second step, the gold seeds are then added to a solution of CTAB with silver nitrate, $HAuCl_4$, and ascorbic acid as the reducing agent.

There are numerous modifications that have been made to this method. For example, to obtain longer rods, and thereby a higher aspect ratio, the rods produced in the second step are used as seeds in a third step.[43] Alternatively, longer nanorods may also be produced by omitting silver nitrate from the growth solution, giving an aspect ratio of up to 25.[44,45] Gold nanorods can also be synthesized by using a cosurfactant of benzyldimethyl-hexadecyl ammonium chloride (BDAC) and CTAB.[42,44,45] Other controlling parameters include the size, shape, and concentration of the seeds; the pH and temperature of the growth solution[46,47]; and the growth time,[42] concentrations of ascorbic acid, surfactant, and initial gold precursor.[48]

3.4.2 Biological Synthesis

Biosynthesis of nanomaterials is gaining popularity as an environmentally friendly alternative to the conventional chemical methods, which may involve high energy consumption, as well as production and use of toxic or hazardous material. Biological synthesis is relatively low cost, energy efficient, and nontoxic.[29,49,50] Successful bioreduction of gold ions has been done using bacteria, fungi, algae, plants, and plant extracts.

Bacteria can deposit the nanoparticles intracellularly or extracellularly.[49] Konishi et al.[51] used the Fe(III)-reducing bacteria *Shewanella algae* to reduce Au(III) with H_2 as the electron donor and under pH conditions ranging from 2.0 to 7.0. The bacteria were cultured under anaerobic conditions, where lactate or H_2 acted as the electron donor and Fe(III)citrate acted as the electron acceptor. The bacteria were incubated for 3–5 days, harvested by centrifugation, and resuspended in a bicarbonate buffer of pH 7.0. This process was repeated several times before the bacteria were ready to be used in bioreduction.

At pH 7.0, AuNPs of sizes between 10 and 20 nm were deposited in the periplasmic space of the bacterial cells. At pH 2.8, AuNPs of different shapes ranging in size from 15 to 200 nm, especially 100–200 nm triangular and hexagonal AuNPs, were deposited on the bacterial cells. At pH 2.0, intracellular deposition of 20 nm AuNPs occurred, along with extracellular deposition of 350 nm AuNPs. Complete reduction of 1 mM $AuCl_4^-$ was achieved within 30 min at 25°C, corresponding to the chemical reduction of 20 mM citric acid at 50°C.

Fungi are generally easy to handle, and most have the advantage of extracellular secretion of reducing enzymes, which makes it easier to retrieve the synthesized AuNPs.[49] Their drawbacks are that they are slow reducers (48–120 h),[52–55] and being eukaryotes, are harder to manipulate genetically.[50] Perhaps the most studied fungus in the biosynthesis of nanoparticles of different kinds is *Fusarium oxysporum*.[49] Using this fungus, Mukherjee et al.[56] observed the extracellular deposition of 20–40 nm stable AuNPs of spherical and triangular morphologies.

Singaravelu et al.[57] were the first to report the use of marine alga in the biosynthesis of AuNPs. *Sargassum wightii* was found to produce extracellular AuNPs in 12 h.

Wang et al.[58] used Barbated Skullcup, a Chinese herbal extract, to produce AuNPs of an average size range of 10–100 nm. Almost complete reduction of $HAuCl_4$ occurred in 3 h at room temperature, and the AuNPs produced were highly stable and of fair monodispersity.

3.4.3 Characterization

Transmission electron microscopy (TEM) is the most common method used to determine the shape and size of the AuNPs, as well as their dispersity. Other characterization methods used to determine the dimensions of AuNPs include scanning electron microscopy, atomic force microscopy, small-angle x-ray scattering (SAXS), and x-ray diffraction. Spectroscopic methods rely on the absorption and scattering properties of AuNPs. The SPR measured, however, is based on the collective oscillations of a group of AuNPs, which may be of various shapes and sizes. It is therefore used as a measure of the heterogeneity of the AuNPs produced.[19] Dark field scattering[59] and absorptive properties[60] of the AuNPs can be measured to determine the spectrum of each nanostructure individually. SAXS can be used to monitor the growth of AuNPs, and the thermodynamics of growth can be measured by monitoring changes in temperature during growth by isothermal titration calorimetry.[61] TEM can also be used to determine the mean number of gold atoms (N_{Au}) in a gold nanoparticle by measuring the mean diameter (d) of the AuNP core and volume of Au atoms (V_{Au}), following Leff's model:[62]

$$N_{Au} = 4 \pi (d/2)^3/V_{Au}$$

3.5 Biomedical Applications

3.5.1 Therapeutics

3.5.1.1 Drug Delivery

Owing to their high surface-to-volume ratio, optical tunability, low cytotoxicity, and ability to interact with organic molecules, AuNPs can serve as excellent vehicles for drug delivery. The high surface-to-volume ratio allows for a large amount of drugs to be loaded onto the same nanoparticle, and the surface properties of the gold nanoparticles can be tuned to alter hydrophobicity and charge, improving cellular uptake. In addition, AuNPs readily form strong linkages with thiols, which can be used to link gold nanoparticles with other molecules. By using AuNPs as drug delivery carriers, a higher concentration of the drug can be delivered to the specific cells or tissues with higher efficiency and less systemic adverse effects.[63]

3.5.1.1.1 Surface Modification

In order for an AuNP to be able to carry a drug, surface modification is first performed to enable its conjugation to the selected molecule, as well as to increase its biocompatibility and stability. AuNPs are modified according to the application in which they will be utilized and to the desired molecule to be attached to the nanoparticles. Surface modification is done to increase water solubility of the AuNPs, prevent their aggregation, as well as prevent or slow down their uptake by the reticuloendothelial system (RES), thereby increasing their circulation lifetime.[63,64] It allows for further functionalization of the AuNPs with conjugates such as antibodies, biosensors, and other ligands, and may also be done to remove or replace original capping agents.[63] For example, the seed-mediated growth method produces gold nanorods capped with CTAB, which is cytotoxic,[65] but can be replaced via thiol exchange.[66]

Surface modification occurs by covalent bond formation or by noncovalent interactions. Though the covalent bonds are stronger, they are also plagued with loss of structural integrity of the conjugate and difficulty in drug release, and while noncovalent interactions offer more flexibility, they may also be less stable. Surface modification strategies include directly linking the biomolecules to the AuNPs, or functionalizing the biomolecules with thiol linkers first, and then allowing them to react with the AuNPs. Other strategies depend on hydrophobic interactions and electrostatic adsorption, in a layer-by-layer manner, to coat the surface of the AuNPs. For example, AuNPs can be coated with poly(sodium-4-stryrene-sulfonate) (PSS). PSS is anionic and changes the surface charge of the AuNPs from positive to negative, allowing their further conjugation to antibodies, or other proteins and biomolecules with positively charged groups[19] (Figure 3.2D). On the other hand, direct electrostatic adsorption can occur at pH values higher than the protein's isoelectric point, causing it to become negatively charged and enabling its attachment to the AuNPs.[19]

In other cases, the surface modification occurs during the synthesis of the AuNPs, as opposed to postsynthesis. For example, the Brust–Schiffrin method produces thiolated

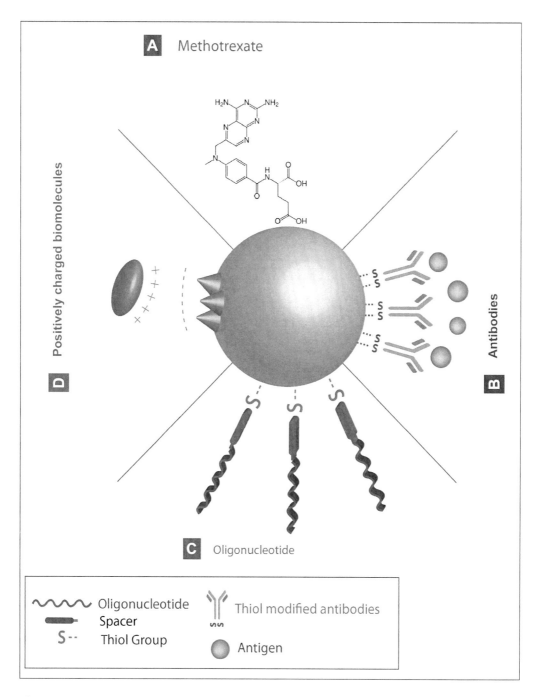

FIGURE 3.2
Functionalization of gold nanoparticles with (A) methotrexate attached by replacing the citrate cap, (B) antibodies, (C) oligonucleotides (with a spacer), and (D) electrostatic interactions between a negatively charged citrate-capped AuNP and a positively charged biomolecule.

AuNPs stabilized by 4-mercaptophenol (Figure 2.3B), with a size range from 1.5 to 6 nm.[35] The citrate reduction method also yields citrate-capped AuNPs that can be directly conjugated to biomolecules (Figure 2.3A). AuNPs can bear a negative or positive charge during synthesis by reduction of $HAuCl_4$ in the presence of, for example, citrate or 2-aminoethanethiol, respectively[34,67]; or postsynthesis by functionalization with ligands having acidic or amine groups, respectively.[68]

Niidome et al.[66] studied the biodistribution of polyethylene glycol (PEG) versus non-PEG AuNPs. After 0.5 h, 54% of the PEG NPs were found in the blood, and after 72 h, 35% were found to have accumulated in the liver; with the non-PEG NPs, 30% accumulation in the liver occurred after only 0.5 h. The length of the PEG chains affects the stabilizing capacity of PEG for AuNPs in circulation, where shorter chains of molecular weight below 2,000 were found to have lower capacity. Another role for PEG is to be used as a spacer between the AuNP and conjugated molecules, improving access to their targets.[69] This, for example, causes a conjugated oligonucleotide to be spaced farther away from the AuNP, increasing the accessibility of the oligonucleotide for hybridization with its target (Figure 3.2C).

Thiolated PEG is also used to replace the CTAB surfactant on gold nanorods.[66] Pegylation bestows higher bioavailability and lower immunogenicity and uptake by RES on AuNPs.[70,71] Surface modification using alkanethiol-terminated PEG chains at low concentrations results in a mushroom-shaped conformation, while high concentrations result in a brush-like conformation[72,73] (Figure 3.3C and D).

3.5.1.1.2 *Types of Targeting*

There are two ways by which cancer cells can be targeted: passive and active targeting, the latter being more tumor cell specific[74,75] (Figure 3.4). Drugs may be delivered via active targeting, passive targeting, or a combination of both, and once the AuNPs have reached their target destination, drug release can be triggered by internal or external stimuli.

In passive targeting, the leaky vasculature surrounding malignant cells[75–77] and dysfunctional lymphatic drainage[78,79] allow particles of sizes up to 400 nm to extravasate and accumulate in the tumor,[79,80] an effect known as enhanced permeability and retention (EPR)[77,78] (Figure 3.4). However, not all tumors show EPR,[81] and passive targeting does not enable the targeting of single cells or small tumors.[82]

Active targeting, on the other hand, uses molecular recognition by conjugating the AuNPs to molecular probes, such as antibodies or other ligands (Figure 3.2), which bind specifically to the targeted cells[83–85] (Figure 3.4).

Folate can be used for cancer cell targeting. Folate is a precursor needed by cells for nucleic acid synthesis and is overexpressed by ovarian, breast, lung, brain, kidney, prostate, colon, and some myeloid cancer cells, among others. Normal cells also express folate receptors, but on the apical surface of their polarized epithelium. This polarization is lost with malignant transformation, and the folate receptors become accessible to blood-borne molecules through the vascular epithelium.[86] Therefore, only the overexpressed folate receptors of malignant cells are targeted, while folate receptors of normal cells are protected.[86]

Bhattacharya et al.[86] modified 5 nm gold nanospheres with polyethylene glycol (PEG)-amines and PEG-thiols and further functionalized them with folic acid through noncovalent interactions to target folate receptors on ovarian and multiple myeloma cancer cell lines. AuNPs modified with PEG-amines had a better binding capacity and drug release profile than those modified with PEG-thiols. Cancer cell lines that overexpressed folate receptors the most showed the highest uptake of the folic acid conjugates.

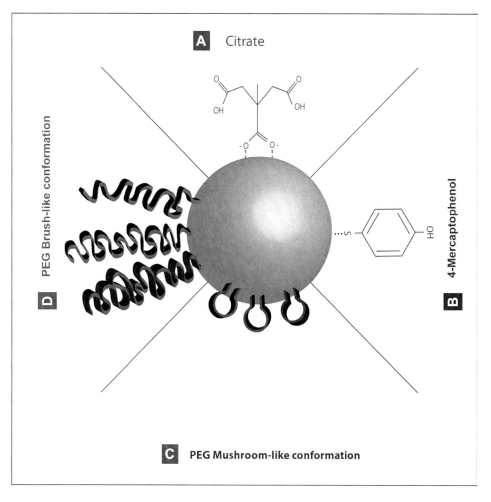

FIGURE 3.3
Surface modification of gold nanoparticles with (A) citrate, (B) 4-mercaptophenol, (C) PEG, mushroom-like conformation, and (D) PEG, brush-like conformation.

Galactose was used by Bergen et al.[87] for active targeting of hepatocytes. Galactose was conjugated to AuNPs of different sizes targeting hepatocytes via recognition by galactose receptor asialoglycoprotein. The AuNPs were synthesized by the reduction of $HAuCl_4$ with sodium borohydride in the presence of 2-aminoethanethiol. The galactose-PEG-AuNPs were prepared by reacting galactose-PEG-thiol (O-pyridyl disulfide) with colloidal gold. Uptake of galactose-PEG-AuNPs was found to be 16-fold that of PEG-AuNPs by hepatocytes.

Kang et al.[88] targeted the cell nucleus by conjugating 30 nm pegylated (PEG) AuNPs to an arginine-glycine-aspartic acid peptide (RGD; the cell attachment site of adhesive extracellular matrix and cell surface proteins), as well as a lysine-lysine-lysine-arginine-lysine peptide, which is a nuclear localization signal (NLS) peptide. The RGD targets cell surface alpha v beta 6 integrins, facilitating the internalization of the AuNPs via endocytosis. The NLS associates with the cytoplasmic karyopherins, and then translocation into the nucleus occurs. AuNPs conjugated only to RGD were compared to AuNP conjugated to both RGD and NLS. The RGD-AuNPs were specific to the cytoplasm of the cancer cells, while the RGD-NLS-AuNPs were specific to the nuclei of the cancer cells.

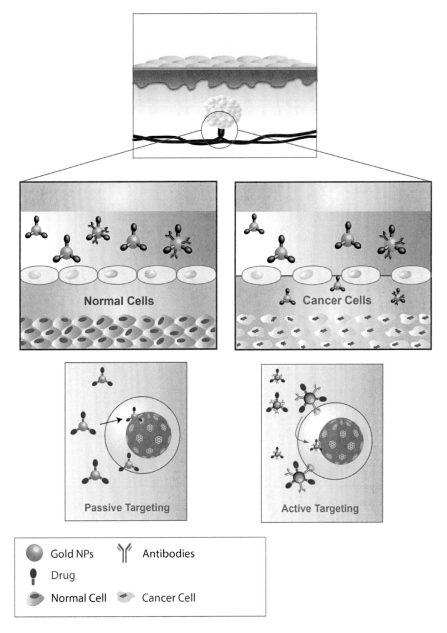

FIGURE 3.4
Different types of targeting.

3.5.1.1.3 Drug Release

Drug release can occur as a result of internal or external stimuli. An example of an internal stimulus for intracellular drug release is glutathione (GSH). The intracellular concentration of GSH is 1–10 mM, while the plasma concentration is 2 μM.[89] This vast concentration difference is utilized for intracellular drug release, via place exchange between GSH and thiols, which does not occur extracellularly due to insufficient GSH concentrations.

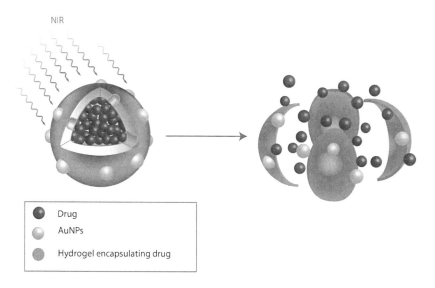

FIGURE 3.5
Drug delivery using AuNPs embedded in a hydrogel. AuNPs are embedded in a hydrogel encapsulating the drug. NIR irradiation leads to hydrogel collapse and drug release.

Hong et al.[89] coated 2 nm AuNPs with a monolayer of a cationic tetra(ethylene glycol)-conjugated ligand and a thiolated hydrophobic dye. The AuNPs were then incubated with human liver cells (Hep G2). The cationic ligand allowed the AuNPs to cross the cell membrane, and the dye functioned as both a model for hydrophobic drugs and a fluorophore probe, allowing the detection of its release by fluorescent spectroscopy. Strong intracellular fluorescence was detected after 96 h of incubation.

Light, on the other hand, can be used as an external stimulus to trigger drug release. Gold nanoshells have been embedded in a temperature-sensitive polymer-hydrogel and irradiated at 808 nm laser wavelength.[90] The nanoshells convert the absorbed radiation into heat, which shrinks the hydrogel reversibly, and the drug is released (Figure 3.5). The same hydrogel principle has been applied using nanorods coated with poly(nisopropylacryamide) (PNIPAM) gel.[91,92] Another way has also been done by Wijaya et al.,[93] where controlled drug release was achieved by using two sets of nanorods of different lengths conjugated to two different oligonucleotides. Irradiation at 800 nm wavelength melted the shorter set of nanorods into nanospheres, causing the release of the conjugated oligonucleotides (Figure 3.6). Melting and release were achieved in the set of longer nanorods under irradiation at 1,100 nm. Ipe et al.[94] delivered amino acids using AuNPs attached to spiropyran. In its open form, which is induced by UV radiation, spiropyran forms complexes with amino acids. These complexes can then be broken by returning the molecule to its closed form by irradiation with visible light or heat.

3.5.1.1.4 Photodynamic Therapy (PDT)

Photodynamic therapy (PDT) is another technique used in cancer therapy, where cytotoxic singlet oxygen is locally generated by light activation of a nontoxic photosensitizer. The photosensitizer absorbs light at suitable wavelengths, the absorbed energy is used to excite oxygen into a singlet state, and the singlet oxygen causes the degradation of cellular components. Photosensitizers are more likely to be taken up by cancer cells than normal ones.[95] Targeting the location of the tumor using light within 600–700 nm in the

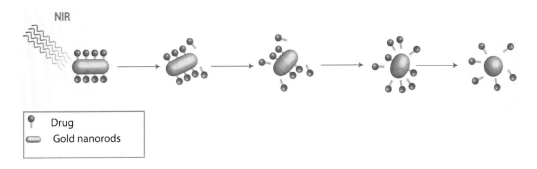

FIGURE 3.6
Drug delivery using gold nanorods. The gold nanorod is conjugated to the drug. Upon NIR irradiation, the nanorod gradually melts into a nanosphere, and the drug is released.

presence of these agents and oxygen (supplied by blood and tissues) leads to the release of cytotoxic singlet oxygen.[95,96] However, full penetration of light into deep tissue tumors is still a challenge,[95] and the absence of oxygen, such as in tissue hypoxia with bacterial infection, renders PDT ineffective.[97] Moreover, because these agents are only relatively specific, widespread burns of the tissues containing the photosensitizers may occur upon exposure of the body to light.[95] Hone et al.[98] used the PDT principle by functionalizing 2–4 nm AuNPs with pthalocyanines (Pc), which act as photosensitizers and release cytotoxic singlet oxygen upon IR light stimulation. The quantum yield of the AuNP-bound Pc was 50% higher than that of free Pc.

3.5.1.1.5 Anticancer Drugs

3.5.1.1.5.1 Tumor Necrosis Factor (TNF)   Tumor necrosis factor (TNF) has both an anticancer and antiangiogenesis (new vessel formation) effect. Paciotti et al.[64] functionalized 33 nm AuNPs with TNF by direct conjugation between the citrate capping on the AuNPs and the TNF at an optimum pH between 8 and 10. Each AuNP was able to carry 400 molecules of TNF, which acted as both a targeting and therapeutic agent.

3.5.1.1.5.2 Paclitaxel   Gibson et al.[99] functionalized AuNPs with paclitaxel, a chemotherapeutic drug. Hexaethylene glycol (HEG) was linked at the C7 position of paclitaxel, acting as a carboxyl linker, and covalently bound to AuNPs modified by 4-mercaptophenol. HEG is expected to increase water solubility of the functionalized AuNPs and reduce their uptake by the RES, thereby increasing circulation lifetime. Each AuNP was able to carry over 70 paclitaxel molecules.

3.5.1.1.5.3 Methotrexate (MTX)   Methotrexate (MTX) is an analogue of folic acid that disrupts cell metabolism by inhibiting dihydrofolate reductase and other enzymes involved in purine synthesis. MTX is used as an anticancer drug-targeting malignant cell that overexpresses folate receptors. Drawbacks of systemic or free MTX therapy include the development of cellular resistance via decreased uptake, as well as increased efflux of MTX. Chen et al.[100] functionalized 13 nm colloidal AuNPs with MTX by direct conjugation to target Lewis lung carcinoma cells. Both the citrate-capped AuNPs and the MTX molecules are negatively charged at neutral pH. At higher concentrations, MTX replaces the citrate capping (Figure 3.2A). The MTX-AuNPs were found to efficiently transport the drug into the cancer cells and exert their cytotoxic activity in a faster and more effective manner than that observed with free MTX.

3.5.1.1.5.4 Combined Drugs (2 in 1)   Priyabrata et al.[101] loaded two anticancer agents on the same AuNPs. Vascular endothelial growth factor (VEGF) is secreted by malignant cells, leading to both neovascularization and increased resistance to apoptosis. AuNPs were successfully functionalized with both VEGF antibody-2C3, as an antiangiogenic agent, and gemcitabine, as an anticancer agent, providing a 2-in-1 system and a model for combined drug therapy.

3.5.1.1.6 Antibiotics

Saha et al.[102] conjugated antibiotics ampicillin, streptomycin, and kanamycin to 14 nm AuNPs. Conjugation was performed during the synthesis of the AuNPs as a result of the combined reducing ability of the antibiotics and sodium borohydride. Targeted bacterial strains were *Escherichia coli* DH5a, *Micrococcus luteus*, and *Staphylococcus aureus*. Conjugated streptomycin and kanamycin proved more effective than ampicillin, and all antibiotics were more heat stable than their free forms. Ampicillin activity, when conjugated to AuNPs, was higher than its free form only at elevated temperatures, and was lower after 2 weeks.

3.5.1.2 Thermal Therapy

3.5.1.2.1 Cancer

Current cancer therapies are notorious for being invasive, toxic, and ridden with adverse effects that may be localized or systemic, physical or psychological. The effective elimination of cancer cells without harming the surrounding healthy tissue has been the ultimate goal and challenge in cancer therapy.

Thermal therapies have been used to treat cancer in a more localized manner. They are minimally invasive and can be used in patients who are not suitable candidates for surgery or have inoperable tumors.[95] Thermal therapies used include cryotherapy,[103] microwaves,[104] focused ultrasound,[105] radiofrequency (RF) ablation,[106] and laser photothermal therapy.[107] Subjecting cells to temperatures between 41 and 47°C induces apoptosis,[108] an organized manner of cell death, which involves normally functioning proteins. At temperatures above 50°C, necrosis occurs, which is a form of cell death based on protein denaturation accompanied by lipid melting.[109] However, when using these nonspecific techniques, it is difficult to accurately target only the cancer cells without also destroying the surrounding normal tissue.

3.5.1.2.1.1 Photothermal Therapy (PTT)   Photothermal therapy (PTT) using AuNPs is similar to combining PDT and thermal therapy techniques, but with higher specificity, uptake, and penetration. It is also less cytotoxic, with little or no injury to normal tissue.[74] Tissues have an "optical transparency window" that corresponds to the NIR, between 750 and 1,100 nm,[81] allowing deeper penetration of light, while minimizing absorption by water and hemoglobin.[19,75] Owing to their optical and physical properties, AuNPs can be tuned so that their SPR falls in the NIR region, while at the same time requiring lower laser energy.[75,95,110]

AuNPs are injected either systemically (intravenous route) or directly into the tumor. Once the AuNPs have reached the target site, whether by passive or active means, short laser pulses are directed at the site and, via the AuNPs' electron-electron collisions followed by electron-phonon interactions, this light energy is converted into heat, leading to thermal ablation of the malignant cells, which are more sensitive to heat than normal cells due to their higher metabolic rates.[33,110] Tong et al.[111] demonstrated that when the AuNPs are on the cytoplasmic membrane, they require 10 times less laser energy to destroy the cell than when they are taken up by the cell. Thus, they proposed that cell death by PTT was mainly due to apoptosis, initiated by membrane disruption, followed by blebbing and damage to the cell's actin filaments.

The tunability of colloidal gold nanospheres is limited, and increasing their size can only slightly shift their SPR into the NIR region,[81,110] though with an increase in size, scattering is favored over absorption.[25] Therefore, while colloidal nanospheres can be used in treatment of superficial tumors, nanocages, nanorods, and nanoshells are preferred due to their high absorption efficiency and their SPR ranges, which can be tuned to extend to the NIR and mid-infrared regions.[19] This can be done by changing the aspect ratio of nanorods (length/diameter) and nanoshells (core/shell).[3,19,76,95,112] Aggregated nanospheres or nanosphere clusters have also been used successfully in PTT, shifting the maximum absorption from 525 nm to 700–1,000 nm.[110]

One of the first applications of AuNPs in PTT was done in 2003 by Hirsch et al.[112] using gold nanoshells and continuous wave NIR laser to target breast carcinomas in mice. Huang et al.[110] targeted human oral squamous carcinoma (HSC) cells using 30 nm sized gold nanospheres conjugated to anti-epidermal growth factor receptor (anti-EGFR) antibodies via noncovalent interactions. EFGRs are overexpressed in many types of cancers, including ovarian, breast, and 80% of oral cancer cells. With the exposure to laser at 800 nm for 2 min, each pulse lasting 100 fs and repeated at a rate of 1 kHz, the active and selective targeting of the oral cancer cells using the anti-EGFR-conjugated nanospheres was achieved at 20 times lower laser energy than that used in the absence of the AuNPs. Moreover, aggregation of the AuNPs further enhanced the photothermal effect. Huang et al.[83] used gold nanorods of a 3.9 aspect ratio, which were also conjugated to anti-EGFR to target oral cancer cells using a laser wavelength of 800 nm. Compared to the normal cells, half the laser energy was required to damage the cancer cells.

O'Neal et al.[76] targeted subcutaneous tumors in mice using gold nanoshells with a silica core coated with thiolated polyethylene glycol (PEG-SH) via covalent bonding to increase the nanoparticles' biocompatibility and avoid their uptake by the reticuloendothelial system (RES). The diameter of the silica core synthesized was 100 nm, and the gold shell thickness 8–10 nm, with peak absorption at 805–810 nm. The PEG nanoshells were injected into the tail vein of mice, and 6 h was allowed for the nanoshells to systemically reach and accumulate in the tumors passively, due to the EPR effect. After exposure to NIR light at 808 nm, complete regression of the tumors was observed.

Chen et al.[113] successfully used gold nanocages to actively target the breast cancer cell line SK-BR-3, which overexpresses EGFR, by conjugating monoclonal anti-HER2 antibodies to the nanocages. The gold nanocages were synthesized using a galvanic replacement reaction between 40 nm silver nanocubes and chloroauric acid. Controlling the molar ratio of the two gives the ability to tune the extinction peak of the nanocages from the visible to the NIR regions. The nanocages in this study were tuned to have an extinction peak at 810 nm. The nanocages were then pegylated and conjugated to the anti-HER2 antibodies, subjected to laser at 810 nm, and thermally ablated the cells at an intensity power of 1.5 W/cm^2, showing a linear increase in effect with irradiation power density.

3.5.1.2.1.2 Radiofrequency Ablation (RFA) The use of AuNPs has also been extended to work in RFA. Thermal ablation using an RF current (10 kHz–900 MHz) causes agitation of tissue ions drawn toward the alternating current, leading to friction, which results in the generation of heat.[114] However, like other thermal therapy technologies currently in use, RFA is nonspecific and causes damage to normal as well as cancer cells.[115] Moreover, recurrence after RFA has been reported to occur at high rates.[116,117] Cardinal et al.[115] used AuNPs exposed to an RF field to thermally ablate liver cancer cells both in vitro and in vivo. To study the effect in vitro, HepG2 liver cancer cells were cultured in the presence of AuNPs synthesized by the citrate reduction method. About 80% of the cells cultured in vitro had been successfully ablated after exposure to 35 W RF power for 7 min. To study the effect in vivo, JM-1 cells were injected into rats and allowed 14 days to form subcutaneous tumors. The AuNPs were then directly injected into the subcutaneous tumors and exposed to RF power of 35 W. Histopathologic assessment of the subcutaneous tumors after RFA using AuNPs also showed signs of widespread thermal injury and apoptosis. In both models, the hyperthermia generated by the AuNPs depended on the exposure time and power of RF current.

3.5.1.2.2 Bacteria

Photothermal technology has also been used in targeting and killing bacterial cells. Zharov et al.[118,119] used PTT and AuNPs to target *Staphylococcus aureus* bacteria. Gold nanospheres of diameters 10–40 nm were conjugated with anti-protein A monoclonal antibodies, targeting protein A, which is attached to the cell wall of Staph isolate UAMS-1. Upon laser radiation, only bacteria labeled with 40 nm AuNPs were killed after 500 pulses at 532 nm laser wavelength. Higher killing efficiency was observed when the nanospheres clustered due to aggregation.

3.5.1.2.3 Wound Closure

Lasers have been used in the closure of wounds as an alternative to suturing.[120] In laser soldering, exogenous chromophores are mixed with proteins or polymers, and the paste is inserted between the wound's margins. Upon NIR radiation, the chromophores convert the radiation to heat, and these proteins or polymers coagulate, acting as clots and closing the wound.[120] These clots are then naturally resorbed by the tissue. Gobin et al.[121] applied this method to seal muscular and cutaneous tears by replacing the chromophores with gold nanoshells in an albumin filler.

In laser welding, chromophores, such as indocyanine green (ICG),[122] are applied to the wound's margins[123] and convert NIR radiation to heat, which leads to reorganization of the tissue by diffusion,[124] leading to minimal scar formation. Such chromophores, however, are highly diffusive and have limited efficiency and stability, as opposed to AuNPs.[28]

Ratto et al.[28] used gold nanorods as NIR chromophores for the welding of connective tissue of the porcine eye lens capsule. The gold nanorods were synthesized using the seed-mediated method[42] to have an aspect ratio of approximately 4, and placed in a sandwich manner between the internal side of the donor capsule and the external side of the recipient capsule, so that adhesion of the capsules can occur while preserving their curvature. The contact welding technique was used with an 810 nm diode laser and 40 ms duration of pulses. Collagen denaturation and fusion of patches were achieved, indicating a rise in temperature beyond 50°C (the temperature at which collagen denaturation occurs).

3.5.2 Diagnostics

AuNPs have gained popularity in the diagnostics arena, due to their unique optical and physical properties and their tunability by variation of shape, size, and compositions, thus enabling their use in the detection of multiple targets simultaneously.[2] AuNPs can be applied in immuno and molecular assays using a wide range of detection techniques, such as colorimetric, fluorescent based, scanometric, and electrochemical. Assays involving

AuNPs are relatively simple, inexpensive, and less time-consuming. Assays based on colorimetric detection can be used in laboratories with limited infrastructure, without the need for sophisticated instruments, as well as at the point of care.

Suspensions of AuNPs are relatively stable and reliable as they are less likely to produce false negative or false positive results caused by nonspecific aggregation.[125] In turbidimetric assays, AuNPs increase sensitivity since their suspensions do not significantly scatter visible light, thereby reducing any background signals that may occur.[126]

Molecular assays include colorimetric techniques, where AuNPs can be modified or unmodified. AuNPs are also used as quenchers in Förster resonance energy transfer (FRET)-based assays. Examples of other assays include the molecular biobarcode assay (BCA), where AuNPs are used in conjunction with magnetic microparticles for the detection of DNA.[127] This method has been extended for multiplexed detection of DNA, as well as by designing different barcodes to suit different targets.[128] In electrochemical techniques, for example, biotinylated AuNPs and magnetic beads coated with streptavidin are used in a process that leads to detection of target DNA.[129] AuNPs have been used in molecular diagnostics for nucleic acid detection, as well as in immunoassays to detect different proteins. In the next section, we will expand upon molecular techniques, in addition to applications in immunoassays.

3.5.2.1 Nucleic Acid Assays

3.5.2.1.1 Modified AuNPs' Colorimetric Method

3.5.2.1.1.1 Cross-Linking (Two Probe) Method   This method can be used for the detection of ssDNA by functionalization of two sets of AuNPs, each with a different oligonucleotide probe, and each complementary to one end of the target DNA (Figure 3.7). This causes them to align in a tail-to-tail manner. Upon hybridization to the target, the close proximity of the AuNPs causes them to aggregate, shifting the solution color from red to blue. By spotting

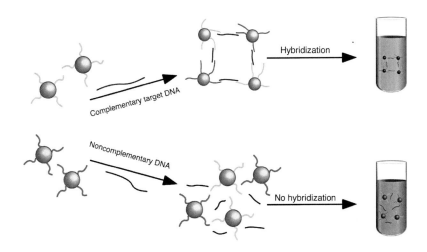

FIGURE 3.7
Modified cross-linking colorimetric detection of nucleic acid targets. Two sets of oligonucleotide probes are conjugated to the AuNPs. If the added DNA is complementary, hybridization occurs, cross-linking the AuNPs together in a tail-to-tail fashion. This brings the AuNPs closer to one another and aggregation occurs, turning the solution color from red to blue. If the added DNA is noncomplementary, no hybridization occurs, the AuNPs do not aggregate, and the solution color remains red.

the solutions onto a white C18 reverse-phase thin-layer chromatography (TLC) plate, we can better observe and record the change in color. This is a relatively fast, easy, and low-cost method compared to others, such as those based on the use of fluorophores.[2]

Multiplexing can be achieved by changing the composition of the AuNPs, such as using gold nanoshells containing a silver, copper, or platinum core.[130] The silver/gold (Ag/Au) nanoshells shift in color from yellow to brown upon aggregation following target hybridization. This has been used by Cao et al.,[131] where spherical AuNPs were functionalized with specific oilgonucleotides to recognize the wild type sequence, while Ag/Au nanoshells were functionalized to recognize a mutant with a single nucleotide polymorphism (SNP). The wild type target was added to the two solutions, each containing one probe type, and the solution temperature was raised from 20 to 70°C. At different temperatures, aliquots of the solutions were then spotted on a C18 reverse-phase TLC plate. The gold solution in the AuNP probe system exhibited a higher melting temperature (T_m), detected by a color change from blue to red, than the Ag/Au core-shell probe system designed to detect the mutant target. In a second experiment, when the mutant target was added to the same two solutions, the opposite results were obtained, where the Ag/Au core-shell probe system showed a higher T_m than the AuNP probe system. In this method, the Ag/Au core-shell system served as a control to ensure accurate results; however, a similar strategy could also be used for the simultaneous detection of two different DNA targets."

3.5.2.1.1.2 Non-Cross-Linking (Single-Probe) Method   Citrate-capped AuNPs possess a negative charge that can be neutralized by the addition of a salt, which allows them to aggregate and change from red to blue. Hybridization of oligonucleotide-functionalized AuNPs increases their negativity, thereby requiring higher salt concentrations for aggregation to occur (Figure 3.8). The single-probe method is easier and faster than the two-probe method.[132]

Chakrabarti and Klibanov[133] used AuNPs functionalized with short, six-base peptide nucleic acids (PNAs), which have a single-base mismatch selectivity five times greater than DNA-functionalized AuNPs. When the PNA-AuNPs hybridize with the target DNA, their resistance to aggregation in the presence of high salt concentrations increases. In the event that the DNA target contains a mutation, the PNA-AuNPs do not hybridize perfectly to the target and aggregate at low salt concentrations, changing in color to blue. Sato et al.[134] used this method with AuNPs functionalized with one type of oligonucleotide. In the presence of NaCl at a concentration higher than 0.5 M, addition of target DNA led to the aggregation of the AuNPs and their color changed to purple.

Multiplex detection was done by Kim and Lee[135] by using gold nanoclusters instead of single AuNPs. The nanoclusters were synthesized in the presence of dithiothreitol (DTT) and monothiol DNA, where each AuNP was 15 nm in diameter and the 19-base DNA sequence used was 6 nm long. As the amount of DTT increased, the size of the gold nanoclusters increased. Four solutions were selected: red, purple, violet, and blue, which contained 35, 42, 49, and 115 nm DNA-AuNP clusters, respectively. Upon DNA hybridization with the target, red turned to purple, purple solution turned to pale violet, violet solution turned to pale blue, and the blue solution turned to pale indigo. A spot test would allow for better differentiation between color changes. Moreover, an even wider variety of colors can be obtained by synthesizing gold nanoclusters with different core compositions. The cluster sizes could be controlled by the DTT and monothiol DNA ratios. Quantitation was done by conducting melting experiments, and the limit of detection of the test was 1 nM, which is five times lower than that using single DNA-AuNPs having the same sequence.

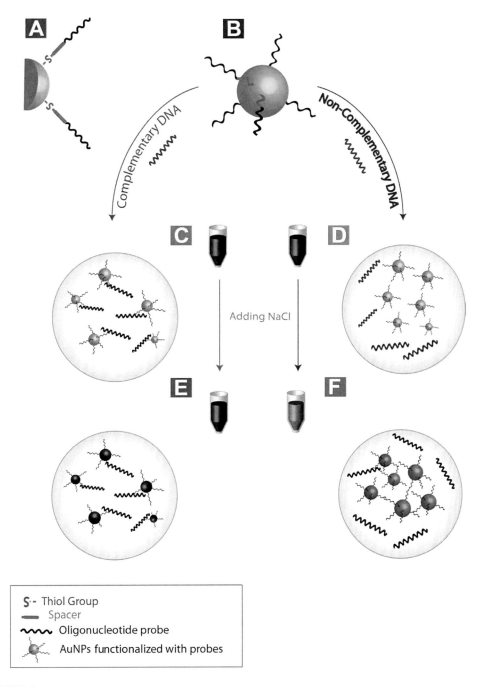

FIGURE 3.8
(See color insert.) Modified non cross-linking colorimetric detection of nucleic acid targets. (A and B) AuNP functionalized with one set of oligonucleotide probe, which is attached to the AuNP via a spacer and thiol bond. (C) In the presence of the complementary target DNA, hybridization occurs. (D) In the absence of complementary target, no hybridization occurs. (E) Upon addition of NaCl, the solution where hybridization occurred remains red. This is because the hybridized oligonucleotides increase the negative charge of AuNPs, thereby increasing their resistance to aggregation. (F) In the solution where no hybridization occurred, addition of NaCl induces aggregation of AuNPs by neutralizing their negative charge, and the solution color changes from red to blue.

3.5.2.1.2 Unmodified AuNPs' Colorimetric Method

As mentioned above, addition of NaCl to citrate-capped AuNPs leads to neutralization of their negative charge, which causes them to aggregate. ssDNA stabilizes AuNPs by increasing their negativity, thereby preventing their aggregation upon addition of salt. Adsorption of ssDNA onto AuNPs occurs as a result of the partial uncoiling of ssDNA, which exposes its bases, and the resulting van der Waals attractions lead to adsorption onto AuNPs. This adsorption is further strengthened due to electrostatic attractions between the ssDNA and AuNPs, and between them and counter-ions in solution. In contrast, the phosphate backbone of dsDNA prevents its adsorption onto AuNPs due to repulsion between the two negatively charged molecules.

This principle can be used in colorimetric nucleic acid detection, where addition of ssDNA to a salt solution containing denatured complementary target DNA leads to hybridization. When AuNPs are added to this mixture, no adsorption or stabilization of AuNPs occurs due to hybridization of ssDNA to the target DNA. The ssDNA is therefore, no longer free to become adsorbed onto the AuNPs and the salt in solution causes the AuNPs to aggregate and their color to change from red to blue. On the other hand, if the target DNA is not present, hybridization does not occur and the ssDNA remains free in solution. The addition of AuNPs leads to their stabilization due to adsorption of ssDNA, thereby preventing the aggregation of the AuNPs, and their color remains red. The main advantage of this method is that no functionalization is required.

A modification of this method can be done by using probes carrying fluorescent tags. In the absence of the target DNA, the probes are adsorbed onto the AuNPs and the fluorescence is quenched, giving no signal. On the other hand, when the target DNA is present, no adsorption occurs and a fluorescent signal indicates the presence of the target. These methods have been used for both DNA and RNA detection.[136]

Our research group[137] reported the first assay for hepatitis C virus (HCV) using unmodified AuNPs targeting HCV RNA performed on clinical samples. In this study, HCV RNA was extracted from patient sera and amplified using reverse transcriptase (RT) PCR. A hybridization buffer was added containing two primers each complementary to the sense and antisense stands of the dsDNA amplicon. Citrate-capped colloidal AuNPs were then added to the mixture. In HCV positive specimens, the AuNPs aggregated and the solution color changed from red to blue. In HCV negative specimens, no change in color was observed due to the stabilization of the AuNPs' surface by the probes. In this method, the use of AuNPs has allowed the detection of HCV after only 15 PCR cycles rather than 40 cycles, thus reducing cost and time. The detection limit was 3 femtomolar.

In another experiment, we used unmodified AuNPs for the detection of unamplified full-length HCV RNA in serum samples[138] (Figure 3.9). In this study, a colloidal solution of AuNPs with a diameter of 15 nm was prepared by citrate reduction of hydrogen tetrachloroaurate (III). The size and extinction spectra of the prepared AuNPs were characterized using field emission scanning electron microscopy and spectrophotometry, respectively. Serum samples were collected from healthy volunteers (n = 20) and chronic HCV patients (n = 25; positive by HCV antibody and RT-PCR tests). The assay was performed by mixing the extracted RNA and hybridization buffer containing phosphate-buffered saline (PBS) and NaCl, and a primer targeting a sequence in the 5'UTR of HCV specific to all genotypes and subtypes. The mixture was denatured at 95°C for 30 s, annealed at 59°C for 30 s, and then cooled to room temperature for 10 min. Colloidal AuNPs were then added to the mixture and the color was observed. Serial dilutions of HCV positive samples with known viral load were used to determine the detection limit of the assay. In HCV positive

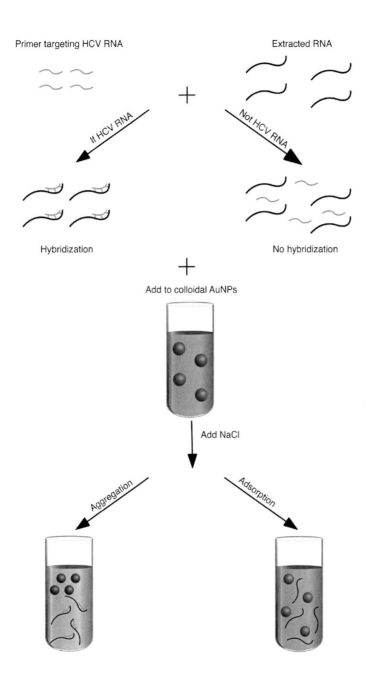

FIGURE 3.9
Unmodified AuNPs' colorimetric detection. A primer targeting HCV RNA (short strands) is mixed with extracted RNA (long strands) in a hybridization buffer solution containing NaCl. If the extracted RNA is HCV RNA, hybridization occurs. Upon addition of colloidal AuNPs to the solution with non-hybridized primer, the primer adsorbs onto the AuNPs, protecting them from aggregation, and the color remains red (bottom-right tube). Upon addition of AuNPs to the solution where hybridization occurred, the AuNPs aggregate, and the color changes to blue (bottom-left tube).

specimens, the color of the solution changed from red to blue within 1 min. Also, a red shift and a broadening of the absorbance peak of AuNPs were observed. The assay has a sensitivity of 92% and a specificity of 90%. The assay detection limit was 50 copies/reaction. The developed assay is highly sensitive, has a turnaround time of 30 min, and eliminates the need for thermal cycling and detection instruments. This assay could be further developed for HCV viral load monitoring and genotyping.

3.5.2.1.3 *Förster Resonance Energy Transfer (FRET)*

FRET is the nonradiative transfer of excitation energy emitted from a donor and absorbed by an acceptor via dipole-dipole interactions. This can only occur if the distance between the donor and acceptor is less than the Förster distance, roughly less than 10 nm, where the energy transfer efficiency between the donor and acceptor is at 50%. Moreover, the donor's emission spectrum must correspond with the acceptor's absorption spectrum.

Due to their large molar extinction coefficient, AuNPs serve as excellent acceptors. When dealing with metal nanoparticles, the FRET phenomenon is referred to as nanometal surface energy transfer (NSET).[2] Metal nanoparticles have higher energy transfer efficiency due to the presence of free conduction bands of electrons, leading to more dipole-dipole interactions, which can occur at longer distances compared to FRET. AuNPs can also accept a wide range of emission frequencies, from visible to NIR, which makes them excellent quenchers.

Compared to conventional fluorescent dyes, AuNPs have stood their ground, as they are not susceptible to photobleaching and have superior absorption and scattering properties. Not only are they more efficient quenchers than fluorescent dyes, but their unique optical properties allow for the detection of zeptomolar nucleic acid concentrations with a sensitivity that is five times higher than that obtained using fluorescent dyes.[139]

Dubertret et al.[140] targeted DNA using ssDNA in a hairpin conformation with an AuNP attached at one end and a fluorescent dye attached at the other (Figure 3.10). Because the AuNP and dye are in close proximity in this conformation, the AuNP is able to quench the dye's fluorescent emission. When the target complementary DNA is present, the hairpin structure opens and hybridization occurs. In this open form, the distance between the AuNP and dye exceeds the NSET distance, and the unquenched fluorescence of the dye can be detected. Compared to the use of organic quenchers, AuNPs increase the sensitivity of the test 100-fold, and increase the ability to detect single-base mismatches 8-fold.

Griffin et al.[141] used fluorophore-labeled ssRNA and AuNPs to detect synthetic HCV RNA targets. In this assay, ssRNA was attached at one end to the AuNP and at the other end to a fluorophore which adsorbed onto the same AuNP, forming a loop. When the complementary target RNA hybridizes with the RNA probe, they form a dsRNA that is released into solution. In this case, two observations can be made: the change of solution color from red to blue due to the aggregation of AuNPs, and the unquenching fluorescent signal of the dye, the strength of which is directly proportional to the concentration of the target. The size of the AuNPs was also found to be of significance, with an increase in sensitivity as well as quenching efficiency up to 1,000-fold, as the AuNP size increased from 5 to 70 nm. At 110 nm AuNP size, the limit of detection (LOD) reached 300 fM of RNA.

3.5.2.2 *Immunoassays*

AuNPs can be coated with proteins/antibodies and aggregate in the presence of the target antigen. Thanh and Rosenzweig[126] used 10 nm AuNPs coated with protein A antigen to detect anti-protein A antibodies in aqueous solutions and serum. The presence of the

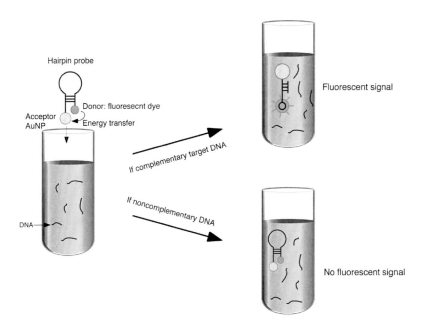

FIGURE 3.10
Detection of nucleic acid targets using FRET-based probes. A hairpin probe with AuNP (energy acceptor) and fluorescent dye (energy donor) attached is added to a solution containing denatured DNA strands. If DNA is complementary, hybridization occurs, which opens the hairpin structure, separates the dye from the AuNP quencher, and a fluorescent signal is detected. If DNA is noncomplementary, no hybridization occurs and no fluorescent signal is emitted by the dye due to quenching by the AuNP.

antibodies caused the solution color to change from red to purple within 30 min. The LOD was 1 µg/ml, with comparable sensitivity to enzyme-linked immunosorbent assays (ELISAs), but with the added advantage of being a simple one-step test, as opposed to the multiple steps required in ELISA.

Tang and Hewlett[142] developed a BCA using AuNPs to detect human immunodeficiency virus type 1 (HIV-1) p24 antigen. The detection sensitivity using this method was superior to that of ELISA, where the LOD (0.1 pg/ml) was 100- to 150-fold that of ELISA (10–15 pg/ml). Moreover, the AuNP-based BCA was able to detect the HIV-1 p24 antigen in patients with serconversion 3 days earlier than ELISA could.

Yang et al.[143] developed a chemiluminescent immunoassay using AuNPs for the detection of serum carcinoembryonic antigen (CEA). AuNPs were labeled with luminol and secondary antibodies, forming antibody-luminol-AuNP conjugates. Magnetic beads labeled with primary antibodies captured the serum CEA, which was then sandwiched by the secondary antibodies attached to the luminol-AuNPs. The chemiluminescent intensity was proportional to CEA concentrations.

Chen et al.[144] developed an immuno-PCR assay using functionalized AuNPs to detect Hantaan virus nucleocapsid protein (HNP). AuNPs were functionalized with monoclonal antibody L13 specific to HNP antigen. The same AuNPs were conjugated to oligonucleotides composed of two strands: one directly attached to the AuNP and acting as the capture strand, and the other attached to the capture strand by being partially complementary, and acting as the signal amplification DNA. An ELISA microplate was coated with a polyclonal antibody used to capture the HNP antigen. Upon addition of the functionalized

AuNPs to the microplate, a sandwich immunocomplex was formed, and heating resulted in the release of the signal DNA. This DNA was then identified by PCR/gel electrophoresis and SYBR-Green real-time PCR. The LOD using this technique was 10 fg/ml, making it about seven times more sensitive than ELISA.

AuNPs have even been used to develop an aggregation-based home pregnancy test,[145] where AuNPs are used in conjunction with latex microparticles and derivatized with antibodies against beta human chorionic gonadotropin (β-HCG) hormone to form a colorless mixture. Addition of urine containing the hormone leads to aggregation of the micro- and nanoparticles, and the color changes to pink.

3.5.3 Imaging

The unique optical properties of AuNPs make them highly attractive contrast agents in imaging applications. Contrast, in this context, is the difference in scattering and absorption of diseased cells compared to healthy cells, which sets them apart and enables their identification. However, this contrast is often weak, presenting difficulty in distinguishing diseased cells from healthy ones, as well as from small tumors and blood vessels.[146] This is where the need for contrast-enhancing agents arises.

Optical imaging includes reflectance-based and absorption-based techniques, such as optical coherence tomography (OCT), x-ray computed tomography (CT), photoacoustic imaging (PAI), and confocal laser scattering microscopy (CLSM). These techniques can best utilize the optical properties of AuNPs. Other attractive characteristics of AuNPs are their size and shape tunability, as well as the ability to functionalize them with biomolecules that can target biomarkers for cell-specific imaging. AuNPs are both biocompatible and of low cytotoxicity, thus enabling their use in vivo.

3.5.3.1 X-Ray Computed Tomography (CT)

X-ray computed tomography (CT) is a technique that depends on the absorbance of x-rays by tissues and their components. The x-rays pass through the tissue and are received by a detector to give cross-sectional images. Though different tissues and cells have different degrees of x-ray absorption, the contrast is often weak and contrast agents are needed.

Commonly used in CT are the iodinated contrast agents, which efficiently absorb x-rays. However, they suffer from short circulation lifetime due to rapid renal clearance, which allows little time for imaging. Additional limitations include the high viscosity of the injectable solutions, their nonspecificity, as well as renal membrane impermeability to these agents, causing them to accumulate in the kidneys, which may lead to renal toxicity and limits their use in renal patients.[147] Moreover, images obtained using iodinated contrast agents still suffer from low resolution, making it difficult to distinguish microtumors from benign or healthy background tissues and small blood vessels.[148] X-ray attenuation depends on the density and atomic number of the element involved. Low molecular weight iodinated agents are therefore either bound or encapsulated in higher molecular weight molecules.[149,150]

Bismuth sulfide nanoparticles have also been successfully used and exhibit greater x-ray attenuation than iodine-based agents.[151] Limitations of these nanoparticles, however, include difficulty in controlling their size and shape and their nonspecificity due to the inability to functionalize their surface with cell-specific biomolecules. AuNPs, on the other hand, are tunable, can be functionalized to target specific cells and tissues, have a longer circulation lifetime, and their absorption coefficient is superior to that of the other contrast agents. They also have higher density and atomic number, thus exhibiting greater attenuation.

Xu et al.[152] tested the effect of size and concentration of AuNPs compared to Omnipaque, which is a commercial iodine-based contrast agent. The sizes of AuNPs used in this study ranged from 4 to 20 nm. Uptake by HeLa S3 cells was found to be dependent on the size and concentration of AuNPs incubated with the cells, where cellular uptake increased with the increase of concentration and decrease in size of the AuNPs. Greater x-ray attenuation was observed with AuNPs of all sizes compared to Omnipaque using the same concentrations. X-ray attenuation also increased with smaller AuNPs compared to larger ones, and at higher concentrations compared to lower ones. This may be due to the increase in surface-to-volume ratio as the AuNPs' size decreased.

Sun et al.[153] compared CT images obtained using eXIA 160 (another commercial iodinated contrast agent) with AuNPs coated with heparin to target liver cells. Functionalization was done by adding heparin conjugated to 3,4-dihydroxyphenylalanine (DOPA) to citrate-capped AuNPs, where each AuNP carried 4×10^3 heparin-DOPA conjugates. The AuNP conjugates showed good stability in PBS containing fetal bovine serum, and the negative charge of heparin prevented their aggregation. The conjugates were also found to be non-cytotoxic and biocompatible in human hepatocellular carcinoma (HepG2) cell lines. The CT images using AuNPs showed well-defined liver tissue delineated from background tissues and blood vessels in mice, with 3.3 times greater attenuation than eXIA 160. The contrast resulting from the AuNP conjugates was maintained for up to 24 h, allowing a prolonged time for imaging. Since the liver is one of the most frequent sites for metastasis, the use of AuNPs in liver CT imaging can significantly aid in early detection and treatment.

Alric et al.[154] used gadolinium-coated AuNPs to enhance both MRI and CT images, making it possible to use AuNPs in dual-modality imaging, while Popovtzer et al.[155] conjugated AuNPs with UM-A9 antibodies for targeted CT imaging of squamous cell carcinoma of the head and neck, where the attenuation coefficient of targeted cells was five times higher than that of untargeted or normal cells. AuNPs have also been used in vascular CT imaging.[148,154]

3.5.3.2 Photoacoustic Imaging (PAI)

PAI, also known as optoacoustic or thermoacoustic imaging, is a technique that combines both optical absorption and ultrasound waves. Tissues are exposed to short pulses in the NIR region, and the absorbed photons are converted into heat, causing thermal expansion, which leads to emission of acoustic or ultrasound waves. These waves travel with minimal disturbance to the surface and can be detected using wideband detectors or acoustic transducers. An inverse algorithm is then used to reconstruct the signals into images, depending on the time of flight, amplitude, and time difference between acoustic wave peaks. The higher the absorption of a tissue component, the stronger the ultrasound signal it emits. For example, due to hemoglobin's high optical absorption in the NIR region compared to other cells and tissue, PAI has been very useful in the imaging of blood vessels. For the same reason, it has been useful in tumor imaging due to the stimulation of angiogenesis by tumors.[156] PAI, therefore, gets the best of both optical and ultrasound worlds, where it depends on the different degrees of light absorption by tissue components for optical contrast, and on the emitted ultrasound waves for determining their location.

Gold nanospheres have been deemed unsuitable for use in PAI due to their limited tunability in the NIR region.[157] Gold nanoshells, nanorods, and nanocages are preferred due to their higher tunability. Hu et al.[158] compared the scattering, absorption, and extinction of AuNPs of different shapes. Engineered so that their SPR peaks correspond to a wavelength of 800 nm, gold nanorods with an aspect ratio of 3.3 (20 nm diameter), gold nanoshells with a 50 nm core and 3.2 shell thickness, and nanocages with a 50 nm inner length and 6 nm thickness were used. The highest absorption and scattering were observed in the nanocages, followed by nanorods, and then by nanoshells. The absorption of AuNPs is five times higher than that of free ICG, and their functionalization enables targeted imaging, as mentioned before. PAI is also used to monitor uptake and excretion of contrast agents.

Copland et al.[159] conjugated 40 nm gold nanospheres to Herceptin® (a monoclonal antibody against HER2) to target SK-BR-3 breast cancer cells in vitro, and PAI images were obtained. At a tissue depth of 6 cm, low concentrations of AuNPs (10^9 AuNPs/ml) showed good sensitivity.

Eghtedari et al.[157] used PEG gold nanorods of 50 nm length and 15 nm diameter, with SPR tuned at 760 nm wavelength, and enhanced PAI images were obtained needing very low concentrations of the nanorods. Nanorods have also been functionalized with anti-EGFR targeting A431 skin cancer cells,[160] with antibodies targeting prostate cancer cells,[161] and with Herceptin antibodies targeting breast cancer cells[162]—all showing enhanced PAI contrast. Nanorods of different aspect ratios have also been conjugated to two monoclonal antibodies simultaneously targeting both HER2 and CXCR4, enabling the determination of oncogene expression levels through imaging.[163]

3.5.3.3 Dark Field Microscopy

The scattering properties of AuNPs can also be harnessed in imaging applications and do not necessarily require sophisticated instruments. AuNPs can be used in vitro in dark field microscopy. Images in dark field microscopy are obtained by capturing the light scattered by a sample using an objective lens with the optional addition of a dark field condenser for higher quality. The sample gives a bright image surrounded by a dark background. Disadvantages include interference of external particles, such as dust, overlapping images, and nonspecificity, which present a challenge for accurate interpretation. AuNPs can be used as contrast agents in dark field microscopy due to their SPR-enhanced absorption and large scattering cross sections, where 80 nm AuNPs have five orders of magnitude greater scattering ability than dyes.[25] AuNPs can strongly scatter the light that corresponds to their SPR frequency,[164] giving clearer images that can be more easily interpreted. In addition, functionalization of AuNPs allows for targeted imaging. Loo et al.[165] used gold nanoshells functionalized to target HER2 on breast carcinoma cells. Dark field microscopy images of the targeted cells with gold nanoshells showed high optical contrast compared to control cells. El-Sayed et al.[166] used anti-EGFR-labeled gold nanospheres to target cancer cells, leading to a dark field image of well-defined cancer cells, compared to untargeted normal cells showing dispersed AuNPs.

3.6 Biodistribution and Toxicity

Biodistribution and cytotoxicity are important factors to consider in biomedical applications when using AuNPs. Reports on the toxicity of AuNPs have been somewhat mixed, with some deeming AuNPs noncytotoxic, and others holding a more reserved view.

Shukla et al.[167] studied macrophage interactions with AuNPs and found no secretion of pro-inflammatory cytokines when the AuNPs were internalized by the macrophages. Internalization of AuNPs also did not disrupt cellular function. The authors concluded that AuNPs are biocompatible, noncytotoxic, and nonimmunogenic, and have an antioxidant effect by reducing reactive oxygen and nitrite production. Connor et al.[168] studied the uptake and toxicity of 18 nm AuNPs in human leukemia cells, with no significant observations of acute toxicity caused by the AuNPs themselves, though toxicity may be caused by AuNP precursors. In addition, in vivo studies by Cho et al.[169] revealed that 13 nm PEG AuNPs caused acute but transient inflammation in the liver cells of mice, marked by increased neutrophil influx, and apoptosis 7 days after a single injection of AuNPs in a dose-dependent manner. AuNPs accumulated most in the liver and spleen due to trapping by Kupffer cells and spleen macrophages, respectively, showing a tendency for accumulation in organs with phagocytic activity. Other organs, such as the kidneys, lungs, and brain, showed gradual clearance of the AuNPs. The authors concluded, however, that AuNPs did not cause acute cytotoxicity.

Balasubramanian et al.[170] attributed similar findings of AuNP accumulation to the discontinuous endothelium of the liver and spleen, allowing the passage of AuNPs of up to 100 nm to enter from circulation, and to the opsonization of the AuNPs, where binding with circulating antibodies leads to their uptake by the RES. Accumulation of AuNPs in membrane-barrier-controlled environments is mainly size dependent. For example, the blood-brain barrier allows only AuNPs of less than 20 nm to pass through.[171] The renal glomerular basement membrane (GBM) also blocks out aggregated AuNPs due to their large size. However, an additional mechanism of the GBM may be dependent on time and charge. In general, the kidney excretes AuNPs in the urine after 4–5 h of intravenous (I.V.) injection,[148,170] but with time, the AuNPs begin to accumulate. The authors hypothesized that the AuNPs become coated with negatively charged proteins that are repelled by the negatively charged GBM.[170] Moreover, redistribution of the AuNPs occurred after 1 month of administration.

The same study reported differences in distribution according to the route of administration, where less accumulation of AuNPs in the liver was observed via the inhalation route vs. I.V. injection.[170] Another study examined the distribution of various sizes of AuNPs administered orally. Widespread distribution occurred with the small 4 nm AuNPs, while the larger 58 nm AuNPs were limited to the gastrointestinal tract.[172]

There are concerns about transplacental distribution of AuNPs, which may affect fetal development either directly or indirectly by causing placental insufficiency. Myllynen et al.[173] examined the effect of 15 and 30 nm sized PEG AuNPs within a 6 h time period. The AuNPs were not found in fetal blood, and therefore did not cross the placental barrier, but accumulated instead in the trophoblastic layer. The fate of these AuNPs after 6 h, however, needs to be monitored to see if their accumulation could lead to placental or fetal adverse effects. Another study compared 1.4 and 18 nm sized AuNPs and reported that AuNPs do cross the placenta in a size-dependent manner.[174]

3.7 Future Prospects and Challenges

The many qualities of AuNPs hailed throughout this chapter make them an attractive and versatile tool for biomedical applications. Healthcare resources are moving toward the development of more cost-effective, rapid, and less invasive options for patients. AuNPs have proven to be useful tools in fulfilling these sought-after benefits, in addition to offering higher sensitivity and short turnaround time of diagnostic tests, enabling point-of-care testing. AuNPs offer a great deal of hope in cancer therapy and imaging. Their high surface-to-volume ratio makes them excellent vehicles for drug delivery, and their unique optical properties, which can be tuned to extend to the NIR (the optical transparency window of tissues), enable their use in photothermal therapy, as well as in diagnostic and imaging applications. AuNPs are therefore expected to remain in the spotlight, though the popularity of gold nanospheres is ebbing in favor of anisotropic gold nanostructures, especially nanorods and nanoshells, due to their higher tunability. Efforts are expected to be further poured into optimization, whether in terms of identifying the best-suited size and shape of AuNPs for each application, or in terms of optimizing the applications themselves. The present unmet needs in healthcare and the advantages offered by AuNPs urge toward their clinical approval.

Among the major challenges facing the use of AuNPs in vivo is the need for more long-term studies on their biodistribution and possible cytotoxic effects, especially with regard to redistribution. Patent issues represent another potential challenge facing the commercialization of AuNPs in biomedical applications. Patents are a double-edged sword, especially in a new and emerging field such as nanomedicine. On the one hand, patents encourage innovation while protecting the rights of inventors. Because nanotechnology and its applications in many scientific disciplines represent a new arena with still much room for growth, inventors and investors are leaving no stone unpatented. In a special report, Bawa[175] elegantly explained that the large number of patents being issued in the field, with scientists, universities, and companies racing to obtain broad patents in upstream processes, is resulting in "patent thickets" and overlaps that may eventually hinder progress, innovation, and commercialization of products.

This challenge was also faced by biotechnology when it emerged as a budding field, and has since been gradually overcome. Therefore, nanomedicine must be geared toward the facilitation of commercialization early on in the process of development so that patients can reap the benefits of nanomedical services and products. Nanomedicine and the use of gold nanoparticles in biomedical applications hold great promise in providing higher-quality healthcare, including reduced cost, earlier detection, and more efficient and targeted treatment, with the use of less or noninvasive methods.

Acknowledgments

We thank Tamer Samir, from the Yousef Jameel Science and Technology Research Center at the American University in Cairo, and Ahmed Teleb for their support and professional help with the artwork.

References

1. Daniel, M.C., and Astruc, D., Gold nanoparticles: assembly, supramolecular chemistry, quantum-size-related properties, and applications toward biology, catalysis, and nanotechnology, *Chem. Rev.*, 104, 293, 2004.
2. Radwan, S.H., and Azzazy, H.M., Gold nanoparticles for molecular diagnosis, *Expert Rev. Mol. Diagn.*, 9, 511, 2009.

3. Loo, C., et al., Nanoshell-enabled photonics-based imaging and therapy of cancer, *Technol. Cancer Res. Treat.*, 3, 33, 2004.
4. Oldenburg, S.J., et al., Infrared extinction properties of gold nanoshells, *J. Appl. Phys. Lett.*, 75, 2897, 1999.
5. Nehl, C.L., et al., Scattering spectra of single gold nanoshells, *Nano Lett.*, 4, 2355, 2004.
6. Rasch, M.R., Sokolov, K.V., and Korgel, B.A., Limitations on the optical tunability of small diameter gold nanoshells, *Langmuir*, 25, 11777, 2009.
7. Kawamura, G., et al., Shape control synthesis of multi-branched gold nanoparticles, *Mater. Chem. Physics*, 115, 229, 2009.
8. Noguez, C., Surface plasmons on metal nanoparticles: the influence of shape and physical environment, *J. Phys. Chem. C*, 111, 3806, 2007.
9. Nehl, C.L., and Hafner, J.H., Shape-dependent plasmon resonances of gold nanoparticles, *J. Mater. Chem.*, 18, 2415, 2008.
10. Song, J.H., et al., Crystal overgrowth on gold nanorods: tuning the shape, facet, aspect ratio, and compositions of the nanorods, *Chem. Euro. J.*, 11, 910, 2005.
11. Sanvicens, N., and Marco, M.P., Multifunctional nanoparticles—properties and prospects for their use in human medicine, *Trends Biotechnol.*, 26, 425, 2008.
12. Hu, M., et al., Ultrafast laser studies of the photothermal properties of gold nanocages, *J. Phys. Chem. B*, 110, 1520, 2006.
13. Shankar, S.S., Bhargava, S., and Sastry, M., Synthesis of gold nanospheres and nanotriangles by the Turkevich approach, *J. Nanosci. Nanotechnol.*, 5, 1721, 2005.
14. Jin, R., et al., Controlling anisotropic nanoparticle growth through plasmon excitation, *Nature*, 425, 487, 2003.
15. Patra, C.R., et al., Fabrication of gold nanoparticles for targeted therapy in pancreatic cancer, *Adv. Drug Deliv. Rev.*, 62, 346, 2009.
16. Sun, Y.G., and Xia, Y.N., Shape-controlled synthesis of gold and silver nanoparticles, *Science*, 298, 2176, 2002.
17. Chen, H., et al., Shape- and size-dependent refractive index sensitivity of gold nanoparticles, *Langmuir*, 24, 5233, 2008.
18. Tsung, C.K., et al., Shape- and orientation-controlled gold nanoparticles formed within mesoporous silica nanofibers, *Adv. Funct. Mater.*, 16, 2225, 2006.
19. Huang, X., Neretina, S., and El-Sayed, M.A., Gold nanorods: from synthesis and properties to biological and biomedical applications. *Adv. Mater.*, 21, 1, 2009.
20. Song, K.H., Kim, C., and Cobley, C.M., Near-infrared gold nanocages as a new class of tracers for photoacoustic sentinel lymph node mapping on a rat model, *Nano Lett.*, 9, 183, 2009.
21. Khlebtsov, N.G., and Dykman, L.A., Optical properties and biomedical applications of plasmonic nanoparticles, *J. Quant. Spectrosc. Radiative Transfer*, 111, 1, 2010.
22. Azzazy, H.M., and Mansour, M.M., *In vitro* diagnostic prospects of nanoparticles, *Clin. Chim. Acta*, 403, 1, 2009.
23. Baptista, P., et al., Gold nanoparticles for the development of clinical diagnosis methods, *Anal. Bioanal. Chem.*, 391, 943, 2008.
24. West, J.L., and Halas, N.J., Engineered nanomaterials for biophotonics applications: improving sensing, imaging, and therapeutics, *Annu. Rev. Biomed. Eng.*, 5, 285, 2003.
25. Jain, P.K., et al., Calculated absorption and scattering properties of gold nanoparticles of different size, shape, and composition: applications in biological imaging and biomedicine, *J. Phys. Chem. B*, 110, 7238, 2006.
26. Kelly, K.L., et al., The optical properties of metal nanoparticles: the influence of size, shape, and dielectric environment, *J. Phys. Chem. B*, 107, 668, 2003.
27. Castaneda, M.T., et al., Electrochemical sensing of DNA using gold nanoparticles, *Electroanalysis*, 19, 743, 2007.
28. Ratto, F., et al., Photothermal effects in connective tissues mediated by laser-activated gold nanorods, *Nanomed. Nanotechnol. Biol. Med.*, 5, 143, 2009.

29. Deplanche, K., and Macaskie, L.E., Biorecovery of gold by *Escherichia coli* and *Desulfovibrio desulfuricans*, *Biotechnol. Bioeng.*, 99, 1055, 2008.

30. Ghosh, S.K., and Pal, T., Interparticle coupling effect on the surface plasmon resonance of gold nanoparticles: from theory to applications, *Chem. Rev.*, 107, 4797, 2007.

31. Azzazy, H.M., Mansour, M.M., and Kazmierczak, S.C., Nanodiagnostics: a new frontier for clinical laboratory medicine, *Clin. Chem.*, 52, 1238, 2006.

32. Jennings, T., and Strouse, G., Past, present, and future of gold nanoparticles, *Adv. Exp. Med. Biol.*, 620, 34, 2007.

33. Huang, X., and El-Sayed, M.A., Gold nanoparticles: optical properties and implementations in cancer diagnosis and photothermal therapy, *J. Adv. Res.*, 1, 13, 2010.

34. Frens, G., Controlled nucleation for the regulation of the particle size in monodisperse gold suspensions, *Nat. Phys. Sci.*, 241, 20, 1973.

35. Brust, M., et al., Synthesis of thiol-derivatised gold nanoparticles in a two-phase liquid–liquid system, *J. Chem. Soc. Chem. Commun.*, 801, 1994.

36. Brust, M., et al., Synthesis and reactions of functionalised gold nanoparticles, *J. Chem. Soc. Chem. Commun.*, 1655, 1995.

37. Oldenburg, S.J., et al., Nanoengineering of optical resonances, *Chem. Phys. Lett.*, 288, 243, 1998.

38. Stober, W., Fink, A., and Bohn, E., Controlled growth of monodisperse silica spheres in the micron size range, *J. Colloid Interface Sci.*, 26, 62, 1968.

39. Duff, D.G, Baiker, A., and Edwards, P.P., A new hydrosol of gold clusters. 1. Formation and particle size variation, *Langmuir*, 9, 2301, 1993.

40. Liu, Z., et al., Fabrication and near-infrared photothermal conversion characteristics of Au nanoshells, *Appl. Phys. Lett.*, 86, 113109, 2005.

41. Lyon, J. L., et al., Synthesis of Fe oxide core/Au shell nanoparticles by iterative hydroxylamine seeding, *Nano Lett.*, 4, 719, 2004.

42. Nikoobakht, B., and El-Sayed, M.A., Preparation and growth mechanism of gold nanorods (NRs) using seed-mediated growth method, *Chem. Mater.*, 15, 1957, 2003.

43. Murphy, C.J., et al., Anisotropic metal nanoparticles: Synthesis, assembly, and optical applications, *J. Phys. Chem. B*, 109, 13857, 2005.

44. Jana, N., Gearheart, L., and Murphy, C., Wet chemical synthesis of high aspect ratio cylindrical gold nanorods, *J. Phys. Chem. B*, 105, 4065, 2001.

45. Gole, A., and Murphy, C.J., Seed-mediated synthesis of gold nanorods: role of size and nature of the seed, *Chem. Mater.*, 16, 3633, 2004.

46. Park, W.M, Huh, Y.S., and Hong, W.H., Aspect-ratio-controlled synthesis of high-aspect-ratio gold nanorods in high-yield, *Curr. Appl. Phys.*, 9, e140, 2009.

47. Park, H.J., Ah, C.S., and Kim, W.-J., Temperature-induced control of aspect ratio of gold nanorods, *J. Vac. Sci. Technol.*, 24, 1323, 2006.

48. Jiang, X.C., and Pileni, M.P., Gold nanorods: influence of various parameters as seeds, solvent, surfactant on shape control, *Colloids Surf. A*, 295, 228, 2007.

49. Narayanan, K.B., and Sakthivel, N., Biological synthesis of metal nanoparticles by microbes, *Adv. Colloid Interface Sci.*, 156, 1, 2010.

50. Thakkar, K.N., et al., Biological synthesis of metallic nanoparticles, *Nanomed. Nanotechnol. Biol. Med.*, 6, 257, 2009.

51. Konishi, Y., et al., Microbial deposition of gold nanoparticles by the metal-reducing bacterium *Shewanella algae*, *Electrochimica Acta*, 53, 186, 2007.

52. Lloyd, J.R., Microbial reduction of metals and radionuclides, *FEMS Microbiol. Rev.*, 27, 411, 2003.

53. Lloyd, J.R., Lovley, D.L., and Macaskie, L.E., Biotechnological application of metal-reducing microorganisms, *Adv. Appl. Microbiol.*, 53, 85, 2003.

54. Kashefi, K., et al., Reductive precipitation of gold by dissimilatory Fe(III)-reducing bacteria and archaea, *Appl. Environ. Microbiol.*, 67, 3275, 2001.

55. Mukherjee, P., et al., Bioreduction of chloroaurate ions by the fungus *Verticillium* and surface trapping of gold nanoparticle thus formed, *Angew. Chem. Int. Ed.*, 40, 3585, 2001.

56. Mukherjee, P., et al., Extracellular synthesis of gold nanoparticles by the fungus *Fusarium oxysporum*, *Chem. Biochem.*, 5, 461, 2002.
57. Singaravelu, G., et al., A novel extracellular synthesis of monodisperse gold nanoparticles using marine alga, *Sargassum wightii* Greville, *Colloids Surf. B Biointerfaces*, 57, 97, 2007.
58. Wang, Y., et al., Barbated Skullcup herb extract-mediated biosynthesis of gold nanoparticles and its primary application in electrochemistry, *Colloids Surf. B Biointerfaces*, 73, 75, 2009.
59. Hu, M., et al., Dark-field microscopy studies of single metal nanoparticles: understanding the factors that influence the linewidth of the localized surface plasmon resonance, *J. Mater. Chem.*, 18, 1949, 2008.
60. Berciaud, S., et al., Observation of intrinsic size effects in the optical response of individual gold nanoparticles, *Nano Lett.*, 5, 515, 2005.
61. Rao, C.N.R., and Biswas, K., Characterization of nanomaterials by physical methods, *Annu. Rev. Anal. Chem.*, 2, 435, 2009.
62. Leff, D.V., et al., Thermodynamic control of gold nanocrystal size: experiment and theory, *J. Phys. Chem.*, 99, 7036, 1995.
63. Pissuwan, D., et al., The forthcoming applications of gold nanoparticles in drug and gene delivery systems, *J. Control. Release*, 149, 65, 2011.
64. Paciotti, G.F., et al., Colloidal gold: a novel nanoparticle vector for tumor directed drug delivery, *Drug Delivery*, 11, 169, 2004.
65. Leonov, A.P., et al., Detoxification of gold nanorods by treatment with polystyrenesulfonate, *ACS Nano*, 2, 2481, 2008.
66. Niidome, T., et al., PEG-modified gold nanorods with a stealth character for *in vivo* applications, *J. Controlled Release*, 114, 343, 2006.
67. Niidome, T., et al., Preparation of primary amine-modified gold nanoparticles and their transfection ability into cultivated cells, *Chem. Commun.*, 1978, 2004.
68. Srivastava, S., et al., Controlled assembly of protein nanoparticle composites through protein surface recognition, *Adv. Mater.*, 17, 617, 2005.
69. Shenoy, D. et al., Surface functionalization of gold nanoparticles using hetero-bifunctional poly(ethylene glycol) spacer for intracellular tracking and delivery, *Int. J. Nanomed.*, 1, 51, 2006.
70. Mishra, S., Webster, P., and Davis, M.E., PEGylation significantly affects cellular uptake and intracellular trafficking of non-viral gene delivery particles, *Eur. J. Cell Biol.*, 83, 97, 2004.
71. Moghimi, S.M., Chemical camouflage of nanospheres with a poorly reactive surface: towards development of stealth and target-specific nanocarriers, *Biochim. Biophys. Acta*, 1590, 131, 2002.
72. Tokumitsu, S., et al., Grafting of alkanethiol-terminated poly-(ethylene glycol) on gold, *Langmuir*, 18, 8862, 2002.
73. Carignano, M.A., and Szleifer, I., Controlling surface interactions with grafted polymers, *Interface Sci.*, 11, 187, 2003, and references therein.
74. Stern, J.M., and Cadeddu, J.A., Emerging use of nanoparticles for the therapeutic ablation of urologic malignancies, *Urol. Oncol. Semin. Orig. Invest.*, 26, 93, 2008.
75. Maksimova, I.L., et al., Near-infrared laser photothermal therapy of cancer by using gold nanoparticles: computer simulations and experiment, *Med. Laser Appl.*, 22, 199, 2007.
76. O'Neal, D.P., et al., Photothermal tumor ablation in mice using near infrared absorbing nanoparticles, *Cancer Lett.*, 209, 171, 2004.
77. Maeda, H., et al., Vascular permeability enhancement in solid tumor: various factors, mechanisms involved and its implications, *Int. Immunopharmacol.*, 3, 319, 2003.
78. Maeda, H., The enhanced permeability and retention (EPR) effect in tumor vasculature: the key role of tumor-selective macromolecular drug targeting, *Adv. Enzyme Regul.*, 41, 189, 2001.
79. Ishida, O., et al., Size-dependent extravasation and interstitial localization of polyethyleneglycol liposomes in solid tumor-bearing mice, *Int. J. Pharm.*, 190, 49, 1999.
80. Kong, G., Braun, R.D., and Dewhirst, M.W., Hyperthermia enables tumor-specific nanoparticle delivery: effect of particle size, *Cancer Res.*, 60, 4440, 2000.

81. Zharov, V.P., et al., Self-assembling nanoclusters in living systems: application for integrated photothermal nanodiagnostics and nanotherapy, *Nanomed. Nanotechnol. Biol. Med.*, 1, 326, 2005.

82. Anderson, R.R., and Parrish, J.A., Selective photothermolysis: precise microsurgery by selective absorption of pulsed radiation, *Science*, 220, 524, 1983.

83. Huang, X., El-Sayed, I.H., and El-Sayed, M.A., Cancer cell imaging and photothermal therapy in near-infrared region by using gold nanorods, *J. Am. Chem. Soc.*, 128, 2115, 2006.

84. Liao, H., and Hafner, J.H., Gold nanorod bioconjugates, *Chem. Mater.*, 17, 4636, 2005.

85. Allen, M., Ligand-targeted therapeutics in anticancer therapy, *Nat. Rev. Cancer*, 2, 750, 2002.

86. Bhattacharya, R., et al., Attaching folic acid on gold nanoparticles using noncovalent interaction via different polyethylene glycol backbones and targeting of cancer cells, *Nanomed. Nanotechnol. Biol. Med.*, 3, 224, 2007.

87. Bergen, J.M., et al., Gold nanoparticles as a versatile platform for optimizing physicochemical parameters for targeted drug delivery, *Macromol. Biosci.*, 6, 506, 2006.

88. Kang, B., Mackey, M.A., and El-Sayed, M.A., Nuclear targeting of gold nanoparticles in cancer cells induces DNA damage, causing cytokinesis arrest and apoptosis, *J. Am. Chem. Soc.*, 132, 1517, 2010.

89. Hong, R., et al., Glutathione-mediated delivery and release using monolayer protected nanoparticle carriers, *J. Am. Chem. Soc.*, 128, 1078, 2006, and references therein.

90. Sershen, S.R., et al., Temperature sensitive polymer-nanoshell composites for photo-thermally modulated drug delivery, *J. Biomed. Mater. Res.*, 51, 293, 2000.

91. Shiotani, A., et al., Stable incorporation of gold nanorods into N-isopropylacrylamide hydrogels and their rapid shrinkage induced by near-infrared laser irradiation, *Langmuir*, 23, 4012, 2007.

92. Kawano, T., et al., PNIPAM gel-coated gold nanorods for targeted delivery responding to a near-infrared laser, *Bioconj. Chem.*, 20, 209, 2009.

93. Wijaya, A., et al., Selective release of multiple DNA oligonucleotides from gold nanorods, *ACS Nano*, 3, 80, 2008.

94. Ipe, B.I., Mahima, S., and Thomas, K.G., Light-induced modulation of self-assembly on spiropyran-capped gold nanoparticles: a potential system for the controlled release of amino acid derivatives, *J. Am. Chem. Soc.*, 125, 7174, 2003.

95. El-Sayed, I.H., Huang, X., and El-Sayed, M.A., Selective laser photo-thermal therapy of epithelial carcinoma using anti-EGFR antibody conjugated gold nanoparticles, *Cancer Lett.*, 239, 129, 2006.

96. Gad, F., et al., Targeted photodynamic therapy of established soft-tissue infections in mice, *Photochem. Photobiol. Sci.*, 3, 451, 2004.

97. Fuchs, J., and Thiele, J., The role of oxygen in cutaneous photodynamic therapy, *Free Radic. Biol. Med.*, 24, 835, 1998.

98. Hone, D.C., et al., Generation of cytotoxic singlet oxygen via phthalocyanine-stabilized gold nanoparticles: a potential delivery vehicle for photodynamic therapy, *Langmuir*, 18, 2985, 2002.

99. Gibson, J.D., Khanal, B.P., and Zubarev, E.R., Paclitaxel-functionalized gold nanoparticles, *J. Am. Chem. Soc.*, 129, 11653, 2007.

100. Chen, Y.-H., et al., Methotrexate conjugated to gold nanoparticles inhibits tumor growth in a syngeneic lung tumor model, *Mol. Pharm.*, 4, 713, 2007.

101. Priyabrata, M., Resham, B., and Debabrata, M., Gold nanoparticles bearing functional anti-cancer drug and anti-angiogenic agent: a "2 in 1" system with potential application in therapeutics, *J. Biomed. Nanotechnol.*, 1, 224, 2005.

102. Saha, B., et al., *In vitro* structural and functional evaluation of gold nanoparticles conjugated antibiotics, *Nanoscale Res. Lett.*, 2, 614, 2007.

103. Akduman, B., Barqawi, A.B., and Crawford, E.D., Minimally invasive surgery in prostate cancer: current and future perspectives, *Cancer J.*, 11, 355, 2005.

104. Mirza, A.N., et al., Radiofrequency ablation of solid tumors, *Cancer J.*, 7, 95, 2001.

105. Uchida, T., et al., Five years' experience of transrectal high-intensity focused ultrasound using the Sonablate device in the treatment of localized prostate cancer, *Int. J. Urol.*, 13, 228, 2006.

106. McDougal, W.S., et al., Long-term followup of patients with renal cell carcinoma treated with radio frequency ablation with curative intent, *J. Urol.*, 174, 61, 2005.

107. Prudhomme, M., et al., Interstitial diode laser hyperthermia in the treatment of subcutaneous tumor, *Lasers Surg. Med.*, 19, 445, 1996.

108. Milleron, R.S., and Bratton, S.B., 'Heated' debates in apoptosis, *Cell. Mol. Life Sci.*, 64, 2329, 2007.

109. Cherukuri, P., et al., Targeted hyperthermia using metal nanoparticles, *Adv. Drug Delivery Rev.*, 62, 339, 2010.

110. Huang, X., et al., The potential use of the enhanced nonlinear properties of gold nanospheres in photothermal cancer therapy, *Lasers Surg. Med.*, 39, 747, 2007.

111. Tong, L., et al., Gold nanorods mediate tumor cell death by compromising membrane integrity, *Adv. Mater.*, 19, 3136, 2007.

112. Hirsch, L.R., et al., Nanoshell mediated near-infrared thermal therapy of tumors under magnetic resonance guidance, *Proc. Natl. Acad. Sci. USA*, 100, 13549, 2003.

113. Chen, J., et al., Immuno gold nanocages with tailored optical properties for targeted photothermal destruction of cancer cells, *Nano Lett.*, 7, 1318, 2007.

114. Goodman, M., and Geller, D.A., Radiofrequency ablation of hepatocellular carcinoma, in *Hepatocellular cancer*, ed. B. Carr, Human Press, Totowa, 2005, p. 171.

115. Cardinal, J., et al. Noninvasive radiofrequency ablation of cancer targeted by gold nanoparticles, *Surgery*, 144, 125, 2008.

116. Abdalla, E.K., et al., Recurrence and outcomes following hepatic resection, radiofrequency ablation, and combined resection/ablation for colorectal liver metastases, *Ann. Surg.*, 239, 818, 2004.

117. Mulier, S., et al., Local recurrence after hepatic radiofrequency coagulation: multivariate meta-analysis and review of contributing factors, *Ann. Surg.*, 242, 158, 2005.

118. Zharov, V.P., Galitovsly, V., and Viegas, M., Photothermal guidance of selective photothermolysis with nanoparticles, *Proc. SPIE*, 5319, 291, 2004.

119. Zharov, V.P., et al., Photothermal nanotherapeutics and nanodiagnostics for selective killing of bacteria targeted with gold nanoparticles, *Biophys. J.*, 90, 619, 2006.

120. McNally, K.M., et al., Photothermal effects of laser tissue soldering, *Phys. Med. Biol.*, 44, 983, 1999.

121. Gobin, A.M., et al., Near infrared laser-tissue welding using nanoshells as an exogenous absorber, *Lasers Surg. Med.*, 37, 123, 2005.

122. Rossi, F., Pini, R., and Menabuoni, L., Experimental and model analysis on the temperature dynamics during diode laser welding of the cornea, *J. Biomed. Opt.*, 12, 0140311, 2007.

123. DeCoste, S.D., et al., Dye-enhanced laser welding for skin closure, *Lasers Surg. Med.*, 12, 25, 1992.

124. Matteini, P., et al., Microscopic characterization of collagen modifications induced by low-temperature diode-laser welding of corneal tissue, *Lasers Surg. Med.*, 39, 597, 2007.

125. Liu, Y. et al., Single chain fragment variable recombinant antibody functionalized gold nanoparticles for a highly sensitive colorimetric immunoassay, *Biosens Bioelectron.*, 24, 2853, 2009.

126. Thanh, N.T.K., and Rosenzweig, Z., Development of an aggregation-based immunoassay for anti-protein A using gold nanoparticles, *Anal. Chem.*, 74, 1624, 2002.

127. Nam, J.M., Stoeva, S.I., and Mirkin, C.A., Bio-bar-code-based DNA detection with PCR-like sensitivity, *J. Am. Chem. Soc.*, 126, 5932, 2004.

128. Stoeva, S.I., et al., Multiplexed DNA detection with biobarcoded nanoparticle probes, *Angew. Chem. Int. Ed. Engl.*, 45, 3303, 2006.

129. Gestwicki, J.E., Strong, L.E., and Kisseling, L.L., Visualization of single multivalent receptor-ligand complexes by transmission electron microscopy, *Angew. Chem. Int. Ed.*, 39, 4567, 2000.

130. Cao, Y., Jin, R., and Mirkin, C.A., DNA-modified core-shell Ag/Au nanoparticles, *J. Am. Chem. Soc.*, 123, 7961, 2001.

131. Cao, Y.C., et al., A two-color-change, nanoparticle-based method for DNA detection, *Talanta*, 67, 449, 2009.

132. Storhoff, J.J., et al., What controls the optical properties of DNA-linked gold nanoparticle assemblies? *J. Am. Chem. Soc.*, 122, 4640, 2000.

133. Chakrabarti, R., and Klibanov, A.M., Nanocrystals modified with peptide nucleic acids (PNAs) for selective self-assembly and DNA detection, *J. Am. Chem. Soc.*, 125, 12531, 2003.

134. Sato, K., Hosokawa, K., and Maeda, M., Rapid aggregation of gold nanoparticles induced by non-cross-linking DNA hybridization, *J. Am. Chem. Soc.*, 125, 8102, 2003.

135. Kim, J.-Y., and Lee, J.-S., Synthesis and thermally reversible assembly of DNA-gold nanoparticle cluster conjugates, *Nano Lett.*, 9, 4564, 2009.

136. Li, H., and Rothberg, L., Detection of specific sequences in RNA using differential adsorption of single-stranded oligonucleotides on gold nanoparticles, *Anal. Chem.*, 77, 6229, 2005.

137. Shawky, S.M., Radwan, S., and Azzazy, H.M., Rapid colorimetric method for detection of HCV RNA using gold nanoparticles, paper presented at Hepatitis C 3rd International Conference, The Third Decade and Beyond, Dublin, Ireland, 2009.

138. Shawky, S.M., and Azzazy, H.M., Rapid detection of unamplified HCV RNA in clinical specimens using unmodified gold nanoparticles, paper presented at Annual Meeting of the American Association for Clinical Chemistry, Anaheim, California, 2010.

139. Thaxton, C.S., et al., A bio-barcode assay based upon dithiotheitol-induced oligonucleotide release, *Anal. Chem.*, 77, 8174, 2005.

140. Dubertret, B., Calame, M., and Libchaber, A.J., Single-mismatch detection using gold-quenched fluorescent oligonucleotides, *Nat. Biotechnol.*, 19, 365, 2001.

141. Griffin, J., et al., Size- and distance-dependent nanoparticle surface-energy transfer (NSET) method for selective sensing of hepatitis C virus RNA, *Chemistry*, 15, 342, 2009.

142. Tang, S., and Hewlett, I., Nanoparticle-based immunoassays for sensitive and early detection of HIV-1 capsid (p24) antigen, *J. Infect. Dis.*, 201 (Suppl. 1), S59, 2010.

143. Yang, X., Guo, Y., and Wang, A., Luminol/antibody labeled gold nanoparticles for chemiluminescence immunoassay of carcinoembryonic antigen, *Anal. Chim. Acta*, 666, 91, 2010.

144. Chen, L., et al., Gold nanoparticle enhanced immuno-PCR for ultrasensitive detection of Hantaan virus nucleocapsid protein, *J. Immunol. Methods*, 346, 64, 2009.

145. Bangs, L.B., New developments in particle-based immunoassays: introduction, *Pure Appl. Chem.*, 68, 1873, 1996.

146. Kim, H.W., et al., Micro-CT imaging with a hepatocyte-selective contrast agent for detecting liver metastasis in living mice, *Acad. Radiol.*, 15, 1282, 2008.

147. Mattrey, R.F., and Aguirre, D.A., Advances in contrast media research, *Acad. Radiol.*, 10, 1450, 2003.

148. Hainfeld, J.F., et al., Gold nanoparticles: a new x-ray contrast agent, *Br. J. Radiol.*, 79, 248, 2006.

149. Kao, C.Y., et al., Long-residence-time nano-scale liposomal iohexol for x-ray-based blood pool imaging, *Acad. Radiol.*, 10, 475, 2003.

150. Torchilin, V.P., PEG-based micelles as carriers of contrast agents for different imaging modalities, *Adv. Drug Delivery Rev.*, 54, 235, 2002.

151. Rabin, O., et al., An x-ray computed tomography imaging agent based on long-circulating bismuth sulphide nanoparticles, *Nat. Mater.*, 5, 118, 2006.

152. Xu, C., Tung, G.A., and Sun, S., Size and concentration effect of gold nanoparticles on x-ray attenuation as measured on computed tomography, *Chem. Mater.*, 20, 4167, 2008.

153. Sun, I.-C., et al., Heparin-coated gold nanoparticles for liver-specific CT imaging, *Chem. Eur. J.*, 15, 13341, 2009.

154. Alric, C., et al., Gadolinium chelate coated gold nanoparticles as contrast agents for both x-ray computed tomography and magnetic resonance imaging, *J. Am. Chem. Soc.*, 130, 5908, 2008.

155. Popovtzer, R., et al., Targeted gold nanoparticles enable molecular CT imaging of cancer, *Nano Lett.*, 8, 4593, 2008.

156. Rayavarapu, R.G., et al., Synthesis and bioconjugation of gold nanoparticles as potential molecular probes for light-based imaging techniques, *Int. J. Biomed. Imaging*, 2007, 29817, 2007.

157. Eghtedari, M., et al., High sensitivity of *in vivo* detection of gold nanorods using a laser optoacoustic imaging system, *Nano Lett.*, 7, 1914, 2007.

158. Hu, M., et al., Gold nanostructures: engineering their plasmonic properties for biomedical applications, *Chem. Soc. Rev.*, 35, 1084, 2006.

159. Copland, J. A., et al., Bioconjugated gold nanoparticles as a molecular based contrast agent: implications for imaging of deep tumors using optoacoustic tomography, *Mol. Imaging Biol.*, 6, 341, 2004.

160. Mallidi, S., et al., Molecular specific optoacoustic imaging with plasmonic nanoparticles, *Opt. Express*, 15, 6583, 2007.

161. Agarwal, A., et al., Targeted gold nanorod contrast agent for prostate cancer detection by photoacoustic imaging, *J. Appl. Phys.*, 102, 064701, 2007.

162. Eghtedari, M., et al., Optoacoustic imaging of nanoparticle labeled breast cancer cells: a molecular based approach for imaging of deep tumors, *Lasers Surg. Med.*, S16 (Suppl.), 52, 2004.

163. Li, P.C., et al., Photoacoustic imaging of multiple targets using gold nanorods, *IEEE Trans. Ultrason. Ferroelectr. Freq. Control*, 54, 1642, 2007.

164. Jain, P.K., et al., Noble metals on the nanoscale: optical and photothermal properties and some applications in imaging, sensing, biology, and medicine, *Accounts Chem. Res.*, 41, 1578, 2008

165. Loo, C., et al., Gold nanoshell bioconjugates for molecular imaging in living cells, *Opt. Lett.*, 30, 1012, 2005.

166. El-Sayed, I.H., Huang, X., and El-Sayed, M.A., Surface plasmon resonance scattering and absorption of anti-EGFR antibody conjugated gold nanoparticles in cancer diagnostics: applications in oral cancer, *Nano Lett.*, 5, 829, 2005.

167. Shukla, R., et al., Biocompatibility of gold nanoparticles and their endocytotic fate inside the cellular compartment: a microscopic overview, *Langmuir*, 21, 10644, 2005.

168. Connor, E.E., et al., Gold nanoparticles are taken up by human cells but do not cause acute cytotoxicity, *Small*, 1, 325, 2005.

169. Cho, W.-S., et al., Acute toxicity and pharmacokinetics of 13 nm-sized PEG-coated gold nanoparticles, *Toxicol. Appl. Pharmacol.*, 236, 16, 2009.

170. Balasubramanian, S.K., et al., Biodistribution of gold nanoparticles and gene expression changes in the liver and spleen after intravenous administration in rats, *Biomaterials*, 31, 2134, 2010.

171. De Jong, W.H., et al., Particle size-dependent organ distribution of gold nanoparticles after intravenous administration, *Biomaterials*, 29, 1912, 2008.

172. Hillyer, J.F., and Albrecht, R.M., Gastrointestinal persorption and tissue distribution of differently sized colloidal gold nanoparticles, *J. Pharm. Sci.*, 90, 1927, 2001.

173. Myllynen, P.K., et al., Kinetics of gold nanoparticles in the human placenta, *Reprod. Toxicol.*, 26, 130, 2008.

174. Saunders, M., Transplacental transport of nanomaterials, *WIREs Nanomed. Nanobiotechnol.*, 1, 671, 2009, and references therein.

175. Bawa, R., Will the nanomedicine "patent land grab" thwart commercialization? *Nanomedicine*, 1, 346, 2005.

4

Theragnostic Nanoparticles with the Power to Diagnose, Treat, and Monitor Diseases

Shelton D. Caruthers, Gregory M. Lanza, and Samuel A. Wickline

School of Medicine, Washington University in St. Louis, St. Louis, Missouri, USA

CONTENTS

4.1 Introduction

As recently as the mid-1990s, a new thrust in medical imaging was mounting. Once again, the promise offered by clinical imaging was focusing on more than just pretty pictures of gross anatomy, but rather on visualizing the underlying physiology, and the dysfunction therein, of pathology. The term *molecular imaging* was adopted to describe this noninvasive quantitative characterization of biologic processes at the molecular (and cellular) level,[1,2] and its progress was driven by contemporaneous developments in multiple disciplines, including genomics, proteomics, molecular biology, medical imaging, computer science, and nanotechnology. The great promise of molecular imaging incorporates early (even presymptomatic) disease detection with patient-specific characterization that could lead to personalized medicine.[3]

While molecular imaging—which often employs site-targeted contrast agents—involves all imaging modalities, nuclear medicine has become synonymous with the term because the other clinical modalities typically lack the sensitivity to the contrast agents and have difficulty detecting the small local concentration of molecularly targeted agents. Where nuclear medicine techniques such as positron emission tomography are exquisitely sensitive to small molecules such as [18]Fluorodeoxyglucose (FDG), this small molecule approach cannot deliver enough contrast agent to rise above the detection limit for other common imaging modalities such as magnetic resonance imaging. This obstacle was obliterated, however, when nanotechnology approaches offered means to deliver tens of thousands of copies of the imaging agent for every single molecular binding site, thereby raising the local concentration well above the detection threshold for imaging.

In the recent decade, there has been an explosion of nanotechnology research and development toward site-targeted imaging contrast agents, therapeutics, and combinations thereof. This chapter is not a thorough review of this very broad field of research, but rather a discourse on the overall clinical goal for this area of nanomedicine focusing on examples from one lab to illustrate that story. For a more extensive scrutiny of the subject, the reader is directed to many helpful publications freely available via the Internet.[4–16]

4.2 Nanoparticle Approach to Targeted Imaging

Whether specifically based upon fullerenes, nanocages, nanotubes, nanocrystals, nanocolloids, dendrimers, liposomes, viruses, copolymers, or many other nano-sized constructs, the general nanoparticle approach to theragnostics (Figure 4.1) is to have a central core or backbone into or onto which other components can be placed for targeting, imaging, drug delivery, and pharmacodynamic control.

4.2.1 Targeting

Passive targeting can be influenced through size, charge, and surface coating of nanoparticles.[17] Accumulation of the nanoparticles occurs based on natural responses of the body to the foreign objects, be that differential permeability, phagocytosis, or other

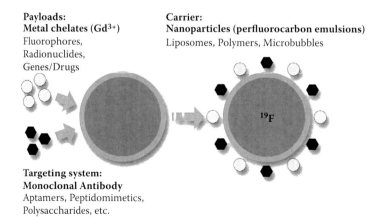

Payloads:
Metal chelates (Gd^{3+})
Fluorophores,
Radionuclides,
Genes/Drugs

Carrier:
Nanoparticles (perfluorocarbon emulsions)
Liposomes, Polymers, Microbubbles

^{19}F

Targeting system:
Monoclonal Antibody
Aptamers, Peptidomimetics,
Polysaccharides, etc.

FIGURE 4.1

Generalized paradigm for nanoparticles as imaging and therapeutic agents. Regardless of the technique employed, the common theme is to have a core (whether solid, liquid, hollow, or polymeric scaffolding) into or onto which "payloads" (such as drugs and imaging agents) are incorporated and delivered to a specific site in the body by targeting ligands (such as antibodies, etc.) attached to the outer surface. The specific nanoparticle emulsion discussed herein is a liquid perfluorocarbon core encapsulated by a phospholipid monolayer that holds drug (e.g., fumagillin) and targeting peptidomimetics (e.g., to $\alpha_v\beta_3$ integrin). (Reprinted with permission from Lanza, G.M., et al., *Medica Mundi*, 47, 34–39, 2003.)

clearance mechanisms. For example, iron oxide nanoparticles (10–100 nm) are cleared by the reticuloendothelial system (RES), and therefore normally accumulate in organs of the RES, namely, the liver, spleen, lymph nodes, and bone marrow. Similarly, macrophages, which are prevalent in areas of interest, such as tumors,[18] lymph nodes, and atherosclerotic plaques,[19] will take up nanoparticles and concentrate them for imaging. Coating the surface of nanoparticles with polyethylene glycol (PEG) chains of various lengths can also affect the pharmacodynamics.[20,21]

Active targeting is frequently achieved via attaching to the outer surface of the particle antibodies, antibody fragments, aptamers, peptides, peptidomimetics, or other small molecules that have an affinity for the binding site of interest.[22] The avidity of the particle—that is, its ability to "stick and stay" at the intended binding site—can be controlled by the way the targeting ligand is put on the nanoparticle, the number of ligands per particle, and whether or not the ligands are free to move along the surface of the particle. The targeting effectiveness of a nanoparticle is determined by the likelihood that a nanoparticle gets to the intended site, that the targeting ligand is presented to the complementary moiety in an orientation compatible with binding, and whether more than one ligand will bind the nanoparticle to the target. Because of this multifaceted nature, the binding affinity of a nanoparticle construct can far exceed that of the individual component targeting ligands. Through phage display[23] and many other novel techniques, more and more unique receptor targeting ligands that can be utilized as the targeting component of multicompartmental nanoparticles are being discovered.

4.2.2 Imaging

Just as the targeting ligand is one component of these multicomponent nanoparticles, so is the imaging agent. As previously stated, a major strength of nanoparticle-based, site-targeted imaging contrast agents over the small molecule approach is that multiple copies of the imaging constituent can be embedded in the core or attached to the surface. From a very simplistic perspective, one might imagine this being akin to the difference between shipping a single lantern to a given postal address and delivering an entire truckload of lanterns to the same address. The latter would indeed be much easier to spot in the dark!

Nanoparticle imaging agents have been developed for every medical imaging technique, including optical, ultrasound, magnetic resonance imaging (MRI), x-ray computed tomography (CT), positron emission tomography (PET), and single-photon emission computed tomography (SPECT). Often, the new agents are constructed by "plugging in" existing imaging agent components, such as chelates of radioisotopes (i.e., for nuclear medicine) or lanthanides (i.e., for MRI).

4.2.3 Therapeutic Delivery

In addition to delivering a truckload of imaging agent to a particular locale, nanoparticles have the ability to deliver drugs. An important benefit of nanotechnology is the repackaging of drugs to control the biodistribution and pharmacodynamics of drug release. More than just time-released capsules, nanotech approaches permit entirely new methods of formulating drugs with the potential for greater safety and efficacy.[24] But also, because drugs can be sequestered into nanoparticles and released only upon binding at the intended site—a concept described as contact-facilitated drug delivery[25]—systemic effects can be reduced while a greater local concentration is achieved. Examples of this are discussed in more depth later in this chapter and are based on a representative therapeutic that is a perfluorocarbon nanoemulsion.

4.2.4 Perfluorocarbon Nanoparticle Emulsion

Perfluorocarbons (PFCs)—hydrocarbons with all the hydrogen atoms replaced with fluorine atoms—exhibit a unique combination of properties, including exceptional inertness, stability, low surface tension, simultaneous hydrophobicity and lipophobicity, and high gas-dissolving capacity.[26] They are therefore of particular interest to biomedical research, including drug delivery, oxygen therapy, and artificial blood substitutes.[27,28] One of the very first successful targeted imaging nanoparticles was a liquid PFC emulsion based on an artificial blood substitute.[29] It followed the generic nanoparticle paradigm, having a liquid core (approximately 250 nm) encapsulated with a lipid monolayer into which various components for imaging and targeting could be inserted. Originally, it was a site-targeted ultrasound agent, relying on the differing speeds of sound between water and the PFC to generate reflections from the layer of nanoparticles bound to the target. Later,[30] the outer membranes of the nanoparticles were decorated with gadolinium diethylene-triamine-pentaacetic acid (Gd-DTPA), a common contrast agent for MRI. Unlike previous targeted MRI contrast agents, this vehicle delivered over 90,000 copies of the gadolinium chelate to each binding site, thereby overcoming the poor sensitivity of MRI to low concentrations of Gd-DTPA. This basic nanoparticle platform has been the backbone from which many similar constructs have been developed for applications in ultrasound,[29] CT,[31] MRI,[32] and even multiple imaging modalities with one agent.[33] By incorporating multiple markers or different perfluorocarbons, this system allows multicolor detection of multiple targets simultaneously.[34]

4.3 Detecting Markers of Disease

The goal of a diagnostic nanoparticle contrast agent is to home to markers of disease and be detected, preferably in a quantitative way, by the clinically relevant imaging technique. The biomarker needs to identify specifically a disease early in its course. The first incarnations of the PFC nanoparticle contrast agent were targeted to fibrin. Detecting exposed fibrin associated with atherosclerotic plaque can help distinguish the difference between lesions that are restricting the lumen but are stable and those at increased risk for disastrous rupture.[35,36] In atherosclerosis, many other markers point to the disease, some at a much earlier stage than plaque rupture. One such early indicator (which is not only specific, but a requirement for disease progression) is angiogenesis within the *vasa vasorum*.[37] Angiogenesis is a complex process with many molecular signatures. One of those, which holds a critical role in new blood vessel formation, the $\alpha_v\beta_3$ integrin, has garnered much attention as a target.[38]

Targeting paramagnetic PFC nanoparticles to the $\alpha_v\beta_3$ integrin, Winter et al. demonstrated the ability to image and quantify with MRI early-stage atherosclerosis.[39] In these experiments, the heterogeneous nature of the disease was clearly visible along the length the aorta. However, in control subjects, in which the disease was not present or else non-targeted contrast agent was administered, no signal enhancement was appreciated. The specificity of nanoparticle binding was further verified through competitive blockade experiments.

Another disease in which angiogenesis plays a critical role is cancer. When the metabolic demands of a growing tumor exceed the supply of nutrients that can be delivered via diffusion, new vessels are recruited to do the job. Again targeting $\alpha_v\beta_3$, PFC nanoparticles have been used to image this neovasculature in models of cancer, e.g., the Vx-2 model, wherein small tumors are implanted into the popliteal fossa of rabbits and allowed to grow for about 2 weeks. In one study[40] using a clinical 1.5 T MRI scanner, the intravenously injected nanoparticles were imaged accumulated at sites of angiogenesis surrounding the tumor, indicating asymmetrical distributions of neovasculature typically at tumor-muscle interfaces and nearby vasculature. The $\alpha_v\beta_3$-targeted paramagnetic nanoparticles provided a signal intensity 126% greater than that at baseline, representing a change twice that due to control, nontargeted nanoparticles (which also accumulated nominally due to passive extravascular leakage). As with the atherosclerosis experiments, in vivo competitive blockade experiments confirmed receptor specificity of the targeted agent. In the same cancer model, indium (^{111}In) radiolabeled nanoparticles were used to image the neovasculature with much greater detection sensitivity using single-pinhole, planar nuclear imaging techniques.[41] Combining the ^{111}In and Gd-DTPA labels on the $\alpha_v\beta_3$-targeted particles provided a unique dual-modality SPECT/MRI contrast agent that could image Vx-2 tumor neovascularity with the increased sensitivity of nuclear images and the high-resolution anatomical images of MRI.[33] Other models of cancer in which these diagnostic particles have been employed successfully are a human melanoma[42] and breast carcinoma[43] in mice. These models, with regard to aggressiveness and angiogenesis, are a bit different from the Vx-2 model, and those differences can be characterized with these targeted nanoparticle imaging agents.

4.4 Characterizing Disease

In the science of medical diagnosis, a tool providing sensitive and specific detection of the presence (or absence) of a disease is invaluable. Nanotechnology offers the potential to extend this one step further by characterizing the disease by the presence or extent of certain biological markers. This can be important not only in understanding the disease type and stage in an individual, but also in determining a priori which therapy will (or will not) be effective. For example, in some, but not all, breast cancers the human epidermal growth factor receptor 2 (HER2) gene is amplified.[44] In the 25–30% overexpressing HER2, the patients have a much poorer prognosis than those not expressing it, but fortunately, they also preferentially respond to treatment with trastuzumab, a recombinant monoclonal antibody-targeted therapy.[45,46] Already mentioned above, for atherosclerosis, the extent of angiogenesis in the *vasa vasorum* is indicative of the severity of the disease,[37,47] and the presence or absence of exposed fibrin in a plaque is useful in predicting its predilection to rupture.[35,36]

The presence and extent of angiogenesis is an important marker for characterizing cancer. Employing MRI molecular imaging with integrin-targeted nanoparticles as before, Lanza et al. described (quantitatively and visually) the amount of angiogenesis associated with MDA-MB-435 breast carcinoma[43] and the more aggressive Vx-2 carcinoma.[48] Not only could the difference in extent of angiogenesis be distinguished between these two tumor types, but by imaging repeatedly over time, this group[49] detected the onset in the Vx-2 tumors of the angiogenic switch[50,51]—the point when the tumor transitions into the angiogenic phenotype, becoming more aggressive and prone to metastasis.

This is an excellent example of noninvasive phenotyping through molecular imaging using a single biomarker. Of course, this could be extended to use a cocktail of agents targeted to a variety of biomarkers. If each uniquely targeted agent had a distinguishable signature (i.e., a unique spectrum on magnetic resonance), then multiparametric imaging could be performed to characterize disease more completely.[52] Caruthers et al. demonstrated the potential of this using multiple perfluorocarbon nanoparticles, each with a unique signature (e.g., color) on MRI, in both in vitro[34] and in vivo[53] experiments.

Early diagnosis and specific characterization of disease through biomarkers such as angiogenesis in cancer have great utility. Phenotyping the disease helps guide the selection of effective therapy. Importantly, it can help avoid applying the wrong therapy, thereby saving the patient the high costs associated with ineffective therapy, namely, exposure to potentially harmful side effects and lost time getting to the successful therapy regimen.

4.5 Targeted Drug Delivery

While early, accurate diagnosis with quantitative characterization of disease could have considerable clinical relevance, this information has far greater impact if it helps guide and monitor earlier, more effective therapies. Nanoparticles designed for site-targeted molecular imaging can also be designed to be multifunctional, incorporating therapeutics for targeted delivery.

Nanotechnology has much to offer in the (re)development of drugs, drug delivery, and pharmacokinetic control.[54] From transdermal systems to controlled-release approaches, there is a vast array of activity in this field.[24,55–69] One example is the liquid perfluorocarbon nanoparticles presented above. The method employed is not based on slow release or targeted cell uptake of the particle, but rather on a process labeled contact facilitated drug delivery[70] (Figure 4.2).

These therapeutic nanoparticles are manufactured with lipophilic drugs incorporated into their outer lipid monolayer. The drugs do not dissolve into the perfluorocarbon core and, due to lipophilicity, do not readily diffuse into the aqueous surrounding (i.e., blood serum). The drug is trapped at the nanoparticle surface in high concentration, but with no release, and therefore no effect. If the nanoparticle binds to the targeted cell surface receptor and remains in close proximity for an extended time, there is an opportunity for interaction between the two lipid membranes. Through these direct interactions, the lipophilic drug can exchange from the nanoparticle membrane to the targeted cell membrane, later to be internalized by the cell's normal processes[71,72] (Figure 4.3). This exchange does not require cell disruption or particle internalization, but it can be augmented by adding energy through various techniques, such as ultrasound.[73] However, without the direct contact afforded by ligand-directed binding, no drug is released and therapeutic effect is not induced.[71,74] As a result, high concentrations of drug can be accumulated at the site of interest, though the blood serum level of the drug is effectively zero (thereby drastically reducing systemic side effects).

Antiangiogenic therapy can be delivered with the same agents used in targeted imaging of angiogenesis. Packaging a potent antiangiogenic compound fumagillin,[75] the

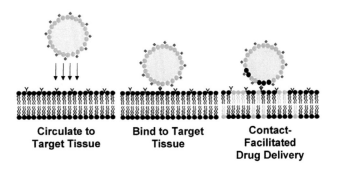

Circulate to Target Tissue   **Bind to Target Tissue**   **Contact-Facilitated Drug Delivery**

FIGURE 4.2

Contact facilitated drug delivery is the concept wherein drug is sequestered as part of the outer membrane of a nanoparticle, unable to escape until binding to the targeted cell occurs. Owing to the close proximity and relatively long duration afforded by binding, the phospholipid monolayer of the nanoparticle and the phospholipid bilayer of the cell can interact, exchanging phospholipids and dissolved molecules, such as drugs. Once the drugs exchange into the outer layer of the cell bilayer, they disperse and can be "flipped" into the inner surface through normal ATP-dependent pathways. (Reprinted with permission from Lanza, G.M., et al., *Curr Pharm Biotechnol*, 5, 495–507, 2004.)

MRI-visible $\alpha_v\beta_3$-targeted particles have been demonstrated effective in both atherosclerosis[74] and cancer[43,48] models. Moulton et al. have shown that the water-soluble form of fumagillin (TNP-470), when administered systemically, has an antiangiogenic effect that reduces the overall plaque growth in atherosclerosis.[76] A model of early atherosclerosis, hypercholesterol rabbits were given angiogenesis-targeted nanoparticles either containing (treatment group) or not containing (control group) fumagillin.[74] MRI signal enhancement averaged over the entire abdominal aorta was identical for both groups, indicating both presence of disease and delivery of theragnostic nanoparticles. After 1 week, the antiangiogenic effect was assessed using diagnostic $\alpha_v\beta_3$-targeted nanoparticles (and corroborated with histology), confirming that the disease was dramatically reduced in those receiving targeted fumagillin nanoparticles, whereas the control group remained the same as before (Figure 4.4). A different control group, which received fumagillin-containing nanoparticles that were not targeted, also showed no therapeutic response. While this study validated the concept that targeting is required for therapeutic response, it further elucidated two other important points. One point was that the MRI signal level at the time of initial dosing not only confirmed the delivery of drug, but also predicted the response to therapy. Second, the therapeutic effect achieved in this study was similar to that by Moulton et al., but the total drug amount used was more than 10,000 times less. So not only are potentially harmful side effects reduced by the sequestration of the drug in the nanoparticles, but the concentrating effect of targeting can administer high concentrations of drug at the intended site even though very small total amounts of drug are used.

These same concepts were repeated in the Vx-2[48] and MDA-MB-435[43] cancer models, the Vx-2 being the larger, more aggressive tumor model. In the Vx-2 model, a single dose of 30 µg/kg of fumagillin (1,000 times less than typical systemic dosing) was administered intravenously via $\alpha_v\beta_3$-targeted MRI nanoparticles. After 16 days, MRI was performed using the diagnostic-only particles to quantify tumor size and assess neovascularity. For those receiving targeted therapy, tumor volume was reduced significantly compared to controls receiving no drug (Figure 4.5). Histology also confirmed the response to therapy, which was not observed in the control tumors. Similarly, Schmieder et al.[43] delivered

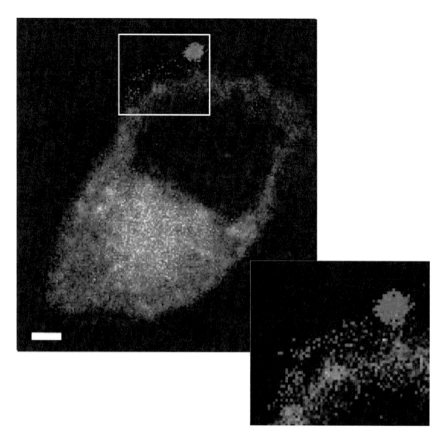

FIGURE 4.3
Fluorescent image of nanoparticle drug delivery. A C32 melanoma cell, transfected to express green fluorescent protein, is targeted by a perfluorocarbon nanoparticle carrying red Rhodamine dye (a fluorescent mimic for a drug). The dye streams from the particle as part of the cell's phospholipid membrane and can be seen over the entire cell. The inset shows a magnified view of the Rhodamine dye melting from the nanoparticle to the cancer cell. (Reprinted with permission from Crowder, K.C., et al., *Ultrasound Med Biol*, 31, 1693–1700, 2005.)

fumagillin to smaller tumors in mice. These tumors, which exhibit a much lower level of angiogenesis, had a measurable, albeit diminutive, effect of therapy. Considering these two studies, one might appreciate the power of these theragnostic nanoparticles in characterizing the amount of angiogenesis prior to initiating therapy, predicting response, and guiding choices.[77] In the cases where only minute amounts of angiogenesis are present, an aggressive and costly regimen of antiangiogenic therapy may not be warranted; another course may be more effective.

   While the previous examples of targeted therapy all involved fumagillin as the drug agent, chemotherapeutics such as paclitaxel,[71] doxorubicin,[71] and rapamycin,[78] and even toxic peptides such as mellitin[79] and thrombolytic enzymes[80] have been effectively incorporated into perfluorocarbon nanoparticles. In many cases, targeted nanoparticles offer a new opportunity to candidate drugs that would otherwise be far too toxic to use systemically. Mellitin, a toxic peptide derived from honeybee venom, nonspecifically attacks lipid membranes, disrupting virtually every cell it encounters. Designing it into a perfluorocarbon nanoparticle (coined nanobee[81]), however, isolates its disruptive effects to only those cells specifically targeted.[79]

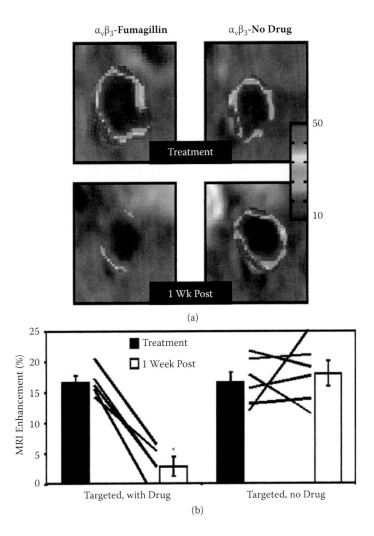

FIGURE 4.4
Targeted drug delivery with image-based monitoring of response. Using an $\alpha_v\beta_3$-targeted nanoparticle to highlight with MRI angiogenesis in these atherosclerotic aortas, we can see significant enhancement at baseline (a, top row). When imaged a week later (second row) using the angiogenesis-targeted nanoparticles, the subject receiving fumagillin via targeted theragnostic nanoparticles shows marked reduction in signal, whereas the control subject, having received no drug, shows a similar amount of angiogenesis. Averaging across all subjects (b, lines indicate individual subjects), we see a significant reduction of atherosclerotic-related angiogenesis in those receiving theragnostic particles, whereas those receiving imaging-only nanoparticles (no drug) showed no change in disease level. (Reprinted with permission from Winter, P.M., et al., *Arterioscler Thromb Vasc Biol*, 26, 2103–2109, 2006.)

4.6 Monitoring Response to Therapy

Just as the diagnostic nanoparticles can be used to detect and characterize markers of disease, so they can be used to evaluate the change in these markers as a result of therapy. This can be true as a method to appraise the effects of traditional drugs, or as in the examples previously cited, it can be to monitor the response to nanoparticle-delivered therapy.

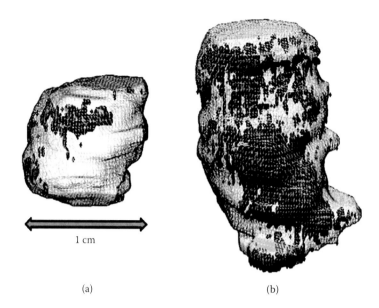

1 cm

(a)   (b)

FIGURE 4.5
(See color insert.) Three-dimensional (3D) neovascular maps in Vx-2 tumors. With MRI and $\alpha_v\beta_3$-targeted nanoparticles, areas of angiogenesis surrounding Vx-2 tumors can be highlighted (dark blue) on 3D mesh models of the tumor (digitally resected from the 3D image data set). These two tumors are the same age, but tumor (a) received antiangiogenic therapy through $\alpha_v\beta_3$-targeted fumagillin-laden theragnostic nanoparticles 16 days prior to this imaging and is much smaller than tumor (b), which received only drug-free particles for imaging. Note the heterogeneous distribution of neovasculature around both the treated and nontreated tumor. (Reprinted with permission from Winter, P.M., et al., *Faseb J*, 22, 2758–2767, 2008.)

The benefits of rapid feedback—vs. a wait-and-see method taking weeks, months, even years—are obvious.

In the examples of both cancer[48] and atherosclerosis,[74] clear evidence of the regression of disease was measured within a week of the initial therapy. Beyond that, serial imaging over multiple weeks perceived the recrudescence of atherosclerosis in fat-fed rabbits after the one-time therapy effects subsided.[82] One week following a single treatment by fumagillin-laden targeted nanoparticles, the amount of angiogenesis had regressed to about one-quarter the value seen in the controls. However, over the following 3 weeks, the amount of atherosclerotic-related neo-vasculature returned to the level measured in the controls. The response was similar independent of the two dosages tested, i.e., 30 and 90 µg/kg. In a similar study, this temporal tool was used to evaluate the combination effects of traditional therapy (i.e., oral atorvastatin) in conjunction with targeted nanotherapeutics. As before, the single dose of nanotherapy induced a significant antiangiogenic response that was monitored weekly by molecular imaging. By the fourth week, when the level of angiogenesis had returned, a second dose of nanotherapeutics was administered, generating the same intermittent response as monitored by diagnostic nanoparticles. However, in those subjects receiving dietary atorvastatin in combination with the theragnostic nanoparticles, the antiangiogenic effect persisted to the end of the 8-week study. In the clinic, this type of swift and repeatable interrogation of drug efficacy can lead to more timely adjustments in therapy, thereby improving patient outcomes.

4.7 Conclusion

Theragnostic nanoparticles are a new class of agents that offer the ability to diagnose and characterize disease in early stages, to deliver therapy specific for that disease phenotype, and to monitor serially the response to that therapy. The ability to characterize disease quantitatively through specific biomarkers (like HER2 or angiogenesis) can help stratify patients by their predicted response to different therapy options. Targeted nanoparticles, which not only safely and stealthily carry highly potent therapeutics, but also are visible via noninvasive clinical imaging, can offer confirmation of drug dosing with a prediction of response at the time of dosing. The actual response, beginning only days posttherapy, may be monitored serially employing the same particles used to diagnose, thus either confirming effect or suggesting an alteration sooner rather than later. All these features coalesce to help usher in the possibility of personalized medicine.

References

1. Service, R.F. Molecular Imaging: new probes open windows on gene expression, and more. *Science* 280, 1010–1011 (1998).
2. Weissleder, R. Molecular imaging: exploring the next frontier. *Radiology* 212, 609–614 (1999).
3. Pettigrew, R.I., Fee, C.A., and Li, K.C. Changes in the world of biomedical research are moving the field of "personalized medicine" from concept to reality. *J Nucl Med* 45, 1427 (2004).
4. Weissleder, R., and Mahmood, U. Molecular imaging. *Radiology* 219, 316–333 (2001).
5. O'Reilly, R.K., Hawker, C.J., and Wooley, K.L. Cross-linked block copolymer micelles: functional nanostructures of great potential and versatility. *Chem Soc Rev* 35, 1068–1083 (2006).
6. Weber, W.A., Czernin, J., Phelps, M.E., and Herschman, H.R. Technology insight: novel imaging of molecular targets is an emerging area crucial to the development of targeted drugs. *Nat Clin Pract Oncol* 5, 44–54 (2008).
7. Tran, T.D., et al. Clinical applications of perfluorocarbon nanoparticles for molecular imaging and targeted therapeutics. *Int J Nanomed* 2, 515–526 (2007).
8. Tomalia, D.A., Reyna, L.A., and Svenson, S. Dendrimers as multi-purpose nanodevices for oncology drug delivery and diagnostic imaging. *Biochem Soc Trans* 35, 61–67 (2007).
9. Michalet, X., et al. Quantum dots for live cells, *in vivo* imaging, and diagnostics. *Science* 307, 538–544 (2005).
10. Kaneda, M.M., Caruthers, S., Lanza, G.M., and Wickline, S.A. Perfluorocarbon nanoemulsions for quantitative molecular imaging and targeted therapeutics. *Ann Biomed Eng* 37, 1922–1933 (2009).
11. Shokeen, M., Fettig, N.M., and Rossin, R. Synthesis, *in vitro* and *in vivo* evaluation of radiolabeled nanoparticles. *Q J Nucl Med Mol Imaging* 52, 267–277 (2008).
12. Surendiran, A., Sandhiya, S., Pradhan, S.C., and Adithan, C. Novel applications of nanotechnology in medicine. *Indian J Med Res* 130, 689–701 (2009).
13. Junghanns, J.U., and Muller, R.H. Nanocrystal technology, drug delivery and clinical applications. *Int J Nanomed* 3, 295–309 (2008).
14. Wickline, S.A., Neubauer, A.M., Winter, P., Caruthers, S., and Lanza, G. Applications of nanotechnology to atherosclerosis, thrombosis, and vascular biology. *Arterioscler Thromb Vasc Biol* 26, 435–441 (2006).

15. Janjic, J.M., and Ahrens, E.T. Fluorine-containing nanoemulsions for MRI cell tracking. *Wiley Interdiscip Rev Nanomed Nanobiotechnol* 1, 492–501 (2009).

16. Lin, W.B., Hyeon, T., Lanza, G.M., Zhang, M.Q., and Meade, T.J. Magnetic nanoparticles for early detection of cancer by magnetic resonance imaging. *Mrs Bulletin* 34, 441–448 (2009).

17. Briley-Saebo, K.C., et al. Clearance of iron oxide particles in rat liver: effect of hydrated particle size and coating material on liver metabolism. *Invest Radiol* 41, 560–571 (2006).

18. Enochs, W.S., Harsh, G., Hochberg, F., and Weissleder, R. Improved delineation of human brain tumors on MR images using a long-circulating, superparamagnetic iron oxide agent. *J Magn Reson Imaging* 9, 228–232 (1999).

19. Ruehm, S.G., Corot, C., Vogt, P., Kolb, S., and Debatin, J.F. Magnetic resonance imaging of atherosclerotic plaque with ultrasmall superparamagnetic particles of iron oxide in hyperlipidemic rabbits. *Circulation* 103, 415–422 (2001).

20. Li, S.-D., and Huang, L. Pharmacokinetics and biodistribution of nanoparticles. *Molecular Pharmaceutics* 5, 496–504 (2008).

21. Zamboni, W.C. Concept and clinical evaluation of carrier-mediated anticancer agents. *Oncologist* 13, 248–260 (2008).

22. Britz-Cunningham, S.H., and Adelstein, S.J. Molecular targeting with radionuclides: state of the science. *J Nucl Med* 44, 1945–1961 (2003).

23. Koivunen, E., Arap, W., Rajotte, D., Lahdenranta, J., and Pasqualini, R. Identification of receptor ligands with phage display peptide libraries. *J Nucl Med* 40, 883–888 (1999).

24. Gradishar, W.J. Albumin-bound paclitaxel: a next-generation taxane. *Expert Opin Pharmacother* 7, 1041–1053 (2006).

25. Lanza, G.M., and Wickline, S.A. Targeted ultrasonic contrast agents for molecular imaging and therapy. *Curr Probl Cardiol* 28, 625–653 (2003).

26. Krafft, M. Fluorocarbons and fluorinated amphiphiles in drug delivery and biomedical research. *Adv Drug Del Rev* 47, 209–228 (2001).

27. Lowe, K.C. Engineering blood: synthetic substitutes from fluorinated compounds. *Tissue Eng* 9, 389–399 (2003).

28. Spiess, B.D. Perfluorocarbon emulsions as a promising technology: a review of tissue and vascular gas dynamics. *J Appl Physiol* 106, 1444–1452 (2009).

29. Lanza, G., et al. A novel site-targeted ultrasonic contrast agent with broad biomedical application. *Circulation* 94, 3334–3340 (1996).

30. Lanza, G., et al. Enhanced detection of thrombi with a novel fibrin-targeted magnetic resonance imaging agent. *Acad Radiol* 5(Suppl 1), s173–s176 (1998).

31. Winter, P.M., et al. Molecular imaging of human thrombus with computed tomography. *J A Coll Cardiol* 43, A10 (2004).

32. Flacke, S., et al. A novel MRI contrast agent for molecular imaging of fibrin: implications for detecting vulnerable plaques. *Circulation* 104, 1280–1285 (2001).

33. Lijowski, M., et al. High sensitivity: high-resolution SPECT-CT/MR molecular imaging of angiogenesis in the Vx2 model. *Invest Radiol* 44, 15–22 (2009).

34. Caruthers, S.D., et al. *In vitro* demonstration using 19F magnetic resonance to augment molecular imaging with paramagnetic perfluorocarbon nanoparticles at 1.5 tesla. *Invest Radiol* 41, 305–312 (2006).

35. Davies, M.J., and Thomas, A.C. Plaque fissuring—the cause of acute myocardial infarction, sudden ischaemic death, and crescendo angina. *Br Heart J* 53, 363–373 (1985).

36. Naghavi, M., et al. From vulnerable plaque to vulnerable patient: a call for new definitions and risk assessment strategies. Part I. *Circulation* 108, 1664–1672 (2003).

37. Moulton, K.S., et al. Inhibition of plaque neovascularization reduces macrophage accumulation and progression of advanced atherosclerosis. *Proc Natl Acad Sci USA* 100, 4736–4741 (2003).

38. Winter, P.M., et al. Molecular imaging of [alpha]v[beta]3-integrin: an opportune biochemical signature for oncologic and cardiovascular diseases. *Acad Radiol* 12, 43 (2005).

39. Winter, P.M., et al. Molecular imaging of angiogenesis in early-stage atherosclerosis with alpha(v)beta3-integrin-targeted nanoparticles. *Circulation* 108, 2270–2274 (2003).

40. Winter, P.M., et al. Molecular imaging of angiogenesis in nascent Vx-2 rabbit tumors using a novel alpha(nu)beta3-targeted nanoparticle and 1.5 tesla magnetic resonance imaging. *Cancer Res* 63, 5838–5843 (2003).

41. Hu, G., et al. Imaging of Vx-2 rabbit tumors with alpha(nu)beta(3)-integrin-targeted (111)In nanoparticles. *Int J Cancer* 120, 1951–1957 (2007).

42. Schmieder, A.H., et al. Molecular MR imaging of melanoma angiogenesis with alphanubeta3-targeted paramagnetic nanoparticles. *Magn Reson Med* 53, 621–627 (2005).

43. Schmieder, A.H., et al. Three-dimensional MR mapping of angiogenesis with {alpha}5 {beta}1 ({alpha}{nu}{beta}3)-targeted theragnostic nanoparticles in the MDA-MB-435 xenograft mouse model. *Faseb J* 22, 4179–4189 (2008).

44. Slamon, D.J., et al. Human breast cancer: correlation of relapse and survival with amplification of the HER-2/neu oncogene. *Science* 235, 177–182 (1987).

45. Vogel, C.L., et al. Efficacy and safety of trastuzumab as a single agent in first-line treatment of HER2-overexpressing metastatic breast cancer. *J Clin Oncol* 20, 719–726 (2002).

46. Carlson, R.W., et al. HER2 testing in breast cancer: NCCN Task Force report and recommendations. *J Natl Compr Canc Netw* 4 (Suppl 3), S1–S22 (2006); quiz, S23–S24.

47. Moulton, K.S. Plaque angiogenesis: its functions and regulation. *Cold Spring Harb Symp Quant Biol* 67, 471–482 (2002).

48. Winter, P.M., et al. Minute dosages of alpha(nu)beta3-targeted fumagillin nanoparticles impair Vx-2 tumor angiogenesis and development in rabbits. *Faseb J* 22, 2758–2767 (2008).

49. Schmieder, A.H., et al. Time-resolved molecular imaging of the "angiogenic switch" in animal models of cancer. *Proc Int Soc Magn Reson Med* 16, 14 (2008).

50. Bergers, G., and Benjamin, L.E. Tumorigenesis and the angiogenic switch. *Nat Rev Cancer* 3, 401–410 (2003).

51. Folkman, J., and Hanahan, D. Switch to the angiogenic phenotype during tumorigenesis. *Princess Takamatsu Symp* 22, 339–347 (1991).

52. Morawski, A.M., et al. Quantitative "magnetic resonance immunohistochemistry" with ligand-targeted (19)F nanoparticles. *Magn Reson Med* 52, 1255–1262 (2004).

53. Yildirim, M., Caruthers, S.D., Nederveen, A.J., Stoker, J., and Lamerichs, R. *In vivo* multicolor imaging of perfluorocarbon emulsions using ultrafast spectroscopic imaging (F-uTSI). *Proc Int Soc Magn Reson* 18, 999 (2010).

54. Caruthers, S.D., Wickline, S.A., and Lanza, G.M. Nanotechnological applications in medicine. *Curr Opin Biotechnol* 18, 26–30 (2007).

55. Banerjee, R. Liposomes: applications in medicine. *J Biomater Appl* 16, 3–21 (2001).

56. Cattel, L., Ceruti, M., and Dosio, F. From conventional to stealth liposomes: a new frontier in cancer chemotherapy. *J Chemother* 16 (Suppl 4), 94–97 (2004).

57. Farokhzad, O.C., and Langer, R. Nanomedicine: developing smarter therapeutic and diagnostic modalities. *Adv Drug Deliv Rev* 58, 1456–1459 (2006).

58. Lanza, G., et al. Nanomedicine opportunities in cardiology. *Ann NY Acad Sci* 1080, 451–465 (2006).

59. Leary, S.P., Liu, C.Y., and Apuzzo, M.L. Toward the emergence of nanoneurosurgery. Part III. Nanomedicine: targeted nanotherapy, nanosurgery, and progress toward the realization of nanoneurosurgery. *Neurosurgery* 58, 1009–1026 (2006); discussion, 1009–1026.

60. Lesniak, M.S., Langer, R., and Brem, H. Drug delivery to tumors of the central nervous system. *Curr Neurol Neurosci Rep* 1, 210–216 (2001).

61. Majoros, I.J., Myc, A., Thomas, T., Mehta, C.B., and Baker, J.R., Jr. PAMAM dendrimer-based multifunctional conjugate for cancer therapy: synthesis, characterization, and functionality. *Biomacromolecules* 7, 572–579 (2006).

62. Pison, U., Welte, T., Giersig, M., and Groneberg, D.A. Nanomedicine for respiratory diseases. *Eur J Pharmacol* 533, 341–350 (2006).

63. Prausnitz, M.R. Microneedles for transdermal drug delivery. *Adv Drug Deliv Rev* 56, 581–587 (2004).
64. Santini, J.T., Jr., Cima, M.J., and Langer, R. A controlled-release microchip. *Nature* 397, 335–338 (1999).
65. Tao, S.L., and Desai, T.A. Microfabricated drug delivery systems: from particles to pores. *Adv Drug Deliv Rev* 55, 315–328 (2003).
66. Wagner, V., Dullaart, A., Bock, A.K., and Zweck, A. The emerging nanomedicine landscape. *Nat Biotechnol* 24, 1211–1217 (2006).
67. Muller, R.H., Radtke, M., and Wissing, S.A. Nanostructured lipid matrices for improved microencapsulation of drugs. *Int J Pharm* 242, 121–128 (2002).
68. Guzman, L.A., et al. Local intraluminal infusion of biodegradable polymeric nanoparticles. A novel approach for prolonged drug delivery after balloon angioplasty. *Circulation* 94, 1441–1448 (1996).
69. Kolodgie, F.D., et al. Sustained reduction of in-stent neointimal growth with the use of a novel systemic nanoparticle paclitaxel. *Circulation* 106, 1195–1198 (2002).
70. Lanza, G.M., et al. Novel paramagnetic contrast agents for molecular imaging and targeted drug delivery. *Curr Pharm Biotechnol* 5, 495–507 (2004).
71. Lanza, G.M., et al. Targeted antiproliferative drug delivery to vascular smooth muscle cells with a magnetic resonance imaging nanoparticle contrast agent: implications for rational therapy of restenosis. *Circulation* 106, 2842–2847 (2002).
72. Kaneda, M.M., Sasaki, Y., Lanza, G.M., Milbrandt, J., and Wickline, S.A. Mechanisms of nucleotide trafficking during siRNA delivery to endothelial cells using perfluorocarbon nanoemulsions. *Biomaterials* 31, 3079–3086 (2010).
73. Crowder, K.C., et al. Sonic activation of molecularly-targeted nanoparticles accelerates transmembrane lipid delivery to cancer cells through contact-mediated mechanisms: implications for enhanced local drug delivery. *Ultrasound Med Biol* 31, 1693–1700 (2005).
74. Winter, P.M., et al. Endothelial alpha(v)beta3 integrin-targeted fumagillin nanoparticles inhibit angiogenesis in atherosclerosis. *Arterioscler Thromb Vasc Biol* 26, 2103–2109 (2006).
75. Ingber, D., et al. Synthetic analogues of fumagillin that inhibit angiogenesis and suppress tumour growth. *Nature* 348, 555–557 (1990).
76. Moulton, K.S., et al. Angiogenesis inhibitors endostatin or TNP-470 reduce intimal neovascularization and plaque growth in apolipoprotein E-deficient mice. *Circulation* 99, 1726–1732 (1999).
77. Schmieder, A.H., et al. MR molecular imaging of neovasculature may predict response to antiangiogenic therapy in animal cancer models. *Proc Int Soc Magn Reson Med* 16, 799 (2008).
78. Cyrus, T., et al. avb3-Integrin-targeted paramagnetic nanoparticles deliver rapamycin into the vascular wall and inhibit stenosis following balloon injury. *Circulation* 114, 217–217 (2006).
79. Soman, N.R., et al. Molecularly targeted nanocarriers deliver the cytolytic peptide melittin specifically to tumor cells in mice, reducing tumor growth. *J Clin Invest* 119, 2830–2842 (2009).
80. Marsh, J.N., et al. Fibrin-targeted perfluorocarbon nanoparticles for targeted thrombolysis. *Nanomed* 2, 533–543 (2007).
81. Landau, E. Ultra-tiny 'bees' target tumors. CNNhealth.com (Cable News Network, 2009).
82. Winter, P.M., et al. Antiangiogenic synergism of integrin-targeted fumagillin nanoparticles and atorvastatin in atherosclerosis. *JACC Cardiovasc Imaging* 1, 624 (2008).
83. Lanza, G.M., Lamerichs, R., Caruthers, S., and Wickline, S.A. Molecular imaging in MR with targeted paramagnetic nanoparticles. *Medica Mundi* 47, 34–39 (2003).

5

Plasmonic Nanostructures in Biosensing

Wenwei Zheng,[1,2] **Heather C. Chiamori,**[1] **Liwei Lin,**[1] **and Fanqing Frank Chen**[2]

[1]*University of California, Berkeley, California, USA*

[2]*Lawrence Berkeley National Laboratory, Berkeley, California, USA*

CONTENTS

5.1 Introduction

In view of basic science, understanding biological systems increasingly depends on our ability to dynamically and quantitatively measure the molecular processes with high spatial and temporal resolution, within the context of a living cell. A living cell responds to its changing environment both inside and outside itself in such a dynamic way that signaling proteins, enzymes, mRNA, and transcription and translation factors are continuously modified or synthesized, transferred from one organelle to another, and finally transported to the locations where they perform appropriate cell function. These intracellular biomolecular complexes are not only heterogeneously distributed in space, but perpetually changed over time with the change of surrounding microenvironments.[1] To quantitatively follow the intracellular biochemical distribution locally and temporarily is vital for understanding intracellular organization and function in cell signaling, growth, differentiation, apoptosis, cell developmental processes, and relevant diseases. Moreover, in biotechnology industry, combinatorial methods are increasingly applied to synthesize new biocatalysts or drugs, demanding simultaneous analysis of thousands of pathogens, mutants, or therapeutic drugs. Furthermore, in personalized medicine, as dictated by economic reasons, mass application of screening and diagnostic tools has to be fast, convenient, and low cost, requiring miniaturization, parallelization, integration, as well as automation of biosensing devices.

To address those major challenges in biosensing research, biosensors based on plasmon-resonant nanoparticle bioconjugates, so-called plasmonic nanobiosensors,[2] are being developed, by synergizing three core techniques: plasmonics that manipulates electromagnetic radiation (light) at dielectric-metal interfaces by tuning properties of nanomaterials, nanofabrication, and controlled synthesis of nanomaterials containing metals (e.g., Au, Ag, Pt, and Cu), and bioconjugate techniques that modify surface of nanomaterials with various bioreceptors (e.g., antibodies, enzymes, aptamers,[3] and molecular imprint polymers[4]). Biomaterials such as proteins, DNA, or RNA have dimensions of 2–20 nm, similar to those of plasmonic nanostructures; thus, the two classes of materials are structurally compatible. Plasmonic nanomaterials exhibit unique electronic, photonic, and catalytic properties, providing electronic or optical transduction of biological phenomena, while biomaterials have unique recognition and catalytic properties. Their integration yields plasmonic nanobiosensors with multiple synergetic advantages over traditional molecular imaging techniques: sensitivity, stability, biocompatibility, selectivity, and spectroscopic imaging capability. Furthermore, recent significant advancements in controlled synthesis and nanofabrication,[5] theory and electrodynamic modeling of optical properties,[6] and surface functionalization[7a] of plasmonic nanomaterials greatly enhance the ability to control and tune the unique optical and electronic properties of plasmonic nanostructures through their sizes, structures, composition, and shapes, building libraries of probes for different analytes. Thus, plasmonic nanobiosensors promise a great potential in the development of high-throughput techniques for the parallel analysis of multiple components in samples.

These recent advancements in nanotechnology and nanoplasmonics also enable hybrid plasmonic nanostructures as subnanometer and nanometer tools to directly interface with intracellular processes. The plasmonic nanobiosensors focus electromagnetic fields to significantly enhance spectral information for localized surface plasmon spectroscopy (LSPR),[8] surface-enhanced Raman spectroscopy (SERS),[9] plasmon resonance energy transfer (PRET),[10] and integrated photoacoustic-photothermal contrast agents.[11] In this way, we can acquire quantitative spectral snapshots of the intracellular biochemical distribution over time as a result of local microenvironmental changes. In this review, we will discuss nanoplasmonic theory, summarize major synthetic and nanofabrication techniques, and describe surface functionalization of plasmonic nanobiosensors. We will then exemplify the applications and summarize the synergetic relationships among three core techniques contained in plasmonic nanobiosensing to address the major aforementioned challenges in biosensing research.

5.2 Theoretical Background

Nanophotonics is defined as "the science and engineering of light matter interactions that take place on wavelength and subwavelength scales where the physical, chemical, or structural nature of natural or artificial nanostructured matter controls the

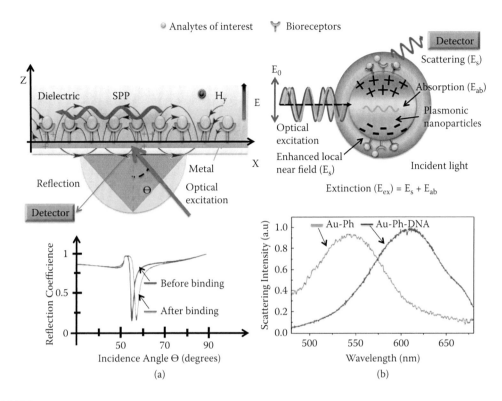

FIGURE 5.1
(a) A typical experiment setup (top) using surface plasmon polaritons (or propagating plasmon) and a resulting reflectivity spectrum (bottom) obtained in an angle-solved mode. (b) A typical measurement setup (top) using localized surface plasmons and a resulting scattering spectrum in biosensing (bottom). Au-Ph: phosphine-coated Au nanoparticles (NPs), in which a surfactant phosphine was used to solubilize naked Au nanoparticles in aqueous solution.

interactions."[12] One of the major subfields of nanophotonics is plasmonics, in which manipulation of light is based on interaction processes between electromagnetic radiation and free-electron plasma (or conduction electrons) at dielectric-metal planar interfaces or nanocurved interfaces, resulting in surface plasmon polaritons (SPPs) or localized surface plasmon polaritons (LSPs),[13] respectively. A plasmon is a quantum quasi-particle representing the elementary excitations, or modes, of the charge density oscillations in a free-electron plasma.[14] As shown in Figure 5.1a, SPPs are propagating dispersive electromagnetic (EM) waves coupled to the electron plasma of a conductor at the planar interface between dielectric and metal materials, having a combined EM wave and surface charge character. The surface charge is generated by the electric field normal to the surface, while the EM waves propagate in the x direction (**H** is in the y direction) for distances on the order of tens to hundreds of microns. This combined character leads to the field component normal to the surface being enhanced near the surface and decaying exponentially with distance in the z direction on the order of 200 nm.[15] The enhancement reaches its maximum when SPPs resonate with the incident exciting laser, which is called surface plasmon resonance (SPR), corresponding to the minimum reflectivity as shown in Figure 5.1a.

Localized surface plasmon polaritons are nonpropagating collective oscillations of the conduction electrons of metallic nanostructures against the background of ionic metal cores[16] coupled to the exciting electromagnetic field (Figure 5.1b). For a particle much smaller than the wavelength of the exciting light, the dipolar plasmon is dominant in the oscillation, which contains an effective restoring force on the driven electrons. When the exciting laser light is in resonance with the dipolar plasmon, the metal particle will radiate light characterized by dipolar radiation,[17] leading to electrical field (E) amplification both inside and in the near-field zone outside the particle. This resonance is called the localized surface plasmon resonance (LSPR). Typical materials for plasmonic applications are noble metals, particularly silver or gold. Silver displays sharper and more intense LSPR bands than gold, while the gold nanostructure is chemically more stable than silver.

The response of both SPR and LSPR sensors to change of refractive index can be described by using the following equation,[18] which was initially developed for propagating SPR.[19]

$$\Delta\lambda_{max} \approx m(n_{adsorbate} - n_{medium}) [1 - \exp(-2d/ld)] \tag{5.1}$$

where $\Delta\lambda_{max}$ is the wavelength shift when the extinction reaches maximum due to resonant absorption, m is the sensitivity factor, $n_{adsorbate}$ and n_{medium} are the refractive indices of the adsorbate and medium surrounding the nanoparticle, respectively, d is the effective thickness of the adsorbate layer, and ld is the electromagnetic field decay length. Therefore, both sensors are highly sensitive to the dielectric properties of the medium adjacent to the dielectric-metal interface. Figure 5.1a shows a typical experimental setup and a resulting reflectivity spectrum obtained in an angle-solved mode (measured against angle of incidence at a fixed wavelength) that demonstrates a resonance angle shift upon binding of analyte in a SPP biosensor, while Figure 5.1b illustrates a typical measurement scheme for LSPR biosensors and a resulting scattering spectrum that shows resonance wavelength shift upon binding of DNA (Au-Ph-DNA, red line) to a phosphine (Ph) surfactant-coated gold nanoparticle (Au-Ph, green line).[20] To selectively bind and detect analytes of interest in biosensing, the metallic surface of both sensors is generally modified using various strategies, such as bioreceptors (Figure 5.1) or hydrophilic coating.[7b,7c]

One of the important driving forces in biosensor research is to develop large-scale biosensor arrays composed of highly miniaturized signal transducer elements that enable the real-time and parallel monitoring of multiple species, especially for high-throughput screening applications such as drug discovery and proteomics research where many thousands of ligand-receptor or protein-protein interactions must be rapidly measured. To meet this need, LSPR sensors stand out as better candidates due to the "nano" advantages over SPR.[18,21] For example, SPR sensors require a minimum of 10×10 μm area for sensing experiments. In contrast, to deliver the same information as do SPR sensors, the spot size for LSPR sensors can be minimized down to the sub-100 nm regime (around the size of a single nanoparticle) using single-nanoparticle LSPR.[22] What's more, because of the large refractive index sensitivity (around 2×10^6 nm RIU^{-1}) SPR sensors require accurate control of temperature and complex optical instrumentation, while LSPR nanosensors do not due to a lower refractive index sensitivity (around 2×10^2 nm RIU^{-1}). Today, the LSPR not only has been recognized as an ultrasensitive method for detecting molecules of both biological and chemical interest, but also plays a major role in all other surface-enhanced biosensing techniques, such as surface-enhanced Raman scattering (SERS),[9b] surface-enhanced hyper-Raman scattering,[23] surface-enhanced infrared spectroscopy,[24] and surface-enhanced fluorescence.[25]

Raman scattering is inelastic scattering of a photon from a molecule in which the frequency change precisely matches the difference in vibrational energy levels.[9b] Raman spectroscopy is a highly specific Raman scattering-based technique that detects and identifies molecules by their unique vibrational fingerprints. However, as a scattering process, Raman scattering is exceedingly weak: typical Raman scattering cross sections per molecule are in the range of 10^{-30} and 10^{-25} cm^2.[23] Thus, Raman spectroscopy usually needs a relatively large volume of molecules to produce detectable scattering signal intensities. This disadvantage has prevented Raman spectroscopy from many important sensing applications, such as surface and in vivo sensing where the number of molecules that produce Raman scattering is small. SERS is about Raman scattering of a single molecule or an ensemble of molecules of interest appearing (e.g., by binding and absorbing) in close proximity (within a few nanometers of the nanoparticle surface) of a plasmonic nanoparticle that will produce an enhancement of the Raman signal. It was discovered in 1974[26] and correctly explained in 1977.[27] Two primary mechanisms are generally thought to be responsible for large SERS signals.[9b,14,23,28] First, chemical enhancement (CE) corresponds to any modification of the Raman polarizability tensors (αR) upon adsorption of the molecule onto the metal surface. It can result from changes in the molecular electronic state or resonant enhancements from either existing molecular excitations or newly formed charge transfer states. Second, both optical excitation and Raman scattering resonate with the LSP modes in the metal NPs simultaneously, resulting in local field electromagnetic (EM) enhancement and radiation EM, respectively.[14,28b] The EM contribution is, in any case, believed to be much larger ($>10^2$) than the CE effect.[8b] More specifically,[28b] if we assume that g is the local field enhancement averaged over the surface of the nanoparticle, the average magnitude of the local field (E_s) around the surface will be $E_s = g \times E_0$, where E_0 is the magnitude of the incident field. The average molecule on the surface will then be excited by the enhanced local field E_s, resulting in the Raman-scattered light near the surface with a field strength $E_R \propto \alpha_R \times E_s \propto \alpha_R \times g \times E_0$. The Raman-scattered (or radiated) fields (E_R) at the Raman-shifted wavelength will be further enhanced by the metal nanoparticle with the radiation enhancement factor g' in the same way as the optical excitation was, giving the magnitude of the SERS-scattered filed: $E_{SERS} = \alpha_R \times g' \times E_s \propto \alpha_R \times g' \times g \times E_0$. As the average light intensity is proportional to the square of its electromagnetic field, we will theoretically have the SERS enhancement factor (assuming that the difference in α_R between SERS and non-SERS ($E_{non\text{-}SERS} = \alpha_R \times E_0$) fields can be ignored compared to EMs):

$$\text{EF} = \frac{I_{SERS}}{I_{non-SERS}} \left| \frac{E_{SERS}}{E_{non-SERS}} \right|^2 = (g')^2 \cdot g^2 \tag{5.2}$$

Assuming that $g \approx g'$, we will have *EF* fourth-power dependence on g, the key to the extraordinate enhancements SERS provides. This approximation takes advantage of the fact that the plasmon width is generally large compared to the Stokes shift, especially the low-frequency one. However, for SERS on isolated homogeneous particles where the plasmon width is small, this assumption leads to an overestimate of the EF by factors of 3 or more.

5.3 Controlled Synthesis and Nanofabrication of Plasmonic Nanostructures

The localized surface plasmon resonance exhibited by plasmonic nanostructures is sensitive to the surrounding environment as well as composition, structure, size, and shape of the nanoparticle.[8b] The ability, then, to tune the morphology of these "sensing" elements is essential for LSPR biosensing applications. Recent advances in control over nanoparticle size and shape allow electromagnetic energy to be concentrated and transported, as well as plasmon energy to be transferred.[29] When chemical and biological molecules are combined with these nanostructured surfaces, both in vitro and in vivo static and dynamic molecular interactions can be detected.

The current design toolbox for engineering nanostructures with different morphologies is based on nanofabrication routes utilizing *bottom up*, *top down*, or both. With bottom-up methods, metal nanocrystals are generally synthesized using solution-based chemical processes, which are also called assay-based or "wet chemistry" methods. The composition, size, and shape of the nanoparticles depend heavily upon control of chemical reactions and conditions during synthesis.[8a] Top-down lithographic techniques provide repeatable and precise nanopattern generation of over large areas, taking advantage of the well-established micro- and nanofabrication tools available.

5.3.1 Bottom-up Chemical Synthesis

Bottom-up chemical synthesis routes include sol-gel, micelles, chemical precipitation from supersaturated conditions, hydrothermal, and pyrolysis.[30] Achieving specific size and shape of nanoparticles is facilitated by either adding organic ligands or capping materials to inhibit further crystal growth or by replacing the growth environment with an inert one.[30] For metal nanoparticle synthesis, chemical reduction of metal salts with a stabilizer is often used, where the stabilizer functions as a growth inhibitor in particular directions, thus controlling shape as well as providing colloidal stability.[8a] A two-step process of metal nanoparticle seeded growth is particularly successful, where diverse shapes such as rods, plates, and pyramids are reproducible.[8a] Typically, these solution-based synthesis routes have three stages: first, nucleation; second, transformation of nuclei to seeds; and finally, seed evolution into nanocrystals.[31] Additional parameters to control nanoparticle morphology include reaction time and temperature as well as concentration of solvents, reagents, and surfactants.[30] For seeded growth, the final shape of the metal nanocrystal is determined by the structure of the seed and the binding affinity of capping meterials

Due to recent significant advances in nanoparticle synthesis,[5a,5c] "wet chemistry" methods can now be used to fabricate plasmonic NPs with controllable sizes, narrow size distributions, and a wide variety of shapes. Nanospheres are the simplest form of nanoparticles and are formed from seed-mediated growth, as described earlier.[32] Due to shape symmetry, the plasmon resonance mode for solid gold colloidal nanospheres typically occurs around 520 nm and can only be tuned around that mode at approximately 50 nm.[33] Nanorods provide additional functionality due to shape anisotropy, which results in two plasmonic resonance modes, dependent on the radial and longitudinal axes of the nanorod,[29] and can be tuned by modifying the aspect ratio.[34] Other fabricated shapes include cubes, tetrahedrons, octahedrons, triangular plates, bipyramids,[35] and prisms.[36] Mixed metallic-alloy

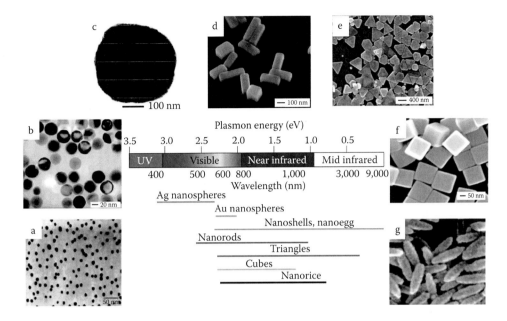

FIGURE 5.2
(See color insert.) Nanoparticle resonance range of plasmon resonances for a variety of particle morphologies. (© 2007 Macmillan Publisher's Ltd. Used with permission.) TEM images of © 2004 (Au spheres (b)), © 1995 (Ag spheres (a)), and © 2001(SiO₂/Ag (core-shell) nanoshells (c)), and SEM images of © 2007 (nanorods (d)), © 2006 (cubes (e) and nanorices (g)), American Chemical Society and © 2006 (triangles (f)). (Wiley-VCH. Used with permission.)

and shell-core structures have also been demonstrated.[33,37] Nanoshells consist of a spherical dielectric core encased in a metal shell. The core-shell structure has tunable plasmonic properties highly sensitive to the inner and outer shell diameters,[38] with the advantage of extending into the near infrared (NIR), a region of relative transparency for living tissues.[39] Nanorice is a hybrid nanostructure combining the advantages of nanorod anisotropy and the tunable plasmon resonances of nanoshells.[40] Plasmon resonance tuning occurs by changing the nanorice length as well as varying the relative size of inner and outer metallic shells. Branched nanoparticles, such as nanostars, are of interest due to their sharp edges and can exhibit higher SERS intensity at the tips.[41] Since the LSPR frequency is dependent on both the shape and the size of the nanostructures, the wide variety of NPs mentioned above can have their LSPR frequency varied from the entire visible to mid-infrared part of the electromagnetic spectrum, as shown in Figure 5.2, demonstrating the highly tunability of plasmonic nanobiosensors, which enables a wide variety of applications in biosensing.

5.3.2 Top-Down Nanofabrication

In top-down fabrication of LSPR substrates, NPs are condensed, deposited, or grown on a supporting substrate such as silicon wafers and glass slides. Some of the important requirements for an ideal LSPR substrate in practical diagnostic applications are that the substrate produces a high enhancement, generates a reproducible and uniform response, has a stable shelf-life, and is easy and simple to fabricate. Currently, there are five major fabrication techniques that could potentially produce the desired nanoplasmonic substrates for either

LSPR or SERS that can meet these requirements: electron beam lithography, nanoimprint lithography (NIL), template-based methods, nanosphere lithography (NSL), and an oblique angle vapor deposition (OAD) or glancing angle deposition (GLAD) method.

5.3.2.1 Electron Beam Lithography

Electron beam lithography (EBL) is a direct-write nanopattern transfer method using a virtual mask for sequential feature exposure on a substrate. High-resolution features on the order of sub-10 nm are obtainable.[42] In addition to nanoscale feature resolution, additional advantages include control over beam energy and dose, elimination of physical masks, high accuracy over small regions on a wafer, reduction in defect densities, and ability for large depth of focus on changing topographies.[42] Although EBL provides an effective method for fabricating reproducible nanoscale features, including nanodots,[43] nanowells,[44] and nanoring antennaes,[45] the long writing times and high costs are the main disadvantages of using EBL for high-throughput, low-cost LSPR applications.

5.3.2.2 Nanoimprint Lithography

Nanoimprint lithography (NIL) has the potential to be a low-cost, mass production method for fabricating LSPR nanostructures. The process is twofold. First, a hard mask, typically made of metal, dielectrics, or semiconductor material, is pressed into a thin layer of polymer heated above its glass transition temperature.[46] When pressed together, the viscous polymer conforms to the mold topography, creating thickness variations in the substrate upon removal of the mold.[47] A reactive ion etch (RIE) is then used to complete the pattern transfer. NIL processes have been used to create gold rectangular, cylindrical, and diamond-shaped nanoblocks based on grating mold orientation,[48] flat, grated, and pillared silver nanostructures for SERS,[49] and gold nanodisks.[50] A potential disadvantage of the NIL process is the embedded cost of fabricating the original mold, which requires access to high-resolution lithographic tools, such as EBL. The mold-making process involves placing photoresist on the mold substrate, nanopattern exposure, hard mask (metal) deposition on the template, followed by RIE for selective etching of the mold.[51] Once the mold is fabricated, though, the reuse of the NIL mold for multiple transfers provides a high level of repeatability for nanoscale features.

5.3.2.3 Template-Based Methods

Template-based methods typically utilize the regular array of nanometer-scale holes in a hard material such as anodic aluminum oxide (AAO) for electrochemical deposition of metals, semiconductors, or polymers.[52] For AAO templates, the pore diameter is controlled by changes in oxidation conditions and feature sizes can range from 5 to 500 nm.[53] Following material deposition, which can include electrochemical deposition thermal vacuum evaporation, or RF sputtering, the AAO template is removed by chemical dissolution.[53,54] The advantages of AAO templates include large working areas (>1 cm^2), compatibility with different materials, and tunable size properties.[53] AAO-templated nanostructures for plasmonic applications include nanorods,[54,55] nanopores,[56] nanopillars,[57] nanowires,[52b] and nanocrescents.[58] Depending on the composition of the electrochemical solution, different surface roughness and size of nanostructures may result.[55] Branching of nanostructures at the base is another issue with the process, and pore diameters of less than 20 nm can exhibit irregular patterns.[53,54] Block copolymers (BCPs) are another technique to fabricate template-based nanostructures for LSPR applications. Nanostructures such as spheres, cylinders, and lamellae are all dependent on the composition and chain structure of the polymers.[59] The features can range from 5 to 50 nm as a function of the BCP molecular weights.[59] The final self-assembled nanoparticle pattern is dependent on the BCP domain symmetry. Generally, micrometer-scale areas are achievable, but defects can exist at grain boundary edges.[60] Examples of defect-free, large-area BCP domains on templated and lithographically defined surfaces have been demonstrated,[60,61] with both methods using a combination of top-down and bottom-up processes. The high cost of extreme ultraviolet interferometric lithography limits wide-scale application of the latter process,[60] while e-beam lithography of grooves combined with plasma etching is required for the former.[61]

5.3.2.4 Nanosphere Lithography

Nanosphere lithography (NSL) is a templating method using monolayer or double-layer colloidal nanoparticles for submicron and nanometer-scale patterns. Preparation techniques include electrostatic deposition, self-assembly, drop casting, spin coating, and evaporation,[53,62] forming a hexagonal, close-packed monolayer on the substrate with controllable size, shape, and interparticle spacing.[63] After metal is deposited onto the nanoparticle array by thermal, e-beam, or pulsed laser deposition, the NSL templates are then removed by burning off at temperatures or chemical dissolution with organic solvents.[53] The advantages of NSL include low cost, high-throughput compatibility with many materials, and the capability of producing well-ordered arrays on different substrates.[64] Nanostructures such as nanopillars,[65] nanohole arrays,[66] nanowire arrays,[67] nanobowls,[68] nanotriangles,[63] nanorings,[69] and nanocrescents[70] have been fabricated successfully for LSPR applications. However, NSL suffers from several disadvantages. First, formation of colloidal particles into a mask has limited geometries due to the hexagonal close-packed formation.[71] Although a modified NSL with varying gaps can be fabricated, the process would involve etching, ion beam techniques, or spin coating prior to pattern transfer.[71,72] Different gap sizes affect the tunability of substrate plasmonic resonances,[73] so it is important to have flexibility with adjusting the gap between particles. Since the size of nanoparticles and gap distance between holes or features are interdependent, control over substrate features has additional constraints.[71] Finally, structural defects such as nanosphere polydispersity, dislocations, vacancies, polycrystalline domains, or local polystyrene (PS) or latex bead disorder are often transferred to the new substrates, leaving limited defect-free areas (10 to 100 μm^2).[74] The NSL method has been modified in multiple ways, including transferring monolayers via submersion in millipore water,[75] liquid-gas interface self-assembly,[74a] angle-resolved NSL,[76] shadow NSL with annealed PS, and fabricating dimers.[8a]

5.3.2.5 Oblique Angle Deposition and Glancing Angle Deposition

The OAD method is a physical vapor deposition at a glancing angle (>75°) of the substrate normal with respect to the incoming vapor direction (Figure 5.3a), causing a geometrical shadowing effect that leads to a preferential growth of nanorods on the substrate in the direction of deposition (Figure 5.3b and c). More specifically, in the initial stages of film

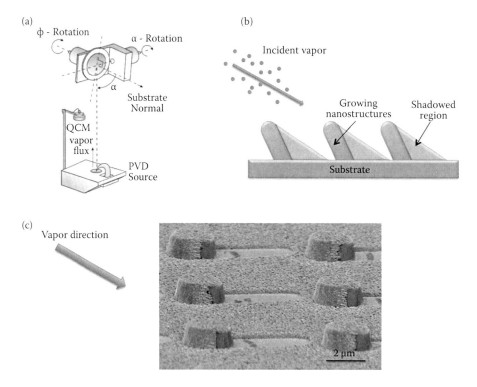

FIGURE 5.3
Schematic of oblique angle deposition (OAD) or glancing angle deposition (GLAD). (a) Typical GLAD apparatus that consists of a rotation stage with substrate angle adjustment. (© 2007 Springer. Used with permission.) (b) Conceptual illustration of column growth. (c) Example of shadowing effect with lithographically patterned substrate (unpublished).

growth atoms condense and form nuclei, resulting in geometrical shadowing of regions of the substrate that prevents film growth in those regions. The resulting film consists of columns that grow off the nuclei and are inclined in the direction of the vapor source. Therefore, the porosity of the film can be controlled by simply changing the incident angle. Since the growth process uses thermal or e-beam evaporation, various materials can be used for GLAD, including materials metals, metal oxides, silicon, silicon oxides, and even their combination.[77] The glancing angle deposition (GLAD) is a combination of OAD and carefully controlled substrate motion (Figure 5.3a) to fabricate a wide variety of nanostructure morphologies, ranging from simple nanostructures like close-packed nanospheres and tilt nanowires to more complex ones such as chevron and helical posts[77c] and nanotubes (Figure 5.4a).[78] GLAD can also be combined with various lithography techniques, including EBL,[79] NSL,[80] and conventional photolithography,[81] to grow highly ordered nanostructure arrays on a large scale, as shown in Figure 5.4b. GLAD techniques can also be used to fabricate heteronanostructures, including gold-coated silicon "matchsticks" for biosensing applications[82] and gold-titanium dioxide-gold sandwich structures for LSPR substrates.[83]

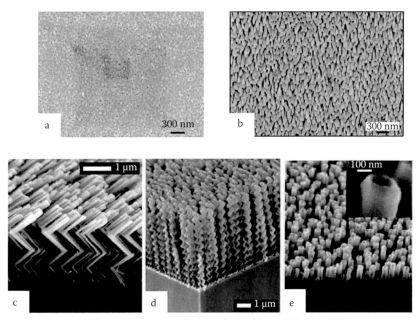

(a) Various nonpatterned morphologies fabricated by GLAD

FIGURE 5.4

Various morphologies fabricated by GLAD. (a) Various nonpatterned morphologies: Ag spheres (a) and tilt nanowires (b) (unpublished data); Si chevrons (b) and helical posts (c) (© 2007 Springer. Used with permission.); nanotubes (d) (© 2007 American Chemical Society. Used with permission.). (b) Various patterned nanostructures fabricated by combining GLAD with e-beam lithography (a, b) (© 2005 IEEE. Used with permission.), nanosphere lithography (c), and photolithography ((d), © 2007 American Physical Society. Used with permission. (e), Unpublished data.).

5.4 Surface Biofunctionalization of Plasmonic Nanomaterials

The significant advancement in the controlled nanofabrication offers researchers the capability to tune plasmonic properties of nanomaterials. However, the surface of plasmonic nanomaterials (normally gold and silver NPs) cannot interact with the biological analyte selectively. To overcome this limit, surface functionalization techniques of nanomaterials by biological recognition elements (bioreceptors or bioprobes) have recently been developed to form hybrid nanomaterials that incorporate the highly selective catalytic and recognition properties of biomaterials such as enzymes and DNA, with the highly sensitive and easily tunable electronic and photonic features of plasmonic nanomaterials.[7a,7c,84]

Often, NPs synthesized in organic medium tend to aggregate to form clusters in an aqueous solution that is generally required by biomedical applications. Therefore, before they can be modified by any bioreceptors, an additional step, called water solubilization, is included, during which colloidally unstable NPs are stabilized in an aqueous solution by conjugating with hydrophilic ligands. The ligands could be ionic, resulting in ionic stabilization of NPs by coulombic repulsion. Another way to stabilize NPs is steric stabilization, during which NP aggregation is prevented by coating with a physical barrier. The barriers include polymeric ligands such as poly(ethylene glycol) (PEG)[85] and small molecular ligands like bis(p-sulfonatophenyl) phenylphosphine surfactants.[20] These ligands for water solubilization are then replaced by phase transfer or ligand exchange or modified by ligand addition with desired biofunctional ligands.[7a] Water solubilization may be performed either as the final stage of the biofunctionalization process of NPs or as an intermediate stage.

Four strategies are generally used to functionalize the surface of plasmonic nanomaterials (mostly Au, Ag, and Cu surfaces) with biomolecules such as bioreceptors. The first is electrostatic adsorption of positively charged biomolecules to negatively charged nanoparticles, or vice versa. For example, gold and silver NPs synthesized by citrate reduction are stabilized by citrate ligands at a pH slightly above their isoelectric point. This results in the anionic citrate coated NPs that can be bound to the positively charged amino acid side chains of immunoglobulin G (IgG) molecules.[86] The second strategy is ligand-like binding to the metallic surface of plasmonic nanomaterials by chemisorption of, e.g., thiol groups. In the former case, metal nanostructures can be functionalized with L-cysteine through the Au-S bonds and then bound to target proteins through peptide bonds with the cysteine moieties.[87] They can be directly bound to thiol derivative analytes such as protein-containing cysteine residues (e.g., serum albumin)[88] or thiolated DNA.[89] The third strategy is covalent binding through biofunctional linkers, exploiting functional groups on both particles and biomolecules. The bifunctional linkers have anchor groups that can be attached to NP surfaces and functional groups that can be further covalently coupled to the target biomolecules. They are extensively used to covalently conjugate biomolecules with various NPs,[90] especially when no linking moieties like thiol groups are available in biomolecules. The common anchor groups include thiols, disulfides, and phosphine ligands that are used to bind the bifunctional linkers to Au, Ag, CdS, and CdSe NPs. The fourth strategy is based on noncovalent, affinity-based receptor-ligand systems. More specifically, nanoparticles are functionalized with bioreceptors (e.g., antibodies) that provide affinity sites for binding of the corresponding ligand (e.g., antigens) or ligand modified proteins and oligonucleotides. The most well-known example in the last several decades is the avidin-biotin system.[91] For example, biotinylated proteins (e.g., immunoglobulins and serum albumins) or biotinylated oligonucleotides (e.g., single-stranded DNA (ssDNA))[89b] have been widely used to modify streptavidin-functionalized Au NPs by affinity binding.[90] Regarding molecular recognition, the system consists of a ligand, the small molecule biotin (vitamin H), and a receptor, the globular protein avidin that is present in, e.g., egg whites. Avidin consists of four identical subunits, yielding four binding pockets that specifically whites recognize and bind to biotin. The dissociation constant is of the order of 10^{15} M, and the affinity bond, though not covalent, is found to be extremely stable, resisting harsh chemical and physical (e.g., elevated temperature) conditions.

5.5 Application

5.5.1 Case 1: Molecular Plasmonic Rulers

Förster resonance energy transfer (FRET) has served as a molecular ruler to monitor conformational changes and measure intramolecular distances of single biomolecules.[92] However, such a ruler sometimes suffers from difficulty in distinguishing changes in relative dye orientation from changes in distance,[92a] limited observation time of a few tens of seconds due to blinking and rapid photobleaching of fluorescence, and an upper distance limit of ~10 nm. Silver and gold NPs have LSPR in the visible range and do not blink or bleach. Alivisatos, Liphardt, and co-workers have exploited them as a new class of molecular ruler to monitor the distance between ssDNA linked single pairs of Au and Ag NPs,[89b] overcoming the limitations of organic fluorophores.[93] They first attached a streptavidin-functionalized NP to the bovine serum (BSA) BSA-biotin-coated glass surface and then introduced a second NP modified by a thiol-ssDNA-biotin bifunctional linker to be attached to the first NP via biotin-streptavidin binding (Figure 5.5a and c). Light scattering was measured by transmission dark field microscopes (Figure 5.5a). For both gold and silver NPs, they observed the significant color change and spectral shift between a single isolated NP and a pair of adjacent NPs (Figure 5.5b–e). The LSPR shift was used to follow the directed assembly of gold and silver nanoparticle dimers in real time. The team also used the ruler to study the kinetics of single DNA hybridization events by monitoring the LSPR shift (Figure 5.5f) due to the resulting 2 nm distance increase between the pairs of the adjacent NP. These "plasmon rulers" make it possible to continuously monitor separations of up to 70 nm for >3,000 s and become an alternative to dye-based FRET for in vitro single-molecule experiments, especially for applications demanding long observation times without dye bleaching.

Chen, Lee, and co-workers demonstrate another molecular ruler in which double-stranded DNA is attached to a 20 nm Au nanoparticle through the thiol-Au chemistry.[20] Instead of monitoring the LSPR between a pair of DNA-linked metal NPs, they monitored the LSPR of individual Au-DNA conjugates cleaved by various endonuclease enzymes. The team found that the LSPR λ_{max} increases with the increased length of the attached double-strand DNA (dsDNA) (Figure 5.6). An average λ_{max} red shift of approximately 1.24 nm is observed per DNA base pair.

They also used this nanoplasmonic molecular ruler to monitor, in real time, DNA being digested by an enzyme. Therefore, this system allows for a label-free, quantitative, real-time measurement of nuclease activity with a time resolution of one second due to high quantum efficiency of Rayleigh scattering compared with fluorescence or Raman scattering. They can also serve as a new DNA footprinting platform that can accurately detect and map the specific binding of a protein to DNA, which is essential to genetic information processing.

5.5.2 Case 2: Multiplexed LSPR Detection

Plasmonic nanobiosensors are potentially ideal biosensors for the high-throughput screening applications in the proteomics and drug discovery. One of the requirements to introduce them to a wider proteomics and drug discovery community is to develop the substrates that are well compatible with current high-throughput platforms. Recently, Endo and co-workers developed a promising multiarray LSPR-based nanochip biosensor suitable for screening biomolecular interactions.[94] This LSPR nanobiosensing array provides rapid, label-free detection of protein concentration in small sample volumes. The plasmonic parts of the biochip are fabricated using nanosphere lithography to form three-layer structures: the bottom layer is a gold film deposited onto a glass substrate, the middle is the silica nanosphere self-assembled monolayer (SAM), and the top is the gold shell coated over silica nanosphere cores to form a SiO_2-Au core-shell array. Subsequently, the bifunctional thiol linker (4,4′-dithiodibutyric acid) SAM is formed over the gold shells.

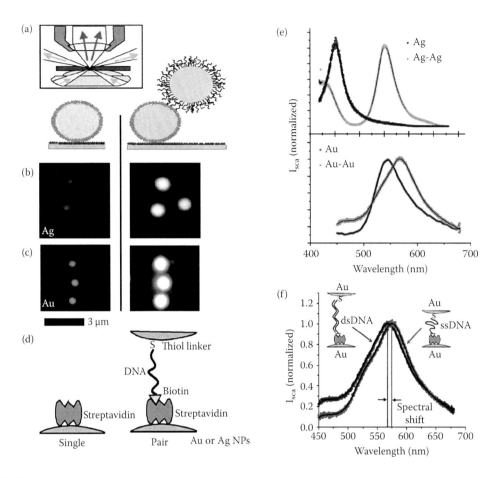

FIGURE 5.5

(See color insert.) DNA-functionalized gold and silver nanoparticles (NPs) as molecular plasmonic rulers. (a) First, streptavidin-functionalized nanoparticles are attached to the BSA-biotin-coated glass surface (a, d left). Then, a second NP is modified by thiol-ssDNA-biotin bifunctional linker and attached to the first particle via biotin-streptavidin affinity binding (a, d right). Inset: Principle of transmission dark field microscopy. (b) Single silver particles appear blue (left) and particle pairs appear blue-green (right). (c) Single gold particles appear green (left) and gold particle pairs appear orange (right). (e) Representative scattering spectra of single particles and particle pairs for silver (top) and gold (bottom). The LSPR scattering spectrum for silver NPs shows a larger LSPR shift (102 nm vs. 23 nm), stronger light scattering, and a smaller half width at half maximum (HWHM) than gold particles. (f) Example of the LSPR spectral shift between a gold particle pair connected with ssDNA (red) and dsDNA (blue), corresponding to the distance increase of 2.1 nm between the particle pair upon DNA hybridization. (© 2005 Macmillan Publisher's Ltd. Used with permission.)

Protein A was then immobilized on the SAM and six different antibodies were attached to protein A using a nanoliter dispensing system, which results in 300 nanospot arrays. A photograph of such a multiarray sensor is shown in Figure 5.7a. Using the same dispensing system, they dispensed different concentrations of specific antigens onto the surface and then measured the change in the absorbance at λ_{max} for each spot that contains analyte with varying concentrations, as shown in Figure 5.7b and c. The limit of detection for this sensor is 100 pg ml$^-$, while the sensor response scales linearly with concentrations up

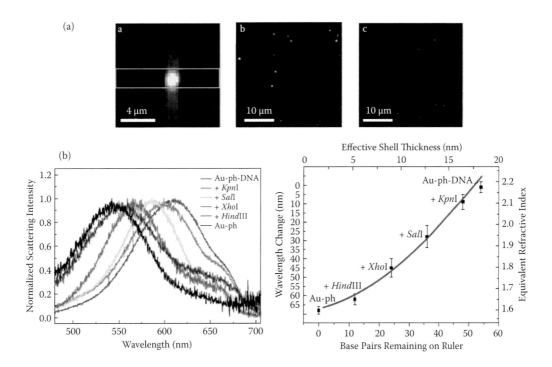

FIGURE 5.6

(See color insert.) Nanoplasmonic molecular rulers for measuring nuclease activity. (a) Dark field scattering images: a, single-target single nanoparticle isolated from other Au or Au-DNA NPs for spectroscopic examination; b, c, true-color images of single Au NP (b) (540 nm peak) and Au-DNA NP (c) (607 nm peak). (b) Typical LSPR scattering spectra and the corresponding maximum scattering wavelengths (λ_{max}) of the Au-DNA nanoconjugates after cleavage reactions with four endonucleases (*Hind*III, *Xho*I, *Sal*I, and *Kpn*I) and as a function of the number of base pairs remaining attached to the Au nanoparticle after the cleavage. The red curve is a fit from a semiempirical model using a Langevin-type dependence of the refractive index vs. dsDNA length. The error bars represent the standard deviation in the measurement of 20 nanoparticles. (© 2006 Macmillan Publisher's Ltd. Used with permission.)

to 1 seml[-u]. This biochip can potentially fit a variety of applications, such as point-of-care of care devices, cancer diagnosis, and microorganism detection for biodefense.

5.5.3 Case 3: Coupling LSPR with SERS

For a particular Raman band, Van Duyne and his team found that the optimal condition is achieved when the LSPR λ_{max} equals the excitation wavelength (in absolute wavenumbers) minus one-half the Stokes shift of the Raman scattering band.[95] Furthermore, the LSPR can be easily tuned by changing the size and shape of NPs. Therefore, LSPR nanobiosensors become ideal candidates for a complementary molecular identification platform to SERS[8b] so that a Raman fingerprint could be used to identify unknown molecules that cannot be achieved by biological recognition elements of LSPR nanobiosensors. The team demonstrated the complementary nature of LSPR spectral shift assays and SERS molecular identification with the antidinitrophenyl immunoassay system.[96] Binding of the antidinitrophenyl ligand to 2,4-dinitrobenzoic acid was quantified by measuring LSPR shift, while SERS was used to verify the identity of the adsorbed molecules.

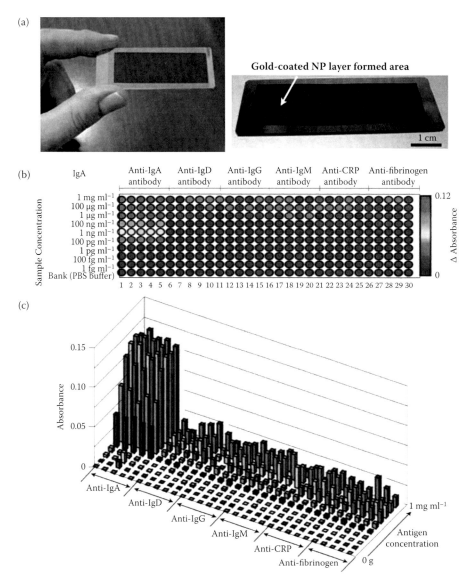

FIGURE 5.7
(See color insert.) LSPR-based nanochips for multiplexed sensing of, e.g., immunoglobulin A (IgA). (a) Photograph of the LSPR-based nanochip (20 mm × 60 mm). The nanochip structure consists of a flat gold film, self-assembled monolayer of silica nanospheres, and a gold coating over the nanospheres. (b and c) Absorbance measurements at each spot of the multiarray nanochip for monitoring the binding of IgA to six different types of antibodies. The antibodies were immobilized on the chip by the nanoliter dispensing system and result in a total of 300 spots separated by 1 mm to prevent cross-contamination. Various concentrations of antigens were incubated for 30 min and LSPR absorption spectra were then taken with a fiber-coupled ultraviolet-visible spectrometer in a reflection configuration. Detection was linear up to 1 μi ml^{-ml} with a limit of detection of 100 pg ml^{-i} for all of the proteins. (© 2006 American Chemical Society. Used with permission.)

5.6 Summary

Plasmonic nanobiosensors are built on the synergetic combination of plasmonic nanostructures, plasmonics, and surface biofunctionalization. The resonant electromagnetic behavior of noble metal NPs, so-called plasmonic behavior, is due to the confinement of the conduction electrons to the small particle volume. Exquisite control synthetic and fabrication of plasmonic nanostructures in combination with advances in theory and the emergence of quantitative electromagnetic modeling tools have provided a better understanding of the optical properties of isolated and electromagnetically coupled nanostructures of various sizes and shapes. This better understanding of plasmonics enables us to more effectively design and synthesize plasmonic nanostructures that fit the need of different applications. Plasmonic nanostructures offer an extremely sensitive response to the dielectric change of the environments within tens of nanometers from the nanomaterial's surface. Surface biofunctionalization integrates unique recognition properties of bioreceptors with unique optical properties of plasmonic nanostructures to make the nonspecific but sensitive dielectric response specific to the analyte of interest. Furthermore, the bioreceptors (e.g., enzymes, antigens, and antibodies) have dimensions in the range of 2–20 nm, similar to those of nanostructures, indicating that the two classes of materials are structurally compatible. The size of the bioreceptors also put the low limit on the minimum size of plasmonic nanomaterials for biosensing. The nanosize of both bioreceptors and nanostructures made highly miniaturized signal transducers possible, while multiplex, real-time, and parallel sensing can be achieved by combining the highly tunable size, shape, structures, and composition of plasmonic nanostructures with a wide variety of available bioreceptors and rapid development of surface biofunctionalization strategies.

This synergetic combination of plasmonic nanostructures, plasmonics, and surface biofunctionalization offers plasmonic nanobiosensors the attractive advantages over other biosensors to better address the major challenges previously mentioned in biosensing. First and foremost, plasmonic nanobiosensors have tunable optical properties to fit a wide variety of applications. The frequency and intensity of the nanoparticle LSPR extinction or scattering bands are highly sensitive to size, shape, orientation, composition, and structure (e.g., shell-core structure) of the NPs, as well as the local dielectric environments, which reduce or eliminate amplification procedures. Therefore, the LSPR can be tuned during fabrication by controlling these parameters with a variety of chemical synthesis and nanofabrication techniques. The nanoparticle array-based biosensors can also be transitioned to single NPs to improve LSPR sensitivity to the level of a single molecule. In addition, plasmonic nanobiosensors possess greater spatial resolution, both lateral and normal, compared with SPR. The ultimate lateral spatial resolution is achieved with single NPs. Plasmonic nanobiosensors have an inherent advantage of the label-free nature over other optical biosensors that require organic fluorescent dyes or inorganic agents such as quantum dots to transduce the binding event because the measured quantity is the optical response of the metal. The label-free nature can significantly reduce sample preparation procedures for biosensing. Moreover, unlike fluorophores, plasmonic NPs do not blink or bleach, providing a virtually unlimited photon budget for observing molecular binding over arbitrarily long time intervals, which makes long-term continuous biosensing possible. What's more, observation of local surface plasmons can be done using simple and inexpensive spectrophotometric equipment in the transmission (or reflection) configuration, which is important for fabricating a large biosensing array that consists of thousands of biosensors.

Acknowledgment

The work was funded by a generous gift from the Agilent Foundation, NIEHS grant 1RC2ES018812-01, National Natural Science Foundation of China grant NSFC-30828010, DOD grant W81XWH-07-1-0663_BC061995, U.S. DOE contract DE-AC03-76SF00098, DARPA, and the MEMS fundamental science program.

References

1. Spiller, D. G., Wood, C. D., Rand, D. A., White, M. R. H. Measurement of single-cell dynamics. *Nature* 2010, 465 (7299), 736–745.
2. (a) Anker, J. N., Hall, W. P., Lyandres, O., Shah, N. C., Zhao, J., Van Duyne, R. P. Biosensing with plasmonic nanosensors. *Nat. Mater.* 2008, 7 (6), 442–453. (b) Lal, S., Link, S., Halas, N. J. Nano-optics from sensing to waveguiding. *Nat Photonics* 2007, 1 (11), 641–648.
3. (a) Nguyen, T., Hilton, J. P., Lin, Q. Emerging applications of aptamers to micro- and nanoscale biosensing. *Microfluid. Nanofluid.* 2009, 6 (3), 347–362. (b) O'Sullivan, C. K. Aptasensors—the future of biosensing. *Anal. Bioanal. Chem.* 2002, 372 (1), 44–48.
4. Ge, Y., Turner, A. P. F. Too large to fit? Recent developments in macromolecular imprinting. *Trends Biotechnol.* 2008, 26 (4), 218–224.
5. (a) Lu, X. M., Rycenga, M., Skrabalak, S. E., Wiley, B., Xia, Y. N. Chemical synthesis of novel plasmonic nanoparticles. *Annu. Rev. Phys. Chem.* 2009, 60, 167–192. (b) Stewart, M. E., Anderton, C. R., Thompson, L. B., Maria, J., Gray, S. K., Rogers, J. A., Nuzzo, R. G. Nanostructured plasmonic sensors. *Chem. Rev.* 2008, 108 (2), 494–521. (c) Xia, Y. N., Halas, N. J. Shape-controlled synthesis and surface plasmonic properties of metallic nanostructures. *MRS Bull.* 2005, 30 (5), 338–344.
6. (a) Zhao, J., Pinchuk, A. O., McMahon, J. M., Li, S. Z., Ausman, L. K., Atkinson, A. L., Schatz, G. C. Methods for describing the electromagnetic properties of silver and gold nanoparticles. *Acc. Chem. Res.* 2008, 41 (12), 1710–1720. (b) Pitarke, J. M., Silkin, V. M., Chulkov, E. V., Echenique, P. M. Theory of surface plasmons and surface-plasmon polaritons. *Rep. Prog. Phys.* 2007, 70 (1), 1–87. (c) Prodan, E., Radloff, C., Halas, N. J., Nordlander, P. A hybridization model for the plasmon response of complex nanostructures. *Science* 2003, 302 (5644), 419–422.
7. (a) Thanh, N. T. K., Green, L. A. W. Functionalisation of nanoparticles for biomedical applications. *Nano Today* 2010, 5 (3), 213–230. (b) Genevieve, M., Vieu, C., Carles, R., Zwick, A., Briere, G., Salome, L., Trevisiol, E. Biofunctionalization of gold nanoparticles and their spectral properties. *Microelectron. Eng.* 2007, 84 (5–8), 1710–1713. (c) Wang, C. G., Ma, Z. F., Wang, T. T., Su, Z. M. Synthesis, assembly, and biofunctionalization of silica-coated gold nanorods for colorimetric biosensing. *Adv. Funct. Mater.* 2006, 16 (13), 1673–1678.
8. (a) Sepulveda, B., Angelome, P. C., Lechuga, L. M., Liz-Marzan, L. M LSPR-based nanobiosensors. *Nano Today* 2009, 4 (3), 244–251. (b) Willets, K. A., Van Duyne, R. P., Localized surface plasmon resonance spectroscopy and sensing. *Annu. Rev. Phys. Chem.* 2007, 58, 267–297.
9. (a) Wachsmann-Hogiu, S., Weeks, T., Huser, T. Chemical analysis *in vivo* and *in vitro* by Raman spectroscopyófrom single cells to humans. *Curr. Opin. Biotechnol.* 2009, 20 (1), 63–73. (b) Stiles, P. L., Dieringer, J. A., Shah, N. C., Van Duyne, R. R. Surface-enhanced Raman spectroscopy. *Annu. Rev. Anal. Chem.* 2008, 1, 601–626.
10. (a) Choi, Y. H., Kang, T., Lee, L. P. Plasmon resonance energy transfer (PRET)-based molecular imaging of cytochrome c in living cells. *Nano Lett.* 2009, 9 (1), 85–90. (b) Choi, Y., Park, Y., Kang, T., Lee, L. P. Selective and sensitive detection of metal ions by plasmonic resonance energy transfer-based nanospectroscopy. *Nat. Nanotechnol.* 2009, 4 (11), 742–746. (c) Liu, G. L., Long, Y. T., Choi, Y., Kang, T., Lee, L. P. Quantized plasmon quenching dips nanospectroscopy via plasmon resonance energy transfer. *Nat. Methods* 2007, 4 (12), 1015–1017.

11. Kim, J. W., Galanzha, E. I., Shashkov, E. V., Moon, H. M., Zharov, V. P. Golden carbon nanotubes as multimodal photoacoustic and photothermal high-contrast molecular agents. *Nat. Nanotechnol.* 2009, 4 (10), 688–694.

12. National Research Council (U.S.), Committee on Nanophotonics Accessibility and Applicability. *Nanophotonics: accessibility and applicability.* Washington, DC: National Academies Press, 2008, p. xviii.

13. Maier, S. A. *Plasmonics: fundamentals and applications.* New York: Springer, 2007, p. xxiv.

14. Le Ru, E. C., Etchegoin, P. G. *Principles of surface-enhanced Raman spectroscopy: and related plasmonic effects.* 1st ed. Amsterdam: Elsevier, 2009, p. xxiii.

15. (a) Brockman, J. M., Nelson, B. P., Corn, R. M. Surface plasmon resonance imaging measurements of ultrathin organic films. *Annu. Rev. Phys. Chem.* 2000, 51, 41–63. (b) Knoll, W. Interfaces and thin films as seen by bound electromagnetic waves. *Annu. Rev. Phys. Chem.* 1998, 49, 569–638. (c) Barnes, W. L., Dereux, A., Ebbesen, T. W. Surface plasmon subwavelength optics. *Nature* 2003, 424 (6950), 824–830.

16. Moskovits, M. Surface-enhanced spectroscopy. *Rev. Mod. Phys.* 1985, 57 (3), 783–826.

17. Jackson, J. D. *Classical electrodynamics.* 3rd ed. New York: Wiley, 1999, p. xxi.

18. Haes, A. J., Van Duyne, R. P. A unified view of propagating and localized surface plasmon resonance biosensors. *Anal. Bioanal. Chem.* 2004, 379 (7–8), 920–930.

19. Jung, L. S., Campbell, C. T., Chinowsky, T. M., Mar, M. N., Yee, S. S. Quantitative interpretation of the response of surface plasmon resonance sensors to adsorbed films. *Langmuir* 1998, 14 (19), 5636–5648.

20. Liu, G. L., Yin, Y. D., Kunchakarra, S., Mukherjee, B., Gerion, D., Jett, S. D., Bear, D. G., Gray, J. W., Alivisatos, A. P., Lee, L. P., Chen, F. Q. F. A nanoplasmonic molecular ruler for measuring nuclease activity and DNA footprinting. *Nat. Nanotechnol.* 2006, 1 (1), 47–52.

21. Haes, A. J., Van Duyne, R. P. A nanoscale optical biosensor: sensitivity and selectivity of an approach based on the localized surface plasmon resonance spectroscopy of triangular silver nanoparticles. *J. Am. Chem. Soc.* 2002, 124 (35), 10596–10604.

22. (a) McFarland, A. D., Van Duyne, R. P. Single silver nanoparticles as real-time optical sensors with zeptomole sensitivity. *Nano Lett.* 2003, 3 (8), 1057–1062. (b) Raschke, G., Kowarik, S., Franzl, T., Sonnichsen, C., Klar, T. A., Feldmann, J., Nichtl, A., Kurzinger, K. Biomolecular recognition based on single gold nanoparticle light scattering. *Nano Lett.* 2003, 3 (7), 935–938.

23. Kneipp, K. Surface-enhanced Raman scattering. *Phys. Today* 2007, 60 (11), 40–46.

24. (a) Ataka, K., Heberle, J. Biochemical applications of surface-enhanced infrared absorption spectroscopy. *Anal. Bioanal. Chem.* 2007, 388 (1), 47–54. (b) Adato, R., Yanik, A. A., Amsden, J. J., Kaplan, D. L., Omenetto, F. G., Hong, M. K., Erramilli, S., Altug, H. Ultra-sensitive vibrational spectroscopy of protein monolayers with plasmonic nanoantenna arrays. *Proc. Natl. Acad. Sci. USA* 2009, 106 (46), 19227–19232.

25. Fort, E., Gresillon, S. Surface enhanced fluorescence. *J. Phys. D Appl. Phys.* 2008, 41 (1) 1–31, .

26. Fleisch, M., Hendra, P. J., Mcquilla, A. J. Raman-spectra of pyridine adsorbed at a silver electrode. *Chem. Phys. Lett.* 1974, 26 (2), 163–166.

27. (a) Jeanmaire, D. L., Vanduyne, R. P. Surface Raman spectroelectrochemistry. 1. Heterocyclic, aromatic, and aliphatic-amines adsorbed on anodized silver electrode. *J. Electroanal. Chem.* 1977, 84 (1), 1–20. (b) Albrecht, M. G., Creighton, J. A. Anomalously intense Raman-spectra of pyridine at a silver electrode. *J. Am. Chem. Soc.* 1977, 99 (15), 5215–5217.

28. (a) Aroca, R. Surface-enhanced vibrational spectroscopy. Chichester, UK: Wiley, 2006, p. xxv. (b) Kneipp, K., Moskovits, M., Kneipp, H. *Surface-enhanced Raman scattering: physics and applications.* Berlin: Springer, 2006, p. xvii.

29. Liu, G. L. Nanoplasmonic-particle-enhanced optical molecular sensing. *IEEE J. Select. Topics Quantum Electron.* 2010, 16 (3), 662–671.

30. Burda, C., Chen, X., Narayanan, R., El-Sayed, M. A. Chemistry and properties of nanocrystals of different shapes. *Chem. Rev.* 2005, 105 (4), 1025.

31. Xia, Y., Xiong, Y., Lim, B., Skrabalak, S. E. Shape-controlled synthesis of metal nanocrystals: simple chemistry meets complex physics. *Angew. Chem. Int. Ed.* 2009, 48 (1), 60–103.

32. Nikoobakht, B., El-Sayed, M. A. Preparation and growth mechanism of gold nanorods (NRs) using seed-mediated growth method. *Chem. Mater.* 2003, 15 (10), 1957–1962.

33. Sun, Y., Mayers, B., Xia, Y. Metal nanostructures with hollow interiors. *Adv. Mater.* 2003, 15 (7–8), 641–646.

34. Stewart, M. E., Anderton, C. R., Thompson, L. B., Maria, J., Gray, S. K., Rogers, J. A., Nuzzo, R. G. Nanostructured plasmonic sensors. *Chem. Rev.* 2008, 108 (2), 494–521.

35. (a) Tao, A. R., Habas, S., Yang, P. Shape control of colloidal metal nanocrystals. *Small* 2008, 4 (3), 310–325. (b) Wiley, B. J., Im, S. H., Li, Z.-Y., McLellan, J., Siekkinen, A., Xia, Y. Maneuvering the surface plasmon resonance of silver nanostructures through shape-controlled synthesis. *J. Phys. Chem. B* 2006, 110 (32), 15666–15675.

36. Bastys, V., Pastoriza-Santos, I., Rodr'guez-González, B., Vaisnoras, R., Liz-Marzán, L. M. Formation of silver nanoprisms with surface plasmons at communication wavelengths. *Adv. Funct. Mater.* 2006, 16 (6), 766–773.

37. Jackson, J. B., Halas, N. J. Silver nanoshells: variations in morphologies and optical properties. *J. Phys. Chem. B* 2001, 105 (14), 2743–2746.

38. Halas, N. Playing with plasmons: tuning the optical resonant properties of metallic nanoshells. *MRS Bull.* 2005, 30, 362–367.

39. Svoboda, K., Block, S. M. Biological applications of optical forces. *Annu. Rev. Biophys. Biomol. Struct.* 1994, 23 (1), 247–285.

40. Wang, H., Brandl, D. W., Le, F., Nordlander, P., Halas, N. J. Nanorice: a hybrid plasmonic nano-structure. *Nano Lett.* 2006, 6 (4), 827–832.

41. (a) Grzelczak, M., Perez-Juste, J., Mulvaney, P., Liz-Marzan, L. M. Shape control in gold nanoparticle synthesis. *Chem. Soc. Rev.* 2008, 37 (9), 1783–1791. (b) Barbosa, S., Agrawal, A., Rodriguez-Lorenzo, L., Pastoriza-Santos, I., Alvarez-Puebla, R. A., Kornowski, A., Weller, H., Liz-Marzan, L. M. Tuning size and sensing properties in colloidal gold nanostars. *Langmuir* 2010, 26 (18), 14943–14950.

42. Madou, M. J. *Fundamentals of microfabrication: the science of miniaturization.* 2nd ed. Boca Raton, FL: CRC Press, 2002.

43. Lin, Y., Zou, Y., Mo, Y., Guo, J., Lindquist, R. G. E-beam patterned gold nanodot arrays on optical fiber tips for localized surface plasmon resonance biochemical sensing. *Sensors-Basel* 2010, 10 (10), 9397–9406.

44. Li, K., Clime, L., Tay, L., Cui, B., Geissler, M., Veres, T. Multiple surface plasmon resonances and near-infrared field enhancement of gold nanowells. *Anal. Chem.* 2008, 80 (13), 4945–4950.

45. Clark, A. W., Cooper, J. M. Nanogap ring antennae as plasmonically coupled SERRS substrates. *Small* 2011, 7 (1), 119–125.

46. Chou, S. Y., Krauss, P. R., Renstrom, P. J. Imprint of sub-5 nm vias and trenches in polymers. *Appl. Phys. Lett.* 1995, 67 (21), 3114–3116.

47. Chou, S. Y., Krauss, P. R., Renstrom, P. J. Nanoimprint lithography. *J. Vac. Sci. Technol. B* 1996, 14 (6), 4129–4133.

48. Lucas, B. D., Kim, J.-S., Chin, C., Guo, L. J. Nanoimprint lithography based approach for the fabrication of large-area, uniformly-oriented plasmonic arrays. *Adv. Mater.* 2008, 20 (6), 1129–1134.

49. Alvarez-Puebla, R., Cui, B., Bravo-Vasquez, J.-P., Veres, T., Fenniri, H. Nanoimprinted SERS-active substrates with tunable surface plasmon resonances. *J. Phys. Chem. C* 2007, 111 (18), 6720–6723.

50. Lee, S.-W., Lee, K.-S., Ahn, J., Lee, J.-J., Kim, M.-G., Shin, Y.-B. Highly sensitive biosensing using arrays of plasmonic Au nanodisks realized by nanoimprint lithography. *ACS Nano* 2011, 5 (2), 897–904.

51. Guo, L. J. Nanoimprint lithography: methods and material requirements. *Adv. Mater.* 2007, 19, 495–513.

52. (a) Masuda, H., Fukuda, K. Ordered metal nanohole arrays made by a two-step replication of honeycomb structures of anodic alumina. *Science* 1995, 268 (5216), 1466–1468. (b) Du, Y., Shi, L., He, T., Sun, X., Mo, Y. SERS enhancement dependence on the diameter and aspect ratio of silver-nanowire array fabricated by anodic aluminium oxide template. *Appl. Surf. Sci.* 2008, 255 (5), 1901–1905.

53. Lei, Y., Yang, S., Wu, M., Wilde, G. Surface patterning using templates: concept, properties and device applications. *Chem. Soc. Rev.* 2011, 40 (3), 1247–1258.

54. Broglin, B. L., Andreu, A., Dhussa, N., Heath, J. A., Gerst, J., Dudley, B., Holland, D., El-Kouedi, M. Investigation of the effects of the local environment on the surface-enhanced Raman spectra of striped gold/silver nanorod arrays. *Langmuir* 2007, 23 (8), 4563–4568.

55. Bok, H.-M., Shuford, K. L., Kim, S., Kim, S. K., Park, S. Multiple surface plasmon modes for a colloidal solution of nanoporous gold nanorods and their comparison to smooth gold nanorods. *Nano Lett.* 2008, 8 (8), 2265–2270.

56. Choi, D., Choi, Y., Hong, S., Kang, T., Lee, L. P. Self-organized hexagonal-nanopore SERS array. *Small* 2010, 6 (16), 1741–1744.

57. Ruan, C., Eres, G., Wang, W., Zhang, Z., Gu, B. Controlled fabrication of nanopillar arrays as active substrates for surface-enhanced Raman spectroscopy. *Langmuir* 2007, 23 (10), 5757–5760.

58. Qin, Y., Pan, A., Liu, L., Moutanabbir, O., Yang, R. B., Knez, M. Atomic layer deposition assisted template approach for electrochemical synthesis of Au crescent-shaped half-nanotubes. *ACS Nano* 2011, 5 (2), 788–794.

59. Bang, J., Jeong, U., Ryu, D. Y., Russell, T. P., Hawker, C. J. Block copolymer nanolithography: translation of molecular level control to nanoscale patterns. *Adv. Mater.* 2009, 21 (47), 4769–4792.

60. Kim, S. O., Solak, H. H., Stoykovich, M. P., Ferrier, N. J., de Pablo, J. J., Nealey, P. F. Epitaxial self-assembly of block copolymers on lithographically defined nanopatterned substrates. *Nature* 2003, 424 (6947), 411–414.

61. Cheng, J. Y., Mayes, A. M., Ross, C. A. Nanostructure engineering by templated self-assembly of block copolymers. *Nat. Mater.* 2004, 3 (11), 823–828.

62. Deckman, H. W., Dunsmuir, J. H. Natural lithography. *Appl. Phys. Lett.* 1982, 41 (4), 377–379.

63. Malinsky, M. D., Kelly, K. L., Schatz, G. C., Van Duyne, R. P. Nanosphere lithography: effect of substrate on the localized surface plasmon resonance spectrum of silver nanoparticles. *J. Phys. Chem. B* 2001, 105 (12), 2343–2350.

64. Hulteen, J. C., Treichel, D. A., Smith, M. T., Duval, M. L., Jensen, T. R., Van Duyne, R. P. Nanosphere lithography: size-tunable silver nanoparticle and surface cluster arrays. *J. Phys. Chem. B* 1999, 103 (19), 3854–3863.

65. Kuo, C.-W., Shiu, J.-Y., Chien, F.-C., Tsai, S.-M., Chueh, D.-Y., Chen, P. Polymeric nanopillar arrays for cell traction force measurements. *Electrophoresis* 2010, 31, 3152–3158.

66. Masson, J.-F., Murray-Methot, M.-P., Live, L. S. Nanohole arrays in chemical analysis: manufacturing methods and applications. *Analyst* 2010, 135, 1483–1489.

67. Peng, K., Zhang, M., Lu, A., Wong, N.-B., Zhang, R., Lee, S.-T. Ordered silicon nanowire arrays via nanosphere lithography and metal-induced etching. *Appl. Phys. Lett.* 2007, 90, 163123.

68. Xu, M., Lu, N., Xu, H., Qi, D., Wang, Y., Chi, L. Fabrication of functional silver nanobowl arrays via sphere lithography. *Langmuir* 2009, 25 (19), 11216–11220.

69. Aizpurua, J., Hanarp, P., Sutherland, D. S., Kall, M., Bryant, G. W., Garcia de Abajo, F. J. Optical properties of gold nanorings. *Phys. Rev. Lett.* 2003, 90 (5), 057401.

70. Lee, S. E., Lee, L. P. Biomolecular plasmonics for quantitative biology and nanomedicine. *Curr. Opin. Biotechnol.* 2010, 21 (4), 489–497.

71. Vossen, D. L. J., Fific, D., Penninkhof, J., van Dillen, T., Polman, A., van Blaaderen, A. Combined optical tweezers/ion beam technique to tune colloidal masks for nanolithography. *Nano Letters* 2005, 5 (6), 1175–1179.

72. Jiang, P., McFarland, M. J. Wafer-scale periodic nanohole arrays templated from two-dimensional nonclose-packed colloidal crystals. *J. Am. Chem. Soc.* 2005, 127, 3710–3711.

73. Masson, J.-F., Gibson, K. F., Provencher-Girard, A. Surface-enhanced Raman spectroscopy amplification with film over etched nanospheres. *J. Phys. Chem. C* 2010, 114 (51), 22406–22412.

74. (a) Rybczynski, J., Ebels, U., Giersig, M. Large-scale, 2D arrays of magnetic nanoparticles. *Colloids Surf. Physicochem. Eng. Aspects* 2003, 219 (1–3), 1–6. (b) Haynes, C. L., Van Duyne, R. P. Nanosphere lithography: a versatile nanofabrication tool for studies of size-dependent nanoparticle optics. *J. Phys. Chem. B* 2001, 105 (24), 5599–5611.

75. Burmeister, F., Schafle, C., Matthes, T., Bohmisch, M., Boneberg, J., Leiderer, P. Colloid monolayers as versatile lithographic masks. *Langmuir* 1997, 13 (11), 2983–2987.

76. Haynes, C. L., McFarland, A. D., Smith, M. T., Hulteen, J. C., Van Duyne, R. P. Angle-resolved nanosphere lithography: manipulation of nanoparticle size, shape, and interparticle spacing. *J. Phys. Chem. B* 2002, 106 (8), 1898–1902.

77. (a) Zhou, C. M., Gall, D. Two-component nanorod arrays by glancing-angle deposition. *Small* 2008, 4 (9), 1351–1354. (b) Zhou, C. M., Li, H. F., Gall, D. Multi-component nanostructure design by atomic shadowing. *Thin Solid Films* 2008, 517 (3), 1214–1218. (c) Steele, J. J., Brett, M. J. Nanostructure engineering in porous columnar thin films: recent advances. *J. Mater. Sci.-Mater. El.* 2007, 18 (4), 367–379.

78. Huang, Z. F., Harris, K. D., Brett, M. J. Morphology control of nanotube arrays. *Adv. Mater.* 2009, 21 (29), 2983.

79. Jensen, M. O., Brett, M. J. Periodically structured glancing angle deposition thin films. *IEEE T. Nanotechnol.* 2005, 4 (2), 269–277.

80. Zhou, C. M., Gall, D. Surface patterning by nanosphere lithography for layer growth with ordered pores. *Thin Solid Films* 2007, 516 (2–4), 433–437.

81. Ye, D. X., Lu, T. M. Ballistic aggregation on two-dimensional arrays of seeds with oblique incident flux: growth model for amorphous Si on Si. *Phys. Rev. B* 2007, 76 (23), 235402-1–8.

82. Fu, J., Park, B., Siragusa, G., Jones, L., Tripp, R., Zhao, Y., Cho, Y.-J. An Au/Si hetero-nanorod-based biosensor for *Salmonella* detection. *Nanotechnology* 2008, 19 (15), 155502-1–7.

83. Fu, J., Zhao, Y. Optical properties of silver/gold nanostructures fabricated by shadowing growth and their sensing applications. In *Nanostructured thin films III*. Vol. 7766. San Diego: SPIE, 2010, p. 77660B1-12.

84. Wang, X., Liu, L. H., Ramstrom, O., Yan, M. D. Engineering nanomaterial surfaces for biomedical applications. *Exp. Biol. Med.* 2009, 234 (10), 1128–1139.

85. Cobley, C. M., Chen, J. Y., Cho, E. C., Wang, L. V., Xia, Y. N. Gold nanostructures: a class of multifunctional materials for biomedical applications. *Chem. Soc. Rev.* 2011, 40 (1), 44–56.

86. Shenton, W., Davis, S. A., Mann, S. Directed self-assembly of nanoparticles into macroscopic materials using antibody-antigen recognition. *Adv. Mater.* 1999, 11 (6), 449.

87. Naka, K., Itoh, H., Tampo, Y., Chujo, Y. Effect of gold nanoparticles as a support for the oligomerization of L-cysteine in an aqueous solution. *Langmuir* 2003, 19 (13), 5546–5549.

88. Hayat, M. A. *Colloidal gold: principles, methods, and applications*. San Diego: Academic Press, 1989, p. v.

89. (a) Mirkin, C. A., Letsinger, R. L., Mucic, R. C., Storhoff, J. J. A DNA-based method for rationally assembling nanoparticles into macroscopic materials. *Nature* 1996, 382 (6592), 607–609. (b) Sonnichsen, C., Reinhard, B. M., Liphardt, J., Alivisatos, A. P. A molecular ruler based on plasmon coupling of single gold and silver nanoparticles. *Nat. Biotechnol.* 2005, 23 (6), 741–745.

90. Niemeyer, C. M. Nanoparticles, proteins, and nucleic acids: biotechnology meets materials science. *Angew Chem. Int. Ed.* 2001, 40 (22), 4128–4158.

91. (a) Green, N. M. Avidin. *Adv. Protein Chem.* 1975, 29, 85–133. (b) Wilchek, M., Bayer, E. A. The avidin biotin complex in bioanalytical applications. *Anal. Biochem.* 1988, 171 (1), 1–32.

92. (a) Weiss, S. Fluorescence spectroscopy of single biomolecules. *Science* 1999, 283 (5408), 1676–1683. (b) Yildiz, A., Forkey, J. N., McKinney, S. A., Ha, T., Goldman, Y. E., Selvin, P. R. Myosin V walks hand-over-hand: single fluorophore imaging with 1.5-nm localization. *Science* 2003, 300 (5628), 2061–2065. (c) Zhuang, X. W., Bartley, L. E., Babcock, H. P., Russell, R., Ha, T. J., Herschlag, D., Chu, S. A single-molecule study of RNA catalysis and folding. *Science* 2000, 288 (5473), 2048.

93. (a) Taton, T. A., Mirkin, C. A., Letsinger, R. L. Scanometric DNA array detection with nano-particle probes. *Science* 2000, 289 (5485), 1757–1760. (b) Yguerabide, J., Yguerabide, E. E. Light-scattering submicroscopic particles as highly fluorescent analogs and their use as tracer labels in clinical and biological applications. II. Experimental characterization. *Anal. Biochem.* 1998, 262 (2), 157–176.

94. (a) Endo, T., Kerman, K., Nagatani, N., Takamura, Y., Tamiya, E. Label-free detection of peptide nucleic acid-DNA hybridization using localized surface plasmon resonance based optical biosensor. *Anal. Chem.* 2005, 77 (21), 6976–6984. (b) Endo, T., Kerman, K., Nagatani, N., Hiepa, H. M., Kim, D. K., Yonezawa, Y., Nakano, K., Tamiya, E. Multiple label-free detection of antigen-antibody reaction using localized surface plasmon resonance-based core-shell structured nanoparticle layer nanochip. *Anal. Chem.* 2006, 78 (18), 6465–6475.

95. McFarland, A. D., Young, M. A., Dieringer, J. A., Van Duyne, R. P. Wavelength-scanned surface-enhanced Raman excitation spectroscopy. *J. Phys. Chem. B* 2005, 109 (22), 11279–11285.

96. Yonzon, C. R., Zhang, X. Y., Van Duyne, R. P. Localized surface plasmon resonance immunoassay and verification using surface-enhanced Raman spectroscopy. *Proc. SPIE* 2003, 5224, 78.

6

Emerging Microscale Technologies for Global Health: CD4⁺ Counts

ShuQi Wang,[1] Feng Xu,[1] Alexander Ip,[1] Hasan Onur Keles,[a] and Utkan Demirci[1,2]

[1]Center for Biomedical Engineering, Department of Medicine, Brigham and Women's Hospital, Harvard Medical School, Boston, Massachusetts, USA

[2]Harvard-MIT Health Sciences and Technology, Cambridge, Massachusetts, USA

CONTENTS

6.1 Introduction

Globally, HIV-1 has caused 25 million deaths and another 33 million infections, mostly occurring in sub-Saharan Africa.[1] To curb the HIV/AIDS epidemic and improve the quality of life of AIDS patients, antiretroviral treatment (ART) has been rapidly expanded in developing countries. Despite these efforts, there were only 4 million AIDS patients on treatment by the end of 2008, accounting for only 42% of AIDS patients who urgently need ART.[3] The main reason for this is the lack of cost-effective and easy-to-use ART monitoring tools in resource-limited settings. In industrialized countries, ART is closely monitored for immunological recovery (CD4⁺ cell count, every 3–4 months), virological failure (two consecutive viral loads > 50 copies/ml), and drug resistance.[4,5] However, these assays require expensive instruments, air conditioning, and skillful operators. In contrast, ART in developing countries is often monitored by the World Health Organization (WHO) clinical staging and CD4⁺ T lymphocyte counts.[6] In rural areas, blood samples are often collected at local clinics and sent to centralized laboratories.

The sample-to-result turnaround can be several weeks.[7] Hence, ART monitoring tools are urgently needed at the point of care (POC) to effectively decrease the turnaround time.

Due to the lack of resources, ART is often initiated in sub-Saharan Africa without testing CD4[+] cell count and viral load.[8,9] This has been a concern since HIV-1 lacks stringent proof-reading in RNA synthesis and results in mutations in the genome. In addition, coinfection of different HIV-1 subtypes facilitates RNA recombination. Thus, HIV-1 can rapidly produce drug-resistant strains and can render the first-line drugs inefficacious. The second-line therapy, though available, can yield annual costs up to $1,037 per person, 2.4 times higher than the first-line therapy.[10] Due to discontinuous logistics or drug toxicity, ART is often interrupted, which breeds more chances for drug resistance to develop, and thus leads to treatment failure.[11] When this scenario occurs, although ART is switched to second-line drugs, AIDS patients most likely experience rapid disease progression. Some studies have shown that viral load monitoring is needed to detect early virological failure.[12,13] However, current commercial viral load assays (e.g., reverse-transcriptase polymerase chain reaction (RT-PCR)) and inexpensive alternatives (e.g., ExaVir reverse transcriptase assay) are still not practical for resource-limited settings.[14] Accordingly, CD4[+] T cell count is an essential standard for monitoring ART, and it can be safely used to determine if the treatment regimens should be changed.[15,16] Thus, the remaining issue transforms to how to develop low-cost, portable, and rapid CD4[+] cell counting systems without the need for highly trained end users at the POC.[17–19]

To address this global health issue, extensive efforts exist to simplify CD4[+] cell counting technologies, where flow cytometry is the "gold standard." However, the implementation of simplified flow cytometers is still challenging due to high cytometer cost ($27,000–$35,000), high cost per test ($5–$20), limited throughput (30–50 samples/day), and need for trained skillful operators. Simplified flow cytometers such as EasyCD4,[20] CyFlow,[21] and Pan-leucogating[22] have been developed, significantly reducing the initial cost on instrument setup. In addition, manual assays such as Dynabeads and Cytosphere have been developed for de-centralized laboratories. However, these two methods are time-consuming and labor-intensive. More recently, microfluidic-based CD4[+] cell count technology has moved rapidly forward, from fluorescence staining to lensless imaging, to suit the need for POC diagnosis.[18,23,24] These microfluidic methods uptake small volumes of whole blood and report CD4[+] cell count within 30 minutes, showing the potential to deliver an affordable solution to scale up ART in resource-limited settings.

In this chapter, we will first present the gold standard (i.e., flow cytometry) and its simplified derivatives. Manual CD4[+] cell counting methods, which were developed for resource-limited settings, will be also reviewed. Most importantly, we will discuss the latest microfluidic and imaging techniques for developing POC CD4[+] assays.

6.2 Flow Cytometry for CD4[+] Cell Counts in Resource-Limited Settings

6.2.1 Gold Standard in Resource-Limited Settings

Traditionally, CD4[+] cell count is performed by using a dual-platform technology; a hematology analyzer is used to quantify the total number of white blood cells, and a flow cytometer is used to measure the percentage of CD4[+] T lymphocytes. To simplify the testing procedure and reduce the running costs, Becton Dickinson developed a portable single-platform-based flow cytometer, i.e., FACSCount.[25,26] This system excludes the hematology analyzer for measurement of total white blood cells. FACSCount starts with whole blood and involves no lysis of red blood cells. Blood cells are first labeled with fluorochrome-conjugated monoclonal antibodies (MAbs) (anti-CD3⁺, CD4⁺, and CD8⁺), and are then introduced into a sheath flow, in which single-cell flow is maintained. Upon exposure to a monochromatic light, fluorescent-labeled blood cells are identified. Microfluorospheres provided by the manufacturer serve as a quantification standard for counting absolute CD4⁺ and CD8⁺ cells, thus eliminating the need for a hematology analyzer.[26]

6.2.2 Simplified Flow Cytometry for CD4⁺ Cell Count in Resource-Limited Settings

To facilitate CD4⁺ cell count in resource-limited settings, other types of single-platform flow cytometers were also developed, including EasyCD4 (Guava), CyFlow (Partec), and Pan-leucogating (Beckman Coulter) (Table 6.1). Compared with FACSCount, they reduce the cost not only on initial instruments, but also on sample testing. However, they can only

TABLE 6.1

Comparison of Commercial CD4 Diagnostics for Resource-Limited Settings

Assay	Company	Methodology	Advantages	Disadvantages
FACSCount	Becton Dickinson	Bead-based single-platform flow cytometry	Automated and well established; as a reference method	Absolute CD4 count only; expensive instrument and high running cost
EasyCD4	Guava Technology	Capillary-based volumetric flow cytometry	Low consumption of samples and reagents; reduced biohazardous waste; portable	Absolute CD4 count only; expensive instrument; daily calibration required
CyFlow	Partec	Volumetric single-platform flow cytometry	Does not require daily calibration; portable	Absolute CD4 count only; does not differentiate between CD4⁺ lymphocytes and monocytes
Pan-leucogate	Beckman Coulter	Dual platform; uses CD45 and side scatter to gate on leukocytes	Can provide CD4%; day-aged blood samples can be used; portable; can be adapted to single platform	Daily quality control; operating skills required
Dynal T4 Quant	Dynal Biotech	Monocyte depletion; isolation of CD4 cells with magnetic beads; microscopic counting	Reduced start-up costs; portable	Absolute CD4 count only; high cost per test; limited throughput; underestimates CD4⁺ counts
Cytospheres	Beckman Coulter	Manual counting CD4 cells label latex beads under a light microscope	Reduced start-up costs; portable	Absolute CD4 count only; expertise required to distinguish monocytes from CD4 cells; limited throughput

be set up in centralized laboratories due to the need for laboratory infrastructure, such as constant power supply and skillful operators.

EasyCD4 is a microcapillary cytometer that relies on the use of a microcapillary to maintain single-cell suspension.[27] This design offers two advantages. First, the use of a capillary excludes the need for sheath flow, and thus reduces the consumption of filtered water. This is particularly important since clean water is not readily available in most resource-limited settings. Second, a small volume of blood (10 μl) is needed for $CD4^+$ cell counting. To achieve specific $CD4^+$ detection, EasyCD4 utilizes two-color conjugated anti-$CD3^+$ and anti-$CD4^+$ MAbs, which are conjugated with phycoerythrin (PE)-Cy5 and PE, respectively. Since anti-$CD3^+$ MAb targets an epitope on the T cell receptor (TCR), $CD3^+$ gating excludes the interference of monocytes ($CD3^-$), which also express $CD4^+$. These two MAbs are first added to 10 μl of EDTA-containing whole blood and mixed gently. Red blood cells in the mixture are then lysed prior to $CD4^+$ counting. Evaluation studies have demonstrated that EasyCD4 has a strong correlation with FACSCount ($R^2 = 0.92 \sim 0.99$) and can be established with quality assurance. Most importantly, EasyCD4 can accurately identify AIDS patients with $CD4^+$ cells below 200 copies/μl.[20,27,28] However, a recent study shows that EasyCD4 gives a consistently higher $CD4^+$ cell count (25 cells/μl) than FACSCount.[29]

CyFlow (Partec) is a single-platform volumetric flow cytometry.[21] To reduce cost, it uses only anti-$CD4^+$ MAb conjugated with PE, and $CD4^+$ cells are counted by a single green solid-state laser (532 nm). The simple testing protocol involves no lyse and no wash, reducing the hands-on time and systematic error. In brief, 100 μl of whole blood is mixed with 10 μl of anti-$CD4^+$ MAb and incubated in darkness for 10–15 minutes, where it is then followed by the addition of 800 μl of no-lyse buffer. For the volumetric measurement, two electrodes are immersed into the mixture at different levels. The actual counting takes place when the fluid level falls between these two electrodes during aspiration. Although CyFlow involves no complex gating strategies or any optical alignment, it still requires substantial technical expertise to differentiate between monocytes and $CD4^+$ T lymphocytes, since they both express $CD4^+$ molecules. Clinical evaluations have been performed in Thailand,[21] Zimbabwe,[30] and Malawi,[31] showing high correlations with FACSCount ($R^2 = 0.92$ to 0.99).

Pan-leucogating (PLG) is a dual-platform flow cytometer. Compared to conventional four-color flow cytometry, it utilizes only two MAbs, anti-$CD45^+$ and anti-$CD4^+$, for $CD4^+$ cell count to reduce cost. $CD45^+$ is a pan-leukocyte marker, and it is used to identify white blood cells. After gating with anti-$CD45^+$, white blood cells are further gated with anti-$CD4^+$ to determine the percentage of $CD4^+$ cells in white blood cells (CD4%). $CD4^+$ cell count is then calculated by multiplying CD4% with white blood cell count, which is obtained from a hemocytometer. The advantage of this approach is that PLG uses less variable white blood cells ($CD45^+$ population), instead of total T lymphocytes, in the gating strategy, which increases reproducibility and enables testing day-old samples.[32] Evaluation studies in Thailand and Barbados have shown that PLG is highly correlated with standard flow cytometry, with an R^2 of 0.95 for absolute $CD4^+$ cell count.[22,33] PLG can also be adapted to single-platform flow cytometry, in which bead count rate (BCR) is introduced for proactive quality control.[34] This simpler flow cytometer proves to be reliable, inexpensive, and has high throughput capability, showing the promise of guiding ART in resource-limited settings.

6.2.3 Manual $CD4^+$ Cell Counting Methods in Resource-Limited Settings

Although modified flow cytometry techniques reduce the initial cost for instrument setup and reagent cost, they can only be established in centralized laboratories due to the requirement for laboratory space, reliable power supply, and refrigerated storage. To be more suitable for rural areas, two manual CD4+ cell counting methods have been developed: Dynabeads and Cytospheres (Table 6.1). However, they are still considered expensive, labor-intensive, and low throughput.

Dynabeads is a manual CD4+ cell counting method that relies on the use of MAb-coated magnetic microbeads.[35] Three steps are involved: depletion of monocytes from whole blood, isolation of CD4+ cells, and staining/counting nuclei in a hemocytometer. In the first step, CD14 MAb-coated Dynabeads are added to fresh EDTA-anticoagulated blood. CD14+ MAb specifically binds to monocytes, which also express CD4+ on the cell surface. When applying a magnetic microbead concentrator, the monocyte-CD14 MAb-Dynabead complex is separated. The supernatant is transferred to a new tube containing Dynabeads coated with anti-CD4+ MAb. After the CD4+ cell separation, CD4+ cells are lysed and the nuclei are stained with acridine orange and counted under a fluorescence microscope. In a modified format, the number of free CD4 Dynabeads decreases, with less Dynabead input and more washing, enabling a reliable CD4+ cell count under a light microscope.[36] Dynabeads CD4 only requires a light microscope, a hemocytometer, and a counter, significantly reducing the initial setup cost, and only costs approximately $5 per assay. Once CD4+ cells are lysed, however, counting needs to be performed within 1 hour, which restricts the batch size to less than six. Therefore, this manual method is suitable for local clinics where throughput is not a concern. In a resource-limited context, Dynabeads CD4 showed a high correlation of 0.89 with flow cytometry. However, Dynabeads CD4 underestimated the CD4+ cell count up to 269 cells when the CD4+ cell count was above 1,000 cells/µl.[35] This was further confirmed in a recent study, in which the median differences between Dynabeads and flow cytometry were 1, 16, 24, and 90 cells/µl with CD4+ cell counts of <200, 200–350, 350–500, and >500, respectively.[37]

The Cytosphere assay is another bead-based manual CD4+ cell counting method.[38] It uses CD4+ MAb (MY4)-conjugated latex spheres to enumerate CD4+ T lymphocytes under a light microscope. MY4, however, cannot differentiate the CD4+ molecules between CD4+ T lymphocytes and monocytes. To reduce the overestimation caused by monocytes, the Cytosphere assay uses anti-CD14+ monoclonal antibody to block monocytes. The blocked monocytes form a golden coloration, which is easily distinguished from CD4+ cells. The following counting is carried out on a hemocytometer, on which cells attached with three or more latex spheres are recognized as CD4+ T lymphocytes. Evaluation in India showed that this FDA-approved assay had high correlation with flow cytometry (R = 0.97), and it was reliable in identifying AIDS patients with CD4+ cell counts below 200 cells/µl.[38] However, a recent large-scale study demonstrated that the Cytosphere assay underestimated CD4+ cells up to 210 cells/µl when it is above 500 cell/µl, with a correlation coefficient of 0.78.[37]

Based on the principle of Cytospheres, a new microbead-based assay was developed, utilizing a different anti-CD4+ monoclonal antibody (MT4).[39] This antibody has a stronger affinity to CD4+ T lymphocytes than monocytes, eliminating the need for monocyte blocking by anti-CD14+ MAb, as in the Cytosphere assay. In this assay, MT4 MAb is coated on latex microbeads and used to capture CD4+ cells from whole blood. Once bound to CD4+ cells, latex microbeads and CD4+ cells form a rosette structure, which can be easily identified under a light microscope. On a hemocytometer, cells with three or more microbeads are recognized as CD4+ cells and manually counted. The evaluation compared with standard flow cytometry showed a high correlation, with $R^2 = 0.941$ among 60 health people and 140 HIV-1 infected individuals. In another strategy, this MT4 MAb is used to deplete CD4+ cells from whole blood to facilitate CD4+ cell counting.[40] With this method, white blood cells are counted by an automatic hematology analyzer with or without depleting CD4+ cells by MT4 MAb, and the CD4+ cell count is obtained by the difference between the two values.

6.3 Advances in Microfluidics and Imaging Technologies for POC CD4+ Testing

Although efforts have been put into simplified and affordable flow cytometers, they remain too complex for district hospitals or POC use in developing countries, as they require expensive reagents, multiple sample preparation steps, and costly maintenance. Current manual CD4+ cell counting methods are restricted by the high cost for a single assay (despite the reduced cost for instrumentation), intensive labor, and low throughput. Thus, on-chip CD4+ cell counting methods have recently been developed to facilitate expanding ART in developing countries.[18,23,24] This is mainly because microchip fabrication is inexpensive and can be easily scaled up with recent advances in microelectromechanical systems.[41] CD4+ cell counting can be realized on-chip with fingerprick whole blood, eliminating the need for sample processing and transport. Furthermore, the actual detection and counting have been made easier, as counting can be performed using a lensless imaging system, in contrast to flow cytometry and manual counting under a microscope.[18] These instruments have the potential to be portable, user-friendly, and do not require costly maintenance. Based on microfluidic chips, other detection methods, such as chemiluminescence,[42] quantum dot-based fluorescence,[17,43] and impedance spectroscopy,[44] are also utilized for on-chip HIV monitoring applications, including CD4+ cell counting. These technologies represent the latest advances in POC CD4+ cell counting.

6.3.1 Design of CD4 Microfluidic Chips

To facilitate CD4+ cell detection from whole blood, CD4+ cells need to be effectively separated from red blood cells (RBCs), monocytes and other lymphocytes. A circular flow cell containing a 3 μm Nuclepore polycarbonate filter[23] was built in a metal case with a poly(methyl methacrylate) (PMMA) top and bottom. There is an inlet and outlet on the PMMA top, allowing flow-through of blood samples. The PMMA bottom serves as a support for the filter, which permits passage of RBCs and retention of lymphocytes, thus eliminating the need for lysis of RBCs. The device only uses microliters of a blood sample; this significantly reduces reagent consumption and biohazardous waste. This device is not disposable and still represents a macroscale device, which requires a technician to operate properly. Later, straight microchambers made of polydimethylsiloxane (PDMS) were used to maintain a constant shear stress.[24,45] In addition to surface chemistry (anti-CD4+ antibody), shear stress helped to differentiate monocytes from CD4+ cells. Due to the differences in cell size and density of CD4+ molecules on the cell surface, the adhesion of monocytes to the chamber surface is kept at a low level (<5 cells/mm²).[45] Recently, Cheng et al. designed a double-stage cascaded microchip. It includes four monocyte depletion chambers, which are coated with anti-CD14+ MAb, to remove monocytes before CD4+ cell capture.[46]

Moon et al. used a combination strategy of surface chemistry and shear stress to selectively capture CD4+ cells in a straight microchannel.[18] To facilitate lensless imaging,

however, PMMA was used instead of PDMS due to its transparence. The selectively captured CD4 cells are counted by their shadow images that fall on a lensless CCD surface
in a few seconds. Basically, this type of microchip is composed of three layers, including
a glass coverslip, double-sided adhesive, and PMMA base with an inlet and an outlet.[17] A
microchannel is formed between the glass coverslip and the PMMA base, with a dimension of $25 \times 4 \times 50$ μm. It is noted that microchip fabrication can be easily scaled up via a
laser cutter. In addition, the number of microchannels can be increased on each microfluidic device to increase the throughput and the multiplexing capability.

6.3.2 Microscopy-Based Detection and Counting

Similar to flow cytometry, microfluidic-based CD4⁺ cell count initially used fluoro-
chrome or quantum dot-labeled anti-CD3⁺ and anti-CD4⁺ MAbs, achieving fluorescence
detection.[23] Captured cells are viewed through different fluorescence filters under a
miscroscope. As shown in Figure 6.1, CD3⁺ cells are shown in red (panel A) and CD4⁺
cells (including monocytes) are shown in green (panel B). In a processed image (panel
C), images viewed with different fluorescence filters are digitally overlapped, with
CD3⁺CD4⁺ T lymphocytes appearing in yellow. For quantification, CD4⁺ T lymphocytes
are counted from processed images, which represent CD4⁺ T lymphocytes in 0.18 μl of

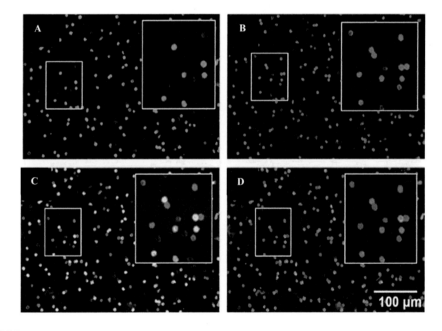

FIGURE 6.1
(See color insert) Fluorescence imaging (10×) of CD4 cells captured in a nanobiochip. (A) Cells are labeled with
anti-CD4 antibody conjugated with Qdot 565 (in green). (B) Cells are labeled with anti-CD3⁺ antibody conjugated
with Qdot 655 (in red). (C) CD3⁺CD4⁺ cells are shown in yellow in the emerged image. (D) Absolute CD4 cell count
(in yellow) is obtained in the processed picture, in which cells in red or green are deleted. A long-pass emission
filter allows for CD4⁺ T lymphocyte counting instead of image processing. Boxes on the right in each panel show
magnified images from the left (2×). (Adapted from Jokerst, J.V., et al., *Lab Chip*, 8(12), 2079–2090, 2008.)

whole blood.[23] In contrast, a colorimetric method has been developed to facilitate differentiation of CD4[+] T lymphocytes from monocytes and other immune cells. However, fluorescence microscopy-based detection is not practical for resource-limited settings, as fluorescence microscopes are expensive and require skilled operation and dedicated laboratory space.

To simplify detection, Cheng et al. used a combination strategy of surface chemistry and shear stress to selectively capture CD4[+] cells from whole blood, which enabled CD4[+] cell counting under a light microscope.[24,45] This offers a great advantage for resource-limited settings since a light microscope is affordable and accessible in most district hospitals. However, these devices still take a long time per test to count manually thousands of captured cells under a microscope. Although this microchip method is less accurate than flow cytometry for the range where CD4[+] cells are more than 800 cells/µl, it can effectively identify clinically relevant thresholds of 200, 350, and 500 cells/µl.[24]

6.3.3 Impedance Sensing

Another label-free CD4[+] counting method can be achieved electrically, as opposed to optical counting.[44,46] In this method, patterned surface electrodes were used to measure the change of conductivity in the microchannel, which is proportional to the number of captured CD4[+] cell count after cell lysis. As with the previous design, CD4[+] cells are captured from whole blood in a microfluidic channel functionalized with anti-CD4[+] MAb. In an ion-free media, CD4[+] cells are passed through the channel and captured. After injecting a low-conductivity cell lysis buffer, intracellular ions are released, increasing the conductivity. The bulk conductance is detected by built-in electrodes and measured by an external impedance spectroscope. As reported, concentrations as low as 20 CD4[+] cells/µl can be detected. This method successfully eliminates the need for cell labeling and fluorescence detection in the previous CD4[+] cell counting system. In addition, the lengthy optical counting process under microscopy is transformed to reading electrical signals, significantly shortening the sample-to-answer time. However, this method assumes identical ion concentration for all cells, which needs to be validated using clinical samples. Also, cells can release ions without lysis and may contribute to measurement errors, which necessitates stringent storage of blood samples.

6.3.4 Lensless Imaging

To further simplify CD4[+] cell counting, a lensless, ultra-wide-field cell array based on shadow imaging (LUCAS) has been developed.[47] LUCAS is designed to provide shadow images for cell counting over an ultra-wide field of view, e.g., a few square centimeters, rather than images with a high spatial resolution (Figure 6.2). LUCAS significantly increases the imaging speed to a few seconds of the whole microchip, in comparison to other imaging technologies.[48,49] In addition to this sensor, an algorithm has also been developed to simultaneously count cells within a few seconds.[51] This automatic counting allows the counting of CD4[+] cells 100 times more efficiently than manual methods under a conventional light microscope. Furthermore, LUCAS uses a charged-coupled device (CCD) as a light source due to the high signal-to-noise ratio. This device can be powered by rechargeable batteries and increases portability. Therefore, the simplified design permits CD4[+] cell capturing and label-free counting in such a portable system, which is suitable for resource-limited settings.

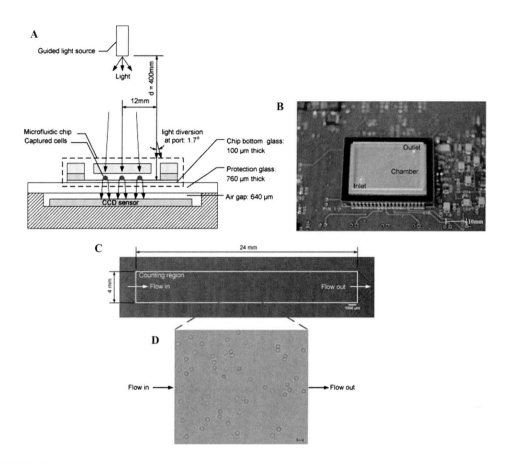

FIGURE 6.2
Schematic illustration of lensless imaging technology. (A) Design of lensless imaging to detect the captured cells in a microchamber. When a light source sheds light on CD4 cells, which are captured in a microchamber, they form shadows on a CCD sensor. (B) Prototype of the microchip and the lensless CCD imaging platform. The field of view of the CCD sensor is 35 × 25 mm. (C) Design of the microchamber. The microchamber has a dimension of 4 × 24 mm. (D) The shadow image of captured CD4 cells in the microchamber is shown, with a scale bar of 100 μm. (Adapted from Moon, S., et al., *Biosens. Bioelectron.*, 24(11), 3208–3214, 2009.)

6.3.5 Single-Platform Image Cytometer (SP ICM)

Different from other simplified flow cytometers, a single-platform image cytometer has been developed.[50–52] This image cytometer consists of two major components: a cell separation chamber and an image detection system. To selectively separate and count CD4⁺ cells from blood, immunomagnetic and fluorescent labeling are both employed. T lymphocytes from whole blood are first separated with immunomagnetic nanoparticles functionalized with CD3⁺ MAb, upon a homogeneous magnetic force. After selective separation with a magnet, T lymphocytes are then labeled with anti-CD4⁺ and anti-CD8⁺ MAbs, which are conjugated to PE and PerCP, respectively. For fluorescence imaging, two light-emitting diodes (LEDs) are placed symmetrically with filters of 595AF60 for PE fluorescence and 695AF55 for PerCP. For rapid imaging, processed cells are mounted in an analysis chamber and pictures are recorded by a CCD camera. Recent studies have shown that this image cytometer has a high correlation (R > 0.96) with single-platform flow cytometers in estimating CD4+ cells.[54,55] However, it shows underestimation of CD8+ cells, which may be caused by the limited sensitivity or cross talk between these two fluorescent colors. Nevertheless, this portable and battery-powered image cytometer is an attractive alternative to flow cytometry for determining absolute CD4+ and CD8+ counts and the CD4+/CD8+ ratio.

6.4 Conclusion and Future Prospects

In the context of the HIV/AIDS epidemic, the WHO is expanding access to ART in resource-limited settings. However, this objective is essentially restricted due to lack of ART monitoring tools (e.g., CD4+ cell count), which need to be inexpensive, rapid, and simple to use. Despite the progress in simplified CD4 automatic and manual counting methods, they are still relatively expensive and require skilled operators. These factors prevent their use as POC diagnostic and monitoring tools. Recent advances in microfluidic technologies have shown promise in developing such POC CD4 assays in terms of cost, simplicity, and turn around time. However, current detection methods, such as fluorescence imaging, impedance sensing, and manual counting under a light microscope, are not practical. Lensless imaging is a promising approach for detection and counting, since it excludes the need for expensive fluorescence microscopes and manual counting, and enables cell counts within seconds. The emerging POC technologies would facilitate expansion of ART in resource-limited settings, especially in rural areas.

Acknowledgments

We acknowledge support from the NIH (grant R01AI081534 and grant RZ1A108710), and a Young Investigator Award from the W. H. Coulter Foundation. This work was also suppoted by the Center for Integration of Medicine and Innovative Technology (CIMIT) under U.S. Army Medical Research Acquisition Activity Cooperation Agreements, DAM017-02.2-0006, W81XWH-07-0011, and W81XWH-09-2-0001. This work was made possible by the grant awarded and administered by the U.S. Army Medical Research & Material Command (USAMRMC) and the Telemedicine & Advanced Technology Research Center (TATRC) at Fort Detrick, MD.

References

1. UNAIDS. (2008). Report on the global AIDS epidemic. http://data.unaids.org/pub/GlobalReport/2008/JC1510_2008GlobalReport_en.zip (accessed November 20, 2009).
2. World Health Organization. (2006). Towards universal access by 2010. http://www.who.int/hiv/toronto2006/towardsuniversalaccess.pdf (accessed November 20, 2009).
3. World Health Organization. (2009). Scaling up priority HIV/AIDS interventions in the health sector—progress report. http://www.who.int/hiv/pub/tuapr_2009_en.pdf (accessed November 20, 2009).

4. Gazzard BG, et al. (2008). British HIV Association guidelines for the treatment of HIV-1-infected adults with antiretroviral therapy 2008. *HIV Med* 9(8):563–608.

5. Hammer SM, et al. (2008). Antiretroviral treatment of adult HIV infection: 2008 recommendations of the International AIDS Society—USA panel. *JAMA* 300(5):555–570.

6. World Health Organization. (2006). Antiretroviral therapy for HIV infection in adults and adolescents. http://www.who.int/hiv/pub/guidelines/artadultguidelines.pdf (accessed November 25, 2009).

7. Calmy A, et al. (2007). HIV viral load monitoring in resource-limited regions: optional or necessary? *Clin Infect Dis* 44(1):128–134.

8. Mugyenyi P, et al. (2010). Routine versus clinically driven laboratory monitoring of HIV antiretroviral therapy in Africa (DART): a randomised non-inferiority trial. *Lancet* 375(9709):123–131.

9. Harries AD, et al. (2010). Diagnosis and management of antiretroviral-therapy failure in resource-limited settings in sub-Saharan Africa: challenges and perspectives. *Lancet Infect Dis* 10(1):60–65.

10. Long L, Fox M, Sanne I, and Rosen S. (2010). The high cost of second-line antiretroviral therapy for HIV/AIDS in South Africa. *AIDS*, 24: 915–919

11. Carr A. (2003). Toxicity of antiretroviral therapy and implications for drug development. *Nat Rev Drug Discov* 2(8):624–634.

12. Mee P, Fielding KL, Charalambous S, Churchyard GJ, and Grant AD. (2008). Evaluation of the WHO criteria for antiretroviral treatment failure among adults in South Africa. *AIDS* 22(15):1971–1977.

13. van Oosterhout JJG, et al. (2009). Diagnosis of antiretroviral therapy failure in Malawi: poor performance of clinical and immunological WHO criteria. *Trop Med Int Health* 14(8):856–861.

14. Fiscus SA, et al. (2006). HIV-1 viral load assays for resource-limited settings. *PLoS Med* 3(10):e417.

15. Brown ER, et al. (2009). Comparison of CD4 cell count, viral load, and other markers for the prediction of mortality among HIV-1-infected Kenyan pregnant women. *J Infect Dis* 199(9):1292–1300.

16. Fowler MG and Owor M. (2009). Monitoring HIV treatment in resource-limited settings: reassuring news on the usefulness of CD4(+) cell counts. *J Infect Dis* 199(9):1255–1257.

17. Kim YG, Moon S, Kuritzkes DR, and Demirci U. (2009). Quantum dot-based HIV capture and imaging in a microfluidic channel. *Biosens Bioelectron* 25(1):253–258.

18. Moon S, et al. (2009). Integrating microfluidics and lensless imaging for point-of-care testing. *Biosens Bioelectron* 24(11):3208–3214.

19. Lee WG, Kim YG, Chung BG, Demirci U, and Khademhosseini A. (2010). Nano/microfluidics for diagnosis of infectious diseases in developing countries. *Adv Drug Deliv Rev* 62(4–5):449–457.

20. Spacek LA, et al. (2006). Evaluation of a low-cost method, the Guava EasyCD4 assay, to enumerate CD4-positive lymphocyte counts in HIV-infected patients in the United States and Uganda. *JAIDS J Acquir Immune Defic Syndr* 41(5):607–610.

21. Pattanapanyasat K, et al. (2005). Evaluation of a new single-parameter volumetric flow cytometer (CyFlow(green)) for enumeration of absolute CD4(+) T lymphocytes in human immunodeficiency virus type 1-infected Thai patients. *Clin Diagn Lab Immunol* 12(12):1416–1424.

22. Pattanapanyasat K, et al. (2005). A multicenter evaluation of the PanLeucogating method and the use of generic monoclonal antibody reagents for CD4 enumeration in HIV-infected patients in Thailand. *Cytometry B Clin Cytom* 65(1):29–36.

23. Rodriguez WR, et al. (2005). A microchip CD4 counting method for HIV monitoring in resource-poor settings. *PLoS Med* 2(7):e182.

24. Cheng X, et al. (2007). A microchip approach for practical label-free CD4+ T-cell counting of HIV-infected subjects in resource-poor settings. *J Acquir Immune Defic Syndr* 45(3):257–261.

25. Lopez A, et al. (1999). Enumeration of CD4(+) T-cells in the peripheral blood of HIV-infected patients: an interlaboratory study of the FACSCount system. *Cytometry* 38(5):231–237.

26. Strauss K, et al. (1996). Performance evaluation of the FACSCount System: a dedicated system for clinical cellular analysis. *Cytometry* 26(1):52–59.

27. Balakrishnan P, et al. (2006). A reliable and inexpensive EasyCD4 assay for monitoring HIV-infected individuals in resource-limited settings. *JAIDS J Acquir Immune Defic Syndr* 43(1):23–26.

28. Thakar MR, Kumar BK, Mahajan BA, Mehendale SM, and Paranjape RS. (2006). Comparison of capillary based microflurometric assay for CD4+ T cell count estimation with dual platform flow cytometry. *AIDS Res Ther* 3:26.

29. Pattanapanyasat K, et al. (2008). Comparison of 5 flow cytometric immunophenotyping systems for absolute CD4(+) T-lymphocyte counts in HIV-1-infected patients living in resource-limited settings. *JAIDS J Acquir Immune Defic Syndr* 49(4):339–347.

30. Manasa J, et al. (2007). Evaluation of the Partec flow cytometer against the BD FACSCalibur system for monitoring immune responses of human immunodeficiency virus-infected patients in Zimbabwe. *Clin Vaccine Immunol* 14(3):293–298.

31. Fryland M, et al. (2006). The Partec CyFlow Counter could provide an option for CD4+ T-cell monitoring in the context of scaling-up antiretroviral treatment at the district level in Malawi. *Trans R Soc Trop Med Hyg* 100(10):980–985.

32. Denny TN, et al. (2008). A North American multilaboratory study of CD4 counts using flow cytometric panLeukogating (PLG): a NIAID-DAIDS Immunology Quality Assessment Program study. *Cytometry B Clin Cytom* 74 (Suppl 1):S52–S64.

33. Sippy-Chatrani N, et al. (2008). Performance of the Panleucogating protocol for CD4+ T cell enumeration in an HIV dedicated laboratory facility in Barbados. *Cytometry B Clin Cytom* 74 (Suppl 1):S65–S68.

34. Glencross DK, et al. (2008). Large-scale affordable PanLeucogated CD4+ testing with proactive internal and external quality assessment: in support of the South African national comprehensive care, treatment and management programme for HIV and AIDS. *Cytometry B Clin Cytom* 74 (Suppl 1):S40–S51.

35. Diagbouga S, et al. (2003). Successful implementation of a low-cost method for enumerating CD4+ T lymphocytes in resource-limited settings: the ANRS 12–26 study. *Aids* 17(15):2201–2208.

36. Bi XQ, et al. (2005). Modified Dynabeads method for enumerating CD4(+) T-lymphocyte count for widespread use in resource-limited situations. *JAIDS Acquir Immune Defic Syndr* 38(1):1–4.

37. Lutwama F, et al. (2008). Evaluation of Dynabeads and Cytospheres compared with flow cytometry to enumerate CD4+ T cells in HIV-infected Ugandans on antiretroviral therapy. *JAIDS Acquir Immune Defic Syndr* 48(3):297–303.

38. Balakrishnan P, et al. (2004). An inexpensive, simple, and manual method of CD4 T-cell quantitation in HIV-infected individuals for use in developing countries. *JAIDS J Acquir Immune Defic Syndr* 36(5):1006–1010.

39. Nouanthong P, Pata S, Sirisanthana T, and Kasinrerk W. (2006). A simple manual rosetting method for absolute CD4+ lymphocyte counting in resource-limited countries. *Clin Vaccine Immunol* 13(5):598–601.

40. Srithanaviboonchai K, et al. (2008). Novel low-cost assay for the monitoring of CD4 counts in HIV-infected individuals. *JAIDS J Acquir Immune Defic Syndr* 47(2):135–139.

41. Ziaie B, Baldi A, Lei M, Gu Y, and Siegel RA. (2004). Hard and soft micromachining for BioMEMS: review of techniques and examples of applications in microfluidics and drug delivery. *Adv Drug Deliv Rev* 56(2):145–172.

42. Wang Z, et al. (2010). Microfluidic CD4+ T-cell counting device using chemiluminescence-based detection. *Anal Chem* 82(1):36–40.

43. Jokerst JV, Floriano PN, Christodoulides N, Simmons GW, and McDevitt JT. (2008). Integration of semiconductor quantum dots into nano-bio-chip systems for enumeration of CD4+ T cell counts at the point-of-need. *Lab Chip* 8(12):2079–2090.

44. Cheng X, et al. (2007). Cell detection and counting through cell lysate impedance spectroscopy in microfluidic devices. *Lab Chip* 7(6):746–755.

45. Cheng X, et al. (2007). A microfluidic device for practical label-free CD4(+) T cell counting of HIV-infected subjects. *Lab Chip* 7(2):170–178.

46. Cheng XH, et al. (2009). Enhancing the performance of a point-of-care CD4+T-cell counting microchip through monocyte depletion for HIV/AIDS diagnostics. *Lab Chip* 9(10):1357–1364.
47. Ozcan A and Demirci U. (2008). Ultra wide-field lens-free monitoring of cells on-chip. *Lab Chip* 8(1):98–106.
48. X. Heng, et al. (2006). Optofluidic microscopy—a method for implementing a high resolution optical microscope on a chip. *Lab Chip* (6):1274–1276.
49. Lange D, Storment CW, Conley CA, and Kovacs GTA. (2005). A microfluidic shadow imaging system for the study of nematode *C. elegans* in space. *Sensors Actuators B* 107:904–914.
50. Alyassin MA, et al. (2009). Rapid automated cell quantification on HIV microfluidic devices. *Lab Chip* 9(23):3364–3369.
51. Li X, Tibbe AG, Droog E, Terstappen LW, and Greve J. (2007). An immunomagnetic single-platform image cytometer for cell enumeration based on antibody specificity. *Clin Vaccine Immunol* 14(4):412–419.
52. Ymeti A, et al. (2007). A single platform image cytometer for resource-poor settings to monitor disease progression in HIV infection. *Cytometry A* 71(3):132–142.
53. Li X, et al. (2007). CD4+ T lymphocytes enumeration by an easy-to-use single platform image cytometer for HIV monitoring in resource-constrained settings. *Cytometry B Clin Cytom* 72(5):397–407.
54. Li X, et al. (2009). CD4 and CD8 enumeration for HIV monitoring in resource-constrained settings. *Cytometry B Clin Cytom* 76(2):118–126.
55. Li X, et al. (2010). Clinical evaluation of a simple image cytometer for CD4 enumeration on HIV-infected patients. *Cytometry B Clin Cytom* 78(1):31–36.

7

Nanoviricides—A Novel Approach to Antiviral Therapeutics

Randall W. Barton, Jayant G. Tatake, and Anil R. Diwan

NanoViricides, Inc., West Haven, Connecticut, USA

CONTENTS

7.1 Introduction

Current medical approaches to viral diseases encompass vaccines, antibodies, and antiviral chemotherapeutic agents. However, many viral diseases lack effective vaccines (e.g., HIV/AIDS, HCV),[1–5] or in the case of influenza A, vaccines suffer from incomplete coverage of all viruses due to antigenic drift and antigenic shift.[6,7] Similarly, many viral diseases lack effective antiviral drugs.[8,9] Nonetheless, antiviral drugs have shown efficacy for certain viral diseases, as both prophylactic and therapeutic agents, and this efficacy has established the importance of antiviral drugs in the absence of vaccines.[8–10] In addition, antiviral therapeutics can be used as adjuncts to vaccines.[11,12]

In general, the current antiviral therapies suffer from limited efficacy, incomplete coverage due to genetic heterogeneity of the virus, rapid emergence of virulent, readily transmissible, drug-resistant mutants, and side effects. This has led to the use of combination antiviral drug regimens to overcome these deficiencies in efficacy and to limit development of drug-resistant mutants.[13,14] While combination drug therapy has shown increased efficacy, there remains a significant unmet medical need for effective, novel, and safe drugs. Equally importantly, there are a number of viruses for which there are no therapeutic agents, even ones with limited efficacy.[8,9]

There are a number of stages in the viral life cycle that represent potential targets for the development of antiviral therapies: (1) viral attachment and entry into the cell, (2) uncoating of the virus, (3) transcription of viral mRNA, (4) translation of viral mRNA, (5) replication of viral DNA or RNA, (6) maturation of viral proteins, (7) assembly of

Cycle of Infection and Replication

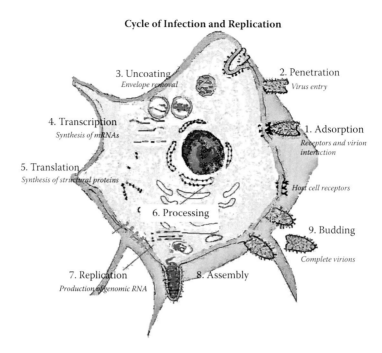

3. Uncoating
Envelope removal

2. Penetration
Virus entry

4. Transcription
Synthesis of mRNAs

1. Adsorption
Receptors and virion interaction

5. Translation
Synthesis of structural proteins

Host cell receptors

6. Processing

9. Budding

Complete virions

7. Replication
Production of genomic RNA

8. Assembly

FIGURE 7.1
Schematic depicting the steps in the cycle of virus infection and replication. (Reproduced from the CDC website: http://www.cdc.gov/rabies/transmission/virus.html.)

viral particles, and (8) budding or release of mature virus. The current antiviral drugs approved by the U.S. FDA and international regulatory agencies exploit many of these targets as their mechanisms of action. As depicted in Figure 7.1, there are a number of mechanisms by which antiviral agents can inhibit viral infection, replication, and maturation and release. Table 7.1 summarizes many of the currently available antiviral

TABLE 7.1

Antiviral Therapeutic Targets

Infectious Event	Molecular Target	Therapeutic Class	Examples
Cell binding and infection	Cell surface receptors Attachment sites	Fusion/entry inhibitors, antibodies	Fuzeon, Marovic, (HIV), XTL/Cubist HepexB, Rimantidine (influenza)
Genome replication	Reverse-transcriptase (RT) Integrase Viral polymerases	Nucleoside (NRTI) Nonnucleoside (NNRTI) Integrase inhibitors Nucleoside analogues	Nevirapine, zidovudine, cytarabine, lamivudine, zalcitabine, stavudine, others
Processing	HIV protease	Protease inhibitors	Atazanavir, fosamprenavir, tipranavir, saquinavir, indinavir, ritonavir, nelfinavir, amprenavir, lopinavir
Budding	HIV Flu neuraminidase	Budding inhibitors N inhibitors	Panacos PA-457, TamiFlu, Relenza, Peramivir

therapeutics and their mechanisms of action; it is not meant to be exhaustive but simply to provide examples. Most of the current antiviral strategies are based on inhibition or modulation of the intracellular biochemical pathways, mediated by both viral and host cell enzymes for the production of new virus particles. For example, in the case of HIV, (1) HIV binds to specific cell surface receptors that are required for viral entry into the cell, (2) the viral genomic RNA (v-RNA) is copied into a complementary DNA molecule (v-cDNA) by the viral reverse transcriptase (v-RT), (3) an enzyme called integrase (v-I) inserts copies of the v-cDNA into the nuclear DNA of the cell, (4) this DNA is copied by the cell machinery to make viral messenger RNA (v-mRNA), and is also copied to make new viral genomic RNA (v-RNA), (5) the v-mRNAs are then translated by the cell to make viral precursor proteins, (6) the precursor proteins are processed by the viral protease (v-P) to form mature viral proteins, and (7) the resulting mature viral proteins and two copies of the newly made viral genomic RNA self-assemble, possibly with assistance from both viral-encoded and host machinery factors, such as molecular chaperones, into new virus particles and are released from the cell by budding.[15]

This chapter introduces the novel NanoViricides nanotechnology that possesses potent antiviral efficacy by targeting the mechanisms of cell attachment or cell binding of viruses. NanoViricides utilizes the technology of proprietary self-assembling amphiphilic polymers that form nanoscopic polymeric micelles that can blunt viruses through multiple mechanisms. Moreover, they have the potential to prevent the development of drug-resistant mutants. Polymeric micelles are amphiphilic polymers that contain hydrophilic and hydrophobic components, so-called block copolymers, allowing self-assembly into a variety of structures depending on the surrounding solvent. In aqueous environments, the hydrophobic components have a tendency to self-assemble in order to avoid contact with water, while the hydrophilic components form a hydrated shell; the result is a thermodynamically stable structure. This hydrophilic deformable shell is exposed to the aqueous environment, while a substantially hydrophobic core or interior that may be shallow or deep allows for encapsulation of guest molecules. Alternating polymeric micelles contain alternating hydrophilic and hydrophobic segments, resulting in a very regular structure that leads to uniform properties. In contrast, we have developed "pendant" polymeric micelles. In these, the hydrophobic segments are attached into a monomer repeat unit as pendants or "tails," while the pendant-free polymer structure is hydrophilic or amphiphilic in nature by itself. By virtue of the free ends, the tails self-assemble in more stable thermodynamic conformations and enable a truly hydrophobic core/hydrophilic shell type of structure in an aqueous milieu, as opposed to alternating block copolymers containing hydrophobic and hydrophilic segments (series or S-type polymeric micelles). We have worked on T-type polymeric micelles, wherein there is a single tail per repeat unit, and later developed and switched over to π-type polymeric micelles possessing two tails per repeat unit, resembling a diacyl-glycerol-like motif of the biological cell membrane. The polymeric micelles formed from these π-polymers are therefore capable of forming structures similar to the biological membrane. Rather than parallel assemblies forming a biological bilayer lipid membrane, a single-polymer chain of a T-type or π-type polymer would form a single layer, globular micelle with a hydrophobic core at a limiting low concentration of the polymer. At higher concentrations, multiple-polymer chains aggregate together to form a larger globular stable micelle, which can range from a few nanometers to a few hundred nanometers in size. At even higher concentrations, liquid crystalline assemblies resembling biological membranes may occur. Because of the significant energy differences reflecting addition of a single-pendant polymeric chain to an existing assembly, the distribution of particle sizes of the ensemble, as well as structural dynamics of the ensemble, are expected to follow thermodynamic phase-change and phase-equilibria-like behaviors, rather than simple aggregation dynamics, resulting in discrete shifts and population equilibria between different sizes and structures.

Polymeric micelles have been reported for use as drug solubilization and drug delivery agents because of potential properties of sustained, controlled drug release and enhanced circulating half-lives of the polymeric micelles.[16–18] Work on specific targeting of drugs and encapsulated active pharmaceutical ingredient (API) has also been performed. However, most of the work involved attachment of a single targeting moiety at either one or both ends of the polymeric chain, resulting in very low densities of the targeting moieties on the polymeric micelle surface.

Herein we show that the NanoViricides nanoscopic polymeric micelle technology serves as a therapeutic agent technology and not simply as a drug delivery technology. In addition to developing T- and π-type polymeric chain structures, we have also incorporated chemical functionally available groups within the monomer itself. We attach targeting moieties at these functional points. This enables a significantly larger density of targeting moieties (one or two per monomer, as opposed to one or two per polymeric chain), with greatly enhanced efficiency of targeting. In fact, in the first generation of NanoViricides technology, which we are working on at present, we have not used the feature of encapsulating any APIs. We plan to utilize this additional feature in future generations of nanoviricides to tackle viruses that cannot be controlled by our current approach alone.

7.2 What Is a Nanoviricide?

When virus-specific ligands are chemically attached to a proprietary flexible polymer structure that mimics a biological cell membrane, a nanoviricide® is created. Our flexible polymer backbone is comprised of a polyethylene glycol (PEG) chain and alkyl pendants. PEG forms the hydrophilic shell and imparts nonimmunogenicity; PEGylation is a familiar technique to minimize immunogenicity of proteins and antibodies. The alkyl chains float together to make a flexible core, like an immobilized oil droplet. The resulting materials are polymeric surfactants that form stable micelles, rather than liposomes that form dynamic micelles. Thus, these nanoviricide micelles are freely soluble and conformationally flexible in bodily fluids. Chemical groups uniformly distributed along the polymer chain allow attachment of virus-specific ligands such as chemical moieties, peptides, antibody fragments, or other proteins. On first contact with a virus particle, a nanoviricide micelle may bind to a virus particle because of specific interaction between a ligand attached to the nanoviricide and the glycoproteins on the virus surface. This may cause the flexible nanoviricide to reach very close to the virus surface, leading to additional ligands binding to additional viral coat proteins, in a mode called cooperative binding. Cooperative binding is a well-known natural process that forms the basis of biological recognition, such as antibody-antigen binding complexes, DNA hybridization, and protein assembly, among others. The attachment of multiple ligands to a single-polymer chain, coupled with the fact that multiple-polymer chains make up a single micelle, leads to a very high avidity for the resulting micelle binding to the virus. This is analogous to the Velcro effect that results from cooperative interactions. This property enables the use of ligands that individually may have low affinity, but the polyvalency can result in very high avidity. This concept has been shown quite convincingly with polymeric vs. monomeric inhibitors.[19–21]

An analogy may be drawn with neutralizing antibodies. Antibodies are an integral part of the natural immune process to bring infections under control, and antibodies can ameliorate or prevent many viral infections, and may also aid in the resolution of viral infections. Some antibodies do this by a process called neutralization. Antibodies can neutralize viral infectivity by causing aggregation of virus particles, inhibiting virus binding to its host cell surface receptor(s), blocking endocytic uptake into cells, and preventing virus uncoating in endosomes. When antiviral antibodies bind to the viruses, the resulting complexes can be lysed upon subsequent binding by complement factors, or virus-antibody complexes can be taken up by phagocytes and destroyed. In contrast, upon binding of a nanoviricide micelle to a virus particle, van der Waals interactions may promote encapsulation of the virus particle by virtue of reassembly of the polymeric chains comprising the nanoviricide, wherein the lipid tails would integrate with the lipid membrane coating the virus (for enveloped viruses). Nonenveloped viruses also exhibit significant lipidic regions on the virus particle surface, with either adsorbed or covalently attached alkyl chains as well as hydrophobic amino-acid-rich regions. Thus, it is likely that similar interactions can happen between a nanoviricide and a nonenveloped virus particle, but possibly to a lesser extent than with an enveloped virus particle. This is akin to a coordinated surfactant attack on the virus particle, and it may lead to dismantling of the virus envelope or stripping off of the viral surface lipids, as well as stripping off of the glycoproteins necessary for viral adsorption and cellular entry.

Many distinctly different viruses use common mechanisms to access tissues, attach to cells, and internalize, for example, sialic acid.[20–22] These steps are primarily where a neutralization strategy such as nanoviricides can be useful. A nanoviricide acts like a decoy of a human cell. When the virus sees the appropriate ligand displayed on a nanoviricide micelle, the virus binds to it. The nanoviricide, being flexible, allows maximization of binding by spreading onto the virus particle, and finally fusing with the lipid-coated virus surface by a phase inversion, wherein the fatty core of the nanoviricide merges with the viral lipid coat and the hydrophilic shell of the nanoviricide becomes the exterior of the particle, thus engulfing the virus. In the process, the coat proteins that the virus uses for binding to cells would be expected to become unavailable. This proposed mechanism of action is shown schematically in Figure 7.2. This highly targeted surfactant attack leads to loss of the coat proteins, and the nanoviricide may further dismantle the engulfed virus capsid. The loss of virus particle integrity renders the virus noninfectious. Support for this mechanism is shown in the electron photomicrographs in Figure 7.3, in which murine cytomegalovirus (CMV) was incubated with a nanoviricide. As can be seen, the binding of the nanoviricide to the CMV results in loss of the viral envelope; the resulting CMV naked capsids are noninfectious. Thus, nanoviricide binding not only renders the CMV inactive, but also appears to result in the disruption of capsid organization. In vivo disrupted capsids would likely be rapidly cleared by host defense mechanisms. Given this mechanism of action, nanoviricides are not expected to interfere with intracellular replication of the virus, or with host effectors to any appreciable extent.

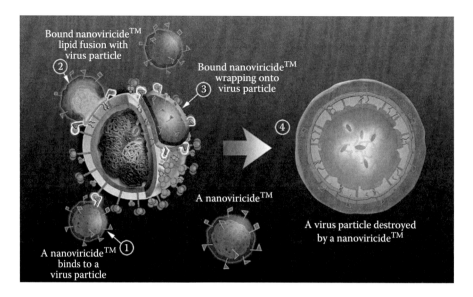

FIGURE 7.2
(See color insert) Graphic representation of the mechanism of action of nanoviricide binding and inactivation of viruses.

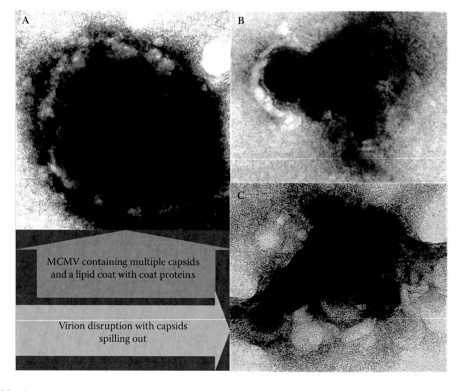

FIGURE 7.3
Effects of two different nanoviricides binding to murine cytomegalovirus (MCMV). (A) Control MCMV virion. (B, C) MCMV virions treated with two different nanoviricides.

7.3 Evaluation of Nanoviricides in a Rabbit Model of EKC

Adenovirus serotypes are the most common etiologic agents of external ocular viral infection in many parts of the world. Although follicular conjunctivitis is the most common, epidemic keratoconjunctivitis (EKC) is the most serious adenoviral ocular disease responsible for global and community epidemics. The disease is very contagious and can cause community and medical facility epidemics resulting in significant patient morbidity, societal losses from worker and student absenteeism, and increased direct medical costs. EKC has a variable course, but the appearance of immune-based subepithelial infiltrates (SEIs) can impair visual acuity for months in some patients.[23,24] Currently, there is no effective topical or systemic antiviral therapy to reduce the symptoms or duration of the disease, reduce viral shedding, or prevent the vision-threatening subepithelial infiltrates.[25,26] In addition, no treatment will protect the second eye or prevent transmission within households, medical offices, and communities. Nonsteroidal anti-inflammatory eye drops, steroid eye drops, and artificial tears have been used as supportive therapy.

The animal model uses New Zealand rabbits utilizing topical and intrastromal ocular inoculation with adenovirus type 5.[27] Viral replication is associated with clinical signs of infection—acute conjunctivitis, blepharitis, and subepithelial infiltrates. Three groups of rabbits per group were compared: an infected control group treated with the vehicle (group A), an infected experimental group treated with NNVC B (group B), and an infected experimental group treated with NNVC C (group C).

On day 0 both eyes of each rabbit were inoculated with adenovirus (Ad) 5.4×10^5 PFU in 100 µl by intrastromal injection and scarification. Fifteen hours after viral inoculation, the rabbit eyes were treated with two drops (50 µl per drop) twice a day with vehicle (group A), NNVC B (group B), or NNVC C (group C) for 10 days. All rabbits were monitored daily from days 1–10, and then at days 21 and 37, until sacrifice, on a grading system of 0–4 for conjunctival injection (redness and dilation or prominence of conjunctival blood vessels), blepharitis (crusting, discharge, and inflammation of eyelid margins), iris hyperemia, and corneal subepithelial infiltrates (SEIs). As shown in Figure 7.4, the vehicle

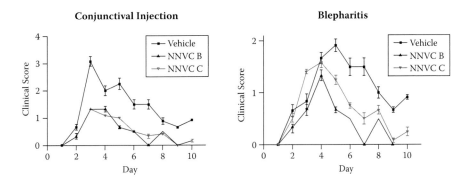

FIGURE 7.4
Time course of the clinical scores for conjunctival injection (redness and dilation or prominence of conjunctival blood vessels) and blepharitis (crusting, discharge, and inflammation of eyelid margins) in vehicle control- and NNVC B- and NNVC C-treated rabbits following adenovirus 5 ocular infection. Treatment (twice daily) with two drops at a time was started 3 hours postinfection.

Clinical Symptoms at 5 Days Post Infection

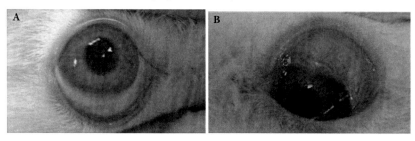

FIGURE 7.5
Clinical symptoms at 5 days postinfection with adenovirus 5. Treatment (twice daily) with two drops at a time was started 3 hours postinfection. A: Vehicle control. B: NNVC B.

group had markedly higher clinical scores for conjunctival injection and blepharitis than the NNVC B- and NNVC C-treated groups. Moreover, both parameters (conjunctivitis and blepharitis) were still significantly elevated at day 10 in the vehicle control group, while in the NNVC B and C treatment groups both parameters had returned to baseline. An example of the clinical symptoms is shown in Figure 7.5, in which a typical vehicle-treated eye and an NNVC B-treated eye are compared at day 5 postinfection. Clearly many of the clinical symptoms of the conjunctiva in the NNVC B-treated eye are significantly reduced compared to the conjunctiva of the vehicle-treated eye.

The subepithelial infiltrates, characteristic of an immune response to the viral infection, were not detected until day 21, and only in a few instances. However, by day 37 (day of sacrifice), as shown in Table 7.2, 83% of the rabbit eyes in the vehicle control group had detectable infiltrates, whereas none of the rabbit eyes in the NNVC B group had detectable infiltrates. In the NNVC C treatment group, 50% of the rabbit eyes had infiltrates. The rabbits were sacrificed on day 37, and the corneas removed for histopathology studies to confirm the cellular composition of the infiltrates.

Thus, nanoviricide treatment had a marked effect on the severity and duration of both conjunctivitis and blepharitis, and perhaps more importantly, nanoviricide treatment dramatically decreased the incidence of the vision-threatening subepithelial infiltrates. Given that, currently, no effective topical or systemic therapeutic agent has been approved by the U.S. FDA, NanoViricides, Inc. can potentially develop a highly effective therapeutic for this disease for which there is a significant unmet medical need.

TABLE 7.2

Incidence of Subepithelial Infiltrates

Treatment	No. of Eyes with Subepithelial Infiltrates
Vehicle	5/6 (83%)
NNVC B	0/6 (0%)
NNVC C	3/6 (50%)

7.4 Evaluation of Nanoviricides in a SCID-hu Thy/ Liv Model of HIV-1 Infection

Antiretroviral agents can, indeed, improve clinical outcomes in HIV-1 infection, and moreover, such therapies have demonstrably reduced the death rate of AIDS in this country and other parts of the world.[13] Once it was confirmed that no one drug could achieve durable viral suppression or clinical benefit, clinical researchers have quickly adopted various combinations of therapeutics as they have become available, comprised of protease and reverse transcriptase inhibitors.[28,29] This has helped facilitate the development of what is called highly active antiretroviral therapy (HAART).[13] However, challenges for HIV therapy still remain, particularly in the areas of long-term complications of therapy, persistent low-level viremia, and viral drug resistance and genetic diversity.[30–32] Resistance to HAART eventually leads to AIDS.

The SCID-hu mouse model, in which human fetal lymphoid tissue is implanted into severe combined immunodeficient (SCID) mice, has been used to study both normal hematopoiesis and the pathogenesis of HIV-1 in human lymphoid tissue.[33,34] The SCID-hu Thy/Liv model is generated by coimplantation of human fetal thymus and liver beneath the kidney capsule. The implanted tissues become vascularized, fuse, and grow to become a Thy/Liv "organ," resembling a histologically normal human thymus. The model has been validated as an in vivo model for evaluation of anti-HIV-1 drug candidates.[35]

For evaluation of the efficacy of nanoviricides, C.B-17 SCID mice were coimplanted with fetal human thymus and liver tissue fragments under the capsule of the left kidney of anesthetized mice. The implants were inoculated with 5,000 viral particles of HIV-1, Ba-L, stock virus (1,250 × $TCID_{50}$) by direct injection after surgical exposure of the implanted kidney. The mice were randomly sorted into groups. Mice treated with nanoviricides received an i.v. injection of 100 μl of a 10 mg/ml solution (~50 mg/kg) at 24, 48, and 72 hours after virus inoculation. A control treatment group received 100 μl of PBS i.v. at 24, 48, and 72 hours. A positive control group received the HAART cocktail (AZT + 3TC + Efavirenz at 40 + 20 + 40 mg/kg, respectively), administered p.o. one time daily for the duration of the study, beginning 24 hours after virus inoculation. The total drug load with NNVC 4 and 5 administered in this study was about 150 mg/kg. In contrast, the total drug load administered for the HAART cocktail was 4,200 mg/kg (1,680 + 840 + 1,680 of AZT + 3TC + Efavirenz, respectively).

Treatment with HAART and NNVC 4 and 5 resulted in a significant reduction in viral load in the Thy/Liv implant, as shown by quantitative PCR (qPCR) (Figure 7.6) and viral particle counts by electron microscopy (EM) (Table 7.3). qPCR analysis showed that HAART and NNVC 4 and 5 treatment reduced the implant viral load by approximately 0.5 $\log_{10}$, compared to the vehicle control group, on day 30. NNVC 4 treatment produced a viral load reduction slightly greater than that of HAART treatment. Similarly, a reduction in viral load was also observed by EM analysis of implant lymphocytes. As shown in Table 7.3, viral particle counts were reduced approximately 20-fold by NNVC 4 and 5 treatment (EM). In contrast, HAART treatment reduced the lymphocyte viral particle count by approximately fivefold. Similar to the reduction in viral load, both HAART and NNVC treatment had long-term effects on thymocyte depletion, as shown by the proportion of CD4+CD8+ thymocytes (double positive) in the fifth week postinfection (Figure 7.6).

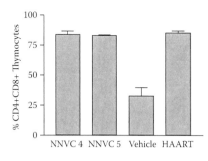
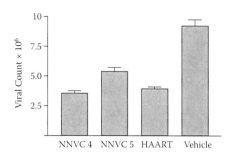

FIGURE 7.6

Effect of NNVC treatment on implant thymocyte depletion and viral load in implant at 30 days postinfection. Left: Thymocyte depletion. The percentage of CD4⁺CD8⁺ thymocytes in the implant are expressed as the mean ± SEM, n = 3. Right: Implant viral load. Results are the mean ± SEM of the qPCR at three different dilutions from three implants per group.

Implants in the HAART and NNVC 4 and 5 treatment groups had 80%–85% CD4⁺ CD8⁺ cells, while the vehicle control groups had approximately 30% CD4⁺CD8⁺ thymocytes.

As shown in Figure 7.7, the control, vehicle-treated mice died at 30 days postinfection, while the mice treated with the HAART cocktail survived until 42 days postinfection, a 40% increase in survival. The groups treated with nanoviricides had increased survival compared to the control vehicle-treated mice. Of significance, the NNVC 4 and 5 groups had survival that was comparable or slightly superior to that of mice treated with HAART, 44.3 and 42.2 days, respectively. There was also a significant body weight loss in the vehicle treatment group. Body weight loss in the HAART treatment group was reduced by more than 40%. Consistent with the survival results, treatment with NNVC 4 and 5 had a similar beneficial effect on body weight loss, approximately 50% less than the vehicle-treated group.

In summary, treatment of SCID-hu mice with NNVC 4 and 5 following HIV-1 Ba-L infection of Thy/Liv implants resulted in significantly reduced viral load and significantly improved double-positive CD4⁺ CD8⁺ thymocyte proportion. These effects appear to have resulted in improved survival and reduced body weight loss. Importantly, comparison with mice treated with the HAART cocktail for the duration of the study revealed that the NNVC 4 and 5 antiviral agents were comparable or slightly superior to HAART treatment for all parameters evaluated. It is important to note that NNVC 4 and 5 were single administrations only at 24, 48, and 72 hours postinfection, while the HAART cocktail was administered daily for the duration of the study. The NNVC total drug load was only 150 mg/kg, as opposed to a total HAART drug load of 4,200 mg/kg; thus, equivalent effects were observed with NNVC drug candidates at ~1/20th of the HAART drug load. It would be important to determine if extended nanoviricide administration shows significantly

TABLE 7.3

Viral Particles in Aspirated Implant Lymphocytes

Treatment	No. of Viral Particles per EM Field
Vehicle	240
NNVC 4	11
NNVC 5	11
HAART	47

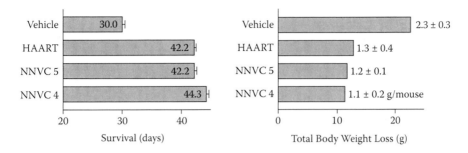

FIGURE 7.7
Effect of nanoviricides treatment on survival and body weight loss in HIV-1 infected SCID-hu Thy/Liv mice. The mortality and body weight of the mice were monitored. In the right-hand graph, the results are expressed as total body weight loss per group, and the mean weight loss + SEM/mouse is shown to the right of the bar, N = 10.

greater efficacy. Additionally, we are not aware of any anti-HIV drug candidates that are superior by themselves to the HAART cocktail.

While published studies of this model report no lethality in HIV-infected, SCID-Hu, Thy/Liv mice, others have found lethality similar to that observed by us (K. Menon, personal communication). Of note in our study, a reduction in lethality was associated with a reduction in thymus depletion. Also, no lethality was observed in Thy/Liv-implanted, but uninfected and untreated, SCID mice (group of six mice observed for 60 days), indicating that lethality was associated with HIV infection. One possibility is a preacquired infection that leads to death upon HIV superinfection in this model (K. Menon, personal communication). Finally, both the biochemical and clinical parameters in the NNVC 4 and 5 treatment groups were comparable to those of the positive control, the HAART treatment group, providing credence to the study results.

The ligands we have designed in the case of HIV-1 are thought to be broadly neutralizing. In silico modeling indicates that our ligands dock to the CD4 binding site of gp120 from the HIV-1 JRFL (HAART-resistant) strain. Further study is needed to determine the binding interactions in silico as well as in vitro.

If the results of this study can be extrapolated to humans, then it is likely that a clinically relevant dosing frequency such as once weekly i.v. injection of a relatively small quantity of an appropriate nanoviricide may be equivalent to the current HAART treatment protocol. Additionally, the putative mechanism of nanoviricides is orthogonal and complementary to the mechanisms of existing agents (other than antibodies and fusion inhibitors). Thus, the existing intracellular HIV replication inhibitors could be combined with nanoviricide therapy with potentially significant advantages. It is not unlikely that such a combination would result in a "functional cure" of HIV, wherein the clinical pathology of HIV is eliminated, other than the reservoirs in memory cells, allowing the patient to lead a normal life, and possibly also minimizing the spread of the disease. It is interesting to note here that the intracellular viral count was significantly diminished by nanoviricide treatments. This is understandable in light of the increased $CD4^+/CD8^+$ double-positive counts. The newly formed cells may have been protected from infection by HIV particles in the presence of nanoviricide treatment. HIV-1 is capable of infecting cells by cell-to-cell direct-spread mechanisms in cell cultures. The significance of this pathway in vivo is not known. However, in this model, such cell-to-cell spread could account for the small amount of virus particles still found inside the human lymphoid cells in nanoviricide-treated animals.

7.5 Additional Properties

Preliminary safety studies, including organ histology, blood pathology, and clinical examination, have shown the various nanoviricides to be safe at the levels tested, which have been both at and above levels shown to be therapeutic. Some of the chemical components of the nanoviricides are currently incorporated into some biological therapeutic agents in order to prolong circulating half-life and reduce immunogenicity. Our nanoviricides exploit this property and should allow a half-life sufficient to enable optimal therapeutic dosing intervals. Our preclinical pharmacodynamic studies indicate that we can achieve efficacy with a medically acceptable dosing frequency.

The nanoviricide technology can be employed to produce distinctly different classes of nanoviricide antiviral therapeutic agents, as described below:

1. Fixed antiviral therapeutics. To make a "fixed" nanoviricide, one would simply develop a specific ligand to which a virus binds and attach the ligand to the backbone.

2. Broad-spectrum therapeutics. If the selected ligand binds to not one but many virus strains, types, or classes, one would obtain a broad-spectrum nanoviricide.

3. Accurate Drug in Field (ADIF™) platform therapeutics. A backbone nanoviricide material can be made with a chemically reactive functional group preattached along the polymer chain. We have developed such groups that enable storage at room temperature, possibly for several years. When a novel, emerging, or advanced pathogen occurs, purified antibodies can be fragmented and the isolated fragments, devoid of the Fc fragment, can simply be mixed in with the backbone nanoviricide material, resulting in a facile reaction leading to covalent attachment of the fragments to the nanoviricide backbone material, to make an accurate, effective, nanoviricide drug that is specific to the threat. Thus, this is an ideal platform for novel, emerging, or advanced agent biothreat scenarios.

As mentioned previously, the development of virulent, readily transmissible, drug-resistant mutants continues to be a significant challenge with antiviral therapeutics and has led to the use of cocktail therapies with their associated side effect and drug interaction concerns. The polyvalent nature of the nanoviricide technology addresses the development of drug-resistant mutants. Polyvalency has been shown to improve the potency of monomeric inhibitors 100- to 1,000-fold.[19–21] In a nanoviricide, this could result in considerable tolerance for mutants that arise with decreased binding affinity in the viral monomeric ligand binding site. Moreover, if multiple ligands that bind to different sites on the virus coat are attached to the same nanoviricide, this can produce synergistic effects, leading to even higher potency than the parent ligands, with the polyvalent multiple different ligands likely precluding the development of mutants resistant to the nanoviricide.

7.6 Summary

NanoViricides, Inc. believes that by employing its technology of proprietary self-assembling amphiphilic polymers that form nanoscopic, targeted polymeric micelles, it has developed antiviral therapeutics that can have a clinically significant impact on the treatment of important viral diseases. We have developed virus-specific, yet broadly neutralizing, nanoviricides, using ligands specific to certain viruses, such as HIV, influenza, and EKC, among others, as well as broad-spectrum nanoviricides that are highly effective against many distinctly different types of viruses, using ligands that mimic common virus cell binding features.

References

1. Von Herrath, M.G., D.P. Berger, D. Homann, T. Tishon, A. Sette, and M.B. Oldstone. Vaccination to treat persistent viral infection. *Virology* 268:411, 2000.
2. Lindenburg, C.E., I. Stolte, M.W. Langendam, F. Miedema, I.G. Williams, R. Colebunders, J.N. Weber, M. Fisher, and R.A. Coutinho. Long-term follow-up: no effect of therapeutic vaccination with HIV-1 p17/p24:Ty virus-like particles on HIV-1 disease progression. *Vaccine* 20:2343, 2002.
3. Dikici, B., A.G. Kalayci, F. Ozgenc, M. Bosnak, M. Davutoglu, A. Ece, T. Ozkan, T. Ozeke, R.V. Yagci, and K. Haspolat. Therapeutic vaccination in the immunotolerant phase of children with chronic hepatitis B infection. *Pediatric Infectious Diseases Journal* 22:345, 2003.
4. Nevens, F., T. Roskams, H. Van Vlierberghe, Y. Horsmans, D. Sprengers, A. Elewaut, V. Desmet, G. Leroux-Roels, E. Quinaux, E. Depla, et al. A pilot study of therapeutic vaccination with envelope protein E1 in 35 patients with chronic hepatitis C. *Hepatology* 38:1289, 2003.
5. Rappuoli, R. Bridging the knowledge gaps in vaccine design. *Nature Biotechnology* 25:1361, 2007.
6. Boni, M.F. Vaccination and antigenic drift in influenza. *Vaccine* 26 (Suppl 3):C8, 2008.
7. Treanor, J. Influenza vaccine—outmaneuvering antigenic shift and drift. *New England Journal of Medicine* 350:218, 2004.
8. Fox, J.L. Antivirals become a broader enterprise. *Nature Biotechnology* 25:1395, 2007.
9. Pauwels, R. Aspects of successful drug discovery and development. *Antiviral Research* 71:77, 2006.
10. De Clerq, E. Three decades of antiviral drugs. *Nature Reviews Drug Discovery* 6:941, 2007.
11. Harper, S.A., J.S. Bradley, J.A. Englund, T.M. File, S. Gravenstein, F.G. Hayden, A.J. McGeer, K.M. Neuzil, A.T. Pavia, M.L. Tapper, T.M. Uyeki, and R.K. Zimmerman. Seasonal influenza in adults and children—diagnosis, treatment, chemoprophylaxis, and institutional outbreak management: clinical practice guidelines of the Infectious Diseases Society of America. *Clinical Infectious Diseases* 48:1003, 2009.
12. WHO Guidelines on the use of vaccines and antivirals during influenza pandemics. WHO/CDS/CSR/RMD/2004.8.
13. Broder, S. The development of antiretroviral therapy and its impact on the HIV-1/AIDS pandemic. *Antiviral Research* 85:1, 2010.
14. Hayden, F.G. Antivirals for influenza: historical perspective and lessons learned. *Antiviral Research* 71:372, 2006.
15. De Clerq, E. The design of drugs for HIV and HCV. *Nature Reviews Drug Discovery* 6:1001, 2007.
16. Kwon, G.S. Polymeric micelles for delivery of poorly water-soluble compounds. *Critical Reviews in Therapeutic Drug Carrier Systems* 20:357, 2003.
17. Aliabadi, H.M., and A. Lavasanifar. Polymeric micelles for drug delivery. *Expert Opinion in Drug Delivery* 3:139, 2006.
18. Uchegbu, I.F. Pharmaceutical nanotechnology: polymeric vesicles for drug and gene delivery. *Expert Opinion in Drug Delivery* 3:629, 2006.
19. Johansson, S.M.C., N. Arnberg, M. Elofsson, G. Wadell, and J. Kihlberg. Multivalent HSA conjugates of 3'-sialyllactose are potent inhibitors of adenoviral cell attachment and infection. *Chem Bio Chem* 6:358, 2005.

20. Johansson, S.M.C., E.C. Nilsson, M. Elofsson, N.Ahlskog, J. Kihlberg, and N. Arnberg. Multivalent sialic acid conjugates inhibit adenovirus type 37 from binding to and infecting human corneal epithelial cells. *Antiviral Research* 73:92, 2007.

21. Honda, T., S. Yoshida, M. Arai, T. Masuda, and M. Yamashita. Synthesis and anti-influenza evaluation of polyvalent sialidase inhibitors bearing 4-guanidino-Neu5Ac2en derivatives. *Bioorganic and Medicinal Chemistry Letters* 12:1929, 2002.

22. Ravindrath, R.M.H., and M.C. Graves. Attenuated murine cytomegalovirus binds to N-acetylglucosamine, and shift to virulence may involve recognition of sialic acids. *Journal of Virology* 64:5430, 1990.

23. Gordon, J.S. Adenoviruses and other nonherpetic viral diseases. In *The cornea*, ed. G. Smolin and R.A. Thoft., 3rd ed. Boston: Little, Brown and Co., 1994, pp. 215–227.

24. Kanski, J.J. *Clinical ophthalmology*. 3rd ed. Boston: Butterworth- Heinemann, 1998, p. 86.

25. Lenaerts, L., and L. Naesens. Antiviral therapy for adenovirus infections. *Antiviral Research* 71:172, 2006.

26. Lenaerts, L., E. De Clercq, and L. Naesens. Clinical features and treatment of adenovirus infections. *Reviews in Medical Virology* 18:357, 2008.

27. Gordon, Y.J., E. Romanowski, and T. Araullo-Cruz. An ocular model of adenovirus type 5 infection in the NZ rabbit. *Investigative Ophthalmology and Visual Science* 33:574, 1992.

28. Vermund, S.H. Millions of life-years saved with potent antiretroviral drugs in the United States: a celebration, with challenges. *Journal of Infectious Disease* 194:1, 2006.

29. Richman, D.D., D.M. Margolis, M. Delaney, W.C. Greene, D. Hazuda, and R.J. Pomerantz. The challenge of finding a cure for HIV infection. *Science* 323:1304, 2009.

30. Menéndez-Arias, L. Molecular basis of human immunodeficiency virus resistance: an update. *Antiviral Research* 85:210, 2010.

31. Estéa, J.A., and T. Cihlarb. Current status and challenges of antiretroviral research and therapy. *Antiviral Research* 85:25, 2010.

32. Sungkanuparph, S., R.K. Groger, E.T. Overton, V.J. Fraser, and W.G. Powderly. Persistent low-level viraemia and virological failure in HIV-1-infected patients treated with highly active antiretroviral therapy. *HIV Medicine* 7:437, 2006.

33. McCune, J.M., R. Namikawa, H. Kaneshima, L.D. Shultz, M. Lieberman, et al. The SCID-hu mouse: murine model for the analysis of human hematolymphoid differentiation and function. *Science* 241:1632, 1988.

34. Namikawa, R., H. Kaneshima, M. Lieberman, I.L. Weissman, and J.M. McCune. Infection of the SCID-hu mouse by HIV-1. *Science* 242:1684, 1988.

35. Stoddart, C.A., C.A. Bales, J.C. Bare, G. Chkhenkeli, S.A. Galkina, A.N. Kinkade, M.E. Moreno, et al. Validation of the SCID-hu Thy/Liv mouse model with four classes of licensed antiretrovirals. *PLoS ONE* 2:655, 2007.

8

Antibody-Conjugated Nanoparticles of Biodegradable Polymers for Targeted Drug Delivery

Bingfeng Sun, Heni Rachmawati, and Si-Shen Feng

Division of Bioengineering, National University of Singapore, Singapore

CONTENTS

8.1 Nanotechnology

Nanotechnology is a multidisciplinary field and can be defined as a science and engineering discipline that involves the design, synthesis, characterization, and application of materials and devices whose smallest functional organization in at least one dimension is on the nanometer scale.[1] Nanotechnology opens the door to a new generation of techniques and devices for cancer diagnosis, treatment, and prevention. For example, intracellular imaging can be achieved with the aid of nanotechnology, in which multicolor fluorescence imaging is realized by attaching imaging agents such as semiconductor quantum dots (QDs) to selected molecules, or coating the QDs with a layer of amphiphilic molecules. Moreover, nanoparticles with a drug encapsulated in a polymer matrix can realize a controlled and sustained manner for drug delivery. Controlling the localization of nanoparticles to the diseased cells by modification of the nanoparticle surface with targeting moieties is now gaining great interest in the field of nanomedicine. Not only can anticancer drugs be incorporated in the nanoparticles to kill the cancer cells, but diagnostic agents as well, which can be utilized to image the presence of cancer cells. The latter will be powerful in detecting the development of cancer at very early stages. This, in turn, promises better and more effective treatment for cancer diseases than ever before.

8.2 Nanoparticles

Recently, biodegradable nanoparticles have attracted great interest due to their advantages over conventional therapeutic strategies. Types of such nanoparticles include polymeric micelles, liposomes, and polymer-based nanoparticles. Nanoparticles of biodegradable polymers can be used as therapeutics containing small-molecule drugs, peptides, proteins, and nucleic acids. Compared to conventional therapeutic strategies, nanoparticles can improve the solubility of poorly soluble drugs and increase the drug half-life, as well as the specificity to the target sites, e.g., tumors. Most nanoparticles preferentially accumulate within tumors via the enhanced permeability and retention (EPR) effect.[2] Thus, polymeric nanoparticles allow for enhancing the intracellular drug concentration in cancer cells while avoiding toxicity in normal cells, resulting in potent therapeutic effects.

There are a variety of nanoparticle systems being developed for cancer diagnosis and therapeutics. The material properties of each nanoparticle system have been developed to enhance delivery to the tumor. For example, hydrophilic surfaces can be used to provide the nanoparticles with stealth properties for longer circulation times, and positively charged surfaces can enhance endocytosis. Nanoparticles can be tailor-made to achieve both controlled drug release and disease-specific localization by altering the polymer characteristics and surface chemistry.[3–5]

8.2.1 Nanoparticle Properties and Emulsifiers

Physicochemical properties of nanoparticles would determine in vitro and in vivo performances of the nanoparticles. These properties include particle size and size distribution, surface morphology, surface chemistry, surface charge, surface adhesion/erosion, drug diffusivity, drug encapsulation efficiency, drug stability, drug release kinetics, etc. Several factors influencing an optimal design of nanoparticles include the polymer type as well as the molecular weight, the copolymer blend ratio, the type of organic solvent, the emulsifier/stabilizer, the oil-to-water phase ratio, the mechanical strength of mixing, temperature, pH, etc. Among these, the polymer type, that is, its molecular structure and molecular weight, and the copolymer blend ratio are the major factors in determining the degradation rate of the nanoparticles, and thus the in vitro and in vivo release of the encapsulated drug.

Emulsifiers, the macromolecules that are added in the emulsification process for nanoparticle preparation, play an important role in stabilizing the colloidal suspension during nanoparticle fabrication. The amphiphilic emulsifier molecules are supposed to stay at the oil-water interface to decrease the interfacial tension, that is, the nanoparticle surface energy per unit area, to facilitate the nanoparticle formation.[6] Poly(vinyl alcohol) (PVA) has been preferentially chosen as an emulsifier in nanoparticle fabrication due to its excellent stabilizing ability to avoid particle aggregation during postpreparative steps (e.g., freeze-drying and purifying), high yield of dry particle powder, and ease to be redispersed in solution after lyophilized.[7] It has also been found, however, that PVA may be adsorbed or tightly associated with the surface layer, and thus cannot be completely removed from the surface of nanoparticles.[8–10] Therefore, PVA should not be used as an emulsifier in the preparation of nanoparticle formulation for intravenous (i.v.) administration, although its safety has been revised.[11] Instead, phospholipids such as dipalmitoyl phosphatidylcholine (DPPC) and TPGS were found to be more effective emulsifiers in nanoparticle fabrication.[12,13] Phospholipids with short and saturated chains showed higher efficiency in emulsion of polymeric micro/nanoparticles of better physical/chemical properties. There is no significant difference in morphology between TPGS- and PVA-emulsified PLGA nanoparticles. However, TPGS has been found to be able to improve the drug encapsulation efficiency up to 100%, compared with PVA-emulsified nanoparticles, at only 59%. Moreover, the amount of TPGS needed in the fabrication process was only 0.015% (w/w), which was far less than the 1% PVA needed in a similar process.[14,15]

8.2.2 Nanoparticle Fabrication

Nanoparticles can be fabricated by various methods, such as polymerization or dispersion of the preformed polymers, solvent extraction/evaporation method, salting-out method, dialysis method, supercritical fluid spray technique, and nanoprecipitation method. Solvent extraction/evaporation is used in current research due to its acceptable drug loading efficiency, ease of processing, and good reproducibility. In this review, we highlight the most useful methods of nanoparticle fabrication: solvent extraction/evaporation, dialysis, and nanoprecipitation.

8.2.2.1 Solvent Extraction/Evaporation Method

The solvent extraction/evaporation method is the most widely used technique for nanoparticle fabrication. In this technique, a selected polymer is first dissolved in an organic solvent such as dichloromethane, chloroform, or ethyl acetate. The hydrophobic drug is then dissolved in the polymer solution. The solution formed is dispersed in an aqueous phase with or without surfactant/stabilizer, such as PVA, gelatin, poloxamer 188, DPPC, TPGS, etc. The mixture is emulsified by either high-speed homogenization or high-voltage sonicator, leading to the formation of an oil-in-water emulsion. After a stable emulsion is formed, the organic solvent is evaporated under increased temperature, reduced pressure, or continuous stirring at room temperature. Further processes, including multiple centrifugation/washing and lyophilization, will finally result in dried nanoparticles.[12,13] Figure 8.1(A) shows the field emission scanning electron microscopy (FESEM) image of PLGA nanoparticles prepared by solvent extraction/evaporation method. The nanoparticles are in spherical shape and in the nanosized range. The pharmaceutical characteristics of the nanoparticles can be influenced by many factors of the fabrication process, such as the polymer concentration in the solvent, ratio of the organic to the aqueous phase, type and concentration of emulsifiers, drug loading ratio, strength of the mixing energy in emulsifying and evaporation, and post-treatment of nanoparticles, including centrifugation, washing, lyophilization, sterilization, pH condition, and temperature.[15]

8.2.2.2 Dialysis Method

Dialysis is another self-assembling method for nanoparticle fabrication, in which nanoparticles can be fabricated with or without surfactant/additive/stabilizer. The drug and polymer are dissolved in a water-miscible organic solvent. The solution is then transferred into a cellulose membrane bag, which is then immersed in a container filled with water for 1 or 2 days. The water is exchanged at a certain interval to maintain the osmotic pressure, which removes the solvent and unloaded drug from the membrane bag. The nanoparticle dispersion is then centrifuged to further eliminate the unloaded drug. In dialysis, the molecular weight cutoff (MWCO) of the membrane is one of the additional controlling factors that may influence the suitability of dialysis for some polymers.[16] In addition, the porosity of the membrane controls the rate at which the organic phase is dialyzed, thus affecting the size of nanoparticles. The main limitations of the dialysis method are that it is time-consuming and a large-volume tank is required. Therefore, large-scale operation may not be feasible.

8.2.2.3 Nanoprecipitation Method

Nanoprecipitation is a nanoparticle synthesis process involving solvent displacement followed by interfacial deposition of preformed polymer. This technique was developed by Fessi et al.[17] The oily phase commonly used is water-soluble organic solvents such as acetone, acetonitrile, and dimethylformamide. The polymer and drug are dissolved in the organic phase. When the organic phase is mixed dropwise with the aqueous phase under gentle magnetic stirring, an emulsification is formed spontaneously. The solvent is evaporated overnight with gently stirring. The particles are collected by filtration to remove the aggregation. The filtrate is then centrifuged and lyophilized. Figure 8.1(B) shows the FESEM image of PLGA nanoparticles prepared by the nanoprecipitiation method, which are more uniform in size than those shown in Figure 8.1(A).

8.2.3 Nanoparticles for Targeted Drug Delivery

Nanoparticle-based drug delivery systems have considerable potential for treatment of various diseases, and are especially attractive for difficult-to-treat diseases like cancer.

Targeted therapy is defined as a type of treatment that uses drugs or other substances to identify and attack specific target cells, e.g., cancer cells (http://www.cancer.gov). Targeted therapy may be more effective and have fewer side effects than general treatments without

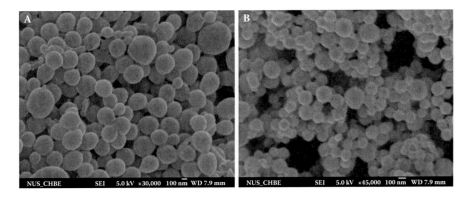

FIGURE 8.1
Field emission scanning electron microscope (FESEM) images of PLGA nanoparticles fabricated by (A) the solvent extraction/evaporation method and (B) the nanoprecipitation method.

targeting effects. Current cancer therapy usually involves intrusive processes to allow chemotherapy to shrink any cancer present or surgery to remove the tumor, if possible. For the majority of patients with an advanced stage of cancer, treatment is limited to chemotherapy or radiotherapy. The purpose of chemotherapy and radiotherapy is to kill the cancer cells, as these cells are more susceptible to the actions of these drugs and therapies because they grow at a much faster rate than healthy cells. The effectiveness of the treatment is directly related to the treatment's ability to target and kill the cancer cells while affecting as few healthy cells as possible. Moreover, the degree of change in the patient's quality of life and eventual life expectancy is directly related to the targeting ability of the treatment.[18,19]

Though there have been several decades of research and development, targeted delivery has not yet fulfilled the initial promise in the treatment of cancer. The targeted delivery of anticancer drugs to solid tumors is a complicated issue because of the impediments to drug delivery.[20,21] The major impediments to drug delivery arise from tumor heterogeneity. Cancer cells typically occupy less than half of the total tumor volume, among which approximately 1 ~ 10% are contributed by tumor vasculature, and the rest are occupied by a collagen-rich interstitium. The heterogeneous distribution of blood vessels, antigen, and receptor expression in tumors is also a problem in affinity-targeted delivery of drugs to solid tumors.[22,23]

The development of a wide spectrum of nanoscale technologies has tremendous potential to make an important contribution in cancer imaging, diagnosis, treatment, and prevention.[24,25] These technological innovations have the potential to turn molecular discoveries arising from genomics and proteomics into widespread benefit for patients. These devices include, but are not limited to, functionalized carbon nanotubes, nanomachines, magnifiers, self-assembling polymeric nanoconstructs, nanomembranes, and nano-sized silicon chips for drug, protein, nucleic acid, or peptide delivery and biosensors and laboratory diagnostics. In terms of biodegradable polymeric nanoparticles, targeting ability can be achieved by functionalizing the surface of drug-loaded nanoparticles with targeting ligands, which can then recognize and bind to certain receptors overexpressed on the membrane of the cancer cells.

Targeted cancer therapies interfere with cancer cell growth and division in different ways and at various points during the development, growth, and spread of cancer. Many of these therapies focus on proteins that are involved in the signaling process. Targeted molecular therapies target tumor antigens to alter the signaling either by monoclonal antibodies (mAbs) or by small molecule drugs that interfere with these target proteins. By blocking the signals that tell cancer cells to grow and divide uncontrollably, the approaches can help to stop the growth and division of cancer cells. This new type of approach has been termed targeted therapy, the goal of which is to provide molecular level-based agents that are more specific for cancer cells.[26] Targeted therapies are also often useful in combination with cytotoxic chemotherapy or radiotherapy to produce additive or synergistic anticancer activity because their toxicity profiles often do not overlap with traditional cytotoxic chemotherapy. Thus, targeted therapies represent a new and promising approach to cancer therapy leading to beneficial clinical effects. There are multiple types of targeted therapies available, including monoclonal antibodies, tyrosine kinase inhibitors, and antisense inhibitors of growth factor receptors.

8.2.3.1 Passive Targeting

Targeted delivery can be achieved in two manners: passive and active. Passive targeting makes use of the tumor microenvironment, which is characterized by a leaky tumor vasculature and a dysfunctional lymphatic drainage system. Most polymeric nanoparticles display the EPR effect, which is a consequence of the increased vasculature permeability and decreased lymphatic function of tumors. It occurs when nanoparticles extravasate out of the tumor microvasculature, leading to an accumulation of drugs in the tumor interstitium.[27] Thus, passive targeting is achieved by incorporating the therapeutic agent into a macromolecule or nanoparticle that passively reaches the target organ through the EPR effect.

Passive targeting also involves the use of other innate characteristics of the nanoparticles that can induce targeting to the tumor, such as surface charge. For example, cationic liposomes have been found to exhibit a tendency to bind through electrostatic interactions to negatively charged phospholipid head groups preferentially expressed on tumor endothelial cells.[28,29] Catheters can be alternatively used to infuse nanoparticles to the target organ or tissues. For example, localized delivery of drug-loaded nanoparticles to sites of vascular restenosis may be helpful for providing sustained drug release at specific sites on the arterial wall.[30]

For passive targeting to be successful, the nanoparticles with chemotherapeutic agents encapsulated have to circulate in the blood for an extended time so that there will be multiple possibilities for the nanoparticles to pass by the target sites. Nanoparticles normally have short-circulation half-lives due to natural defense mechanisms of the body to eliminate them after opsonization by the mononuclear phagocytic system (MPS).[31] Therefore, the particle surfaces need to be modified to be invisible to opsonization. A number of studies applied various surface modification approaches to classical nanocarriers to increase their circulation half-lives for effective passive targeting or sustained drug effect. These approaches include incorporation of linear dextrans, sialic acid-containing gangliosides, and lipid derivatives of hydrophilic polymers such as polyethylene glycol (PEG), to provide steric stabilization around the liposomes for protection from the MPS uptake.[32,33] For example, PEG, a hydrophilic polymer, is commonly used to lengthen the circulation time. This is because PEG has desirable attributes, such as a low degree of immunogenicity and antigenicity, chemical inertness of the polymer backbone, and availability of the terminal primary hydroxyl groups for derivatization.[34]

Passive targeting results in high drug concentrations in tumors and reduced drug toxicities to the normal tissues. However, the biggest limitation of passive targeting lies in its inability to deliver a sufficiently high level of drug concentration to the tumor site.[35] Because this technology lacks tumor specificity and has less ability to control the release of the entrapped agents, the focus has gradually shifted from passive to active targeting nanoparticles.

8.2.3.2 Active Targeting

In comparison with passive targeting, exploiting the characteristic features of tumor biology that allow nanocarriers to accumulate in the tumor by the EPR effect, active targeting is achieved by delivering drug-encapsulated nanoparticles to uniquely identified sites while having minimal side effects.[36] There are still several limitations in passive targeting approaches, although they form the basis of clinical therapy. Certain tumors do not exhibit the EPR effect, and the permeability of vessels may not be the same through a single tumor.[22] It is difficult to control the passive targeting process due to the inefficient diffusion of some drugs as well as the random nature of the approach. Thus, multiple-drug resistance (MDR) might occur, which is a situation where chemotherapy treatments fail patients owing to the resistance of cancer cells to one or multiple different drugs. MDR happens when transporter proteins that expel drugs from cells are overexpressed on the surface of cancer cells. This drug resistance, at its worst when the toxic side effects are still in full force, will kill healthy cells in the body but has no harmful effects on the cancer cells.[37]

Active targeting attempts to take advantage of overexpressed tumor-associated antigens or receptors to selectively target the drug to the tumor. In general, active targeting is achieved through the administration of nanoparticles with cell-specific ligands conjugated on their surface. These ligands can recognize and then bind to specific receptors that are uniquely expressed on cancer cells. In the case of local drug delivery, the cytotoxic drug encapsulated in the nanoparticles can be delivered directly to cancer cells while minimizing harmful toxicity to healthy cells adjacent to the target tissue.[38]

8.3 Antibody-Conjugated Nanoparticles for Targeting

8.3.1 Functionalization of Nanoparticles for Targeting

As described above, tyrosine kinase inhibitors, monoclonal antibodies, and antisense inhibitors of growth factor receptors can function as a cell-specific homing device.

8.3.1.1 Tyrosine Kinase Inhibitors

Protein tyrosine kinases (PTKs) are enzymes that catalyze the phosphorylation of tyrosine residues and are especially important targets, as they play an important role in the modulation of growth factor signaling and oncogenic transformation of cells.[39] There are two main classes of PTKs: receptor PTKs and nonreceptor PTKs. Receptor tyrosine kinases are multidomain proteins. Several approaches to target PTKs have been developed, and classification of such inhibitors is based on the mode of action. It has been found that activation of protein phosphorylation-related pathways in tumors can occur through overexpression of this protein, compared to normal cells.[40] Therefore, the targeted therapeutics ascribed to therapeutic agents are as close to being monospecific as possible to avoid the harmful side effects that sometimes arise with traditional cancer therapies. For this reason, small molecule inhibitors of protein kinases have emerged as essential for studying targeted therapy, and these drugs are also called signal transduction inhibitors.[41]

Among the tyrosine kinases, the epidermal growth factor receptor (EGFR) family is the most widely investigated. EGFR is the cell surface receptor for members of the EGFR family of extracellular protein ligands. EGFR is a member of the ErbB family of receptors, a subfamily of four closely related receptor tyrosine kinases: EGFR (ErbB-1), HER2/c-neu (ErbB-2), HER3 (ErbB-3), and HER4 (ErbB-4). Mutations affecting EGFR expression or activity could result in cancer. EGFR exists on the cell surface and is activated by binding of its specific ligands, including epidermal growth factor and transforming growth factor α (TGFα). Overexpression of EGFR has been associated with a poor prognosis in a variety of solid tumors, and thus represents an attractive therapeutic target.[42]

8.3.1.2 Monoclonal Antibody

Monoclonal antibodies (mAbs) were first described in 1975 by Kohler and Milstein, who shared the Nobel Prize in Physiology or Medicine in 1984 for the discovery. Monoclonal antibodies target specific molecules and are used as passive immunotherapy to treat various diseases, including certain types of cancer. These mAbs selectively target tumor tissues and have been safely administered in cancer patients. It was in the 1980s that the development of blocking mAbs to EGFR as a cancer therapy was proposed by Mendelsohn for the first time.[43] Since then, various antibodies have become valuable therapeutic agents for targeting of extracellular proteins in various diseases, including cancer, autoimmunity, and cardiovascular disorders.

Development of mAbs of human origin has proven to be a hard task. The first generation of humanized mAbs was simply chimeric antibodies, in which the variable regions of murine mAbs were linked to the constant region of a human immunoglobulin (Ig) G.[44] As expected, these humanized molecules should lack the immunogenicity of a murine mAb and interact more efficiently with the human immune system.[45] The second generation of humanized mAbs was the reshaped antibodies, in which the antigen-binding loops of the murine mAbs were built into a human IgG. The main problem was related to the specificity. It is usually not enough to maintain the original binding specificity by combining the complementarity-determining regions (CDRs) from the murine mAb with the framework of a human IgG, since several specific residues in the original murine framework contribute to maintaining the CDR conformations. Molecular modeling techniques allowed this problem to be resolved, and thus reshaped antibodies have successfully been produced and used for clinical purposes.[46]

The differences between mAbs and small molecule drugs lie in several pharmacological properties. Antibodies are administered intravenously and act only on the receptors expressed on the cell surface.[47] Small molecule tyrosine kinase inhibitors are orally available small, membrane-permeable synthetic compounds that block or compete with adenosine triphosphate (ATP) binding, thus inhibiting the intracellular, downstream signaling cascade stimulated by a receptor or several receptors. The half-life of many tyrosine kinase inhibitors, such as gefitinib, is approximately 24–48 h, whereas the half-life of monoclonal antibodies such as bevacizumab is about 3–4 weeks. mAbs cannot cross the blood-brain barrier efficiently due to their large size, whereas current evidence suggests that small molecule drugs can successfully cross the blood-brain barrier.[48] Small molecules are generally less specific than therapeutic monoclonal antibodies and, therefore, a higher risk of toxicity potential comes along with small molecule drugs.[49]

8.3.2 Trastuzumab-Conjugated Nanoparticles

8.3.2.1 HER2-Targeted Therapy

The EGFR family is thought to play a primary role in the control of epithelial cell proliferation and mutations affecting EGFR expression or activities that could result in cancer.

The human epidermal growth factor receptor 2 (HER2) is a member of the EGFR family, which is a receptor tyrosine-specific protein kinase family consisting of four semihomologous receptors: EGFR (ErbB1), HER2/neu (ErbB2), HER3 (ErbB3), and HER4 (ErbB4). These receptors can interact with several ligands and generate intracellular signals as homodimer or heterodimer pairs.[50] HER2 is a cell membrane surface-bound receptor tyrosine kinase and is normally involved in the signal transduction pathways leading to cell growth, survival, and differentiation in a complex manner. HER2 levels correlate strongly with the pathogenesis and prognosis of breast cancer. The level of HER2 in human cancer cells with gene amplification is much higher than that in normal adult tissues, which potentially reduces the toxicity of HER2-targeting drugs. It is well known that HER2 is overexpressed in 25%–30% of invasive breast cancers, but in normal tissues its expression is at a much lower level.[51] HER2 overexpression is found in both the primary tumor and metastatic sites, indicating that anti-HER2 therapy may be effective in all disease sites.[52] Thus, HER2 is of great interest as an important therapeutic target in breast cancer.

8.3.2.2 Assessment of HER2 Status

The HER2 status of a tumor is the critical determinant of response to trastuzumab as it provides prognostic information. Thus, accurate assessment of HER2 expression levels is essential for identifying breast cancer patients who will benefit from trastuzumab.[53] There are several methods to test the HER2 levels, and testing can be done at the same time as initial breast cancer surgery, or samples of cancer cells from previous biopsies or surgery may be used. Two of the most widely used methods of measuring HER2 levels in the clinical setting are immunohistochemistry (IHC) and fluorescence in situ hybridization (FISH). Both methods have been approved by the FDA for selecting patients for trastuzumab-based therapy.

IHC is widely used, as it entails staining paraffin-embedded tissue with a HER2-specific antibody. When using commercially available kits such as HercepTest (Dako, Carpinteria, California) and Pathway HER2 (Ventana, Tucson, Arizona), staining is graded semiquantitatively on a scale from 0 (no detectable HER2) to 3+ (high HER2 expression) on the basis of comparison with cell lines of known HER2 receptor density. Tumors with a staining score of 3+ are the most responsive to trastuzumab.[54] The disadvantages of IHC include the subjective interpretation and semiquantitative nature of the results. Currently available IHC kits provide control slides against which samples are compared. Such standardization is essential to ensuring accurate assessment of HER2 status.

FISH detects HER2 gene amplification and is more specific and sensitive than IHC.[55] Importantly, FISH offers quantitative results, possibly eliminating subjectivity and variability among different laboratories. Furthermore, FISH predicts prognosis and response to trastuzumab more accurately than IHC does, as the subset of patients whose tumors overexpress HER2 in the absence of gene amplification is less likely to respond to trastuzumab-based therapy. In general, a concordance rate of approximately 80% can be achieved by IHC and FISH.[56]

8.3.2.3 Trastuzumab (Herceptin)

Trastuzumab (Herceptin) is a humanized monoclonal antibody that can specifically bind to the membrane region of HER2/neu with a high affinity and inhibit signal transduction as well as cell proliferation, which offers an excellent strategy for drug targeting due to the easy accessibility of HER2. Trastuzumab was the first HER2-targeted therapy approved by the FDA for the treatment of metastatic breast cancer (MBC), either in the first-line setting in combination with paclitaxel, or as a monotherapy for patients who had received at least one prior chemotherapy regimen.[57]

8.3.2.4 Mechanism of Action of Trastuzumab

Trastuzumab consists of two antigen-specific sites binding to the HER2 receptor, which prevent the activation of its intracellular tyrosine kinase. The mechanism of action of trastuzumab has not been fully characterized and appears to be complex. Trastuzumab exerts its antitumor therapeutic effects against tumor cells with HER2 overexpressed by several possible mechanisms.

Herceptin reduces signaling from pathways that are activated by HER2, and thus promotes cell cycle arrest and apoptosis. These pathways include the PI3 kinase (PI3K) and MAP kinase (MAPK) cascades. Research shows that HER2 remains at the same level after trastuzumab-based treatment; therefore, it is still unclear whether trastuzumab downregulates HER2.[58] Trastuzumab therapy also increases membrane localization and activity of PTEN. Nagata et al. demonstrated that the interaction between HER2 and the Src tyrosine kinase is disrupted in response to trastuzumab treatment, leading to inactivation of Src with subsequent activation of the PI3K inhibitor phosphatase and tensin homolog (PTEN) the protein product of the phosphatase and tensin homolog deleted on the chromosome 10 gene.[59] Thus, Herceptin activates the PTEN phosphatase, resulting in rapid Akt dephosphorylation, and inhibits cell proliferation.

Trastuzumab possesses not only targeting ability but also therapeutic effect, compared to many targeted agents. The cytotoxic property enables trastuzumab to block proliferation and promote cell death, which may be related in part to induction of an immune response. Trastuzumab activated an antibody-dependent cellular cytotoxicity (ADCC) response in multiple breast cancer cell lines.[60,61] Natural killer (NK) cells, expressing the Fc gamma receptor, are a principal immune cell type involved in ADCC. NK-mediated cell lysis is activated after trastuzumab binds to the Fc gamma receptor. Trastuzumab recruits immune effective cells that are responsible for antibody-dependent cytotoxicity, which has been put forward in preclinical models. Results showed that mice bearing BT474 HER-2-overexpressing xenografts demonstrated a tumor regression rate of 96% when treated with Herceptin.[62] However, mice lacking the Fc receptor lost much of the protective effect of Herceptin, with only 29% tumor growth inhibition observed. Therefore, NK cells and ADCC are important contributors to the cytotoxic activity of Herceptin. However, it should be noted that patients with advanced metastatic breast cancer are immunosuppressed and may not be the optimal population to study. Additional studies are needed to better understand the importance of ADCC in mediating the response to Herceptin.

Overexpression of HER2 in human cancer cells is associated with increased angiogenesis.[63] Trastuzumab might also play a role as an antiangiogenic agent, as it has been shown to induce normalization and regression of the vasculature in a human breast tumor model which is HER2 positive.[64] Trastuzumab suppresses angiogenesis by both induction of antiangiogenic factors and repression of proangiogenic factors. Expression of multiple proangiogenic factors was reduced, while expression of antiangiogenic factors was increased in tumors with therapy of trastuzumab, compared with control-treated tumors in vivo.[65] Moreover, the combined use of trastuzumab and paclitaxel more effectively inhibited HER2-mediated angiogenesis than either treatment alone, which reflects more pronounced antitumor effects.[64] Trastuzumab may also prevent ligation of HER2 with its ligand, by which apoptosis may be induced.[66]

8.3.2.5 Clinical Efficacy of Trastuzumab

The therapeutic efficacy and tolerability of trastuzumab have been investigated in various studies. The initial clinical trials investigating the safety of trastuzumab were performed on women with metastatic breast tumors overexpressing HER2. Results from phase I and II trials in patients demonstrated that Herceptin has an acceptable tolerability profile and promising clinical efficacy in patients.[67] A response rate of 15–35% was reported for trastuzumab monotherapy in MBC, showing that Herceptin has significant biostatic activity as a single agent.[57] Cobleigh et al. reported a study in which the benefit of Herceptin as a single agent for MBC was evaluated.[68] In the study, 222 women were enrolled, of whom all were extensively pretreated, and a quarter had received more than two prior therapies and two-thirds had received paclitaxel. Two hundred and thirteen patients received Herceptin, and at 11 months the overall response rate was found to be 15% by an independent response evaluation committee. The median response duration was 8.4 months, with an estimated median survival of 13 months. Treatment was well tolerated, with only two patients discontinuing therapy due to toxicity. In addition, the side effects attributable to Herceptin include fever, chills, pain, asthenia, nausea, vomiting, and cardiac dysfunction.

Herceptin has been approved for first-line use in combination with chemotherapy. It has been found that Herceptin helps to increase the clinical benefits, such as response rate, time to disease progression, and overall survival of first-line chemotherapies, such as paclitaxel, docetaxel, doxorubicin, and cisplatin in patients with HER2-overexpressed MBC.[54,69,70] Several studies have examined different trastuzumab-taxane combinations. For example, higher response rates have been reported for docetaxel, which yielded a 73% response rate in three phase II trials.[71] An increased response rate of 50%–80% was reported when trastuzumab was combined with single-agent cytotoxic chemotherapy in first-line MBC.[72] Several studies have evaluated the role of Herceptin as a preoperative therapy in patients with early-stage breast cancer. The greatest effect in vitro and in vivo was seen with the combination of paclitaxel and Herceptin.[73] The approved combination therapy indication for trastuzumab is with 3-weekly paclitaxel, and after that, both weekly paclitaxel and docetaxel regimens are widely used. A phase II study of preoperative Herceptin in combination with paclitaxel was reported by Burstein et al., and a pathologic complete response rate of 18% was found.[74] Seidman et al. reported that women with MBC of HER2 positive or -negative responded differently to Herceptin.[75] In the study, patients were treated with standard-dose trastuzumab plus weekly paclitaxel (90 mg/m^2). As expected, toxic effects were typical of single-agent paclitaxel, and cardiac function was preserved for at least a year. The response rate was 81% in patients who were HER2 positive by IHC evaluation and 43% in those who were HER2 negative.

The success with Herceptin in the treatment of metastatic breast cancer in combination with chemotherapy has inspired intensive studies on developing effective targeted drug delivery systems. The combination of trastuzumab and chemotherapeutic agents such as paclitaxel and docetaxel for the treatment of HER2-overexpressed cancer has been suggested as a promising means of targeted chemotherapy.

8.3.3 Trastuzumab-Conjugated Nanoparticles

Nanoparticles of biodegradable polymers as a drug delivery system have aroused continuous interest in recent years. Drug-loaded nanoparticles have considerable potential to provide an ideal solution for the major problems encountered in chemotherapy. It has been established that nanoparticles can become concentrated preferentially in tumors by virtue of the EPR effect of the vasculature. Once accumulated at the target site, biodegradable polymeric nanoparticles can act as a local drug deposit and provide a source for a continuous supply of encapsulated therapeutic agent. The advantages of nanoparticle-based drug delivery systems include sustained and controlled drug release, improved bioavailability, reduced systemic side effects, and high capability to cross various physiological barriers.[29,76] Nevertheless, the low selectivity of NPs toward cancer cells hinders the advantages of the nanoparticle formulation for efficient chemotherapy, as the therapeutic agents also possess high toxicity to healthy cells. Therefore, it is necessary to develop effective therapies with a specific effect on cancer cells.

Effective drug targeting can be realized by the functionalization of nanoparticles with small molecule ligands such as folate and thiamine, peptides such as arginine-glycine-aspartic acid (RGD), sugar residues such as galactose, antibodies and antibody fragments such as anti-HER2, nucleic acid aptamers such as anti-PSMA aptamer, and proteins such as transferrin.[76–83] Among various ligands, trastuzumb, the monoclonal antibody directed against HER2, is of great interest because it has synergistic therapeutic effects with chemotherapy, in addition to its targeting ability.[84]

Synergistic antitumor effects were found when trastuzumab was administered in combination with chemotherapeutic agents such as paclitaxel, docetaxel, and doxorubicin.[74,85] Two different targeting approaches have been reported by Nobs et al. for immunotargeting with trastuzumab conjugated to nanoparticles.[86] One is a direct method using trastuzumab-labeled NPs, and the other is a pretargeting method using the avidin-biotin technology. Specific trastuzumab-labeled NPs binding to tumor cells produced a mean 10-fold or higher signal increase over the control. The two-step method was evaluated in vitro by incubating SKOV-3 cells first with biotinylated mAbs, followed by NPs. The relative fluorescence associated with the specific binding of NPs produced a sixfold increase in the flow cytometry signal, compared to nonspecific binding, which suggested that trastuzumab-functionalized NPs might be a useful drug carrier for tumor targeting.

Steinhauser et al. reported a time-dependent cell uptake study, in which a significant enrichment of trastuzumab-modified nanoparticles in 64.23% of HER2 positive SK-BR-3 cells was achieved, compared to 3.59% of the nude NPs with 30 min incubation. However, after 24 h incubation the nonspecific uptake of PEGylated nanoparticles without trastuzumab increased dramatically (39.62%) compared to trastuzumab-modified nanoparticles (51.22%).[87] Research on doxorubicin-loaded, trastuzumab-modified human serum albumin nanoparticles was reported by Anhorn et al., which was the first demonstration of doxorubicin-loaded nanoparticles with a specific trastuzumab-based targeting of HER2-overexpressing breast cancer cells.[88] It was also reported that there was specific targeting with trastuzumab-functionalized doxorubicin-loaded nanoparticles with a cellular binding of 73.80%, whereas doxorubicin-loaded nanoparticles without trastuzumab had a marginal cellular binding of 5.57%. With such a targeting system, with trastuzumab as the targeting ligand binding the HER2 receptor, higher drug levels in tumor tissue were achieved. Lee et al. employed cationic P(MDS-co-CES) micelles to delivery paclitaxel and Herceptin simultaneously, which is convenient for adjusting the dosage of paclitaxel and Herceptin by simply altering the initial loading.[89] Their results showed that the cytotoxicity of paclitaxel was increased with codelivery of Herceptin, and the degree of increment in cytotoxicity depends on the level of HER2 expression. All these results indicate that trastuzumab-functionalized nanoparticles can significantly promote targeted delivery of the drug to the corresponding cancer cells, and therefore greatly enhance its therapeutic effects and reduce its side effects.

8.4 Trastuzumab-Conjugated Nanoparticles of PLA-TPGS/TPGS-COOH Copolymer Blend

In this section, we report a strategy developed in our earlier research on how to pre-pare nanoparticles of biodegradable novel copolymers for targeted chemotherapy, by which the targeting effect can be controlled. The NP formulation consisted of a blend of two-component copolymers. One is poly(lactide)-D-α-tocopheryl polyethylene glycol suc-cinate (PLA-TPGS), which is of a desired hydrophobic-lipophilic balance (HLB) and can thus result in high drug encapsulation efficiency and high cellular adhesion/adsorption. Another is carboxyl group-terminated TPGS (TPGS-COOH), which plays the role of linker molecules when appearing on the NP surface. This combination provides a simple tech-nique for conjugation of the molecular probes on the NP surface. Moreover, the targeting effect can be quantitatively controlled by adjusting the copolymer blend ratio. Docetaxel was used as a prototype drug; it is a hydrophobic anticancer drug and has excellent thera-peutic effects against a wide spectrum of cancers, such as breast cancer, non-small-cell lung cancer, ovarian cancer, and head and neck cancer.[90] Docetaxel is semisynthetic analog of paclitaxel, but more effective as an inhibitor of microtubule depolymerization due to its ability to alter tubulin processing within the cells. Moreover, there is a synergistic antitu-mor effect when trastuzumab is combined with docetaxel.[85]

Docetaxel-loaded nanoparticles of a PLA-TPGS and TPGS-COOH blend were prepared by a modified solvent extraction/evaporation method. Four levels of the mass ratio between PLA-TPGS and TPGS-COOH were used in this research (1:0, 9:1, 4:1, and 2:1), for which the weight amounts of TPGS-COOH in the blend were 0, 10, 20, and 33.3%. In brief, 10 mg of docetaxel and 100 mg of the PLA-TPGS/TPGS-COOH blend at various weight ratios were dissolved in 8 ml of dichloromethane (DCM). The solution formed was poured into 120 ml of aqueous phase solution containing 0.03% (w/v) TPGS as the emulsifier under gentle stirring, and sonicated at 25 W output for 120 s. The emulsion was evaporated overnight and then centrifuged at 10,500 rpm for 20 min. The pellet was resuspended in water and freeze-dried for 2 days. Hereinafter, the docetaxel-loaded PLA-TPGS/TPGS-COOH NPs of 0, 10, 20, and 33.3% TPGS-COOH are termed NP0, NP10, NP20, and NP33, respectively, and the NPs with no trastuzumab conjugated are called nude NPs.

Before trastuzumab conjugation, the docetaxel-loaded PLA-TPGS/TPGS-COOH NPs were first activated according to a procedure similar to that described by Wuang et al.,[91] where polypyrrole nanoparticles were synthesized to formulate iron oxides (IOs) for tar-geted MRI. In brief, 10 mg of the PLA-TPGS/TPGS-COOH NPs were dispersed in 4 ml of deionized (DI) water, followed by adding 20 mg of 1-ethyl-3-(3-dimethylamino)-propyl carbodiimide (EDC) and 8 mg of N-hydroxysuccinimide (NHS). One milligram of N-Boc ethylenediamine was added to introduce amine groups to the NPs. The pH of the reaction mixture was adjusted to 8 with triethylamine. After 4 h of reaction at room temperature, the mixture was centrifuged, washed with DI water, and dried under reduced pressure. Then the product was treated with trifluoroacetic acid (TFA) at 0°C for 30 min to remove N-Boc protection. After evaporation of TFA, the product was washed with DI water and dried. Trastuzumab was then conjugated to the NPs by using carbodiimide chemistry. The NPs were dispersed in 2 ml of 0.1 M PBS and placed in an ultrasonic bath for 30 min. Then 10 mg of NHS and 50 mg of EDC were added. After that, 500 μl of trastuzumab (10 mg/ml) was added for reaction and triethylamine was used to adjust the pH of the reaction mixture to 8. After 4 h reaction at room temperature, the resulting product was

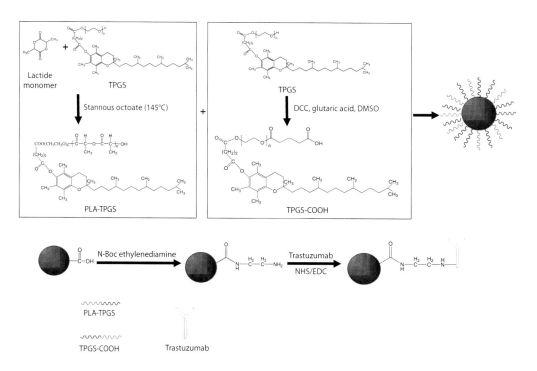

FIGURE 8.2
Schematic illustration for preparation of the trastuzumab-functionalized, docetaxel-loaded PLA-TPGS/TPGS-COOH nanoparticles. (Reproduced with permission from Sun and Feng, *Nanomedicine* 4 (4): 431–445, 2009.)

centrifuged, washed with DI water, and dried under reduced pressure. Hereinafter, the docetaxel-loaded, trastuzumab-conjugated NPs of 10, 20, and 33.3% TPGS-COOH are termed NP10-HER, NP20-HER, and NP33-HER, respectively. The detailed scheme for preparation of nanoparticles is shown in Figure 8.2.

Coumarin-6-loaded PLA-TPGS/TPGS-COOH NPs with and without trastuzumab conjugation were prepared in the same way as the docetaxel-loaded NPs, except the drug was replaced by 0.05% (w/v) coumarin-6. Coumarin-6-loaded NPs were used for the in vitro cellular uptake experiment.

8.5 Characterization of Trastuzumab-Conjugated Nanoparticles

8.5.1 Size and Size Distribution

Particle size, polydispersity, zeta potential, and drug encapsulation efficiency of the docetaxel-loaded PLA-TPGS/TPGS-COOH NPs of 0, 10, 20, and 33.3% TPGS-COOH with or without trastuzumab surface modification are listed in Table 8.1, from which it can be seen that the NPs of all formulations in this study were around 300 nm in diameter, with a polydispersity of around 0.15. Surface modification of the NPs by trastuzumab slightly increased the particle size and polydispersity, but this did not seem to be affected by the amount of trastuzumab on the NP surface. From our point of view, NPs of around 200 nm

TABLE 8.1

Characteristics of Docetaxel-Loaded, Trastuzumab-Functionalized Nanoparticles of PLA-TPGS and TPGS-COOH Copolymer Blend at Various Component Ratios

Nanoparticles	Size (nm)	Polydispersity	Zeta Potential (mV)	EE (%)
NP0	301.8 ± 9.5	0.143 ± 0.02	–42.73 ± 3.88	54.7 ± 4.73
NP10	294.7 ± 9.2	0.151 ± 0.03	–41.68 ± 4.12	55.8 ± 4.62
NP10-HER	308.5 ± 9.0	0.202 ± 0.03	–32.25 ± 2.95	54.5 ± 6.25
NP20	286.2 ± 10.2	0.156 ± 0.03	–40.33 ± 4.01	57.4 ± 4.19
NP20-HER	305.7 ± 11.8	0.175 ± 0.05	–25.08 ± 1.15	55.7 ± 5.75
NP33	292.5 ± 9.8	0.121 ± 0.03	–41.50 ± 3.28	48.2 ± 4.03
NP33-HER	311.2 ± 9.4	0.155 ± 0.04	–23.94 ± 2.09	47.5 ± 4.58

Source: Reproduced with permission from Sun et al., *Nanomedicine* 4(4): 431–445, 2009.

in diameter would be appropriate to be used as drug carriers to achieve high cellular uptake through a mechanism called endocytosis, in which the individual nanoparticle would first adhere to the cell surface and then bend a small piece of the lipid bilayer membrane. The nanoparticles would then be enveloped and engulfed into the cytoplasm. In this process, the surface energy of the NPs is sacrificed to provide the bending energy. Too small NPs would not have enough surface energy to complete the cell membrane bending process. Moreover, too small NPs would result in too small drug encapsulation efficiency and cause too fast drug release. We found that placebo nanoparticles have similar sizes and size distributions as their corresponding drug-loaded nanoparticles. Other physicochemical properties, such as surface morphology, surface chemistry, and surface charge, were also not effected by the drug loading since, as the XPS investigation shows in Section 8.5.5, no drug molecules appear on the nanoparticle surfaces due to their high hydrophobicity.

8.5.2 Surface Morphology

The surface morphology of the docetaxel-loaded PLA-TPGS/TPGS-COOH NPs was investigated by a field emission scanning electron microscopy system (FESEM; JSM-6700F, JEOL, Tokyo, Japan). The samples were prepared by dripping a drop of the NP suspension onto the copper tape placed on the surface of the sample stub and drying it overnight. The stub was coated with a platinum layer by the Auto Fine Platinum Coater (JEOL) for 40 s before measuring.

Figure 8.3 shows the surface morphology of the docetaxel-loaded PLA-TPGS/TPGS-COOH NPs, and the images were studied by FESEM. The docetaxel-loaded NPs without trastuzumab modification (Figure 8.3A) were spherical in shape with a smooth surface within the FESEM resolution level. The image of NPs with trastuzumab conjugation had a blurry surface, and there seemed to be adhesion between the NPs, compared with NP20, which was probably caused by the trastuzumab on the nanoparticle surface.

8.5.3 Surface Charge

Zeta potential is an important factor to determine the stability of the NPs in dispersion. Surface charge also plays an important role in the interaction between the cell membrane

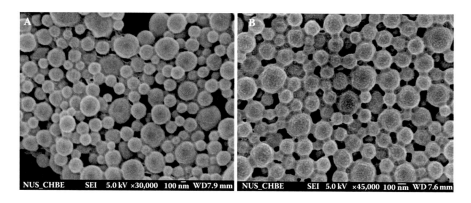

FIGURE 8.3
FESEM images of docetaxel-loaded nanoparticles of PLA-TPGS (67%) and TPGS-COOH (33%) copolymer blend: (A) NP33 and (B) NP33-HER. (Reproduced with permission from Sun and Feng, *Nanomedicine*, 4(4): 431–445, , 2009.)

and the NPs. A low absolute value of the zeta potential indicates colloidal instability, which could lead to aggregation of the NPs in dispersion. A high absolute value of the zeta potential indicates a high surface charge of the NPs, which leads to strong repellent interactions among the NPs in dispersion, and thus high stability. It can be seen from Table 8.1 that the zeta potential of docetaxel-loaded PLA-TPGS/TPGS-COOH NPs without trastuzumab modification was found to be around −40 mV, indicating that the nanoparticle dispersion is stable. The absolute values of zeta potential for trastuzumab-functionalized NPs are lower than those without trastuzumab on the surface, and the more trastuzumab on the surface, the lower the absolute values were. This is because of the positive charge of trastuzumab.

8.5.4 Drug Encapsulation Efficiency

The drug encapsulation efficiency (EE) of the docetaxel-loaded PLA-TPGS/TPGS-COOH NPs is also included in Table 8.1, from which it can be found that (1) the EE data have no significant difference among all types of NPs, and (2) the slight difference in EEs among all types of NPs is caused by the component ratio between PLA-TPGS and TPGS-COOH. For example, the EE of NP20, which contains 20% TPGS-COOH, is the highest among all the NPs. Either lower or higher content of TPGS-COOH in the polymeric matrix caused lower EE. Surface functionalization should increase the EE in general since it would decrease the drug diffusion from the NPs into the aqueous phase in the preparation process. However, drug diffusion from the NPs should also depend on the porosity of the NP matrix. The porosity of the polymeric matrix would be high if too much TPGS-COOH was included, which would speed the drug diffusion from the NPs into the aqueous phase in the preparation process.

8.5.5 Surface Chemistry

X-ray photoelectron spectroscopy (XPS) was applied to study the surface chemistry of the docetaxel-loaded PLA-TPGS/TPGS-COOH NPs with or without surface modification. Figure 8.4 shows the XPS wide-scan spectrum of the trastuzumab-functionalized,

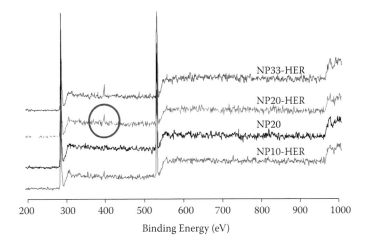

FIGURE 8.4
X-ray photoelectron spectroscopy (XPS) wide-scan spectra of the docetaxel-loaded PLA-TPGS/TPGS-COOH nanoparticles. (Reproduced with permission from Sun et al., *Nanomedicine* 4(4): 431–445, 2009.)

docetaxel-loaded NPs of various component copolymer blend ratios. From their chemical structure, it can be seen that PLA-TPGS and TPGS-COOH do not contain any nitrogen atoms, while docetaxel and trastuzumab contain nitrogen atoms. Therefore, nitrogen could be used as a marker to detect the presence of docetaxel and trastuzumab on the NP surfaces. It can be seen from the third spectrum from the top (for NP20) in Figure 8.4 that there is no N 1s signal at the 397.6 eV bond energy position, which means there are no docetaxel molecules on the NP surface. This agrees with our earlier research on paclitaxel-loaded polymeric nanoparticles prepared with either phospholipids or PVA or TPGS as emulsifier. The second spectrum from the top (for NP20-HER), however, the nitrogen peak at the 397.6 eV bond energy position (as shown in the circle in Figure 8.4) must come from trastuzumab, which confirms the successful conjugation of trastuzumab on the NP surface. It can be seen from a close comparison of the XPS spectra among NP10-HER, NP20-HER, and NP33-HER that there was an increase in the percentage of the N 1s region; as the more TPGS-COOH is included in the polymeric matrix, the more trastuzumab can be conjugated on the NP surface.

8.5.6 In Vitro Docetaxel-Release Kinetics

Figure 8.5 shows the 30-day in vitro release profiles of docetaxel from the drug-loaded PLA-TPGS/TPGS-COOH NPs with and without trastuzumab conjugation. Docetaxel was released from the NPs in a pH 7.4 PBS buffer at 37°C. From Figure 8.5, it can be seen that docetaxel was released in a biphasic style with an initial burst and subsequent accumulative release. It is obvious that the drug release of NPs with trastuzumab conjugation is significantly faster than that without trastuzumab conjugation, which means modification of the NP surface significantly speeds up the drug release. This can be explained by effects of the surface modification on the surface morphology, which makes the NPs blurry and thus increases the surface area, which increases the speed at which the drug is released from the NPs.

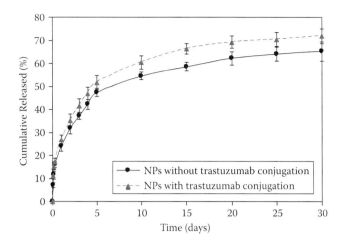

FIGURE 8.5
In vitro drug release profiles of docetaxel-loaded nanoparticles of PLA-TPGS (67%) and TPGS-COOH (33%) copolymer blend within 30 days. (Reproduced with permission from Sun et al., *Nanomedicine* 4(4): 431–435, 2009.)

8.6 In Vitro Evaluation of Trastuzumab-Conjugated Nanoparticles

8.6.1 In Vitro Cellular Uptake

MCF-7 breast adenocarcinoma cells have a moderate level of HER2 overexpression on their surface, and SK-BR-3 breast adenocarcinoma cells have high overexpression of HER2. The targeting effect of the trastuzumab-functionalized nanoparticles can thus be evaluated by employing these two cell lines to investigate the cellular uptake of coumarin-6-loaded nanoparticles. Figure 8.6 shows the cellular uptake efficiency of coumarin-6-loaded NP20 and NP20-HER by (A) SK-BR-3 breast cancer cells and (B) MCF-7 breast cancer cells at the same 125 µg/ml NP concentration for 0.5, 1.0, 2.0, and 4.0 h, respectively. It can be seen from

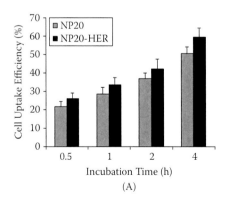
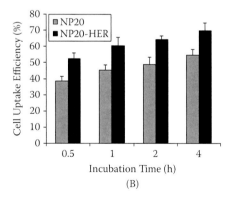

FIGURE 8.6
Cellular uptake efficiency of coumarin-6-loaded nanoparticles after 0.5, 1, 2, and 4 h incubation at 125 µg/ml nanoparticle concentration by (A) MCF-7 breast cancer cells and (B) SK-BR-3 breast cancer cells. Each point represents mean ± SD (n = 6, $p < 0.05$). (Reproduced with permission from Sun et al., *Nanomedicine* 4(4): 431–445, 2009.)

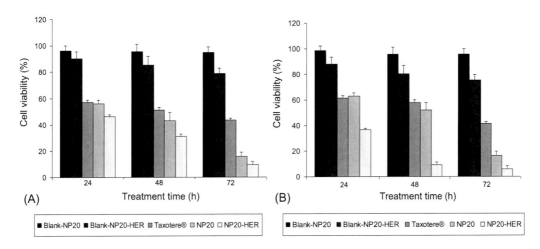

FIGURE 8.7
In vitro viability of (A) MCF-7 breast cancer cells and (B) SK-BR-3 breast cancer cells treated with placebo nanoparticles (Blank-NP20 and Blank-NP20-HER), Taxotere, and docetaxel-loaded nanoparticles (NP20 and NP20-HER) at 25 µg/ml docetaxel concentration after 24 h (left group), 48 h (middle group), and 72 h (right group) treatment, respectively (n = 6). (Reproduced with permission from Sun et al., *Nanomedicine* 4(4): 431–445, 2009.)

Figure 8.6 that NP20-HER demonstrated much higher cellular uptake efficiency for both MCF-7 and SK-BR-3 cells than NP20. For example, it can be found from Figure 8.6A that NP20-HER achieved 1.36-, 1.33-, 1.31-, and 1.28-fold higher cellular uptake efficiencies than NP20 after 0.5, 1.0, 2, and 4 h incubation with SK-BR-3 cells, respectively. It can be seen in Figure 8.7B, instead, that NP20-HER showed 1.21-, 1.17-, 1.15-, and 1.17-fold higher cellular uptake efficiencies than NP20 after 0.5, 1.0, 2, and 4 h incubation with MCF-7 cells, respectively. NP20-HER thus showed 1.12-, 1.34-, 1.14-, and 1.09-fold higher targeting effects for SK-BR-3 cancer cells (of high HER 2 overexpression) than for MCF-7 breast cancer cells (of moderate HER2 overexpression) after 0.5, 1.0, 2, and 4 h cell culture, respectively.

It can be seen from the first group in Figure 8.6B that the cellular uptake efficiencies for a 30 min incubation of naked nanoparticles and trastuzumab-modified nanoparticles are 38.6 and 52.5% for SK-BR-3 cells, respectively. This demonstrates that in such a short incubation period, the uptake of trastuzumab-modified nanoparticles was enhanced in comparison with the cellular uptake of the naked nanoparticles, which is in accordance with the literature. Moreover, it is straightforward from both Figure 8.6A and B that the cellular uptake depends on the incubation time, and the longer the incubation time, the higher the cellular uptake efficiency achieved. Similar trends were also found in our earlier work, where trastuzumab was physically attached to the PLGA-MMT NPs for targeted delivery of paclitaxel.

8.6.2 In Vitro Cytotoxicity

Figure 8.7 shows the in vitro viability of (A) MCF-7 breast cancer cells (of moderate HER2 overexpression) and (B) SK-BR-3 breast cancer cells (of high HER2 overexpression) treated with placebo nanoparticles with (Blank-NP20-HER) and without (Blank-NP20) trastuzumab conjugation, Taxotere®, and docetaxel-loaded PLA-TPGS/TPGS-COOH nanoparticles of 20% TPGS-COOH with (NP20-HER) and without (NP20) 25 µg/ml docetaxel concentration after 24, 48, and 72 h treatment, respectively (n = 6). The results are shown in the form of cell viability. The sum of viability and mortality is always 100%. The reason we chose 25 μg/ml concentration of docetaxel is that the IC50 of the MCF-7 breast cancer cells has been found to be 13.2 μM, i.e., 10.7 μg/ml (the molecular weight of docetaxel is 808). From Figure 8.7, the following conclusions can be obtained.

The viability for both cell lines was decreased with an increase of the incubation time, which means the longer the culture time, the higher the cytotoxicity. This is straightforward.

From the first column of each group for Blank-NP20, it can be found from Figure 8.7A for MCF-7 cells that the viabilities for placebo PLA-TPGS/TPGS-COOH NPs without trastuzumab functionalization were 96.1 ± 2.6%, 95.5 ± 3.8%, and 94.8 ± 5.2% after 24, 48, and 72 h incubation, respectively. In comparison, similar results can be found in Figure 8.7B; that is, the viabilities of SK-BR-3 cells for placebo PLA-TPGS/TPGS-COOH NPs without trastuzumab functionalization were 98.4 ± 4.0%, 95.5 ± 5.7%, and 95.7 ± 4.2% after 24, 48, and 72 h incubation, respectively. This means that the cytotoxicities of the placebo PLA-TPGS/TPGS-COOH NPs have high biocompatibility with fractional cytotoxicity.

The second column of each group in Figure 8.7A and B stands for the cell viability of placebo PLA-TPGS/TPGS-COOH NPs with trastuzumab surface modification (Blank-NP20-HER). In Figure 8.7A, the viabilities of MCF-7 cells are 90.3 ± 3.3%, 85.3 ± 4.8%, and 78.7 ± 5.2% after 24, 48, and 72 h cell culture, respectively. In Figure 8.7B, however, the viabilities of SK-BR-3 cells are slightly lower: 88.2 ± 5.1%, 80.3 ± 6.6%, and 75.3 ± 4.3% for 24, 48, and 72 h, respectively. The results clearly show the targeting effects of the trastuzumab-functionalized NPs for the breast cancer cells of high HER2 overexpression.

The third column of every group in Figure 8.7 shows the viabilities of the two breast cancer cell lives after 24, 48, and 72 h treatment with Taxotere, which are 56.9 ± 2.6%, 51.3 ± 4.3%, and 43.2 ± 3.6% for MCF-7 cells and 61.3 ± 1.9%, 57.8 ± 2.1%, and 41.4 ± 1.4% for SK-BR-3 cells, respectively. This means that Taxotere may be more effective for MCF-7 breast cancer cells than for SK-BR-3 breast cancer cells in a short time (24 and 48 h), and the effect may become inverse in a longer time (72 h). In comparison, the in vitro viabilities shown in the fourth column of each group in Figure 8.7A and B are 55.8 ± 3.1%, 43.2 ± 3.0%, and 15.6 ± 3.4% for MCF-7 cells and 62.7 ± 2.8%, 51.9 ± 6.1%, and 16.4 ± 3.2% for SK-BR-3 cells after 24, 48, and 72 h treatment, respectively. This means that evaluated by cellular mortality, the NP formulation has an in vitro therapeutic effect (1.025-fold for MCF-7 cells and 0.964-fold for SK-BR-3 cells) comparable to that of Taxotere after 24 h cell culture. However, the NP formulation of docetaxel would be 1.166- and 1.486-fold for MCF-7 cells and 1.140- and 1.427-fold for SK-BR-3 cells, and thus more effective than that of Taxotere, after 48 and 72 h treatment, respectively. This may be attributed to the sustainable drug release manner of the NP formulation.

The fifth column at the right of every group in Figure 8.7A and B shows the viabilities of the MCF-7 and SK-BR-3 breast cancer cell lines after 24, 48, and 72 h treatment with NP20-HER, which are 46.3 ± 2.7%, 31.0 ± 3.2%, and 9.4 ± 3.1% for MCF-7 cells and 36.7 ± 1.3%, 9.2 ± 2.0%, and 5.9 ± 2.5% for SK-BR-3 cells, respectively. It can thus be concluded that, judged by cellular mortality, the trastuzumab-functionalized NP formulation can be 1.215-, 1.215-, and 1.073-fold for MCF-7 cells and 1.697-, 1.886-, and 1.126-fold for SK-BR-3 cells more effective than the NP formulation with no trastuzumab functionalization after 24, 48, and 72 h treatment, respectively. The targeting effects of trastuzumab functionalization thus showed 39.7, 55.2, and 4.9% higher cancer cell mortality for SK-BR-3 cells of high HER2 overexpression than for MCF-7 cells of moderate HER2 overexpression after 24, 48, and 72 h treatment, respectively.

Combined with the advantages of the PLA-TPGS/TPGS-COOH NP formulation and the trastuzumab functionalization, the trastuzumab-functionalized NP formulation can be 1.245-, 1.417-, and 1.594-fold for MCF-7 cells and 1.635-, 2.150-, and 1.606-fold for SK-BR-3 cells more effective than Taxotere after 24, 48, and 72 h treatment, respectively.

8.7 Summary

Nanoparticle formulation has great advantages over conventional drug formulations, resulting in a sustained and controlled manner for drug release, thus increasing the therapeutic effect and reducing side effects. Size and surface modifications are the two key factors determining the performance of biodegradable polymeric NPs for cancer therapy. Targeted chemotherapy can be realized by functionalization of the NP surface by targeting ligands such as monoclonal antibodies. A copolymer strategy can be further employed to synthesize nanoparticles of two component copolymers: one component copolymer has ideal HLB for high drug encapsulation efficiency, high cellular adhesion, and internalization, and another copolymer facilitates the conjugation of the ligand molecules on the NP surface. The targeting effects can thus be quantitatively controlled by adjusting the blend ratio between the two copolymers. Development of multifunctional NPs is also feasible by coencapsulation of the imaging, therapeutic, and reporting agents in the polymeric nanoparticles, with their surface functionalized by targeting ligands. In this chapter, we provided a preliminary, proof-of-concept investigation for a targeted drug delivery system of molecular probe-conjugated nanoparticles of a blend of two biodegradable copolymers, PLA-TPGS and TPGS-COOH. It was found that judged by in vitro cellular mortality, the trastuzumab-functionalized PLA-TPGS/TPGS-COOH NP formulations of docetaxel can be 1.215-, 1.215-, and 1.073-fold for MCF-7 cells and 1.697-, 1.886-, and 1.126-fold for SK-BR-3 cells more effective than the PLA-TPGS/TPGS-COOH NP formulation with no trastuzumab functionalization after 24, 48, and 72 h treatment, respectively. The targeting effects of trastuzumab functionalization thus showed 39.7, 55.2, and 4.9% higher cancer cell mortalities for SK-BR-3 cells of high HER2 overexpression than for MCF-7 cells of moderate HER2 overexpression after 24, 48, and 72 h treatment, respectively. The strategy greatly simplified the existing nanoparticle technologies for targeted drug delivery.

References

1. Emerich, D. F., and Thanos, C. G., Nanotechnology and medicine, *Expert Opinion on Biological Therapy* 3 (4), 655–663, 2003.
2. Shenoy, D., Little, S., Langer, R., and Amiji, M., Poly(ethylene oxide)-modified poly(beta-amino ester) nanoparticles as a pH-sensitive system for tumor-targeted delivery of hydrophobic drugs. Part 2. *In vivo* distribution and tumor localization studies, *Pharmaceutical Research* 22 (12), 2107–2114, 2005.
3. Kreuter, J., Drug targeting with nanoparticles, in *1st International Meeting on the Scientific Basis of Modern Pharmacy Medecine Et Hygiene*, Athens, Greece, 1994, pp. 253–256.
4. Moghimi, S. M., Hunter, A. C., and Murray, J. C., Long-circulating and target-specific nanoparticles: theory to practice, *Pharmacological Reviews* 53 (2), 283–318, 2001.
5. Panyam, J., and Labhasetwar, V., Biodegradable nanoparticles for drug and gene delivery to cells and tissue, *Advanced Drug Delivery Reviews* 55 (3), 329–347, 2003.
6. Feng, S. S., and Huang, G. F., Effects of emulsifiers on the controlled release of paclitaxel (Taxol [(R)]) from nanospheres of biodegradable polymers, *Journal of Controlled Release* 71 (1), 53–69, 2001.
7. Quintanar-Guerrero, D., Allemann, E., Fessi, H., and Doelker, E., Preparation techniques and mechanisms of formation of biodegradable nanoparticles from preformed polymers, *Drug Development and Industrial Pharmacy* 24 (12), 1113–1128, 1998.

8. Lee, S. C., Oh, J. T., Jang, M. H., and Chung, S. I., Quantitative analysis of polyvinyl alcohol on the surface of poly(D,L-lactide-co-glycolide) microparticles prepared by solvent evaporation method: effect of particle size and PVA concentration, *Journal of Controlled Release* 59 (2), 123–132, 1999.

9. Shakesheff, K. M., Evora, C., Soriano, I., and Langer, R., The adsorption of poly(vinyl alcohol) to biodegradable microparticles studied by x-ray photoelectron spectroscopy (XPS), *Journal of Colloid and Interface Science* 185 (2), 538–547, 1997.

10. Scholes, P. D., Coombes, A. G. A., Illum, L., Davis, S. S., Watts, J. F., Ustariz, C., Vert, M., and Davies, M. C., Detection and determination of surface levels of poloxamer and PVA surfactant on biodegradable nanospheres using SSIMS and XPS, *Journal of Controlled Release* 59 (3), 261–278, 1999.

11. Yamaoka, T., Tabata, Y., and Ikada, Y., Comparison of body distribution of poly(vinyl alcohol) with other water-soluble polymers after intravenous administration, *Journal of Pharmacy and Pharmacology* 47 (6), 479–486, 1995.

12. Mu, L., and Feng, S. S., A novel controlled release formulation for the anticancer drug paclitaxel (Taxol[(R)]): PLGA nanoparticles containing vitamin E TPGS, *Journal of Controlled Release* 86 (1), 33–48, 2003.

13. Mu, L., Seow, P. H., Ang, S. N., and Feng, S. S., Study on surfactant coating of polymeric nanoparticles for controlled delivery of anticancer drug, *Colloid and Polymer Science* 283 (1), 58–65, 2004.

14. Mu, L., and Feng, S. S., Vitamin E TPGS used as emulsifier in the solvent evaporation/extraction technique for fabrication of polymeric nanospheres for controlled release of paclitaxel (Taxol[(R)]), *Journal of Controlled Release* 80 (1–3), 129–144, 2002.

15. Mu, L., and Feng, S. S., PLGA/TPGS nanoparticles for controlled release of paclitaxel: effects of the emulsifier and drug loading ratio, *Pharmaceutical Research* 20 (11), 1864–1872, 2003.

16. Vangeyte, P., Gautier, S., and Jerome, R., About the methods of preparation of poly(ethylene oxide)-b-poly(epsilon-caprolactone) nanoparticles in water analysis by dynamic light scattering, *Colloids and Surfaces A—Physicochemical and Engineering Aspects* 242 (1–3), 203–211, 2004.

17. Fessi, H., Puisieux, F., Devissaguet, J. P., Ammoury, N., and Benita, S., Nanocapsule formation by interfacial polymer deposition following solvent displacement, *International Journal of Pharmaceutics* 55 (1), R1–R4, 1989.

18. Fitzgerald, D., and Pastan, I., Targeted toxin therapy for the treatment of cancer, *Journal of the National Cancer Institute* 81 (19), 1455–1463, 1989.

19. Allen, T. M., Ligand-targeted therapeutics in anticancer therapy, *Nature Reviews Cancer* 2 (10), 750–763, 2002.

20. Poste, G., and Kirsh, R., Site-specific (targeted) drug delivery in cancer-therapy, *Bio-Technology* 1 (10), 869–878, 1983.

21. Brigger, I., Dubernet, C., and Couvreur, P., Nanoparticles in cancer therapy and diagnosis, *Advanced Drug Delivery Reviews* 54 (5), 631–651, 2002.

22. Jain, R. K., Barriers to drug-delivery in solid tumors, *Scientific American* 271 (1), 58–65, 1994.

23. Jain, R. K., Transport of molecules, particles, and cells in solid tumors, *Annual Review of Biomedical Engineering* 1, 241–263, 1999.

24. Balshaw, D. M., Philbert, M., and Suk, W. A., Research strategies for safety evaluation of nanomaterials. Part III. Nanoscale technologies for assessing risk and improving public health, *Toxicological Sciences* 88 (2), 298–306, 2005.

25. Nie, S. M., Xing, Y., Kim, G. J., and Simons, J. W., Nanotechnology applications in cancer, *Annual Review of Biomedical Engineering* 9, 257–288, 2007.

26. Mothersill, C., and Seymour, C., Radiation-induced bystander and other non-targeted effects: novel intervention points in cancer therapy? *Current Cancer Drug Targets* 6 (5), 447–454, 2006.

27. Seymour, L. W., Passive tumor targeting of soluble macromolecules and drug conjugates, *Critical Reviews in Therapeutic Drug Carrier Systems* 9 (2), 135–187, 1992.

28. Thurston, G., McLean, J. W., Rizen, M., Baluk, P., Haskell, A., Murphy, T. J., Hanahan, D., and McDonald, D. M., Cationic liposomes target angiogenic endothelial cells in tumors and chronic inflammation in mice, *Journal of Clinical Investigation* 101 (7), 1401–1413, 1998.

29. Krasnici, S., Werner, A., Eichhorn, M. E., Schmitt-Sody, M., Pahernik, S. A., Sauer, B., Schulze, B., Teifel, M., Michaelis, U., Naujoks, K., and Dellian, M., Effect of the surface charge of liposomes on their uptake by angiogenic tumor vessels, *International Journal of Cancer* 105 (4), 561–567, 2003.

30. Maeda, H., The enhanced permeability and retention (EPR) effect in tumor vasculature: the key role of tumor-selective macromolecular drug targeting, in *41st International Symposium on Regulation of Enzyme Activity and Synthesis in Normal and Neoplastic Tissues*, G. Weber, Indianapolis, IN, 2000, pp. 189–207.

31. Owens, D. E., and Peppas, N. A., Opsonization, biodistribution, and pharmacokinetics of polymeric nanoparticles, *International Journal of Pharmaceutics* 307 (1), 93–102, 2006.

32. Allen, T. M., and Chonn, A., Large unilamellar liposomes with low uptake into the reticuloendothelial system, *FEBS Letters* 223 (1), 42–46, 1987.

33. Lee, V. H. L., Nanotechnology: challenging the limit of creativity in targeted drug delivery, *Advanced Drug Delivery Reviews* 56 (11), 1527–1528, 2004.

34. Rossi, J., Giasson, S., Khalid, M. N., Delmas, P., Allen, C., and Leroux, J. C., Long-circulating poly(ethylene glycol)-coated emulsions to target solid tumors, *European Journal of Pharmaceutics and Biopharmaceutics* 67 (2), 329–338, 2007.

35. Boghaert, E. R., Khandke, K., Sridharan, L., Armellino, D., Dougher, M., DiJoseph, J. F., Kunz, A., Hamann, P. R., Sridharan, A., Jones, S., Discafani, C., and Damle, N. K., Tumoricidal effect of calicheamicin immuno-conjugates using a passive targeting strategy, *International Journal of Oncology* 28 (3), 675–684, 2006.

36. Peer, D., Karp, J. M., Hong, S., FaroKhzad, O. C., Margalit, R., and Langer, R., Nanocarriers as an emerging platform for cancer therapy, *Nature Nanotechnology* 2 (12), 751–760, 2007.

37. Gottesman, M. M., Mechanisms of cancer drug resistance, *Annual Review of Medicine* 53, 615–627, 2002.

38. Ciardiello, F., and Tortora, G., A novel approach in the treatment of cancer: targeting the epidermal growth factor receptor, *Clinical Cancer Research* 7 (10), 2958–2970, 2001.

39. Blume-Jensen, P., and Hunter, T., Oncogenic kinase signalling, *Nature* 411 (6835), 355–365, 2001.

40. Longati, P., Comoglio, P. M., and Bardelli, A., Receptor tyrosine kinases as therapeutic targets: the model of the MET oncogene, *Current Drug Targets* 2 (1), 41–55, 2001.

41. Smith, J. K., Mamoon, N. M., and Duhe, R. J., Emerging roles of targeted small molecule protein-tyrosine kinase inhibitors in cancer therapy, *Oncology Research* 14 (4–5), 175–225, 2004.

42. Mendelsohn, J., and Baselga, J., The EGF receptor family as targets for cancer therapy, *Oncogene* 19 (56), 6550–6565, 2000.

43. Mendelsohn, J., Epidermal growth factor receptor inhibition by a monoclonal antibody as anticancer therapy, in *Symposium on Foundations of Clinical Cancer Research—Perspective for the 21st-Century*, Houston, TX, 1997, pp. 2703–2707.

44. Morrison, S. L., Johnson, M. J., Herzenberg, L. A., and Oi, V. T., Chimeric human-antibody molecules—mouse antigen-binding domains with human constant region domains, *Proceedings of the National Academy of Sciences of the United States of America—Biological Sciences* 81 (21), 6851–6855, 1984.

45. Lobuglio, A. F., Wheeler, R. H., Trang, J., Haynes, A., Rogers, K., Harvey, E. B., Sun, L., Ghrayeb, J., and Khazaeli, M. B., Mouse human chimeric monoclonal-antibody in man—kinetics and immune response, *Proceedings of the National Academy of Sciences of the United States of America* 86 (11), 4220–4224, 1989.

46. Hale, G., Clark, M. R., Marcus, R., Winter, G., Dyer, M. J. S., Phillips, J. M., Riechmann, L., and Waldmann, H., Remission induction in non-Hodgkin lymphoma with reshaped human monoclonal-antibody CAMPATH-1H, *Lancet* 2 (8625), 1394–1399, 1988.

47. Iannello, A., and Ahmad, A., Role of antibody-dependent cell-mediated cytotoxicity in the efficacy of therapeutic anti-cancer monoclonal antibodies, *Cancer and Metastasis Reviews* 24 (4), 487–499, 2005.

48. Imai, K., and Takaoka, A., Comparing antibody and small-molecule therapies for cancer, *Nature Reviews Cancer* 6 (9), 714–727, 2006.

49. Baselga, J., Targeting the epidermal growth factor receptor with tyrosine kinase inhibitors: small molecules, big hopes, *Journal of Clinical Oncology* 20 (9), 2217–2219, 2002.

50. Spivakkroizman, T., Rotin, D., Pinchasi, D., Ullrich, A., Schlessinger, J., and Lax, I., Heterodimerization of C-ERBB2 with different epidermal growth-factor receptor mutants elicits stimulatory or inhibitory responses, *Journal of Biological Chemistry* 267 (12), 8056–8063, 1992.

51. Olayioye, M. A., Update on HER-2 as a target for cancer therapy—intracellular signaling pathways of ErbB2/HER-2 and family members, *Breast Cancer Research* 3 (6), 385–389, 2001.

52. Tanner, M., Jarvinen, P., and Isola, J., Amplification of HER-2/neu and topoisomerase II alpha in primary and metastatic breast cancer, *Cancer Research* 61 (14), 5345–5348, 2001.

53. Nahta, R., and Esteva, F. J., HER-2-targeted therapy: lessons learned and future directions, *Clinical Cancer Research* 9 (14), 5078–5084, 2003.

54. Esteva, F. J., Valero, V., Booser, D., Guerra, L. T., Murray, J. L., Pusztai, L., Cristofanilli, M., Arun, B., Esmaeli, B., Fritsche, H. A., Sneige, N., Smith, T. L., and Hortobagyi, G. N., Phase II study of weekly docetaxel and trastuzumab for patients with HER-2-overexpressing metastatic breast cancer, *Journal of Clinical Oncology* 20 (7), 1800–1808, 2002.

55. Persons, D. L., Bui, M. M., Lowery, M. C., Mark, H. F. L., Yung, J. F., Birkmeier, J. M., Wong, E. Y., Yang, S. J., and Masood, S., Fluorescence *in situ* hybridization (FISH) for detection of HER-2/neu amplification in breast cancer: a multicenter portability study, *Annals of Clinical and Laboratory Science* 30 (1), 41–48, 2000.

56. Fornier, M., Risio, M., Van Poznak, C., and Seidman, A., HER2 testing and correlation with efficacy of trastuzumab therapy, *Oncology—New York* 16 (10), 1340, 2002.

57. Vogel, C. L., Cobleigh, M. A., Tripathy, D., Gutheil, J. C., Harris, L. N., Fehrenbacher, L., Slamon, D. J., Murphy, M., Novotny, W. F., Burchmore, M., Shak, S., Stewart, S. J., and Press, M., Efficacy and safety of trastuzumab as a single agent in first-line treatment of HER2-overexpressing metastatic breast cancer, *Journal of Clinical Oncology* 20 (3), 719–726, 2002.

58. Nahta, R., Takahashi, T., Ueno, N. T., Hung, M. C., and Esteva, F. J., P27(kip1) down-regulation is associated with trastuzumab resistance in breast cancer cells, *Cancer Research* 64 (11), 3981–3986, 2004.

59. Nagata, Y., Lan, K. H., Zhou, X. Y., Tan, M., Esteva, F. J., Sahin, A. A., Klos, K. S., Li, P., Monia, B. P., Nguyen, N. T., Hortobagyi, G. N., Hung, M. C., and Yu, D. H., PTEN activation contributes to tumor inhibition by trastuzumab, and loss of PTEN predicts trastuzumab resistance in patients, *Cancer Cell* 6 (2), 117–127, 2004.

60. Lewis, G. D., Figari, I., Fendly, B., Wong, W. L., Carter, P., Gorman, C., and Shepard, H. M., Differential responses of human tumor-cell lines to anti-P185(HER2) monoclonal-antibodies, *Cancer Immunology Immunotherapy* 37 (4), 255–263, 1993.

61. Cooley, S., Burns, L. J., Repka, T., and Miller, J. S., Natural killer cell cytotoxicity of breast cancer targets is enhanced by two distinct mechanisms of antibody-dependent cellular cytotoxicity against LFA-3 and HER2/neu, *Experimental Hematology* 27 (10), 1533–1541, 1999.

62. Clynes, R. A., Towers, T. L., Presta, L. G., and Ravetch, J. V., Inhibitory Fc receptors modulate *in vivo* cytoxicity against tumor targets, *Nature Medicine* 6(4), 443–446, 2000.

63. Laughner, E., Taghavi, P., Chiles, K., Mahon, P. C., and Semenza, G. L., HER2 (neu) signaling increases the rate of hypoxia-inducible factor 1 alpha (HIF-1 alpha) synthesis: novel mechanism for HIF-1-mediated vascular endothelial growth factor expression, *Molecular and Cellular Biology* 21 (12), 3995–4004, 2001.

64. Izumi, Y., Xu, L., di Tomaso, E., Fukumura, D., and Jain, R. K., Tumor biology—Herceptin acts as an anti-angiogenic cocktail, *Nature* 416 (6878), 279–280, 2002.

65. Kos, K. S., Zhou, X. Y., Lee, S., Zhang, L. L., Yang, W. T., Nagata, Y., and Yu, D. H., Combined trastuzumab and paclitaxel treatment better inhibits ErbB-2-mediated angiogenesis in breast carcinoma through a more effective inhibition of Akt than either treatment alone, *Cancer* 98 (7), 1377–1385, 2003.

66. Cuello, M., Ettenberg, S. A., Clark, A. S., Keane, M. M., Posner, R. H., Nau, M. M., Dennis, P. A., and Lipkowitz, S., Down-regulation of the erbB-2 receptor by trastuzumab (Herceptin) enhances tumor necrosis factor-related apoptosis-inducing ligand-mediated apoptosis in breast and ovarian cancer cell lines that overexpress erbB-2, *Cancer Research* 61 (12), 4892–4900, 2001.

67. Smith, I. E., Efficacy and safety of Herceptin (R) in women with metastatic breast cancer: results from pivotal clinical studies, in *25th Congress of the European Society of Medical Oncology*, Lippincott Williams & Wilkins, Hamburg, Germany, 2000, pp. S3–S10.

68. Cobleigh, M. A., Vogel, C. L., Tripathy, D., Robert, N. J., Scholl, S., Fehrenbacher, L., Wolter, J. M., Paton, V., Shak, S., Lieberman, G., and Slamon, D. J., Multinational study of the efficacy and safety of humanized anti-HER2 monoclonal antibody in women who have HER2-overexpressing metastatic breast cancer that has progressed after chemotherapy for metastatic disease, *Journal of Clinical Oncology* 17 (9), 2639–2648, 1999.

69. Romond, E. H., Perez, E. A., Bryant, J., Suman, V. J., Geyer, C. E., Davidson, N. E., Tan-Chiu, E., Martino, S., Paik, S., Kaufman, P. A., Swain, S. M., Pisansky, T. M., Fehrenbacher, L., Kutteh, L. A., Vogel, V. G., Visscher, D. W., Yothers, G., Jenkins, R. B., Brown, A. M., Dakhil, S. R., Mamounas, E. P., Lingle, W. L., Klein, P. M., Ingle, J. N., and Wolmark, N., Trastuzumab plus adjuvant chemotherapy for operable HER2 positive breast cancer, *New England Journal of Medicine* 353 (16), 1673–1684, 2005.

70. Hussain, M. H. A., MacVicar, G. R., Petrylak, D. P., Dunn, R. L., Vaishampayan, U., Lara, P. N., Chatta, G. S., Nanus, D. M., Glode, L. M., Trump, D. L., Chen, H., and Smith, D. C., Trastuzumab, paclitaxel, carboplatin, and gemcitabine in advanced human epidermal growth factor receptor-2/neu-positive urothelial carcinoma: results of a multicenter phase II National Cancer Institute trial, *Journal of Clinical Oncology* 25 (16), 2218–2224, 2007.

71. Coudert, B. P., Arnould, L., Moreau, L., Chollet, P., Weber, B., Vanlemmens, L., Molucon, C., Tubiana, N., Causeret, S., Misset, J. L., Feutray, S., Mery-Mignard, D., Garnier, J., and Fumoleau, P., Pre-operative systemic (neo-adjuvant) therapy with trastuzumab and docetaxel for HER2-overexpressing stage II or III breast cancer: results of a multicenter phase II trial, *Annals of Oncology* 17 (3), 409–414, 2006.

72. Piccart-Gebhart, M. J., Procter, M., Leyland-Jones, B., Goldhirsch, A., Untch, M., Smith, I., Gianni, L., Baselga, J., Bell, R., Jackisch, C., Cameron, D., Dowsett, M., Barrios, C. H., Steger, G., Huang, C. S., Andersson, M., Inbar, M., Lichinitser, M., Lang, I., Nitz, U., Iwata, H., Thomssen, C., Lohrisch, C., Suter, T. M., Ruschoff, J., Suto, T., Greatorex, V., Ward, C., Straehle, C., McFadden, E., Dolci, M. S., Gelber, R. D., and Team, H. T. S., Trastuzumab after adjuvant chemotherapy in HER2 positive breast cancer, *New England Journal of Medicine* 353 (16), 1659–1672, 2005.

73. Baselga, J., Norton, L., Albanell, J., Kim, Y. M., and Mendelsohn, J., Recombinant humanized anti-HER2 antibody (Herceptin(TM)) enhances the antitumor activity of paclitaxel and doxorubicin against HER2/neu overexpressing human breast cancer xenografts, *Cancer Research* 58 (13), 2825–2831, 1998.

74. Burstein, H. J., Harris, L. N., Gelman, R., Lester, S. C., Nunes, R. A., Kaelin, C. M., Parker, L. M., Ellisen, L. W., Kuter, I., Gadd, M. A., Christian, R. L., Kennedy, P. R., Borges, V. F., Bunnell, C. A., Younger, J., Smith, B. L., and Winer, E. P., Preoperative therapy with trastuzumab and paclitaxel followed by sequential adjuvant doxorubicin/cyclophosphamide for HER2 overexpressing stage II or III breast cancer: a pilot study, *Journal of Clinical Oncology* 21 (1), 46–53, 2003.

75. Seidman, A. D., Fornier, M. N., Esteva, F. J., Tan, L., Kaptain, S., Bach, A., Panageas, K. S., Arroyo, C., Valero, V., Currie, V., Gilewski, T., Theodoulou, M., Moynahan, M. E., Moasser, M., Sklarin, N., Dickler, M., D'Andrea, G., Cristofanilli, M., Rivera, E., Hortobagyi, G. N., Norton, L., and Hudis, C. A., Weekly trastuzumab and paclitaxel therapy for metastatic breast cancer with analysis of efficacy by HER2 immunophenotype and gene amplification, *Journal of Clinical Oncology* 19 (10), 2587–2595, 2001.

76. Lucarini, M., Franchi, P., Pedulli, G. F., Pengo, P., Scrimin, P., and Pasquato, L., EPR study of dialkyl nitroxides as probes to investigate the exchange of solutes between the ligand shell of monolayers of protected gold nanoparticles and aqueous solutions, *Journal of the American Chemical Society* 126 (30), 9326–9329, 2004.

77. Kim, S. H., Jeong, J. H., Chun, K. W., and Park, T. G., Target-specific cellular uptake of PLGA nanoparticles coated with poly(L-lysine)-poly(ethylene glycol)-folate conjugate, *Langmuir* 21 (19), 8852–8857, 2005.

78. Oyewumi, M. O., Liu, S. Q., Moscow, J. A., and Mumper, R. J., Specific association of thiamine-coated gadolinium nanoparticles with human breast cancer cells expressing thiamine transporters, *Bioconjugate Chemistry* 14 (2), 404–411, 2003.

79. Meyer, A., Auemheimer, J., Modlinger, A., and Kessler, H., Targeting RGD recognizing integrins: drug development, biomaterial research, tumor imaging and targeting, *Current Pharmaceutical Design* 12 (22), 2723–2747, 2006.

80. Jeong, Y. I., Seo, S. J., Park, I. K., Lee, H. C., Kang, I. C., Akaike, T., and Cho, C. S., Cellular recognition of paclitaxel-loaded polymeric nanoparticles composed of poly(gamma-benzul L-glutamate) and poly(ethylene glycol) diblock copolymer endcapped with galactose moiety, *International Journal of Pharmaceutics* 296 (1–2), 151–161, 2005.

81. Cirstoiu-Hapca, A., Bossy-Nobs, L., Buchegger, F., Gurny, R., and Delie, F., Differential tumor cell targeting of anti-HER2 (Herceptin[(R)]) and anti-CD20 (Mabthera[(R)]) coupled nanoparticles, in *6th European Workshop on Particulate Systems*, Elsevier Science Bv, Geneva, Switzerland, 2006, pp. 190–196.

82. Smith, J. E., Medley, C. D., Tang, Z. W., Shangguan, D., Lofton, C., and Tan, W. H., Aptamer-conjugated nanoparticles for the collection and detection of multiple cancer cells, *Analytical Chemistry* 79 (8), 3075–3082, 2007.

83. Bellocq, N. C., Pun, S. H., Jensen, G. S., and Davis, M. E., Transferrin-containing, cyclodextrin polymer-based particles for tumor-targeted gene delivery, *Bioconjugate Chemistry* 14 (6), 1122–1132, 2003.

84. Moasser, M. M., Targeting the function of the HER2 oncogene in human cancer therapeutics, *Oncogene* 26 (46), 6577–6592, 2007.

85. Slamon, D. J., Leyland-Jones, B., Shak, S., Fuchs, H., Paton, V., Bajamonde, A., Fleming, T., Eiermann, W., Wolter, J., Pegram, M., Baselga, J., and Norton, L., Use of chemotherapy plus a monoclonal antibody against HER2 for metastatic breast cancer that overexpresses HER2, *New England Journal of Medicine* 344 (11), 783–792, 2001.

86. Nobs, L., Buchegger, F., Gurny, R., and Allemann, E., Biodegradable nanoparticles for direct or two-step tumor immunotargeting, *Bioconjugate Chemistry* 17 (1), 139–145, 2006.

87. Steinhauser, I., Spankuch, B., Strebhardt, K., and Langer, K., Trastuzumab-modified nanoparticles: optimisation of preparation and uptake in cancer cells, *Biomaterials* 27 (28), 4975–4983, 2006.

88. Anhorn, M. G., Wagner, S., Kreuter, J., Langer, K., and von Briesen, H., Specific targeting of HER2 overexpressing breast cancer cells with doxorubicin-loaded trastuzumab-modified human serum albumin nanoparticles, *Bioconjugate Chemistry* 19 (12), 2321–2331, 2008.

89. Lee, A. L. Z., Wang, Y., Cheng, H. Y., Pervaiz, S., and Yang, Y. Y., The co-delivery of paclitaxel and Herceptin using cationic micellar nanoparticles, *Biomaterials* 30 (5), 919–927, 2009.

90. Piccart, M. J., Gore, M., Huinink, W. T. B., Vanoosterom, A., Verweij, J., Wanders, J., Franklin, H., Bayssas, M., and Kaye, S., Docetaxel—An active new drug for treatment of advanced epithelial ovarian-cancer, *Journal of the National Cancer Institute* 87 (9), 676–681, 1995.

91. Wuang, S. C., Neoh, K. G., Kang, E. T., Pack, D. W., and Leckband, D. E., HER-2-mediated endocytosis of magnetic nanospheres and the implications in cell targeting and particle magnetization, *Biomaterials* 29 (14), 2270–2279, 2008.

9

The Biosynthesis of Anisotropic Nanoparticles

Mariekie Gericke

Biotechnology Division, Mintek, Randburg, South Africa

CONTENTS

9.1 Introduction

The synthesis of anisotropic nanoparticles is a rapidly developing research field and has received considerable attention over the past few years. These nanoparticles, which include various shapes, such as rods, discs, triangles, hexagons, cubes, and nanoshells, exhibit unique properties, which either strongly differ from or are more pronounced than those of symmetric, spherical nanoparticles. Their unusual optical and electronic properties, improved mechanical properties, and specific surface-enhanced spectroscopies make them ideal structures for applications in photonics, optical sensing and imaging, biomedical labeling and sensing, catalysis, and electronic devices, among others. The size and morphology of these nanoparticles are critical in the determination of these novel properties, and with an ever-broadening range of applications, the controlled synthesis of anisotropic materials has never been more important.[1–4]

Various strategies have been employed to synthesize anisotropic nanoparticles, and although some success has been achieved in developing chemical and physical procedures for the synthesis of anisotropic nanoparticles, there are still a number of bottlenecks and challenges involved in the controlled production of these materials, and to understanding the shape-guiding mechanisms.[2,3]

Over the last decade the utilization of biological systems such as yeasts, fungi, bacteria, and plants to produce nanoparticles has been demonstrated and has emerged as a promising potential method for the synthesis of anisotropic metal nanoparticles. This is not unexpected since many organisms, both unicellular and multicellular, are known to produce inorganic nanomaterials, either intracellularly or extracellularly. Some well-known examples include the synthesis of magnetic nanoparticles by magnetotactic bacteria,[5] siliceous materials produced by diatoms,[6] and production of gypsum and calcium carbonate layers by S-layer bacteria.[7] In addition, the interactions between microorganisms and metals have been well documented,[8] and the ability of microorganisms to extract and accumulate metals is already employed in biotechnological processes such as bioleaching and bioremediation, and it is well known that microorganisms such as bacteria, yeast, and fungi play an important role in remediation of toxic metals through reduction of the metal ions.[9]

Even though the biological generation of anisotropic nanoparticles could be demonstrated, the development of reliable approaches for stable production of nanoparticles of different materials, and understanding the detailed mechanisms involved in controlling the shape of the nanoparticles, are still a challenge. A brief overview of the current research on the use of biological systems as biofactories for the synthesis of metal nanoparticles is provided in this chapter. The production of nanoparticles using bacteria, fungi, yeast, actinomycetes, and biotemplates, the potential to manipulate key reaction conditions to achieve controlled size and shape of the particles, and progress on gaining a better understanding of the biological mechanisms involved in the process will be reviewed.

9.2 Biosynthesis of Nanoparticles

A wealth of different microbes have been evaluated for their ability to produce anisotropic nanoparticles. A brief selection of some of these organisms and the characteristics of the particles produced will be discussed.

9.2.1 Bacterial Production of Nanomaterials

Several bacterial species have been successfully used for intracellular and extracellular synthesis of nanoparticles. In early studies, Beveridge[10] demonstrated the formation of gold particles of nanoscale dimensions inside the cell walls of *Bacillus subtilis* after exposure to Au^{3+} ions under ambient temperature and pressure conditions. More recently, the synthesis of gold nanoplates by *Rhodopseudomonas capsulata* was demonstrated by He et al.[11] The group also demonstrated the extracellular production of gold nanowires (50–60 nm) with a network structure after exposure of cell-free extract of *R. capsulata* to gold.[12] Further process developments include the extracellular synthesis of gold nanoparticles (15–30 nm) by, e.g., *Pseudomonas aeroginosa*,[13] as well as the use of bacterial cell supernatant for the reduction of gold ions, resulting in extracellular biosynthesis of gold nanoparticles.[14] This could be an advantage from a process point of view, since it will eliminate the complex process of harvesting and recovery of the particles.

Although silver is toxic to most microorganisms, Klaus et al.[15] reported on a metal-accumulating bacterium, *Pseudomonas stutzeri* AG259, capable of producing silver-based single crystals in the size range of a few nanometers up to 200 nm. Growing these bacteria in a solution of $AgNO_3$ resulted in the reduction of the Ag^+ ions and the formation of silver nanoparticles of well-defined size and distinct morphology within the periplasmic space of the bacteria.

The exact mechanism leading to the formation of silver nanoparticles is yet to be elucidated, but it is speculated that the nanoparticle formation might result from specific mechanisms of resistance.[16] Furthermore, Joerger and colleagues[17] have shown that biocomposites of nanocrystalline silver and the bacteria, when thermally treated, yield a carbonaceous material with interesting optical properties for potential application in functional thin-film coatings.[17]

Even bacteria normally not exposed to metal ions could be used to synthesize nanoparticles. Nair and Pradeep[18] demonstrated the ability of *Lactobacillus* (present in buttermilk) to produce large amounts of gold and silver nanoparticles within the cells. In addition, the formation of alloy nanoparticles of silver and gold could be shown after exposure of the organisms to mixtures of silver and gold. Similarly, Jha and colleagues[19] and Jha and Prasad[20] demonstrated the use of *Lactobacillus* sp. to produce TiO_2 (8–35 nm) and $BaTiO_3$ (20–80 nm) particles.

In addition to gold and silver nanoparticles, attention has also been focused on the development of methods for the synthesis of semiconductors, so-called quantum dots, such as CdS, ZnS, and PbS. Holmes and co-workers[21] demonstrated that exposure of *Klebsiella aerogenes* to Cd^{2+} ions resulted in the formation of CdS nanoparticles in the nanometer range (20–200 nm).

Although much less is known about the bioreduction of PGMs, the sulfate-reducing bacterium *Desulfovibrio desulfuricans* has been shown to produce palladium nanoparticles in the presence of an electron donor.[22] Similarly, *Shwanella algae* produced platinum nanoparticles, around 5 nm in diameter, located in the periplasm when exposed to $PtCl_6^{2-}$ in the presence of lactate as electron donor.[23]

One of the most intriguing examples of biologically controlled mineralization is the formation of magnetic nanocrystals in magnetosomes within magnetotactic bacteria. Magnetosomes consist of magnetic iron mineral particles enclosed within membrane vesicles. In most cases the magnetosomes are organized in a chain or chains (Figure 9.1), and in many cases the iron mineral particles consist of magnetite (Fe_3O_4) or greigite (Fe_3S_4) characterized by narrow size distributions and uniform morphology.[24]

9.2.2 The Use of Yeast in Nanoparticle Synthesis

Yeasts have been successfully explored for the production of mainly cadmium nanoparticles. Dameron and co-workers[25] were the first to demonstrate that yeasts such as *Schizosaccharomyces pombe* and *Candida glabrata* were capable of intracellular production of CdS nanoparticles when challenged with cadmium salt in solution. These cadmium sulfide nanocrystals can potentially be used in quantum semiconductor crystallites.[25,26] Conditions have also been standardized for the synthesis of silver nanoparticles using a silver-tolerant yeast strain MKY3.[27] More recently, the yeast strain *Pichia jadinii* has been identified for its ability to produce gold nanoparticles, which included spheres, triangles, and hexagons,[28] and Agnihotri et al.[29] demonstrated the use of *Yarrowia lipolytica* for the synthesis of gold nanoparticles.

9.2.3 Nanoparticle Formation by Actinomycetes

Sastry et al.[30] and Ahmad et al.[31] described the extracellular formation of gold nanoparticles by the extremophilic actinomycete *Thermomonospora* when exposed to gold ions. Gold

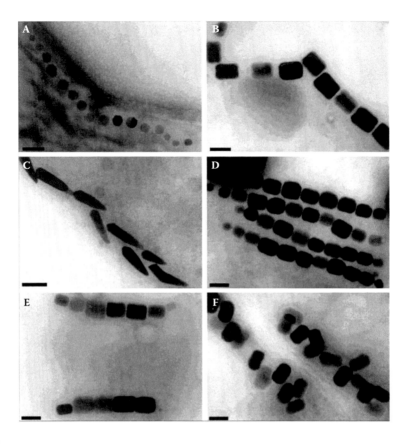

FIGURE 9.1
Electron micrographs of crystal morphologies and intracellular organization of magnetosomes found in various magnetotactic bacteria. Shapes of magnetic crystals include cubo-octahedral (A), elongated hexagonal pris-
matic (B, D–F,) and bullet-shaped morphologies (C). The particles are arranged in one (A–C), two (E), or multiple
chains (D) or irregularly (F). (Bar equivalent to 100 nm). (From Schuler, D., and Frankel, R.B., *Appl. Microbiol.
Biotechnol.*, 52, 464, 1999.) With permission.

nanoparticles with very good monodispersity have been produced, and the nanoparticles
were stable, even after up to 4 months incubation at 25°C. This is an important aspect of
nanoparticle synthesis, since the lack of sufficient stability of many nanoparticle preparations has to some extent impeded the development of applications of nanomaterials.
Similarly, Ahmad and co-workers[32] used the alkalotolerant *Rhodococcus* sp. for intracellular synthesis of good quality monodisperse gold nanoparticles.

9.2.4 The Use of Fungi in Nanoparticle Synthesis

One of the first works defining the use of fungi for nanoparticle synthesis was carried out by
Mukherjee and colleagues for the production of gold nanoparticles using *Verticillium* sp.[33]
Gold nanoparticles of around 20 to 25 nm in diameter were formed intracellularly. The
particles were mostly spherical, but there were also a few triangular and hexagonal particles present. They concluded that compared to bacteria, fungi could be a good source

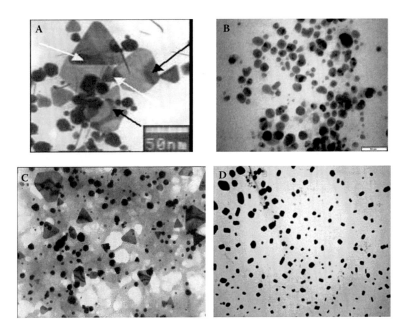

FIGURE 9.2
TEM photographs. (A) Au nanoparticles formed with *Trichothecium* sp. (From Ahmad, A., et al., *J. Biomed. Nanotechnol.*, 1, 47, 2005. With permission) (B) Ag particles synthesized by *A. fumigatus*. (From Bhainsa, K.C., and D'Souza, S.F., *Colloids Surf. B Biointerfaces*, 47, 160, 2006. With permission) (C) Au nanoparticles formed after exposing cell-free extract from *Verticillium luteoalbum* to AuCl₄⁻. (From Gericke, M., and Pinches, A., *Hydrometallurgy*, 83, 132, 2006. With permission) (D) Pt nanoparticles formed with *Fusarium oxysporum*. (From Riddin, T.L., et al., *Nanotechnology*, 17, 3482, 2006. With permission.)

of nanoparticles, since fungi are known to secrete much higher amounts of proteins and could potentially produce nanoparticles at a higher rate.

Further screening showed that *Fusarium oxysporum* behaved very differently when exposed to gold and silver ions. The reduction of the metal ions occurred extracellularly, resulting in the formation of stable gold and silver nanoparticles of 20–50 nm, an advantage from a downstream point of view.[34] During intracellular nanoparticle formation, an additional process step is needed to release the nanoparticles from the biomass, whereas extracellular synthesis results in a much cheaper and simpler process.[9]

Much attention has therefore been focused on the development of processes for the extracellular production of nanoparticles, and a number of reports describing the successful synthesis of extracellular nanoparticles can be found in the literature. Ahmad et al.[35] found that the fungus *Trichothecium* sp. can synthesize gold nanoparticles of different morphologies, both intra- and extracellularly, by changing its growing conditions. Gericke and Pinches[28] demonstrated the formation of gold nanoparticles, including triangles, hexagons, and spheres, when exposing cell-free extracts obtained from *Verticillium luteoalbum* to gold ions. The ability of *Fusarium oxysporum* to form Pt nanoparticles with interesting morphologies was reported by Riddin et al.[36]

It was shown that silver nanoparticles could also be produced extracellularly by a number of fungal species, including *Aspergillus flavus*,[37] *Fusarium semitectum*,[38] and *Aspergillus fumigatis*.[39] A few examples of nanoparticles produced by fungi are shown in Figure 9.2.

Recently Sanghi and Verma[40] demonstrated long-term operation with an immobilized fungus, *Coriolus versicolor*, in continuous column mode. The immobilized fungus served a dual purpose of both bioremediating cadmium and extracellularly synthesizing stable CdS nanoparticles in aqueous conditions. The fungus immobilized in the column could remove 98% cadmium within 2 h. The research highlighted the potential of this approach to remediate both waste streams and promotes large-scale growth of CdS nanoparticles.

9.3 Control of Growth Parameters to Manipulate Nanoparticle Characteristics

Although the screening results are very promising and show that a wide variety of organisms have the ability to produce a range of nanoparticles with interesting shapes, it is realized that for most applications, nanoparticles of well-defined size and shape are required. In addition, for a biological process to successfully compete with chemical nanoparticle synthesis, very strict control over average particle size in a specific size range and uniform particle morphology is required. The cellular mechanism leading to the reduction of the metal ions and formation of nanoparticles is not well understood, but it has been suggested that proteins and enzymes play a role in regulating gold crystal morphology.[33,41]

In an attempt to achieve better size and shape control, the effect of a number of growth parameters on gold nanoparticle synthesis using the fungal culture *V. luteoalbum* was investigated.[28] It was demonstrated that the rate of particle formation, and therefore the size of the nanoparticles, could, to an extent, be manipulated by controlling parameters such as pH (Figure 9.3), temperature, gold concentration, and exposure time to gold ions.

In similar studies, the effect of pH on the shape of gold nanoparticles produced in *Rhodopseudomonas capsulata* was established by He et al.[11] The results from TEM analyses shown in Figure 9.4 indicate the various sizes and shapes formed with pH changes ranging from 4 to 7.

Sanghi and Verma[42] investigated the ability of the fungus *Coriolus versicolor* to produce Ag nanoparticles. The resulting nanoparticles displayed controllable structural and optical properties, depending on experimental parameters such as pH and reaction temperature.

In another study, Ahmad et al.[35] showed that incubation of the fungus *Trichothecium* sp. in the presence of gold ions under stationary conditions resulted in the forma-

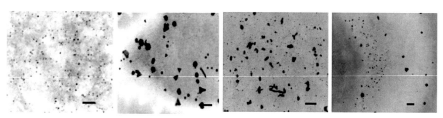

pH 3 (Scale bar = 100 nm)   pH 5 (Scale bar = 100 nm)   pH 7 (Scale bar = 100 nm)   pH 9 (Scale bar = 100 nm)

FIGURE 9.3
TEM images showing the effect of pH on gold nanoparticle morphology in *V. luteoalbum*. (From Gericke, M., and Pinches, A., *Hydrometallurgy*, 83, 132, 2006. With permission.)

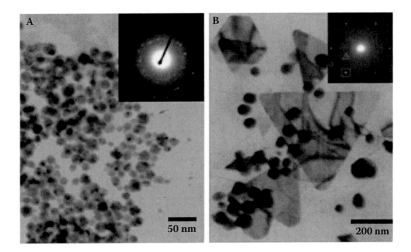

FIGURE 9.4
TEM analyses of the gold nanoparticles formed by *R. capsulata* at (A) pH 7 and (B) pH 4. (From He, S., et al., *Mater. Lett.*, 61, 3984, 2007. With permission.)

tion of extracellular nanoparticles, whereas incubation under shaking conditions led to intracellular gold nanoparticle formation.

9.4 Mechanism of Biological Nanoparticle Synthesis

While the findings discussed above give an indication of the potential for cellular manipulation to achieve control of the process within the cells, it is acknowledged that much research still remains to be done, especially in understanding the detailed mechanism of the reaction on a molecular and cellular level, including isolation and identification of the compounds responsible for the process.

So far, little is known about the interaction between biomolecules and nanoparticles, though various models for the enzymatic production of nanoparticles have been hypothesized. NADH, a coenzyme occurring in most cells could serve as a reductant in various metabolic processes. There are indications that NADH- and NADH-dependent enzymes are important factors in the biosynthesis of metal nanoparticles.[9,43–46] The models proposed are very similar, as they all point toward the possible role of an NADH-dependent reductase or hydrogenase being responsible for nanoparticle production.

An example is the bacterium *R. capsulate*, known to secrete cofactor NADH and NADH-dependent enzymes, which may be responsible for the bioreduction of Au^{3+} to Au^0 and the subsequent formation of gold nanoparticles (Figure 9.5A). The reduction seems to be initiated by electron transfer from the NADH by NADH-dependent reductase as an electron carrier; the gold ions obtain electrons and are reduced to Au^0.[11]

The same observation was reported with a strain of *F. oxysporum*, where it was demonstrated that a nitrate reductase and anthraquinone are involved in the in vitro production of silver nanoparticles.[9] The authors were able to purify the nitrate reductase and utilized the enzyme for the synthesis of nanoparticles. It is thought that the Ag^0 reduction was mainly due to a conjugation between the electron shuttle systems, with the reductase participating as shown in Figure 9.5B.

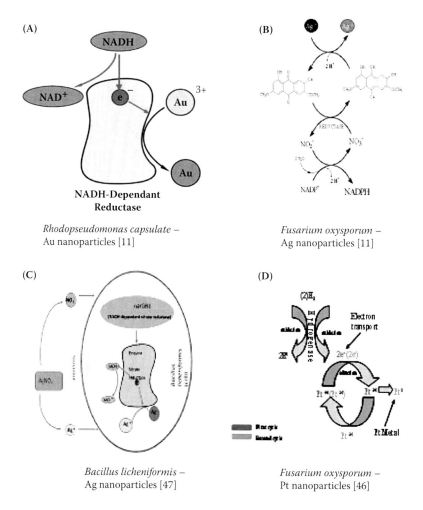

FIGURE 9.5
(See color insert.) Summary of typical models proposed for the mechanism of nanoparticle synthesis. (A) (From He, S., et al., *Mater. Lett.*, 61, 3984, 2007.) With permission (B) From Duran, N. et al., J. Nanobiotechnology, 3,8, 2005.With permission (C) From Kalimuthy, K. et al., Celloids Surf. B Biointes faces 65, 150, 2008. With permission (D) From Govender., Y. et al. Biotechnol. Lett, 31, 95, 2008. With permission.

In the case of *B. licheniformis* it is hypothesized that the enzyme involved in the synthesis of nanoparticles is induced by nitrate ions to reduce silver ions to metallic silver.[47] *B. licheniformis* is known to secrete the cofactor NADH and NADH-dependent enzymes, especially nitrate reductase, which are important factors in the biosynthesis of nanoparticles. The mechanism is the electron shuttle enzymatic metal reduction process (Figure 9.5c), earlier proposed for gold and silver nanoparticles. The mechanism of silver nanoparticles' formation was expanded by Kumar et al.[48] by demonstrating the synthesis of silver nanoparticles stabilized by the capping peptide in vitro using the enzyme nitrate reductase purified from *Fusarium oxysporum* and phytochelatin in the presence of a cofactor (α-NADPH).

Most models have been proposed for the synthesis of silver and gold nanoparticles. Govender et al.,[46] however, investigated Pt nanoparticle biosynthesis and suggested that platinum salts may act as an electron acceptor during the redox mechanism of a hydrogenase.

The authors proposed that a network of hydrophobic channels exists between the active site and the molecular surface in the Ni-Fe hydrogenase, and that these channels serve as a passage for metal ions. It is believed that at conditions suitable for platinum nanoparticle formation, the H_2PtCl_6 platinum salt is rapidly reduced by a passive two-electron transfer process into a species at a remote site on the molecular surface of the enzyme (Figure 9.5D). The Pt^{2+} migrates to the active region through the network of channels and is reduced, during conditions favorable for hydrogenase activity, by a second, slower two-electron reduction into Pt^0.

Another recent example is the work performed by Jha et al.,[19] who suggested that the synthesis of TiO_2 nanoparticles produced by *Lactobacillus* sp. and *S. cerevisiae* might have resulted due to pH-sensitive membrane-bound oxido-reductases, as well as the partial pressure of gaseous hydrogen of the culture solution, which seemed to play an important role in the process.

9.5 The Use of Templates for Nanoparticle Synthesis

Biological materials naturally display an astonishing variety of sophisticated nanostructures that are difficult to obtain even with the most technologically advanced synthetic methodologies. As the need for nano-engineered materials with improved performance characteristics is becoming increasingly important, the potential of biological scaffolds for the fabrication of novel types of nanostructures is being actively explored and recognized as a unique approach for the synthesis and organization of inorganic nanostructures into well-defined architectures.[49] These synthesis techniques utilize the molecular recognition properties of the biological molecules to nucleate and control growth of the nanoscale structure.[50]

A range of biotemplates has been evaluated and described in the literature, and a few examples are presented in Figure 9.6. A brief overview of advances in the use of protein-mediated biotemplating and trends involved in its applications for the nanostructuring of inorganic material is discussed here.

Certain proteins, such as ferritin and ferritin-like protein (FLP), have cavities in the center (Figure 9.6A), and these protein cavities can be used as templates for the growth of nanoparticles.[51] Yoshimura[51] showed that the protein shells served as a template to restrain particle growth and as a coating to prevent coagulation between nanoparticles, enabling the preparation of nanoparticles with uniform size and shape. Zhang et al.[52] also demonstrated that gold nanoparticles could form on the surface of apoferritin, and in turn become stabilized in solution by the protein cage.

Crystalline bacterial cell surface layers (S-layers) provide an outer cell envelope in many eubacteria and archaebacteria. These layers can be readily isolated and reconstructed in vitro to generate a two-dimensional monomolecular protein array with a pore size in the 2–6 nm range (Figure 9.6B).[7] Reconstructed S-layers have, for example, been used as templates for the in situ nucleation of ordered two-dimensional arrays of cadmium sulfide nanocrystals and gold nanoparticles, 5 nm in size,[54,55] and for core-shell quantum dot arrays.[56]

The synthesis of nanowires, 25–30 nm in diameter, templated using naturally occurring nano-sized tubes, e.g., rhapidosomes[57] and microtubuli,[58,59] has been reported (Figure 9.6C). Improved process control has enabled biotemplating directed to the formation of arrays of metallic silver grains, approximately 5.2 nm in size.[60]

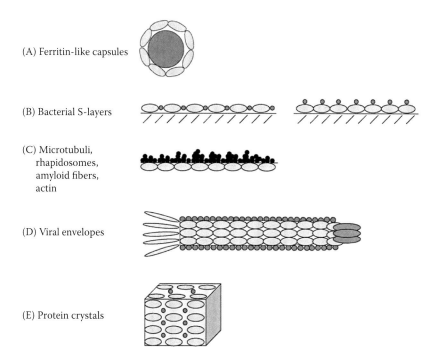

(A) Ferritin-like capsules

(B) Bacterial S-layers

(C) Microtubuli,
     rhapidosomes,
     amyloid fibers,
     actin

(D) Viral envelopes

(E) Protein crystals

FIGURE 9.6

The main configurations of protein biotemplates. (A) The apoferritin protein shell on the outside is loaded with magnetoferritin to produce magnetic particles. (B) Bacterial S-layers have been used as templates for the construction of two-dimensional arrays of nanoparticles. Protein molecules are shown as ovals and nanoparticles as spheres. (C) Microtubuli, rhapidosomes, amyloid fibers, and actin have all been employed as biotemplates for the construction of nanowires. The deposited metal grains are shown as black spheres. (D) The regularity of viral envelopes has been exploited for biotemplating. (E) The three-dimensional array of a protein crystal structure was used to synthesize a nanostructured protein-hydrogel complex. Ovals depict the regular arrangement of protein molecules; hydrogel molecules are in gray and fit within the voids. (From Lagziel-Simis, S., et al., *Curr. Opinion Biotechnol.*, 17, 569, 2006. With permission.)

Similarly to the protein ferritin, viruses have been used in the template-directed synthesis of nanostructures. A typical example of the use of virus envelopes (Figure 9.6D) is the use of the tobacco mosaic virus as a biotemplate for the synthesis of nickel and cobalt nanowires a few atoms in diameter.[61] Viral capsids have also been used for the synthesis of magnetic virus particles,[62] and gold cobalt hybrid nanowires have recently been synthesized on the envelope protein of the M13 phage. These hybrid nanowires were successfully used as highly effective anodes in a lithium ion battery.[53]

Using the protein-mediated control of crystal growth and the crystallization of gold as a model system, Brown et al.[41] found that polypeptides could control the morphology of the resulting gold crystals (Figure 9.6E). Similarly, Naik et al.[63] randomly identified silver binding peptides and demonstrated the reduction of silver ions by the peptides, Slocik et al.[64] described the successful formation of gold nanoparticles by a tyrosine tripeptide, and Braun et al.[65] utilized gold binding peptides of 14–30 amino acid repeats.

Nucleic acids are also ideal structural templates, and a variety of secondary structures can be achieved using DNA and RNA; e.g., duplexes, hairpins, triplexes, cruciforms, and tetraplexes can be exploited to engineer nanoparticle organization.[66]

9.6 Conclusions

Undoubtedly, nanomaterials will play a key role in many technologies of the future. One of the main challenges in this area is the development of reliable experimental protocols for the synthesis of nanoparticles over a range of chemical compositions, sizes, shapes, and high monodispersity.

In the last decade the biosynthesis of nanoparticles has received considerable attention, and the utilization of biological systems such as yeasts, fungi, and bacteria has been demonstrated for the synthesis of a wide range of precious and base metal nanoparticles. Although some success has been achieved in the biological generation of anisotropic nanoparticles, there are still big challenges, including the development of reliable approaches for stable production of nanoparticles of different materials, and understanding the detailed mechanisms involved in controlling the size and shape of these particles. Further efforts should now focus on demonstrating the reproducibility, scale-up, and control of size and shape of the process, as well as the properties on a larger scale, addressing mechanistic issues and determining the potential applications of these particles.

References

1. Chili, M.M., and Revaprasadu, N., Synthesis of anisotropic gold nanoparticles in water-soluble polymer, *Mater. Lett.*, 62, 3896, 2008.
2. Hall, S.R., Biotemplated syntheses of anisotropic nanoparticles, *Proc. R. Soc. A*, 465, 335, 2009.
3. Hao, E., Schatz, G.C., and Hupp, J.T., Synthesis and optical properties of anisotropic metal nanoparticles, *J. Fluoresc.*, 14, 331, 2004.
4. Tréguer-Delapierre, M., et al., Synthesis of non-spherical gold nanoparticles, *Gold Bull.*, 41, 195, 2008.
5. Roh, Y., et al., Microbial synthesis and the characterization of metal-substituted magnetites, *Solid State Commun.*, 118, 529, 2001.
6. Sarikaya, M., Biomimetics: materials fabrication through biology, *Proc. Natl. Acad. Sci. USA*, 96, 14183, 1999.
7. Sleytr, U.B., et al., Crystalline bacterial cell surface layers (S layers): from supramolecular cell structure to biomimetics and nanotechnology, *Angew. Chem. Int. Ed.*, 38, 1034, 1999.
8. Beveridge, T.J., et al., Metal–microbe interactions: contemporary approaches, *Microb. Phys.*, 38, 177, 1997.
9. Duran, N., et al., Mechanistic aspects of biosynthesis of silver nanoparticles by several *Fusarium oxysporum* strains, *J. Nanobiotechnol.*, 3, 8, 2005.
10. Beveridge, T.J., Role of cellular design in bacterial metal accumulation and mineralisation, *Ann. Rev. Microbiol.*, 43, 147, 1989.
11. He, S., et al., Biosynthesis of gold nanoparticles using the bacteria *Rhodopseudomonas capsulata*, *Mater. Lett.*, 61, 3984, 2007.
12. He. S., et al., Biological synthesis of gold nanowires using extract of *Rhodopseudomonas capsulata*, *Biotechnol. Progr.*, 24, 476, 2008.
13. Husseiny, M.I., et al., Biosynthesis of gold nanoparticles using *Pseudomonas aeruginosa*, *Spectrochim. Acta A*, 67, 1003, 2007.
14. Shahverdi, A.R., et al., Rapid synthesis of silver nanoparticles using culture supernatants of enterobacteria: a novel biological approach, *Process Biochem.*, 42, 919, 2007.

15. Klaus, T., et al., Silver-based crystalline nanoparticles, microbially fabricated, *Proc. Natl. Acad. Sci.*, 96, 13611, 1999.

16. Mandal, D., et al., The use of microorganisms for the formation of metal nanoparticles and their application, *Appl. Microbiol. Biotechnol.*, 69, 485, 2006.

17. Joerger, R., Klaus, T., and Granqvist, C.-G., Biologically produced silver-carbon composite materials for optically functional thin film coatings, *Adv. Mater.*, 12, 407, 2000.

18. Nair, B., and Pradeep, T., Coalescence of nanoclusters and formation of submicron crystallites assisted by *Lactobacillus* strains, *Crystal Growth Design*, 4, 295, 2002.

19. Jha, A.K., Prasad, K., and Kulkarni, A.R., Synthesis of TiO$_2$ nanoparticles using microorganisms, *Colloids Surf. B Biointerfaces*, 71, 226, 2009.

20. Jha, A.K., and Prasad, K., Ferroelectric BaTiO$_3$ nanoparticles: biosynthesis and characterization, *Colloids Surf. B Biointerfaces*, 75, 330, 2010.

21. Holmes, J.D., et al., Energy-dispersive x-ray analysis of the extracellular cadmium sulphide crystallites of *Klebsiella aerogenes*, *Arch. Microbiol.*, 163, 143, 1995.

22. Yong, P., et al., Bioreduction and biocrystallization of palladium by *Desulfovibrio desulfuricans* NCIMB 8307, *Biotechnol. Bioeng.*, 80, 369, 2002.

23. Konishi, Y., et al., Bioreductive deposition of platinum nanoparticles on the bacterium *Shewanella algae*, *J. Biotechnol.*, 128, 648, 2007.

24. Schuler, D., and Frankel, R.B., Bacterial magnetosomes: microbiology, biomineralization and biotechnological applications, *Appl. Microbiol. Biotechnol.*, 52, 464, 1999.

25. Dameron, C.T., Biosynthesis of cadmium sulphide quantum semiconductor crystallites, *Nature*, 338, 596, 1989.

26. Kowshik, M., et al., Microbial synthesis of semiconductor CdS nanoparticles, their characterization and their use in the fabrication of an ideal diode, *Biotechnol. Bioeng.*, 78, 583, 2002.

27. Kowshik, M., et al., Extracellular synthesis of silver nanoparticles by a silver-tolerant yeast strain MKY3, *Nanotechnology*, 14, 95, 2003.

28. Gericke, M., and Pinches, A., Biological synthesis of metal nanoparticles, *Hydrometallurgy*, 83, 132, 2006.

29. Agnihotri, M., et al., Biosynthesis of gold nanoparticles by the tropical marine yeast *Yarrowia lipolytica* NCIM 3589, *Mater. Lett.*, 63, 1231, 2009.

30. Sastry, M., et al., Biosynthesis of metal nanoparticles using fungi and actinomycete, *Curr. Sci.*, 85, 162, 2003.

31. Ahmad, A., et al., Extracellular biosynthesis of silver nanoparticles using the fungus *Fusarium oxysporum*, *Colloids Surf. B Biointerfaces*, 28, 313, 2003.

32. Ahmad, A., et al., Intracellular synthesis of gold nanoparticles by a novel alkalotolerant actinomycete, *Rhodococcus* species, *Nanotechnology*, 14, 824, 2003.

33. Mukherjee, P., et al., Bioreduction of AuCl$_4^-$ ions by the fungus, *Verticillium* sp. and surface trapping of the gold nanoparticles formed, *Angew. Chem. Int. Ed.*, 40, 3585, 2001.

34. Mukherjee, P., et al., Extracellular synthesis of gold nanoparticles by the fungus *Fusarium oxysporum*, *ChemBioChem*, 5, 461, 2002.

35. Ahmad, A., et al., Extra-/intracellular biosynthesis of gold nanoparticles by an alkalotolerant fungus, *Trichothecium* sp., *J. Biomed. Nanotechnol.*, 1, 47, 2005.

36. Riddin, T.L., Gericke, M., and Whiteley, C.G., Analysis of the inter- and extracellular formation of platinum nanoparticles by *Fusarium oxysporum* f. sp. *Lycopersici* using response surface methodology, *Nanotechnology*, 17, 3482, 2006.

37. Vigneshwaran, N., et al., Biological synthesis of silver nanoparticles using the fungus *Aspergillus flavus*, *Mater. Lett.*, 61, 1413, 2007.

38. Basavaraja, S., et al., Extracellular biosynthesis of silver nanoparticles using the fungus *Fusarium semitectum*, *Mater. Bull.*, 43, 1164, 2007.

39. Bhainsa, K.C., and D'Souza, S.F., Extracellular biosynthesis of silver nanoparticles using the fungus *Aspergillus fumigatus*, *Colloids Surf. B Biointerfaces*, 47, 160, 2006.

40. Sanghi, R., and Verma, P., A facile green extracellular biosynthesis of CdS nanoparticles by immobilized fungus, *Chem. Eng. J.*, 155, 886, 2009.

41. Brown, S., Sarikaya, M., and Johnson, E., A genetic analysis of crystal growth, *J. Mol. Biol.*, 299, 725, 2000.

42. Sanghi, R., and Verma, P., Biomimetic synthesis and characterisation of protein capped silver nanoparticles, *Bioresource Technol.*, 100, 1, 2009.

43. Huang, Y., Programmable assembly of nanoarchitectures using genetically engineered viruses, *Nanotechnol. Lett.*, 5, 1429, 2005.

44. Willner, I., Baron, R., and Willner, B., Growing metal nanoparticles by enzymes, *Adv. Mater.*, 18, 1109, 2006.

45. Willner, I., Basnar, B., and Willner, B., Nanoparticle enzyme hybrid systems for nanobiotechnology, *FEBS J.*, 274, 302, 2007.

46. Govender, Y., et al., Bioreduction of platinum salts into nanoparticles: a mechanistic perspective, *Biotechnol. Lett.*, 31, 95, 2008.

47. Kalimuthu, K., Biosynthesis of silver nanocrystals by *Bacillus licheniformis, Colloids Surf. B Biointerfaces*, 65, 150, 2008.

48. Kumar, S.A., et al., Nitrate reductase-mediated synthesis of silver nanoparticles from AgNO3, *Biotechnol. Lett.*, 29, 439, 2007.

49. Sotiropoulou, S., et al., Biotemplated nanostructured materials, *Chem. Mater.*, 20, 821, 2009.

50. Kriplani, U., and Kay, B.K., Selecting peptides for use in nanoscale materials using phage-displayed combinatorial peptide libraries, *Curr. Opinion. Biotechnol.*, 16, 470, 2005.

51. Yoshimura, H., Protein-assisted nanoparticle synthesis, *Colloids Surf. A Physicochem. Eng. Aspects,* 283, 464, 2006.

52. Zhang, L., Structure and activity of appoferritin stabilized gold nanoparticles, *J. Inorg. Biochem.*, 101, 1719, 2007.

53. Lagziel-Simis, S., et al., Protein mediated nanoscale biotemplating, *Curr. Opinion Biotechnol.*, 17, 569, 2006.

54. Shenton, W., et al., Synthesis of cadmium sulfide superlattices using self-assembled bacterial S-layers, *Nature*, 389, 585, 1997.

55. Dieluweit, S., Pum, D., and Sleytr, U.B., Formation of a gold superlattice on an S-layer with square lattice symmetry, *Supramol. Sci.*, 5, 15, 1998.

56. Mark, S.S., et al., Bionanofabrication of metallic and semiconductor nanoparticle arrays using S-layer protein lattices with different lateral spacings and geometries, *Langmuir*, 22, 3763, 2006.

57. Pazirandeh, M., Bural, S., and Campbell, J.R., Metallized nanotubules derived from bacteria, *Biomimetics*, 1, 41, 1992.

58. Kirsch, R., et al., Three dimensional metallization of microtubules, *Thin Solid Films*, 305, 248 1997.

59. Mertig, M., Kirsch, R., and Pompe, W., Biomolecular approach to nanotube fabrication, *Appl. Physics A*, 66, S723, 1998.

60. Behrens, S., et al., Silver nanoparticle and nanowire formation by microtule templates, *Chem. Mater.*, 16, 3085, 2004.

61. Knez, M., et al., Biotemplate synthesis of 3nm nickel and cobalt nanowires, *Nanotechnol. Lett.*, 3,1079, 2003.

62. Liu, C., et al., Magnetic viruses via nano-capsid template, *J. Magnetism Magn. Mater.*, 302, 47, 2006.

63. Naik, R.R., et al., Biomimetic synthesis and patterning of silver nanoparticles, *Nature Mater.*, 1, 169, 2002.

64. Slocik, J.M., et al., Viral templates for gold nanoparticles synthesis, *J. Mater. Chem.*, 15, 749, 2005.

65. Braun, R., Mehmet, S., and Klaus, S., Genetically engineered gold-binding polypeptides: structure prediction and molecular dynamics, *J. Biomater. Sci. Polym. E*, 13, 747, 2002.

66. Gourishankar, A., et al., DNA and RNA as templates for building nanoassemblies via electrostatic complexation with gold nanoparticles, *Curr. Appl. Phys.*, 5, 102, 2004.

10

Nanotechnology in Biomaterials: Nanofibers in Tissue Engineering

Deniz Yucel, Halime Kenar, Albana Ndreu, Tugba Endoğan, Nesrin Hasirci, and Vasif Hasirci

BIOMATEN, Center of Excellence in Biomaterials and Tissus Engineering, and Departments of Biological Sciences and Chemistry, Middle East Technical University, Ankara, Turkey

CONTENTS

10.1 Biomaterials

Devices made of synthetic or natural materials have been used in the medical area and introduced into the human body to improve human health since ancient times. Romans, Chinese, and Aztecs used gold wires in dentistry more than 2,000 years ago. Ancient Egyptians and Greeks sutured wounds with plant fibers and animal-derived materials and used wood for prosthetic limbs. Since then, scientists have continued their research to find bioactive agents to cure illnesses and to improve the quality of human life by using artificial prostheses.

There are various definitions of biomaterials expressed in different ways given in the literature, but more or less they have the same meaning. One of the very early definitions came from the Clemson University Advisory Board for Biomaterials (1976), which described a biomaterial as "a systemically and pharmacologically inert substance designed for implantation within or incorporation with living systems" (Park, 1981). Another definition of biomaterials is "materials of synthetic as well as of natural origin in contact with tissue, blood, and biological fluids, and intended for use for prosthetic, diagnostic, therapeutic, and storage applications without adversely affecting the living organism and its components" (Bruck, 1980). Biomaterials can be defined as materials to be used to substitute a tissue or organ of a living system that is not fully functioning, or to support the biological system in its function while in intimate contact with the tissues in a safe, reliable, economic, and physiologically acceptable manner (Park and Bronzino, 2003). Black (1999) defined biomaterial as "a nonviable material used in a medical device, intended to interact with biological systems." In addition, during the Second Consensus Conference on Definitions in Biomaterials Science, a biomaterial was defined as "a material intended to interface with biological systems to evaluate, treat, augment, or replace any tissue, organ or function of the body" (Williams et al., 1987; Williams, 1992). The same definition is found in the *Williams Dictionary of Biomaterials* (Liverpool University Press, 1999).

Biomaterials are used in many application areas, such as orthopedic (total joint replacements, bone plates and screws, etc.), dental (tooth fillings, crowns, implants, etc.), cardiovascular (cardiac pacemaker, artificial heart valve, blood vessels, etc.), ophthalmic (contact and intraocular lenses), wound healing (sutures, wound dressing, tissue adhesive, ligating clips, staples, etc.), and drug delivery (oral, inhalation injectable delivery systems, micro- and nanodevices, sheets, fibers, and sponges, etc.) systems. Biomaterials can be found in different forms, such as tubes (artificial veins, artificial trachea), fibers (sutures), films and membranes (wound dressings, drug delivery systems, artificial kidney), gels (burn covers, topical drug delivery systems), and porous scaffolds (tissue engineering), depending on the site and type of use. Metals (titanium, stainless steel, cobalt-chromium alloys, gold, silver, platinum), polymers (nylon, silicones, Teflon, dacron, polylactides), ceramics (aluminum oxide, carbon, hydroxyapatite), natural materials (collagen, hyaluronic acid, reconstituted tissue such as porcine heart valves), and composites (carbon or ceramic wires, fiber-reinforced polymers) have been used as biomaterials in the design and production of biomedical devices. Material to be used as a biomaterial must fulfill some requirements. First, it should be compatible with the biological environment and should not cause any adverse carcinogenic or allergic tissue reaction; i.e., it should be biocompatible. Degradation products of a biomaterial must also be nontoxic and noncarcinogenic, and have the desired bioactivity toward the cells and tissue. Physical, chemical, and mechanical properties of the biomaterial should match those of the target tissue.

10.2 Tissue Engineering

Tissue engineering was once considered a subfield of biomaterials, but since it has gained too much importance in the last decades, it is now considered a field in its own right. Tissue engineering is the process of creating living three-dimensional (3D) tissues and organs by using engineering and materials methods, and specific combinations of cells and cell signals, both chemical and mechanical (Griffith, 2002). This approach aims to reproduce both the form and function of the tissue. Different from nonliving biomaterials, tissue engineering is based on the formation and regeneration of a new functional tissue by combining living components (cells, tissue fragments) and carrier scaffolds in a construct to be implanted into the body (Kneser et al., 2002; Guelcher et al., 2006).

There are three approaches in repairing tissues (Langer et al., 1993; Ma, 2004b). The first one is the injection of tissue-specific viable cells directly into the damaged site of the tissue. The second approach is the delivery of tissue-inducing substances, such as growth and differentiation factors, to target locations. The third one is the tissue engineering approach, in which cells are grown in 3D scaffolds (Langer et al., 1993). The use of cell suspensions or tissue-inducing substances is considered when the defects are small, well contained, or not suitable for implanting a 3D structure. To engineer tissues with practical size and predetermined shapes, the first two approaches are not appropriate. The main principle of the tissue engineering strategy is to seed a porous, biodegradable 3D cell carrier (scaffold, matrix) with cells, add signaling molecules like growth factors, culture the construct in a medium for growing and proliferation of cells, and then implant it into the defect area to grow a new tissue. It is expected that the cells of the natural tissue would also attach to the scaffold, proliferate, differentiate, and get organized, forming a normal, healthy tissue while the scaffold itself degrades. At the end, healthy tissue fills the defect site and no artificial matrix remains. Scaffolds, cells, and signaling biomolecules are the main components of tissue engineering devices. All of these components play very important roles during tissue regeneration. The extracellular matrix (ECM) is a microenvironment composed of parenchymal cells (functional cells), mesenchymal cells (support cells), and structural materials. Every tissue and organ in our body is formed in these microenvironments. The body serves as a bioreactor, which applies biomechanical forces and provides biochemical signals to the cells and the ECM to develop and maintain tissue and organs (Barnes et al., 2007). In the tissue engineering approach, the biological way of tissue formation in the body is mimicked by taking place in a porous matrix.

A scaffold is one of the essential components of tissue engineering since it provides 3D support for the cells, as ECM does. In vitro, cells do not grow in a way to fill a volume if there is no 3D support. Porous scaffolds provide some place for cell attachment and viability through transport of nutrients and metabolites, and allow cell expansion and tissue organization. There are some requirements for proper scaffolds. First, they should be biocompatible, and therefore should not cause any adverse responses, such as allergic and carcinogenic reactions in the body. They must have an interconnected, highly porous structure with a high surface area for cell adhesion and reorganization, for transport of nutrients and removal of metabolic waste, and for homogeneous cell distribution throughout the scaffold. They should have adequate mechanical strength to withstand the pressures and mechanical forces applied by the biological system. The degradation rate of the scaffold should be in accordance with the tissue growth; when the tissue is totally regenerated, the scaffold should be almost fully degraded. A suitable surface chemistry is extremely important for cell attachment, proliferation, and differentiation (Ma, 2004a; Stevens et al., 2008.) The material of the scaffold should be easily processed to achieve the desired shapes and sizes. Finally, the scaffold should have some specific properties for specific tissue engineering applications, such as osteoconductivity for bone, electrical conductivity for nerve, and contraction for myocardial tissue engineering.

Various materials have been studied as potential scaffolds. Some metals are good choices for load-bearing implants since they have superior mechanical properties (Catledge et al., 2004). Recently, porous titanium (Ti) and titanium alloys were used as scaffold biomaterials for bone tissue engineering, but they are not degradable in the body. Inorganic materials are also used in bone tissue engineering studies (Hench and Polak, 2002). Ceramics such as hydroxyapatite (HAp), Bioglass, A-W glass ceramics, and β-tricalcium phosphate (β-TCP) closely mimic bone tissues. They are osteoconductive or osteoinductive, and enhance the absorption of bone ECM proteins, resulting in improved osteoblast adhesion (Ducheyne and Qiu, 1999; Dieudonné et al., 2002; Smith et al., 2004).

However, ceramics have some drawbacks, like fragility, brittleness, low tensile strength, and being mostly not biodegradable. Biological and synthetic polymeric materials have been widely used in various tissue engineering applications due to their degradability, biocompatibility, mechanical properties suitable for many applications, and ease of processability. Some examples of natural biodegradable polymers are collagen, fibrinogen, gelatin, chitosan, cellulose, alginate, starch, silk, and hyaluronic acid. The main advantages of these materials are their low immunogenicity, bioactivity, capability of interacting with the host tissue, chemical versatility, and availability from various sources, as in starch and chitosan. Of these, collagen, fibrin, hyaluronic acid, agarose, chitosan, and alginate are materials especially used in the production of scaffolds for engineering of bone and cartilage (Sabir et al., 2009). The most widely used synthetic polymers are a family of linear aliphatic polyesters, poly(α-hydroxy acid) (PHA), such as poly(lactic acid) (PLA), poly(glycolic acid) (PGA), and their copolymer poly(lactic acid-co-glycolic acid) (PLGA) (Mooney et al., 1996; Seal et al., 2001). For example, PGA has been used in the engineering of cartilage, tendon, ureter, intestine, blood vessel, heart valve, and other tissues. Poly(ε-caprolactone) (PCL), poly(propylene fumarate) (PPF), poly(carbonate), poly(phosphazenes), poly(phosphoesters) and poly(anhydrides), tyrosine-derived polymers, and biodegradable urethane-based polymers are other important synthetic, biodegradable polymers used in tissue engineering applications. A problem with synthetic polymers is their degradation products, which may reduce the local pH, induce an inflammatory response, or create some adverse reactions and being nonbiocompatible.

In the past few years, polymer-ceramic composites have gained increased importance in the engineering of several types of tissues, such as bone, cartilage, tendon, and ligament (Marra et al., 1999; Zhang and Ma, 1999; Ma et al., 2001; Zhao et al., 2002; Liu and Ma, 2004). The composite is expected to have improved compressive mechanical properties compared to the polymer, and better structural integrity and flexibility than ceramics, and thus the combination of ceramic and polymer created reinforced porous scaffolds with enhanced bioactivity and controlled resorption rates (Zhang and Ma, 1999; Ma et al., 2001).

During the selection of a scaffold material, an important step is to decide on the fabrication technique. The processing technique is expected not to change certain material properties, like biocompatibility and surface chemistry. The technique should be reproducible in terms of scaffold porosity, pore size, pore distribution, and interconnectivity (Leong et al., 2003). There are many techniques that can be used to process materials into scaffolds, such as solvent casting, phase inversion, fiber bonding, melt-based technologies, high-pressure-based methods, freeze drying, and rapid prototyping (Ma and Langer, 1999; Kinikoglu et al., 2009; Sangsanoh et al., 2009; Yilgor et al., 2008; Vrana et al., 2007; Zorlutuna et al., 2007; Mooney et al., 1996; Thomson et al., 1995; Mikos et al., 1993). Each production technique has some disadvantages and advantages over the others and may not be applicable to some polymers. Therefore, appropriate processing techniques should be chosen to produce a scaffold with the desired properties for the specific application.

To date, scaffolds have been produced in our group in various forms, like foam (Zorlutuna et al., 2008; Kinikoglu et al., 2009, Dogan et al., 2009, Ulubayram et al., 2001), film (Vrana et al., 2007; Zorlutuna et al., 2007; Kenar et al., 2008; Zorlutuna et al., 2009), and fibers (Ndreu et al., 2008; Kenar et al., 2009; Yucel et al., 2010), each having different chemistries, and a number of studies have been carried out with them either in vitro or in vivo. In the last decades, foam and film-based scaffolds were popular for 3D and 2D applications; however, in recent years there has been an increase in the use of fibrous scaffolds, especially made by electrospinning.

10.3 Importance of Nano/Microfibrous Scaffolds

Many extracellular proteins have a fibrous structure with diameters on the low or submicrometer scales. For example, collagen is the most abundant ECM protein in the human body, and it has a fibrous structure with fiber bundle diameters varying from 50 to 500 nm, depending on the site and species (Hay, 1991). In the native tissues, the structural ECM proteins are one to two orders of magnitude smaller than the cells, and therefore, the cells are in direct contact with many ECM fibers. It is now well known that many biologically active molecules, extracellular matrix components, and cells interact with each other using functional groups on the nanoscale. It is therefore a useful approach to mimic the fibrous nature of the extracellular matrix when fabricating scaffolds (Barnes et al., 2007). Nanofibers have a high surface-to-volume ratio, which enhances the number of cells that adhere to the surface (Ma and Langer, 1999). Cell migration, proliferation, and differentiation follow adhesion (Palecek et al., 1997), and it is expected that these reactions will be much higher on nanofibrous scaffolds than on foams or films. As a result, nanofibers have attracted the attention of many researchers who considered them as potential scaffolds for applications such as tissue engineered vasculature (Xu et al., 2004; Lee et al., 2007), bone (Sui et al., 2007; Cui et al., 2007), nerve (Yang et al., 2005; Schnell et al., 2007), and tendon and ligament (Lee et al., 2005; Sahoo et al., 2006).

10.4 Fabrication of Nano/Microfibrous Scaffolds

Since fibrous scaffolds with nanometric fiber diameters have been found to be satisfactory for various tissue engineering applications, extensive research toward developing processes for the fabrications of these fibrous structures is being conducted. The main methods used in the fabrication of fibrous scaffolds are self-assembly, template synthesis, drawing and extrusion, phase separation, wet spinning, and electrospinning. Each of these procedures leads to fibers with different dimensions and organizations.

In the case of self-assembly, atoms and molecules arrange themselves through weak and noncovalent interactions, and it is known as a bottom-up method since the built-up nanoscale fibers are formed starting from smaller molecules. It yields fibers with small dimensions (less than 100 nm wide and up to a few micrometers long), and it offers novel properties and functionalities that cannot be achieved by conventional organic synthesis (Hasseinkhan et al., 2006). The main disadvantage of this method is its complexity, in addition to being a long and extremely elaborate process with low productivity (Ma et al., 2005).

Another fabrication method is template synthesis. As the name implies, a nanoporous membrane is used as a mold or template in order to obtain the material of interest in the form of solid nanofibrous or hollow-shaped tubules. This method uses a polymer solution to be extruded through a porous membrane by application of pressure. Fibers are formed when the extruded polymer comes into contact with a precipitating solution (Feng et al., 2002). Fiber diameter depends on the pore size of the template used, and it varies from a few to a hundred nanometers. Many materials, like metals, polymers, and carbon, can be utilized as templates for fiber fabrication; however, this method is limited in that the length of nanofibers can only be a few micrometers.

Drawing is a method to produce single nanofibers. It requires a minimum amount of equipment and is a discontinuous process. A micropipette is dipped into a droplet near the solution-solid surface contact line via a micromanipulator, and then the micropipette is withdrawn from the liquid at a certain speed, yielding a nanofiber. These steps are repeated many times, each time with a different droplet. The solution viscosity, however, increases with solvent evaporation, and some fiber breaking occurs due to instabilities that occur during the process (Ondarcuhu and Joachim, 1998). The main limitation of this method is that it can only use solutions of viscoelastic materials, which can undergo strong deformations that are cohesive enough to withstand the stresses developed during the pull. Again, the fiber diameter is dependent on the orifice size used. It is difficult to obtain fiber diameters of less than 100 nm.

Phase separation is another process utilized in nanofiber production. The main principle of this method is that a polymer dissolved in an appropriate solvent is stirred at a certain gelation temperature for a period of time until a homogeneous solution is obtained. Then, the formed gel is immersed in water several times for solvent exchange to occur. The final gel is lyophilized. Phase separation occurs due to chemical incompatibility and results in nanofibers, but the problem is that a long time is needed to convert the polymer into a nanoporous foam. The fiber diameter ranges from 50 to 500 nm with a length of a few micrometers (Zhang et al., 2005), which is a disadvantage for some applications. Moreover, only polymers that have gelation capability can be used to obtain nanofibrous structures.

Wet spinning is a process similar to drawing, with the difference being that a polymeric solution is precipitated or coagulated by dilution, adding into a nonsolvent or a chemical reaction. Fibers produced by this method are continuous and in microscale (from 10 to 100 μm). Polymer solution concentration and the spinneret diameter are the two crucial parameters that influence the resulting fiber diameter.

The last, and one of the most preferred, methods used to prepare fibers in nano- and low microscale is the electrospinning process. High electrostatic forces are applied to draw continuous fibers from a polymeric solution through a syringe. Basically, four components are required to accomplish the process: a capillary tube ending in a needle of small diameter, a high-voltage supplier, a syringe pump, and a metallic component. One of the electrodes is connected to the needle of the syringe containing the polymer, and the other one is connected to the collector. Leakage of the polymer solution is prevented by surface tension; however, as the applied potential increases, the surface tension contribution starts to decrease as a result of charge repulsion, and the fluid shape at the needle tip becomes like a cone, what is called the Taylor cone. The voltage is increased until the repulsive electrostatic forces overcome the influence of solution surface tension. Meanwhile, the solvent evaporates and the charged jet of the solution is ejected from the needle tip in the shape of a long, continuous, thin fiber. Therefore, fibers with various diameters are collected on either a stationary or rotating collector. Rotating collectors are used to produce aligned nanofibers, which are more preferable in certain applications. A variety of polymers have been used until now to produce fibers from tens of nanometers to a few micrometers (Jin et al., 2004; Li et al., 2005; Ma et al., 2005; Ndreu et al., 2008; Erisken et al., 2008; Jeong et al., 2009; Kenar et al., 2009; Yucel et al., 2010).

Certain properties are expected from a polymer that is proven to be spinnable. The fibers should be consistent, defect-free, and have controllable diameters. The main parameters to control are process conditions such as polymer type, its concentration, solvents, applied potential, distance between the needle tip and the collector, and the collector type. These parameters are modified to control fiber thickness, porosity and pore size (or rather the spacing between the fibers or fiber density), surface chemistry, and surface topography

(Deitzel et al., 2001; Huang et al., 2003; Katti et al., 2004; Barhate et al., 2006; Ndreu et al., 2008; Wang J et al., 2009). Therefore, optimization of the process conditions in order to get the best results is challenging with so many parameters.

Finally, it can be said that compared to the other nano/microfiber-production methods, electrospinning is more economical, simpler, yields continuous fibers, and is versatile enough to be used for the production of nanofibers from a variety of materials. Therefore, electrospinning represents an attractive technique to be used for producing nanofiber-based scaffolds for tissue engineering.

10.5 Nano/Microfibers in Tissue Engineering Applications

The use of polymeric nano/microfibers in tissue engineering applications is rapidly growing. The main areas include the engineering of skin (Venugopal and Ramakrishna, 2005; Noh et al., 2006; Yang X et al., 2009; Jeong et al., 2009; Torres et al., 2009), ligament and tendon (Petrigliano et al., 2006; Teh et al., 2007, 2008; Sahoo et al., 2009; Fan et al., 2009; Yin et al., 2009), skeletal muscle (Riboldi et al., 2005; Choi et al., 2008), cartilage (Li WJ et al., 2005; Li et al., 2006; Janjanin et al., 2008; Lu et al., 2008; Hu et al., 2009), bone (Santos et al., 2008; Zhang et al., 2008; Reed et al., 2009; Wang J et al., 2011; Prabhakaran et al., 2009), cardiovascular (Kenar et al., 2009; Rockwood et al., 2008; Li WJ et al., 2006), and nerve (Yucel et al., 2010; Neal et al., 2009; Nisbet et al., 2008; Corey et al., 2007) tissues. The increase in the number of review articles (Ashammakhi et al., 2007a, 2007b, 2008; Agarwal et al., 2009; Jang et al., 2009) and publications related to the application of nano/microfibers in these fields shows the interest in electrospinning for tissue engineering. In this review we will concentrate more on the studies carried out by our research group, and also by some of the leading researchers in the fields of bone, cardiovascular, and nerve tissue engineering.

10.5.1 Bone Tissue Engineering

Prior to designing a bone substitute or a tissue-engineered bone, it is important to know well the structure and function of the natural bone so that the material type chosen is the most suitable to regenerate the function of the damaged bone. Bone is a complex, physically hard, rigid, and highly organized tissue. This tissue is composed of an intracellular matrix in a fibrous form made mainly of collagen types I and V, and some inorganic materials. Apart from the major collagen matrix, some other noncollageneous proteins, like osteocalcin and osteopontin, are also present. There are three cell types present in the bone structure: osteoblasts, osteocytes, and osteoclasts. Osteoblasts, derived from mesenchymal precursor cells in the marrow, are the bone cells responsible for formation and structural organization of the bone matrix, and also bone mineralization. When osteoblasts become trapped in the matrix, they secrete and become osteocytes. Osteocytes are the most abundant cells found in compact bone. They are networked to each other via long cytoplasmic extensions that occupy tiny canals called canaliculi, which are used for exchange of nutrients and waste. The space that an osteocyte occupies is called a lacuna. Unlike osteoblasts, these cells have reduced synthetic activity and are not capable of mitotic division. They are actively involved in the turnover of bony matrix, through various mechanosensory mechanisms. They destroy bone through a mechanism called osteocytic osteolysis. The third bone cell type, osteoclasts, are polarized cells and have a crucial role in bone resorption, which includes an initial mineral dissolution and then an organic phase degradation (Jang et al., 2009).

As mentioned above, the main approach in the tissue engineering process is mimicking the ECM. Therefore, for bone tissue engineering it is crucial to make a design whose properties would mimic best the structure and function of bone ECM. It is important to choose a material that interacts well with bone cells in the bone formation processes. Moreover, it is favorable that materials with good mechanical properties and can withstand various external stresses applied on bone are incorporated into the material design. Since the organization of all bone compartments is in nanoscale or at the molecular level, a nanoscale-based design would be optimal to mimic the natural bone. Various strategies have been tried in designing a tissue engineered bone.

Both synthetic and natural materials have been utilized in this field. Among all polymeric materials, a group of polyesters, such as PLA, PGA, PCL, poly(3-hydroxybutyric acid) (PHB), and their copolymers, has been the most extensively studied polymers to produce the nanofibrous system for the regeneration of bone tissues (Burg et al., 2000). A summary of nano/microfibrous scaffolds used in bone tissue engineering studies is given in Table 10.1. Yoshimoto et al. (2003) used electrospun PCL nanofibers, and microscopic analysis of mesenchymal stem cell-seeded scaffolds revealed a good cell penetration in addition to a high level of mineralization and collagen type I production after 4 weeks of culture (Yoshimoto et al., 2003). Sombatmankhong et al. produced mats from PHB, poly(3-hydroxybutyrate-co-3-hydroxyvalerate) (PHBV) and their 50/50 w/w blends and studied the biocompatibility of these mats by a human osteosarcoma cell line (Saos-2) and mouse fibroblasts (L929). Both cells adhered well on all types of fibrous scaffolds; however, the highest alkaline phosphatase (ALP) activity was seen on the PHB/PHBV blend (Sombatmankhong et al., 2007).

In another study, PLA and poly(ethylene glycol) (PEG)-PLA diblock polymers with controlled topographies and chemistries were prepared by electrospinning in order to study the influence of surface topography and fiber diameter on cell behavior. In the presence of osteogenic factors the cell numbers on fibrous scaffolds were reported to be equal to or higher than those on smooth surfaces. Furthermore, the authors indicated an increase in cell density with an increase in fiber diameter and no influence of surface substratum on ALP activity (Badami et al., 2006). The influence of fiber diameter on cell behavior was also studied by other researchers. Ndreu et al. (2008) studied the influence of different types of nano/microfibers (diameter range: 300 nm–1.5 μm) on Saos-2 cells. Good cell adherence, proliferation, and infiltration were observed on all types of scaffolds (PHBV, PHBV-PLGA, PHBV-P(L,DL)LA, and PHBV-PLLA); however, the highest number of cell attachments was observed on the fibers with the lowest diameter (PHBV-PLLA fiber diameter: ~350 nm) when the scaffolds were examined with a scanning electron microscope (SEM) and a confocal laser scanning microscope (CLSM) (Figure 10.1).

Reed and his co-workers (2009) carried out a study on the reconstruction of a compound tissue based on the fact that healing does not occur as a single tissue, but as a compound one, such as bone-periosteum-skin. In order to perform this study, they seeded human foreskin fibroblasts, murine keratinocytes, and periosteal cells on PCL nanofibers and showed that these scaffolds supported the growth of all these cell types. Even though keratinocytes grew randomly throughout the scaffolds, they reported improved longevity of the cocultured cells and a strong degree of osteoinduction. As a consequence, construction of a composite tissue has been reported by the authors (Reed et al., 2009).

Compared to the synthetic materials, natural materials have better biocompatibility, and more importantly, they possess the main cues that are necessary for a cell to attach and grow. Natural polymers are also advantageous in that they can be degraded by naturally

TABLE 10.1

Fibrous Scaffolds Fabricated by Electrospinning for Bone Tissue Engineering

Polymer Type	Polymer Concentration/Composition/Solvent	Fiber Diameter	In Vitro Studies: Cell Type	Reference
PLA/BG nanofiller	3.5%, w/v; 90:10, w:w CHL	Hundreds of nanometers	MC3T3-E1	Noh et al., 2010
Gelatin/siloxane	15–25% w/w Gelatin:GPMS:CaNO₃ formic acid	Nanometer scale	BMSCs	Ren et al., 2010
PVA/COL/n-HAp	7% w/w PVA:COL (55:45) n-HAp: 5–10% w/w of PVA deionized water, HAc	~320 nm	—	Asran et al., 2010
PAN/β-TCP	10% w/w DMF	~300 nm	MG-63	Liu et al., 2010
CHI	7% w/w TFA/DCM 70:30 v/v	126 ± 20 nm	MC3T3-E1	Sangsanoh et al., 2009
PCL/nano-apatite (nAp)	80:20% w/w 80% TFE in H₂O	320–430 nm	RBM	Yang F. et al., 2009
COL-HAp	80:20% w/w Sol concentration (0.023–0.040 g/ml) HFP	75–160 nm	MC3T3-E1	Song et al., 2008
CHI/PVA and CECS/PVA	CHI (7% w/w): PVA (10% w/w) (80:20 v/v) Aq. HAc	100–700 nm	L929	Yang et al., 2008
CHI-HAp	HAp/CHI (30:70) + 10% w/w UHMWPEO HAc:DMSO (10:1, w/w)	214 ± 25 nm	hFOB	Zhang et al., 2008
PCL-β-TCP	12% w/w DCM 0–15% w/w β-TCP	200–2,000 nm	MC3T3-E1	Erisken et al., 2008
PLGA/COL	8 g/ml (50:50 (w:w)) HFP	382 ± 125 nm	BMHSCs	Ma et al., 2008
PLA/HAp	6% w/w DCM and 1,4-dioxane	313 nm	MG63	Sui et al., 2007

(*continued*)

TABLE 10.1 (CONTINUED)

Fibrous Scaffolds Fabricated by Electrospinning for Bone Tissue Engineering

Polymer Type	Polymer Concentration/Composition/Solvent	Fiber Diameter	In Vitro Studies: Cell Type	Reference
PHB, PHBV, PHB - PHBV (50:50 w/w)	14% w/v CHL	2,300–4,000 nm	SaOS-2 and L929	Sombatmankhong et al., 2007
Silk fibroin (Nang-Lai and DOAE-7 type)	10–40% (w/v) 85% FA	217–610 nm (Nang-Lai) 183–810 nm (DOAE-7)	MC3T3-E1	Meechaisue et al., 2007
PLA (L- and DL-type)/PEG	8.2% w/w (L-type), 5% w/w (DL-type), 26% w/w (PEG-PLLA), 26% w/w (PEG-PDLLA) HFP	246 ± 79 nm (L-type) 141 ± 39 nm (DL-type) 171 ± 66 nm (PEG-PLLA) 889 ± 446 nm (PEG-PDLLA)	MC3T3-E1	Badami et al., 2006
Silk/PEO/nHAp/BMP-2	7.86% (w/v), silk:PEO (82:18 w/w) 5 g nHAP/100 g silk 3 mg BMP-2/mg silk fibroin	520 ± 55 nm	hMSCs	Li et al., 2006
Gelatin/HAp	80:20 (% w/w) HFP	200–400 nm	MG63	Kim et al., 2005
PCL-CaCO$_3$	PCL:CaCO$_3$ = 75:25 and 25:75% w/w CHL:MeOH (75:25% w/w)	760–900 nm	hFOB	Fujihara et al., 2005
Silk fibroin/PEO	7.5% w/w (80/20 w/w) in H$_2$O	700 ± 50 nm	Bone marrow stromal cells (BMSCs)	Jin et al., 2004
PCL, gelatin, PCL/gelatin	Gelatin: 2.5–12.5% w/v PCL: 10% w/v 50:50 blend: 10% w/v TFE	Tens of nm–1 μm	BMSC	Zhang et al., 2005
PCL	10% w/w CHL	400 ± 200 nm	BMSC	Yoshimoto et al., 2003

Abbreviations: BG, bioactive glass; BMSCs, bone marrow-derived mesenchymal stem cells; CECS, N-carboxyethyl chitosan; CHI, chitosan; CHL, chloroform; COL, collagen; DCM, dichloromethane; DMF, N,N-dimethyl formamide; DMSO, dimethyl sulfoxide; FA, formic acid; GPMS, 3-(glycidoxypropyl) trimethoxysilane; HAc, acetic acid; HAp, hydroxyapatite; hFOB, human fetal osteoblast; HFP, 1,1,1,3,3,3-hexafluoro-2-propanol; hMSC, human bone marrow–derived mesenchymal stem cell; L929, mouse fibroblasts; MeOH, methanol; MC3T3-E1, mouse calvaria-derived murine preosteoblast cells; MG63, human osteoblasts cells; PAN, poly(acrylonitrile); PCL, poly(ε-caprolactone); PEG, poly(ethylene glycol); PEO, poly(ethylene oxide); PHB, poly(hydroxybutyric acid); PHBV, poly(3-hydroxybutyric-co-3-hydroxyvaleric acid); PLA, poly(lactic acid); PLGA, poly(lactic acid-co-glycolic acid); RBM, rat bone marrow cells; TFA, trifluoroacetic acid; TFE, 2,2,2-trifluoroethanol; β-TCP, β-tricalcium phosphate.

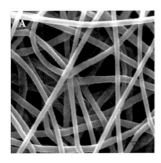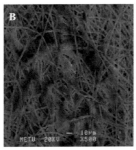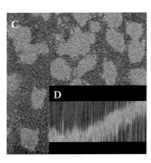

FIGURE 10.1
SEM micrographs of (A) unseeded and (B) Saos-2 cell-seeded PHBV-P(L,DL-LA) (70:30) scaffolds 14 days after incubation. (C) CLSM images of Saos-2-seeded PHBV10 scaffolds 7 days after incubation (top view showing cell-matrix interaction). (D) CLSM of cell infiltration within the scaffold as viewed in cross section (Z-axis direction). (From Ndreu et al., *Nanomed*, 3, 45–60, 2008.)

occurring enzymes, and therefore, they can be used as biodegradable scaffolds. The most commonly used natural polymers are proteins (collagen, gelatin, fibrinogen, silk, casein, etc.) and polysaccharides (chitosan, cellulose, hyaluronic acid, etc.). Sangsanoh et al. (2009) studied biological evaluation of electrospun chitosan nanofibers and films in vitro with four different cell types: Schwann cells, osteoblast-like cells, keratinocytes, and fibroblasts. Both substrate forms were seen to support cell attachment and especially promoted proliferation of keratinocytes. While Schwann cells attached well on nanofibrous structures, they did not adhere on films. On the other hand, even though osteoblast-like cells and fibroblasts showed good attachment on both nanofibrous and film surfaces they did not proliferate well (Sangsanoh et al., 2009). Jin et al. (2004) produced silk fibroin/poly(ethylene oxide) (PEO) nanofibrous matrices (700–750 nm), examined the behavior of human bone marrow stromal cells (BMSCs), and reported that the matrices supported BMSC attachment and proliferation during 14 days of incubation with extensive matrix coverage (Jin et al., 2004).

A blend of natural and synthetic polymers is advantageous in that they combine the better mechanical properties of synthetic polymers with the biofunctionality of natural ones. In this approach the properties of the scaffold can be changed easily and can be arranged as desired by changing the composition and the ratio of the blend. Ngiam et al. (2009) prepared nano-hydroxyapatite (n-HAp) containing electrospun PLGA and PLGA/collagen (COL) blend nanofibers. Biomineralization of n-HAp on electrospun nanofibers was achieved by using an alternate calcium and phosphate solution dipping method. n-HAp crystallite deposition was predominant on PLGA/COL nanofibers compared to on PLGA, which shows that collagen is a good template for n-HAp nucleation. It was shown that the amount of n-HAp present had a greater influence on early cell behavior, such as attachment and proliferation, than in the intermediate stages because their ALP and total protein expression results for mineralized PLGA and PLGA/COL were comparable during the culture period (Ngiam et al., 2009).

An important requirement in bone tissue engineering is that the scaffold should be highly porous so that cells can be homogeneously distributed throughout the structure. Poor cell infiltration into the structure is reported when electrospun scaffolds are used; the cells are observed more on the surface than the inside of the scaffold. Therefore, it is of great importance to find an approach to produce nanofibrous scaffolds with a higher degree of cell penetration. Ekaputra et al. (2008) used three strategies to solve this problem: selective leaching of a water-soluble fiber phase (PEO or gelatin), the use of micron-sized fibers as the scaffold, and a combination of micron-sized fibers with codeposition of a hyaluronic acid (HA)-derived hydrogel, Heprasil. In the first procedure, PCL/COL solution was electrospun in the presence of PEO, which was then removed selectively in order to increase the scaffold porosity. In the second approach, PCL/COL was electrospun in micrometer-sized PCL/COL (µmPCL/COL). The third method involved the cospraying of µmPCL/COL matrix with Heprasil. All the scaffold types supported attachment and proliferation of human fetal osteoblasts (hFOB). While selective leaching improved cellular infiltration to a low degree compared to meshes obtained by conventional electrospinning, µmPCL/COL microfibers allowed better penetration. This effect was more pronounced when Heprasil regions were present in the structure.

A fabrication method that can be used to solve the low penetration problem is the use of electrospun nano/microfibers in combination with 3D macrofiber deposition. In this design, macrofibers are expected to play a role in the structural integrity and mechanical properties of the scaffolds, whereas the nano/microfibers can act as a cell entrapment system and a cell adhesion promoter. The influence of this nano/microfibrous network on the behavior of endothelial cells (EC) seeded onto a scaffold based on a starch and PCL blend was studied by Santos et al. (2008). They have shown that the nanofibers influenced cell morphology and the maintenance of the physiological expression pattern. They also showed these EC-seeded scaffolds to be sensitive to pro-inflammatory stimulus; that is, the cells could migrate and form capillary-like structures within a 3D gel of collagen in response to some angiogenesis-stimulating factors (Santos et al., 2008).

A procedure that has been recently reported is on-site layer-by-layer (LBL) tissue regeneration, which involves electrospinning one layer of the polymeric solution, then seeding the cells onto this layer and repeating these steps to create a 3D scaffold with the cells already inside. This method is advantageous in that a highly specific 3D environment can be created by varying the composition of the fibrous layers, fiber thickness, diameter, and orientation as desired, and the neotissue formation can be unambiguously reduced (Agarwal et al., 2009).

Another way to improve cell carrier properties is surface modification of the fibers. A higher degree of spreading and growth of Saos-2 on PHBV and some of its blends has been shown by our research group by treating with oxygen plasma in order to increase hydrophilicity by adding some functional groups on the scaffold surface (Ndreu et al., 2008).

An appropriate biomechanical property is another challenge in bone tissue engineering, since the engineered tissue must be able to retain its shape and withstand the forces applied after implantation. It is well known that a significant part of the bone is constituted of calcium phosphate mineral phase. The presence of bone-inducing inorganic components within bone substitutes enhances calcium phosphate deposition after the osteogenic differentiation process. The apatite minerals and proteins/peptides, as well as drugs encapsulated within the nanofibers, are a promising strategy used for nanofibrous materials to achieve adequate mechanical properties and bone-specific bioactivity (Jang et al., 2009).

HAp plays a crucial role in the biomechanics of bone tissue, especially in increasing the compressive strength. Therefore, an increase in the number of papers reporting the addition of this material to nanofibrous scaffolds is being seen. Gelatin-HAp electrospun nanofibrous scaffolds were produced by Kim et al. (2005) to mimic the human bone matrix, and higher cellular activity was reported with these scaffolds, compared to those prepared with only gelatin. Zhang et al. (2008) have reported the use of electrospinning in combination with an in situ coprecipitation synthesis in the preparation of nanocomposite fibers of HAp/chitosan (CHI). Production of continuous nanofibers was possible after addition of a fiber-forming polymer, ultrahigh molecular weight PEO. Continuous HAp/CHI nanofibers with a diameter of 214 ± 25 nm were produced successfully, and the HAp nanoparticles were incorporated into the electrospun nanofibers. In vitro cell culture with hFOB showed that the incorporation of HAp nanoparticles into CHI nanofibrous scaffolds led to significantly more bone formation than that with electrospun pure CHI scaffolds (Zhang et al., 2008). In another study, COL and HAp were combined in PLLA, forming three different scaffold types: PLLA, PLLA/HA, and PLLA/COL/HAp. In these scaffolds COL and HAp were expected to provide cell recognition sites and introduce osteoconductivity, an important issue in bone mineralization, respectively. Fiber diameter was found to decrease upon the addition of HAp and COL to PLLA. Human fetal osteoblasts were found to adhere and grow actively on PLLA/COL/HAp nanofibers, with mineral deposition increased by 57% in comparison to PLLA/HAp nanofibers (Prabhakaran et al., 2009).

Some other ingredients introduced to polymers to improve the biomechanical properties and material-cell interaction for bone tissue engineering applications are calcium carbonate ($CaCO_3$) and β-TCP nanoparticles. Fujihara and his co-workers (2005) used a guided bone regeneration (GBR) approach with the aim of producing a calcium-rich and mechanically stable membrane. This was achieved by the preparation of PCL nanofibers containing $CaCO_3$ nanoparticles. A PCL-based membrane was compared with that based on PCL/$CaCO_3$, and a higher cell adherence was observed in the latter. The tensile strength of PCL nanofibrous membranes decreased when the amount of calcium carbonate nanoparticles was increased; that is, an increase in the amount of $CaCO_3$ resulted in more brittle membranes. However, it was stated that these membranes have sufficient tensile strength for guided bone regeneration (Fujihara et al., 2005).

Functionally graded, nonwoven mats of PCL with β-TCP nanoparticles were produced by using a hybrid, twin-screw extrusion/electrospinning (TSEE) process (Erisken et al., 2008). This procedure allows feeding of various liquid and solid materials, achieving melting, dispersion, pressurization, and electrospinning within a single process, with the composition changing in a time-dependent manner. The potential of these scaffolds for tissue engineering and bone-cartilage interface formation was demonstrated by using mouse calvaria-derived murine preosteoblast (MC3T3-E1) cells. The ability to control the incorporation of β-TCP resulted in a better mimicking of the compositional and structural characteristics of bone tissue (Erisken et al., 2008).

10.5.2 Nerve Tissue Engineering

The structure of the nervous system is important for its functionality; therefore, the architecture of the nerve tissue should be well known in order to be able to design the most suitable nerve substitute for nerve regeneration. In nervous tissue the axons are bundled into cables that provide long-distance communication between the brain and spinal cord and the rest of the body. The nerves in the peripheral nervous system (PNS) are composed of bundled motor and sensory axons in parallel arrays in connective tissue and are called the nerve trunk (Schmidt and Leach, 2003). The individual axons and their Schwann cell sheaths are surrounded by endoneurium, which is composed predominantly of oriented collagen fibers. Groups of axons are wrapped by the perinerium, which is composed of many layers of flattened cells (i.e., fibroblasts) and collagen to form fascicles. An outer sheath of loose fibrocollagenous tissue, the epineurium, assembles individual nerve fascicles into a nerve trunk. The vascularization in these nerves is provided by capillaries within the support tissue of the nerve trunk, or by vessels penetrating into the nerve from surrounding arteries and veins. Within the central nervous system (CNS), the brain and spinal cord, such bundled together axons are called nerve tracts. Typically, the neuron consists of dendrites, an axon, and a cell body (Schmidt and Leach, 2003). A gray matter localized in the center of the spinal cord is composed of the cell bodies of excitatory neurons, glial cells, and blood vessels. The gray matter is surrounded by white matter, which provides protection and insulation of the spinal cord. White matter consists of axons and glial cells, such as oligodendrocytes, astrocytes, and microglia. Oligodendrocytes myelinate the axons in the CNS, while astrocytes are active in the blood-nerve barrier, separating the CNS from blood proteins and cells. Fascicles, axons in bundles, come out from the white matter and then exit the encasing bone of the spinal column. Following travel through the PNS-CNS transition zone, nerves enter the PNS.

Nervous system injuries may result in permanent functional disabilities, and thus have a significant impact on the patient's quality of life. In severe damage of the peripheral nervous system, self-regeneration of the tissue is difficult, and it is almost impossible in the central nervous system (Meek et al., 2002; Schmidt and Leach, 2003). For that reason, the repair of the damaged tissue in the nervous system is complicated. Surgical interventions (end-to-end anastomosis) are used for small nerve gaps (Sanghvi et al., 2004); however, for long gaps (longer than 5 mm) the common treatment is to use autografts. The inherent drawbacks of autografts, such as limited supply, permanent loss of the nerve function at the donor site, and the requirement of a second surgery, are great challenges for researchers and compel them to investigate alternative treatments (Millesi, 1991; Terzis et al., 1997).

The rapidly developing alternative, nerve tissue engineering, aims to restore the function of damaged neural tissue through use of native cells and cell carriers. The cells commonly used in nerve tissue engineering are mature neurons, stem cells, and supportive cells, such as Schwann cells, astrocytes, and oligodendrocytes. The construct developed by nerve tissue engineering should imitate the anatomic structure of the nervous system for proper performance. When the native architecture of the nerve tissue is considered, it becomes apparent that a directional cell growth is a prerequisite for functional nerve regeneration. The desired physical properties of a cell carrier are biodegradability, high porosity, and also a cell guiding tool, such as an internal oriented matrix with appropriate chemical structure (ECM proteins, especially laminin, for nerve tissue). Topographical cues for cell migration could serve to mimic the native nerve fascicles (Hudson et al., 1999). The guidance needed for the tissue-engineered nerve construct may be achieved by the use of films with patterned surfaces, or longitudinally oriented microchannels and fibrous scaffolds.

Fibrous scaffolds to be used in nerve regeneration could be produced by electrospinning and self-assembly (Table 10.2). Since they are inherently porous and have a high surface area-to-volume ratio, these fibrous scaffolds promote attachment, survival, and growth of cells (Chiu et al., 2005). An architecture similar to ECM makes fibrous scaffolds a promising tool for nerve tissue engineering. The ideal scaffold should be biocompatible to avoid inflammatory response, be biodegradable, and have appropriate mechanical properties to provide the support required for proper healing (Hudson et al., 1999). The material of the cell carrier is a very important determinant in achieving a scaffold suitable for nerve tissue engineering. The degradation rate and mechanical properties of synthetic polymers such PLLA, PLGA, PHBV, and PCL can be controlled by chosing the appropriate compound, and as such, they are very suitable for tissue engineering. Natural materials such as PHBV, collagen, gelatin, chitosan, and especially laminin have specific physical properties and biomolecular recognition sites as in native tissue and, therefore, they have been increasingly investigated as fibrous scaffolds for nerve tissue regeneration (Cao et al., 2009). Blends of polymers have also been extensively used in nerve tissue engineering to achieve the desired scaffold properties. For instance, natural polymers lack proper

TABLE 10.2
Nano/Microfibrous Textures Used for Nerve Tissue Engineering

Polymer Type	Polymer Concentration/ Composition/Solvent	Fiber Diameter	In Vitro Studies: Cell Type	In Vivo Studies: Nerve Injury Model	Reference
Fibers Fabricated by Electrospinning					
Laminin	3–8% w/v HFP	90–300 nm (RF)	Human adipose stem cells and PC12 cells	—	Neal et al., 2009
Laminin-PLLA blend	10% w/v; 1:250, w:w HFP	100–500 nm (RF)	Rat PC12 cells	—	Koh et al., 2008
PCL	10% w/v CHL:MeOH 3:1, v:v	750 ± 100 nm (RF)	Brain-derived neural stem cells	—	Nisbet et al., 2008
PCL-pp*	16% w/w DCM:MeOH 8:2, v:v	2.26 ± 0.08 μm (RF) 1.03 ± 0.03 μm (AF)	Human Schwann cells	—	Chew et al., 2008
	CHL:MeOH 3:1, v:v	100–500 nm (RF)	Mouse (129 strain) embryonic stem cells	—	Thouas et al., 2008
PCL-PP	20% w/v DCM: DMF 80:20, v:v	250 nm (RF/AF)	Mouse embryonic stem cells CE3 and RW4	—	Xie et al., 2009b
PCL-PP	10–20% w/v DCM:DMF	220 ± 36 nm (50 ± 3 nm shell thickness)	Chick embryonic (E8) dorsal root ganglia explant	—	Xie et al., 2009a
PCL/GEL	10% w/v; 1:1, w:w TFE	232 ± 194 nm (RF) 160 ± 86 nm (AF)	Rat Schwann cell line (RT4-D6P2T)	—	Gupta et al., 2009
PCL/CHI	6% w/w; 70:30, w:w HFP	189 ± 56 nm (RF/AF)	Neonatal mouse cerebellum C17.2 stem cells	—	Ghasemi-Mobarakeh et al., 2008
	PCL:CHI, 75:25, w:w PCL in HFIP, CHI in TFA/DCM	190 ± 26 nm (RF)	Rat Schwann cell line RT4-D6P2T	—	Prabhakaran et al., 2008

* PCL-PP: PCL coated with polypyrrole

(*continued*)

TABLE 10.2 (CONTINUED)

Nano/Microfibrous Textures Used for Nerve Tissue Engineering

Polymer Type	Polymer Concentration/ Composition/Solvent	Fiber Diameter	In Vitro Studies: Cell Type	In Vivo Studies: Nerve Injury Model	Reference
PLLA	3 w/w% CHL	524 ± 305 nm (RF/AF)	Dorsal root ganglia explant	—	Corey et al., 2007
PLLA	1% w/w 3% w/w 2% w/w 5% w/w DCM:DMF 70:30, v:v	250 nm (RF) 1.25 μm (RF) 300 nm (AF) 1.5 μm (AF)	Neonatal mouse cerebellum C17.2 stem cells	—	Yang et al., 2005 Ekaputra et al., 2008
PLLA	8% w/w CHL	1.2–1.6 μm (AF)	Embryonic (E9) chick dorsal root ganglia explants and rat Schwann cells	—	Wang et al., 2009
PLLA/ PLGA	10% w/v THF and DMF	510 ± 380 nm (PLLA, RF) 760 ± 300 nm (PLGA, RF)	Mouse embryonic cortical neurons	—	Nisbet et al., 2007
PLGA, 75/25 (coated with polypyrrole)	6.5% w/w (RF) 7.0% w/w (AF) HFIP	520 ± 150 nm (RF) 430 ± 180 nm (AF) (85 ± 41 nm shell thickness)	PC12 cells and rat embryonic hippocampal neurons	—	Lee et al., 2009
Poly(ε-caprolactone-co-ethyl ethylene phosphate) (PCLEEP)	12% w/w DCM	5.08 ± 0.05 μm (AF)	—	15 mm nerve lesion gap in rat sciatic nerve (cell-free)	Chew et al., 2007
Blend of PLGA 75:25/ PCL	5.5% w/w PCL: 4% w/w PLGA CHL:MeOH 3:1	279 ± 87 nm (RF)	—	10 mm nerve gap in rat sciatic nerve (cell-free)	Panseri et al., 2008
Fibers Fabricated by Self-Assembly					
RAD16-I* (Arg-Ala-Asp) and RAD16-II**	Cell culture media	10–20 nm	Rat PC12 cells Mouse hippocampal and cerebellar granule neurons Rat hippocampal neurons	Injection of self-assembled (20 μl) in rat right leg (cell-free)	Holmes et al., 2000

IKVAV (Ile–Lys–Val–Ala–Val)	Cell culture media	5–8 nm	Mouse (E13) neural progenitor cells	Spinal cord injection (T10) at a depth of 1.5 mm in rats (cell free)	Silva et al., 2004
RADA 16-I (Arg–Ala–Asp–Ala)	Milli-Q water	10 nm	—	1.5–2 mm deep cut wound in the optic tract, superior colliculus in hamster brain (cell free)	Ellis-Behnke et al., 2006
RADA16-I peptide	Cell culture media	10 nm	Rat neural progenitor cells and Schwann cells	1 mm tissue removal by a transection in spinal cord dorsal column (between C6 and C7) in rats	Guo et al., 2007

Abbreviations: RF, random fiber; AF, aligned fiber; HFP, 1,1,1,3,3,3-hexafluoro-2-propanol; PCL, poly(ε-caprolactone); CHL, chloroform; MeOH, methanol; DCM, dichloromethane; DMF, dimethylformamide; GEL, gelatin; TFE, 2,2,2-trifluoroethanol; CHI, chitosan; HFIP, hexafluoro-2-propanol; THF, tetrahydrofuran; TFA, trifluoroacetic acid; PLLA, poly(L-lactic acid); PLGA, poly(lactic acid-co-glycolic acid).

mechanical properties and also lose mechanical strength very quickly due to degradation, while synthetic polymers alone are generally too hydrophobic and lack binding sites for cell adhesion (Gupta et al., 2009). Consequently, a blend or a graft of natural and synthetic polymers is an effective way to obtain scaffolds with desirable properties. A summary of nano/microfibrous scaffolds used in vitro and in vivo in nerve regeneration studies is given in Table 10.2.

Electrospinning is a straightforward technique to obtain random and aligned nano/microfibers. Even though the randomly oriented fibers are easier to obtain and are more frequently used in nerve tissue engineering, the use of the aligned fibers is a challenging but rewarding approach in mimicking the unidirectional orientation to maintain function of the nerve tissue (Table 10.2). Fibers aligned longitudinally (parallel to nerve axis) may provide a suitable environment that encourages contact guidance and the directional neuronal/axonal growth desired in nerve tissue engineering. Yang et al. (2005) observed the positive effect of contact guidance by aligned fibrous scaffolds (diameters of aligned micro- and nanofibers: 1.5 μm and 300 nm, respectively) on neural stem cells that had been induced to elongate neurites parallel to the fiber axis. The cell elongation and the neurite outgrowth were random on the randomly oriented mats (diameters of random micro- and nanofibers: 1.25 μm and 250 nm, respectively), proving the significance of guidance. This guidance was observed on aligned nano/microfibers regardless of fiber diameter. The effect of fiber arrangement was also observed in our studies carried out with neural stem cells seeded on fibrous mats. Electrospun PHBV-PLGA fibers were produced under optimized conditions (PHBV-PLGA solution: 1:1 w/w, 15% w/v, in chloroform: dimethylformamide (CHL:DMF)) with a diameter of ca. 1.5 μm (Yucel et al., 2010). The stem cells seeded on the aligned fibers were well oriented along the axis of the fibers, while the cells on the random fibers appeared to be in clusters distributed in all directions (Figure 10.2). Similar results were also obtained with dorsal root ganglion (DRG) explants cultured on PLLA nanofibers (Patel et al., 2007). Significant extension of neurites was observed on aligned fibers, while there was no neurite outgrowth on randomly oriented fibers. Like other cells, human Schwann cells responded to the fiber orientation with elongation along the fibers

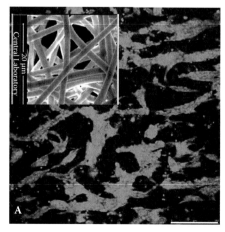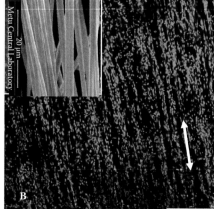

FIGURE 10.2
CLSM images of neural stem cells on fibronectin-coated PHBV-PLGA scaffolds. (A) Randomly oriented and (B) aligned fibers (100×; scale bar, 250 μm; two-sided arrow, direction of fiber axis; Acridine orange stain). Insets show the SEM images of the fibrous mats.

(Chew et al., 2008). In addition, aligned electrospun fibers appeared to enhance Schwann cell maturation more than the randomly oriented fibers did, as was shown by the upregulation of the myelin-specific gene. In vitro results were confirmed by the in vivo studies in which anatomical and functional measurements showed that constructs with aligned polymer fibers significantly enhanced nerve regeneration in 17 mm peripheral nerve gaps in rats over those of randomly oriented fibers (Kim et al., 2008).

Micro- and nanoscale topography was shown to play an important role in neural stem cell differentiation. Yang et al. (2005) observed that the mouse neural stem cell differentiation rate might depend on the fiber dimension. The differentiation rate on nanofibers 300 nm in diameter was twice as high as on microfibers with a 1.5 μm diameter. However, these results on the effect of fiber diameter on neural cell differentiation are controversial. Christopherson and his colleagues (2009) found that under differentiation conditions, rat neural stem cells stretched multidirectionally and preferentially differentiated into oligodendrocytes on fibers with a diameter of 283 nm, but they elongated along a single fiber axis and preferentially differentiated into neuronal lineage on fibers 749 nm and 1,452 nm thick. Despite these conflicting results, topographical and biological cues seem to be important in the control of stem cell behavior.

Electrospinning approaches involving blending, coaxial extrusion, and emulsion spinning are used in the incorporation of bioactive agents like proteins and growth factors to enhance cell attachment, growth, differentiation, and survival of cells in nerve tissue engineering applications. Incorporation of laminin, a neurite-promoting basement membrane protein, to a synthetic polymer by blending was reported as a more efficient method to produce biomimetic scaffolds than covalent immobilization and physical adsorption of the bioactive agents (Koh et al., 2008). In a recent study, nerve growth factor (NGF), a common neurotrophic factor involved in peripheral nerve regeneration, was incorporated in poly(L-lactide-co-ε-caprolactone) fibers by emulsion electrospinning (Li et al., 2010). Pheochromocytoma 12 (PC12) cells, which differentiate into a neuronal phenotype in the presence of NGF, were used in a study where NGF was released from fibrous mats for a sustained period, revealing its bioactivity. A glial cell-derived neurotrophic factor (GDNF), important in motor neuron survival, was entrapped in aligned biodegradable polymeric fibers (a copolymer of ε-caprolactone and ethyl ethylene phosphate (PCLEEP)) (Chew et al., 2007). An in vitro release profile of GDNF showed that after an initial burst release of about 30%, the remaining GDNF was released in a fairly sustained manner for almost 2 months.

Electrical stimulation is another tool that affects neurite extension. Electrical stimulation using an external electrical field and an electroconductive polypyrrole (PPy) film was shown to enhance neurite extension in vitro (Schmidt et al., 1997). Xie et al. (2009a) cultured DRG explants on conductive (PPy) core-sheath nanofibers electrospun with PCL, and results showed that upon electrical stimulation, the rate of neurite extension was enhanced on both random and aligned nanofibers without any change in the direction of the neurites' extension. In another study, PPy was deposited on electrospun PLGA fibers to investigate the response of PC12 cells (Lee et al., 2009). Upon application of 10 mV/cm on PPy-PLGA scaffolds, PC12 cells exhibited 40–50% longer neurites and 40–90% more neurite formation than unstimulated cells on the same scaffolds.

In conclusion, the topography of nano/microfibrous scaffolds closely resembles that of the natural ECM, and this is an important factor in tissue engineering. Considering the architecture of longitudinally oriented nerve tissue, uniaxially aligned fibers could create an appropriate environment for guided nerve regeneration. The use of the aforementioned tools with aligned fibrous scaffolds could be a promising approach in the design of a biomimetic substitute for nerve regeneration.

10.5.3 Cardiac Muscle Tissue Engineering

The heart is the center of the circulation system and distributes blood throughout the body. The function of the heart is vital to supply oxygen and nutrients to, and remove waste products from, the body via the blood in order to maintain the balance that is necessary to sustain life (Vander et al., 1994). Strong muscular contractions in the ventricles pump blood out of the heart and into the circulatory system. These muscular contractions are produced by the muscle tissue that makes up the walls of the ventricles. Healthy heart muscle wall is composed of three layers: a muscular sheet, the myocardium, lined on either side by two collagenous membranes (containing type I and type III collagen and elastin); the endocardium, which is populated by endothelial cells; and the epicardium, also called visceral pericardium (van de Graaff, 1998). The myocardium is the layer of functional beating muscle that consists of fibroblasts and highly oriented cardiomyocytes (muscle fibers) in a matrix of collagen. The cardiomyocytes are connected end to end in the longitudinal direction and side by side in the transverse direction (Parker and Ingber, 2007). Collagen types I and III are the predominant interstitial collagens in the myocardium that generate structural integrity for the adjoining cardiomyocytes, providing the means by which myocyte shortening is translated into overall ventricular pump function. Basement membrane components include laminin, entactin, fibronectin, collagen type IV, and fibrillin, and proteoglycans include chondroitin sulfate, dermatan sulfate, and heparan sulfate (Kassiri and Khokha, 2005). The cardiomyocytes facilitate conduction of the electrical signals needed to initiate contractile movement in order to pump blood out of the ventricles (Vander et al., 1994). The elementary myocardial functional units are not lone cardiomyocytes, but rather are multicellular assemblies of these highly oriented cells. The cardiomyocytes are connected with intercalated discs that integrate individual electrical activation and contraction into a pumping action. Gap junctions in these intercalated discs allow the action potential to travel through the membranes of the myocytes, thus facilitating signal propagation and a synchronized contractile pulse (Vander et al., 1994).

  Heart failure, stemming from cardiovascular diseases, is the number one cause of death in industrialized countries. About 5 million people in the United States have heart failure, and the number is growing; heart failure contributes to or causes about 300,000 deaths each year. The inability of the heart to deliver sufficient blood to meet the body's metabolic requirements leads to heart failure. The most common causes are coronary artery disease and hypertension (high blood pressure), which damage the myocardium. Damage to any part of the intricate structure of the heart, though, can impair cardiac performance and result in heart failure; examples include diseases of the heart valves, the electrical conduction system, or external pressure around the heart, due to constriction of the pericardial sac in which the heart is located (Jawad et al., 2007). Congenital heart disease is also a considerable problem worldwide, affecting approximately 1% of infants (Wu et al., 2006). Congenital heart defects such as atrial septal defect, ventricular septal defect, double outlet ventricles, and hypoplastic left heart syndrome are associated with aplastic, defective, or necrotic myocardial structures, where patch closure, correction of the defect, or revascularization is usually required (Kofidis et al., 2002). Although the currently available prostheses are adequate for restoring ventricular geometry and maintaining ventricular pressure, and thus may be life saving, they do not actively adapt to the physiological environment and mechanical demands, as they are nonliving materials. When implanted into the immature heart of a child, these materials do not grow with the pediatric patient. Therefore, these prostheses have limited durability and are prone to infection, immunologic reactivity, and thrombosis, which often requires repeated operations (Mirensky and

Breuer, 2008). These constructs are also subject to obstructive tissue ingrowths and fibrotic responses with shrinkage and calcification leading to graft failure (Endo et al., 2001). At that point, tissue engineering provides a new experimental approach aimed at vital, semi-autologous replacement structures with the capacity of growth and adaptability.

The negligible ability of adult cardiomyocytes to proliferate has triggered an intense search for progenitor cells that can be used in repairing damaged myocardium. It is clear that natural cardiomyocyte regeneration, including differentiation of progenitors residing in the myocardium or the recruitment of stem cells from outside (e.g., from endothelial cells or from the bone marrow), is insufficient to overcome cardiomyocyte death in the acutely or chronically damaged heart (Nakamura and Schneider, 2003). The potential of cardiac stem cells resident in the heart, embryonic stem cells, bone marrow stem cells, and fetal stem cells (from tissues of umbilical cord) to differentiate into cardiomyogenic lineage either in vivo or in vitro in the presence of differentiation factors, conditioned media, or cocultures has been investigated, but there is still a need for elaboration and improvement in the procedures to be able to obtain high numbers of fully differentiated cardiomyocytes. Since the main concern in myocardial tissue engineering is the functionality of the final construct, the majority of in vitro studies use rat neonatal cardiomyocytes or cardiomyocytes derived from embryonic stem cells that are already able to contract.

Fibrous scaffolds whose fiber diameter ranges from several microns to submicron sizes (Table 10.3) are increasingly gaining importance over spongy scaffolds with random pores used to obtain tissue-engineered cardiac muscle. The main obstacle in overcoming the issues involved in the engineering of cardiovascular grafts is production of substitutes that can withstand the pulsation (i.e., have adequate shape recovery), the high pressure, and the flow rate of the blood (Nisbet et al., 2009). An additional challenge in cardiovascular tissue engineering is the necessity for mechanical and biological compliance. Polymers naturally found in the cardiac tissue, like collagen and elastin, are good choices as scaffold materials, since they carry attachment sites for cells, and especially elastin has elastic properties, but they have poor mechanical properties. On the other hand, synthetic polymers have higher mechanical strength, but have problems, such as being too hydrophobic and rigid, lacking cell binding functional groups, or having inappropriate degradation rates (rapidly degrading or nondegrading). Thus, as mentioned earlier, blends or copolymers of synthetic or natural polymers are employed often in the production of scaffolds with the desired properties (Zong et al., 2005, Li et al., 2006a, 2006b). Currently cell survival in 3D cardiac grafts of spongy design is a critical issue; generally the cells are concentrated in the outer regions of the grafts and are scarce in the inside due to restricted transportation of nutrients and waste in and out of the scaffold, resulting in an undesired core effect (Bursac et al., 1999). Electrospinning enables us to fabricate highly porous scaffolds with extensively interconnected pores that allow cell-to-cell contact and migration in all directions, facilitate the transportation of nutrients and metabolites, and encourage blood vessel formation. All of these combined help cell survival throughout the scaffold, including the central portion. Electrospun mats of PCL with random, unaligned fibers were seeded with cardiomyocytes by Ishii et al. (2005), and the mats were stacked to generate a 3D cardiac graft. Five layers of these mats were successfully stacked without any indication of core ischemia. The individual layers adhered intimately, morphological and electrical communication between the layers was established, and synchronized beating was also observed. This study demonstrated the feasibility of obtaining a functional 3D cardiac graft from cell sheets supported by nanofibers mimicking the native structure of cardiac ECM. Fiber density on the electrospun mats could be tailored depending on the intended use. For example, electrospun mats with tight fiber spacing may act as a sieve to prevent blood cell

TABLE 10.3

Nano/Microfibrous Textures Used in Cardiac Muscle Tissue Engineering

Polymer Type	Polymer Concentration/ Composition/Solvent	Fiber Diameter	In Vitro Studies: Cell Type	Reference
Extrusion				
PGA	—	13 μm	Neonatal rat cardiomyocytes	Bursac et al., 1999 Papadaki et al., 2001
Collagen type I	25 mg/ml DI water-HCl (pH 2.5–3.0)	~ <250 nm (AF)	Rat embryonic cardiac myocytes	Evans et al., 2003 Yost et al., 2004
Electrospinning				
PCL	10% (w/w) CHL:MeOH (1:1)	100 nm–5 μm (RF)	Neonatal rat cardiomyocytes	Shin et al., 2004 Ishii et al., 2005
PLLA Blend of PGLA (90:10)/PLLA Blend of PLGA (75:25)/PEG-PDLLA	30–50% (w/w) DMF	0.9–1 μm (RF/AF)	Neonatal rat cardiomyocytes	Zong et al., 2005
Blend of PLGA/gelatin/α-elastin	10% PLGA:8% gelatin: 20% elastin (w/v) HFP	380 ± 80 nm (RF)	H9c2 rat cardiac myoblasts Neonatal rat bone marrow stromal cells	Li et al., 2006a
Blends of PANi/gelatin	3% (w/v) CPSA-PANi 8% (w/v) gelatin HFP	From 803 ± 121 nm to 61 ± 13nm (RF)	H9c2 rat cardiac myoblasts	Li et al., 2006b
Biodegradable polyurethane	15% (w/v) DCM	2–10 μm (RF)	Mouse ESC-derived cardiomyocytes C2C12 skeletal myoblast line HL-1 cardiomyocyte cell line	Fromstein et al., 2008 Zhang et al., 2005
Biodegradable polyurethane	18% (w/v) DCM	1.8 ± 1.26 μm (AF) 2.0 ± 1.92 μm (RF)	Neonatal rat cardiomyocytes	Rockwood et al., 2008
Blend of PHBV/P(L-D,L)LA/PGS	5% (w/v) PHBV:P(L-D,L)LA:PGS 49:49:2 (w/w) CHL:DMF (10:1)	1.10–1.25 μm (AF)	Human Wharton's jelly MSCs	Kenar et al., 2009

Abbreviations: PGA, poly(glycolic acid); PCL, poly(ε-caprolactone); PLLA, poly(L-lactide); PGLA, poly(glycolide-co-lactide); PLGA, poly(lactide-co-glycolide); PEG, poly(ethylene glycol); PDLLA, poly(D,L-lactide); PHBV, poly(3-hydroxybutyrate-co-3-hydroxyvalerate); P(L-D,L)LA, poly(L-D,L-lactic acid); PGS, poly(glycerol sebacate); CHL, chloroform; MeOH, methanol; DI, deionized; DMF, N,N-dimethyl formamide; HFP, 1,1,1,3,3,3-hexafluoro-2-propanol; DCM, dichloromethane; PANi, polyaniline (conductive polymer).

leakage or provide a homogeneous surface to obtain intact endothelial cell sheets, or act as a supporting outer surface in cardiovascular scaffolds (Hong et al., 2009; Soletti et al., 2010). Conversely, fiber spacing and density may also be adjusted and loosened to promote cell infiltration in the myocardial patches.

Anisotropy in the architecture of the constructs plays an important role in achieving compliance and functionality in cardiovascular tissue engineering. Parallel submicron fibers may be produced to have circumferential orientations similar to those of the cells and fibrils of the medial layer of the native artery (Xu et al., 2004). It was demonstrated that anisotropic poly(ester urethane)urea scaffolds exhibit mechanical properties closely resembling those of the native pulmonary heart valve leaflets (Courtney et al., 2006). Cardiac muscle contracts like a syncytium owing to the multicellular assembly of myofibers oriented parallel to each other, connected end to end via intercalated discs in the longitudinal direction and connected side to side in the transverse direction. It is this cellular arrangement that integrates individual contraction into a synchronous pumping action, whereas myofiber disarrays are known to cause arrhythmia. It was shown by Black et al. (2009) that cardiomyocyte alignment augments the twitch force. Prior to emergence of techniques to produce aligned nano/microfibrous scaffolds, anisotropy had been introduced to the cardiomyocytes on the polymer scaffolds in three different ways: (1) by micropatterning (e.g., using biodegradable, elastomeric polyurethane films patterned by microcontact printing of laminin (McDevitt et al., 2003)), (2) by cyclic stretching (e.g., stretching of cardiomyocyte-seeded scaffolds (Akhyari et al., 2002; Zimmermann et al., 2002, 2004; Zimmermann and Eschenhagen, 2003; Fedak et al., 2003)), and (3) by applying electrical signals designed to mimic those in the native heart to the tissue engineered constructs (e.g., rat cardiomyoctes resuspended in Matrigel were seeded onto collagen sponges (Radisic et al., 2004)). Electrospinning, on the other hand, offers a much simpler way to achieve the anisotropy and allows greater control over composition, mechanical properties, and structure of a graft, thus making it easier to match the properties of the scaffold and the native tissue. It was shown by Zong et al. (2005) that cardiomyocytes can align parallel to micron-size fibers. Their electrospun mats with oriented fibers of PLLA and PLGA were used to assess the influence of scaffolds on primary cardiomyocyte attachment, structure, and function. The primary cardiomyocytes showed a preference for relatively hydrophobic surfaces (PLLA), where they aligned along the direction of fibers and developed mature contractile machinery (sarcomers). The only drawback of these aligned fiber scaffolds was the inability of the cardiomyocytes to penetrate into the scaffold due to the inadequate space among the fibers. A very similar phenomenon was reported by Evans et al. (2003), who observed that rat embryonic cardiac myocytes were unable to penetrate the 3D aligned collagen fiber scaffold with a pore size of 1–10 microns on the outer surface of a tubular scaffold. Aligned fiber scaffolds with a fiber density suitable for cell penetration were produced by our group by using a frame collector to obtain aligned fibers (Kenar et al., 2010). At least eight or nine Wharton's jelly (WJ) MSC layers could be obtained in a single mat of average thickness of $12 \pm 3\ \mu m$. Both cytoskeletal and nuclear alignments of the cells were observed (Figure 10.3). A wrapping method was used to obtain a thicker homogeneous construct, which maintained its original anisotropic property. Another technique, microintegration, was developed by Stankus et al. (2006), where simultaneous electrospraying of cells and electrospinning of poly(ester urethane)urea (PEUU) was utilized in order to fabricate thicker constructs with more uniform cell incorporation for use in cardiovascular tissue engineering applications. Smooth muscle cell-microintegrated PEUU was strong, flexible, and anisotropic, and Trypan blue staining revealed no significant decrease in cell viability as a result of the fabrication process.

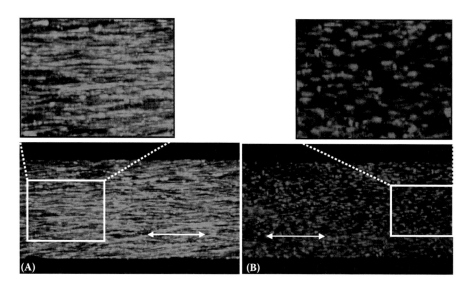

FIGURE 10.3
(See color insert.) Cross-sectional (X-axis ~ 60° tilted) CLSM images of FITC-Phalloidin and propidium iodide-stained WJ MSCs on a single, electrospun, aligned fiber mat of PHBV:P(L-D,L)LA:PGS (poly(glycerol sebacate)) (100×) showing (A) F-actin and (B) nuclear alignment (two-sided arrow, direction of aligned fiber axis). Top images are the magnified versions of insets.

Studies carried out with aligned fiber mats with fiber diameters ranging from 0.25 to 3.00 μm show that cells grown on them align equally well along the direction of the fibers (Evans et al., 2003; Yost et al., 2004; Zong et al., 2005; Rockwood et al., 2008; Kenar et al., 2010). The single study carried out on cardiomyocyte function on electrospun random and aligned polymer fibers showed that cellular alignment enhances cardiomyocyte maturation; cells grown on aligned electrospun polyurethanes had significantly lower steady-state levels of atrial natriuretic peptide (ANP) and, as a result, released less ANP over time than cardiomyocytes on random fibers (Rockwood et al., 2008).

In conclusion, nanofibers are novel scaffold materials for tissue engineering, and the accumulating data show them to be superior to most other designs due to the versatility of the approach. They are especially suited to the highly organized nature of biological tissues. With developments in the processing systems, they will become the major scaffold type.

References

Agarwal S, Wendorff JH, Greiner A. 2009. Progress in the field of electrospinning for tissue engineering applications. *Adv Mater* 21: 3343–3351.

Akhyari P, Fedak PW, Weisel RD, Lee TY, Verma S, Mickle DA, Li RK. 2002. Mechanical stretch regimen enhances the formation of bioengineered autologous cardiac muscle grafts. *Circulation* 106: I137–I142.

Ashammakhi N, Ndreu A, Nikkola L, Wimpenny I, Yang Y. 2008. Advancing tissue engineering by using electrospun nanofibers. *Regen Med* 3: 547–574.

Ashammakhi N, Ndreu A, Piras AM, Nikkola L, Sindelar T, Ylikauppila H, Harlin A, Gomes ME, Neves NM, Chiellini E, Chiellini F, Hasirci V, Redl H, Reis RL. 2007a. Biodegradable nanomats produced by electrospinning: expanding multifunctionality and potential for tissue engineering. *J Nanosci Nanotechnol* 7: 862–882.

Ashammakhi N, Ndreu A, Yang Y, Ylikauppila H, Nikkola L, Hasirci V. 2007b. Tissue engineering: a new take-off using nanofiber-based scaffolds. *J Craniofac Surg* 18: 3–17.

Asran AS, Henning S, Michler GH. 2010. Polyvinyl alcohol-collagen-hydroxyapatite biocomposite nanofibrous scaffold: mimicking the key features of natural bone at the nanoscale level. *Polymer* 51: 868–876.

Badami AS, Kreke MR, Thompson MS, Riffle JS, Goldstein AS. 2006. Effect of fiber diameter on spreading, proliferation, and differentiation of osteoblastic cells on electrospun poly(lactic acid) substrates. *Biomaterials* 27: 596–606.

Barhate RS, Loong CK, Ramakrishna S. 2006. Preparation and characterization of nanofibrous filtering media. *J Membr Sci* 283: 209–218.

Barnes CP, Sell SA, Boland ED, Simpson DG, Bowlin GL. 2007. Nanofiber technology: designing the next generation of tissue engineering scaffold. *Adv Drug Delivery Rev* 59: 1413–1433.

Black J. 1999. *Biological performance of materials*, 3rd ed. M. Dekker, New York, p. 6.

Black LD, Meyers JD, Weinbaum JS, Shvelidze YA, Tranquillo RT. 2009. Cell-induced alignment augments twitch force in fibrin gel-based engineered myocardium via gap junction modification. *Tissue Eng A* 15(10): 3099–3108.

Bruck SD. 1980. *Properties of biomaterials in the physiological environment*. CRC Press, Boca Raton, FL.

Burg KJL, Porter S, Kellam JF. 2000. Biomaterial developments for bone tissue engineering. *Biomaterials* 21: 2347–2359.

Bursac N, Papadaki M, Cohen RJ, Schoen FJ, Eisenberg SR, Carrier RL, Vunjak-Novakovic G, Freed LE. 1999. Cardiac muscle tissue engineering: toward an *in vitro* model for electrophysiological studies. *Am J Physiol Heart Circ Physiol* 277: H433–H444.

Cao H, Liu T, Chew SY. 2009. The application of nanofibrous scaffolds in neural tissue engineering. *Adv Drug Deliv Rev* 61(12): 1055–1064.

Catledge SA, Vohra YK, Bellis SL, Sawyer AA. 2004. Mesenchymal stem cell adhesion and spreading on nanostructured biomaterials. *J Nanosci Nanotechnol* 4: 986–989.

Chew SY, Mi R, Hoke A, Leong KW. 2007. Aligned protein-polymer composite fibers enhance nerve regeneration: a potential tissue-engineering platform. *Adv Funct Mater* 17(8): 1288–1296.

Chew SY, Mi R, Hoke A, Leong KW. 2008. The effect of the alignment of electrospun fibrous scaffolds on Schwann cell maturation. *Biomaterials* 29(6): 653–661.

Chiu JB, Luu YK, Fang D, Hsiao B, Chu B, Hadjiargyrou M. 2005. Electrospun nanofibrous scaffolds for biomedical applications. *J Biomed Nanotechnol* 1: 115–132.

Choi JS, Lee SJ, Christ GJ, Atala A, Yoo JJ. 2008. The influence of electrospun aligned poly(ε capro-lactone)/collagen nanofiber meshes on the formation of self-aligned skeletal muscle myotubes. *Biomaterials* 29: 2899–2906.

Christopherson GT, Song H, Mao HQ. 2009. The influence of fiber diameter of electrospun substrates on neural stem cell differentiation and proliferation. *Biomaterials* 30(4): 556–564.

Corey JM, Lin DY, Mycek KB, Chen Q, Samuel S, Feldman EL, Martin DC. 2007. Aligned electrospun nanofibers specify the direction of dorsal root ganglia neurite growth. *J Biomed Mater Res A* 83(3): 636–645.

Courtney T, Sacks MS, Stankus J, Guan J, Wagner WR. 2006. Design and analysis of tissue engineering scaffolds that mimic soft tissue mechanical anisotropy. *Biomaterials* 27: 3631–3638.

Cui W, Li X, Zhou S, Weng J. 2007. *In situ* growth of hydroxyapatite within electrospun poly(DL-lactide) fibers. *J Biomed Mater Res A* 82(4): 831–841.

Deitzel JM, Kleinmeyer J, Harris D, Tan NCB. 2001. The effect of processing variables on the morphology of electrospun nanofibres and textiles. *Polymer* 42: 261–272.

Dieudonné SC, van den Dolder J, de Ruijter JE, Paldan H, Peltola T, van't Hof MA, Happonen RP, Jansen JA. 2002. Osteoblast differentiation of bone marrow stromal cells cultured on silica gel and sol–gel-derived titania. *Biomaterials* 23: 3041–3051.

Dogan S, Demirer S, Kepenekci I, Erkek B, Kiziltay A, Hasirci N, Müftüoğlu S, Nazikoglu A, Renda N, Dincer UD, Elhan A, Kuterdem E. 2009. Epidermal growth factor-containing wound closure enhances wound healing in non-diabetic and diabetic rats. *Int Wound J* 6: 107–115.

Ducheyne P, Qiu Q. 1999. Bioactive ceramics: the effect of surface reactivity on bone formation and bone cell function. *Biomaterials* 20: 2287–2303.

Ekaputra AK, Prestwich GD, Cool SM, Hutmacher DW. 2008. Combining electrospun scaffolds with electrosprayed hydrogels leads to three-dimensional cellularization of hybrid constructs. *Biomacromolecules* 9: 2097–2103.

Ellis-Behnke RG, Liang YX, You SW, Tay DK, Zhang S, So KF, Schneider GE. 2006. Nano neuro knitting: peptide nanofiber scaffold for brain repair and axon regeneration with functional return of vision. *Proc Natl Acad Sci USA* 103(13): 5054–5059.

Endo S, Saito N, Misawa Y, Sohara Y. 2001. Late pericarditis secondary to pericardial patch implantation 25 years prior. *Eur J Cardiothorac Surg* 20: 1059–1060.

Erisken C, Kalyon DM, Wang H. 2008. Functionally graded electrospun polycaprolactone and β-tricalcium phosphate nanocomposites for tissue engineering applications. *Biomaterials* 29: 4065–4073.

Evans HJ, Sweet JK, Price RL, Yost M, Goodwin RL. 2003. Novel 3D culture system for study of cardiac myocyte development. *Am J Physiol Heart Circ Physiol* 285: H570–H578.

Fan H, Liu H, Toh SL, Goh JCH. 2009. Anterior cruciate ligament regeneration using mesenchymal stem cells and silk scaffold in large animal model. *Biomaterials* 30: 4967–4977.

Fedak PWM, Weisel RD, Verma S, Mickle DAG, Li R-K. 2003. Restoration and regeneration of failing myocardium with cell transplantation and tissue engineering. *Semin Thorac Cardiovasc Surg* 15(3): 277–286.

Feng L, Li S, Li K, Zhai J, Song Y, Yiang L. 2002. Superhydrophobic surface of aligned polyacrylonitrile nanofibers. *Angew Chem Int Ed* 41: 1221–1223.

Fromstein JD, Zandstra PW, Alperin C, Rockwood D, Rabolt JF, Woodhouse KA. 2008. Seeding bioreactor-produced embryonic stem cell-derived cardiomyocytes on different porous, degradable, polyurethane scaffolds reveals the effect of scaffold architecture on cell morphology. *Tissue Eng A* 14(3): 369–378.

Fujihara K, Kotaki M, Ramakrishna S. 2005. Guided bone regeneration membrane made of polycaprolactone/calcium carbonate composite nano-fibers, *Biomaterials* 26: 4139–4147.

Ghasemi-Mobarakeh L, Prabhakaran MP, Morshed M, Nasr-Esfahani MH, Ramakrishna S. 2008. Electrospun poly(epsilon-caprolactone)/gelatin nanofibrous scaffolds for nerve tissue engineering. *Biomaterials* 29(34): 4532–4539.

Griffith LG. 2002. Emerging design principles in biomaterials and scaffolds for tissue engineering. *Ann NY Acad Sci USA* 961: 83–95.

Guelcher SA, Hollinger JO. 2006. *An introduction to biomaterials.* Taylor & Francis Group, CRC Press, Boca Raton, FL.

Guo J, Su H, Zeng Y, Liang YX, Wong WM, Ellis-Behnke RG, So KF, Wu W. 2007. Reknitting the injured spinal cord by self-assembling peptide nanofiber scaffold. *Nanomedicine* 3(4): 311–321.

Gupta D, Venugopal J, Prabhakaran MP, Dev VR, Low S, Choon AT, Ramakrishna S. 2009. Aligned and random nanofibrous substrate for the *in vitro* culture of Schwann cells for neural tissue engineering. *Acta Biomater* 5(7): 2560–2569.

Hay ED. 1991. *Cell biology of extracellular matrix*, 2nd ed. Plenum Press, New York.

Hench LL, Erthridge EC. 1982. *Biomaterials—an interfacial approach,* Vol. 4, A. Noordergraaf, ed. Academic Press, New York.

Hench LL, Polak JM. 2002. Third-generation biomedical materials. *Science* 295: 1014–1017.

Holmes TC, de Lacalle S, Su X, Liu G, Rich A, Zhang S. 2000. Extensive neurite outgrowth and active synapse formation on self-assembling peptide scaffolds. *Proc Natl Acad Sci USA* 97(12): 6728–6733.

Hong Y, Ye S-H, Nieponice A, Soletti L, Vorp DA, Wagner WR. 2009. A small diameter, fibrous vascular conduit generated from a poly(ester urethane)urea and phospholipid polymer blend. *Biomaterials* 30: 2457–2467.

Hosseinkhani H, Hosseinkhani M, Tian F, Kobayashi H, Tabata Y. 2006. Ectopic bone formation in collagen sponge self-assembled peptide-amphiphile nanofibers hybrid scaffold in a perfusion culture bioreactor. *Biomaterials* 27: 5089–5098.

Hu J, Feng K, Liu X, Ma PX. 2009. Chondrogenic and osteogenic differentiations of human bone marrow-derived mesenchymal stem cells on a nanofibrous scaffold with designed pore network. *Biomaterials* 30: 5061–5067.

Huang ZM, Zhang YZ, Kotaki M, Ramakrishna S. 2003. A review on polymer nanofibers by electrospinning and their application in nanocomposites. *Compos Sci Technol* 63: 2223–2253.

Hudson TW, Evans GR, Schmidt CE. 1999. Engineering strategies for peripheral nerve repair. *Clin Plast Surg* 26: 617–628.

Ishii O, Shin M, Sueda T, Vacanti JP. 2005. *In vitro* tissue engineering of a cardiac graft using a degradable scaffold with an extracellular matrix–like topography. *J Thorac Cardiovasc Surg* 130: 1358–1363.

Jang JH, Castano O, Kim HW. 2009. Electrospun materials as potential platforms for bone tissue engineering. *Adv Drug Deliv Rev* 61: 1065–1083.

Janjanin S, Li WJ, Morgan MT, Shanti RM, Tuan RS. 2008. Mold-shaped, nanofiber scaffold-based cartilage engineering using human mesenchymal stem cells and bioreactor. *J Surg Res* 149: 47–56.

Jawad H, Ali NN, Lyon AR, Chen QZ, Harding SE, Boccaccini AR. 2007. Myocardial tissue engineering: a review. *J Tissue Eng Regen Med* 1(5): 327–342.

Jeong L, Yeo IS, Kim HN, Yoon YI, Jang DH, Yung SY, Min BM, Park WH. 2009. Plasma-treated silk fibroin nanofibers for skin regeneration. *Int J Biol Macromol* 44: 222–228.

Jin HJ, Chen J, Karageorgiou V, Altman G.H, Kaplan DL. 2004. Human bone marrow stromal cell responses on electrospun silk fibroin mats. *Biomaterials* 25(6): 1039–1047.

Kassiri Z, Khokha R. 2005. Myocardial extra-cellular matrix and its regulation by metalloproteinases and their inhibitors. *Thromb Haemost* 93: 212–219.

Katti DS, Robinson KW, Ko FK, Laurencin CT. 2004. Bioresorbable nanofiber-based systems for wound healing and drug delivery: optimization of fabrication parameters. *J Biomed Mater Sci* 70B: 286–296.

Kenar H, Kocabas A, Aydinli A, Hasirci V. 2008. Chemical and topographical modification of PHBV surface to promote osteoblast alignment and confinement. *J Biomed Mater Res A* 85: 1001–1010.

Kenar H, Kose GT, Hasirci V. 2009. Design of a 3D aligned myocardial tissue construct from biodegradable polyesters. *J Mater Sci Mater Med* 21(3): 989–997.

Kim HW, Song JH, Kim HE. 2005. Nanofiber generation of gelatin-hydroxyapatite biomimetics for guided tissue regeneration. *Adv Funct Mater* 15: 1988–1994.

Kim YT, Haftel VK, Kumar S, Bellamkonda RV. 2008. The role of aligned polymer fiber-based constructs in the bridging of long peripheral nerve gaps. *Biomaterials* 29(21): 3117–3127.

Kinikoglu B, Auxenfans C, Pierrillas P, Justin V, Breton P, Burillon C, Hasirci V, Damour O. 2009. Reconstruction of a full-thickness collagen-based human oral mucosal equivalent. *Biomaterials* 30: 6418–6425.

Kneser U, Schaefer DJ, Munder B, Klemt C, Andree C, Stark GB. 2002. Tissue engineering of bone. *Min Invas Ther Allied Technol* 11: 107–116.

Kofidis T, Akhyari P, Wachsmann B, Boublik J, Mueller-Stahl K, Leyh R, Fischer S, Haverich A. 2002. A novel bioartificial myocardial tissue and its prospective use in cardiac surgery. *Eur J Cardiothorac Surg* 22(2): 238–243.

Koh HS, Yong T, Chan CK, Ramakrishna S. 2008. Enhancement of neurite outgrowth using nanostructured scaffolds coupled with laminin. *Biomaterials* 29(26): 3574–3582.

Langer R, Vacanti JP. 1993. Tisue engineering. *Science* 260: 920–926.

Lee CH, Shin HJ, Cho IH, Kang YM, Kim IA, Park KD, Shin JW. 2005. Nanofiber alignment and direction of mechanical strain affect the ECM production of human ACL fibroblast. *Biomaterials* 26(11): 1261–1270.

Lee JY, Bashur CA, Goldstein AS, Schmidt CE. 2009. Polypyrrole-coated electrospun PLGA nanofibers for neural tissue applications. *Biomaterials* 30(26): 4325–4335.

Lee SJ, Yoo JJ, Lim GJ, Atala A, Stitzel J. 2007. *In vitro* evaluation of electrospun nanofiber scaffolds for vascular graft application. *J Biomed Mater Res A* 83(4): 999–1008.

Leong KF, Cheah CM, Chua CK. 2003. Solid free form fabrication of three dimensional scaffolds for engineering replacement tissues and organs. *Biomaterials* 24: 3262–3278.

Li M, Guo Y, Wei Y, MacDiarmid AG, Lelkes PI. 2006b. Electrospinning polyaniline-contained gelatin nanofibers for tissue engineering applications. *Biomaterials* 27: 2705–2715.

Li M, Mondrinos MJ, Chen X, Gandhi MR, Ko FK, Lelkes PI. 2006a. Co-electrospun poly(lactide-co-glycolide), gelatin, and elastin blends for tissue engineering scaffolds. *J Biomed Mater Res A* 79: 963–973.

Li WJ, Jiang YJ, Tuan RS. 2006. Chondrocyte phenotype in engineered fibrous matrix is regulated by fiber size. *Tissue Eng* 12: 1775–1785.

Li WJ, Tuli R, Okafor C, Derfoul A, Danielson KG, Hall DJ, Tuan RS. 2005. A three dimensional nanofibrous scaffold for cartilage tissue engineering using human mesenchymal stem cells. *Biomaterials* 26: 599–609.

Li X, Su Y, Liu S, Tan L, Mo X, Ramakrishna S. 2010. Encapsulation of proteins in poly(L-lactide-co-caprolactone) fibers by emulsion electrospinning. *Colloids Surf B Biointerfaces* 75(2): 418–424.

Liu H, Cai Q, Lian P, Fang Z, Duan S, Yang X, Deng X, Ryu S. 2010. β-Tricalcium phosphate nanoparticles adhered carbon nanofibrous membrane for human osteoblasts cell culture. *Mater Lett* 64: 725–728.

Liu X, Ma PX. 2004. Polymeric scaffolds for bone tissue engineering. *Ann Biomed Eng* 32(3): 477–486.

Lu G, Sheng B, Wei Y, Wang G, Zhang L, Ao Q, Gong Y, Zhang X. 2008. Collagen nanofiber-covered porous biodegradable carboxymethyl chitosan microcarriers for tissue engineering cartilage. *Eur Polym J* 44: 2820–2829.

Ma K, Chan CK, Liuo S, Hwang WYK, Feng O, Ramakrishna S. 2008. Electrospan nano fiber scaffolds for rapid and rich capture of bone marrow derived hematopoietic stem cells. *Biomaterials*. 29: 2096–2103.

Ma PX. 2004a. Scaffolds for tissue fabrication. *Mater Today Rev* 30–41.

Ma PX. 2004b. Tissue engineering. In: Kroschwitz JI, ed., *Encyclopedia of polymer science and technology*, 3rd ed. John Wiley & Sons, Hobokan, NJ.

Ma PX, Langer R. 1999. In: Yarmush M, Morgan, J, eds., *Tissue engineering methods and protocols*. Humana Press, Totowa, NJ, pp. 47–56.

Ma PX, Zhang R, Xiao G, Franceschi R. 2001. Engineering new bone tissue *in vitro* on highly porous poly(α-hydroxyl acids)/hydroxyapatite composite scaffolds. *J Biomed Mater Res* 54: 284–293.

Ma Z, Kotaki M, Inai R, Ramakrishna S. 2005. Potential of nanofiber matrix as tissue engineering scaffolds. *Tissue Eng* 11: 101–109.

Marra KG, Szem JW, Kumta PN, DiMilla PA, Weiss LE. 1999. *In vitro* analysis of biodegradable polymer blend/hydroxyapatite composites for bone tissue engineering. *J Biomed Mater Res* 47: 324–335.

McDevitt TC, Woodhouse KA, Hauschka SD, Murry CE, Stayton PS. 2003. Spatially organized layers of cardiomyocytes on biodegradable polyurethane films for myocardial repair. *J Biomed Mater Res A* 66(3): 586–595.

Meechaisue C, Wutticharoenmongkol P, Waraput R, Huangjing T, Ketbumrung N, Pavasant P, Supaphol P. 2007. Preparation of electrospun silk fibroin fiber mats as bone scaffolds: a preliminary study. *Biomed Mater* 2: 181–188.

Meek MF, Varejao AS, Geuna S. 2002. Muscle grafts and alternatives for nerve repair. *J Oral Maxillofac Surg* 60: 1095–1096.

Mikos AG, Bao Y, Linda LG, Ingber DE, Vacanti JP, Langer R. 1993. Preparation of poly(glycolic acid) bonded fiber structures for cell attachment and transplantation. *J Biomed Mater Res* 27: 183–189.

Millesi H. 1991. Indications and techniques of nerve grafting. In: Gelbertman RH, ed., *Operative nerve repair and reconstruction*. Lippincott JB, Philadelphia, PA, pp. 525–544.

Mirensky TL, Breuer CK. 2008. The development of tissue-engineered grafts for reconstructive cardiothoracic surgical applications. *Pediatr Res* 63(5): 559–568.

Mooney DJ, Baldwin DF, Suh NP, Vacanti JP, Langer R. 1996. Novel approach to fabricate porous sponges of poly(D,L-lacto-co-glycolic acid) without the use of organic solvents. *Biomaterials* 17: 1417–1422.

Nakamura T, Schneider MD. 2003. The way to a human's heart is through the stomach: visceral endoderm-like cells drive human embryonic stem cells to a cardiac fate. *Circulation* 107: 2638–2639.

Ndreu A, Nikkola L, Ylikauppila H, Ashammakhi N, Hasirci V. 2008. Electrospun biodegradable nanofibrous mats for tissue engineering. *Nanomed* 3: 45–60.

Neal RA, McClugage SG, Link MC, Sefcik LS, Ogle RC, Botchwey EA. 2009. Laminin nanofiber meshes that mimic morphological properties and bioactivity of basement membranes. *Tissue Eng Part C Methods* 15(1): 11–21.

Ngiam M, Liao S, Patil AJ, Cheng Z, Chan CK, Ramakrishna S. 2009. The fabrication of nano-hydroxyapatite on PLGA and PLGA/collagen nanofibrous composite scaffolds and their effects in osteoblastic behavior for bone tissue engineering. *Bone* 45: 4–16.

Nisbet DR, Forsythe JS, Shen W, Finkelstein DI, Horne MK. 2009. Review paper: a review of the cellular response on electrospun nanofibers for tissue engineering. *J Biomater Appl* 24: 7–29.

Nisbet DR, Pattanawong S, Ritchie NE, Shen W, Finkelstein DI, Horne MK, Forsythe JS. 2007. Interaction of embryonic cortical neurons on nanofibrous scaffolds for neural tissue engineering. *J Neural Eng* 4(2): 35–41.

Nisbet DR, Yu LM, Zahir T, Forsythe JS, Shoichet MS. 2008. Characterization of neural stem cells on electrospun poly(epsilon-caprolactone) submicron scaffolds: evaluating their potential in neural tissue engineering. *J Biomater Sci Polym Ed* 19(5): 623–634.

Noh HK, Lee SW, Kim JM, Oh JE, Kim KH, Chung CP, Choi SC, Park WH, Min BM. 2006. Electrospinning of chitin nanofibers: degradation behavior and cellular response to normal human keratinocytes and fibroblasts. *Biomaterials* 27: 3934–3944.

Noh K, Lee H, Shin U, Kim H. 2010. Composite nanofiber of bioactive glass nanofiller incorporated poly(lactic acid) for bone regeneration. *Mater Lett* 64: 802–805.

Ondarcuhu T, Joachim C. 1998. Drawing a single nanofiber over hundreds of microns. *Europhys Lett* 42: 215–220.

Palecek SP, Loftus JC, Ginsberg MH, Lauffenburger DA, Horwitz AF. 1997. Integrin-ligand binding properties govern cell migration speed through cell-substratum adhesiveness. *Nature* 385: 537–540.

Panseri S, Cunha C, Lowery J, Del Carro U, Taraballi F, Amadio S, Vescovi A, Gelain F. 2008. Electrospun micro- and nanofiber tubes for functional nervous regeneration in sciatic nerve transections. *BMC Biotechnol* 11: 8–39.

Papadaki M, Bursac N, Langer R, Merok J, Vunjak-Novakovic G, Freed LE. 2001. Tissue engineering of functional cardiac muscle: molecular, structural, and electrophysiological studies. *Am J Physiol Heart Circ Physiol* 280: H168–H178.

Park JB. 1981. *Biomaterials science and engineering*. Plenum Press, New York, p. 1.

Park J. B, Bronzino JD. 2003. *Biomaterials: principles and applications*, CRC Press, Boca Raton, FL, p. 1.

Parker KK, Ingber DE. 2007. Extracellular matrix, mechanotransduction and structural hierarchies in heart tissue engineering. *Phil Trans R Soc B* 362: 1267–1279.

Patel S, Kurpinski K, Quigley R, Gao H, Hsiao BS, Poo MM, Li S. 2007. Bioactive nanofibers: synergistic effects of nanotopography and chemical signaling on cell guidance. *Nano Lett* 7: 2122–2128.

Petrigliano FA, McAllister DR, Wu BM. 2006. Tissue engineering for anterior cruciate ligament reconstruction: a review of current strategies. *Arthroscopy J Arthroscopic Related Surg* 22: 441–451.

Prabhakaran MP, Venugopal J, Ramakrishna S. 2009. Electrospun nanostructured scaffolds for bone tissue engineering. *Acta Biomater* 5: 2884–2893.

Prabhakaran MP, Venugopal JR, Chyan TT, Hai LB, Chan CK, Lim AY, Ramakrishna S. 2008. Electrospun biocomposite nanofibrous scaffolds for neural tissue engineering. *Tissue Eng Part A* 14(11): 1787–1797.

Radisic M, Park H, Shing H, Consi T, Schoen FJ, Langer R, Freed LE, Vunjak-Novakovic G. 2004. Functional assembly of engineered myocardium by electrical stimulation of cardiac myocytes cultured on scaffolds. *Proc Natl Acad Sci USA* 101(52): 18129–18134.

Reed CR, Han L, Andrady A, Caballero M, Jack MC, Collins JB, Saba SC, Loboa EG, Cairns BA, van Aalst JA. 2009. Composite tissue engineering on polycaprolactone nanofiber scaffolds. *Ann Plast Surg* 62: 505–512.

Ren L, Wang J, Yang FY, Wang L, Wang D, Wang TX, Tian MM. 2010. Fabrication of gelatin-siloxane fibrous mats via sol-gel and electrospinning procedure and its application for bone tissue engineering. *Mater Sci Eng C* 30: 437–444.

Riboldi SA, Sampaolesi M, Neuenschwander P, Cossu G, Mantero S. 2005. Electrospun degradable polyesterurethane membranes: potential scaffolds for skeletal muscle tissue engineering. *Biomaterials* 26: 4606–4615.

Rockwood DN, Akins RE, Parrag IC, Woodhouse KA, Rabolt JF. 2008. Culture on electrospun polyurethane scaffolds decreases atrial natriuretic peptide expression by cardiomyocytes *in vitro*. *Biomaterials* 29: 4783–4791.

Sabir MI, Xu X, Li L. 2009. A review on biodegradable polymeric materials for bone tissue engineering applications. *J Mater Sci* 44: 5713–5724.

Sahoo S, Ang LT, Goh JCH, Toh SL. 2009. Bioactive nanofibers for fibroblastic differentiation of mesenchymal precursor cells for ligament/tendon tissue engineering applications. *Differentiation* 79: 102–110.

Sahoo S, Ouyang H, Goh JC, Tay TE, Toh SL. 2006. Characterization of a novel polymeric scaffold for potential application in tendon/ligament tissue engineering. *Tissue Eng* 12(1): 91–99.

Sanghvi AB, Archit B, Murray JL, Schmidt CE. 2004. Tissue engineering of peripheral nerve. *Encyclopedia Biomater Biomed Eng* 1(1): 1613–1621.

Sangsanoh P, Suwantong O, Neamnark A, Cheepsunthorn P, Pavasant P, Supaphol P. 2009. *In vitro* biocompatibility of electrospun and solvent-cast chitosan substrata towards Schwann, osteoblast, keratinocyte and fibroblast cells. *Eur Polym J* 46: 428–440.

Santos MI, Tuzlakoglu K, Fuchs S, Gomes ME, Peters KE, Unger RE, Piskin E, Reis RL, Kirkpatrick CJ. 2008. Endothelial cell colonization and angiogenic potential of combined nano- and microfibrous scaffolds for bone tissue engineering. *Biomaterials* 29: 4306–4313.

Schmidt CE, Leach JB. 2003. Neural tissue engineering: strategies for repair and regeneration. *Annu Rev Biomed Eng* 5: 293–347.

Schmidt CE, Shastri VR, Vacanti JP, Langer R. 1997. Stimulation of neurite outgrowth using an electrically conducting polymer. *Proc Natl Acad Sci USA* 94(17): 8948–8953.

Schnell E, Klinkhammer K, Balzer S, Brook G, Klee D, Dalton P, Mey J. 2007. Guidance of glial cell migration and axonal growth on electrospun nanofibers of poly-epsilon-caprolactone and a collagen/poly-epsilon-caprolactone blend. *Biomaterials* 28(19): 3012–3025.

Seal BL, Otero TC, Panitch A. 2001. Polymeric biomaterials for tissue and organ regeneration. *Mater Sci Eng* 34(4–5): 147–230.

Shin M, Ishii O, Sueda T, Vacanti JP. 2004. Contractile cardiac grafts using a novel nanofibrous mesh. *Biomaterials* 25: 3717–3723.

Silva GA, Czeisler C, Niece KL, Beniash E, Harrington DA, Kessler JA, Stupp SI. 2004. Selective differentiation of neural progenitor cells by high-epitope density nanofibers. *Science* 303(5662): 1352–1355.

Smith IO, Baumann MJ, McCabe LR. 2004. Electrostatic interactions as a predictor for osteoblast attachment to biomaterials. *J Biomed Mater Res A* 70: 436–441.

Soletti L, Hong Y, Guan J, Stankus JJ, El-Kurdi MS, Wagner WR, Vorp DA. 2010. A bilayered elastomeric scaffold for tissue engineering of small diameter vascular grafts. *Acta Biomaterialia* 6: 110–122.

Sombatmankhong K, Sanchavanakit N, Pavasant P, Supaphol P. 2007. Bone scaffolds from electrospun fiber mats of poly(3-hydroxybutyrate), poly(3-hydroxybutyrate-co-3-hydroxyvalerate) and their blend. *Polymer* 48: 1419–1427.

Song JH, Kim HE, Kim HW. 2008. Electrospun fibrous web of collagen–apatite precipitated nanocomposite for bone regeneration. *J Mater Sci Mater Med* 19: 2925–2932.

Stankus JJ, Guan J, Fujimoto K, Wagner WR. 2006. Microintegrating smooth muscle cells into a biodegradable, elastomeric fiber matrix. *Biomaterials* 27: 735–744.

Stevens B, Yang Y, Mohandas A, Stucker B, Nguyen KT. 2008. A review of materials, fabrication methods, and strategies used to enhance bone regeneration in engineered bone tissues. *J Biomed Mater Res B Appl Biomater* 85B: 573–582.

Sui G, Yang X, Mei F, Hu X, Chen G, Deng X, Ryu S. 2007. Poly-L-lactic acid/hydroxyapatite hybrid membrane for bone tissue regeneration. *J Biomed Mater Res A* 82(2): 445–454.

Teh TKH, Goh JCH, Toh SL. 2007. Advanced micro-nano fibrous silk scaffold for tendon/ligament tissue engineering. *J Biomech* 40: 119–120.

Teh TKH, Goh JCH, Toh SL. 2008. Comparative study of random and aligned nanofibrous scaffolds for tendon/ligament tissue engineering. *J Biomech* 41: 527.

Terzis JK, Sun DD, Thanos PK. 1997. History and basic science review: past, present, and future of nerve repair. *J Reconstr Microsurg* 13: 215–225.

Thomson RC, Yaszemski MJ, Powers JM, Mikes AG. 1995. Fabrication of biodegradable polymer scaffolds to engineer trabecular bone. *J Biomater Sci* 7: 23–38.

Thouas GA, Contreras KG, Bernard CC, Sun GZ, Tsang K, Zhou K, Nisbet DR, Forsythe JS. 2008. Biomaterials for spinal cord regeneration: outgrowth of presumptive neuronal precursors on electrospun poly(epsilon)-caprolactone scaffolds microlayered with alternating polyelectrolytes. *Conf Proc IEEE Eng Med Biol Soc* 2008: 1825–1828.

Ulubayram K, Cakar AN, Korkusuz P, Ertan C, Hasirci N. 2001. EGF containing gelatin-based wound dressings. *Biomaterials* 22: 1345–1356.

Vargas EA, de Vale BNC, de Brito J, de Queiroz AA. 2010. Hyperbranched polyglycerol electrospun nanofibers for wound dressing applications. *Acta Biomater* 6: 1069–1078.

van de Graaff KM. 1998. *Human anatomy.* 5th ed. Weber State University, McGraw-Hill Companies.

Vander AJ, Sherman JH, Luciano DS. 1994. *Human physiology.* 6th ed. McGraw-Hill Companies.

Venugopal J, Ramakrishna S. 2005. Biocompatible nanofiber matrices for the engineering of a dermal substitute for skin regeneration. *Tissue Eng* 11: 847–854.

Vrana NE, Elsheikh A, Builles N, Damour O, Hasirci V. 2007. Effect of human corneal keratocytes and retinal pigment epithelial cells on the mechanical properties of micropatterned collagen films. *Biomaterials* 28: 4303–4310.

Wang HB, Mullins ME, Cregg JM, Hurtado A, Oudega M, Trombley MT, Gilbert RJ. 2009. Creation of highly aligned electrospun poly-L-lactic acid fibers for nerve regeneration applications. *Neural Eng* 6(1): 016001.

Wang J, Shah A, Yu X. 2011. The influence of fiber thickness, wall thickness and gap distance on the spiral nanofibrous scaffolds for bone tissue engineering. *Mater Sci Eng C* 31: 50–56.

Williams DF. 1987. *Definition in biomaterials.* In: *Progress in biomedical engineering.* Elsevier, Amsterdam, p. 67.

Williams DF, Black J, Doherty PJ. 1992. Second consensus conference on definitions in biomaterials, Chester, England. In: Doherty PJ, Williams RF, Williams DF, Lee AJC, eds., *Biomaterial–tissue interfaces. Advances in biomaterials*, Vol. 10. Elsevier, Amsterdam.

Wu KH, Lu XD, Liu YL. 2006. Recent progress of pediatric cardiac surgery in China. *Chin Med J (Engl)* 119(23): 2005–2012.

Xie J, Macewan MR, Willerth SM, Li X, Moran DW, Sakiyama-Elbert SE, Xia Y. 2009a. Conductive core-sheath nanofibers and their potential application in neural tissue engineering. *Adv Funct Mater* 19(14): 2312–2318.

Xie J, Willerth SM, Li X, Macewan MR, Rader A, Sakiyama-Elbert SE, Xia Y. 2009b. The differentiation of embryonic stem cells seeded on electrospun nanofibers into neural lineages. *Biomaterials* 30(3): 354–362.

Xu CY, Inai R, Kotaki M, Ramakrishna S. 2004. Aligned biodegradable nanofibrous structure: a potential scaffold for blood vessel engineering. *Biomaterials* 25(5): 877–886.

Yang D, Jin Y, Zhou Y, Ma G, Chen X, Lu F, Nie J. 2008. *In situ* mineralization of hydroxyapatite on electrospun chitosan-based nanofibrous scaffolds. *Macromol Biosci* 8: 239–246.

Yang F, Both SK, Yang X, Walboomers XF, Jansen JA. 2009. Development of an electrospun nano-apatite/PCL composite membrane for GTR/GBR application. *Acta Biomater* 5(9): 3295–3304.

Yang F, Murugan R, Wang S, Ramakrishna S. 2005. Electrospinning of nano/microscale poly(L-lactic acid) aligned fibers and their potential in neural tissue engineering. *Biomaterials* 26(15): 2603–2610.

Yang X, Shah JD, Wang H. 2009. Nanofiber enabled layer-by-layer approach toward three dimensional tissue formation. *Tissue Eng A* 15: 945–956.

Yilgor P, Sousa RA, Reis RL, Hasirci N, Hasrici V. 2008. 3D plotted PCL scaffolds for stem cell based bone tissue engineering. *Macromol Symp* 269: 92–99.

Yin Z, Chen X, Chen JL, Shen WL, Nguyen TMH, Gao L, Ouyang HW. 2010. The regulation of tendon stem cell differentiation by the alignment of nanofibers. *Biomaterials* 31: 2163–2175.

Yoshimoto H, Shin YM, Terai H, Vacanti JP. 2003. A biodegradable nanofiber scaffold by electrospinning and its potential for bone tissue engineering. *Biomaterials* 24: 2077–2082.

Yost MJ, Baicu CF, Stonerock CE, Goodwin RL, Price RL, Davis JM, Evans H, Watson PD, Gore CM, Sweet J, Creech L, Zile MR, Terracio L. 2004. A novel tubular scaffold for cardiovascular tissue engineering. *Tissue Eng* 10(1/2): 273–284.

Yucel D, Kose GT, Hasirci V. 2010. Polyester based nerve guidance conduit design. *Biomaterials* 31: 1596–1603.

Zhang R, Ma PX. 1999. Poly(alpha-hydroxy acids)/hydroxyapatite porous composites for bone tissue engineering. 1. Preparation and morphology. *J Biomed Mater Res* 44: 446–455.

Zhang Y, Lim CT, Ramakrishna S, Huang ZM. 2005. Recent development of polymer nanofibers for biomedical and biotechnological applications. *J Mater Sci Mater Med* 16: 933–946.

Zhang Y, Venugopal JR, El-Turki A, Ramakrishna S, Su B, Lim CT. 2008. Electrospun biomimetic nanocomposite nanofibers of hydroxyapatite/chitosan for bone tissue engineering. *Biomaterials* 29: 4314–4322.

Zhang Y, Ouyang H, Lim CT, Ramakrishna S, Huang ZM. 2005. Electrospinning of gelatin fibers and gelatin/PCL composite fibrous scaffolds. *J Biomed* Mater Res Part B: *Appl Biomater* 72B: 156–165.

Zhao F, Yin YJ, Lu WW, Leong JC, Zhang WJ, Zhang JY, Zhang M. F, Yao KD. 2002. Preparation and histological evaluation of biomimetic three-dimensional hydroxyapatite/chitosan-gelatin network composite scaffolds. *Biomaterials* 23: 3227–3234.

Zimmermann WH, Eschenhagen T. 2003. Cardiac tissue engineering for replacement therapy. *Heart Fail Rev* 8: 259–269.

Zimmermann WH, Melnychenko I, Eschenhagen T. 2004. Engineered heart tissue for regeneration of diseased hearts. *Biomaterials* 25: 1639–1647.

Zimmermann WH, Schneiderbanger K, Schubert P, Didie M, Munzel F, Heubach JF, Kostin S, Neuhuber WL, Eschenhagen T. 2002. Tissue engineering of a differentiated cardiac muscle construct. *Circ Res* 90: 223–230.

Zong X, Bien H, Chung C-Y, Yin L, Fang D, Hsiao BS, Chu B, Entcheva E. 2005. Electrospun fine-textured scaffolds for heart tissue constructs. *Biomaterials* 26: 5330–5338.

Zorlutuna P, Builles N, Damour O, Elsheikh A, Hasirci V. 2007. Influence of keratocytes and retinal pigment epithelial cells on the mechanical properties of polyester-based tissue engineering micropatterned films. *Biomaterials* 28: 3489–3496.

Zorlutuna P, Rong Z, Vadgama P, Hasirci V. 2009. Influence of nanopatterns on endothelial cell adhesion: enhanced cell retention under shear stress. *Acta Biomater* 5: 2451–2459.

Zorlutuna P, Tezcaner A, Hasirci V. 2008. A novel construct as a cell carrier for tissue engineering. *J Biomater Sci Polym Ed* 19: 399–410.

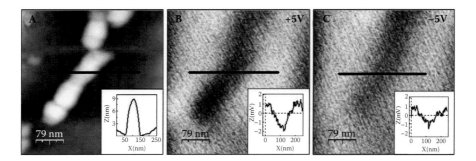

COLOR FIGURE 1.8
(A) Topographic AFM image of 6His-SP1-GNP chain formation on mica (insets shows a cross-section on the chain). (B and C) EFM at positive and negative bias voltage of 6His-SP1-GNP chain on mica, showing clear polarization of the embedded GNPs (no signal is observed at 0 V). The insets show the signal strength. (From Medalsy et al. 2008)

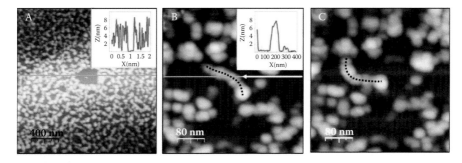

COLOR FIGURE 1.9
6His-SP1-GNP tube structures on HOPG. (A) Large-area topographic imaging shows multiple tubes. (B) A single tube with a height of ~9 nm (dashed line). (C) With contact mode AFM, a single line was scratched, thus applying lateral sheer force on the tube (the arrow in B). The tube was moved and bent but not torn, indicating the physical strength of the structure.

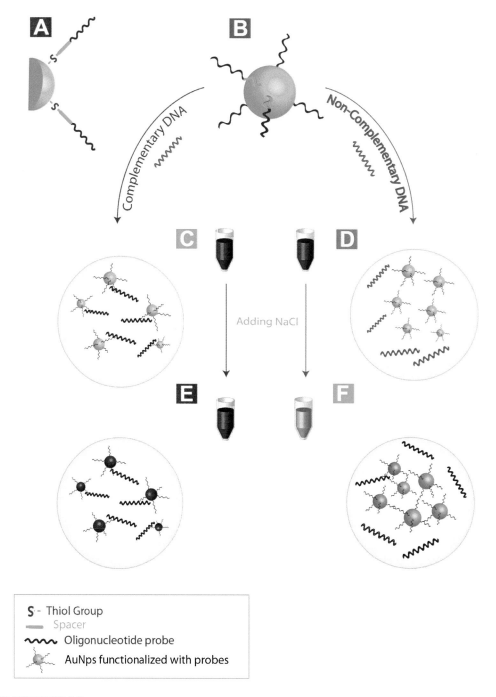

COLOR FIGURE 3.8

Modified non cross-linking colorimetric detection of nucleic acid targets. (A and B) AuNP functionalized with one set of oligonucleotide probe, which is attached to the AuNP via a spacer and thiol bond. (C) In the presence of the complementary target DNA, hybridization occurs. (D) If the absence of complementary target, no hybridization occurs. (E) Upon addition of NaCl, the solution where hybridization occurred remains red. This is because the hybridized oligonucleotides increase the negative charge of AuNPs, thereby increasing their resistance to aggregation. (F) In the solution where no hybridization occurred, addition of NaCl induces aggregation of AuNPs by neutralizing their negative charge, and the solution color changes from red to blue.

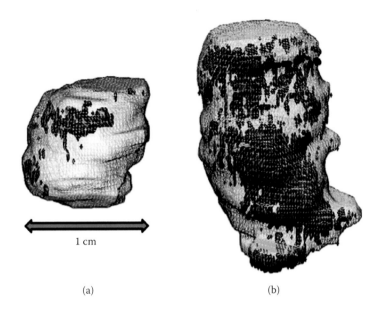

(a)   (b)

COLOR FIGURE 4.5
Three-dimensional (3D) neovascular maps in Vx-2 tumors. With MRI and $\alpha_v\beta_3$-targeted nanoparticles, areas of angiogenesis surrounding Vx-2 tumors can be highlighted (dark blue) on 3D mesh models of the tumor (digitally resected from the 3D image data set). These two tumors are the same age, but tumor (a) received antiangiogenic therapy through $\alpha_v\beta_3$-targeted fumagillin-laden theragnostic nanoparticles 16 days prior to this imaging and is much smaller than tumor (b), which received only drug-free particles for imaging. Note the heterogeneous distribution of neovasculature around both the treated and nontreated tumor. (Reprinted with permission from Winter, P.M., et al., *Faseb J*, 22, 2758–2767, 2008.)

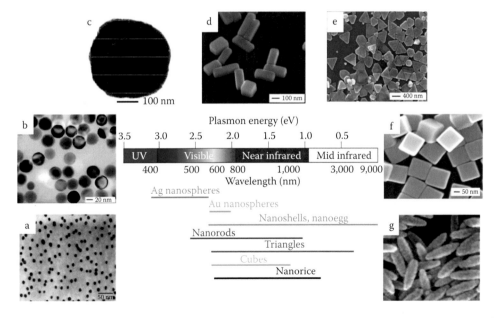

COLOR FIGURE 5.2
Nanoparticle resonance range of plasmon resonances for a variety of particle morphologies. (© 2007 Macmillan Publisher's Ltd. Used with permission.) TEM images of © 2004 (Au spheres (b)), © 1995 (Ag spheres (a)), and © 2001(SiO$_2$/Ag (core-shell) nanoshells (c)), and SEM images of © 2007 (nanorods (d)), © 2006 (cubes (e) and nanorices (g)), American Chemical Society and © 2006 (triangles (f)). (Wiley-VCH. Used with permission.)

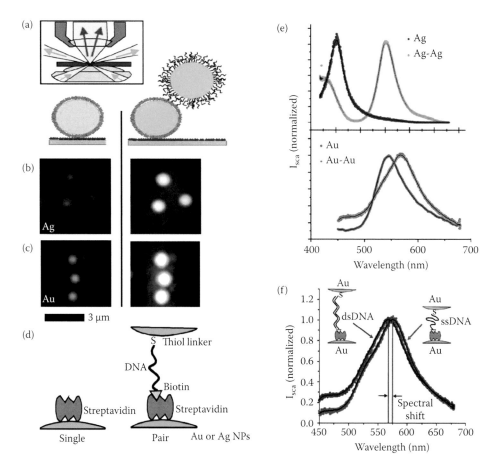

COLOR FIGURE 5.5

DNA-functionalized gold and silver nanoparticles (NPs) as molecular plasmonic rulers. (a) First, streptavidin-functionalized nanoparticles are attached to the BSA-biotin-coated glass surface (a, d left). Then, a second NP is modified by thiol-ssDNA-biotin bifunctional linker and attached to the first particle via biotin-streptavidin affinity binding (a, d right). Inset: Principle of transmission dark field microscopy. (b) Single silver particles appear blue (left) and particle pairs appear blue-green (right). (c) Single gold particles appear green (left) and gold particle pairs appear orange (right). (e) Representative scattering spectra of single particles and particle pairs for silver (top) and gold (bottom). The LSPR scattering spectrum for silver NPs shows a larger LSPR shift (102 nm vs. 23 nm), stronger light scattering, and a smaller half width at half maximum (HWHM) than gold particles. (f) Example of the LSPR spectral shift between a gold particle pair connected with ssDNA (red) and dsDNA (blue), corresponding to the distance increase of 2.1 nm between the particle pair upon DNA hybridization. (© 2005 Macmillan Publisher's Ltd. Used with permission.)

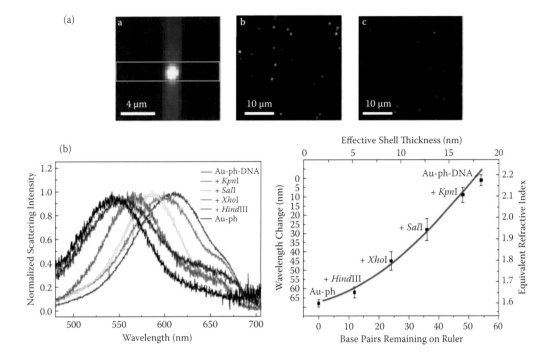

COLOR FIGURE 5.6
Nanoplasmonic molecular rulers for measuring nuclease activity. (a) Dark field scattering images: a, single-target single nanoparticle isolated from other Au or Au-DNA NPs for spectroscopic examination; b, c, true-color images of single Au NP (b) (540 nm peak) and Au-DNA NP (c) (607 nm peak). (b) Typical LSPR scattering spectra and the corresponding maximum scattering wavelengths (λ_{max}) of the Au-DNA nanoconjugates after cleavage reactions with four endonucleases (*Hind*III, *Xho*I, *Sal*I, and *Kpn*I) and as a function of the number of base pairs remaining attached to the Au nanoparticle after the cleavage. The red curve is a fit from a semiempirical model using a Langevin-type dependence of the refractive index vs. dsDNA length. The error bars represent the standard deviation in the measurement of 20 nanoparticles. (© 2006 Macmillan Publisher's Ltd. Used with permission.)

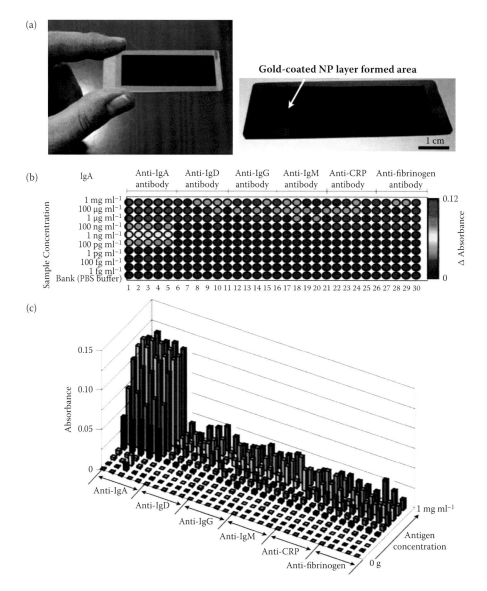

COLOR FIGURE 5.7
LSPR-based nanochips for multiplexed sensing of, e.g., immunoglobulin A (IgA). (a) Photograph of the LSPR-based nanochip (20 mm × 60 mm). The nanochip structure consists of a flat gold film, self-assembled monolayer of silica nanospheres, and a gold coating over the nanospheres. (b and c) Absorbance measurements at each spot of the multiarray nanochip for monitoring the binding of IgA to six different types of antibodies. The antibodies were immobilized on the chip by the nanoliter dispensing system and result in a total of 300 spots separated by 1 mm to prevent cross-contamination. Various concentrations of antigens were incubated for 30 min and LSPR absorption spectra were then taken with a fiber-coupled ultraviolet-visible spectrometer in a reflection configuration. Detection was linear up to 1 μi ml^{-ml} with a limit of detection of 100 pg ml^{-i} for all of the proteins. (© 2006 American Chemical Society. Used with permission.)

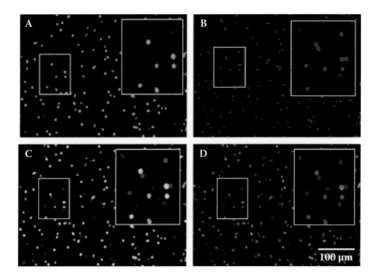

COLOR FIGURE 6.1
Fluorescence imaging (10×) of CD4 cells captured in a nanobiochip. (A) Cells are labeled with anti-CD4 antibody conjugated with Qdot 565 (in green). (B) Cells are labeled with anti-CD3$^+$ antibody conjugated with Qdot 655 (in red). (C) CD3$^+$CD4$^+$ cells are shown in yellow in the emerged image. (d) Absolute CD4 cell count (in yellow) is obtained in the processed picture, in which cells in red or green are deleted. A long-pass emission filter allows for CD4$^+$ T lymphocyte counting instead of image processing. Boxes on the right in each panel show magnified images from the left (2×). (Adapted from Jokerst, J.V., et al., *Lab Chip*, 8(12), 2079–2090, 2008.)

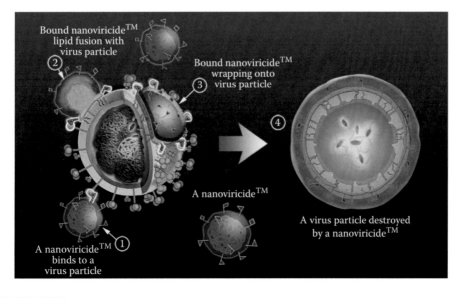

COLOR FIGURE 7.2
Graphic representation of the mechanism of action of nanoviricide binding and inactivation of viruses.

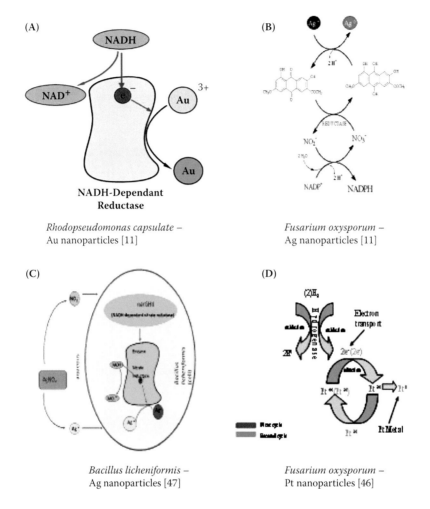

(A)

NADH

NAD⁺

e⁻

Au 3+

Au

**NADH-Dependant
Reductase**

Rhodopseudomonas capsulate –
Au nanoparticles [11]

(B)

Fusarium oxysporum –
Ag nanoparticles [11]

(C)

Bacillus licheniformis –
Ag nanoparticles [47]

(D)

Fusarium oxysporum –
Pt nanoparticles [46]

COLOR FIGURE 9.5
Summary of typical models proposed for the mechanism of nanoparticle synthesis (A)-(D). ((A) From He, S., et al., *Mater. Lett.*, 61, 3984, 2007.) (B) From Duran, N. et al., J. Nanobiotechnology, 3,8, 2005. With permission.) ((C) From Kalimuthy, K. et al., Celloids Surf. B Biointer faces, 65, 150, 2008. With permission) ((D) From Govender, Y. et al. Biotechnol. Lett., 31, 95, 2008. With permission.)

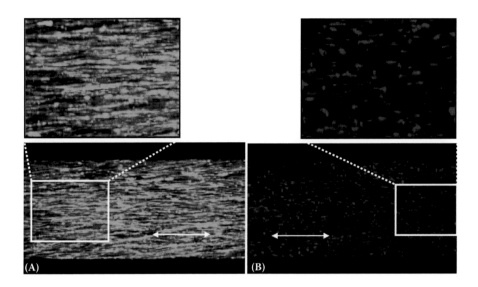

COLOR FIGURE 10.3
Cross-sectional (X-axis ~ 60° tilted) CLSM images of FITC-Phalloidin and propidium iodide-stained WJ MSCs on a single, electrospun, aligned fiber mat of PHBV:P(L-D,L)LA:PGS (poly(glycerol sebacate)) (100×) showing (A) F-actin and (B) nuclear alignment (two-sided arrow, direction of aligned fiber axis). Top images are the magnified versions of insets.

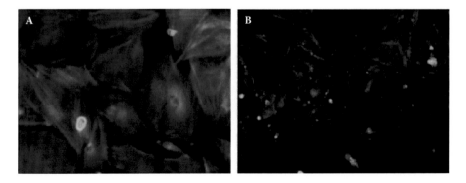

COLOR FIGURE 11.10
Fluorescence micrographs of nano-microcapsules taken up by Saos-2 cells. (A) PHBV, 40×. (B) PLGA, 20×. Cell nuclei were stained with DAPI (blue), cell cytoskeletons were stained with FITC-labeled phalloidin (green), and nano-microcapsules were stained with Nile red (red).

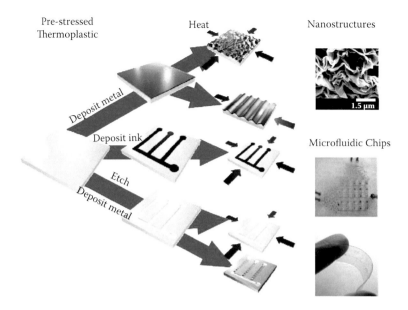

COLOR FIGURE 13.1
Ultra-rapid, low-cost manufacturing process of nano-integrated microsystems.

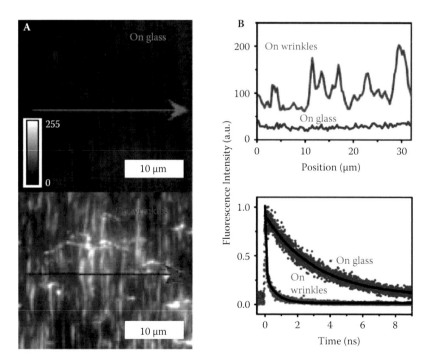

COLOR FIGURE 13.8
(A) Wide-field epifluorescence images of dyes on glass plate (top) and uniaxial wrinkles (bottom). (B) The corresponding intensity profiles along the arrows in (A) (top) and fluorescence lifetime measurements of dyes on a glass plate (top line) and wrinkles (bottom line) (bottom).

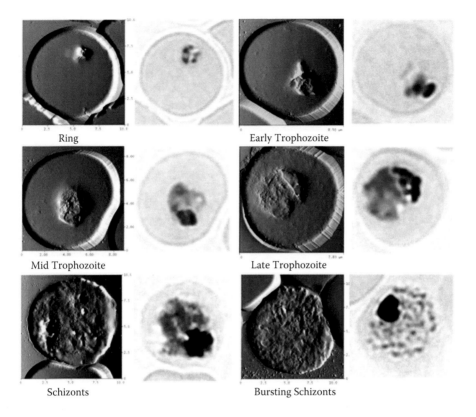

COLOR FIGURE 15.2

Corresponding AFM imaging of external surface and Giemsa staining of interior parasite structure of lab-cultured *P. falciparum* IRBCs. (Details shown in Li, A., et al., *J. Microbiol. Methods*, 66, 434–439, 2006.)

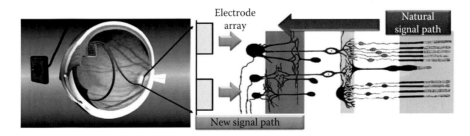

COLOR FIGURE 16.5

Illustration of epiretinal prosthesis. An array of electrodes are implanted inside the eye and hatched on the surface of the retina. The natural signal path of the photoreceptor to the retinal ganglion neuron is blocked. A new signal path is created by electrically stimulating the surviving neurons.

COLOR FIGURE 16.9
Die photograph and testing results. (a) Die photograph. (b) Chip test environment. (c) Measured stimulator output waveforms.

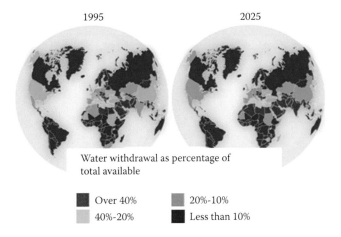

COLOR FIGURE 19.1
Water withdrawal rate as a percentage of total available water, projected until 2025. (From Service, R.F., *Science*, 313, 1088–1090, 2006. With permission.)

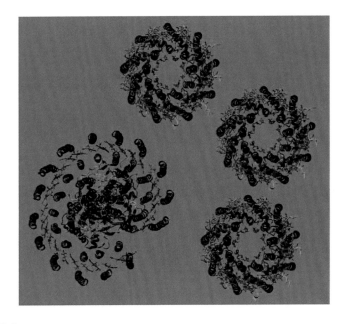

COLOR FIGURE 22.6
Structure of antenna protein system LH1-RC complex surrounded by LH2.

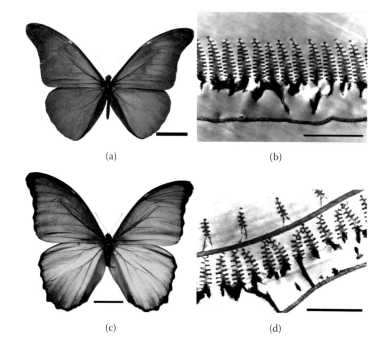

(a)   (b)

(c)   (d)

COLOR FIGURE 25.4

Optical and TEM images of the butterflies *Morpho rhetenor* (a, b) and *Morpho didius* (c, d). Both optical images (a, c) show the brilliant blue that is synonymous with this family of butterflies. The two TEM images (b. and d.) show the substructure of the butterfly scales. TEM image of a section from a *M. rhetenor* scale (b.) shows the multilayer within the ridges responsible for the brilliant colour, while the image from a section of *M. didius* (d.) shows the two discretely configured multilayers. Scale bars a 20 mm b. 1.8 μm c. 20mm d. 1.3 μm.

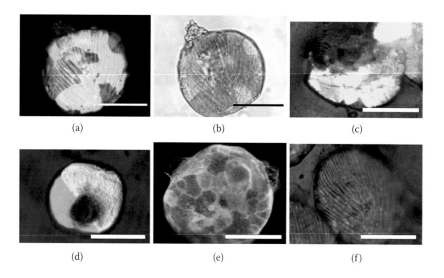

(a)   (b)   (c)

(d)   (e)   (f)

COLOR FIGURE 25.7

White illuminated microscope images to show the (a) reflection and (b) transmission colors of a single scale from the blue region of *E. magnificus*. This scale was selected to show the large size of the domains within these scales, and the consistency of the hue that they produce. By comparison, the scales from other regions demonstrate multiple domains and colors (see Figure 25.9). (c–f) Four images of domained scales from other regions: (c) yellow region, *E. schoenherri pettiti*; (d) green region, *E. schoenherri pettiti*; (e) blue region, *E. magnificus* (dark field image); (f) green region, *E. magnificus*. Scale bars: 30 μm.

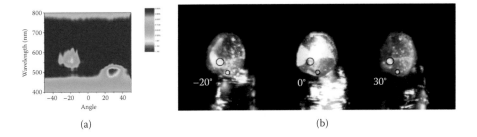

(a)　　　　　　　　　　　　　　　　　　　　　(b)

COLOR FIGURE 25.10

(a) Graph showing the reflection measured when a scale from *E. schoenherri pettiti* is rotated around the azimuthal θ axis, with the reflection spectra recorded by a microspectrometer. (b) Image of the scale examined in (Figure 25.10a) at three angles of incidence: –20°, 0°, and 30°. The larger black circle encompasses the region that the spectra in Figure 25.10a was measured on. The smaller black circle shows another domain that displays a similar color progression. Scale bars: 80 μm (b) (With permission).

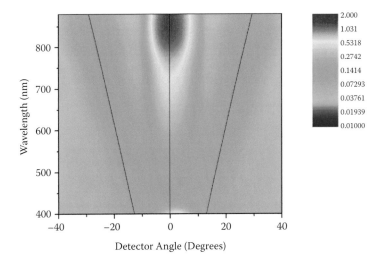

COLOR FIGURE 25.14

Angle-dependent transmission data for a single valve of a *C. wailesii* diatom using the equipment shown in Fgure 25.2. The black lines superimposed on this image correspond to the theoretical zero- and first-order diffraction lines for the 2 μm periodic structure within the valve. Intensity in arbitrary units.

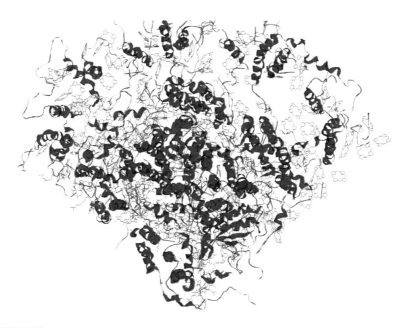

COLOR FIGURE 26.2
The x-ray structure plant of photosystem I (PSI) at 4.4 A° from the protein data bank (code 1qzv). The protein backbone is in red, and the chlorophyll pigments are shown in blue. The four light-harvesting proteins (Lhca1–4) are at the top, the region where light is absorbed.

Control   Carbon nanotubes

COLOR FIGURE 26.5
Tomato seedlings grown with and without ~40 mg/L MWNT additive. (Reprinted from Khodakouskaya et al. *ACS Nano* 3(10): 3221(2009). With permission of the American Chemical Society.)

11

Nanotechnology in Biomaterials: Nanoparticulates as Drug Delivery Systems

Birsen Demirbag, Sinem Kardesler, Arda Buyuksungur, Aysu Kucukturhan, Gozde Eke, Nesrin Hasirci, and Vasif Hasirci

BIOMATEN, Center of Excellence in Biomaterials and Tissue Engineering, and Departments of Biological Sciences and Chemistry Middle East Technical University, Ankara, Turkey

CONTENTS

11.1 Drug Delivery Systems

A drug delivery system (DDS) can be defined as the system that achieves the administration of a therapeutic agent to the patient and improves the drug's efficacy and safety by controlling the concentration, rate, time, and place of release of drugs in the body.[1] The primary purpose of drug delivery systems is to deliver the drug efficiently and precisely to a targeted site in an appropriate period of time, while maintaining a high concentration of the drug in the diseased site and as low as possible in the healthy tissue.[2]

In conventional systems plasma drug levels increase after administration to a patient, and then decrease to an ineffective level; however, the concentration should be in the therapeutic window. When a new administration is made, the same rise and fall happens. To overcome this low dose problem, higher drug concentrations can be applied, but this increases the risk of toxic effects of the drug and the treatment cost. For many diseases, the therapeutic agent should reach a certain concentration and remain constant at the site of action. In conventional systems, however, the plasma drug levels do not remain constant, but fluctuate because of distribution, metabolism and excretion. Drug delivery systems try to overcome this problem by releasing the drug with zero-order kinetics. These systems attain desired plasma levels and sustain the release of the drug for a given period of time. Zero-order release has the advantage of maintaining constant plasma levels, and therefore decreases the risks of toxicity and ineffectiveness. Another problem associated with the conventional systems is exemplified in cancer therapy. Most conventional chemotherapies are relatively nonspecific; for example, any chemotherapeutic drug that is administered intravenously enters systemic distribution, and this results in the exposure of healthy cells to the toxic chemical alongside the target cells.[3] Targeted DDS have therefore been developed to solve these problems by providing drugs with the predetermined concentration mainly at the target site.

Another reason for constructing a drug delivery system is that sometimes the bioactive agent is very unstable and has the risk of losing its bioactivity due to the environmental conditions.

The therapeutic strategy of drug delivery systems has to take into account many factors, such as the disease type, the route of administration, the properties of the drug, and the target site to achieve the highest efficacy with the therapeutic agent. The main aims for a drug delivery system are to[4]

- Protect the drug from the host by minimizing its degradation, prolonging its bio-availability after administration
- Protect the host from the toxic effects and side effects of the drug due to accumulation at the desired site via specific or nonspecific interactions
- Use responsiveness to local stimuli, such as abnormal local pH values or local temperatures, to achieve the kinetics needed

Delivery can be made more effective by targeting actively or passively. For active targeting, the carrier system or the therapeutic agent can be conjugated with a ligand specific to the target tissue or target cell type. One of the mechanisms of passive targeting to a tumor site utilizes the properties of vasculature at the tumor. Tumor vasculature has a chaotic structure, which consists of various loops, dead ends, and openings, and has increased leakiness. Additionally, because the lymphatic drainage is hampered, the drug carrier construct stays longer at the tumor site.[5] These two factors are combined and result in an enhanced permeability and retention (EPR) effect.[6] Macromolecules and nanoparticles take advantage of the EPR effect and can passively be targeted to the tumor site. Targeting of the therapeutic agent reduces systemic levels and, thus, systemic toxicity and achieves high drug levels at the desired site.[1]

Responsive drug delivery systems use the advantage of on-demand or passive triggering of the release of therapeutic agents using a local characteristic of tissues (low pH, high temperature) or an external stimulus (heat, light, temperature, or magnetic field).[4] For example, in cancer therapy the tumor microenvironment has a reduced pH,[7] increased temperature, and increased oxidative potential,[8] and these environmental factors are used to trigger the release of drug at the tumor site.

Drug delivery systems need to take into account the route of administration in order to maximize the amount of drug at the site of action. Different routes of administration require different conditions for successful drug delivery. Table 11.1 shows the routes for the administration of drug delivery systems and the biological systems they interact with.

TABLE 11.1

Routes of Drug Administration

Administration Route	Interacting Biological Systems
Nasal	Nose
Topical and dermal	Skin
Intramuscular (im)	Muscle
Intravenous (iv)	Circulatory system
Parenteral (subcutaneous: sc)	Skin
Ocular	Eye
Pulmonary	Lung
Vaginal	Vaginal mucosa
Rectal	Rectum/colon
Buccal and oral	Oral cavity and gastrointestinal tract
Intraperitoneal (ip)	Peritoneum
Intrathecal (it)	Central nervous system (vertebra)

Although there are many disadvantages associated with the oral route, it still is a very successful, convenient, and therefore preferred route of administration for normal drugs. In this application the drug applied has to pass through the highly variable gastrointestinal system. Low pH of the stomach tends to affect both stability and solubility of drugs. Also, the intestine has some rather demanding requirements, like motility, mucus barriers, high metabolic activity, and relative impermeability of the epithelium.[9] For a drug delivery system, however, most of these routes, including the oral one, have certain limitations because of the special demands of the target site, and therefore case-by-case solutions have to be provided.

In the following sections, nanomaterial-based drug delivery systems using lipoid (liposomes) or polymeric nanoparticles will be discussed.

11.2 Nanoparticulates for Drug Delivery Systems

Nanosystems have been in the spotlight in recent years, especially with their applications in biotechnology. Nanotechnology is focused on processes and materials at the atomic and molecular level. The main concern of this area is on the 1–100 nm scale. The first use of the word *nanotechnology* was during a talk; "There is plenty of room at the bottom," said Nobel laureate Richard Feynman.[10] Nanotechnology has a great range of applications in fibers and textiles,[11] agriculture,[12,13] electronics,[14] forensic science,[15] and medical therapeutics,[16–21] among others. The interaction of nanotechnology with other areas of engineering and science has opened up new opportunities, especially in nanobiotechnology, which is described as the application of principles of nanotechnology in biological systems. In this new field, nanotechnology is the provider of devices and technology, while biological systems are the inspiration and models for the technology.[22]

Nanoparticles are very important tools for nanobiotechnology applications, especially in medicine. They can be utilized in various cell and tissue imaging techniques, e.g., quantum dots,[23] as growth factor delivery materials for tissue engineering,[24,25] and for gene[26] and drug[27] delivery and transport purposes.

In drug delivery applications, nanoparticles are preferred mainly due to their improved bioavailability, solubility, and retention time.[28] Encapsulation of drugs in nanoparticles increases the efficiency of the drug due to improved localization, especially after extravasation, a capability that the microparticles do not have. Moreover, an encapsulated drug is protected from early degradation while in the circulatory system, before reaching the target organ or tissue. Like microparticles, nanoparticles can be designed to target a particular organ or tissue so that the drug accumulates at the predetermined area without excessive dilution along the route.[29] In addition to targeted release, controlled or responsive release can also be achieved with nanoparticles. The control factors can be pH,[30] temperature,[31] and light exposure.[32] In brief, utilization of nanoparticles in drug delivery increases the therapeutic value of the drug, which then can lead to lower risks and hospital costs.

Although nanoparticles are widely utilized in many areas, they may also pose risks to both human health and the environment. These risks have led to the development of a research area called nanotoxicology, which deals with the toxic effects of nanomaterials on living organisms. Although nanoparticles have been found to be dispersed everywhere (air, soil, and water), no exhaustive study has been carried out concerning the long-term detrimental effects of nanoparticles on the environment. However, it is known that there are various adverse health effects of nanoparticles. For example, it is reported that nanoparticles can easily enter the body through inhalation, injection, dermal route, and ingestion, and these particles can be accumulated in the tissues upon entering the body.[33] They can also migrate from these entry points to the circulatory and lymphatic systems and ultimately reach organs and tissues. The deposition of these nanoparticles might lead to inflammation or severe neurological disorders, such as Parkinson's or Alzheimer's disease.[34] Nanoparticles can also have other effects as a result of accumulation in organs and tissues. It would be ideal if the nanoparticles that have accomplished their function were removed from the body. Some can be removed without any degradation process (exhalation, excretion), while others are metabolized in the body. As a consequence of the degradation of these nanoparticles, several by-products, which can also lead to numerous complications, are formed.[33]

Depending on their application areas, nanosystems can be found in different forms and shapes. They are classified in terms of dimension, morphology, composition, and preparation approach. Table 11.2 presents a schematic summary of nanoparticles classified according to these three categories.

The nanoparticles used in nanobiotechnology can also be classified according to the materials they are made of, such as polymeric, lipid based, and inorganic:

- **Polymeric nanoparticles:** These nanoparticles can be made of either homo- or copolymers, and their blends in different shapes (e.g., rod, fiber, sphere, etc.). They can be made of degradable or nondegradable, hydrophilic or hydrophobic polymers, and can constitute the bulk or the coat of the nanoparticles. Due to their nanosize, they have a high surface area-to-volume ratio, indicating that there are more molecules on the surface of a nanoparticle than in the core. Since the surface is part of the interface with the tissue, these molecules have the potential to react with the surrounding molecules, which makes nanoparticles proper tools for

TABLE 11.2

Classification of Nanoparticles

Dimension and Morphology		Composition	
1D	Nanowires	Organic	Synthetic polymeric
		Organic	Biological polymeric
1D	Nanotubes	Organic	Liposomal
1D	Nanofibers	Inorganic	Ceramic
		Inorganic	Carbon
2D	Films and surface coatings	Inorganic	Metallic
		Preparation Approach	
2D	Thick membranes with nanopores	Top-down	Molded structures
		Top-down	Etched structures
3D	Nanospheres	Bottom-up	Layer by layer
		Bottom-up	SAMs
3D	Nanocapsules	Bottom-up	Liposomes
		Bottom-up	Surface coats
3D	Irregular nanoparticles		

* SAMs: self assembled monolayers

targeted delivery after appropriate surface modifications.[35] Well-known examples of polymeric nanoparticles used in nanobiotechnology are made of poly(lactide) and poly(glycolide) and their copolymers.[36,37] In addition, poly(ε-caprolactone) (PCL), gelatin, and chitosan are also widely used in nanobiotechnology.[28] The major advantage of these degradable polymers is that they are mostly biocompatible and are metabolized and cleared from the body via regular pathways.

There are two main methods of production of polymeric nanoparticles:[38]

- Dispersion of preformed polymers
  - Solvent evaporation method
  - Spontaneous emulsification/solvent diffusion method
  - Salting out/emulsification-diffusion method
  - Supercritical fluid technology
  - Ionic gelation
- Polymerization of monomers
  - Anionic polymerization
  - Cationic polymerization

- Bulk polymerization
- Emulsion polymerization
- Suspension polymerization

By using these routes, polymeric nanoparticles carrying bioactive or imaging agents could be prepared.

- **Lipid-based nanoparticles:** Lipid-based nanoparticles are useful tools for drug delivery and gene transfer since they protect their content from being cleared by the circulatory system while also protecting the nontarget tissues from adverse effects of the bioactive agent. More importantly, they can be metabolized and removed from the body without leaving any by-products behind. Nanoparticles made of phospholipids can be spontaneously self-assembled or prepared by similar processes. Depending on their preparation method, lipid nanoparticles can take different shapes and forms, such as micelles, bilayers, and hexagonal vesicles.[39] Lipid-based nanoparticles can have a wide range of sizes (i.e., 20–1000 nm), because of the different methods used in their preparation. Liposomes are a special category of lipid nanoparticles, and their field of use in biotechnology is determined by physicochemical characteristics such as composition, size, loading, and stability, in addition to their interaction with cells. For drug delivery purposes liposomes can be formulated in a suspension, as an aerosol, or in a semisolid form, such as cream, gel, or dry powder.[40] In addition to different methods of lipid-based nanoparticle preparation, there also are large numbers of methods for active and passive loading of drugs in these nanoparticles. For example, hydrophilic drugs can be encapsulated in the aqueous core while hydrophobic ones can be embedded in the bilayer of the phospholipids. For specific purposes, conventional liposomes can be modified; they can be PEGylated (poly(ethylene glycol) modified) or targeted or designed to be charged through the use of functional groups or molecules. These modifications alter their interactions and control their distribution in the body.

- **Inorganic nanoparticles:** Inorganic nanoparticles are preferred in nanobiotechnology due to their durability, magnetic properties, fluorescence, and radio opaqueness.[41] Calcium phosphate, gold, carbon materials, silicon oxide, iron oxide, cadmium selenide, and zinc selenide are some examples of inorganic nanoparticles. Among the metallic nanoparticles, gold nanoparticles are very commonly used for delivery purposes since they can be shaped and brought to the desired size, and their surface can be modified by coating with a variety of molecules. All inorganic nanoparticles can be surface modified to be multifunctional and used in targeting a tissue in drug delivery, to avoid the immune system, and also to fluoresce for imaging.[42]

11.2.1 Liposomes as Drug Delivery Systems

Liposomes form spontaneously when phospholipids are hydrated. Bangham et al. in 1965 demonstrated that phospholipids form closed structures when dispersed in water, and these structures are relatively impermeable to entrapped material.[43] After Bangham et al., there was a great expansion in the use of liposomes as models for biomembranes, for gene delivery, and as drug delivery systems.[44] The main constituents of liposomes are phospholipids, which have an amphiphilic structure; they have polar and nonpolar regions. The polar regions contact with the aqueous phase inside and outside the liposomes, and the nonpolar region is oriented away from the aqueous phase. Liposomes are biocompatible, mask the toxicity of the drug, and increase its time in the circulation, and thus the efficacy. Since their discovery, a number of them have been tested in clinical trials, and some of them were approved by the U.S. Food and Drug Administration (FDA).

The properties of liposomes (i.e., size, charge, release kinetics, etc.) depend on their formulation and the preparation methods. The spontaneous (and uncontrolled) formation of liposomes leads to a population of vesicles ranging from 10 nm to 10 μm in diameter. Liposomes can entrap hydrophilic and hydrophobic molecules in their aqueous compartment and within their membrane, respectively. One has to consider the phase transition temperature, chain length, and degree of saturation of the phospholipids, and the permeability of the vesicles toward various compounds. Although liposomes are formed spontaneously, there are several methods to control their properties and obtain a population with predetermined features.

Liposomes can be designed to be neutral or charged. The major constituents of neutral liposomes are uncharged phosphatidyl cholines (PCs) such as dipalmitoyl phosphatidyl glycerol (DPPG), dipalmitoyl phosphatidyl choline (DPPC), and cholesterol. Phosphatidyl ethanolamine (PE) is used for positively and dipalmitoyl phosphatidic acid (DPPA) for negatively charged liposomes. Sterols, and especially cholesterol, are very important for liposome stability. They themselves cannot form the bilayer structures, but they can be incorporated into the liposomes up to 1:1 molar ratio. Cholesterol presence in the lipid bilayer has a significant effect on the phase transition temperature. As the concentration of cholesterol in the bilayer membrane increases, the phase transition temperature also increases, up to a certain degree. This indicates that cholesterol incorporation increases the stability (and rigidity) of the bilayer. In addition, permeability and fluidity of the membrane are also affected by cholesterol.

Liposomes are classified according to the number of bilayers and size (Figure 11.1):

- Small unilamellar vesicles (SUVs): Their diameters are generally in the range of 20–100 nm, and they have a single bilayer (Figure 11.1A).
- Multilamellar vesicles (MLVs): They are generally larger than 500 nm in diameter and have multiple bilayers (Figure 11.1B).
- Large unilamellar vesicles (LUVs): Liposomes of a single bilayer larger than 1,000 nm in diameter are classified as large unilamellar vesicles (Figure 11.1C).
- Multivesicular vesicles (MVVs): They consist of several vesicles within one large common liposome. They are generally larger than 1000 nm (Figure 11.1D).

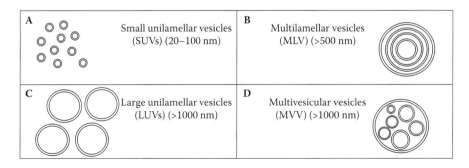

FIGURE 11.1
Classification of liposomes according to their size and bilayer number.

Liposomes are prepared with different methods used in various preparation steps. Following is a list of methods with parameters that could change the type of liposomes obtained as a result.[45]

11.2.1.1 Methods of Liposome Preparation Based on Variations in the Replacement of Organic Solvent(s) by the Aqueous Media

11.2.1.1.1 Removal of Organic Solvent(s) before Hydration

During liposome preparation all lipid-soluble membrane ingredients are dissolved in an organic solvent in order to ensure a homogeneous membrane. In the removal of this organic solvent a rotary evaporator is the most common device used.[46] The organic solvent is evaporated and the lipid is dried on the walls of the glass flask. Then this dried lipid is hydrated and used in liposome production. Compounds to be carried in the lipoid membrane are added during the solution phase, and those to be carried in the aqueous part of the liposomes are added in the hydration step. This leads to small unilamellar vesicles (SUVs) and multilamellar vesicles (MLVs).

11.2.1.1.2 Reverse Phase Evaporation

In this type of liposome preparation, the first step is removal of the organic solvent, as above with a rotary evaporator. The lipids are then redissolved in the organic phase, the aqueous phase is added, and the resulting two-phase system is sonicated until the mixture becomes a homogeneous suspension of oil in water. The organic solvent is removed under vacuum in a rotary evaporator, and the aqueous suspension obtained at the end is called the reverse phase evaporation vesicle (REV).[47] This method leads to very high encapsulation efficiencies and large unilamellar vesicles (LUVs).

11.2.1.1.3 Use of Water-Immiscible Solvents

This technique involves injection of an immiscible organic solution of the phospholipid through a needle into the aqueous phase in a very slow fashion while the organic phase is removed by vaporization. Large vesicles are formed as a result of this process. Ether injection is a common example for this method.[48] This method has very little risk of oxidative degradation of the phospholipids.

11.2.1.1.4 Use of Water-Miscible Solvents

Ethanol injection is a good example for this method. In ethanol injection lipid solution is prepared in ethanol and injected into an aqueous medium through a fine needle. The force of injection also achieves mixing, resulting in an evenly dispersed phospholipid solution.[49] With this technique small unilamellar vesicles (SUVs) are obtained.

Thus, even with a single parameter change, a large variety of liposomes are produced, proving the versatility of liposomes.

11.2.1.2 Applications of Liposomes

Liposomes are widely used in cosmetic products and pharmaceuticals, and they are also used as analytical tools in many disciplines of science[50] (Table 11.3).

Liposomes are used in drug delivery, as controlled and sustained release systems, and in medical diagnostics and gene therapy. They are also used as a model in cell recognition and interaction studies, and in the investigation of the mode of action of active substances in pharmacology and medicine.

TABLE 11.3

Applications of Liposomes in Science

Disciplines	Applications	References
Physics	Aggregation behavior, fractals, soft and high strength materials	51
Biophysics	Permeability, phase transitions in two dimensions, photophysics	52
Physical chemistry	Colloid behavior in a system of well defined physical characteristics, inter- and intra-aggregate forces	52
Biochemistry	Reconstitution of membrane proteins into artificial membranes to investigate their role in transport mechanisms	53
Biology	Model biological membranes, cell function, fusion, recognition	54

11.2.1.2.1 In Medicine and Pharmaceuticals

Bioactive substances are encapsulated either within the lipid bilayer or in the aqueous core of the liposome according to their properties, and form liposome-drug products. Use of appropriate drug carriers reduces the toxic effect of drugs and modifies their absorption, distribution, and release profile. These are important because they can change the drug's pharmacokinetics (the rate of transport of substances administered externally to a living organism) and biodistribution in the body. Liposomal drug products seem to meet these features, which are expected of suitable drug carriers.

The advantages of loading the drug into the liposome, which can be applied as (colloidal) solution, aerosol, or in (semi)solid forms, such as creams and gels, are listed below:[55]

- Increased solubility of lipophilic (amphotericin B, minoxidil) and amphiphilic (anticancer agent doxorubicin or acyclovir) drugs
- Sustained release of systemically or locally drug-administered liposomes (doxorubicin, cytosine arabinose, cortisones, proteins, or peptides such as vasopressin)
- Decrease toxicity, such as nephrotoxicity, cardiotoxicity, and neurotoxicity (amphotericin B has reduced nephrotoxicity, and doxorubicin has reduced cardiotoxicity)
- Site-specific targeting by ligands attached to the liposome surface (anticancer, antimicrobial, and anti-inflammatory drugs)
- Improved transfer of hydrophilic, charged molecules (chelators, antibiotics, plasmids, and genes into cells)
- Enchanced penetration into tissues, especially in the case of topically applied liposomal forms (anesthetics, corticosteroids, and insulin)

Today, five liposomes and liposome-like forms for intravenous administration have been approved and are being marketed for clinical use in the European Community. Four of them (Abelcet, Amphotec, DaunoXome, and Doxil) are marketed as approved drugs in the United States. Table 11.4 shows characteristics of commercially available liposomal products.[56,57]

In addition to these marketed liposomal products, several liposomal drugs, such as liposomal cyclosporin A, liposomal nystatin, liposomal p-ethoxy growth receptor bound protein-2 antisense product, liposomal prostaglandin E1 injection, and liposome-encapsulated recombinant interleukin-2 (1), which are intended to treat frequently occurring diseases such as cancer, have been registered in the United States as orphan drugs (FDA).[56–64]

TABLE 11.4

Characteristics of Commercially Available Liposomal Products

Trade Name, Bioactive Agent, Type of Dosage Form, Company	Phospholipids and Drug/Lipid Ratio	Liposome Type	Particle Size	References
AmBisome®, amphotericin B, freeze-dried liposomes, Gilead Sciences/Fujisawa Healthcare	HSPC DSPG CHOL α-Tocopherol Drug:lipid 1:6 (w:w)	SUV	<100 nm	58, 59
Albelcet®, amphotericin B, liposomal suspension, Enzon	DMPC DMPG Drug:lipid 1:1 molar	MLV	<5 μm	60
DaunoXome®, daunorubicin, liposomal suspension, Gilead Sciences	DSPC CHOL Drug:lipid 18.7:1 (w:w)	SUV	ca. 45 nm	61
Doxil®, doxorubicin, liposomal suspension, J&J ALZA	MPEG-DSPE HSPC Drug:lipid 1:6 (w:w)	LUV	100 nm	62
Myocet®, doxorubicin, liposomal suspension, Elan	Egg-PC CHOL Drug:lipid 1:4 (w:w)	MLV	180 nm	63
Visudyne™, Verteporfin®, benzoporphyrin, freeze-dried liposomes, QLT/Novartis	Egg-DMPC Ascorbyl palmitate BHT Drug:lipid 1:7.5–15 (w:w)	SUV	<100 nm	64

Abbreviations: SUV, small unilamellar vesicle; MLV, multilamellar vesicle; LUV, large unilamellar vesicle; HSPC, hydrogenated soy phosphatidyl choline; PC, phosphatidyl choline; DSPG, distearoylphosphatidyl glycerol; CHOL, cholesterol; DSPC, distearoylphosphatidyl choline; PG, phosphatidyl glycerol; DMPC, dimyristoylphosphatidyl choline; MPEG-DSPE, N-(carbonylmethoxypoly(ethylene glycol) 2000)-1, 2-distearoyl-sn-glycero-3-phosphoethanolamine; BHT, butylated hydroxytoluene.

11.2.1.2.2 In Cosmetics

In cosmetics, liposomes' moisturizing and restoring properties are generally exploited. The first liposomal cosmetic products, Capture (C. Dior) and Niosomes (L'Oréal), were introduced in 1987. Today, many more liposome-containing products, such as Hydra-Radiance Liposome Cream (Avon), Revision Liposomal Cream (Revision SkinCare), and Lipoflow Forte plus C (Lipoflow), have been developed and are sold in the marketplace. Liposomal cosmetic products are used for hydrating skin, reducing skin dryness (the main reason for aging and wrinkles), and replacing skin lipids.[56] Liposomes are also known to contain recombinant proteins for wound or sunburn healing. Not only antiaging skin creams, but also unrinsable sunscreens, long-lasting perfumes, hair conditioners, aftershaves, and similar products have liposomes in their formulations. Table 11.5 shows some of these products.[66]

Liposomes provide many advantages in cosmetic products, such as shorter application periods, decreased side effects, delayed washing out (water resistant), increased skin hydration, and increased penetration into the deeper layers of the epidermis that are absorbed in, etc.

At our research center (BIOMATEN), different types of liposomes are being developed recently. In one study dehydration-rehydration vesicles (DRVs) were developed for use as a drug delivery system to achieve personalized treatment of patients with dermal ailments such as sunburn and psoriasis.[67] DRVs were selected because they have a specific

TABLE 11.5

Some Cosmetic Products Containing Liposomes

Manufacturer	Product	Key Ingredients of Liposomes
Christian Dior	Capture	Liposomes in gel with ingredients
L'Oréal	Efect du Soleil	Tanning agents
Lancome (L'Oréal)	Niosomes, Nactosomes	Glyceropolyether with moisturizer
Payot (Ferdinand Muehlens)	Formule Liposome Gel	Thymoxin, hyaluronic acid
Estee Lauder	Future Perfect Skin Gel	TMF, (Methyl trimethicone) vitamins E and A palmitate, cerebroside ceramide, phospholipids

advantage over conventional liposomes; they are stored in a freeze-dried form (in appropriate conditions) until use. This substantially increases the shelf life of the liposomes from 1 week to several months. DRVs are widely used in liposomal products and drugs such as Visudyne™ and AmBisome®.

Another study involving liposomes attempted to develop a UV-responsive drug delivery system using UV-sensitive liposomes. UV responsiveness was provided by Suprofen incorporated into the liposomal membrane.[68] This achieved on-demand delivery of a particular drug in a site of the body accessible to electromagnetic radiation. Liposomes are also studied as both drug delivery and imaging systems to track their route inside the tissue and for detection of disease sites.[69,70]

11.2.2 Polymeric Nanoparticles as Drug Delivery Systems

11.2.2.1 Some Polymers Used in Nanoparticle Preparation

At present, the use of polymeric materials in dosage forms that are known as controlled release systems is a very promising area of biotechnology, biomedical devices, and pharmaceuticals.[71] As stated earlier, the use of nanoscale materials has significant advantages, such as injectable biocompatible and biodegradable polymeric particles, reducing the inconvenience of surgical insertion of larger implants and also avoiding macrophages when introduced intravenously and accumulating at solid tumor sites after extravasation.[72] Synthetic and natural polymers have been investigated extensively in the last few decades as carriers for controlled drug release and targeting. Among the natural polymers used are proteins like human serum albumin (HSA), collagen, gelatin, and hemoglobin, all of which are both biodegradable and biocompatible. One limitation in their use is high cost and impurities that might induce undesirable effects. Synthetic polymers, on the other hand, have significant advantages over natural ones, such as high purity, ease of production and reproducibility, nontoxic by-products, and controllable biodegradability and biocompatibility.[72–74] Synthetic polymers such as poly(α-hydroxy acids), poly(β-hydroxy acids), poly(ϵ-caprolactone), poly(amides), poly(phosphazenes), poly(orthoesters), poly(anhydrides), and poly(phosphate esters) are commonly used in the construction of controlled release systems.[75] Among them, poly(lactides) and poly(hydroxyalkanoates) are of special interest.

The most widely studied and used biodegradable polymers are the poly(α-esters), such as poly(lactic acid) (PLA), poly(glycolic acid) (PGA), and their copolymers (Figure 11.2), and the poly(hydroxyalkanoates) (PHAs) (Fig 11.3), such as poly(3-hydroxybutyrate) (P3HB), poly(3-hydroxyvalerate) (P3HV), and their copolymers (Figure 11.4). They have a range of pharmaceutical and biomedical uses as a result of these characteristics and their physicochemical

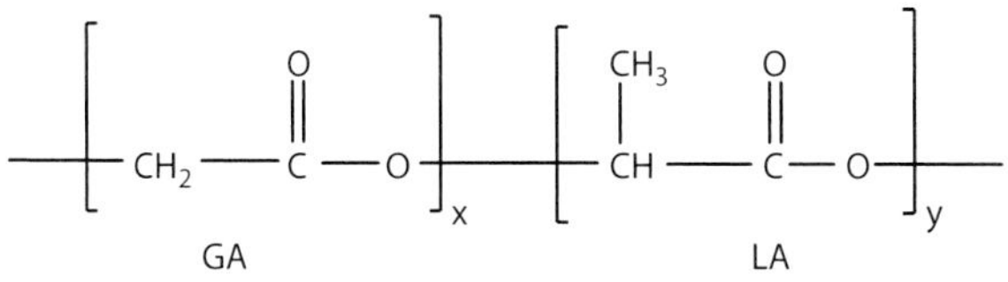

FIGURE 11.2
Molecular structure of poly(lactic acid-co-glycolic acid) block copolymer. x and y can be in the range of several thousands, and the x:y ratio could vary between 100:0 and 0:100.

properties. These polymers and their degradation products have been shown to be nontoxic and biocompatible, and products made using these polyesters have been awarded FDA approval.

For more than three decades, linear polymers of lactide and glycolide have been used in various medical applications. Use of these polymers as carriers for controlled delivery of a wide range of bioactive agents for human and animal use has been an attractive research area.[75] The properties of these linear polyesters have been modified by making copolymers and by varying the ratios of the monomers in the copolymer with respect to each other. Therefore, copolymers such as poly(lactide-co-glycolide) (PLGA) (Figure 11.2) have been extensively investigated for various medical applications, including drug delivery devices, scaffolds for tissue engineering, surgical sutures, bone pins, and stents.[76] PLGA was shown to be biocompatible in many applications and degrades into the nontoxic and natural products lactic acid and glycolic acid, which can be eliminated from the body as they are or after conversion into CO_2 and H_2O.[77]

PHAs (Figure 11.3), on the other hand are the most commonly used natural biotechnological polymers, because they are made by bacteria as well as by other organisms, some of which are transgenic.[78] They are rapidly hydrolyzed in the environment, but more slowly in the human body, and this rate could be regulated by varying the composition, form, and size of the biomedical device.

The best known polymer of the PHA family is poly(β-hydroxybutyrate) (P3HB). The mechanical properties, crystallinity, and solubility of the polyesters can also be controlled by their composition, molecular weight, and heterogeneity index (the molecular weight distribution in a given batch).[74]

11.2.2.2 Nanoparticles

Nanoparticles have a diameter in the range of 5–1000 nm. They can be irregular in shape or spherical. Spherical nanoparticles can be full in the interior with the matrix material or empty. Nanospheres are nanoparticles that are spherical and full, and hollow spherical systems are called nanocapsules, where a core is surrounded by a membrane (Figure 11.5).[71]

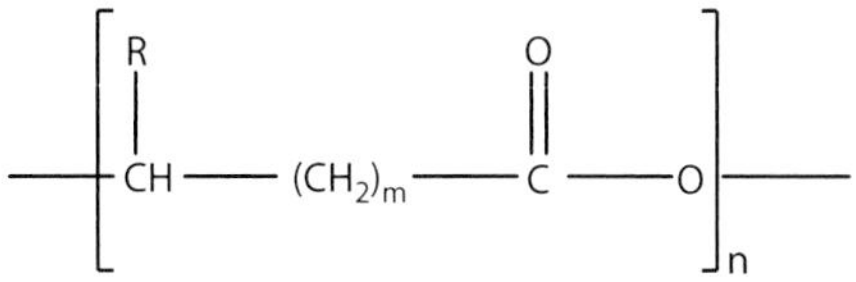

FIGURE 11.3
Molecular structure of poly(hydroxyalkanoates). When m = 1 and R = CH_3, the product is poly(3-hydroxybutyrate) (P3HB).

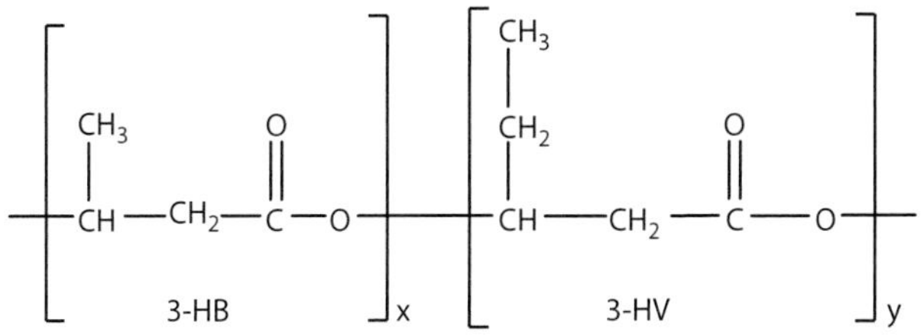

$$\left[\underset{\text{3-HB}}{\underset{|}{\overset{CH_3}{\underset{|}{CH}}} - CH_2 - \overset{O}{\overset{\|}{C}} - O} \right]_x \left[\underset{\text{3-HV}}{\underset{|}{\overset{CH_3}{\underset{|}{\overset{CH_2}{\underset{|}{CH}}}}} - CH_2 - \overset{O}{\overset{\|}{C}} - O} \right]_y$$

FIGURE 11.4
Molecular structure of PHBV. x and y can be in the range of several thousands.

In the current research activities in many biomaterials laboratories, both types of these important biomedical polyesters are used to construct carriers for various bioactive agents. The followings are results from studies carried out in our laboratories.

11.2.2.2.1 Preparation and Characterization of Nano-Microspheres Designed for Controlled Release Purposes

11.2.2.2.1.1 Preparation PHBV and PLGA nano-microspheres were prepared by an oil-in-water (o/w) emulsion technique. These polymers are dissolved in an organic solvent such as dichloromethane, which is immiscible with water. Bioactive agents are dissolved in the organic solvent, and this organic solution is added to an aqueous solution to form an o/w emulsion, and thus the droplets. The aqueous phase generally contains an emulsifier such as poly(vinylalcohol) (PVA), Tween, or Span to provide stabilization of the droplets. Emulsification is achieved by sonication, and continuous mixing prevents cluster formation before the solvent evaporates at room temperature. Ultracentrifugation is generally used to collect the nano-microspheres. Excess aqueous and organic solvents are removed by washing the nano-microparticles, and they are dried in a vacuum oven or by lyophilization.

11.2.2.2.1.2 Characterization of PLGA and PHBV Nano-Microspheres The various nano-microspheres of PHBV and PLGA were characterized for their size, size distribution, and surface topography using various methods such as a particle size analyzer, scanning electron microscopy (SEM), and an image analysis program (e.g., Scion Image of National Institutes of Health (NIH)), and fluorescence microscopy for fluorescent agent-loaded nano-microspheres.

In Figure 11.6, the influence of the drying method on the quality of recovery of PLGA (50:50) nano-microspheres is presented. It can be seen that even a simple stage of the process has a significant influence on the nano-microspheres obtained.

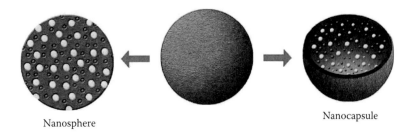

Nanosphere

Nanocapsule

FIGURE 11.5
Schematic representation of nanospheres and nanocapsules carrying drugs in their structure.

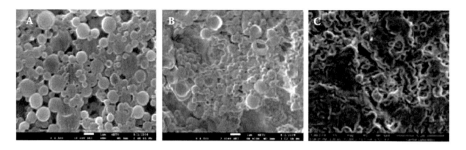

FIGURE 11.6
Changes in PLGA nano-microsphere morphology with the drying technique used. (A) Air drying (8500×).
(B) Lyophilization (8500×). (C) In vacuum oven at 50°C (3000×).

In this specific case air drying at room temperature where the nano-microspheres were
obtained in a dichloromethane-aqueous phase (5% PVA) and dried at room temperature
(RT) without any special devices was determined to be the best drying technique. This
method prevented clustering and the deformation of colloidal particles (Figure 11.6A).

The size of the nano-microparticles was examined with dynamic light scattering and
found to be in the 200 nm–9 µm range.

PHBV nano-microspheres revealed a completely different morphology than PLGA
(Figure 11.7), even though the size range was very similar, in the 350 nm–10 µm range.

It is observed that the surface of the PHBV nano-microspheres is mesoporous, as can be seen
in the SEM. This normally is not a desirable feature and could be the major route for the leakage
of entrapped drug from the nano-microspheres. On the other hand, it could be an advantage in
that water-soluble macromolecules can be loaded onto these micro-nanospheres without putting them through the micro-nanosphere formation process. It was observed that the extent of
the porosity of the sphere surfaces depended on the HV content and also most probably on the
source of the PHBV (company, microorganism type, purification approach, etc.).

11.2.2.2.2 Preparation and Characterization of Nano-Microcapsules
Designed for Controlled Release Purposes

Nano-microcapsules are different from the spheres because the capsules are hollow in the
inside, while the spheres are full.

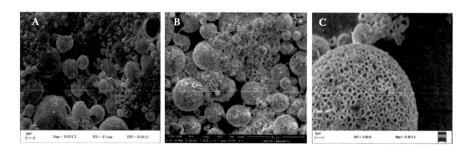

FIGURE 11.7
SEM micrographs of spheres prepared from PHBV copolymers with 5 and 8% HV. They were also of different
sources. (A) PHBV5 (HV 5%) (10,000×). (B) PHBV8 (HV 8%) (3,000×). (C) Magnification of B.

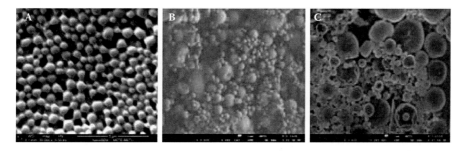

FIGURE 11.8
Influence of drying method on the morphology of PLGA nano-microcapsules. (A) Air dried (30,000×).
(B) Lyophilization (3,000×). (C) Vacuum oven dried at room temperature (3000×).

11.2.2.2.2.1 Preparation   Nanocapsules are prepared in a different way than the spheres
to create the wall. In the studies presented, the capsules were loaded with water-soluble
drugs in the core and hydrophobic drugs in the shell component.

PLGA and PHBV nano-microcapsules were prepared by a double-emulsion (w_1/o/
w_2) technique. Briefly, after the polymer is dissolved in an organic solvent (e.g., dichloromethane DCM), a smaller volume of aqueous phase carrying the drug is added into this
solution and agitated to form an emulsion of the water phase in the organic phase, the size
of the droplets of which is controlled by the agitation (sonication) level. This emulsion is
then added into a much larger volume of the aqueous phase containing the emulsifier to
form the nano-microcapsules. Evaporation of the organic solvent, separation of the capsules, and drying were carried out as mentioned in Section 11.2.2.2.1.1.

11.2.2.2.2.2 Characterization of PLGA and PHBV Nano-Microcapsules   Particle size, SEM,
and fluorescence microscopy are used to examine the physical properties (morphology,
size, size distribution) of the PLGA nano-microcapsules, and this information is then used
to optimize the procedure. In Figure 11.8 the effect of the drying method on capsule topography is presented.

In this study, the nano-microcapsules were obtained in a DCM-water medium, as was
mentioned in Section 11.1.2.2.2.1, and again drying at room temperature (RT) without
any special devices was found to be the best drying method in terms of obtaining round,
smooth, and unagglomerated capsules (Figure 11.8A).

Determination of the size of the PLGA nano-microcapsules by dynamic light scattering
revealed that the particle sizes were in the range 200 nm–10 µm, and the average diameter
was 2 µm.

The morphologies of PHBV nano-microcapsules loaded with calcein (a model drug fluorescent molecule) were also evaluated by fluorescence microscopy and SEM (Figure 11.9).
Their particle size range was 300 nm–20 µm, and the average diameter was found to be
8 µm. This indicates that the loading process led to an increase in size. The use of fluorescent molecule had an advantage; it made the nano-microcapsule easily detectable with
fluorescence microscope (Figure 11.9A).

11.2.2.2.3 Uptake of PLGA and PHBV Nano-Microcapsules by Saos-2 Cells
The uptake of nano-microcapsules of PLGA and PHBV was studied in vitro with osteosarcoma cells (Saos-2). The results were evaluated after incubation of the capsules for 4 h in
the cell culture medium.

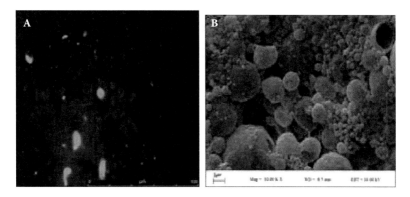

FIGURE 11.9
Fluorescence and SEM micrographs of calcein-loaded PHBV nano-microcapsules. (A) Scale bar is 100 μm.
(B) 30,000×.

   After the nano-microcapsules of PHBV and PLGA were stained with Nile red 0.1% (v/v),
which has a red emission under fluorescence microscopy, Saos-2 cells were incubated
with the nano-microparticles for 4 h. Before examination, the cells were stained with FITC
(fluorescein isothiocyanate)-labeled phalloidin and 4′,6-diamidino-2-phenylindole (DAPI),
respectively, dyes that stain the cell nuclei blue and the cell cytoskeleton green. The cells
were then examined under fluorescence microscope (Figure 11.10). It has to be remembered
that the sizes of most capsules are lower than the resolution of the fluorescent microscope
(in the nano-range), so some of the red regions probably represent clusters of nanocapsules, while the others that are seen as individual specks could be micron sized. It can be
observed that PHBV and PLGA nano-microcapsules are taken up by Saos-2 cells. An interesting finding is that the capsules are generally located near or on the cell nuclei, implying
that they could serve as carriers of agents for gene therapy because they seem to be able to
avoid lysosomes and accumulate in the vicinity of the nuclei.

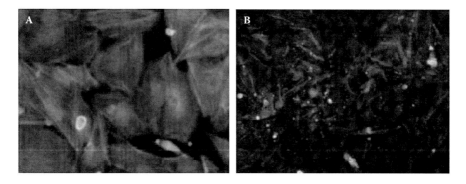

FIGURE 11.10
(See color insert.) Fluorescence micrographs of nano-microcapsules taken up by Saos-2 cells. (A) PHBV, 40×. (B)
PLGA, 20×. Cell nuclei were stained with DAPI (blue), cell cytoskeletons were stained with FITC-labeled phal-
loidin (green), and nano-microcapsules were stained with Nile red (red).

11.3 Conclusion

Nano- and low-micron-sized particles, whether polymeric or lipid based, are very versatile tools for drug delivery applications. They can be made in various sizes, from various materials, can carry molecules of different chemistries, and can be modified for targeting purposes. As such, they are indispensable as drug carriers. And with the advent of nanotechnology, which provides tools such as quantum dots, nanowires, and nanorods, the field has become more exciting than ever.

Acknowledgments

We gratefully acknowledge the support by METU (BAP 07.02.2009.00.01) through which part of this study was conducted.

References

1. Jain, K.K., *Drug delivery systems methods in molecular biology*, Vol. 437, Humana Press, New York, 2008, chap. 1.
2. Uekama, K., Hirayama, F., and Irie, T., Cyclodextrin drug carrier systems, *Chem. Rev.*, 98, 2045, 1998.
3. Pankhurst, Q.A., et al., Applications of magnetic nanoparticles in biomedicine, *J. Phys. D Appl. Phys.*, 36, R167, 2003.
4. Torchillin, V., *Multifunctional pharmaceutical nanocarriers*, Vol. 4, Springer, New York, 2008.
5. Carmeliet, P., and Jain, R.K., Angiogenesis in cancer and other diseases, *Nature*, 407, 249, 2000.
6. Maeda, H., et al., Tumor vascular permeability and the EPR effect in macromolecular therapeutics: a review, *J. Control. Release*, 65, 271, 2000.
7. Gerweck, L.E., and Seetharaman, K., Cellular pH gradient in tumor versus normal tissue: potential exploitation for the treatment of cancer, *Cancer Res.*, 56, 1194, 1996.
8. Cook, J.A., et al., Oxidative stress, redox, and the tumor microenvironment, *Semin. Radiat. Oncol.*, 14, 259, 2004.
9. Malmsten, M., *Surfactants and polymers in drug delivery*, Marcel Dekker, New York, 2002.
10. Feynman R.P., There is plenty of room at the bottom, paper presented at Annual Meeting of the American Physical Society, California Institute of Technology, December 29, 1959.
11. Dubas, S.T., Kumlangdudsana, P., and Potiyaraj, P., Layer-by-layer deposition of antimicrobial silver nanoparticles on textile fibers, *Colloids Surf. A*, 289, 105, 2006.
12. Navrotsky, A., Nanomaterials in the environment, agriculture, and technology (NEAT), *J. Nanoparticle Res.*, 2, 321, 2000.
13. González-Melendi, P., et al., Nanoparticles as smart treatment-delivery systems in plants: assessment of different techniques of microscopy for their visualization in plant tissues, *Ann. Bot.*, 101, 187, 2008.
14. Kruis, F.E., Fissan, H., and Peled, A., Synthesis of nanoparticles in the gas phase for electronic, optical and magnetic applications—a review, *J. Aerosol Sci.*, 29, 511, 1998.
15. Choi, M.J., et al., Metal-containing nanoparticles and nano-structured particles in fingermark detection, *Forensic Sci. Int.*, 179, 87, 2008.

16. Bonduelle, S., et al., Association of cyclosporin to isohexylcyanoacrylate nanospheres and subsequent release in human plasma *in vitro*, *J. Microencapsul.*, 9, 173, 1991.

17. Bender, A.R., et al., Efficiency of nanoparticles as a carrier system for antiviral agents in human immunodeficiency virus-infected human monocytes/macrophages *in vitro*, *Antimicrob. Agents Chemother.*, 40, 1467, 1996.

18. Kawashima, Y., et al., Mucoadhesive D,L-lactide/glycolide copolymer nanospheres coated with chitosan to improve oral delivery of elcatonin, *Pharm. Dev. Technol.*, 5, 77, 2000.

19. Salata, O.V., Applications of nanoparticles in biology and medicine, *J. Nanobiotechnol.*, 2, 2004.

20. Rieux, A.D., et al., Nanoparticles as potential oral delivery systems of proteins and vaccines: a mechanistic approach, *J. Control. Release*, 116, 1, 2006.

21. Jahanshahi, M., and Babaei, Z., Protein nanoparticle: a unique system as drug delivery vehicles, *Afr. J. Biotechnol.*, 7, 4926, 2008.

22. Roco, M.C., Nanotechnology: convergence with modern biology and medicine, *Curr. Opin.*, 14, 337, 2003.

23. Michalet, X., et al., Quantum dots for live cells, *in vivo* imaging, and diagnostics, *Science*, 307, 538, 2005.

24. Basmanav, F.B, Kose, G.T., and Hasirci, V., Sequential growth factor delivery from complexed microspheres for bone tissue engineering, *Biomaterials*, 29, 4195, 2008.

25. Yilgor, P., et al., Incorporation of a sequential BMP-2/BMP-7 delivery system into chitosan-based scaffolds for bone tissue engineering, *Biomaterials*, 30, 3551, 2009.

26. Mao, H.-Q., et al., Chitosan-DNA nanoparticles as gene carriers: synthesis, characterization and transfection efficiency, *J. Control. Release*, 70, 399, 2001.

27. Panyam, J., and Labhasetwar, V., Biodegradable nanoparticles for drug and gene delivery to cells and tissue, *Adv. Drug Deliver. Rev.*, 55, 329, 2003.

28. Kumari, A., Yadav, S.K., and Yadav, S.C., Biodegradable polymeric nanoparticles based drug delivery systems, *Colloid Surf. B*, 75, 1, 2010.

29. Alexis, F., et al., Factors affecting the clearance and biodistribution of polymeric nanoparticles, *Mol. Pharm.*, 5, 505, 2008.

30. Potineni, A., et al., Poly(ethylene oxide)-modified poly(β-amino ester) nanoparticles as a pH-sensitive biodegradable system for paclitaxel delivery, *J. Control. Release*, 86, 223, 2003.

31. Sershen, S.R., et al., Temperature-sensitive polymer-nanoshell composites for photothermally modulated drug delivery, *J. Biomed. Mater. Res.*, 51, 293, 2000.

32. Cirli, O.O., and Hasirci, V., UV-induced drug release from photoactive REV sensitized by suprofen, *J. Control. Release*, 96, 85, 2004.

33. Oberdorster, G., Oberdorster, E., and Oberdorster, J., Nanotoxicology: an emerging discipline evolving from studies of ultrafine particles, *Environ. Health Perspect.*, 113, 823, 2005.

34. Peters, A., et al., Translocation and potential neurological effects of fine and ultrafine particles a critical update, *Particle Fibre Toxicol.*, 3, 2006.

35. Hans, M.L., and Lowman, A.M., Biodegradable nanoparticles for drug delivery and targeting, *Curr. Opin. Solid State Mater.*, 6, 319, 2002.

36. Stolnik, S., et al., Surface modification of poly(lactide-co-glycolide) nanospheres by biodegradable poly(lactide)-poly(ethylene glycol) copolymers, *Pharmaceut. Res.*, 11, 1800, 1994.

37. Katanec, D., Paveli, B., and Ivasovi, Z., Efficiency of polylactide/polyglycolide copolymers bone replacements in bone defects healing measured by densitometry, *Collegium Antropol.*, 28, 331, 2004.

38. Soppimath, K.S., et al., Biodegradable polymeric nanoparticles as drug delivery devices, *J. Control. Release*, 70, 1, 2001.

39. Israelachvili, J.N., Marcelja, S., and Horn, R.G., Physical principles of membrane organization, *Q. Rev. Biophys.*, 13, 121, 1980.

40. Zhang, Y.P., Ceh, B., and Lasic, D.D., Liposomes in drug delivery, in *Polymeric biomaterials*, 2nd ed., Bumitriu, S., ed., Marcel Dekkar Inc., New York, 2002, p. 783.

41. Xu, Z.P., et al., Inorganic nanoparticles as carriers for efficient cellular delivery, *Chem. Eng. Sci.*, 61, 1027, 2006.

42. Liong, M., et al., Multifunctional inorganic nanoparticles for imaging, targeting, and drug delivery, *Am. Chem. Soc.*, 2, 889, 2008.

43. Bangham, A.D., Standish, M.M., and Watkins J.C., The first description of liposomes, *J. Mol. Biol.*, 13, 238, 1965.

44. Wang, B., Siahaan, T., and Soltero, R., *Drug delivery: principles and applications*, 1st ed., John Wiley & Sons, New Jersey, 2005.

45. Verma, A.M.L., et al., Liposomes as carrier systems, *InPharm Communique*, 2, 20, 2009.

46. Torchillin, V., and Weissig, V., *Liposomes: a practical approach*, 1st ed., Oxford University Press, Oxford, 1990.

47. Szoka, F., Jr., and Papahadjopoulos, D., Procedure for preparation of liposomes with large internal aqueous space and high capture by reverse-phase evaporation, *Proc. Natl. Acad. Sci. USA*, 9, 4194, 1978.

48. Deamer, D., and Bangham, A.D., Large volume liposomes by an ether vaporization method, *BBA Biomembranes*, 443, 629, 1976.

49. Batzri, S., and Korn, E.D., Single bilayer liposomes prepared without sonication, *BBA Biomembranes*, 298, 1015, 1973.

50. Barenholz, Y., Liposome application: problems and prospects, *Curr. Opin. Colloid Interface*, 6, 66, 2001.

51. Pajic-Lijakovic, I., et al., Rheological quantification of liposomes aggregation, *Minerva Biotechnol.*, 17, 245, 2005.

52. Ulrich, S.A., Biophysical aspects of using liposomes as delivery vehicles, *Bioscience Rep.*, 22, 129, 2002.

53. Jemioat-Rzeminska, M., Latowski, B., and Strzalka, K., Incorporation of plastoquinone and ubiquinone into liposome membranes studied by HPLC analysis. The effect of side chain length and redox state of quinine. *Chem. Phys. Lipids*, 110, 85, 2001.

54. Yuba, E., et al., pH-sensitive fusogenic polymer-modified liposomes as a carrier of antigenic proteins for activation of cellular immunity, *Biomaterials*, 31, 943, 2010.

55. Lasic, D.D., Liposomes, *Am. Sci.*, 80, 20, 1992.

56. Janoff, S.A., *Liposome: rational design*, 1st ed., Marcel Dekker, New York, 1999.

57. Barenholz, Y., Relevancy of drug loading to liposomal formulation therapeutic efficacy, *J. Liposome Res.*, 13, 1, 2003.

58. Linden, P.K., Amphotericin B lipid complex for the treatment of invasive fungal infections, *Expert Opin. Pharmacother.*, 4, 2099, 2003.

59. Adler-Moore, J., and Proffitt, R.T., AmBisome: liposomal formulation, structure, mechanism of action and pre-clinical experience, *J. Antimicrob. Chemother.*, 49, 21, 2002.

60. Veerareddy, P.R., and Vobalaboina, V., Lipid-based formulations of amphotericin B, *Drugs Today*, 40, 33, 2004.

61. Hempel, G., et al., Population pharmacokinetics of liposomal daunorubicin in children. *Br. J. Clin. Pharmacol.*, 56, 370, 2003.

62. Waterhouse, D.N., et al., A comparison of liposomal formulations of doxorubicin with drug administered in free form: changing toxicity profiles, *Drug Safety*, 24, 903, 2001.

63. Allen, T.M., and Martin, F.J., Advantages of liposomal delivery systems for anthracyclines, *Semin. Oncol.*, 31, 5, 2004.

64. Chowdhary, R.K., Shariff, I., and Dolphin, D., Drug release characteristics of lipid based benzoporphyrin derivative, *J. Pharm. Pharm. Sci.*, 6, 13, 2003.

65. www.fda.gov/orphan/designat/list.htm.

66. Lasic, D.D., and Barenholz, Y., *Handbook of nonmedical applications of liposomes: theory and basic sciences*, Vol. 1, CRC Press, Boca Raton, FL, 1996.

67. Kardesler, S., et al., Topical delivery systems for the treatment of skin diseases, in *International Symposium on Biotechnology: development and Trends*, Ankara, Turkey, September 27–30, 2009, p. 112.

68. Demirbag, B., and Hasirci, V., UV responsive drug delivery from suprofen incorporated liposomes, in *15th International Biomedical Science and Technology Symposium*, Guzelyurt, TRNC, August 16–19, 2009, p. 84.

69. Buyuksungur, A., and Hasirci, V., Bioactive agent carrying quantum dot labeled liposomes, in *15th International Biomedical Science and Technology Symposium*, Guzelyurt, TRNC, August 16–19, 2009, p. 112.

70. Buyuksungur, A., Padeste, C., and Hasirci V., Entrapment in liposome prevents toxicity of quantum dots, in *International Symposium on Biotechnology: Development and Trends*, Ankara, Turkey, September 27–30, 2009, p. 113.

71. Barratt, G., et al., Polymeric micro and nanoparticles as drug carriers, in *Polymeric biomaterials*, 2nd ed., Dumitriu, D., ed., Marcel Dekker, New York, 2002, chap. 28.

72. Jain, R.A., The manufacturing techniques of various drug loaded biodegradable poly(lactide-co-glycolide) (PLGA) devices, *Biomaterials*, 21, 2475, 2000.

73. Astete, C.E., and Sabliov, C.M., Synthesis and characterization of PLGA nanoparticles, *J. Biomater. Sci. Pol. Ed.*, 17, 247, 2006.

74. Nair, L.S., and Laurencin, C.T., Polymers as biomaterials for tissue engineering and controlled drug delivery, *Adv. Biochem. Eng./Biotechnol.*, 102, 47, 2006.

75. Domb, J.A., et al., Biodegradable polymers as drug carrier systems, in *Polymeric biomaterials*, 2nd ed., Dumitriu, D., ed., Marcel Dekker, New York, 2002, chap. 4.

76. Jain, R.A., et al., Controlled delivery of drugs from a novel injectable *in situ* formed biodegradable PLGA microsphere system, *J. Microencapsul.*, 17, 343, 2000.

77. Stevanovic, M., and Uskokovic, D., Poly(lactide-co-glycolide)-based micro and nanoparticles for the controlled drug delivery of vitamins, *Curr. Nanosci.*, 5, 1, 2009.

78. Errico, C., et al., Poly(hydroxyalkanoates)-based polymeric nanoparticles for drug delivery, *J. Biomed. Biotechnol* Vol 2009, p. 1–10, 2009.

12

Microscale Technologies for Controlling the Cellular Microenvironment in Time and Space

Hirokazu Kaji, Yun-Ho Jang, Hojae Bae, and Ali Khademhosseini
Department of Medicine, Harvard Medical School, Boston, Massachusetts, USA

CONTENTS

12.1 Introduction

The ability to control the behavior of living cells in vitro is critical in a wide range of research fields, including fundamental cell biology, tissue engineering, and drug screening.[1-3] In general, cellular behavior is a consequence of intrinsic and extrinsic factors. The extrinsic factors are largely derived from the surrounding microenvironment, along with cell-secreted factors, and are comprised of a milieu of biochemical, biomechanical, and bioelectrical signals generated from the surrounding cells, the extracellular matrix (ECM), and the soluble factors. These components vary in both time and space and are integral to the regulation of cellular behaviors.

Recent studies have confirmed the existence of cells residing in niches, unique to the tissues and organs in which they reside, that contain highly ordered microarchitechtures, cellular compartmentalization, and arrangement.[4] Cells sense each other by direct contact in a process in which cell surface receptors (e.g., cadherins) detect the cell surface receptors of neighboring cells. Cell to cell contact can be used to regulate a variety of biological processes, such as cell proliferation, differentiation, and migration. Also, cells signal each other through small molecules such as hormones as well as larger molecules, which are mostly made of proteins such as chemokines, cytokines, and growth factors. Such molecules typically interact with cells through surface receptors, and different cells exhibit unique types of surface receptors, thereby interacting with specific signaling molecules. The ECM that provides three-dimensional (3D) scaffold or 2D surface with which cells interact is also important in signaling cells. The ECM is often comprised of proteins such as fibronectin and collagen, as well as polysaccharides such as hyaluronic acid (HA). Through interacting with the ECM, cells are provided with an environment to which they can anchor and generate tissues.

Recently, emerging technologies at the interface of engineering and materials science have resulted in a number of new approaches to control the various aspects of the cellular microenvironment.[3,5–7] The merger of microfabrication and advanced biomaterials may be useful to recreate many of the complex features to create in vitro microenvironments with the ability to effectively direct cell behavior. For example, surface patterning can be used to control cell-substrate and cell-cell interactions. To control the interaction with substrates, techniques such as lithography and surface modification have been integrated. Through these modifications, properties such as cell adhesion can be controlled by depositing cells on adhesive regions or controlling the topography of surfaces. Furthermore, to control the interactions of cells with surrounding cells, various patterned coculture techniques have been utilized. These techniques have been successful with controlling the homotypic and heterotypic cell-cell interactions in vitro. Also, microfluidic devices have been used to control the presentation of soluble molecules to cells. Such devices provide a number of advantages, including reduced sample volumes and costs compared to conventional macroscale systems. Moreover, microfluidic channels can be used to perform high-throughput experimentation and to control cell-soluble factor interactions, enabling the presentation of soluble factors in a highly controllable manner.

This chapter highlights the recent advances of microfabrication technologies that can be used to control the cellular microenvironment in vitro. These applications of microscale engineering may be of great benefit to the future of regenerative medicine, and success will be highly dependent on successes in both tissue engineering and stem cell differentiation. These suggest that the development of technologies to achieve the desired outcome will hopefully shorten the time required for the translation of these techniques to the clinic.

12.2 Surface Patterning for Controlling Cell Adhesion

Most of the techniques used in the generation of cellular microenvironments have evolved from semiconductor fabrication processes. Thus, planar patterning of cellular microenvironments has been studied more in comparison with 3D systems. Much of in vitro cell biology studies have been performed on planar surfaces by using microwell plates, culture flasks, or Petri dishes, since they are convenient for investigating cellular responses to biological stimuli. These 2D methods can be miniaturized using selective surface modification through microfabrication techniques and enable the control of cell-cell contacts on geometrically defined 2D surfaces. Controlling the degree of cell-cell contact is useful in cell biology studies and tissue engineering because cell-cell communications in aggregates play a significant role in controlling the cell behavior. One example is in embryonic stem cell (ESC) proliferation and differentition.[3,8] Variety of technologies, such as microtopography,[9–11] microfabricated stencils,[12] microcontact printing,[13] and layer-by-layer deposition,[14] have been established for ESC patterning on 2D surfaces. One approach uses poly(ethylene glycol) (PEG) hydrogel microarrays for uniform cell seeding and embryoid body (EB) formation,[15] and this approach was further used to generate spatially and temporally synchronized beating human EBs.[16] Moreover, homogeneous EBs obtained from the microwell can be used to direct EB-mediated differentiation.[13] Another approach used microfabricated parylene-C stencils to pattern cocultures in both static and dynamic conditions by using several stencils sequentially with different surface properties.[12]

12.2.1 Photolithography

Photolithography is a well-established and widely used method for patterning cells on a surface.[17] In this process, materials that interact with cells are patterned selectively by the transfer of patterns from a photomask to photosensitive polymers prepared on a substrate. Typically, photoresist is used as photosensitive materials patterned by ultraviolet (UV) exposure and selectively removed on the exposed (or unexposed) area. After depositing cell-adhesive (or repellent) materials on the substrate, the patterned photoresist is removed to leave the material of interest selectively. This so-called lift-off process is an effective way to selectively pattern cells on a surface. Recently, direct patterning of cell interacting materials has been investigated to overcome low biocompatibility of the lift-off process using photoresist. Patterning of vitronectin,[18] hexamethyldisilazane (HMDS),[19] cell-adhesive proteins,[20] and PEG[21,22] can eliminate photoresist during the selective surface modification for cell studies. Karp et al. demonstrated cell patterning using photocrosslinkable chitosan.[23] They prepared the polymer films on a slide glass and exposed UV light to crosslink chitosan, then washed out uncrosslinked chitosan to leave patterns of crosslinked polymer. Using this approach, cardiomyocytes were cultured on the glass surface between chitosan lane patterns, and osteoblasts were localized within the patterns. In another interesting example, Yamazoe et al. selectively converted albumin surfaces to cell adhesive through UV exposure (Figure 12.1).[24] UV irradiation changed the contact angle of the albumin from 85° to 23°, which successfully applied to selective attachment of mouse fibroblast cells on the UV-irradiated region.

12.2.2 Microcontact Printing

Compared with the photolithography technique, microcontact printing (μCP) can transfer self-assembled monolayers (SAMs) from a soft elastomeric stamp, typically made of polydimethylsiloxane (PDMS), onto another substrate without any exposure steps.[25,26] For example, methyl-terminated alkanethiol SAMs facilitated protein adsorption, leading to increased cell adhesion, while PEG-modified alkanethiol SAMs prevented cell adhesion. Other mechanisms can also be adapted to use with microstamps for creating structures. Khademhosseini and colleagues used the direct contact of a PDMS stamp on a thin polymer film to generate patterns by using capillary forces (Figure 12.2).[27] Wet polymers coated on a substrate and pressed by a PDMS stamp move spontaneously into a void space, which

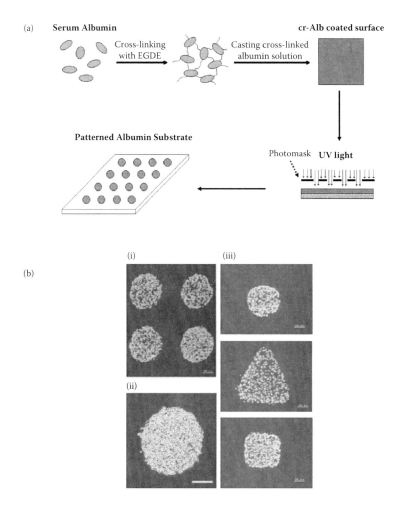

FIGURE 12.1

(a) Selective surface modification on albumin using UV exposure. (b) The selective surface patterning was used to pattern mouse fibroblast cells on an albumin coated substrate. Different shapes (circle, rectangle, and triangle) and culture time (2 h – 5 days) were tested to demonstrate the feasibility. Adapted with permission from reference 24. Copyright 2008 American Chemical Society.

makes negative mold structures without any transfer procedures employed by conventional μCP. It was shown in this experiment that 3T3 cells in microstructures had better stability than the cells in polymeric films. Also, PDMS stamps can be used to print living microbes on an agarose gel.[28] A monolayer of *E. coli* prepared on the surface of agarose gel was transferred to another agarose substrate using a PDMS stamp without damage by exposure to air. During the process, the agarose concentration for both inkpad and substrate is crucial for generating patterns. It was found that the agarose concentrations of 3% and 4% were appropriate for the inkpad and substrate, respectively. Coq et al. studied a self-supporting hydrogel stamp that has a high water/buffer holding capacity for the hosting of sensitive biological molecules.[29] An optimized hydrogel concentration (10% water, 18% PEGDA, and 72% HEA prepolymer solution) in terms of rigidity and high water absorption made it possible to transfer antibodies on various surfaces, including bare Au, with minimal damage under ambient conditions. Recently, other researchers have studied

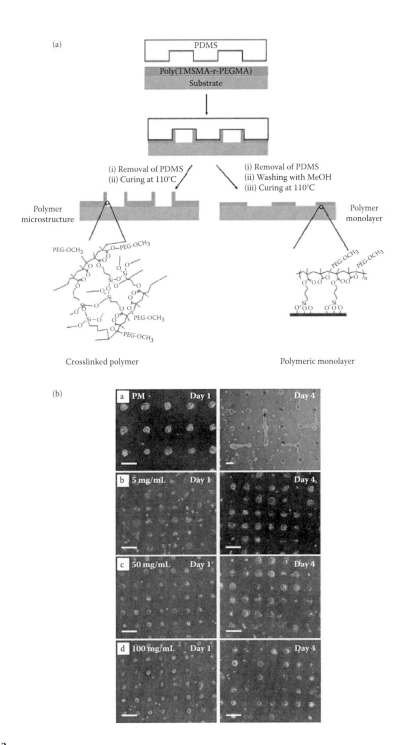

FIGURE 12.2
Micropatterning surfaces using a capillary force lithography approach. (a) PDMS mold is placed on a spin-coated polymeric surface to induce a capillary force and selectively form patterned microstructures or polymeric monolayers. (b) Cell patterns on microstructures showed more stability than cells in polymeric monolayers after 4 days of culture. Adapted with permission from reference 27. Copyright 2003 Wiley-VHC Verlag GmbH & Co. KGaA.

a porous stamp to obtain a permanent hydrophilic surface to print high molecular weight materials.[30] They fabricated hydrophilic porous stamps using a phase separation micromolding process, and demonstrated µCP by printing proteins several times without reinking. Foley et al. developed an interesting technique to print proteins inside recessed areas, such as bottom surfaces of microchannels, which was possible by bending a PDMS stamp itself.[31] Microcontact printing has been employed for differentiation study, where gene and protein analysis of micropatterned human ESCs revealed that endodermal and neuronal expression increased inversely with colony size.[13]

12.2.3 Other Techniques

There are many other techniques to make selective biological patterns on substrates. For example, with the use of microfluidic channels, fluids can be confined on a surface to control cell interaction with materials selectively.[32] By expanding this technique, Takayama and co-workers studied how laminar flow can establish a single stream consisting of two or more fluids, which can establish a subcellular microenvironment by positioning a single mammalian cell at the interface between two different streams.[33] Also, Ostuni et al. developed a PDMS stencil patterning that is straightforward in that it requires no chemical treatment of the surface.[34] They fabricated a cell-repellent PDMS stencil by coating the stencil with bovine serum albumin (BSA), and then attached the PDMS stencil on a substrate to culture bovine capillary endothelial (BCE) cells selectively. Direct forming of liquid phase cell patterns by dispensing nanoliter volumes was developed by Tavana et al. to sustain aqueous environments during pattern formation (Figure 12.3).[35] Aqueous dextran was dispensed

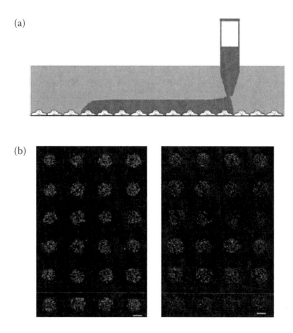

FIGURE 12.3
Nanoliter liquid patterning in an aqueous environment (b). Dispensing reagent of interest in the aqueous environment ensures high reagent activity and cell viability as well as patterning flexibility for investigating different microenvironments between adjacent cells (a). Adapted with permission from reference 35. Copyright 2009 Macmillan Publishers Ltd: Nature Materials.

on the mammalian cells covered with an aqueous PEG phase to deliver genetic materials selectively.

12.3 Dynamic Substrates for Controlling Cell Adhesion and Generating Cocultures

Recent advances in the ability to engineer surface properties of substrates have allowed researchers to dynamically modulate the interactions between cells and the substrate surface in real time using an external trigger such as voltage, heat, or light. These techniques can be used for controlling the adhesion and motility of cells and for the sequential patterning of multiple cell types.

12.3.1 Voltage-Responsive Substrates

Electrochemical transformations available on SAMs can alter the physicochemical properties of the surface in situ, thereby being utilized to dynamically control cell adhesion and migration. For example, Yousaf et al. developed SAMs of which the terminal groups can be electrically switched to allow the immobilization of cell-adhesive molecules.[36] In this approach, an inert and nonadhesive monolayer of a hydroquinone group against a background of tri(ethylene glycol) groups on a gold substrate was used. Electrochemical oxidation of a hydroquinone group to the corresponding benzoquinone, which is reactive to cyclopentadiene by the Diels-Alder reaction, was used to selectively immobilize a cell-adhesive peptide ligand conjugated with the diene. This dynamic substrate was used to prepare cocultures of two different cell types by allowing a first population of cells to attach to a patterned monolayer, and then activating a second pattern for attachment of a second population, as well as to create a cell migration-based assay. Another example, based on electrochemical transformation of SAMs, has been carried out by Jiang et al.[37] They adopted the electrochemical desorption of an ethylene glycol-terminated SAM on a gold substrate to induce cell migration onto the SAM-desorbed area. Recently, they extended this approach to pattern multiple types of cells within microfluidic channels.[38]

Kaji et al. developed a surface patterning technique based on an electrochemical method that enables the localized immobilization of cells under physiological conditions.[39-41] This technique used a microelectrode to electrochemically generate an oxidizing agent HBrO, which acts on a heparin- or albumin-coated substrate, initially antibiofouling, to render these regions cell adhesive. Since this technique can be conducted under cell cultivation conditions, it facilitated the stepwise immobilization of multiphenotype cell arrays and the in situ directional navigation of cell migration (Figure 12.4). Also, since the patterning procedure requires only small numbers of electrodes and a small dry battery, it was readily applicable to miniaturized and semiclosed systems such as microfluidic devices and tubing scaffolds.[42–45]

12.3.2 Light-Responsive Substrates

Photochemistry is also useful for dynamically modulating interactions between cells and the substrate surface. For example, Nakanishi et al. used a SAM of silanes having a photocleavable 2-nitro-benzyl group.[46,47] In this approach, the entire surface of the SAM was

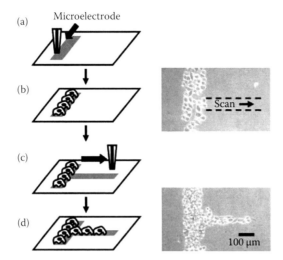

FIGURE 12.4
Microelectrode patterning of a surface for directing cell migration in (a)-(d). The microelectrode can be used to desorb the antibiofouling layer, first allowing cells to attach and then allowing cells to migrate from the initial pattern. Adapted with permission from reference 39. Copyright 2004 American Chemical Society.

initially covered with albumin molecules to prevent cell adhesion. When light was irradiated to cleave the terminal group of the SAMs, albumin desorbed from the surface, allowing fibronectin deposition and subsequent cell attachment corresponding to the irradiated pattern. They used this method to fabricate an array composed of different cell types and to induce cell migration. In another example, poly(N-isopropylacrylamide) (PIPAAm) modified with a photochromic spiropyran has been used to control cell adhesion.[48] The cell adhesiveness of the polymer surface enhanced by the UV irradiation was reset by the visible light irradiation and annealing at 37°C.

12.3.3 Temperature-Responsive Substrates

Okano and colleagues utilized a thermo-responsive polymer, PIPAAm, for controlling cell-substrate interaction.[49–52] In this approach, PIPAAm was covalently grafted as a thin layer onto tissue culture-grade polystyrene dishes by electron beam radiation. Above the lower critical solution temperature (LCST, 32°C) of PIPAAm, the polymer network collapses, making the polymer dehydrated and relatively hydrophobic, thereby becoming cell adhesive. Under its LCST of 32°C, the polymer is hydrated, and cell attachment is highly suppressed. PIPAAm copolymerized with other monomers can be designed to vary the LCST at which the polymer becomes cell adhesive. By using these features, patterned cocultures were generated in which a patterned surface of PIPAAm and its copolymer was prepared to seed two cell types at different temperatures. Since the entire surface became cell repellent at lower temperatures, patterned cells could be removed from the temperature-responsive surface in the form of a cell sheet.

Recently, Kim et al. reported a dynamic display of biomolecular micropatterns based on an elastic creasing instability of thermo-responsible hydrogels (Figure 12.5).[53] In the swelled state of the hydrogel at room temperature, the hydrogel surface exhibited an elastic creasing instability that sequesters functionalized regions within tight folds in the surface. When the hydrogel was deswelled at 37°C, the gel surface was unfolded to

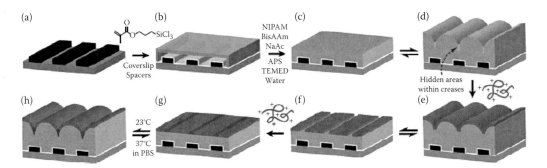

FIGURE 12.5
Dynamic scaffolds based on a thermo-responsible hydrogel. (a) First, a topographically patterned substrate is prepared and chemically modified. (b), (c) Then, a hydrogel is loaded and polymerized while spacers at the side and a coverslip on top define the final thickness of the hydrogel layer. (d) After detachment of the coverslip, the hydrogel is swelled, thereby forming areas with creases. (e), (f) The displayed surface of the hydrogel is then coated, while the hidden areas remain uncoated. (g), (h) Deswelling of the hydrogel at a higher temperature enables the modification of the previously hidden areas. Adapted with permission from reference 53. Copyright 2010 Macmillan Publishers Ltd: Nature Materials.

regenerate the biomolecular patterns. Topographic features on the underlying substrate determined the crease positions and enabled the creation of arbitrary 2D surface chemical patterns through selective deposition of biochemically functionalized polyelectrolytes. This method could provide a valuable component for lab-on-a-chip devices as well as a dynamic substrate for cell biology.

12.3.4 Other Systems

Various approaches not categorized in the above sections have been developed to control cell-substrate and cell-cell interactions. For example, electrostatic interactions have been applied for generating patterned cocultures. To do this, Khademhosseini et al. developed a method that used layer-by-layer deposition of ionic biomolecules to pattern cellular cocultures.[14] HA, a biocompatible and biodegradable material, was patterned on a substrate by capillary force lithography, followed by fibronectin adsorption onto the HA-free region. Then, the first cell type was seeded and only attached to the fibronectin-coated region. Subsequent ionic adsorption of poly-L-lysine to the negatively charged HA pattern was used to change its surface from cell repulsive to cell adhesive. Finally, the second cell type was seeded and attached to the poly-L-lysine pattern.

Hui and Bhatia developed a technique for the dynamic control of cell-cell adhesion that could affect cellular phenotype.[54] In this system, a microfabricated silicon substrate consisting of two interlocking parts was manually manipulated to bring cells in close proximity to each other. The two parts could be joined in discrete configurations, such that different types of cells are adjacent to one another or are separated by a micron-scale gap. By culturing hepatocytes and supportive stromal cells on these substrates, and by adjusting the placement of the interlocking components, the effects of paracrine and juxtacrine signals could be examined to derive important insight into the nature of these interactions.

Takayama et al. developed a chemomechanical technique that allows the reversible control of cell adhesion and spreading.[55] In this approach, a parallel array of cracks in which cell-adhesive proteins are selectively adsorbed was fabricated on a PDMS substrate. The widths of the cracks could be modulated by adjusting the mechanical strain applied to the substrate, thereby controlling the amount of cell-adhesive proteins accessible to the cells. Through in situ adjustment of crack widths, morphology of cells could be reversibly manipulated.

12.4 Microfluidic Systems for Controlling Cell-Soluble Factor Interactions

Microfluidic systems play a key role in cell-related studies since they consume small quantities of samples and provide a rapid analysis in a high-throughput manner within small channels that are typically in the range of 10–100 µm.[56] They offer several advantages, such as reduced analysis time, lower sample volume, and ability to run multiple assays on a single device.[57,58] With the advent of microfluidic systems, cell research has entered into a new era in which the soluble factors can be controlled within microfluidic networks to generate sophisticated conditions, as well as handle cellular molecules precisely for transportation of cells to the desired locations. Soluble factors, such as growth factors, are proteins that bind to receptors on the cell surface and activate specific series of cellular mechanisms, leading to events like cellular proliferation or differentiation. Based on the ability to control cell behavior, effective usage of growth factors has been investigated for directing the lineage-specific differentiation of ESCs using microfluidic systems. For example, activin A was shown to mediate dorsoanterior mesoderm differentiation, while bone morphogenic protein 4 (BMP-4) was shown to mediate formation of hematopoietic cells.[59–61] These benefits have accelerated the application of microfluidic systems to investigate cell-soluble factor interactions.

12.4.1 Microfluidic Techniques for Cell Manipulation

Microfluidic-based cell manipulation approaches have generally focused on how to handle a single cell or a few cells inside microfluidic systems where magnetic, optical, electrical, or mechanical forces are employed. We briefly summarize the description of each manipulation method.

Manipulation using magnetic forces generally utilizes magnetic particles to identify specific cells. The size of the magnetic particles is 10–100 nm, so that it is possible to uptake the magnetic particles into cells with little effect on cell function and viability. Lee et al. implemented an embedded 8×8 microcoil array for the generation of magnetic forces in order to control the location of a single cell on the microcoil surface.[62] Using the magnetic force via magnetic particles is a noninvasive method to control cells inside microfluidics, but the development of a new conjugating method of particles into ligands and the simple integration of magnetic force generators with microfluidics remain a challenge.

Optical tweezers, due to an electric field gradient around a focal point of laser light, are commonly used in cell manipulation since they are easily combined with microfluidics systems and are noncontact methods. Live mammalian cells are successfully sorted using the optical method within a short time (2–4 ms).[63] Though this optical method requires only transparent windows to allow light passing through a chip, its applications in cell manipulation have limitations due to the external complex optical setups.

Dielectrophoresis (DEP) is a popular form of electric force application in microfluidic systems. DEP can easily separate cells based on electrical properties in electric gradients based on a principle similar to optical tweezers. Use of this technique inside a photopatternable hydrogel showed a new opportunity for cell manipulation and immobilization.[64] Though DEP is effective to separate cells with different sizes in fluids, biological damages due to high electric field and the additional process for electrode formation restrict its widespread application in microfluidic systems.

Mechanical manipulation of cells is typically applied to cell sorting by implementing mechanical constriction structures inside microfluidic systems to separate cells of interest based on their mechanical properties, such as size. Contrary to other methods, this approach uses passive structures inside microfluidic systems, leading to numerous advantages, such as simple cell manipulation fabrication, easy procedures, and high throughput. The modified application of mechanical manipulation is cell trapping and docking inside microwell arrays to investigate cell-soluble factors' interaction with different conditions or cell types. C-shaped microwell arrays showed successful cell docking with a uniform number of cells.[65] Double grooves underneath microfluidic channels also can effectively capture cells by microcirculation of fluids.[66]

12.4.2 Microfluidic Systems for Gradient Generation of Soluble Factors

Generation of gradient is essential in cell-related studies implemented for investigation of cellular responses according to different combinations and concentrations of soluble factors. Microfluidic systems make it possible to generate different types of static or dynamic gradients on chips.

The simplest way to create gradients in microfluidic channels is by combining two or more input channels into one gradient channel to generate combinatorial gradients of different biological solutions.[67] This method is reproducible and quantitative, but consumes a considerable amount of expensive reagents, and the cells inside channels inevitably experience shear stress caused by the flow. Complex shapes of gradients can be obtained by using branched structures that control diffusive mixing of liquids, depending on branch configurations.[68] This technique provides flexible and fast generation of gradients, but drawbacks include large dead volumes, single axis gradient, and rapid clearance of cell-secreted factors before gradient stabilization. The parallel flow gradient generator, having physical walls between adjacent streams, uses orthogonal diffusion at disconnected points of the walls in order to generate various types of stream profiles (Figure 12.6).[69] Like the branch-structured generator, this technique may consume a large volume of reagents that are possibly expensive. Recently, gradient range and linearity in a passive diffusive gradient were improved by the addition of backward flow inside a microfluidic channel,[70,71] which is a simple and effective approach to obtain a long-range gradient up to a few centimeters inside a single channel.

The stabilization of generated gradients is also important to analyze cell-soluble factor interactions inside a localized and stabilized microenvironment. Many of the papers about gradient generation mention that cells may be able to interact with soluble factors inside microfluidic channels where chemical or biological gradients are dynamically generated. Kim et al. showed that microfluidic devices can produce a logarithmic scale of flow rates and a logarithmatic concentration of gradients.[72] Based on these gradients, it has been shown that the morphology and proliferation of ESCs varies.[72] These approaches are easy and convenient for testing cellular response, but a number of drawbacks exist, such as large consumption of soluble factors, exposure of cells to shear stress, leading to false response, and cross-contamination between combinatorial conditions. One solution for this purpose is to implement different microfluidic pathways for cell loading and media perfusion.[73] Individual addressability to chambers by bridge and underpass architecture has made this possible. Upadhyaya and Selvaganapathy utilized a nanoporous membrane

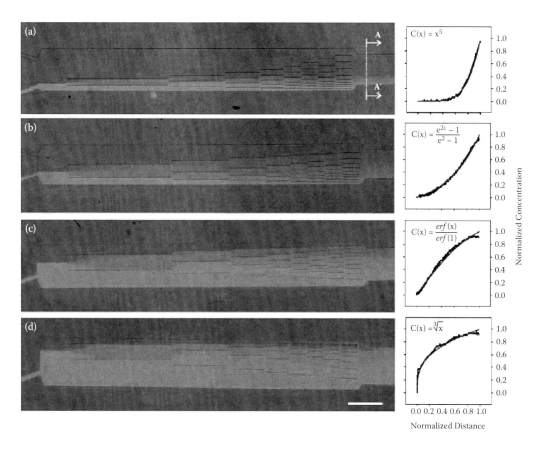

FIGURE 12.6

The locations of physical walls between two liquids can make different kinds of steady state gradient profiles including fifth power, exponential, error, and cubic root functions. Adapted with permission from reference 69. Copyright 2004 American Chemical Society.

at the interface between the gel layer and the microfluidic channels as an electrically controllable valve.[74] Although the response time to allow chemicals to transfer through the nanoporous membrane is low (~20 s), it is effective to stabilize gradients for further application in high-throughput screening analysis. An interesting approach to obtain a static gradient in channel structures was performed by Wu and co-workers.[75] The chemical gradient was formed statically inside an agarose gel layer that connects two reservoirs of different concentrations. The stabilized gradient can be transferred to the microfluidic channels formed under the agarose gel layer. This method makes it possible to generate random gradient patterns by exploiting well-characterized gradient profiles inside the gel layer, as well as by designing channel shapes for certain gradient profiles. However, it uses diffusion inside the gel layer, which takes several hours to reach equilibrium.

12.4.3 Integrated Microfluidic Systems

Integration of functions between cell manipulations and the gradient generation of soluble factors is a critical step to obtain successfully integrated microfluidic systems for assays of cell-soluble factor interaction. This integration process may increase device complexity, which in turn leads to low reproducibility and poor reliability. Moreover, final analysis could become complicated due to unexpected problems, such as interwell communications or flushing out of cell-secreted factors at gradient generation. Thus, device simplicity and reliability should be considered at an integration stage.

An integrated microfluidic system was reported to analyze DNA or mRNA from as little as one cell, which could eliminate blurring by multicell analysis for the typical mRNA procedure.[76] The device is composed of well-organized microfluidic blocks such as microvalves to regulate fluids inside channels and rotary mixers to circulate three different fluids. The same research group developed thousands of individually addressable chambers that were successfully implemented and driven to demonstrate a microfluidic memory device using column and row multiplexers.[77] These techniques utilize a simple fabrication process and highly integrated fluidic circuits to enable high-throughput analysis of biological experiments, but the complexity of plumbing to control numerous valves should be simplified to be widely used with biologists. Hung et al. developed a simplified bioreactor array with the capability of different gradient generation inside bioreactors and continuous perfusion for long-term cell culture.[78,79] The fabricated device has versatile functions, including generation of different conditions into bioreactors and postprocessing after cell culture, but nonuniform cell loading and cross-contamination could be drawbacks. Their later research improved the uniformity of cell loading by adopting barrier structures to increase fluidic resistance inside bioreactors.[65] King et al. created a real-time gene expression array by implementing nano-liter-scale microfluidic bioreactors.[80] A matrix of experiments consisting of row-seeded cells and column-introduced stimuli can provide thousands of single time experiments through combining automated image capture and analysis, but interwell contamination should be verified and improved by devising proper fluidic pathways for perfusion.

12.5 Three-Dimensional Structuring of Cellular Microenvironment

Studying cell behavior on a planar surface is convenient for the experiment and analysis, but it can possibly make different microenvironments on cells, since all the tissue cells in the body interact with ECM or adjacent cells in 3D space. This may lead to the failure to mimic the tissue microenvironment. Therefore, microengineering techniques can be employed to regulate the cell's interactions with its 3D microenvironment.[81] To date, many studies have emphasized the 3D structuring of cellular microenvironments to overcome the limitations of cell studies on the surface. For example, tuned microenvironments were created to investigate specific cell behaviors by restricting the cellular geometry through patterned substrates.[82–84] Other microengineering approaches, such as microscale channels,[85] microwells,[9] and cell-laden hydrogels, have been used to control the cellular microenvironment.[81,86,87] Furthermore, biomaterials have been engineered with the capability of controlling the cell behavior.[81,88,89]

12.5.1 Cell-Laden Hydrogels

Cell-laden hydrogels can be used to generate 3D structures. To engineer cell-laden hydrogels, biocompatible and biodegradable polymer scaffolds which provide structural support and physical cues for cell attachment, orientation, alignment, and spreading can be used. The

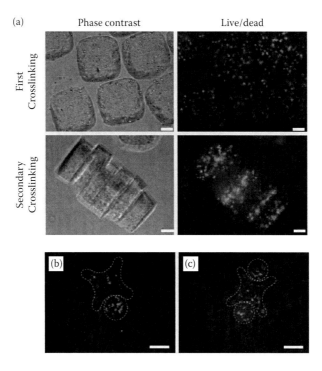

FIGURE 12.7
Individual cell-laden microgels are first patterned by crosslinking with a photomask, and then the second crosslinking can lock up the self-assembled microgels (a). Surface tension drives microgels to minimize the exposed surfaces to hydrophobic fluid, leading to self assembly according to shapes (b) and (c). (Scale bars, 100 μm.) Stained cells encapsulated in microgels can be assembled to implement designed structures. Adapted with permission from reference 95. Copyright 2008 National Academy of Sciences, U.S.A.

hydrogel prepolymer solution is mixed homogeneously with cell suspension and gelled rapidly to construct 3D structures. Among available polymers, PEG-based hydrogels have been used widely because of its biocompatibility and photo patternability to generate customized pattern shapes. PEG has been proven to be useful in biomedicine field due to its ability to prolong the circulating time of drugs in vivo.[92–93] PEG does not initiate cytotoxicity and immunogenicity and is FDA approved for specific applications. Tsang and colleagues developed a method to pattern cell-laden hydrogels layer by layer, which can be used to construct a hepatocyte tissue.[94] Hepatic functions including cumulative albumin secretion and urea synthesis increased 100%–260% rather than unperfused plug with same cell density. Khademhosseini and co-workers combined self-assembly technique with photopatterned PEG building blocks to provide a highly scalable approach for the formation of 3D cell microenvironments (Figures 12.7).[95] They employed shape directed self-assembly to increase the possibility of desired assembly between two different types of microgels. Dielectrophoretic (DEP) forces can localize single cells within a fluid; thus live cells suspended in prepolymer solution were photoimmobilzed after relocation of cells using this electropatterning method.[64] Compared with other methods with homogeneously dispersed cells inside PEG solutions, the proposed way in this research can control cell-cell interactions within a small environment. Even though PEG based hydrogels provide an easy method to construct 3D cell ECM structures, other types of hydrogels are being researched to overcome the drawbacks of PEG hydrogels including non-natural and non-biodegradable properties leading to failure to recreate the complexity of native ECM. Dang et al. developed a method to fabricate cell laden alginate microcapsules patterned by commercially available polypropylene meshes.[96] The meshes having different shapes and sizes are dipped into cell suspended alginate solution and $CaCl_2$ solution sequentially to form alginate microcapsules. This method provides a cheap and convenient way to create non-photopatternable hydrogels, but the size and shape of fabricated hydrogels are limited by available meshes and further assembly of hydrogels should be proposed to build complex 3D cell structures. Liu and Chan-Park studied modified dextran and gelatin to investigate the viability, proliferation and spreading of human umbilical artery smooth muscle cells (SMCs) under different synthetic conditions.[73] The mechanical stiffness of hydrogels was adjusted by changing relative concentrations of dextran and gelatin, or precursor solutions.

12.5.2 Porous Polymers

An important aspect of generating engineered tissue is to direct the 3D organization of cells by biomaterial construct and to establish the conditions necessary for the cells to reconstruct a functional tissue structure. Porous polymer structures may provide a suitable chance to construct 3D cellular microenvironment when cells are cultured inside the polymers and the porous space is used as a channel to supply oxygen and nutrients and allow for removing waste. This approach may assist to overcome ongoing major challenge in tissue engineering which is to create functional, integrated microvasculature. Fibrous structure is one example of porous polymers typically fabricated by electrospinning process.[97] The polymer stream ejected from an electrically biased needle forms a thin fibrous structure on a target plate which is electrically grounded. A high surface to volume ratio and fiber network similar to natural ECM is beneficial to mimic 3D cellular environment, but cell seeding and migration can be difficult due to the small pore size of the fibrous structures. Acceptable pore sizes for cell seeding and migration in the porous structures have been obtained by mixing polymers with porogens (or water to be removed by freeze dry) and removing porogens selectively by appropriate solvents to form pores where the porogens existed. The pore size can be controlled from sub-micron to a few hundred microns by chemical composition and processing condition. Akay et al. reported a high internal phase emulsion (HIPE) polymerization process to construct different pore sizes of 40, 60, and 100 μm with the varied porosity from 70% to 97%.[98] They built bone-like matrix by culturing osteoblasts for 35 days to allow the cells to migrate to a maximum depth of 1.4mm. Cheng et al. created 3D porous structures by removing porogens through temperature induced phase separations without applying chemicals.[99] Sieved ammonium bicarbonate particles were mixed with high viscous polymer solution and removed by decomposition into ammonia over 36°C leaving void spaces they occupied.

12.5.3 Computer-Aided Structuring

Recently, computer aided structuring has emerged to take advantages such as rapid prototyping, customization of 3D structures and automation of manufacturing process. This concept has become an attractive approach since it has a great potential of providing delicate 3D structures.[100,101] By dispensing polymers mixed with cells using a bioplotter, 3D scaffolds were fabricated with a particle size of 30–75 μm.[102] Concentration, viscosity, and dispensing condition should be optimized to deposit polymers properly. Two photon polymerization uses photopolymerization of the polymers only within the focal volume of a laser, leading to easy construction of complex 3D structures via computer controls.[103] The research showed 3D porous structure allows higher cell speed than 2D substrate and the cell speed depends on the pore size. Mutilayered 3D cell-laden hydrogel structures (16.2 μm per each layer) also can be constructed using a mechanical valve for dispensing hydrogel droplets and computer controlled moving stages.[104]

12.6 Summary and Future Directions

In this chapter, we have reviewed some of the microfabrication technologies available for controlling the cellular microenvironment. The widespread uses of lithographic technologies as well as the development of novel biomaterials have improved methods of controlling cell-cell, cell-ECM, and cell-soluble factor interactions spatially and temporally. Much insight has been gained from these studies for directing stem cell fates as a cell source either individually or incorporation into tissue engineering scaffolds and biomimetic cell-based devices. However, in order to develop sophisticated platforms that can control cellular functions more accurately, it will be indispensable to design new materials and to incorporate them into improved fabrication strategies. On the other hand, tissues are more complex than those generated with the technologies that have been developed so far. Thus, integration of microfabrication technologies to engineer the complexities of the cell microenvironment will yield more promising systems for research and therapeutic purposes.

References

1. Niemeyer, C.M. and Mirkin, C.A., Nanobiotechnology, Wiley-VCH, 2004.
2. Willner, I. and Katz, E., Bioelectronics, Wiley-VCH, 2005.
3. Khademhosseini, A. et al., Microscale technologies for tissue engineering and biology, *Proc. Natl. Acad. Sci. USA* 103, 2480, 2006.
4. Murtuza, B., Nichol, J.W., and Khademhosseini, A., Micro- and nanoscale control of the cardiac stem cell niche for tissue fabrication, *Tissue Eng. Part B Rev.* 15, 443, 2009.
5. Kane, R.S. et al., Patterning proteins and cells using soft lithography, *Biomaterials* 20, 2363, 1999.
6. Khetani, S.R. and Bhatia, S.N., Engineering tissues for *in vitro* applications, *Curr. Opin. Biotechnol.* 17, 524, 2006.
7. Liu, W.F. and Chen, C.S., Cellular and multicellular form and function, *Adv. Drug Deliv. Rev.* 59, 1319, 2007.
8. Chung B.G,. Kang L., and Khademhosseini A., Micro- and nanoscale approaches for tissue engineering and drug discovery, *Expt. Opin. Drug Dis.* 2, 1653, 2007.
9. Karp, J.M. et al., Controlling size, shape and homogeneity of embryoid bodies using poly(ethylene glycol) microwells, *Lab Chip* 7, 786, 2007.
10. Khademhosseini A. et al., Micromolding of photocrosslinkable hyaluronic acid for cell encapsulation and entrapment, *J. Biomed. Mater. Res. A* 79, 522, 2006.

11. Moeller, H.C. et al., A microwell array system for stem cell culture, *Biomaterials* 29, 752, 2008.
12. Wright D. et al., Generation of static and dynamic patterned co-cultures using microfabricated parylene-C stencils, *Lab chip* 7, 1272, 2007.
13. Bauwens, C.L. et al., Control of human embryonic stem cell colony and aggregate size heterogeneity influences differentiation trajectories, *Stem Cells* 26, 2300, 2008.
14. Khademhosseini, A. et al., Layer-by-layer deposition of hyaluronic acid and poly-L-lysine for patterned cell co-cultures, *Biomaterials* 25, 3583, 2004.
15. Kang, L. et al., Cell confinement in patterned nanoliter droplets in a microwell array by wiping, *J. Biomed. Mater. Res. A* 2009.
16. Ungrin M.D. et al., Reproducible, ultra high-throughput formation of multicellular organization from single cell suspension-derived human embryonic stem cell aggregates, *PLoS ONE* 3, e1565, 2008.
17. Bhatia, S., Yarmush, M., and Toner, M., Controlling cell interactions by micropatterning in co-cultures: hepatocytes and 3T3 fibroblasts, *J. Biomed. Mater. Res.* 34, 189, 1997.
18. Molnar, P. et al., Photolithographic patterning of C2C12 myotubes using vitronectin as growth substrate in serum-free medium, *Biotechnol. Prog.* 23, 265, 2007.
19. Li, N. and Ho, C., Photolithographic patterning of organosilane monolayer for generating large area two-dimensional B lymphocyte arrays, *Lab Chip* 8, 2105, 2008.
20. Carrico, I. et al., Lithographic patterning of photoreactive cell-adhesive proteins, *J. Am. Chem. Soc.* 129, 4874, 2007.
21. Revzin, A., Tompkins, R., and Toner, M., Surface engineering with poly (ethylene glycol) photolithography to create high-density cell arrays on glass, *Langmuir* 19, 9855, 2003.
22. Hahn, M. et al., Photolithographic patterning of polyethylene glycol hydrogels, *Biomaterials* 27, 2519, 2006.
23. Karp, J. et al., A photolithographic method to create cellular micropatterns, *Biomaterials* 27, 4755, 2006.
24. Yamazoe, H., Uemura, T. and Tanabe, T., Facile cell patterning on an albumin-coated surface, *Langmuir* 24, 8402, 2008.
25. Singhvi, R. et al., Engineering cell shape and function, *Science* 264, 696, 1994.
26. Chen, C. et al., Geometric control of cell life and death, *Science* 276, 1425, 1997.
27. Khademhosseini, A. et al., Direct patterning of protein-and cell-resistant polymeric monolayers and microstructures, *Adv. Mater.* 15, 1995, 2003.
28. Xu, L. et al., Microcontact printing of living bacteria arrays with cellular resolution, *Nano Lett.* 7, 2068, 2007.
29. Coq, N. et al., Self-supporting hydrogel stamps for the microcontact printing of proteins, *Langmuir* 23, 5154, 2007.
30. Xu, H. et al., Microcontact printing of dendrimers, proteins, and nanoparticles by porous stamps, *J. Am. Chem. Soc.* 131, 797, 2009.
31. Foley, J. et al., Microcontact printing of proteins inside microstructures, *Langmuir* 21, 11296, 2005.
32. Takayama, S. et al., Topographical micropatterning of poly (dimethylsiloxane) using laminar flows of liquids in capillaries, *Appl. Phys. Lett* 76, 2946, 2000.
33. Takayama, S. et al., Subcellular positioning of small molecules, *Nature* 411, 1016, 2001.
34. Ostuni, E. et al., Patterning mammalian cells using elastomeric membranes, *Langmuir* 16, 7811, 2000.
35. Tavana, H. et al., Nanolitre liquid patterning in aqueous environments for spatially defined reagent delivery to mammalian cells, *Nat. Mater.* 8, 736, 2009.
36. Yousaf, M.N., Houseman, B.T. and Mrksich, M., Turning on cell migration with electroactive substrates, *Angew. Chem., Int. Ed.* 40, 1093, 2001.
37. Jiang, X. et al., Electrochemical desorption of self-assembled monolayers noninvasively releases patterned cells from geometrical confinements, *J. Am. Chem. Soc.* 125, 2366, 2003.
38. Li, Y. et al., A method for patterning multiple types of cells by using electrochemical desorption of self-assembled monolayers within microfluidic channels, *Angew. Chem. Int. Ed.* 46, 1094, 2007.

39. Kaji, H. et al., *In situ* control of cellular growth and migration on substrates using microelectrodes, *J. Am. Chem. Soc.* 126, 15026, 2004.

40. Kaji, H., Kawashima, T., and Nishizawa, M., Patterning cellular motility using an electrochemical technique and a geometrically confined environment, *Langmuir* 22, 10784, 2006.

41. Kaji, H. et al., Patterning the surface cytophobicity of an albumin-physisorbed substrate by electrochemical means, *Langmuir* 21, 6966, 2005.

42. Hashimoto, M., Kaji, H. and Nishizawa, M., Selective capture of a specific cell type from mixed leucocytes in an electrode-integrated microfluidic device, *Biosens. Bioelectron.* 24, 2892, 2009.

43. Hashimoto, M. et al., Localized immobilization of proteins onto microstructures within a preassembled microfluidic device, *Sens. Actuators B* 128, 545, 2008.

44. Kaji, H. et al., Stepwise formation of patterned cell co-cultures in silicone tubing, *Biotechnol. Bioeng.* 98, 919, 2007.

45. Kaji, H., Hashimoto, M., and Nishizawa, M., On-demand patterning of protein matrixes inside a microfluidic device, *Anal. Chem.* 78, 5469, 2006.

46. Nakanishi, J. et al., Photoactivation of a substrate for cell adhesion under standard fluorescence microscopes, *J. Am. Chem. Soc.* 126, 16314, 2004.

47. Nakanishi, J. et al., Spatiotemporal control of migration of single cells on a photoactivatable cell microarray, *J. Am. Chem. Soc.* 129, 6694, 2007.

48. Edahiro, J. et al., *In situ* control of cell adhesion using photoresponsive culture surface, *Biomacromol.* 6, 970, 2005.

49. Tsuda, Y. et al., Heterotypic cell interactions on a dually patterned surface, *Biochem. Biophys. Res. Commun.* 348, 937, 2006.

50. Tsuda, Y. et al., The use of patterned dual thermoresponsive surfaces for the collective recovery as co-cultured cell sheets, *Biomaterials* 26, 1885, 2005.

51. Yamato, M. et al., Thermally responsive polymer-grafted surfaces facilitate patterned cell seeding and co-culture, *Biomaterials* 23, 561, 2002.

52. Yamato, M. et al., Novel patterned cell coculture utilizing thermally responsive grafted polymer surfaces, *J. Biomed. Mater. Res.* 55, 137, 2001.

53. Kim, J., Yoon, J., and Hayward, R.C., Dynamic display of biomolecular patterns through an elastic creasing instability of stimuli-responsive hydrogels, *Nat. Mater.* 9, 159, 2010.

54. Hui, E.E. and Bhatia, S.N., Micromechanical control of cell-cell interactions, *Proc. Natl. Acad. Sci. USA* 104, 5722, 2007.

55. Zhu, X. et al., Fabrication of reconfigurable protein matrices by cracking, *Nat. Mater.* 4, 403, 2005.

56. Whitesides, G.M., The origins and the future of microfluidics, *Nature* 442, 368, 2006.

57. McDonald, J.C. et al., Fabrication of microfluidic systems in poly(dimethylsiloxane), *Electrophoresis* 21, 27, 2000.

58. Park, T.H. and Shuler, M.L., Integration of cell culture and microfabrication technology, *Biotechnol Prog* 19, 243, 2003.

59. Huber, T.L. et al., Cooperative effects of growth factors involved in the induction of hematopoietic mesoderm, *Blood* 92, 4128, 1998.

60. Johansson, B.M. and Wiles, M.V., Evidence for involvement of activin A and bone morphogenetic protein 4 in mammalian mesoderm and hematopoietic development, *Mol. Cell Bio.* 15, 141, 1995.

61. Valdimarsdotter, G. and Mummery, C., Functions of the TGF- superfamily in human embryonic stem cells, *APMIS* 113, 773, 2005.

62. Lee, H. et al., Integrated cell manipulation system-CMOS/microfluidic hybrid, *Lab Chip* 7, 331, 2007.

63. Wang, M. et al., Microfluidic sorting of mammalian cells by optical force switching, *Nat. Biotechnol.* 23, 83, 2005.

64. Albrecht, D. et al., Photo-and electropatterning of hydrogel-encapsulated living cell arrays, *Lab Chip* 5, 111, 2005.

65. Lee, P. et al., Nanoliter scale microbioreactor array for quantitative cell biology, *Biotechnol. Bioeng.* 94, 5, 2006.

66. Khabiry, M. et al., Cell docking in double grooves in a microfluidic channel, *Small* 5, 1186, 2009.

67. Kamholz, A. et al., Quantitative analysis of molecular interaction in a microfluidic channel: the T-sensor, *Anal. Chem.* 71, 5340, 1999.

68. Dertinger, S. et al., Generation of gradients having complex shapes using microfluidic networks, *Anal. Chem.* 73, 1240, 2001.

69. Irimia, D., Geba, D.A., and Toner, M., Universal microfluidic gradient generator, *Anal. Chem.* 78, 3472, 2006.

70. Du, Y. et al., Convection-driven generation of long-range material gradients, *Biomaterials* 31, 2686, 2010.

71. Du, Y. et al., Rapid generation of spatially and temporally controllable long-range concentration gradients in a microfluidic device, *Lab Chip* 9, 761, 2009.

72. Kim, L. et al., Microfluidic arrays for logarithmically perfused embryonic stem cell culture, *Lab Chip* 6, 394, 2006.

73. Liu, Y. and Chan-Park, M., A biomimetic hydrogel based on methacrylated dextran-graft-lysine and gelatin for 3D smooth muscle cell culture, *Biomaterials* 31, 1158, 2010.

74. Upadhyaya, S. and Selvaganapathy, P., Microfluidic devices for cell based high throughput screening, *Lab Chip* 10, 341, 2010.

75. Wu, H., Huang, B., and Zare, R., Generation of complex, static solution gradients in microfluidic channels, *J. Am. Chem. Soc* 128, 4194, 2006.

76. Hong, J.W. and Quake, S.R., Integrated nanoliter systems, *Nat. Biotechnol.* 21, 1179, 2003.

77. Thorsen, T., Maerkl, S.J., and Quake, S.R., Microfluidic large-scale integration, *Science* 298, 580, 2002.

78. Hung, P.J. et al., Continuous perfusion microfluidic cell culture array for high-throughput cell-based assays, *Biotechnol. Bioeng.* 89, 1, 2005.

79. Hung, P.J. et al., A novel high aspect ratio microfluidic design to provide a stable and uniform microenvironment for cell growth in a high throughput mammalian cell culture array, *Lab Chip* 5, 44, 2005.

80. King, K. et al., A high-throughput microfluidic real-time gene expression living cell array, *Lab Chip* 7, 77, 2007.

81. Fukuda, J. et al., Micromolding of photocrosslinkable chitosan hydrogel for spheroid microarray and co-cultures, *Biomaterials* 2006.

82. Tien, J. and Chen, C.S., Patterning the cellular microenvironment, *IEEE Eng. Med. Biol. Mag.* 21, 95, 2002.

83. Tien, J., Nelson, C.M. and Chen, C.S., Fabrication of aligned microstructures with a single elastomeric stamp, *Proc. Natl. Acad. Sci. USA* 99, 1758, 2002.

84. Nelson, C.M. and Chen, C.S., VE-cadherin simultaneously stimulates and inhibits cell proliferation by altering cytoskeletal structure and tension, *J. Cell Sci.* 116, 3571, 2003.

85. Khademhosseini, A. et al., Molded polyethylene glycol microstructures for capturing cells within microfluidic channels, *Lab Chip* 4, 425, 2004.

86. Khademhosseini, A. et al., Micromolding of photocrosslinkable hyaluronic acid for cell encapsulation and entrapment, *J. Biomed. Mater. Res. A* 79, 522, 2006.

87. Napolitano, A.P. et al., Dynamics of the self-assembly of complex cellular aggregates on micromolded nonadhesive hydrogels, *Tissue Eng. Part A* 13, 2087, 2007.

88. Nichol, J.W. and Khademhosseini, A., Modular tissue engineering: engineering biological tissues from the bottom up, *Soft Matter.* 5, 1312, 2009.

89. Peppas, N.A. et al., Hydrogels in biology and medicine: from molecular principles to bionanotechnology, *Adv. Mater.* 18, 1345, 2006.

90. Burdick, J.A. and Vunjak-Novakovic, G., Engineered microenvironments for controlled stem cell differentiation, *Tissue Eng. Part A* 15, 205, 2009.

91. Khademhosseini, A. et al., Microscale technologies for tissue engineering and biology, *Proc. Natl. Acad. Sci. USA* 103, 2480, 2006.
92. Maruyama, K. et al., Targetability of novel immunoliposomes modified with amphipathic poly(ethylene glycol)s conjugated at their distal terminals to monoclonal-antibodies, *Biochim. Biophys. Acta-Biomembranes* 1234, 74, 1995.
93. Akerman, M.E. et al., Nanocrystal targeting *in vivo*, *Proc. Natl. Acad. Sci. USA* 99, 12617, 2002.
94. Tsang, V. et al., Fabrication of 3D hepatic tissues by additive photopatterning of cellular hydrogels, *FASEB J.* 21, 790, 2007.
95. Du, Y. et al., Directed assembly of cell-laden microgels for fabrication of 3D tissue constructs, *Proc. Natl. Acad. Sci. USA* 105, 9522, 2008.
96. Dang, T. et al., Microfabrication of homogenous, asymmetric cell-laden hydrogel capsules, *Biomaterials* 30, 6896, 2009.
97. Greiner, A. and Wendorff, J., Electrospinning: a fascinating method for the preparation of ultra-thin fibers, *Angew. Chem., Int. Ed.* 46, 5670, 2007.
98. Akay, G., Birch, M., and Bokhari, M., Microcellular polyHIPE polymer supports osteoblast growth and bone formation *in vitro*, *Biomaterials* 25, 3991, 2004.
99. Cheng, K., Lai, Y. and Kisaalita, W., Three-dimensional polymer scaffolds for high throughput cell-based assay systems, *Biomaterials* 29, 2802, 2008.
100. Mironov, V. et al., Organ printing: tissue spheroids as building blocks, *Biomaterials* 30, 2164, 2009.
101. Neagu, A. and Forgac, G., Fusion of cell aggregates: a mathematical model., In Biomed. Eng. Recent Development, Medical and Engineering Publishers, 2002.
102. El-Ayoubi, R. et al., Design and fabrication of 3D porous scaffolds to facilitate cell-based gene therapy, *Tissue Eng. Part A* 14, 1037, 2008.
103. Tayalia, P. et al., 3D cell-migration studies using two-photon engineered polymer scaffolds, *Adv. Mater.* 20, 4494, 2008.
104. Moon, S. et al., Layer by layer three-dimensional tissue epitaxy by cell-laden hydrogel droplets, *Tissue Eng. Part C* 920, 2010.

13

Think Big, Then Shrink: Low-Cost, Ultra-Rapid Tunable Nanowrinkles from Shape Memory Polymers

Aaron Chen, Himanshu Sharma, Diep Nguyen, Soroush M. Mirzaei Zarandi, and Michelle Khine

Department of Chemical Engineering, University of California, Irvine, California, USA

CONTENTS

13.1 Introduction

To translate technological advances from the laboratory into deployable point of care (POC) diagnostics to detect infectious diseases in developing countries, it is necessary to develop a robust, low-cost, and highly sensitive quantitative bioassay platform. The persistent chasm between academic prototyping and real-world solutions must be bridged by developing a versatile and field-rugged platform technology that is widely deployable throughout the poorest areas of the world. The technology has to be sensitive and robust enough to be efficacious in detecting specific infections at their very onset. The entire system must be simple to use and extremely inexpensive. Detection strategies adequate for the laboratory, including expensive high-powered lasers and detector systems, have no place in the field. One of the popular methods to reduce the need for costly instruments and maintain accuracy and sensitivity in diagnostics is the advance in microfluidics and lab-on-a-chip (LOC) technologies.

The analytical field of LOC has been widely applied to medical diagnostics.[1-6] This field is based on the premise of shrinking the entire laboratory onto a miniaturized and integrated micro total analysis system (μTAS). Due to the size of the chip and often microscaled channels embedded in the chip, the advantages of this technology include faster reaction time, less reagent consumption, and lower capital equipment costs. Applications of these technologies include, to name a few, continuously fractionating components of blood,[7-11] nucleic acid extraction upstream of polymerase chain reaction (PCR),[12] and purification of small molecules upstream of an immunoassay.[13,14] Furthermore, areas such as genomics, proteomics, drug discovery, and POC medical diagnostics can also benefit from these technologies.[1–15] To fully realize the potential of this field, rapid prototyping of new and intricate chips without compromising important material properties must first be developed.

Microfabrication of complex patterns into silicon, glass, or quartz is traditionally resource intensive. In addition, it typically requires costly investments in large specialized equipment, a clean room, and consumables.[5] However, the introduction of poly(dimethylsiloxane) (PDMS) has enabled the accelerated progress of rapid prototyping in the field of microfluidics through soft lithography.[16] Soft lithography is generally used to construct features measured in the micro- to nanometer scale, and it drastically reduces the chip fabrication time. PDMS is one of the most popular materials used in microfluidics because it is optically transparent (from 240 to 1,100 nm), flexible, gas permeable (for cell-based assays), and low cost.[17] However, the inherent material properties of this polymer present significant limitations. The disadvantages of PDMS include its hydrophobicity and absorption of organic solvent and small molecules. Hydrophobic and small molecules absorb easily into the porous PDMS matrix. The absorption of molecules could potentially affect results, which is unacceptable for many analytical applications.[18] PDMS also swells in certain organic solvents and has inherently unstable surface properties.[19] Due to its high flexibility, channels fabricated using PDMS tend to expand and contract with different pressures. Although this property could potentially benefit some applications, it could be detrimental to others. These limitations have prevented the industrial use of PDMS for applications in drug discovery and other sensitive assays.[17] Thus, PDMS has been mainly used in academic prototyping.

Instead of using PDMS, industry prefers to work with more rigid and chemical-resistant materials, such as polystyrene (PS). PS is optically transparent, biocompatible, inert, rigid, and its surface can be easily functionalized. In addition, it is economical, recyclable, and can be molded into desired shapes. The applications of PS range from daily items, such as CD cases and cups, to laboratory items, such as cell culture petri dishes, test tubes, and microtiter plates.[20] Its hydrophobic surface can be easily made hydrophilic by various means, through corona discharge, gas plasma, irradiation, and surface chemistry. However, the cost microfabricating intricate patterns on such plastics is limited to academic pursuit.

To overcome this persistent chasm between academic prototyping and industry standards, various approaches have been investigated. For example, hot embossing is used in polymer microfabrication to rapidly produce identical devices using a stamp to create patterns in polymeric materials. However, creating a precise stamp with microfeatures is often time- and expense-intensive. To obviate this issue, an epoxy stamp, which is durable, relatively inexpensive, and rapid to produce, has been developed.[21] Although the epoxy stamp addresses the resource problem for creating a stamp, it does not help reduce the time and expense required for fabricating the mold. Others have attempted to address this issue by developing PS-based thermoplastic elastomer gels to create viscoelastic three-dimensional (3D) stacked chips using a molding approach.[22] Perfluoropolyether (PFPE), which is resistant to organic solvents, was molded to develop a DNA synthesizer.[23] Thermoset polyester was used to develop microfluidic devices.[24] Shrinkable PS films were also used previously to demonstrate high-aspect-ratio microstructures; however, a complete microfluidic chip was not developed heretofore with this approach.[25,26]

In this chapter, we will describe a new manufacturing process that obviates the most expensive and difficult aspect of micro- and nanofabrication, which is patterning at high

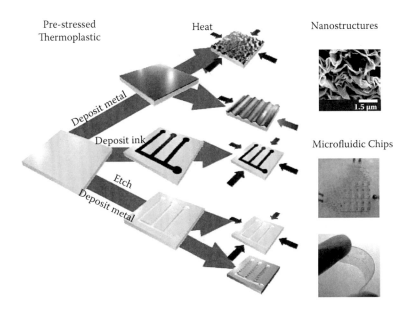

FIGURE 13.1
(See color insert.) Ultra-rapid, low-cost manufacturing process of nano-integrated microsystems.

resolution. This approach eliminates the need for expensive capital equipment and reduces the cost of each chip to pennies. We simply pattern at the large scale, which is inexpensive and easy, and subsequently shrink down to the small scale. We use prestressed polymer sheets and leverage the inherent retraction properties, which can controllably shrink features down to 40%–95% of their patterned sizes, depending on the polymer chain length and amount of extrusion.[27–30] With these tunable shape memory polymers, we can create nanostructures integrated directly into our microsystems (Figure 13.1). This ultra-rapid fabrication approach results in plastic-based microfluidic systems with integrated metallic nanostructures for surface-enhanced sensing based on surface plasmon resonance (SPR). We have demonstrated that our integrated nanostructures can significantly enhance the signal output of the bioassay, therefore also obviating the need for high-powered excitation lasers and detector systems.[31]

13.2 Fabrication and Characterization of Nanowrinkles

Metal wrinkles are thin films of metal on polymer substrates that are formed after the metal buckles during the heating process. The ability to fabricate metal wrinkles into plastics has promising applications in molecular detection, optical devices, electronics, biological assays, filters, and sorters.[32–36] Previous processes for manufacturing the wrinkles have been tedious, time-consuming, and sometimes utilize expensive microfabricated molds or ion/electron beam equipment. Researchers such as Bowden et al. were able to achieve feature sizes of 30 μm after depositing metal onto a thermally expanded PDMS and cooling it down to ambient temperature.[34] Huck et al. utilized this technique to fabricate wrinkles for forming aligned patterns of buckles on planar surfaces that were patterned by a photolithography process.[37] Watanabe and Hirai improved this approach by simply prestretching the PDMS sheet to improve their feature sizes to 6–20 μm striped patterns.[38] Efimenko et al. utilized wrinkled substrates for separating particles based on size by fabricating uniaxial wrinkles by stretching PDMS.[33] Utilizing the wrinkles, two different size silica particles (3 and 10 μm) and 67 nm PS latex in water flowed over the buckled surfaces in a microfluidic cell. After the particles flowed, they began to segregate. The biggest silica particles, 10 μm, were entrapped between the grooves of the buckles, while the smaller silica particles, 3 μm, rested on top of the smaller buckles. The smallest PS latex particles aligned themselves in the smaller wrinkle wavelengths, 1–5 μm. Yoo et al. was able to achieve higher-resolution wrinkles approximately 2 μm in periodicity.[39] Our lab has demonstrated that we could fabricate nanowrinkles by using a two-step approach and tailor the scale range for the nanometer metal wrinkles by adjusting the thickness of the metal deposited on PS sheets.[31] Furthermore, we were able to easily orient the metallic nanowrinkles in a uniaxial or biaxial fashion.

The overall fabrication procedure, shown in Figure 13.2, consisted of two main steps. First, a 10 nm thick gold film was deposited on PS sheets via sputtering. In order to prevent the sample from shrinking or inducing any preheating, the sputtering was divided into four cycles, where 2.5 nm of gold is deposited per cycle. After deposition, the substrate was heated at 160°C for 6 minutes in order to induce 60% shrinkage and force the metal film to collapse, forming bimetallic wrinkles. In order to form uniaxial wrinkles, two clips were placed on the edges of each of the sides of the gold film and heated at 160°C for 6 minutes.

The wrinkles can be further integrated into shrink-induced polymer microfluidic devices. First, a microchannel was patterned by manual scribing on the PS sheet. Then clips were attached on the ends of the PS sheet and the substrate was shrunk by heating it to 150°C. Once the substrate was cool, a diamond saw was used to cut out the center of the sample. Chips were masked by placing adhesive tape on the surface of the sample adjacent to the channel previously scribed into the surface. Using a dissection microscope, tape was

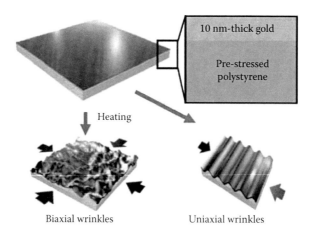

10 nm-thick gold

Pre-stressed polystyrene

Heating

Biaxial wrinkles   Uniaxial wrinkles

FIGURE 13.2
Schematic of the fabrication of biaxial and uniaxial metal wrinkles. Ten nanometers of gold is sputtered on a PS sheet and then heated at 160°C in an oven to induce the buckling of the metal.

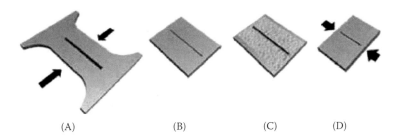

(A)   (B)   (C)   (D)

FIGURE 13.3
Schematic of the fabrication of uniaxial wrinkles inside the polystyrene-based microchannel. (A) A channel was patterned on the polystyrene sheet. The sheet was shrunk unidirectionally. (B) The edges of the chip were cut off. (C) A piece of tape was used to cover the surface on the chip except for the channel. Silver was deposited into the channel by sputter coating. (D) The tape was removed and the chip was then shrunk to form uniaxial wrinkles inside the channel.

aligned until everything except for the channel was covered. The chip was then sputter coated with 45 nm of silver. After deposition, the tape was removed from the surface and the chip was then shrunk without any constraints to form uniaxial wrinkles inside the channel (Figure 13.3).

The characterization of the wrinkles was done using a scanning electron microscope (SEM). SEM images were taken for the biaxial and uniaxial samples to confirm the wrinkle features (Figure 13.4).

Two-dimensional fast Fourier transform (FFT) of the SEM images was conducted to determine the distribution of the wrinkle wavelengths as a probability function. The insets of Figure 13.5A and B indicate the 2D FFT patterns. The wrinkle wavelengths ranged from 200 nm to 1 μm, with a majority of the nanowrinkles peaking around 400 nm. To better understand the tuning of the wavelength of the wrinkles, it was important to first understand how the length scale of the wrinkles depends on the thickness of the metal film, along with material properties. The thickness of deposited gold was varied from 10 to 50 nm, and Figure 13.6 illustrates how the wrinkles' wavelength shifted in hundreds of nanometers, with increasing thickness for the biaxial samples. For uniaxial samples, a similar trend was observed, except two peaks were observed for the wrinkle wavelengths 300 and 800 nm (Figure 13.7).

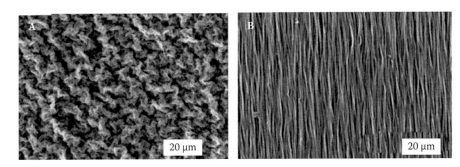

FIGURE 13.4
SEM images of gold (A) biaxial and (B) uniaxial nanowrinkles on shrunk polystyrene sheets.

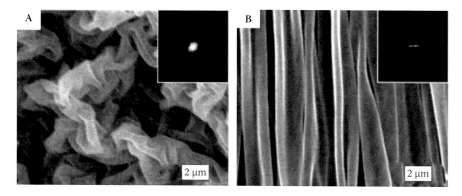

FIGURE 13.5
SEM images of gold nanowrinkles on shrunk polystyrene sheets. (A) Biaxial wrinkles, inset: 2D FFT patterns. (B) Uniaxial wrinkles, inset: 2D FFT patterns.

13.3 Immunoassay Development Using Nanowrinkles

In recent years, there has been a growing interest in metal-enhanced fluorescence (MEF) technology as an emerging tool in the field of immunoassay development. The effect and mechanisms of MEF have been discussed extensively elsewhere.[40–67] Briefly, MEF is the method of using metal structures or particles in close proximity to the fluorophore to increase the brightness and photostability of the fluorophore. Depending on the distance between the fluorophore and metal surfaces or particles, the fluorescence can be quenched or enhanced several thousand-fold. The fluorescence signal enhancement can be explained by at least two mechanisms. First, the metal surfaces and colloids are able to amplify the incident field by concentrating the local excitation intensity. Second, the close proximity of metal and fluorophores can increase the radiative decay rate of the fluorophores.

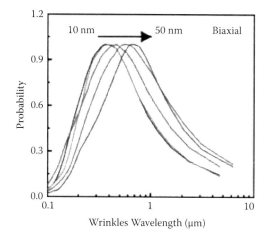

FIGURE 13.6
Wavelength distributions of biaxial wrinkles for varying gold thicknesses (10–50 nm).

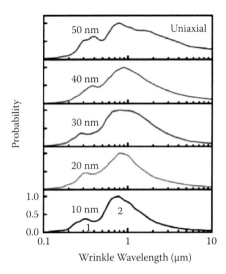

FIGURE 13.7
Wavelength distributions of uniaxial wrinkles for varying gold thicknesses (10–50 nm).

Furthermore, the increase in radiative decay rate is typically accompanied by an increase in quantum yields and a decrease in the lifetime of the fluorophores.[40–67]

The feasibility of immunoassay development utilizing MEF has been demonstrated by several research groups. Many of these studies utilized metal colloids, metal surfaces, metal mirrors, or metal islands. Among all these metal substrates, the highest-intensity enhancement has been observed for metal islands or colloids, rather than for continuous metal surfaces.[62,63] One popular approach is the use of silver island films (SIFs) on glass or metal mirror surfaces. Matveeva et al. were able to demonstrate 9.5- and 41.4-fold fluorescence intensity increases for SIFs on glass and SIFs on gold mirrors, respectively, using rabbit IgG immunoassay. They have also shown that with the same SIF substrates, the fluorescence intensity increases were 9.9- and 50.3-fold using myoglobin immunoassay.[66] In addition, with MEF technology, it is possible to detect myoglobin concentrations below 50 ng/ml, which is lower than the clinical cutoff for myoglobin in healthy patients.[67]

Although various types of metal surfaces have shown significant fluorescence intensity enhancement, there are limitations with these surfaces. Surface plasmons can be created on a continuous metal surface with p-polarized light. The plasmons are free to migrate in the direction parallel to the metal and dielectric interface. On a perfectly smooth metal surface, these plasmons will not be reemitted. On the other hand, if the propagating plasmon wave hits a metal particle, part of the energy can be reemitted as light.[62,63] Therefore, the fluorescence intensity increase is generally smaller for the continuous metal surface than that from a surface with irregularities. In addition, a dielectric-constant material is necessary to create surface plasmons on a continuous metal surface. However, the requirement of dielectric-constant material limits the choices for functionalizing the surface. Although SIF provides necessary topographical features for reemitting plasmons, silver has low chemical stability and limitations on its surface chemistry, which restrict its potential in biomedical applications.[41]

The ability to achieve nanometer-scale wrinkles is particularly useful in biomedical diagnostic devices, and by combining the metallic wrinkled nanostructures with microfluidics,

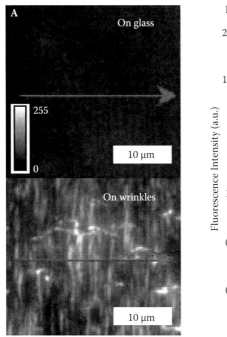

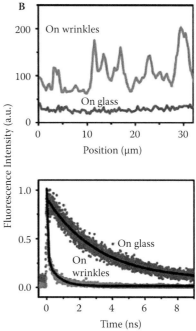

FIGURE 13.8

(See color insert.) (A) Wide-field epifluorescence images of dyes on glass plate (top) and uniaxial wrinkles (bottom). (B) The corresponding intensity profiles along the arrows in (A) (top) and fluorescence lifetime measurements of dyes on a glass plate (top line) and wrinkles (bottom line) (bottom).

LOC devices can be rapidly created. In particular, the nanowrinkles provide the ability for the metallic nanostructures to be used for MEF-based sensing applications.

To demonstrate the utility of the wrinkles for POC technology, dye molecules were dissolved in polymer solution and spin coated on the uniaxial gold wrinkles. Furthermore, parallel to the direction of the wrinkles, there were many hot spots, which indicated a five- to sevenfold enhancement relative to glass (Figure 13.8A).

In SPR, the fluorescence enhancement is accompanied by a decrease in the fluorescence lifetime. In order to confirm that the enhancement was due to plasmon effects as opposed to aggregation of the dyes, fluorescence lifetime measurements were performed with a confocal microscope. With a homemade confocal microscope, the average lifetime of the fluorescence dye was measured at 1/10 less than the average lifetime of the glass, suggesting that the enhancement resulted from strong interactions between the fluorophore and surface plasmons (Figure 13.8B).

Although the gold wrinkles exhibited only a moderate enhancement of fluorescent signal over glass, the enhancement could be improved in several ways, such as the utilization of bimetallic film instead of gold film alone. During the past decade, most of the MEF-based immunoassays have been developed on a single metal film such as gold or silver. However, evidence has shown that bimetallic nanostructures, comprised of gold and silver, demonstrate stronger surface plasmon resonance (SPR) effects than single pure metal.[41,68] The multimetallic alloy and core-shell nanostructures have remarkable catalytic, electronic, and magnetic properties.[69] Yoo et al. demonstrated that the bifrustum nanocrystals made of a gold core and a silver shell showed stronger SPR than pure gold nanoprisms. In addition to extremely strong SPR effects, Au/Ag alloy nanostructures hold a number of advantages over a single pure metal, including better plasmon tenability and a broader surface plasmon band for multiple-probe enhancement.[68]

Since the fluorescent signal intensity of MEF is partly based on the surface irregularity, it is conceivable that the enhancement could be amplified by using biaxial and bimetallic wrinkles. Our lab has recently demonstrated methods to create sharp bimetallic wrinkles for molecular detection using fluorescein. During the biaxial shrinkage of the PS sheet, the metal buckled and formed nanogaps. The edges of these nanogaps created hot spots, and bursts of intensity are emitted when a fluorescein molecule diffuses into a hot spot, and then the intensity drops right back to background. The fluorescent signal enhancement of such a mechanism was in the range of several thousand-fold over glass substrate (data not shown).

13.4 Better Shrinkage Than Polystyrene Sheet

With advances in polymer science, novel polymeric blends have become available for potential application in POC technology. Polyolefin (PO), a thermoplastic commonly used for the manufacturing of shrink-wrap films, possesses excellent shape memory properties that can be recovered through the application of heat. Based upon the blend of the polymer and processing condition, this elastomeric property can be tuned. Furthermore, PO films exhibit attractive optical properties, such as transmission over a large range of wavelengths, in addition to low autofluorescence.[70] Combined, these features make PO an attractive polymer for applications in microfluidics.

By leveraging the shrinkage property of the PO sheets, it was demonstrated that high-resolution microstructures can be generated, obviating the need for photolithography and advanced tooling. In addition, unlike PS sheets, which yielded a modest 40–60% increase in resolution postshrinkage, PO sheets achieved 95% reduction in surface area (Figure 13.9). By simple printing on the polymer sheets, defined ink patterns can be heated to generate high-aspect micromolds. In addition to the physical contraction of the polymer sheet, the patterned features are induced to form well-rounded features, which are difficult to achieve with standard photolithography. This process, which is analogous to the reflow of photoresist in standard photolithography, allows for the reflow of toner ink, creating well-rounded features. Further, the improvement in surface area reduction facilitated the formation of better aspect ratios for the microstructures (Figure 13.10).

The ability to integrate microarray technology and fluorescent protein analysis in microfluidic devices can greatly contribute to the µTAS for POC diagnostics. Recently, it has been shown that by printing protein patterns onto thermoplastics, it is possible to increase array density as well as protein concentration through the heat-induced shrinking of thermoplastics.[71] This method can be demonstrated through micro contact printing (µCP) of primary fluorescently labeled antibodies onto the polymer sheets, or through the sequential attachment of the primary antibody and secondary fluorophore-conjugated antibody.

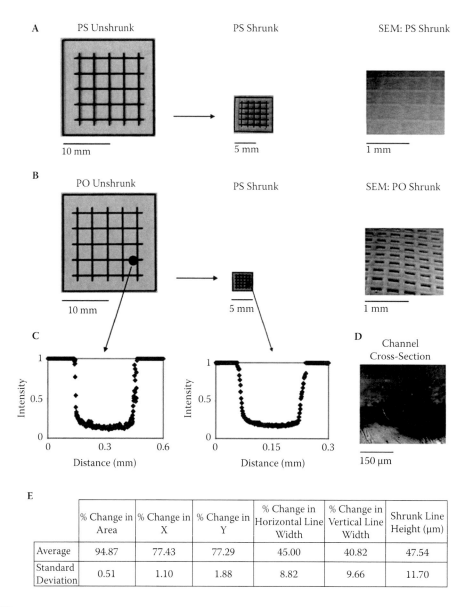

FIGURE 13.9
Characterization of thermoplastic shrinkage. It is apparent that the shrinkage of PO (B) is much greater than that of PS (A). The resulting enhancement in shrinkage of PO allowed for an improved aspect ratio, as shown in the SEM of the PO surface vs. that of PS. Postshrinkage, these patterned features can be used for the fabrication of microfluidic channels (C) with rounded channel shapes (D). The average surface area reduction of PO was approximately 95% in both the x and y directions (E).

Due to the heat-induced shrinking of the polymer, the protein arrays are pulled together, resulting in increased array density as well as antibody concentration. Thus, to fully utilize the size reduction of PO films, fluorescently tagged polyclonal antibodies were deposited onto PO films using standard µCP, whereby PDMS stamps containing the proteins were stamped onto the PO surface. To facilitate the uniform shrinking of the array, the PO

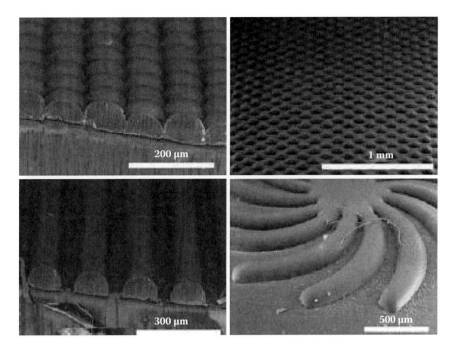

FIGURE 13.10
SEM of ink structures on PO films. Microstructures can be generated on PO films simply by printing (single print) onto the polymer sheets. These microstructures can be for the fabrication of microfluidic devices, negating the need for standard photolithography.

films were laminated onto nonshrinkable polyester backing using a high-tack, low-bond strength latex adhesive. This lamination process allowed for the PO film to slide along the backing and prevented contortion of the film. In addition, slowly applying heat, first at 115°C, then at 135°C, and finally at 155°C, prevented the delimitation of the PO sheet. From this process, we observe that the protein array exhibited a significant reduction in size as well as a dramatic increase in fluorescence (Figure 13.11).

The ability to generate self-assembled micro- and nanostructures plays an important role in the integration of biochips and POC diagnostics. An example of such an application is the regulation of a cellular microenvironment; specifically, substrate modification has been shown capable of eliciting specific cellular responses such as cell alignment.[72,73] Further, as a potential application of such structures in the μTAS, it has been shown that by altering these structures, it is possible to tune the surface plasmon effect, leading to the enhancement of fluorescent signals.[31] Thus, by utilizing the mismatch in the metallic film and the PO, it is possible to generate metallic nanostructures that are smaller than the previous reported wrinkles utilizing PS sheets[31] (Figure 13.12).

With the use of commercially available PO thermoplastic, it was demonstrated that the patterning of microstructures can be achieved, negating the use of standard hot embossing and photolithographic methods. These POs exhibited a 95% reduction in feature size upon the induction of heat and can be utilized for microfluidic applications. Further, by applying standard protein patterning and self-assembled nanostructures, PO films can be used to generate high-density microarrays, which can be applied to biochip designs for point of care diagnostics.

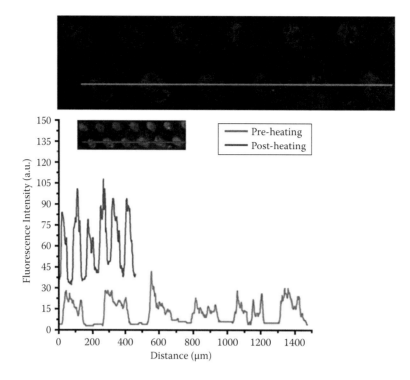

FIGURE 13.11
Fluorescent images of a 6 × 2 microarray of Alexa 555 conjugated anti-rabbit IgG. Through heat-induced shrinking, protein spots were concentrated and exhibited increased fluorescent signal.

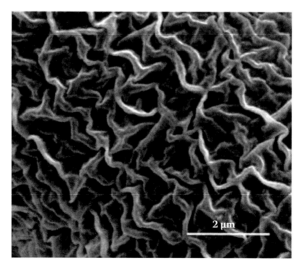

FIGURE 13.12
SEM image of self-assembled gold nanostructures on PO films.

References

1. Schulte, T.H., R.L. Bardell, and B.H. Weigl, Microfluidic technologies in clinical diagnostics, *Clin Chim Acta*, 2002, 321(1–2): 1–10.
2. Verpoorte, E., Microfluidic chips for clinical and forensic analysis, *Electrophoresis*, 2002, 23(5): 677–712.
3. Chin, C.D., V. Linder, and S.K. Sia, Lab-on-a-chip devices for global health: past studies and future opportunities, *Lab Chip*, 2007, 7(1): 41–57.
4. Dittrich, P.S., K. Tachikawa, and A. Manz, Micro total analysis systems. Latest advancements and trends, *Anal Chem*, 2006, 78(12): 3887–908.
5. Voldman, J., M.L. Gray, and M.A. Schmidt, Microfabrication in biology and medicine, *Annu Rev Biomed Eng*, 1999, 1: 401–25.
6. Lee, S.J., and S.Y. Lee, Micro total analysis system (micro-TAS) in biotechnology, *Appl Microbiol Biotechnol*, 2004, 64(3): 289–99.
7. Cheng, X., et al., A microchip approach for practical label-free CD4+ T-cell counting of HIV-infected subjects in resource-poor settings, *J Acquir Immune Defic Syndr*, 2007, 45(3): 257–61.
8. Lancaster, C., et al., Rare cancer cell analyzer for whole blood applications: microcytometer cell counting and sorting subcircuits, *Methods*, 2005, 37(1): 120–27.
9. VanDelinder, V., and A. Groisman, Perfusion in microfluidic cross-flow: separation of white blood cells from whole blood and exchange of medium in a continuous flow, *Anal Chem*, 2007, 79(5): 2023–30.
10. Chen, X., et al., Continuous flow microfluidic device for cell separation, cell lysis and DNA purification, *Anal Chim Acta*, 2007, 584(2): 237–43.
11. Panaro, N.J., et al., Micropillar array chip for integrated white blood cell isolation and PCR, *Biomol Eng*, 2005, 21(6): 157–62.
12. Kokoris, M., et al., Rare cancer cell analyzer for whole blood applications: automated nucleic acid purification in a microfluidic disposable card, *Methods*, 2005, 37(1): 114–19.
13. Helton, K.L., and P. Yager, Interfacial instabilities affect microfluidic extraction of small molecules from non-Newtonian fluids, *Lab Chip*, 2007, 7(11): 1581–88.
14. Yager, P., et al., Microfluidic diagnostic technologies for global public health, *Nature*, 2006, 442(7101): 412–18.
15. Yager, P., G.J. Domingo, and J. Gerdes, Point-of-care diagnostics for global health. *Annu Rev Biomed Eng*, 2008, 10: 107–44.
16. Xia, Y., and G.M. Whitesides, Soft lithography, *Annu Rev Mater Sci*, 1998, 28: 153–84.
17. Mukhopadhyay, R., When PDMS isn't the best. What are its weaknesses, and which other polymers can researchers add to their toolboxes? *Anal Chem*, 2007, 79(9): 3248–53.
18. Toepke, M.W., and D.J. Beebe, PDMS absorption of small molecules and consequences in microfluidic applications, *Lab Chip*, 2006, 6(12): 1484–6.
19. Duffy, D.C., J.C. McDonald, O.J.A. Schueller, and G.M. Whitesides, Rapid prototyping of microfluidic systems in poly(dimethylsiloxane), *Anal Chem*, 1998, 70: 4974–84.
20. Kaur, J., et al., Direct hapten coated immunoassay format for the detection of atrazine and 2,4-dichlorophenoxyacetic acid herbicides, *Anal Chim Acta*, 2008, 607(1): 92–99.
21. Koerner, T., L. Brown, R. Xie, and R.D. Oleschuk, Epoxy resins as stamps for hot embossing of microstructures and microfluidic channels, *Sensors Actuators B Chem*, 2005, 107(2): 632–39.
22. Sudarsan, A.P., J. Wang, and V.M. Ugaz, Thermoplastic elastomer gels: an advanced substrate for microfluidic chemical analysis systems, *Anal Chem*, 2005, 77(16): 5167–73.
23. Huang, Y., et al., Solvent resistant microfluidic DNA synthesizer, *Lab Chip*, 2007, 7(1): 24–26.
24. Fiorini, G.S., et al., Fabrication improvements for thermoset polyester (TPE) microfluidic devices, *Lab Chip*, 2007, 7(7): 923–26.

25. Zhao, X.M., Y. Xia, O.J.A. Schuller, D. Qin, and G.M. Whitesides, Fabrication of polymeric microstructures with high aspect ratios using shrinkable polystyrene films, *Adv Mater*, 1997, 9(3): 251–54.

26. Zhao, X.M., Y. Xia, O.J.A. Schuller, D. Qin, and G.M. Whitesides, Fabrication of microstructures using shrinkable polystyrene films, *Sensors Actuators A Phys*, 1998, 65: 209–17.

27. Chen, C.S., et al., Shrinky-Dink microfluidics: 3D polystyrene chips, *Lab Chip*, 2008, 8(4): 622–24.

28. Grimes, A., et al., Shrinky-Dink microfluidics: rapid generation of deep and rounded patterns, *Lab Chip*, 2008, 8(1): 170–72.

29. Long, M., M.A. Sprague, A. Grimes, B.D. Rich, M. Khine, A simple three-dimensional vortex micromixer, *Appl Phys Lett*, 2009, 94: 133501–3.

30. Nguyen, D., D. Taylor, K. Qian, N. Norouzi, J. Rasmussen, S. Botzet, M. Lehmann, K. Halverson, and M. Khine, Better shrinkage than Shrinky-Dinks, *Lab Chip*, 2010, DOI(10.1039/c001082k).

31. Fu, C.C., A. Grimes, M. Long, C.G.L. Ferri, B.D. Rich, S. Ghosh, L.P. Lee, A. Gopinathan, and M. Khine, Tunable nanowrinkles on shape memory polymer sheets, *Adv Mater*, 2009, 21(44): 4472.

32. Stafford, C.M., et al., A buckling-based metrology for measuring the elastic moduli of polymeric thin films, *Nat Mater*, 2004, 3(8): 545–50.

33. Efimenko, K., et al., Nested self-similar wrinkling patterns in skins, *Nat Mater*, 2005, 4(4): 293–97.

34. Bowden, N., S. Brittain, A.G. Evans, J.W. Hutchinson, and G.M. Whitesides, Spontaneous formation of ordered structures in thin films of metals supported on an elastomeric polymer, *Nature*, 1998, 393: 146.

35. Lacour, S.P., S. Wagner, Z.Y. Huang, and Z. Suo, Stretchable gold conductors on elastomeric substrates, *Appl Phys Lett*, 2003, 82: 2404.

36. Urdaneta, M.G., R. Delille, and E. Smela, Stretchable electrodes with high conductivity and photo-patternability. *Adv Mater*, 2007, 19: 2629.

37. Huck, W.T.S., N. Bowden, P. Onck, T. Pardoen, J. W. Hutchinson, and G. M. Whitesides, Ordering of spontaneously formed buckles on planar surfaces, *Langmuir*, 2000, 16: 3497.

38. Watanabe, M., and T. Hirai, Polypyrrole film with striped pattern, *Polym Sci B Polym Phys*, 2004, 42: 2460.

39. Yoo, P.J., K.Y. Suh, S.Y. Park, and H.H. Lee, Physical self-assembly of microstructures by anisotropic buckling, *Adv Mater*, 2002, 14: 1383.

40. Matveeva, E.G., Z. Gryczynski, et al., Myoglobin immunoassay based on metal particle-enhanced fluorescence, *J Immunol Methods*, 2005, 302(1–2): 26–35.

41. Zhang, J., and J.R. Lakowicz, Metal-enhanced fluorescence of an organic fluorophore using gold particles, *Opt Express*, 2007, 15: 2598–606.

42. Aslan, K., et al., Metal-enhanced fluorescence from plastic substrates, *J Fluoresc*, 2005, 15(2): 99–104.

43. Aslan, K., and C.D. Geddes, Microwave-accelerated metal-enhanced fluorescence: platform technology for ultrafast and ultrabright assays, *Anal Chem*, 2005, 77(24): 8057–67.

44. Aslan, K., and C.D. Geddes, Microwave-accelerated and metal-enhanced fluorescence myoglobin detection on silvered surfaces: potential application to myocardial infarction diagnosis, *Plasmonics*, 2006, 1(1): 53–59.

45. Aslan, K., and C.D. Geddes, Microwave-accelerated metal-enhanced fluorescence (MAMEF): application to ultra fast and sensitive clinical assays, *J Fluoresc*, 2006, 16(1): 3–8.

46. Aslan, K., et al., Metal-enhanced fluorescence: an emerging tool in biotechnology, *Curr Opin Biotechnol*, 2005, 16(1): 55–62.

47. Aslan, K., K.E. Holley, and C.D. Geddes, Microwave-accelerated metal-enhanced fluorescence (MAMEF) with silver colloids in 96-well plates: application to ultra fast and sensitive immunoassays, high throughput screening and drug discovery, *J Immunol Methods*, 2006, 312(1–2): 137–47.

48. Aslan, K., et al., Metal-enhanced fluorescence-based RNA sensing, *J Am Chem Soc*, 2006, 128(13): 4206–7.

49. Aslan, K., J.R. Lakowicz, and C.D. Geddes, Rapid deposition of triangular silver nanoplates on planar surfaces: application to metal-enhanced fluorescence, *J Phys Chem B*, 2005, 109(13): 6247–51.

50. Aslan, K., J.R. Lakowicz, and C.D. Geddes, Metal-enhanced fluorescence using anisotropic silver nanostructures: critical progress to date, *Anal Bioanal Chem*, 2005, 382(4): 926–33.

51. Aslan, K., et al., Fast and slow deposition of silver nanorods on planar surfaces: application to metal-enhanced fluorescence, *J Phys Chem B*, 2005, 109(8): 3157–62.

52. Aslan, K., et al., Annealed silver-island films for applications in metal-enhanced fluorescence: interpretation in terms of radiating plasmons, *J Fluoresc*, 2005, 15(5): 643–54.

53. Aslan, K., et al., Microwave-accelerated metal-enhanced fluorescence: an ultra-fast and sensitive DNA sensing platform, *Analyst*, 2007, 132(11): 1122–29.

54. Aslan, K., S.N. Malyn, and C.D. Geddes, Fast and sensitive DNA hybridization assays using microwave-accelerated metal-enhanced fluorescence, *Biochem Biophys Res Commun*, 2006, 348(2): 612–17.

55. Aslan, K., S.N. Malyn, and C.D. Geddes, Metal-enhanced fluorescence from gold surfaces: angular dependent emission, *J Fluoresc*, 2007, 17(1): 7–13.

56. Aslan, K., S.N. Malyn, and C.D. Geddes, Angular-dependent metal-enhanced fluorescence from silver island films, *Chem Phys Lett*, 2008, 453(4–6): 222–28.

57. Aslan, K., Y. Zhang, and C.D. Geddes, Sonication-assisted metal-enhanced fluorescence-based bioassays, *Anal Chem*, 2009, 81(12): 4713–19.

58. Aslan, K., et al., Microwave-accelerated metal-enhanced fluorescence: application to detection of genomic and exosporium anthrax DNA in <30 seconds, *Analyst*, 2007, 132(11): 1130–38.

59. Geddes, C.D., et al., Metal-enhanced fluorescence: potential applications in HTS, *Comb Chem High Throughput Screen*, 2003, 6(2), 109–17.

60. Geddes, C.D., et al., Roughened silver electrodes for use in metal-enhanced fluorescence, *Spectrochim Acta A Mol Biomol Spectrosc*, 2004, 60(8–9): 1977–83.

61. Gryczynski, Z., et al., Metal-enhanced fluorescence: a novel approach to ultra-sensitive fluorescence sensing assay platforms, *Proc Soc Photo Opt Instrum Eng*, 2004, 5321(275): 275–282.

62. Lakowicz, J.R., Radiative decay engineering: biophysical and biomedical applications, *Anal Biochem*, 2001, 298(1): 1–24.

63. Lakowicz, J.R., Radiative decay engineering 5: metal-enhanced fluorescence and plasmon emission, *Anal Biochem*, 2005, 337(2): 171–94.

64. Lakowicz, J.R., et al., Advances in surface-enhanced fluorescence, *J Fluoresc*, 2004, 14(4): 425–41.

65. Pribik, R., et al., Metal-enhanced fluorescence (MEF): physical characterization of silver-island films and exploring sample geometries, *Chem Phys Lett*, 2009, 478(1–3): 70–74.

66. Matveeva, E.G., et al., Metal particle-enhanced fluorescent immunoassays on metal mirrors, *Anal Biochem*, 2007, 363(2): 239–45.

67. Matveeva, E.G., Z. Gryczynski, and J.R. Lakowicz, Myoglobin immunoassay based on metal particle-enhanced fluorescence, *J Immunol Methods*, 2005, 302(1–2): 26–35.

68. Yoo, H.J., J. E. Millstone, et al., Core-shell triangular bifrustums, *Nano Lett*, 2009, 9(8): 3038–41.

69. Ferrando, R., J. Jellinek, et al., Nanoalloys: from theory to applications of alloy clusters and nanoparticles, *Chem Rev*, 2008, 108(3): 845–910.

70. Torres, A., N. Colls, and F. Mendez, Properties predictor for HDPE/LDPE/LLDPE blends for shrink film applications, *J Plastic Film Sheeting*, 2006, 22(1): 29–37.

71. Sollier, K., et al., "Print-n-shrink" technology for the rapid production of microfluidic chips and protein microarrays, *Lab Chip*, 2009, 9(24): 3489–94.

72. Vernon, R.B., et al., Microgrooved fibrillar collagen membranes as scaffolds for cell support and alignment, *Biomaterials*, 2005, 26(16): 3131–40.

73. Manbachi, A., S. Shrivastava, M. Cioffi, B. G. Chung, M. Moretti, U. Demirci, M. Yliperttulaa, and A. Khademhosseini, Microcirculation within grooved substrates regulates cell positioning and cell docking inside microfluidic channels, *Lab Chip*, 2008, (8): 747.

14

Electric Fields for Vaccine Delivery: How in Vivo *Electroporation Could Fulfill the Promise of DNA Vaccines*

Andrew Y. Choo, Amir S. Khan, Niranjan Y. Sardesai, and J. Joseph Kim
Inovio Pharmaceuticals, Blue Bell, Pennsylvania, USA

CONTENTS

14.1 Immunization: A Revolutionary Approach to Treating Diseases

One of the greatest achievements in modern medicine has been the development of vaccines that provide immunity against pathogens by stimulating the body's immune system to recognize an agent as foreign and by developing "memory" immune responses to reduce the severity of or prevent subsequent infections. This concept of introducing the immune system to agents that mimic specific pathogens was demonstrated over

200 years ago by Edward Jenner. In 1776, Jenner tested the hypothesis that prior cowpox infection could spare a person from the ravages of smallpox. The origin of the hypothesis is from the observation that milkmaids, who were highly exposed to cowpox, often did not develop smallpox infection. By showing that exposure to cowpox protected an 8-year-old boy from smallpox, Jenner revolutionized medicine by demonstrating a legitimate method to prevent diseases.[1] Eventually, Jenner's work paved the way for the eradication of smallpox in 1979. Currently, a number of different vaccines have been approved for immunization and distribution in the United States, including those for diphtheria, hepatitis A and B, human papillomavirus (HPV), influenza, polio, rabies, and others (Table 14.1). Therefore, the consequence of vaccine development is one of the most impactful medical and economical discoveries of the modern era. Not surprisingly, vaccine development has become an intense area of investigation in both academia and industry with the rise in prevalence of infectious diseases such as HIV; the increase in the incidence of newly emergent strains of pathogens, such as dengue, chikungunya, and novel influenza; the increase in the incidence of cancer; and the heightened threat of bioterrorism attacks.

14.1.1 Types of Vaccines: Advantages and Disadvantages

Jenner's early work on the smallpox vaccine utilized cowpox because its infection is attenuated in humans and because the immune response generated from cowpox exposure is sufficient to protect against infection from smallpox. However, not all pathogens can be so easily targeted, in terms of both safety and immunogenicity. Some types of virus may be safe and well tolerated, but may not generate an immune response that can sufficiently prevent infection. This latter aspect is predicated on a number of factors, including the ability of a vaccine to generate different arms of the immune system (antibody vs. T cell) and the expression level of the different types of antigen. Since the ultimate goal of vaccines is to mimic the targeted pathogen—at least in terms of inducing an immune response—several different approaches have been used to achieve this effect.

Most of the vaccines that are currently approved for use in humans are live, attenuated organisms and killed or inactivated organisms (see Table 14.1). Since attenuated organisms are no longer pernicious or infectious, injection into a host usually induces immune responses in the form of antigen-specific antibodies or antigen-specific T cells. However, a concern with live, attenuated organisms, which are generated under conditions that disable their virulent properties, is that the organism may revert back to a virulent form upon introduction into the host, or that ineffective attenuation of the virus may result in transmission of the disease. For example, an attenuated or killed HIV is unlikely to be pragmatic, considering reversion to a virulent form via mutations or extreme sensitivity of the host to attenuated viruses via differences in genetics would be catastrophic.[2] Such a scenario has been observed with the use of early smallpox vaccines, and this would limit its use for patients with immunodeficiency syndromes or conditions.[3]

However, there are benefits to vaccines with such potential pitfalls, most notably the ability of attenuated vaccines to stimulate both cellular and humoral arms of the immune system. Because a live vaccine, like the pathogen of interest, can replicate, albeit ineffectively, in the host, similar types of immune responses are often induced. So antigens produced from attenuated viruses and natural infection present peptides for MHC class I and II presentation to activate CD4+ and CD8+ T cells. Therefore, both the type and magnitude

TABLE 14.1

Vaccines Approved for Use in the United States

Disease/Pathogen	Vaccines
Anthrax	AVA (BioThrax®)
Chicken pox (varicella)	VAR (Varivax®)
	MMRV (ProQuad®)
Diphtheria	DTaP (Daptacel,® Infanrix,® Tripedia®)
	Td (Decavac,® generic®)
	DT (generic®)
	Tdap (Boostrix,® Adacel®)
	DTaP-IPV (Kinrix®)
	DTaP-HepB-IPV (Pediarix®)
	DTaP-IPV/Hib (Pentacel®)
	DTaP/Hib (TriHIBit®)
Hepatitis A	HepA (Havrix,® Vaqta®)
	HepA-HepB (Twinrix®)
Hepatitis B	HepB (Engerix-B,® Recombivax HB®)
	Hib-HepB (Comvax®)
	DTaP-HepB-IPV (Pediarix®)
	HepA-HepB (Twinrix®)
HIB	Hib (ActHIB, PedvaxHIB,® Hiberix®)
	Hib-HepB (Comvax®)
	DTaP/Hib (TriHIBit®)
	DTap-IPV/Hib® (Pentacel®)
HPV	HPV4 (Gardasil®)
	HPV2 (Cervarix®)
Influenza (seasonal)	TIV (Afluria,® Agriflu,® FluLaval,® Fluarix,® Fluvirin,® Fluzone®)
	LAIV (FluMist®)
Japanese encephalitis	JE (Ixiaro,® JE-Vax®)
Measles	MMR (M-M-R® II)
	MMRV (ProQuad®)
Meningitis	MCV4 (Menactra®)
	MPSV4 (Menomune®)
Mumps	MMR® (M-M-R® II)
	MMRV (ProQuad®)
Pertussis	DTaP (Daptacel,® Infanrix,® Tripedia®)
	Tdap (Adacel,® Boostrix®)
	DTaP-IPV (Kinrix®)
	DTaP-HepB-IPV (Pediarix®)
	DTaP-IPV/Hib (Pentacel®)
	DTaP/Hib (TriHIBit®)
Pneumococcal	PCV7 (Prevnar®)
	PPSV23 (Pneumovax® 23)

(*continued*)

TABLE 14.1 (CONTINUED)

Vaccines Approved for Use in the United States

Disease/Pathogen	Vaccines
Polio	Polio (Ipol®)
	DTaP-IPV (Kinrix®)
	DTaP-HepB-IPV (Pediarix®)
	DTaP-Hep-IPV (Pentacel®)
Rabies	Rabies (Imovax® Rabies,® RabAvert®)
Rotavirus	RV1 (Rotarix®)
	RV5 (RotaTeq®)
Rubella	MMR (M-M-R® II)
	MMRV (ProQuad®)
Shingles	ZOS (Zostavax®)
Smallpox	Vaccinia (ACAM2000,® Dryvax®)
Tetanus	DTaP (Daptacel,® Infanrix,® Tripedia®),
	Td (Decavac,® generic®)
	DT (generic®)
	TT (generic®)
	Tdap (Boostrix,® Adacel®)
	DTaP-IPV (Kinrix®)
	DTaP-HepB-IPV (Pediarix®)
	DTaP-IPV/Hib (Pentacel®)
	DTaP/Hib (TriHIBit®)
Tuberculosis	BCG (TICE® BCG,® Mycobax®)
Typhoid	Typhoid oral (Vivotif®)
	Typhoid polysaccharide (Typhim Vi®)
Yellow fever	YF (YF-Vax®)

of immune response induced by the vaccine could closely resemble those of the natural infection by generating antigen-specific antibodies and T cells.[4] Currently, live, attenuated vaccines are available in the United States for pathogens such as yellow fever, rotavirus, and varicella (chicken pox).

The other type of vaccine is killed/subunit vaccines, which are clearly advantageous over attenuated vaccines in terms of safety. These vaccines are either killed/inactivated organisms or a specific protein from a pathogen with antigenic properties, and therefore cannot replicate in the inoculated host. The obvious benefit is safety, as concerns for reversion to a virulent form are minimal with killed vaccines and impossible with subunit vaccines.[4] However, the disadvantage is that these vaccines induce minimal cellular responses in the form of CD8+ T cells because the antigens are extracellular; these antigens are taken up by antigen-presenting cells (APCs) by phagocytosis or endocytosis to be degraded and presented as peptides on MHC class II molecules, which do not stimulate CD8+ T cells (see Vyas et al.[5] for details). While killed/subunit vaccines often stimulate strong antigen-specific antibodies, the ability to activate only one arm of the immune system limits their efficacy; for example, pathogens that require CD8+ T cells to clear infection, such as chronic infections like human immunodeficiency virus (HIV) and hepatitis C virus (HCV), would not be effectively cleared with killed/subunit vaccines.[6]

14.2 DNA Vaccines

14.2.1 The Birth of DNA Vaccination

In the early 1990s, reports from a number of different groups paved the foundation for DNA vaccines. Stephen Johnston and colleagues described the ability of DNA expression plasmids coated onto gold beads and delivered into mice to drive the expression of a foreign protein and stimulate an antibody response.[7] Concurrently, Margaret Liu and colleagues reported on the induction of antibodies and cellular immune responses to influenza A in mice[8]; Harriet Robinson and her group demonstrated protection of mice from influenza by delivering influenza DNA[9]; and David Weiner and colleagues reported that intramuscular injection of HIV antigen-encoding plasmids induced antigen-specific cellular and humoral immune responses.[10] Although a number of previous reports demonstrated the ability of DNA plasmids to be transfected in vivo, these reports were the first to demonstrate an immunological consequence, and have become the foundation for the field of DNA vaccines, which stimulate pathogen-specific immune responses by using the host's cells to express the antigen (see Chattergoon et al.[11] for more details).

14.2.2 Differences between Traditional Vaccines and DNA Vaccines

There are several advantages with DNA vaccines when compared to the aforementioned traditional vaccines. Since DNA vaccines are simply expression plasmids that encode specific antigens, there are no fears of infection from the vaccine. Gene expression from these plasmids is typically driven by the cytomegalovirus (CMV) immediate promoter, and the plasmid includes other elements, such as the transcription termination signal and a prokaryotic antibiotic resistance gene.[4] Moreover, safety studies in humans suggest minimal evidence for plasmid integration into the host's genome, and almost all of the plasmids are cleared from the body, with only a small amount remaining at the site of injection. Over a 2-month experiment conducted by Sheets et al., only a small number of the immunized animals retained the plasmids at the site of injection (10%–20%), with only about ~100 copies of plasmid per site.[12] In addition, there is limited concern for biodistribution, as nontransfected plasmids are rapidly degraded by highly active nucleases, reducing agents, and charged molecules in the blood; in fact, the half-life of DNA plasmids in blood is approximately 2–3 minutes.[13] Therefore, DNA vaccines provide minimal toxicity because the excess DNA, or nontransfected DNA, is degraded and eliminated rapidly.

A distinction between DNA vaccines and killed/subunit vaccines, which are also noninfectious, is that DNA vaccines can mimic the immunological effects of attenuated vaccines. Since DNA vaccines directly transfect the host's cells, gene expression occurs by using the host's machinery—transcription factors, ribosomes, mRNP, etc.—allowing for antigen presentation through both the MHC class I and II pathways.[4] As a result, DNA vaccines induce both antigen-specific antibodies and T cells, unlike killed/subunit vaccines. Consequently, DNA vaccines, like attenuated vaccines, are not limited by the type of immune response needed for the host to clear infection. For example, subunit vaccines rarely generate sufficient cellular immune response required for clearing viral infections such as HCV and HPV. This is one likely reason why HPV preventative vaccines, Gardasil® (Merck & Co.) and Cervarix® (GlaxoSmithKline), fail to clear preexisting infections, as these vaccines work primarily through generating antibodies (see Figure 14.1 for mechanism; see Table 14.2 for differences between vaccines).

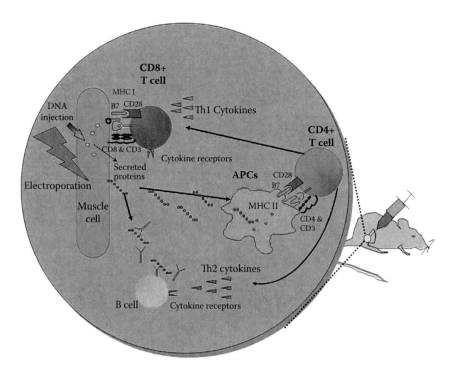

FIGURE 14.1

DNA vaccines activate both Th1 and Th2 immune responses. Injection of DNA and subsequent electroporation increases plasmid transfection into muscle cells. The plasmids, which express an antigen of interest, are
expressed in the host's muscle cells, and the antigen peptides are presented via MHC class I presentation,
leading to CD8$^+$ T cell activation. The antigens can also be secreted by the muscle cells and taken up by antigen-
presenting cells, leading to MHC class II presentation and CD4$^+$ T cell activation. In addition, coexpression of
cytokine plasmids can drive Th1 or Th2 immune responses.

Another advantage with DNA vaccines is that they allow for the engineering of immune
responses with adjuvants. By simply coimmunizing the DNA vaccine with plasmids that
encode certain cytokines, one can skew the vaccine-induced response by altering the
microenvironment of antigen presentation. For example, coexpressing Th1-driving cytokines such as interleuken (IL)-12 plasmids dramatically increases the magnitude of the
antigen-specific T cell response. As shown in Figure 14.1, coexpression of cytokines can

TABLE 14.2

Comparison of Live Attenuated, Subunit/Killed, and DNA Vaccines

	Live Attenuated	Subunit/Killed	DNA
Humoral response	Yes	Yes	Yes
Cellular response	Yes	Minimal to none	Yes
Antigen presentation	Class I and II	Class II	Class I and II
Safety (potential infection?)	Yes	No, but inactivation must be complete	No
Production (cost and ease?)	Expensive and difficult (virus must be attenuated)	Easier to develop than live attenuated (protein, thus strict storage conditions)	Relatively inexpensive and easy to manufacture (DNA is stable)

influence not only the magnitude of response, but also the type of immune response. (See Kim and Weiner[14] for a more detailed analysis.)

14.2.3 Early Evaluation of DNA Vaccines in Human

In mice and smaller organisms, DNA vaccines are highly effective and generate strong and often protective immune responses against various pathogens. Consequently, much effort has been made in expanding the use of DNA vaccines in humans. The earliest DNA vaccine trials in humans investigated the safety of intramuscularly injecting patients with plasmid DNA, and these studies all suggested that plasmids were well tolerated with no measurable anti-DNA immune response.[15,16] However, unlike the magnitude of immune response that was generated in mice, the magnitude in humans was minimal at best. In one study, HIV-negative volunteers were immunized with an HIV *Env/Rev* DNA vaccine, and the authors reported that patients receiving the highest dose did exhibit vaccine-induced immune responses. However, the effects were often transient, and the magnitude of the CD8[+] T cell response was negligible, with fewer than 35 spot-forming units (SFU) in an interferon (IFN)-γ ELISpot assay.[15] These early studies all suggested that DNA immunization may stimulate antigen-specific immune responses in humans, but the magnitude of response was far from that of attenuated vaccines, and definitely not within range to provide protection, suggesting the need for more potent and effective second-generation vaccines.

Although many factors could affect the efficacy of DNA vaccines in humans, the primary reason appears to be inefficient transfection of muscles cells. Upon DNA injection into the muscle, the plasmids are rapidly taken up by the muscle cells, which then express the antigens. Although the plasmids may also directly transfect APCs, this mechanism is not the primary one for priming the immune system. Several pieces of evidence support this notion; first, upon intramuscular injection, the injected DNA is detected primarily at the site of injection, and any DNA detected at other sites, such as the draining lymph nodes, does not confer gene expression in terms of both mRNA and protein expression.[17] Therefore, it's more likely that DNA detected in these regions is likely from APCs that have taken up dead, plasmid-containing cells or degraded extracellular DNA. A more cogent piece of evidence is that DNA plasmids whose promoters are muscle specific are equally as effective as plasmids whose promoters are ubiquitously active in stimulating the immune system.[18] Therefore, even if plasmids do get transfected directly into APCs, the ability of the APCs to actually express the protein is not critical for stimulating the immune system.

So if DNA transfection of muscle cells is important, is there a difference in the mode or efficiency of transfection in smaller animals versus larger animals? Dupuis et al. investigated this exact question by performing two distinct experiments. First, the authors altered the total volume of the vaccine while maintaining the absolute DNA amount; the difference in volume would control DNA dispersion in the tissue following injection and alter the hydrostatic pressure caused by the injection. This experiment would address whether vaccine volume could play a role in determining vaccine efficiency, considering the volume used in mice could never be scaled up to be used in humans and macaques. Second, the authors also electroporated the tissue following injection to increase DNA uptake. Interestingly, these authors demonstrated that injecting DNA at a lower concentration, or higher volume, increased the efficiency of plasmid transfection and vaccine-induced immune response.[17,19] The exact mechanism(s) for increased DNA uptake by the muscle cells is not known, and the importance of this phenomenon in the context of DNA immunization in humans remains unknown. Confounding to the volume hypothesis is that a number of different factors, including salt concentration, pH, and temperature of the vaccine, may all affect the efficiency of DNA transfection.[20] Nevertheless, this study raised concerns and gave credence to the argument that inefficient DNA uptake is limiting vaccine immunogenicity in humans. But perhaps the more important observation that Dupuis et al. made was that electroporation of the tissue immediately following injection could ameliorate the decrease in immune response observed with DNA injection alone. Electroporation, which increases DNA uptake, improved gene expression of the vaccine and augmented both cellular and humoral immune responses.

14.3 Electroporation of DNA Vaccines

14.3.1 Mechanism of Electroporation

In 1982, Neumann et al. demonstrated that pulsing murine cells with electric impulses (8 kB/cm, 5 microseconds) dramatically increased the uptake of linear or circular plasmids in culture.[21] This protocol was termed electroporation. Neumann and colleagues proposed the development of large hydrophobic pores via alterations to the transmembrane voltage induced by the electric pulse allowed for DNA to be taken up with increased efficiency. The enlarged pores become hydrophilic by rotation of lipids between the aqueous medium and the lipid bilayer, allowing for passage of normally impermeable macromolecules such as DNA. A primary limitation with this model, however, is that evidence for these stable hydrophilic pores is lacking. To date, there is not a proven mechanism for electroporation, although the process likely involves the reversible change in the impermeability of the cell membrane to allow objects, such as drugs, nucleic acids, and antibodies, to enter the cell. A point of highlight, however, is that electroporation is a reversible process inasmuch as electric pulse-induced alternations to the cell, such as permeability of ions, are transient. This is clearly distinct from irreversible electroporation, which leads to cell death following permanent changes to membrane physiology.[22]

Although more details need to be filled in, a working model for electroporation has been proposed; Teissie et al. have described a series of events or steps that are involved in electropermeabilization.[23] The first is the trigger step, whereby the external field induces an increase in the transmembrane potential up to the critical permeabilizing threshold. This is followed by the expansion step, where a time-dependent membrane transition takes place as long as the field is maintained at an overcritical value. In the stabilization phase, the membrane organization is recovered as soon as the field is subcritical, leading to the resealing and memory steps. In these steps, membrane permeability recovers to its prior state, commencing the reversing process. The recovery and reversion back to a prior state are critical to induce homeostasis and prevent cell death. (See Teissie et al.[23] for more thorough and mechanistic insight into this process.).

14.3.2 Increased Immunogenicity with Electroporation

In 1990, Wolff et al. made the seminal observation that injection of naked DNA was sufficient to induce gene expression in vivo.[24] As mentioned previously, this observation then paved the way for DNA plasmids that encode various antigens to induce immune responses.

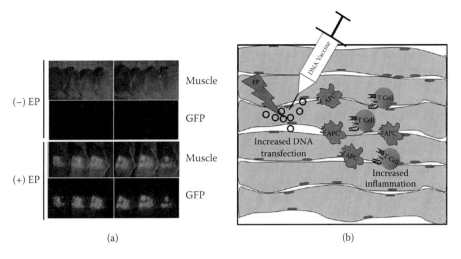

FIGURE 14.2
Electroporation increases DNA transfection and inflammation. (a) GFP expression plasmids were injected into the quadriceps muscle of mice with or without electroporation. (b) Cartoon showing the dual effect of electroporation: increased DNA transfection and inflammation. Electroporation of DNA plasmids leads to an increase in the recruitment of antigen-presenting cells, B cells, and T cells.

However, a clear limitation in the early clinical evaluation of DNA vaccines was limited gene expression. So how was the field going to evolve and overcome this limitation.

Neumann et al. made the groundbreaking observation that pulsing cells in culture with electric currents dramatically improved DNA uptake, leading to increased gene expression.[21] And applying this further, in 1999, Mir et al. demonstrated that electric pulses also improved plasmid uptake in vivo.[25] As shown in Figure 14.2a, gene delivery via electroporation leads to increased DNA expression by several orders of magnitude in skeletal muscles of mice, as well as rats, rabbits, and monkeys. These observations, coupled with poor immunogenicity data in larger animals, transitioned the DNA vaccine field into the electroporation platform.

The initial study that illuminated the connection between electroporation and limited DNA transfection in vivo was from Dupuis et al.[17] These authors provided lucid data showing increased gene expression and immune responses with electroporation. One of the factors that contributed to increased immune response is the increase in protein expression; clearly, in most cases, increased antigen translates to increased immune response as a result of increased antigen presentation and T cell activation. But another consequence of pulsing skeletal muscles with electricity may be involved; upon electroporation, some of the muscle cells undergo cell death, likely from irreversible alterations to membrane physiology. The consequence of this type of tissue damage is inflammation, likely via sending "danger" signals to the immune system. Dying cells, typically through necrotic means, release various proinflammatory inducers, such as HMGB1, leading to maturation of dendritic cells and increased antigen presentation.[26] As a result, increased muscle damage from electroporation may augment ancillary factors to promote inflammation. In fact, this has been observed experimentally: Ahlen et al. reported a clear increase in the number of CD3+ cells at the site of injection[27]; Liu et al. reported an amazing 45-fold increase in the number of macrophages, a 77-fold increase in the number of dendritic cells, and a 2- to 6-fold increase in the number of B and T cells at the site of injection,[28] supporting the hypothesis that electroporation itself induces inflammation.

However, there is some synergy, specifically between DNA injection and electroporation, which may aid in the overall induction of the immune system. For example, electroporation following peptide immunization did not increase vaccine-specific immune responses, suggesting that electroporation alone is not sufficient to augment vaccine potency.[19,29] The exact nature of this synergy is not completely understood. The vaccine plasmids, which are bacterial expression vectors, contain unmethylated CpG sequences, which activate immune cells through Toll-like receptor 9 (TLR9).[30] However, mice deficient in TLR9 are equally as effective as the wild-type control mice are in responding to DNA immunizations; thus, DNA vaccination with electroporation can stimulate the immune system through TLR9-independent mechanisms.[30] Another reason for the DNA/electroporation synergy may be that the kinetics of DNA immunization and electroporation cooperate more effectively than do the kinetics of peptide and electroporation. Antigen expression from plasmids takes a few hours, allowing electroporation-induced inflammation and antigen expression to overlap, while peptide injection provides antigen in a ready form.[29] Although a clear mechanism is preliminary, a combination of these effects—increased transfection and synergy of gene expression and inflammation—is likely important for increasing the immunogenicity of DNA vaccines via electroporation (Figure 14.2b).

14.3.3 Electroporation of DNA Vaccines Increases Immunogenicity in Larger Animals

Working with the observation that electroporation could dramatically increase DNA vaccine efficacy in mice, Babiuk et al. tested this vaccine modality in pigs weighing between 9 and 18 kg.[31] The authors reported dramatic increases in both antibody and cellular responses, especially in pigs that were not boosted with recombinant protein. The total antigen-specific T cell response with DNA and electroporation was comparable to that observed with DNA prime and protein boost.[31] Subsequent work from a number of groups investigated the effect of DNA vaccination followed by electroporation in nonhuman primates. In one report, Luckay et al. reported a 50- to 200-fold increase in DNA vaccine potency with electroporation, and even increased immune responses against less immunogenic antigens.[32] The effect of electroporation is also immediate; Hirao et al. reported that even after a single immunization, the addition of electroporation induced a sixfold increase in T cell responses against HIV antigens.[33] After the second immunization, the increase with electroporation was nearly 20-fold (Figure 14.3a). The overall increase in cellular immune response was associated with not only more antigen-specific T cells, but also more T cells with increased capacity, such as production of IL-2 and tumor necrosis factor (TNF)-α. These increases also had a lasting effect, as antigen-specific T cells were present even 5 months after the last immunization, suggesting a boost in the memory T cell response. But perhaps the most interesting consequence of electroporation was the induction of strong antibody responses. A historical weakness of DNA vaccine technology has been its inability to induce clear antibody responses in nonhuman primates and in human clinical studies.[34] While immunization with DNA alone induced negligible to no antigen-specific antibody responses, the addition of electroporation increased antibody titers tremendously.[33] The overall consequence was not only an increase in total cellular immune responses, but also a balance in the response by stimulating both arms of the immune system. As will be discussed further, this ability to activate both arms of the immune system will have numerous benefits in targeting various diseases and conditions.

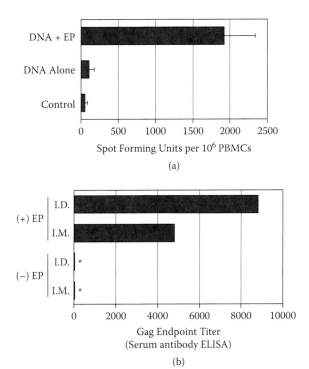

FIGURE 14.3

Electroporation of DNA vaccines increases both cellular and humoral immune responses. (a) Cellular immune response (IFN-γ spot-forming units) was measured from macaque peripheral blood mononuclear (PBMCs) after the second immunization. Rhesus macaques were immunized with 1.0 mg of DNA encoding the HIV Gag and Env antigens. The cellular responses were measured via stimulation with anti-Gag and anti-Env peptides overnight using a standardized ELISpot assay. (b) Serum antibody responses in rhesus macaques with or without electroporation delivered via the intramuscular or intradermal route. Shown is the response after two immunizations. *, ≤50.

14.3.4 Clinical Trials with DNA Vaccines Delivered with Electroporation

There have been several DNA vaccine/electroporation studies in the clinic evaluating its efficacy against a variety of diseases, including HIV, HCV, HPV, and cancer, and probably a number more will enter the clinic in the next few years (see Table 14.3). A number of factors have been evaluated in these studies, most notably safety, tolerability of electroporation, and induction of immune responses. Thus far, no adverse reports with either DNA vaccination or electroporation have been reported, including anti-DNA immune responses and plasmid integration.[15,16,35] However, the primary limitation with previous DNA vaccine trials has not been issues with safety, but rather with immunogenicity and efficacy. Can DNA vaccines, especially in combination with electroporation, induce sufficient immune responses to allow the host to fight infections effectively by eliminating preexisting infections or preventing new infections? Clearly, previous clinical studies with DNA immunization alone suggest otherwise, but can the increased effect with electroporation overcome this hurdle?

To begin to answer this question, several research groups are now investigating electroporation-delivered DNA vaccines in humans. Inovio Pharmaceuticals has designed a therapeutic DNA vaccine against cervical cancer, which is the second most common cancer among women worldwide and the most common cancer in developing countries.[36] The

TABLE 14.3

Electroporation-Based DNA Vaccines in Clinical Evaluation

NCT or GT ID	Phase	Sponsor/Collaborator	Condition	Intervention
NCT0047113	I	Ichor Medical Systems, Inc. Memorial Sloan-Kettering	Melanoma	Xenogenic tyrosinase DNA
NCT0545987	I	Ichor Medial Systems, Inc. Aaron Diamond AIDS Res. Cent. Bill and Melinda Gates Foundation International AIDS Vaccine Initiative Rockefeller University	HIV prevention	ADVAX HIV DNA vaccine
NCT00991354	I	NIAID	HIV prevention	PENNVAX-B with and without IL-12
NCT00563173	I	Inovio Pharmaceuticals Tripep AB	HCV	CHRONVAC-C®
NCT00685412	I	Inovio Pharmaceuticals	HPV	VGX-3100
NCT00859729	I	Uppsala University Karolinska Institutet Cyto Pulse Sciences, Inc.	Prostate cancer	pVAXrcPSAv531 (DNA encoding rhesus prostate-specific antigen (PSA))
NCT01064375	I	Karolinska University Hospital Karolinska Institutet Swedish Institute for Infect. Dis. Con. Cyto Pulse Sciences, Inc.	Colorectal cancer	TetwtCEA DNA (wtCEA with tetanus toxoid Th epitope)
NCT01045915	I	BioAlliance Pharma SA	Melanoma	Plasmid encoding protein AMEP
NCT0753415	I	Merck, Inc. Inovio Pharmaceuticals	Breast, lung, prostate, and other cancers	V934/V935 (hTERT)
NCT01082692	I	Inovio Pharmaceuticals University of Pennsylvania	HIV therapy	PENNVAX-B
N/A	I	VGX International Inovio Pharmaceuticals	Avian influenza (H5N1)	VGX-3400
N/A	I/II	University of Southampton Inovio Pharmaceuticals	Prostate cancer	PSMA27/pDom fusion

development of cervical cancer is closely associated with infection by high-risk types of human papillomaviruses (HPVs).[37] Studies have shown that HPV16 is the predominant subtype of both cancers, accounting for 46%–63% of cervical cancer cases and 90% of head and neck cancer cases.[38,39] In addition, HPV-associated head and neck cancer affects as many as 115,000 people each year globally.[38,40] The rationale for this vaccine is to prevent the progression of HPV-infected cells to malignant cancer. Although preventative vaccines (Gardasil and Cervarix) are sufficient to prevent about 70%–80% of all cervical cancers, they are not effective in controlling existing HPV infections or HPV-associated lesions.[41]

Inovio constructed DNA vaccine plasmids that express either the HPV16 E6 and E7 proteins or the HPV18 E6 and E7 proteins, which are responsible for the transformation of cells and are required for the maintenance of HPV-associated malignancies, making them ideal immunotherapeutic targets.[42,43] To target both HPV16 and 18, this group formulated VGX™-3100, which is a combination of the two aforementioned plasmids and targets both the E6 and E7 proteins. As part of an ongoing phase I clinical trial in the United States, Inovio immunized adult females with postsurgical or ablative treatment of grade 2 or 3 cervical intraepithelial neoplasia (CIN) with varying concentrations of VGX-3100 via intramuscular injection followed by electroporation. Although the study is currently ongoing, Inovio has thus far observed no adverse effects with the vaccine and is currently completing the analysis of the immune responses from VGX-3100 immunization; the initial analyses show that the vaccination scheme can generate significant antigen-specific immune responses in these patients.

There have several other studies that also delivered DNA vaccines with electroporation. ChronVac-C, a HCV NS3/4A DNA vaccine, has been evaluated in humans; this study was conducted by Tripep AB and was the world's first clinical trial of an HCV DNA vaccine delivered by in vivo electroporation.[44] Although no direct comparisons between DNA and DNA + electroporation were conducted, two of three patients in the 0.5 mg and 1.5 mg vaccine groups responded to the vaccine as measured by an increase in HCV-specific cellular immune responses and a reduction in viral load.

Two other studies, however, directly compared the delivery of DNA alone and DNA plus electroporation. Low et al. immunized HLA-A2+ patients with known prostate cancer with the p.DOM-PSMA$_{27}$ vaccine, which encodes a domain (DOM) of fragment C of tetanus toxin fused to the prostate-specific membrane antigen (PSMA).[45] This study delivered the p.DOM-PSMA$_{27}$ vaccine intramuscularly with or without electroporation, and measured anti-DOM antibody responses. Importantly, while DNA alone increased the anti-DOM antibody response by 1.7-fold, DNA plus electroporation increased the response by 24.5-fold, and the response with electroporation persisted to 18 months of follow-up, suggesting a dramatic increase in antibody responses in humans with electroporation. The second study was from the Aaron Diamond AIDS Research Center (ADARC) and the International AIDS Vaccine Initiative (IAVI), and evaluated the cellular immune response induced from immunization with ADVAX, an HIV DNA vaccine.[46] This study also compared the responses from patients immunized with ADVAX alone or with a combination of ADVAX and electroporation. Importantly, the authors reported that while none of the subjects who received ADVAX alone generated detectable cellular immune responses after three injections, all of the subjects who received ADVAX with electroporation generated a modest level of antigen-specific immune responses. Therefore, both of these studies strongly suggest the importance of delivering DNA vaccines with electroporation, and increase the anticipation of the results of other studies involving DNA vaccines delivered via electroporation. A list of such studies that are in clinical evaluation is shown in Table 14.3.

14.4 Emerging Trends with DNA Vaccination with Electroporation and Its Potential Implications

14.4.1 Maximizing Both the Prophylactic and Therapeutic Properties of DNA Vaccines

The limitation of DNA vaccines, at least in their first generation form, is obvious: limited ability to generate antibody responses in large animal models and in humans. This limits the use of DNA vaccines for prophylactic purposes, as many preventative vaccine strategies require a strong antibody response. However, much of this view has changed due to the auspicious immune effects of DNA vaccines when coupled to electroporation.[47] In contrast with DNA alone, DNA followed by electroporation increases both cellular and antibody responses, and this effect can be observed in both large animals (Figure 14.3a and b) and humans (data not shown; mentioned above). Although more clinical evaluation is required, we have observed that the mode of delivery of DNA immunization followed by electroporation dramatically alters the vaccine-induced response. While intramuscular immunization followed by electroporation raises strong T cell responses and modest to strong antibody responses, intradermal immunization followed by electroporation dramatically augments vaccine-induced antibody responses.[33] As shown in Figure 14.3b, electroporation following either intramuscular or intradermal immunization greatly increased antibody responses as measured by antigen-specific endpoint titer. Moreover, the effect was even more demonstrable when the vaccines were delivered intradermally with electroporation. This trend was demonstrated in nonhuman primates, and we are currently studying this phenomenon in humans with a specific emphasis on developing strong antibody-driving vaccines against pathogens such as influenza (H1N1 and H5N1). Unlike currently approved attenuated vaccines, electroporated DNA vaccines could be more specific and efficient by inducing only immune responses that are critical for the host to clear the pathogen, and unlike subunit vaccines, electroporated DNA vaccines could be used against almost all pathogens because of their ability to raise both cellular and humoral immune responses. Much like the next generation of novel cancer drugs, which inhibit cancer-specific pathways, DNA vaccination with electroporation could also provide specificity by inducing only the immune response required for clearing the virus and not inducing unnecessary responses.

14.4.2 Developing Multicomponent and Universal Vaccines

Although the concept of DNA vaccines has been around since the early 1990s, their full potential is only now being realized. First, the magnitude of cellular and humoral immune responses with DNA vaccines plus electroporation in humans and larger animals is significantly greater than that observed in other previous DNA-alone studies. Thus, clinical trials with DNA and electroporation will provide the first realistic opportunity to evaluate the efficacy of DNA vaccines in humans. Second, use of cytokine- and chemokine-encoding plasmids will be the foundation for the next generation of electroporation-delivered vaccines by targeting pathogens more specifically and potently. Finally, the flexibility of DNA vaccines—in terms of modifying gene sequences—has allowed for increased gene expression and the induction of broader immune responses. For example, codon optimization has been used to increase gene expression in host target tissues; codons that are preferentially used in human muscle cells can be incorporated to maximize gene expression in the host's tissue. In addition, conserved epitopes of antigens can be incorporated to increase the broadness of immune responses. For example, Inovio has employed SynCon™ technology, which integrates multiple antigen sequences to formulate a consensus sequence, and increased the coverage or recognition of a greater number of antigenic variations. Consequently, the SynCon™-based influenza vaccines induce antibodies that provide cross-clade protection in a variety of animals.[48] An obvious benefit from these advances is the potential to develop pandemic vaccines more rapidly, as DNA plasmids are safe and easy to produce in large scales. This opens up options not previously available to many government and health agencies, especially during this era of increased bioterrorism threats.

14.5 Where Do We Go from Here?

14.5.1 Improvements in the Electroporation Process

A critical element in efficient electroporation of DNA into tissues is the design of the applicators and electrodes. The applicator consists of an applicator with an attached power cord with the electrodes inserted into the applicator that also serves as the DNA injection device.[49] The injection/electroporation combination may be controlled by nonintegrated or integrated injection systems. The nonintegrated system has two distinct and separate steps: the DNA is first injected, and then the needle-electrode arrays are inserted in the DNA-injected area to deliver the electric pulses. Conversely, the integrated injection system combines the injection and electroporation, although two distinct steps are still required to deliver both the DNA and the electric pulse. The advantage of the latter is that both the DNA and the electrodes are aligned, reducing inconsistencies from injection to injection and misalignment. However, the design of the needle, applicators, and electrodes will require constant evolution with consideration of several factors, including ease of use, patient tolerability, and optimization of immune responses.

While the current generation of applicators has been sufficient for use in animal experiments and other small preclinical and clinical studies, improvements in the equipment's ease of operation may be needed to vaccinate a large number of patients in clinical studies. A movement to a fully integrated system has reduced some of the issues with usage, and the trend is toward increased automation and miniaturization.

14.5.2 Increasing Patient Tolerability of Electroporation

An important criterion with electroporation modalities is patient acceptability and tolerability. Insertion of the needle-electrodes into muscle plays a role in triggering pain, and this depends on myriad factors, including needle thickness and penetration. In addition to the needle-electrode itself, the strength of the electrical voltage also contributes to the pain. Pain increases with greater contact area and higher electrical field intensity; thus, deeper injection of the needle-electrode invariably causes more pain.

One method to reduce the pain would be to make the process less invasive by having smaller needle-electrodes, reduced tissue penetration, and fewer electric pulses, while maintaining transfection efficiency. One example of such a system is the CELLECTRA® (Inovio Pharmaceuticals)-ID device (Figure 14.4). Such optimizations have been made with a number of applicators, although the route of vaccine delivery may limit several

FIGURE 14.4
CELLECTRA®-ID electroporation device. CELLECTRA®-ID is used for intradermal injections, contains three 26-gauge needle-electrodes to generate the electric field (0.2 Amp constant current), and penetrates 3 mm into the site of injection.

factors, including depth of needle-electrode penetration. Intramuscular DNA injections will require needle-electrodes to conduct voltage and create an electrical field around the site of DNA injection. One potential solution for this problem is to decrease the area of DNA transfection. Although the intensity of the electric field must be maintained, decreasing the area of transfection would reduce the voltage and, consequently, overall pain. In this regard, improved manufacturing methods, which have allowed for the production of more concentrated vaccines, have been especially important.[50] The more concentrated DNA solution will ostensibly decrease the total volume of injected vaccine, and therefore decrease the area of needle insertion without altering the efficiency of DNA transfection.

14.5.3 A Fundamental Change in DNA Vaccines: Electroporation and Vaccine Development Go Hand in Hand

Recent advances of DNA delivery via electroporation have illuminated a significant relationship between DNA vaccines and electroporation. While these two fields have been traditionally distinct, the development of future vaccines will entail careful consideration of both aspects, and therefore, these two fields should not be considered mutually exclusive. For example, some pathogens, such as influenza, respiratory syncytial virus (RSV) and HBV, may require primarily an antibody response to prevent infection, thus favoring intradermal delivery of vaccines. Conversely, other pathogens, such as HCV, require a strong T cell response, thus favoring intramuscular delivery of vaccines. Therefore, the development of specific vaccines must be coupled with not only the design of DNA plasmids, but also the choice of the electroporation system.

Inovio currently uses two different electroporation applicators. The CELLECTRA® Adaptive Constant Current Electroporation Device is currently in use in phase I clinical trials to deliver HPV and HIV DNA vaccines. It consists of three components: (1) a pulse generator that is enclosed in a water-resistant case for housing the circuitry and software, (2) an ergonomic applicator that is connected to the electroporation system unit with an attached power cord, and (3) a sterile disposable array with the needle-electrodes and an injection port to deliver the plasmids. One of the CELLECTRA® devices is designed for intramuscular injections and contains five 21-gauge needle-electrodes to generate the electric field (0.5 amps constant current) that will penetrate between 1 and 2 cm at the site of injection. The other device, CELLECTRA®-ID, which is newer and used for intradermal injections, employs a pulse generator identical to that of the intramuscular version, but uses a separate applicator and array. The intradermal version contains three 26-gauge needle-electrodes to generate the electric field (0.2 Amp constant current) and penetrates 3 mm into the site of injection. These interchangeable components add to the versatility of DNA vaccines delivered via electroporation by providing both therapeutic and prophylactic options and by providing pathogen-specific options.

Another area of intense development is to make all electroporation devices completely portable and cordless. In these next-generation devices, the pulse generator that is normally tethered to an applicator in the current generation of electroporation devices becomes integrated into the cordless, handheld applicator, which is powered by a rechargeable battery. Several obvious advantages from having such a system over the current design include portability, better ergonomics, and ease of use.

In summary, the ultimate goal is to develop an effective device with minimal invasiveness to suit the patient, easier handling to suit the user, and decreased size and ancillary components to suit usage in settings that require mass vaccinations. Recent advances have begun to meet these demands, including the incorporation of a fully automated injection-insertion applicator to ensure colocalization of the vaccine and the electric field. The automated injection makes it easier for the user to administer the vaccine more consistently. Currently, a number of other advances are in development, including prototypes that will incorporate additional automation and cordless devices.

Although the field of DNA vaccination has experienced a vacillating pattern of success and failure, the latest clinical data suggest that electroporation-mediated delivery of DNA vaccines may provide the immunogenicity required to have a clinical impact. Therefore, this is a truly exciting period for the vaccine community, as potential therapies or cures for infections such as HPV, HCV, tuberculosis, malaria, dengue, and HIV, as well as cancer, may be just around the corner. In addition, as more studies that utilize electroporation enter the clinic, it will also be imperative that vaccine and device development coevolve, allowing for usage in developing countries and meeting the demand for mass vaccinations.

References

1. Levine, M.M., Lagos, R., and Esparza, J., Vaccines and vaccination in historical perspective, in *New generation vaccines*, 4th ed., ed. Levine, M.M., Informa Healthcare, New York, 2010, chap. 1.
2. Baba, T.W., et al., Live attenuated, multiply deleted simian immunodeficiency virus causes AIDS in infant and adult macaques, *Nat. Med.*, 5, 194, 1999.
3. Belongia, E.A., and Naleway, A.L., Smallpox vaccine: the good, the bad, and the ugly, *Clin. Med. Res.*, 1, 87, 2003.
4. Gurunathan, S., Klinman, D.M., and Seder, R.A., DNA vaccines: immunology, application, and optimization, *Annu. Rev. Immunol.*, 18, 927, 2000.
5. Vyas, J.M., Van der Veen, A.G., and Ploegh, H.L., The known unknowns of antigen processing and presentation, *Nat. Rev. Immunol.*, 8, 607, 2008.
6. Miller, J.D., et al., Differentiation of CD8 T cells in response to acute and chronic viral infections: implications for HIV vaccine development, *Curr. Drug Targets Infect. Disord.*, 5, 121, 2005.
7. Tang, D.C., DeVit, M., and Johnston, S.A., Genetic immunization is a simple method for eliciting an immune response, *Nature*, 356, 152, 1992.
8. Ulmer, J.B., et al., Heterologous protection against influenza by injection of DNA encoding a viral protein, *Science*, 259, 1745, 1993.
9. Robinson, H.L., Hunt, L.A., and Webster, R.G., Protection against a lethal influenza virus challenge by immunization with a haemagglutinin-expressing plasmid DNA, *Vaccine*, 11, 957, 1993.

10. Wang, B., et al., Gene inoculation generates immune responses against human immunodeficiency virus type 1, *Proc. Natl. Acad. Sci. USA*, 90, 4156, 1993.

11. Chattergoon, M., Boyer, J., and Weiner, D.B., Genetic immunization: a new era in vaccines and immune therapeutics, *FASEB J.*, 11, 753, 1997.

12. Sheets, R.L., et al., Biodistribution of DNA plasmid vaccines against HIV-1, Ebola, severe acute respiratory syndrome, or West Nile virus is similar, without integration, despite differing plasmid backbones or gene inserts, *Toxicol. Sci.*, 91, 610, 2006.

13. Zhou, Q.H., et al., Evaluation of pharmacokinetics of bioreducible gene delivery vectors by real-time PCR, *Pharm. Res.*, 26, 1581, 2009.

14. Kim, J.J., and Weiner, D.B., Development of multicomponent DNA vaccination strategies against HIV, *Curr. Opin. Mol. Ther.*, 1, 43, 1999.

15. MacGregor, R.R., et al., First human trial of a DNA-based vaccine for treatment of human immunodeficiency virus type 1 infection: safety and host response, *J. Infect. Dis.*, 178, 92, 1998.

16. Calarota, S., et al., Cellular cytotoxic response induced by DNA vaccination in HIV-1-infected patients, *Lancet*, 351, 1320, 1998.

17. Dupuis, M., et al., Distribution of DNA vaccines determines their immunogenicity after intramuscular injection in mice, *J. Immunol.*, 165, 2850.

18. Loirat, D., et al., Muscle-specific expression of hepatitis B surface antigen: no effect on DNA-raised immune responses, *Virology*, 260, 74, 1999, 2000.

19. Widera, G., et al., Increased DNA vaccine delivery and immunogenicity by electroporation *in vivo*, *J. Immunol.*, 164, 4635, 2000.

20. Wolff, J.A., et al., Conditions affecting direct gene transfer into rodent muscle *in vivo*, *Biotechniques*, 11, 474, 1991.

21. Neumann, E., et al., Gene transfer into mouse lyoma cells by electroporation in high electric fields, *EMBO J.*, 1, 841, 1982.

22. Rols, M.P., Mechanism by which electroporation mediates DNA migration and entry into cells and targeted tissues, *Methods Mol. Biol.*, 423, 19, 2008.

23. Teissie, J., Golzio, M., and Rols, M.P., Mechanisms of cell membrane electropermeabilization: a minireview of our present (lack of?) knowledge, *Biochim. Biophys. Acta*, 1724, 270, 2005.

24. Wolff, J.A., et al., Direct gene transfer into mouse muscle *in vivo*, *Science*, 247, 1465, 1990.

25. Mir, L.M., et al., High-efficiency gene transfer into skeletal muscle mediated by electric pulses, *Proc. Natl. Acad. Sci. USA*, 96, 4262, 1999.

26. Rovere-Querini, P., et al., HMGB1 is an endogenous immune adjuvant released by necrotic cells, *EMBO Rep.*, 5, 825, 2004.

27. Ahlen, G., et al., *In vivo* electroporation enhances the immunogenicity of hepatitis C virus non-structural 3/4A DNA by increased local DNA uptake, protein expression, inflammation, and infiltration of CD3+ T cells, *J. Immunol.*, 179, 4741, 2007.

28. Liu, J., Kjeken, R., Mathiesen, I., and Barouch, D.H., Recruitment of antigen-presenting cells to the site of inoculation and augmentation of human immunodeficiency virus type 1 DNA vaccine immunogenicity by *in vivo* electroporation, *J. Virol.*, 82, 5643, 2008.

29. Dai, Y., et al., DNA vaccination by electroporation and boosting with recombinant proteins enhances the efficacy of DNA vaccines for *Schistosomiasis japonica*, *Clin. Vaccine Immunol.*, 16, 1796, 2009.

30. Spies, B., et al., Vaccination with plasmid DNA activates dendritic cells via Toll-like receptor 9 (TLR9) but functions in TLR9-deficient mice, *J. Immunol.*, 171, 5908, 2003.

31. Babiuk, S., et al., Electroporation improves the efficacy of DNA vaccines in large animals, *Vaccine*, 20, 3399, 2002.

32. Luckay, A., et al., Effect of plasmid DNA vaccine design and *in vivo* electroporation on the resulting vaccine-specific immune responses in rhesus macaques, *J. Virol.*, 81, 5257, 2007.

33. Hirao, L.A., et al., Combined effects of IL-12 and electroporation enhances the potency of DNA vaccination in macaques, *Vaccine*, 26, 3112, 2008.

34. Le, T.P., et al., Safety, tolerability and humoral immune responses after intramuscular administration of a malaria DNA vaccine to healthy adult volunteers, *Vaccine*, 18, 1893, 2000.

35. Robertson, J.S., and Griffiths, E., Assuring the quality, safety, and efficacy of DNA vaccines, *Methods Mol. Med.*, 127, 363, 2006.

36. Franceschi, S., et al., Human papillomavirus and risk factors for cervical cancer in Chennai, India: a case-control study, *Int. J. Cancer*, 107, 127, 2003.

37. Bosch, F.X., et al., Prevalence of human papillomavirus in cervical cancer: a worldwide perspective. International Biological Study on Cervical Cancer (IBSCC) Study Group, *J. Natl. Cancer Inst.*, **87**, 796, 1995.

38. Munoz, N., et al., Against which human papillomavirus types shall we vaccinate and screen? The international perspective, *Int. J. Cancer*, 111, 278, 2004.

39. Hildesheim, A., et al., Risk factors for rapid-onset cervical cancer, *Am. J. Obstet. Gynecol.*, 180, 517, 1999.

40. de Villiers, E.M., Taxonomic classification of papillomaviruses, *Papillomavirus Rep.*, 12, 57, 2001.

41. FUTURE II Study Group, Prophylactic efficacy of a quadrivalent human papillomavirus (HPV) vaccine in women with virological evidence of HPV infection, *J. Infect. Dis.*, 196, 1438, 2007.

42. Peng, S., et al., Characterization of HLA-A2-restricted HPV-16 E7-specific CD8(+) T-cell immune responses induced by DNA vaccines in HLA-A2 transgenic mice, *Gene Ther.*, 13, 67, 2006.

43. Lin, C.T., et al., A DNA vaccine encoding a codon-optimized human papillomavirus type 16 E6 gene enhances CTL response and anti-tumor activity, *J. Biomed. Sci.*, 13, 481, 2006.

44. Sallberg, M., Frelin, L., and Weiland, O., DNA vaccine therapy for chronic hepatitis C virus (HCV) infection: immune control of a moving target, *Expert Opin. Biol. Ther.*, 9, 805, 2009.

45. Low, L., et al., DNA vaccination with electroporation induces increased antibody responses in patients with prostate cancer, *Hum. Gene Ther.*, 20, 1269, 2009.

46. De las Alas, M., and Steinhardt, R., Phase I study of preventive HIV vaccine shows improved immune responses when the vaccine is delivered by electrical impulses, ICHOR Medical Systems, November 30, 2009, www.ichorms.com/images/Release11-30-09.pdf (accessed March 17, 2010).

47. Rice, J., Ottensmeier, C.H., and Stevenson, F.K., DNA vaccines: precision tools for activating effective immunity against cancer, *Nat. Rev. Cancer*, 8, 108, 2008.

48. Laddy, D.J., et al., Electroporation of synthetic DNA antigens offers protection in nonhuman primates challenged with highly pathogenic avian influenza virus, *J. Virol.*, 83, 4624, 2009.

49. Rabussay, D., Applicator and electrode design for *in vivo* DNA delivery by electroporation, *Methods Mol. Biol.*, 423, 35, 2008.

50. Cai, Y., et al., Production of pharmaceutical-grade plasmids at high concentration and high supercoiled percentage, *Vaccine*, 28, 2046, 2010.

15

Atomic Force Microscopy as a Nanotool to Investigate Malaria-Infected Erythrocytes

Bruce Russell, Ang Li, and Chwee Teck Lim
*Division of Bioengineering and Department of Mechanical Engineering,
National University of Singapore, Singapore*

CONTENTS

15.1 The Threat of Malaria to Humanity

Malaria is a life-threatening disease endemic in over 100 countries. Although conservative estimates suggest that 200 million people suffer from malaria, recent data put the figure closer to 600 million.[1,2] Malaria is caused by an intracellular protozoan of the genus *Plasmodium* and is transmitted by the *Anopheles* mosquito. Human malaria is caused by five species of *Plasmodium* spp. listed in order of importance: *Plasmodium falciparum*, *Plasmodium vivax*, *Plasmodium malariae*, *Plasmodium ovale*, and *Plasmodium knowlesi*. The first two species, *P. falciparum* and *P. vivax*, account for almost all malaria cases, and the former is primarily responsible for the 1–2 million malaria-related deaths in sub-Saharan Africa.

15.2 How and Why Malaria Parasites Alter the Host Red Blood Cell

Plasmodium spp. have a complex life cycle in the human host (liver and blood stages) and mosquito vector; however, the symptoms of malaria are entirely due to the life cycle of the parasite within the erythrocyte or red blood cell. Soon after the merozoite enters the host red blood cell, the parasite begins to modify the structural, and correspondingly the biomechanical, properties of the cell. The growing trophozoite stage exports a range of proteins to the erythrocytic cytoskeleton and membrane. In the case of the deadly *P. falciparum*, a protein called knob-associated histidine rich protein (KAHRP) is exported to the red cell cytoskeleton within 22 h of the parasite invading the host cell.[3] KAHRP cross-links the actin and spectrin of the cytoskeleton, both reducing elasticity and causing knob-like excrescences to appear on the cell surface. The knobs of *P. falciparum* and their associated proteins anchor and expose an array of important adhesion proteins known as *P. falciparum* erythrocyte membrane protein 1 (PfEMP1). The PfEMP1-covered knobs of the *P. falciparum* infected red blood cell (IRBC) interact with a range of surface proteins expressed on host endothelial cells, which stick the infected cells to the blood vessel wall to avoid splenic clearance, under considerable shear flow stresses (up to 1 Pa).[4] Unfortunately, the rigid sticky IRBCs of *P. falciparum* readily block the microvasculature of important organs such as the brain, leading to severe diseases. Despite numerous reports of severe diseases in *P. vivax* and *P. malariae,* few studies have investigated nanostructures on the surface of these non-*P. falciparum* species. Another important gap in our knowledge results from most structural studies on the knobs of *P. falciparum* IRBCs relying on culture-adapted laboratory clones from in vitro or primate models. Little has been done to describe these features in field isolates or correlate them with a carefully defined clinical phenotype.

15.3 Traditional Methods to Investigate the Surface Morphology of Malaria-Infected Red Cells

Since the 1950s, scanning electron microscopy (SEM) and transmission electron microscopy (TEM) have been used by numerous investigators to examine morphological changes to the malaria IRBCs.[5] Both of these electron gun-based methods provide an unparalleled tool to investigate in high resolution subnanometer changes to the surface of the red cell (Figure 15.1).

Although TEM has the advantage of observing nano-ultrastructural details inside the cell at extremely high resolution, since the cells have to be sliced into ultra-thin pieces, it does not allow us to completely study the morphological features on the cell surface, such as the knob's full size, density, etc., in three dimensions. Alternatively, SEM is able to produce impressive three-dimensional (3D) images of the surface of the IRBC membrane; however, it does not allow accurate quantitative measurements of the nanofeature, such as knob height and density, and more importantly, SEM could not see through the surface to identify the parasite developmental stages. Both TEM and SEM methodologies require extensive sample preparation and involve numerous fixing and slicing/coating procedures that may severely alter the shape of the IRBC surface, obscuring fine changes or creating artifacts. This has led to some disagreements in the literature regarding the presence of knobs and other nanofeatures in non-*P. falciparum* spp. For example, data from past studies on *P. malariae* IRBCs using TEM disagree: one study by Mackenstedt et al.[7] suggested no knobs but numerous caveolae (caveolae are pleomorphic invaginations on the surface membranes of cells), while Atkinson et al.[8] showed the presence of knobs. Another important problem confronting past studies is that most of them were carried out on clinical isolates that were not confirmed by molecular speciation techniques. As optical

FIGURE 15.1
Upper panel: TEM micrograph of knob-like excrescences (arrow highlights a clear example) from an ex vivo matured *Plasmodium falciparum* clinical isolate from the Thai-Burmese border (SMRU laboratory). Lower panel: SEM scan of a red blood cell-infected laboratory clone of *Plasmodium falciparum*. The entire surface of the infected red cell is covered with numerous knobs, giving the impression of a rough-textured cell surface. Arrow indicates an example of one of these many knobs. Note the smooth uninfected red blood cells in the background.

microscopy has poor sensitivity and specificity in the diagnosis of mixed infections, it is possible that past samples examined by SEM and TEM contain a minority species that was inadvertently reported.

15.4 An Emerging Tool for the Investigation of Malaria: Atomic Force Microscopy

Atomic force microscopy (AFM) is a highly sensitive surface scanning method that provides data on nanometer-scale resolution and at physiological conditions that would be difficult to perform using electron microscopy. Different from the concept of detecting deflected or scattered light or electrons from the sample that light microscopy and electron microscopy use for imaging, AFM relies on a flexible cantilever with a sharp tip at the end to directly sense the sample surface. By means of monitoring the tip-sample interaction from cantilever bending during scanning across the sample surface, AFM

can collect true 3D topographical information and measure surface mechanical properties at nanometer- and piconewton (pN)-scale resolution. Thus, AFM enables the investigation of fresh and air-dried erythrocytes without the need for any other treatments,
such as fixation and surface coating, that may alter delicate nanostructures on the surface of IRBCs. In the late 1990s Aikawa was the first to use AFM to study malaria IRBCs.[9]
Despite some early challenges (such as artifactual AFM images showing double-headed
knobs), an optimized protocol now exists to allow accurate and repeatable AFM imaging of IRBCs.[10]

One of the key advantages of AFM for the examination of malaria IRBCs is that the
same sample can be examined separately or simultaneously by other imaging methods,
especially optical imaging methods. For example, an individual IRBC can be located using
an indexed copper grid on the underside of the slide. The targeted cell is initially scanned
using the AFM; subsequently, the same cell can be fixed for an indirect fluorescent assay
(fluorescently linked antibodies binding to a protein of interest), and finally, the cell can
be stained with a standard Giemsa malaria stain to accurately stage the cell under investigation (Figure 15.2). Infected erythrocytes splutter coated for SEM examination cannot
be stained afterward, and it is almost impossible to determine the erythrocytic stage of a
parasite, or even if a cell is really infected. Table 15.1 outlines the major advantages and
disadvantages of AFM in comparison with SEM and TEM.

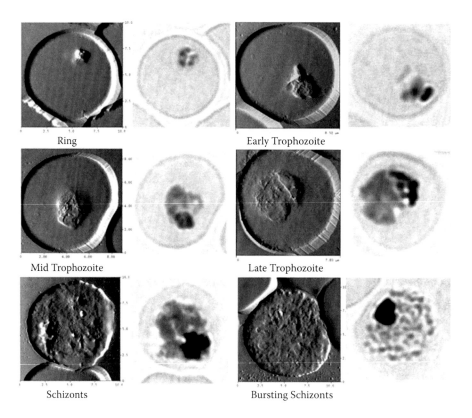

Ring   Early Trophozoite

Mid Trophozoite   Late Trophozoite

Schizonts   Bursting Schizonts

FIGURE 15.2
(See color insert.) Corresponding AFM imaging of external surface and Giemsa staining of interior parasite
structure of lab-cultured *P. falciparum* IRBCs. (Details shown in Li, A., et al., *J. Microbiol. Methods*, 66, 434–439,
2006.)

TABLE 15.1

Comparison of AFM, SEM, and TEM Imaging Methods

	AFM	TEM	SEM
Max. resolution	Subnanometer	Atomic	Nanometer
Typical cost (US$)	100,000–300,000	500,000 or higher	200,000–400,000
Imaging environments	Air, fluid, vacuum, controlled temperature	Vacuum	Vacuum
In situ imaging	Yes	No	No
Simultaneous optical imaging	Yes	No	No
Sample preparation	Easy	Difficult	Medium

15.5 Key Methodological Considerations for AFM Scanning of *Plasmodium* spp. IRBCs

One of the most important considerations before conducting AFM scanning is how best to prepare the sample. On the one hand, RBCs are a highly deformable and nonadherent cell type. It is very difficult to achieve high resolution and stable images directly from liquid environment without any treatments on the cells and substrates. Thus, it will be helpful to immobilize the cells on certain pretreated substrates and reduce the cells' deformability to minimize the scanning-induced movement of the sample. Previously developed protocols involved using a range of fixatives to fix the cells to withstand the scanning force from the AFM tip.[11] However, such fixation methods may alter the surface morphology of the sample significantly without the operator's awareness. A recently developed nonfixation smearing method helped to maintain the cells' structures at low cost and labor.[10]

On the other hand, there are some considerations one should be aware of when preparing clinical samples. First, blood collected from patients should be conserved in anticoagulant such as lithium heparin or citrate, not in ethylenediaminetetraacetic acid (EDTA) which has been observed to cause significant apoptotic-like changes to the IRBC, such as membrane detachment. As we wish to focus on the erythrocytes, it is also better to remove the leukocytes contaminating the sample, thus eliminating a messy artifact-filled background. This can be achieved with commercial filters or simple self-made cellulose filters at a fraction of the cost.[12] It is also important to note that the erythrocytic life cycle of *Plasmodium* spp. has numerous distinct stages (for example, asexual stages have ring, trophozoite, and schizont forms), and depending on the species of the isolate examined, the proportion of these stages may be significantly skewed. For example, most of the stages present in *P. falciparum* isolates collected from the peripheral blood are at the ring or early trophozoite stage, as the mature cytoadherent knob-covered IRBCs at the trophozoite and schizont stages have sequestered to the microvasculature. Therefore, if one wishes to examine the knobs present on the mature asexual form, the isolate must be cultured in a nutrient-rich media at 37°C under microaerophillic conditions for 20 to 40 h, depending on the stage required. Once the sample has the required stage present, it should be washed twice in an isotonic solution before smearing onto a good quality microscopic-grade glass slide. To more efficiently identify both the stage and location of IRBCs, the sample could be supravitally stained with a fluorescent DNA dye (such as 4',6-diamidino-2-phenylindole (DAPI)) or a very low concentration of Giemsa (less than 2%). The smear should not be too dilute, as the isotonic solution supporting the cells will form large salt crystals on the surface of IRBCs, which will impede the scanning.

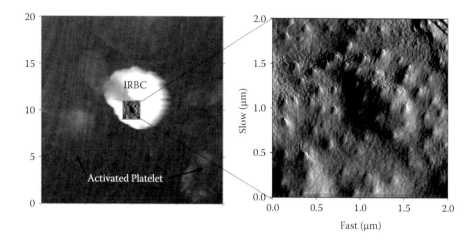

FIGURE 15.3
AFM images of one gluteraldehyde-fixed *Plasmodium falciparum* IRBC initially bound under flow conditions on platelet-coated surface.

After smearing, the samples should be immediately moved into a sealed box or glass jar containing a desiccant such as anhydrous silica. As these samples have not been fixed, it is vital that samples are always transported in a sealed container with plenty of desiccant; otherwise, water droplets condensing on the erythrocytes will cause cell lysis or significant membrane damage.

In certain circumstances, it is advantageous to fix the IRBCs of interest prior to AFM scanning. For example, if one wishes to scan *Plasmodium* IRBCs that have bound to a substrate coated with a particular receptor, it is preferable to fix the adherent cells with fixatives to ensure that they retain their shapes in situ (Figure 15.3). However, when fixing IRBCs with chemicals, one should carefully choose the proper fixative and stick to the optimized fixing protocols in terms of fixing time and fixative concentration. We have found that 1% glutaraldehyde in phosphate-buffered saline (PBS) fixing for 1 h works fine to preserve the cells' surface structures, but a higher concentration and overfixing induce artificial excrescence structures on the cell surface. Also, methanol and formaldehyde/paraformadehyde may not work as well as glutaraldehyde and would alter the cell morphology significantly.

After the sample has been properly prepared, one will need to choose a suitable scanning mode to image the sample surface. The commonly used contact mode and tapping mode in air are both able to reveal the surface structures at fairly high resolution; however, the tapping mode is preferred since the lateral scanning force is significantly reduced in the tapping mode, and thus minimizes the possibility of scanning-induced distortion or removal of the cells. It is also important to check the tip geometry and sharpness prior to scanning to get rid of any possible tip shape-induced artifacts, such as dual knobs.

15.6 The Nanotopography of Red Cells Infected with *Plasmodium* spp.

As previously mentioned, *Plasmodium* spp. cause significant changes to the morphological and biomechanical properties of the host red blood cell. While most studies focus on laboratory clones of *P. falciparum*, it is important to realize that these long-term cultures

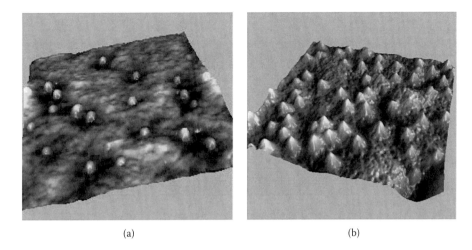

(a)   (b)

FIGURE 15.4
Surface of *Plasmodium falciparum* infected red blood cell. (a) lab clone vs. (b) field isolate.

(some of which are over 30 years old, post isolation) bear little resemblance to the parental cell line both genetically and phenotypically. For example, the density of knobs found on the IRBCs of laboratory clones is relatively low compared with that found on newly isolated parasites from malaria patients. Furthermore, the morphology of knobs on laboratory clones appears more round and smoother than that on field isolates (Figure 15.4).

Aside from *P. falciparum*, other species of human malaria parasites also cause significant changes to the surface nanotopograpy of host RBCs. The poorly understood cause of quartan malaria, *Plasmodium malariae* induces the host cell to form small, spiky knob-like excrescences (which we refer to as spikes) on the surface of all sexual and asexual erythrocytic stages (Figure 15.5c). The function of these spikes is not yet understood; however, it is unlikely they assist the IRBCs to cytoadhere in a *P. falciparum*-like manner, as all stages of this parasite are found circulating in the peripheral blood system.[13] The IRBCs of the second most important cause of human malaria, *P. vivax*, are covered in cavolae-like depressions, as opposed to the raised nanofeatures of *P. falciparum* (knobs) and *P. malariae* (spikes) (Figure 15.4). As with spikes of *P. malariae*, the genesis and function of the caveolae on the surface of *P. vivax* IRBCs are still unknown.

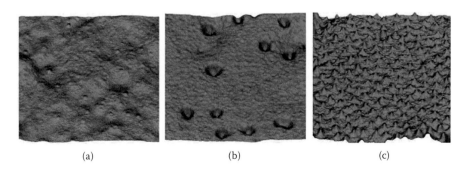

(a)   (b)   (c)

FIGURE 15.5
Three-dimensional surface plots of (a) *P. falciparum*, (b) *P. vivax*, and (c) *P. malariae* IRBCs. Scan size = 2 µm.

15.7 Conclusions and Future Outlook

Despite over 100 years of *Plasmodium* spp. research, we still know little about the pathophysiology of this major cause of human illness. A better understanding of how and why malaria parasites change the nanotopography of host cells will certainly help researchers identify how IRBCs interact with the host endothelia and immune system.

Current rapid instrumental developments of AFM to combine with other microscopy and spectroscopy techniques make it possible to simultaneously monitor the surface physical changes and track different biomarkers, such as fluorescently labeled molecules. The immune-AFM method can even map the spatial and dynamic distribution of certain molecules with nanometer resolution, with the help of the quantum dot labeling technique.[14,15] Furthermore, another exciting aspect of AFM is its ability to measure the force of adhesion between parasite-derived proteins on the surface of the IRBC membrane and their corresponding receptors on the host endothelium. With further development in this nanotechnological technique, a better understanding of the parasitology and pathophysiology of this disease can hopefully be achieved.

Acknowledgments

The authors acknowledge support provided by the Global Enterprise for Micro Mechanics and Molecular Medicine (GEM4) Laboratory at the National University of Singapore and the Singapore-MIT Alliance. We are grateful to the staff and patients attending the Mae Sod Malaria Clinic Shoklo Malaria Research Unit (SMRU). Many thanks to the director of SMRU, Prof. Francois Nosten, and the head of the Malaria Immunobiology Laboratory, Singapore Immunology Network (SIgN, A*STAR, Biopolis), Prof. Laurent Renia.

References

1. Hay SI, Guerra CA, Tatem AJ, Noor AM, Snow RW. (2004). The global distribution and population at risk of malaria: past, present, and future. *Lancet Infect Dis* 4: 327–336.
2. Snow RW, Guerra CA, Noor AM, Myint HY, Hay SI. (2005). The global distribution of clinical episodes of *Plasmodium falciparum* malaria. *Nature* 434: 214–217.
3. Cooke BM, Mohandas N, Coppel RL. (2004). Malaria and the red blood cell membrane. *Semin Hematol* 41: 173–188.
4. Crabb BS, Cooke BM, Reeder JC, Waller RF, Caruana SR, et al. (1997). Targeted gene disruption shows that knobs enable malaria-infected red cells to cytoadhere under physiological shear stress. *Cell* 89: 287–296.
5. Rudzinska MA, Trager W. (1958). [The mechanism of feeding of *Plasmodium malariae*; electron-microscopic studies]. *Wiad Parazytol* 4: 617–618; English translation, 618–619.
6. Aikawa M, Rabbege JR, Udeinya I, Miller LH. (1983). Electron microscopy of knobs in *Plasmodium falciparum*—infected erythrocytes. *J Parasitol* 69: 435–437.

7. Mackenstedt U, Brockelman CR, Mehlhorn H, Raether W. (1989). Comparative morphology of human and animal malaria parasites. I. Host-parasite interface. *Parasitol Res* 75: 528–535.

8. Atkinson CT, Aikawa M, Rock EP, Marsh K, Andrysiak PM, et al. (1987). Ultrastructure of the erythrocytic stages of *Plasmodium malariae. J Protozool* 34: 267–274.

9. Aikawa M. (1997). Studies on falciparum malaria with atomic-force and surface-potential microscopes. *Ann Trop Med Parasitol* 91: 689–692.

10. Li A, Mansoor AH, Tan KS, Lim CT. (2006). Observations on the internal and surface morphology of malaria infected blood cells using optical and atomic force microscopy. *J Microbiol Methods* 66: 434–439.

11. Nagao E, Kaneko O, Dvorak JA. (2000). *Plasmodium falciparum*-infected erythrocytes: qualitative and quantitative analyses of parasite-induced knobs by atomic force microscopy. *J Struct Biol* 130: 34–44.

12. Sriprawat K, Kaewpongsri S, Suwanarusk R, Leimanis ML, Lek-Uthai U, et al. (2009). Effective and cheap removal of leukocytes and platelets from *Plasmodium vivax* infected blood. *Malar J* 8: 115.

13. Boyd MF. (1940). Observations on naturally and artificially induced quartan malaria. *Am J Trop Med* s1–s20: 749–798.

14. Murakoshi M, Iida K, Kumano S, Wada H. (2009). Immune atomic force microscopy of prestin-transfected CHO cells using quantum dots. *Pflügers Archiv Eur J Physiol* 457: 885–898.

15. Wang Y, Chen Y, Cai J, Zhong L. (2009). QD as a bifunctional cell-surface marker for both fluorescence and atomic force microscopy. *Ultramicroscopy* 109: 268–274.

16

Neural Prosthesis Facilitated by Nanotechnology

Wentai Liu, Zhi Yang, and Linh Hoang

University of California, Santa Cruz, Santa Cruz, California, USA

CONTENTS

16.1 Introduction

Research in neural prosthesis has progressed rapidly in recent years, fueled by the unique interdisciplinary efforts fusing engineering, medicine, and biology. This interdisciplinary research requires addressing aspects of humanity and societal impacts, technical challenges and barriers, targeting a wide range of applications. These applications span understanding highly complex biological systems to treating restoring/repairing the lost biological functions, such as deafness, blindness, and paralysis, to building a human-machine interface for performance enhancement (super-person). Neural prosthetic systems will offer viable solutions to neurodisorder diseases, which potentially affect a very large population of people worldwide, and thus occupy the largest market share in healthcare.

The major enabling technology for implantable neural prosthetic systems includes biological recording, stimulation, biosignal processing, wireless communication, sensing, electrodes, hermetic packaging, and powering, where the implants must deal with critical constraints of size, power, reliability, safety, and technology. The additional heterogeneous system testing/measurements under the regulatory and compliance guidelines are critically different from the conventional electronic system designs and accordingly require new design methodology at every design level. Clearly, integration and miniaturization of the implants become essential and require solutions from many fronts—device, circuit, architecture, system, algorithm, design, testing, packaging, and technology.

Nanotechnology plays a critical role in the integration and miniaturization of the implants, especially in the areas of electrode and packaging designs. Both recording and stimulation functions critically depend on the efficacy, efficiency, safety, and reliability of the electrodes. The electrodes provide the contact to the tissues of nerves or muscles in the fashion of "rubber meeting the road" in car driving. The physical properties and dimensions strongly correlate the characteristics of an electrode and its applications accordingly. The core requirements for stimulation and recording lie in the high efficiency of charge transferring, biocompatibility, and durability. Signal-to-noise ratio (SNR) is a critical factor in recording electrode design. The factor that most affects the charge transferring mechanism is the area that, in turn, impacts on selectivity. The electrode is made of metal, which directly affects biocompatibility. The prosthesis systems are often operated in continuously dynamic excitation (stimulation or recording) via electrodes by current, and thus charge, whose long-term reliability and safety effect in terms of electrochemical properties must be carefully studied. One of the great challenges is the inflammatory response resulting in the generation of glial cells around the implanted electrodes, which, in turn, degrades the effectiveness of the recording and stimulation operations.[1] Nanotechnology provides ways of surface modifications to precisely control both chemical and biological reactions, as well as the transition mechanisms of physical property. For example, an anti-inflammatory surface modification could attenuate the formation of glial cells.[2]

In order to improve an electrode's SNR, and thus the sensitivity, surface modification techniques using nanotechnology, to increase the surface area of an electrode and reduce the electrode impedance accordingly, have been reported,[3,4] as shown in Figure 16.1. The report shows that a carbon nanotube (CNT)-coated electrode achieves 25 times the impedance reduction, a 45-fold charge transfer rate, and 65% noise reduction.[5] This performance improvement has a direct impact on the future neural implant if it requires a high density of electrodes, which in turn reduce the geometry size of each electrode.

FIGURE 16.1
Electrodes after surface modification techniques. Such techniques reduce the impedance, interface noise, and increase the charging capacity. (From Keefer, E., et al., *Nat. Nanotech.*, 3, 434–439, 2008.)

In this chapter, visual prosthesis (a biological implant device that restores vision in blind patients) is used as an example to illustrate the design, analysis, and realization of neural prosthesis, which is based on electrical stimulation of neurons/tissues through an array of electrodes powered by integrated microelectronics. The rest of the chapter is organized as follows. Section 16.2 provides a list of neural prostheses and their common system characteristics. Section 16.3 gives an overview of visual prosthesis. Section 16.4 lists the challenges in realizing a high-density retinal implant and describes the fundamentals of neural stimulation. Section 16.5 describes power and data telemetry. Section 16.6 presents an electrical retinal stimulator using a mixed-voltage process. Section 16.7 shows the characterization and measured results of this particular design. Section 16.8 presents a bench-top testing apparatus and measurement results of an epiretinat chip.

16.2 Neural Prosthesis System Characteristics

Brain/neural-related illnesses affect a large population of people worldwide and create an economic burden and opportunity on the order of trillions of dollars. In an aging society, this burden gets worse and consumes a significant portion of the GDP. Consequently, governments, private foundations, and industrial companies have invested a large amount of funding to support relevant research and development to deal with the cure and treatment of brain/neural-related illnesses. Many neural disorder diseases cannot be cured by current medicine; however, treatment using neural prosthetic devices to subside or reduce symptom is possible. Neural prosthesis in general has the benefits of treatment of neural disorders, repair/restoration of the lost biological systems, and enhancement of the biological system performance. The most prominent examples include cochlear implant,[6–8] retinal implant,[9–22] and Parkinson's implant.[23–25] Other candidates include the following:

- Vision
  - Blindness—retinal prosthetics
  - Denervated eyelid (Bell's palsy)
  - Presbyopia—lens implantation for 50+
- Neural disorders and deep brain stimulation
  - Epilepsy, Parkinson's, Compulsory Disorders, Alzheimer's
  - Prostate cancer and impotence
  - Stroke and dementia
- Spinal cord injury—stand and walk, bladder control
- Pain relief—invasive or noninvasive devices
- Antidepression
- Obesity
  - Diabetes—implantable drug pumps
  - Heart disease
- Intelligent artificial upper or lower limbs
- Deaf—cochlear implant
- Musculoskeletal—orthopedic implants for osteoarthritis

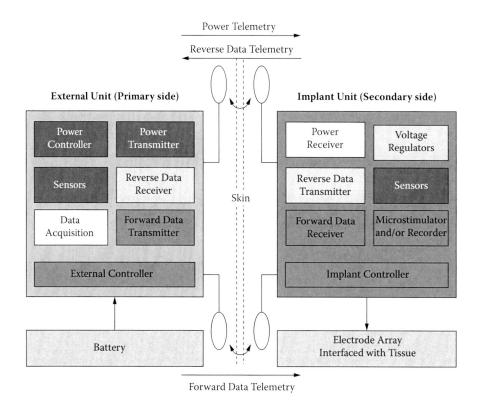

FIGURE 16.2

A generic neural implant architecture. (From Keefer E, Booterman B, Romero M, Rossi A, Gross G. Carbon nanotube coating improves neurenal recording, *Nat Nanotech*; 3: 434–439.)

Neural prosthesis deals with the degeneration of nerve or muscle tissues that have undergone pathological changes due to diseases. Major technology to ensure high efficacy of the neural prosthetic systems involves electrodes, sensors, wireless technology, micro/nanoelectronics, energy sources, signal processing, and packaging. A generic neural prosthetic system is shown in Figure 16.2. It consists of a telemetry system and peripheral accessories to provide power transmission, bidirectional data transmission, stimulation, and recording via electrodes.

Functionally, a telemetry system consists of two links: a power telemetry link to deliver energy for powering electronics and a data telemetry link to communicate information between the external units and the neural implant. To deliver power and data wirelessly for implants, both links have two components of electronics. Those located externally or physically detached from the objects are referred to as primary side units, e.g., external battery, power transmitter, signal processing station, forward data transmitter, reverse data receiver, etc. Those located under the skin (implanted electronics) are typically referred to as secondary units, including the power receiver, forward data demodulator, stimulator,[26] signal sensor and processor,[27] reverse data transmitter, etc. To accomplish the functions of powering and communicating with secondary electronics, the telemetry system should achieve a high power efficiency to avoid heating tissue, a low bit error rate (BER) to ensure reliable data communication, and a small size and light weight to target the highest level of system integration and miniaturization.

16.3 Overview of Visual Prosthesis

Over the past two decades, the engineering and science research communities have witnessed the fast advancement of retinal prosthesis to restore vision in blind patients with retinitis pigmentosa (RP) and age-related macular degeneration (AMD).[9] In a healthy retina, hundreds of millions of photoreceptors initiate neural responses to light and contribute the information used by the visual system to form a representation of the visual world. Photoreceptors are severely absent in blind patients suffering from RP and AMD; however, the remaining neuron circuitries in the signal path can survive at high rates and be used to form input for the visual cortex by applying controlled electrical stimuli.[10] Retinal implant is a prosthetic device that maps visual images to control signals, based on which it stimulates the surviving retinal circuitry. In early studies reported by teams led by Humayun et al.[11] and Liu et al.,[12] simple forms, such as an English character or a matchbox, have been perceived by human subjects with a 4×4 retinal implant. Several approaches for retinal prosthesis[12–21] target different stimulation regions. This chapter focuses on epiretinal prosthesis, where electronics and electrodes are placed directly inside the eye, stimulating the ganglion and bipolar neurons. This approach is preferable from surgical considerations and allows the vitreous to act as a heat sink medium to dissipate heat from the implant.[10,22]

An epiretinal prosthesis system consists of external and implant units. The components of the external unit are a camera, a battery, and power and data transmitters. At the external side, a digital camera captures images that are encoded to bit stream as data input. Transmitters generate radio frequency (RF) electromagnetic fields containing both energy and the encoded image data for the implant. At the implant side, the implant unit includes power and data receivers, a digital controller, and a stimulator array. The power receiver harvests electromagnetic energy and outputs DC supplies powering the remaining electronics. The data receiver demodulates bit stream data, which are further decoded and then used to command the operations of the stimulator array. In this chapter, the design principles, optimization techniques, and realization of a retinal prosthesis microchip are introduced.

16.4 High-Density Neural Stimulation

16.4.1 Challenges of High-Density Retinal Stimulation

A BBC news[28] review of previous generations of epiretinal prosthesis devices that included 16 and 60 electrodes reported the following experience from patients: "When I am walking along the street I can avoid low-hanging branches—I can see the edges of the branches," and "I can't recognize faces but I see them like a dark shadow." There is a clear motivation to improve vision quality by increasing the number of pixels. Studies on pixelized vision suggest that 60 pixels are needed for navigation through an office environment,[29] 100 pixels enable subjects to start recognizing a face,[30] 256 pixels give the subject the ability to read slowly, and 1,000+ pixels are required for a normal reading speed.[9] However, the realization of a high-resolution retinal implant is challenging from several engineering perspectives.

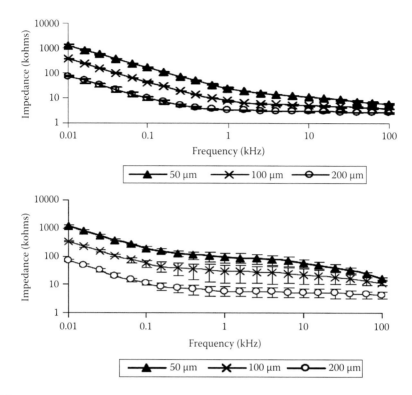

FIGURE 16.3

Electrode impedance vs. electrode size in the vitreous (left) and on the retina (right). The error bars represent 1 standard deviation. The X axis is 1/(pulse duration) in Hz, and the Y axis is the measured impedance in ohms. (From Shah S, Hines A, Zhou D, Greenberg RJ, Humayun MS, Weiland JD. Electrical properties of retinal-electrode interface. *J Neural Eng,* 2007; 4(1): S24–S29.)

First, the size of the electrode array is limited by a central 6 mm diameter area of the human retina.[22] The increase of the number of pixels reduces the area available for an electrode and its corresponding stimulator circuitry. As shown in Figure 16.3, the reduction of electrode size results in higher electrode impedance. Given the same stimulation current, the required compliance voltage to stimulate neurons is proportional to the electrode impedance. This leads to the adoption of high-voltage semiconductor technology, which has transistors much larger in size due to the isolation techniques involved. Since the size available for each stimulator pixel is reduced due to the increased number of pixels, area-efficient circuit architecture is critical to accomplish high-density stimulator design.

Second, because the time to accomplish stimulation of the entire electrode array is fixed, the increase in the number of pixels inevitably reduces the pulse width available for stimulating each pixel. As shown in Figure 16.4, a reduced pulse width significantly increases the current threshold and, consequently, the required compliance voltage. The high compliance voltage resulting from the smaller electrode size and shorter pulse width demands a high power supply voltage. This increases implant power dissipation and causes tissue heating. Consider a simplified case of a disk-shaped implant with diameter D surrounded by liquid with a thermal conductivity λ and dissipated power P_{heat}. The temperature increment of the implant is $\Delta T_{\text{temp}} = P_{\text{heat}}/(4\pi\lambda D)$, where T_{temp} is the temperature elevation at the implant surface. Assuming $D = 6$ mm and $\lambda = 0.58$ W m^{-1} K^{-1}, 50 mW power dissipation

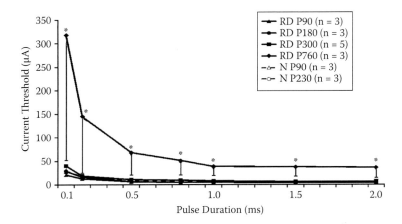

FIGURE 16.4
Strength duration curves for rat retina. Two control groups and four degenerate groups. (From Chan et al. 2008.)

corresponds to 1.2°C temperature increment, bringing potential damage to the retina tissues. To balance the trade-off between the high compliance voltage and tissue temperature increment, a high-efficiency power link that outputs multiple supply voltages for mixed-voltage circuit design is used.

Results showed that for the postnatal (P) day 90 and 180 degenerate groups, threshold currents were similar to those for the normal postnatal (NP) control group (NP90 and NP230). For the P300 and P760 degenerate group, the threshold currents increased significantly.[32]

Third, the power telemetry design must take the coil size, tissue magnetic field exposure, and power efficiency into consideration. Also, the data telemetry should provide a high data rate that controls a large number of stimulation pixels in real time. In a single-band approach, where power and data are transmitted through the same carrier, high power efficiency and high data rate are two contradicting specifications.[12] Dual-band telemetry is a better approach[33,34] because it transmits power and data at different frequencies. However, the power to data interference in the dual-band scheme can potentially corrupt the data demodulation, which should be carefully examined. In addition, a data rate with a minimum of a few megabits per second (Mbps) is required to control a large number of stimulation pixels in real time.

16.4.2 Mechanism of Neural Stimulation

Figure 16.5 illustrates the epiretinal stimulation. The array of electrodes is driven by wirelessly powered stimulators and surgically hatched on the retina. Electrical current stimuli injected through electrodes induce electrical fields to the surrounding environment, which activate nearby neurons. An extensive body of theories describing the response of neurons to an external electric field distribution has been established in the literature.[35–39] Among these theories, the cable equation is widely acknowledged and described as

$$\lambda_m^2 \frac{\partial^2 V_e(x,t)}{\partial x^2} = \lambda_m^{-2} \frac{\partial^2 V_m(x,t)}{\partial x^2} + \tau \frac{\partial V_m(x,t)}{\partial t} + V_m(x,t), \quad \lambda_m = \sqrt{r_m/r_i}, \quad \tau_m = c_m r_m \quad (16.1)$$

where V_m is the transmembrane voltage, r_m is the membrane resistance times the unit length, r_j is the intracellular resistance, c_m is the membrane resistance per unit length, and

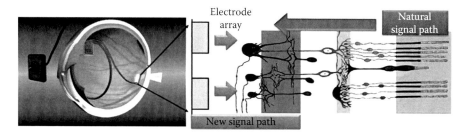

FIGURE 16.5
(See color insert.) Illustration of epiretinal prosthesis. An array of electrodes is implanted inside the eye and hatched on the surface of the retina. The natural signal path of the photoreceptor to the retinal ganglion neuron is blocked. A new signal path is created by electrically stimulating the surviving neurons.

V_e is the extracellular voltage along the axon membrane. Numeric simulations based on cable equations with parameters extracted from retinal neurons,[36] as well as stimulation experiments using retinal ganglion neurons,[40,41] point out that the compliance voltage on the electrode ($V_0(t)$) required to activate neurons dramatically increases as the pulse width or electrode size decreases.

16.4.3 A 256-Channel Epiretinal Implant

Figure 16.6 shows an example of an epiretinal implant system. It includes a power link delivering up to 100 mW to the implant, a data link providing 2 Mbps data, and 256-channel stimulators performing parallel stimulation at a rate of 50 Hz or more. The power regulators, data receiver, digital blocks, and stimulators are integrated onto a single microchip with total area of 5.1×5.3 mm^2.

16.5 Wireless Power Telemetry

16.5.1 Coil Model and Coil Design

Power efficiency, defined as the ratio of power delivered to the implant over the power drained from the battery, is an important parameter for epiretinal implants. The upper bound of the efficiency is analytically derived:[42]

$$\eta_p = \frac{k^2 Q_1 Q_2}{2 + k^2 Q_1 Q_2 + 2\sqrt{1 + k^2 Q_1 Q_2}} \approx \frac{k^2 Q_1 Q_2}{4 + 2k^2 Q_1 Q_2} \tag{16.2}$$

where η_p is the efficiency of the power telemetry, Q_1 and Q_2 are the Q factors of coils, and k is the coupling coefficient between coil pairs. In the case where coils are weakly coupled due to a large separation (e.g., 2%–3% in this system), the expression can be simplified to $0.25k^2 Q_1 Q_2$, suggesting that coils are the most critical components for power telemetry. Given the restrictions on coil size and separation, the achievable coupling coefficient is predetermined. Therefore, the main focus of coil design is improving the power efficiency by increasing the coil's Q factor.

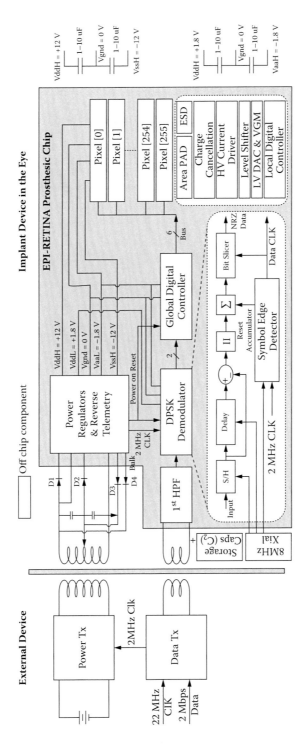

FIGURE 16.6
An example of epiretinal implant architecture.

When coils operate at low frequency, their Q factor is defined as $Q = \omega\, L/R$, where ω is frequency, L is inductance, and R is series resistance. As frequency increases, frequency-related effects, including skin effects, proximity effects, and self-resonance, modify both L and R, dramatically degenerating the Q factor. According to Yang et al.,[43] a formula predicting the Q factor is

$$Q(f) = \frac{2\pi f L}{R} \frac{1 - f^2/f_{self}^2}{1 + f^2/f_h^2} \tag{16.3}$$

where f_h is a parameter to quantify the impact from proximity effect (skin effect) and f_{self} is the coil's self-resonant frequency. f_h and f_{self} are expressed by geometry and physical parameters, whose expressions are shown in Yang et al.[43] For any given coil, there is an optimal frequency f_{peak} that has the maximal Q. To maximize the power efficiency, f_{peak} of the coil pair should be designed in accordance with the power carrier frequency. Based on Equation 16.3, an analytical form of f_{peak} is

$$\frac{1}{f_{peak}^2} \approx \frac{1}{f_h^2} + \frac{3}{f_{self}^2} \tag{16.4}$$

Equation 16.4 represents a key design equation and a closed-form analytical solution for f_{peak}. With this single equation of merit, the maximum Q and the maximum efficiency of the telemetry system can be determined. By changing the design parameters, one may tune f_{peak} close to the target frequency and maximize the power efficiency of the telemetry.

16.5.2 Multiple Power Supplies

A high supply voltage specification of the retinal implant system is derived from the compliance voltage of the stimulator. However, a majority of circuit blocks, e.g., digital controllers, data demodulator, oscillator, and part of the stimulator, can operate at a much lower voltage level to reduce the overall power consumption. As shown in Figure 16.7a, a high-efficiency DC/DC buck converter is a possible solution to convert a higher supply voltage to a lower supply voltage.

However, this approach requires additional off-chip components (inductors), imposing burdens on system integration and miniaturization. Through use of a cascaded resonant tank that reuses the power coil, it is feasible to provide four rectified and regulated voltages (±12V, ±1.8V) with a shared return path, as illustrated in Figure 16.7b. Compared to

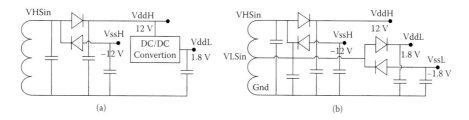

(a)   (b)

FIGURE 16.7
Rectifier comparison. (a) A conventional resonant tank provides four supplies (VddH, VssH, Vddl, and Vssl) through DC/DC conversion. (b) A cascoded resonant tank provides the same four supplies, but without DC/DC conversion. For illustration, regulators and their required voltage overheads are ignored in the picture.

the standard approach of using two regulated power supplies, the cascoded resonant tank saves roughly 40% of power without introducing additional off-chip components.

16.6 Wireless Data Telemetry

Two pairs of coils for transferring power and data are placed coaxially to increase coupling and reduce constraints of surgical operation. However, this configuration causes severe interference between power and data and may corrupt the data demodulation. A noncoherent differential phase shift keying (DPSK) scheme, where 1 is coded as a phase shift of 180° and 0 is coded as no phase shift,[44] can be used to reduce the interference via cancelation. With the assignment of a data rate equal to the power carrier frequency, two consecutive symbols are under the same power interference, which is eliminated by differentiation.

The implemented DPSK receiver uses the subsampling scheme,[45] where the sampling frequency is lower than the data carrier frequency. The adoption of this subsampling scheme is based on two reasons. First, a low-frequency operation reduces demodulator power. Second, it samples the band pass signal without downmixing, which involves a phase-locked loop circuitry. Equation 16.5 gives the relationship among the sampling frequency (f_s), data carrier frequency (f_c), and data rate (f_d), ensuring that nonzero crossing points of the data carrier are sampled:

$$f_s = 4f_c/(2i+1) = 4jf_d, \quad i = 1,2,3,\ldots; j = 1,2,3,\ldots \tag{16.5}$$

Using $S_{i,j}$ to denote the clock sequence, where i is the symbol index and j the sample index, at the end of each symbol, the comparator output is

$$V_{i,\text{out}} = \sum_j |S_{i,j} - S_{i+1,j}| \tag{16.6}$$

When there is no phase shift, $S_{i,j} - S_{i+1,j} \approx 0$ and $V_{i,\text{out}}$ is a small value. When there is a 180° phase shift and the sampled points are not the zero crossing ones, $|S_{i,j} - S_{i+1,j}| \gg 0$ and $V_{i,\text{out}}$ becomes a large positive value. By comparing $V_{i,\text{out}}$ to a predefined threshold, the phase shifts are identified.

16.7 Stimulator and Digital Controller

An example stimulator pixel architecture is shown in Figure 16.8. It consists of a local digital controller, level shifter, digital-to-analog converter (DAC), variable-gain current mirror (VGCM), protection circuitry, driver stage, electrostatic discharge (ESD) protection circuitry, and area pad. Functionally, the LDC generates stimulation waveform patterns based on the decoded bit stream by the demodulator and GDC. The level shifter, DAC, and VGCM interpret the waveform patterns and control the high-voltage (HV) output stage that injects stimulus through the electrode. Any residual charge on the electrode interface due to current mismatch is periodically removed to prevent potential tissue damage. A part of the stimulator circuitry area is shared by the area pad that is composed of the top three metal layers. To further reduce the die area usage of a stimulation pixel, the LDC, DAC, and VGCM are designed using low-voltage (LV) transistors, which have 10 times smaller feature size than HV transistors. To reduce power consumption, only the output current driver and the charge cancelation circuit use the power supply level of ±12 V; the rest of the circuits operate between 0 and ±1.8 V. To properly interface circuit blocks operating at different power supply levels, a protection circuit that is composed of a high-voltage transistor with a gate bias of 1 V is employed.

Imperfections such as process variations, imbalanced power supplies, and leakage from adjacent stimulus sites may lead to a mismatch between the anodic and cathodic pulses and can cause charge residues across the electrode interface, which should be avoided for safety. In Singh et al.,[46] a switch-activated resistor consisting of low-voltage transistors was used to remove the charge. In Ortmanns et al.,[18] a complex charge balancer circuitry was developed that injects counterpulses for removing the charge. Because of the rigid area constraint, an HV positive-channel metal oxide semiconductor (PMOS) transistor can be used, operating in the deep triode region for charge cancelation. This HV PMOS switch is controlled by the LDC through a level shifter operating at ±12 V.

The digital controllers decode the bit stream recovered by the data demodulator and control the stimulator array to generate current waveforms. Functional blocks, which can

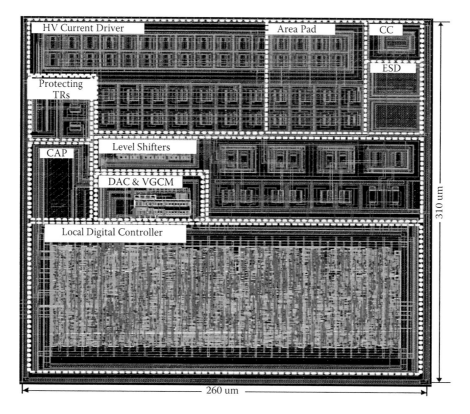

FIGURE 16.8
Stimulator pixel: Layout of a stimulator pixel providing the physical size of each module.

be shared globally, such as packet header detector, packet parser, error detector, and clock divider, are implemented in a global digital controller (GDC). An on-chip network generates pixel addresses and distributes the data from the GDC to individual pixels. Each pixel has a copy of the local digital controller (LDC), which generates stimulation waveform patterns according to the data input. To reduce the power consumption of LDCs, a programmable low-frequency clock is utilized to reduce unnecessary transitions.

A large number of bonding pads bridging circuitry and electrodes is required for a high-density stimulator array. This imposes additional difficulties in system integration and realization. A circuit under pad (CUP) structure, which is compatible with flip-chip bonding technology, is used to facilitate system integration. In the CUP structure, an area pad is made of the top three metal layers, under which an ESD protection circuit, a charge cancelation switch, and part of a current driver are inserted. As a result, the size of a pixel with the CUP structure is reduced by 13.5%, compared with one without the CUP structure.

16.8 Benchtop Testing

The 256-pixel epiretinal chip is fabricated in a 0.18 μm, 32 V complementary metal-oxide-semiconductor (CMOS) process and tested on the benchtop. Figure 16.9a shows the die photograph. The chip includes power regulators, data receiver, digital controllers, and 256-channel stimulation pixels occupying an area of 5.1×5.3 mm^2. For a full-path demonstration, as shown in Figure 16.9b, a commercial camera captures images and encodes them to a 2 Mbps DPSK data packet passing through a data acquisition board made by National Instrument (NI). A field-programable gate array followed by an NI card generates a 2 MHz clock and inputs 2 Mbps data to the data transmitter. The 2 MHz clock is also used to drive the class E power amplifier to generate a 2 MHz sinusoidal RF power signal coupled to the implant side through coils. In Figure 16.9c, the measured data stream recovered from the data demodulator/global digital controller (cyan trace) and output voltage patterns of two randomly picked stimulation channels (magenta and green traces) are displayed. This demonstrates the functionality and flexibility of the system.

16.9 Conclusion

Research in neural prosthesis has progressed rapidly in recent years, fueled by the unique interdisciplinary efforts fusing engineering, medicine, and biology. This chapter presented an extensive outline in developing a retinal prosthesis implant system through its design, analysis, and realization. The major design challenges, such as implant heat dissipation, high-compliance stimulation voltage, and system integration and miniaturization, are analyzed and addressed using the different techniques presented in this chapter. The corresponding realization of the epiretinal implant includes a power telemetry system that delivers 100 mW to the implant, a data telemetry system that provides 2 Mbps data for controlling the implant, and 256-channel stimulators enabled by digital modules to perform parallel stimulation matching arbitrary patterns at a rate of 50 Hz or more.

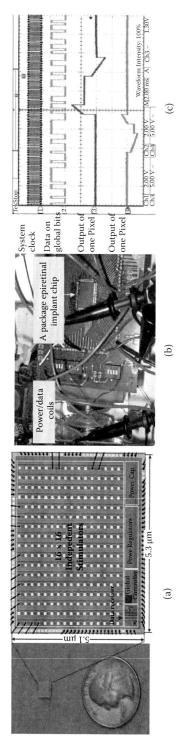

FIGURE 16.9
(See color insert.) Die photograph and testing results. (a) Die photograph. (b) Chip test environment. (c) Measured stimulator output waveforms.

Acknowledgments

This work is partially supported by the National Science Foundation through BMES-ERC and the University of California Lab Fee Research Program. The authors are thankful for the technical support by members of the Integrated Bioelectronics Laboratory at University of California Santa Cruz (UCSC), as well as the chip fabrication service provided by Taiwan Semiconductor Manufacturing Co. (TSMC) and digital standard cells library from ARM.[R]

References

1. Capadona JR, Shanmuganathan K, Tyler DJ, Rowan SJ, Weder C. Stimuli responsive polymer nanocomposites inspired by the sea cucumber dermis. *Science*, 2008; 319(5869):1370–1374.
2. He W, McConnell G, Schneider TM, Bellamkonda R. A novel anti-inflammatory surface for neural electrodes. *Adv Mater*, 2007; 19:3529–3533.
3. Keefer E, Bootterman B, Romero M, Rossi A, Gross G. Carbon nanotube coating improves neuronal recordings. *Nat Nanotech*, 2008; 3:434–439.
4. Ludwig K, Uram J, Yang J, Martin D, Kipke D. Chronic neural recordings using silicon microelectrode arrays electrochemically deposited with a poly(3,4 ethylenedioxythiophene) (PEDOT) film. *J Neural Eng*, 2006; 3:59–70.
5. Plexon. http://www.plexon.com/index.html.
6. Buchegger T, Oberger G, Reisenzahn A, Hochmair E, Stelzer A, Springer A. Ultra-wideband transceivers for cochlear implants. *EURASIP J Appl Signal Process*, 2005; 18:3069–3075.
7. Spelman FA. The past, present, and future of cochlear prostheses. *IEEE Eng Med Biol Mag*, May 1999; 27–33.
8. McDermott H. An advanced multiple channel cochlear implant. *IEEE Trans Biomed Eng*, 1989; 36(7):789–797.
9. Weiland JD, Humayun MS. Visual prosthesis. *Proc IEEE*, 2008; 96(7):1076–1084.
10. Weiland JD, Liu W, Humayun MS. Retinal prosthesis. *Annu Rev Biomed Eng*, 2005; 7:361–401.
11. Humayun MS, Juan Ed, Weiland JD, Dagnelie G, Katona S, Greenberg R, et al. Pattern electrical stimulation of the human retina. *Vision Res*, 1999; 39:2569–2576.
12. Liu W, Vichienchom K, Clements M, DeMarco SC, Hughes C, McGucken E, et al. A neuro-stimulus chip with telemetry unit for retinal prosthetic device. *IEEE J Solid-State Circuits*, 2000; 35(10):1487–1497.
13. Veraart C, Wanet-Defalque MC, Gerard B, Vanlierde A, Delbeke J. Pattern recognition with the optic nerve visual prosthesis. *Artif Organs*, 2003; 11:996–1004.
14. Liu W, Humayun MS. Retinal prosthesis. *IEEE ISSCC Dig Technol Papers*, 2004; 12(1):218–219.
15. Theogarajan L, Wyatt J, Rizzo J, Drohan B, Markova M, Kelly S, et al. Minimally invasive retinal prosthesis. *IEEE ISSCC Dig Technol Papers*, 2006; 2(5):99–100.
16. Ohta J, Tokuda T, Kagawa K, Furumiya T, Uehara A, Terasawa Y, et al. Silicon LSI-based smart stimulators for retinal prosthesis. *IEEE Eng Med Biol Mag*, 2006; 25:47–59.
17. Coulombe J, Sawan M, Gervais JF. A highly flexible system for microstimulation of the visual cortex: design and implementation. *IEEE Trans Biomed Circuits Syst*, 2007; 1(4):258–269.
18. Ortmanns M, Rocke A, Gehrke M, Tiedtke HJ. A 232-channel epiretinal stimulator ASIC. *IEEE J Solid-State Circuits*, 2007; 42(12):2946–2959.
19. Burghartz JN, Engelhardt T, Graf H, Harendt C, Richter H, Scherjon C, et al. CMOS imager technologies for biomedical applications. *IEEE ISSCC Dig Technol Papers*. 2008; 7(4):142–143.

20. Rothermel A, Liu L, Aryan NP, Fischer M, Wuenschmann J, Kibbel S, et al. A CMOS chip with active pixel array and specific test features for subretinal implantation. *IEEE J Solid-State Circuits*, 2009; 44(1):290–300.

21. Butterwick A, Huie P, Jones BW, Marc RE, Marmor M, Palanker D. Effect of shape and coating of a subretinal prosthesis on its integration with the retina. *Exp Eye Res*, 2009; 88(1):22–29.

22. Ameri H, Weiland JD, Humayun MS. Biological considerations for an intraocular retinal prosthesis. *Biological and Medical Physics, Biomedical Engineering*, 2008; 1–29.

23. Le Jeune F, Drapier D, Bourguignon A, Peron J, Mesbah H, Drapier S, Sauleau P, Haegelen C, Travers D, Garin E, Malbert CH, Millet B, Verin M. Subthalamic nucleus stimulation in Parkinson disease induces apathy: a PET study. *Neurology*, 2009; 73:1746–1751.

24. Groiss SJ, Wojtecki L, Sudmeyer M, Schnitzler A. Review: deep brain stimulation in Parkinson's disease. *Ther Adv Neurol Disord*, 2009; 2:379–391.

25. Maks CB, Butson CR, Walter BL, Vitek JL, McIntyre CC. Deep brain stimulation activation volumes and their association with neurophysiological mapping and therapeutic outcomes. *J Neurol Neurosurg Psychiatry*, 2009; 80:659–666.

26. Kelly SK, Shire DB, Doyle P, Gingerich MD, Drohan WA, Rizzo JF, Theogarajan LS, Chen J, Cogan SF, Wyatt JL. The Boston retinal prosthesis: a 15-channel hermetic wireless neural stimulator. In: *2nd International Symposium on Applied Sciences in Biomedical and Communication Technologies*, 2009; 1-6.

27. Sivaprakasam M, Liu W, Wang G, Zhou M, Weiland JD, Humayun MS. Challenges in system and circuit design for high density retinal prosthesis. In: *IEEE/NLM Life Science Systems and Applications Workshop (LSSA)*, 2006; 1–2.

28. Fildes J. Trials for 'bionic' eye implants. *BBC News*, February 2007.

29. Dagnelie G, Keane V, Narla V, Yang L, Weiland JD, Humayun MS. Real and virtual mobility performance in simulated prosthetic vision. *J Neural Eng*, 2007; 4:2092–2101.

30. Thompson RW, Barnett GD, Humayun MS. Facial recognition using simulated prosthetic pixelized vision. *Invest Ophthalmol Vis Sci*, 2003; 44:5035–5042.

31. Shah S, Hines A, Zhou D, Greenberg RJ, Humayun MS, Weiland JD. Electrical properties of retinal-electrode interface. *J Neural Eng*, 2007; 4(1):S24–S29.

32. Chan, LH, Ray, A, Thomas, BB, Humayun, MS Weiland, JD. *In vivo* study of response threshold in retinal degenerate model at different degenerate stages. In: *30th Annu Int Conf Proc IEEE Engineering in Medicine & Biology Society*, 2008; 20–24.

33. Wang G, Liu W, Sivaprakasam M, Kendir A. Design and analysis of an adaptive transcutaneous power telemetry for biomedical implants. *IEEE Trans Circuits Syst*, 2005; 52(10):2109–2117.

34. Ghovanloo M, Atluri S. A wide-band power-efficient inductive wireless link for implantable microelectronic devices using multiple carriers. *IEEE Trans Circuits Syst*, 2007; 54(10):2211–2221.

35. Plonsey R, Barr RC. Electric field stimulation of excitable tissue. *IEEE Trans Biomed Eng*, 1995; 42(4):329–336.

36. Greenberg RJ, Velte TJ, Humanyun MS, Scarlatis GN, De Juan EJ. A computational model of electrical stimulation of the retinal ganglion cell. *IEEE Trans Biomed Eng*, 1999; 46(5):505–514.

37. Holt GR, Koch C. Electrical interactions via the extracellular potential near cell bodies. *J Comp Neurosci*, 1999; 6(2):169–184.

38. Rattay F, Lutter P, Felix H. A model of the electrically excited human cochlear neuron. I. Contribution of neural substructures to the generation and propagation of spikes. *Hearing Res* 2001; 153:43–63.

39. Fall C, Marland E, Wagner J, Tyson J. *Computational cell biology*. New York: Springer Verlag, 2002.

40. Jensen RJ, Rizzo JF, Ziv OR, Grumet A, Wyatt J. Thresholds for activation of rabbit retinal ganglion cells with an ultrafine, extracellular microelectrode. *Invest Ophthalmol Vis Sci*, 2003; 44:3533–3543.

41. Jensen RJ, Ziv OR, Rizzo JF. Thresholds for activation of rabbit retinal ganglion cells with relatively large, extracellular microelectrodes. *Invest Ophthalmol Vis Sci*, 2005; 46:1486–1496.

42. Ko WH, Liang SP, Fung CD. Design of radio-frequency powered coils for implant instruments. *Med Biol Eng Comput*, 1977; 15(6):634–640.

43. Yang Z, Liu W, Basham E. Inductor modeling in wireless links for implantable electronics. *IEEE Trans Magn*, 2007; 43:3851–3860.

44. Proakis JG. *Digital communications*. Singapore: McGraw Hill, 1995.

45. Vaughan RG, Scott NL, White DR. The theory of bandpass sampling. *IEEE Trans Signal Process*, 1991; 39(9):1973–1984.

46. Singh P, Liu W, Sivaprakasam M, Humayun M, Weiland J. A matched biphasic microstimulator for an implantable retinal prosthetic device. *IEEE Int Symp Circuits Syst*, 2004; 1–4.

17

Production of Bare Multiple-Element Magnetic Nanoparticles and Their Use in Fast Detection and Removal of Pathogenic Bacteria from Water Resources

Khaled N. Elshuraydeh, Hanan I. Malkawi, and Mona Hassuneh

Higher Council for Science and Technology, Department of Biological Sciences, Yarmouk University, Irbid, Jordan

CONTENTS

17.1 Introduction and Background

Nanotechnology involves the study, control, and manipulation of materials at the nanoscale, typically having dimensions on the order of 0.1 to 100 nm, exhibiting novel and significantly enhanced physical, chemical, and biological properties, functions, phenomena, and processes due to their nanoscale size.

The large interest in nanostructures results from their numerous potential applications in various areas, such as materials, biomedical sciences, electronics, optics, magnetism, environment, biotechnology, energy storage, and electricity.

Nanoparticles, even from the same material, can be synthesized utilizing a variety of methods. Different methods are used in order to optimize specific properties of materials. These properties include, but are not limited to, size (diameter, length, volume), size distribution, symmetry, surface properties, surface coating, purity, ease of manipulation, yield, and suitability of scaling up.

Magnetic nanoparticles such as magnetite and maghemite have been in use in areas of environment, bioscience, and medicine. Their unique magnetic properties allow them to move in high magnetic field gradients, making them useful in areas such as drug targeting and drug separation. Potential applications for biosensing techniques may also employ magnetic particles by selectively binding the particles to species of interest and mobilizing them under a magnetic field. In almost all applications of nanoparticles, the preparation method of nanomaterials represents one of the most important challenges that will determine the particle size and shape, the size distribution, the surface chemistry of the particles, and consequently their magnetic properties. Due to their magnetic characters, magnetite (Fe_3O_4) nanoparticles can be attracted by a magnetic field and are easily separable in a solution. Magnetite is also one of the most commonly used magnetic materials because it has a strong magnetic property and low toxicity.

To date, many approaches have been developed for the preparation of magnetic nanoparticles; they usually lead to complicated processes or require relatively high temperature.

Several applications of nanomaterials to biology, environment, and medicine have been suggested and are being studied, such as fluorescent biological labels, drug and gene therapy, biodetection of pathogens, detection of proteins, probing of DNA structure, tissue engineering, tumor destruction via healing, separation, and purification of biological molecules and cells, and other applications.

One application of nanoparticles will be the detection of a small number of bacteria in different environments, such as blood, urine, stool, water, and food. The current method for detection of microbial pathogens, such as bacteria in water, food, and clinical samples, is based on conventional microbiological methods, which require at least 3–10 days, and on molecular biology methods, such as polymerase chain reaction (PCR), which requires 24–48 h and detects specific genes for specific microorganisms, but remains limited because of the potential presence of PCR inhibitors, lack of information on the viability of cells, and low sensitivity for quantification of cells.

Thousands of types of bacteria are naturally present in our environment. Some bacteria cause disease (pathogens) in humans; when certain pathogens enter the food and water supply, they can cause foodborne illness or disease. Foodborne disease is caused by consuming contaminated foods, water, or beverages. The analysis of pathogenic bacteria is very important and vital for food and water safety, clinical diagnosis and therapies, potable water, and prevention strategies to combat bioterrorism agents. Many tests currently in use require a high concentration of pathogen cells, typically from 1 million to 10 million cells per milliliter of fluid. The tests also rely on a process known as enrichment, which occurs when a sample believed to be contaminated grows for a period of time in a nutrient broth to allow any pathogen cells present to multiply. Other tests rely on DNA markers, but these also can take days to process, which is a problem because by the time test results come back, food products may already be in food supply warehouses or on store shelves.

Nanosized iron oxides, especially Fe_2O_3 and Fe_3O_4, have been of intense interest due to their characteristic features, such as electric, optical, magnetic, and photosensitive properties. In addition to their inherent characteristics, their amenability for surface modification has led to a great extension of their utilization in biological, medical, and environmental applications. Despite intense efforts in the study of magnetic nanoparticles, their applications in biomedicine and the environment are just emerging.

In the current study, we present a simple sodium chloride-assisted method to prepare multiple-element magnetic nanoparticles (MMNs) and use of those MMNs for fast detection and removal of bacterial pathogens from water resources.

17.2 Materials and Methods

17.2.1 Preparation of Multiple-Element Magnetic Nanoparticles (MMNs)

Fly ash collected from chimneys of iron and steel factories was used for the production of multiple-element magnetic nanoparticles (MMNs).

Wet sieving of fly ash through a 25 μm sieve was performed, and then dried in an oven (temperature of 100–110°C for 55 min). Magnetites in the dry product were separated by a magnetic field (3.8×10^4 G). These constituted 95% of the original product.

The dried magnetic product was then well mixed with sodium chloride salt (NaCl) (ratio: 1:10) in a thin, semispherical ceramic pot and stirred thoroughly with a teaspoon to provide for good dispersion. The contents of the pot were emptied into another, similar pot. The magnetites in the ash attached to the inner surface of the original pot were harvested by applying a permanent magnet of 3,000 G at the outer surface of the pot, and attached dust was blown away by jet air using a small air pump.

The magnet was removed and magnetites were collected in a test tube (50 ml capacity). This process was repeated several times until the required sample size of magnetites was produced.

The test tube was filled with double-distilled water (50 ml) and vortexed for 5 min to provide for good dispersion of the magnetites in the solution. The tube with the content was centrifuged for 20 min at 3,400 rpm, and the supernatant (~30 ml) of the water fraction was poured into another test tube.

Characterization of the MMN in terms of sizes and of chemical components was conducted using transmission electron microscope (Fig. 1), scanning electron microscope and atomic absorption.

17.2.2 Characterization of the MMNs Using Atomic Absorption

MMNs were dissolved in 30 ml of 35% HCL and were subjected to mineral analysis using atomic absorption.

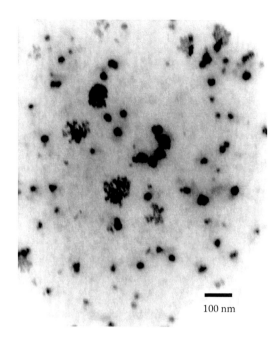

FIGURE 17.1
Image of control of dispersed MMNs with particle size dimensions (≤100 nm) as viewed under a transmission electron microscope (TEM).

17.2.3 Preparation of MMNs for Examination and Imaging under Transmission Electron Microscope (TEM)

MMNs (0.3 g) were dissolved in 50 ml of deionoized water (6 mg/ml) in a test tube and vortexed for 5 min. The MMNs were centrifuged for 20 min at 3,400 rpm, and the supernatant (~30 ml) was transferred to another tube. Then a drop of the supernatant (15 µl) was placed onto a copper grid and dried in a vacuum desiccator. It was then viewed under TEM and photographed (Figure 17.1).

17.2.4 Preparation of MMNs for Examination under Scanning Electron Microscope (SEM)

One drop of the same sample above was placed on a grid; the grid was mounted on a stub and coated with gold by a Polaran E6100 vacuum coater using the sputtering method at 1,200 V and 20 mA for 45 s. Then the samples were studied by SEM (FEI Quanta 200, Netherlands, equipped with EDAX, USA).

The study included morphology and chemical composition using an energy-dispersive x-ray analyzer.

17.2.5 Examining the Effect (Antibacterial Activity) of MMNs on Bacterial Growth

The possibility of antibacterial activity of MMNs on bacteria was examined following the disc diffusion method of Kirby and Bauer (Barry et al., 1979); briefly, a 10-fold serial dilution of MMNs water suspensions was made. An overnight bacterial culture broth was spread evenly on agar plates, and then sterile filter discs were dipped in each MMNs dilution tube (10^{-2}, 10^{-4}, and 10^{-6}) and applied firmly to the surface of the inoculated agar plates. Control

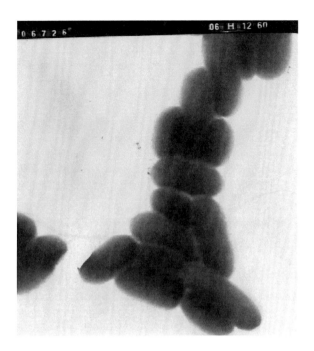

FIGURE 17.2
Intact rod shapes of *Salmonella* bacterial cells (control) before treatment with MMNs as revealed under a TEM.

antibiotics discs were also applied to the agar plates, and the plates were incubated at 37°C for 24–48 h. The inhibition zones around the antibiotic discs were measured if observed.

17.2.6 Preparation and Purification of Bacterial Suspension for Examination and Imaging under Transmission Electron Microscope

Bacterial cells were grown overnight in 10 ml of nutrient broth in a shaker incubator at 37°C, and then 1.5 ml of the bacterial broth was added an Eppendorf tube and centrifuged at 10,000 rpm for 5 min. The supernatant was discarded, and the pellet was washed twice with sterile distilled H_2O and resuspended using vortex. This bacterial suspension was viewed and examined as the control under TEM (Figure 17.2).

17.2.7 Capturing and/or Binding of Bacterial Cells with Unconjugated (Bare) MMNs

- The bacterial suspension (200 µl of *Salmonella typhimurium*) and the MMN suspension (1 ml) were mixed in a sterile Eppendorf tube.
- Three replicates of the previous reaction mixture (bacteria-MMN suspension) were used and incubated in a shaker incubator at 37°C, where the first tube was incubated for 5–10 min, the second tube for 30 min, and the third tube (the third replicate) for 1 h. Then, 15 µl of each tube was placed onto the copper grid and then dried in a vacuum desiccator without using any stain, and then TEM images were performed (Figure 17.3). Figure 17.4 shows this bacterial-MMN aggregate under a TEM with (a) and without (b) stain.
- Two negative controls were also used: 200 µl of bacterial suspension (suspended in deionized water) and 1 ml of MMNs (suspended in deionized water).

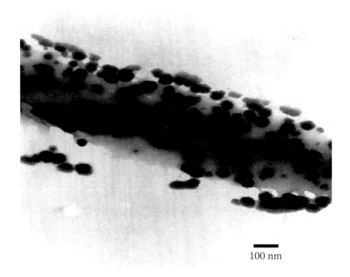

100 nm

FIGURE 17.3
MMNs are attracted and captured by *Salmonella* bacterial cells when mixed for 5–10 min with bacterial suspension as viewed under a TEM.

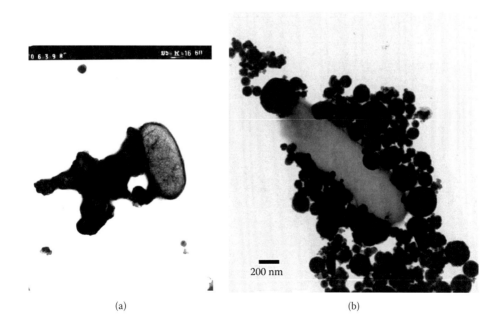

(a)   (b)

FIGURE 17.4
(a) MMNs and *Salmonella* bacterial mixture viewed when stained under TEM. (b) MMN and *Salmonella* bacterial mixture when viewed under a TEM without using any type of staining.

TABLE 17.1

Mineral Composition of MMNs as Revealed by a SEM

Element	Wt %	At %	K-Ratio[a]	Z[b]	A[c]	F[d]
OK	**29.34**	58.41	0.0889	1.0941	0.2767	1.0013
MgK	2.22	2.91	0.0038	1.0542	0.1604	1.0008
AlK	1.27	1.49	0.0028	1.0241	0.2182	1.0015
SiK	3.26	3.69	0.0102	1.0548	0.2966	1.0014
PbM	5.27	0.81	0.0402	0.8284	0.922	1.0003
ClK	2.44	2.19	0.0137	1.0036	0.5545	1.005
KK	0.28	0.23	0.0021	1.0152	0.7168	1.0129
CaK	1.57	1.24	0.013	1.0382	0.7847	1.0206
MnK	2.76	1.6	0.0256	0.9334	0.9626	1.0298
FeK	**27.28**	15.56	0.2651	0.9529	0.9777	1.0431
ZnK	**24.32**	11.85	0.2158	0.9305	0.9464	1.0076
Total	100	100				

Element	Net Intensity of Peak	Background Intensity	Intensity Error	Peak/ Background
OK	38.9	0.95	1.18	40.81
MgK	7.12	2.81	3.59	2.53
AlK	5.76	3.48	4.43	1.65
SiK	22.24	4.72	1.81	4.71
PbM	17.92	6.34	2.21	2.83
ClK	25.56	7.7	1.79	3.32
KK	3.58	9.09	9.33	0.39
CaK	21.04	9.28	2.14	2.27
MnK	26.75	7.63	1.73	3.51
FeK	**248.35**	7.46	0.47	33.29
ZnK	**122.39**	5.57	0.68	21.97

[a] Ratio of intensity of x-ray measure for element in sample/Intensity of a standard for the same element in the database.

[b] Factor for atomic number.

[c] Absorption A.

[d] Fluorescence.

17.2.8 Use of MMN Preparations to Detect Water Samples Artificially Contaminated with Bacteria

- Sterile water samples (negative for the presence of bacteria) were inoculated with the indicated bacterial culture separately; the fresh bacterial cell growth of *Salmonella* was inoculated at different dilutions, and 200 μl of each bacterial suspension dilution was mixed with 1 ml of MMN suspension.

- Three replicates of each of the previous mixtures were used at different incubation times (10 min, 30 min, and 1 h) to evaluate the sensitivity of the assay period. Each tube containing the mixture was incubated with a shaking incubator at 37°C.

- A magnet (400 G) was used for the attraction and removal of the MMN-bacterial mixture aggregate in each tube by placing the magnet on the side walls of the

TABLE 17.2

Total Coliform (TC) cfu/ml of Wastewater Samples before and after Treatment with MMN

Time of Incubation of Wastewater Sample with MMN	TC in Original Sample[a]	TC in Supernatant[b]	TC in Pellet[b]
10 min	$1.2 * 10^8$ cfu/ml	$1.1 * 10^5$ cfu/ml	$1.6 * 10^5$ cfu/ml
20 min	$1.2 * 10^8$ cfu/ml	$4.0 * 10^4$ cfu/ml	$4.4* 10^5$ cfu/ml
30 min	$1.2 * 10^8$ cfu/ml	$7.0 * 10^4$ cfu/ml	$1.4 * 10^5$ cfu/ml
40 min	$1.2 * 10^8$ cfu/ml	$9.0 * 10^4$ cfu/ml	$3.6 * 10^5$ cfu/ml

[a] Before treatment with MMN.
[b] After treatment with MMN.

tubes for 20 min. The supernatant was removed (while the magnet was still on) and transferred into another sterile tube, and the aggregated pellet was washed with sterile deinoized water. For every sample both the supernatant and aggregated pellet were examined under a TEM.

- Serial dilutions for each tube, including the supernatant and pellet, were done; 0.1 ml from each of the different dilutions was inoculated in nutrient agar and EMB agar plates and incubated for 24 h at 37°C. Then the total heterotrophic aerobic bacterial counts and total coliform counts (cfu/ml) were recorded.

17.2.9 Treatment of Wastewater: Using MMNs for Detection, Enumeration, and Removal of Bacterial Contamination

Wastewater samples were collected in sterile containers from different wastewater treatment plants in Jordan. The samples were brought on ice to the laboratory for examination.

- Five milliliters of each wastewater sample was placed in a sterile tube.
- Serial dilutions were done, and 0.1 ml from each dilution was inoculated into nutrient agar medium plates and 0.1 ml onto an EMB agar plate medium and incubated for 24 h at 37°C. The total heterotrophic bacterial and coliform counts (cfu/ml) on each type of media were recorded.
- The MMNs preparation (200 mg) was suspended in 50 ml of sterile deionized water (4 mg/ml), and 200 µl of this MMN suspension was added to the wastewater tube. Four replicates of the previous mixtures were used at different incubation times (10, 20, 30, and 40 min) to evaluate the sensitivity of time in this assay. These four replicates of mixtures were incubated with shaking at 37°C.
- A magnet (400 G) was placed on the side wall of each of the four tubes for 10–20 min, the supernatant was removed (while the magnet was still on) and transferred into another sterile tube, and the pellet was washed by sterile deionized water. Both the supernatant and pellet were examined under a TEM (Figure 17.5a and b).
- Serial dilutions for each tube (for both the supernatant and the pellet) were done; 0.1 ml from each dilution was inoculated in nutrient agar and EMB agar plates and incubated for 24 h at 37°C. Then the total heterotrophic aerobic bacterial counts and the total coliform counts (cfu/ml) were recorded.

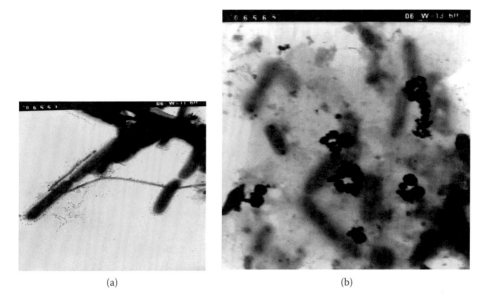

FIGURE 17.5
(a) TEM pictures of wastewater sample before treatment with MMN. (b) Wastewater sample after treatment with MMNs and use of magnet.

17.2.10 Use of MMNs for Detection, Enumeration, and Removal of Bacterial Contamination in Grey Water

Grey water samples were brought on ice in sterile containers to the laboratory.

- Five milliliters of grey water was placed in a sterile tube.
- Serial dilutions were done, and 0.1 ml from each dilution was inoculated into nutrient medium agar plates and EMB medium agar plates and incubated for 24 h at 37°C. Then the total heterotrophic bacterial and coliform counts (cfu/ml) were recorded (Table 17.3).

TABLE 17.3

Total Coliform (TC) cfu/ml of Grey Water Samples before and after Treatment with MMN

Time of Incubation of Grey Water Sample with MMN	TC in Original Sample[a]	TC in Supernatant[b]	TC in Pellet[b]
10 min	$1.2 * 10^7$ cfu/ml	$1.0 * 10^4$ cfu/ml	$1.8 * 10^5$ cfu/ml
20 min	$1.2 * 10^7$ cfu/ml	$3.1 * 10^4$ cfu/ml	$4.0* 10^5$ cfu/ml
30 min	$1.2 * 10^7$ cfu/ml	$5.1 * 10^4$ cfu/ml	$2.2 * 10^5$ cfu/ml
40 min	$1.2 * 10^7$ cfu/ml	$7.3 * 10^4$ cfu/ml	$2.5 * 10^5$ cfu/ml

[a] Before treatment with MMN.
[b] After treatment with MMN.

- MMNs (200 mg) were suspended in 50 ml of sterile deionized water (4 mg/ml), and 200 µl of the MMN suspension was added to each diluted grey water tube. Then four replicates of the previous mixtures were prepared at different incubation times (10, 20, 30, and 40 min) to evaluate the sensitivity of the time in this assay, and incubated with shaking incubator at 37°C.

- A magnet (400 G) was placed on the bottom side of each tube for 20 min, the supernatant was removed (while the magnet was still on) and transferred into other sterile tubes, and the pellet was washed by sterile deionized water.

- Serial dilution for each tube of supernatant and pellet was done; 0.1 ml from different dilutions was inoculated in nutrient agar and EMB agar plates and incubated for 24 h at 37°C. Then the total heterotrophic aerobic bacterial and total coliform counts (cfu/ml) were recorded as before (see Table 17.3).

17.2.11 Transmission Electron Microscopy Sectioning and Imaging

- One milliliter of diluted MMN suspension was placed into 200 µl of freshly prepared *Salmonella* bacterial suspension and incubated for 30 min in a shaker incubator at 37°C.

- Also, a mixture of 1 ml of the MMN suspension and 200 µl of the *Salmonella* suspension was incubated overnight in a shaker incubator at 37°C.

- A freshly prepared bacterial sample (alone) was used as a negative control.

- The three samples above were fixed overnight using special fixative, and dehydrated in ethanol or acetone before being embedded in EPON resin overnight.

- The prepared samples were sectioned using an ultramicrotome, and examined under a transmission electron microscope to reveal the inside of bacterial cells (Figure 17.6).

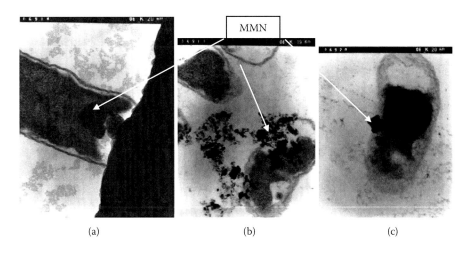

(a)　　　　　　　　(b)　　　　　　　　(c)

FIGURE 17.6

Sections of *Salmonella* bacterial cells when mixed with MMNs as viewed under a TEM. Arrows indicate the penetration of MMNs inside bacterial cells.

17.3 Results and Discussion

A method for the production of multiple-element magnetic nanoparticles (MMNs) from fly ash (waste product in chimneys of iron and steel factories) was achieved successfully through the following steps:

1. Wet sieving of fly ash through a 25 µm sieve and drying in an oven (temperature of 100–110°C for 55 min)

2. Separation of magnetites in the dry product through a magnetic field (3.8×10^4 G)

3. Mixing of the dried magnetic product with sodium chloride salt (NaCl) (ratio: 1:10) in a thin, semispherical ceramic pot and stirring thoroughly to provide for good dispersion

4. Emptying the content of the pot into another, similar pot and harvesting the magnetites in the attached ash to the inner surface of the original ceramic pot by applying a permanent magnet of 3,000 G to the outer surface of the pot

The positive aspects of the method were that it did not use water for the production of nanoparticles, no chemical synthesis was conducted for the preparation of MMNs, the attached dust was blown away by jet air using a small air pump, the raw materials needed for preparation of MMNs were only table salt and fly ash, which are available in abundance, a permanent magnet block (3,000 G) was sufficient to harvest the MMNs, and sodium chloride (NaCl) was the only compound added to the preparation procedures, as a dispersion and dust adsorption agent.

TEM, SEM, and atomic absorption were used for characterization of the sizes and chemical components of MMNs. TEM revealed MMNs dispersed in nanosize dimensions ($\leq$100 nm diameter) (Figure 17.1). The elements in the MMN product were composed mainly of Fe (55.738 ppm) and zinc (11.05 ppm) as revealed by atomic absorption results. Scanning electron microscope results verify the composition of the MMNs of the following oxides: Fe (27.28%), Zn (24.32%), and O_2 (29.34%), which represent the total amount of Fe components (248.35) and Zn (122.39) (Table 17.1).

This method of the production of MMN product proves that MMNs can be preserved for long periods of time in dry conditions without any additives, preservatives, or modifications. We could also conclude that the current method is an efficient way of recycling waste material and cleaning up the environment. In addition, it is an easy and fast, and cheap sources of ingredients are used.

The MMNs were used for detecting and removing bacterial pathogens in water samples when mixing MMNs with pathogen-contaminated water samples to form bacterial pathogen-MMN complexes. Then a magnet was used to aggregate the bacterial pathogen-MMN complexes. A distinguished, intact, rod-shaped bacterial appearance was revealed when control growth bacterial suspension was viewed under a TEM (Figure 17.2). MMNs were attracted and captured by bacterial cells when mixed with bacterial suspension as viewed under TEM (Figure 17.3).

It was also obvious that MMNs improved focusing and removed staining artifacts when viewing bacterial cells under TEM, compared to the regularly used staining methods of bacterial cells (Figure 17.4a and b), which indicates that MMN functions by itself as a staining means. Huge MMN-bacterial aggregates were revealed by TEM when taken from the pellets while placing the magnet on the side wall of the tubes, and few bacterial cells remained in the supernatant, while the rest were captured in the MMN-bacterial aggregates toward the magnet.

When MMNs were used to capture bacterial contaminants, especially coliforms, in the wastewater samples while applying the magnet to the side wall of the tubes, both the supernatant and pellet were examined under a TEM (Figure 17.5a and b). The supernatant showed a reduction of a factor of 10^3–10^4 of bacterial counts (cfu/ml), while the magnet was applied to the side walls of the tubes (Table 17.2), and a reduction of bacterial counts was observed: a factor of 10^2–10^3 for grey water (Table 17.3). This indicates that MMNs are used for fast detection and capturing of bacterial cells (10–20 min maximum), compared to the currently available and used methodologies (2–7 days), and applying MMNs serves as an ideal method of choice for water quality and rapid detection and removal of bacterial contaminants.

When a mixture of MMNs and *Salmonella* cell samples was sectioned using an ultramicrotome and examined under a TEM to reveal the inside of bacterial cells, the results clearly showed that MMNs did enter inside bacterial cells (Figure 17.6). The real mechanism(s) by which MMNs entered inside bacterial cells through the cell wall and cell membrane barriers is not really known, and was not investigated in this study. The suggestion is that it may be through protein-like siderephore compounds secreted by bacterial cells to capture MMNs, or receptor-like compounds on the surface of the bacterial cells that capture the MMNs and engulf them inside, or by invagination of the cell membrane.

The MMNs, when tested, proved to have no effect on bacterial growth and viability when the growth inhibition zone method was applied. In fact, the MMNs were composed of nutrient elements (mainly Fe and Zn), which promoted bacterial growth and had no killing effect. It was also indicated in this study that MMNs penetrated inside bacterial cells as revealed by thin-section imaging under a TEM.

In conclusion, in the application of MMNs, wherein minimal facilities and infrastructures are needed, minimal amounts of MMNs (4–6 mg/ml) are required to detect and decontaminate bacterial pathogens, and inexpensive procedures for both detection and decontamination of microbial pathogens are used. Furthermore, no staining is needed to detect the presence of bacterial cells.

References

Barry AL, Coyle MB, Thornsberry C, Gerlach EH, and Howkinson RW. 1979. Methods of measuring zones of inhibition with the Bauer-Kirby disk susceptibility test. *J. Clin. Microbiol.* 10(6): 885–889.

Cao ZG, Zhou SW, Sun K, Lu XB, Luo G, and Liu JH. 2004. Preparation and feasibility of superparamagnetic dextran iron oxide nanopartciles as gene carrier. *Aizheng* 23: 1105–1109.

Duan H-L, Shon Z-Q, Wang X-W, Chao F-H, and Li J-W. 2005. Preperation of immunomagnetic iron-dextran nanoparticles and application in rapid isolation of *E. coli* O157:H7 from foods. *World J. Gastroenterol.* 11(24): 3360–33664.

El-Boubbou K, Grudea C, and Huang X. 2007. Magnetic glycol-nanoparticles: a unique tool for rapid pathogen detection, decontamination, and strain differentiation. *J. Am. Chem. Soc.* 129: 13392–13393.

Gupta AK and Gupta M. 2005. Synthesis and surface engineering of iron oxide nanoparticles for biomedical applications. *Biomaterials* 26: 3995–4021.

Heo J and Hua SZ. 2009. An overview of recent strategies in pathogen sensing. *Sensors* 9: 4483–4502.

Khaled N, Elshuraydeh KN, Malkawi HI, and Hassuneh M. 2008. Method for the production of bare (non-functionalized) multiple element magnetic nanoparticles and their use in fast detection and removal of pathogenic bacteria from water resources. Registered Patent 2450, Ministry of Industry and Trade, Amman, Jordan. Approved May 2008.

Liu L-H, Dietsch H, Schurtenberger P, and Yan M. 2009. Photoinitiated coupling of unmodified monosaccharides to iron oxides nanoparticles for sensing proteins and bacteria. *Bioconjugate Chem.* 20(7): 1349–1355.

Ravindranath SR, Mauer LJ, Deb-Roy C, and Irudayaraj J. 2009. Biofunctionalized magnetic nanoparticle integrated mid-infrared pathogen sensor for food matrixes. *Anal. Chem.* 91: 2840–2846.

Safarik I and Safarikova M. 2009. Magnetically modified microbial cells: a new type of magnetic adsorbents. *China Particuol.* 5: 19–25.

Tiwari DK, Behari J, and Sen P. 2008. Application of nanoparticles in wastewater treatment. *World Appl. Sci. J.* 3(3): 417–433.

Yang H, Qu L, Wimbrow AN, Jiang X, and Sun Y. 2007. Rapid detection of *Listeria monocytogenes* by nanoparticle-based immunomagnetic separation and real time PCR *Int. J. Food Microbiol.* 118: 132–138.

18

Antimicrobial Nanomaterials in the Textile Industry

Sukon Phanichphant and Suriya Ounnunkad

Materials Science Research Center and Department of Chemistry, Faculty of Science, Chiang Mai University, Chiang Mai, Thailand

CONTENTS

18.1 Introduction

Nanotechnology is an emerging interdisciplinary technology that has prospered in materials science, mechanics, electronics, and aerospace in the past decade. Its profound societal impact has been considered a huge advance, ushering humanity into a more enlightened era.[1–3]

The fundamentals of nanotechnology lie in the fact that properties of substances dramatically change when their size is reduced to the nanometer range. When a bulk material is divided into small size particles with one or more dimensions (length, width, or thickness) in the nanometer range, or even smaller, the individual particles exhibit unexpected properties different from those of the bulk material. It is known that atoms and molecules possess totally different behaviors than those of bulk materials: while the properties of the former are described by quantum mechanics, the properties of the latter are governed by classical mechanics. Between these two distinct domains, the nanometer range is a murky threshold for the transition of a material's behavior. For example, ceramics, which are normally brittle, can easily be made deformable when their grain size is reduced to the low-nanometer range. A gold particle with 1 nm diameter shows red color. Moreover, a small amount of nanosized species can interface with a matrix polymer that is usually in a similar size range, resulting in a structural system with unprecedented performance. These are the reasons why nanotechnology has attracted a large amount of federal funding, research activity, and media attention.

The textile industry has already seen many influences from nanotechnology. Research involving nanotechnology to create unique functions of textile materials is flourishing. These research endeavors mainly focused on using nanosized substances and generating nanostructures during manufacturing and finishing processes.

This chapter emphasizes the use of nanomaterials such as inorganic metals and metal oxide nanoparticles, polymer composites, and dyes in the textile industry as antimicrobial agents. The key advantage of these inorganic antimicrobial agents is improved safety and stability, compared to organic antimicrobial agents.[4] Nanoparticles have a larger surface area and hence higher efficiency than larger-sized particles. Besides, nanoparticles are transparent and do not blur color and brightness of the textile substrates. Research involving the nanoparticles of metal oxide has been focusing on antimicrobial, self-decontaminating, and UV-blocking functions for both military protection gear and civilian health products.[5] For example, nylon fiber filled with ZnO nanoparticles can provide UV shielding and reduce static electricity of nylon fiber. A composite fiber with nanoparticles TiO_2 and MgO now replaces fabrics with active carbon, previously used as chemical and biological protective materials. The photocatalytic activity of TiO_2 and MgO nanoparticles can break up harmful and toxic chemicals and biological agents. These nanoparticles can also be preengineered to adhere to textile substrates by using spray coating or electrostatic methods.[6]

18.2 Antimicrobial Agents

18.2.1 Silver Nanoparticles

Several investigations have been reported on the antimicrobial activity of silver on the textile surface. The polyester-polyamide Ag-loaded textile was carried out by radio frequency (RF)-plasma and vacuum-UV (V-UV) surface activation, followed by chemical reduction of silver salts.[7] The rate of bacterial inactivation by the silver-loaded textile was tested on *E. coli* K-12 and showed a long-lasting residual effect. In an investigation of antimicrobial activity of silver nanoparticles against *E. coli* by Sondi and Salopek-Sondi,[8] the silver hydrosols were prepared by the reduction of $AgNO_3$ with ascorbic acid using Daxad 19 as surfactant. Bacteriological tests were performed on a Luria-Bertani (LB) agar plate supplemented with nanosized silver particles in concentrations of 10–100 µg cm^{-3}. Figure 18.1 shows scanning electron microscopy (SEM) micrographs of native *E. coli* cells and cells treated with 50 µg cm^{-3} of silver nanoparticles in liquid LB medium for 4 h. The treated bacteria cells showed major damage, which was characterized by the formation of "pits" in their cell walls.

Other antimicrobial activities of silver nanoparticles (Ag-NPs) have been performed by Cho and co-workers.[9] In their investigation, Ag-NPs prepared using different stabilizers such as sodium dodecylsulfate (SDS) and poly-(N-vinyl-2-pyrrolidone) (PVP) were tested against *S. aureus* and *E. coli*. The minimum inhibitory concentrations of Ag-NPs for *S. aureus* and *E. coli* were 5 and 10 ppm, respectively. Figure 18.2 shows the antibacterial activity of (A) platinum nanoparticles (Pt-NPs) solution with PVP, (B) Ag-NPs solution stabilized with

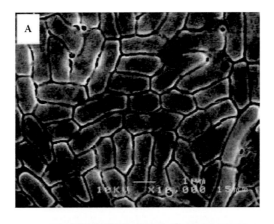

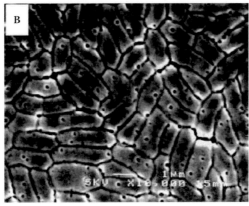

FIGURE 18.1
SEM micrographs of native *E. coli* cells (A) and cells treated with 50 µg cm^{-3} of silver nanoparticles in liquid Lb medium for 4 h. (B) (Reproduced with permission from Sondi, I., and Salopek-Sondi, B., *J. Colloid Interface Sci.*, 275, 177, 2004. Copyright Elsevier.)

PVP, and (C) Ag-NPs solution stabilized with SDS against *S. aureus* (KCTC 1928) (a) and *E. coli* (KCTC 1041) (b). All concentrations of Pt-NPs and Ag-NPs were 10 µl (5.4 ppm).

Son and co-workers[10] reported on their experiments on an antimicrobial cellulose acetate nanofiber containing Ag-NPs using UV irradiation at 245 nm of the polymer nanofibers electrospun with small amounts of silver nitrate. The Ag-NPs generated on the surface of cellulose acetate with an average size of 21 nm exhibited strong antimicrobial activity against Gram-positive *S. aureus* and Gram-negative *E. coli, K. pneumoniae,* and *P. aeruginosa*.

Alternatively, antimicrobial Ag-NPs were immobilized on nylon and silk fibers by following the layer-by-layer deposition method by Dubas and co-workers.[11] The deposition of 20 layers of polydiallydimethylammonium chloride and Ag-NPs capped with polymethacrylic acid onto the fibers resulted in 80% *S. aureus* reduction for the silk fiber and 50% for the nylon fiber. Hydrogel-silver nanocomposites synthesized from Ag-NPs within swollen polyacrylamide-co-acrylic acid hydrogel demonstrated excellent antibacterial effects on *E. coli,* as reported by Thomas et al.[12] The bacterial action depended on the

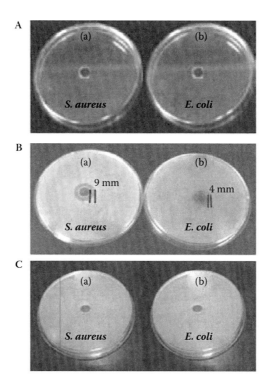

FIGURE 18.2
Antibacterial activity of (A) Pt-NPs solution with PVP, (B) Ag-NPs solution stabilized with PVP, and (C) Ag-NPs solution stabilized with SDS against *S. aureus* (KCTC 1928) (a) and *E. coli* (KCTC 1041) (b). All concentrations of Pt-NPs and Ag-NPs are 10 µl (5.4 ppm). (Reproduced with permission from Cho, K.H., et al., *Electrochem. Acta*, 51, 956, 2005. Copyright Elsevier.)

size of the particles, the amount of Ag-NPs within the hydrogel, and amount of monomer acid in the hydrogel-silver nanocomposites. Figure 18.3 shows the formation of Ag-NPs within the swollen copolymeric network.

When Petica and co-workers[13] investigated the antibacterial and antifungal properties of colloidal silver solutions (CSSs), they discovered that CSSs showed good activities against *Staphylococcus aureus* (ATCC) (Gram-positive cocci), *Pseudomonas aeruginosa* (ATTC), *E. coli* (ATCC), and *Acinetobacter* spp. (Gram-negative coccobacillus). In another, similar investigation by Llić and co-workers,[14] it was found that cotton fabric loaded with silver nanoparticles from 10 and 50 ppm colloids exhibited excellent antimicrobial activity against Gram-negative bacterium *E. coli*, Gram-positive bacterium *S. aureus*, and fungus *C. albicans*. Thus, we can assume that silver nanoparticles exhibited excellent antibacterial and antifungal activities.

Rai and co-workers,[15] in an article on Ag-NPs as a new generation of antimicrobials, described the action mechanism of silver, silver ion–AgNO$_3$, silver zeolite, and Ag-NPs. They also went on to describe the effect of size and shape on the antimicrobial activity of nanoparticles, as well as the applications of silver nanoparticles. Sharma and co-workers[16] reported in their review that Ag-NPs prepared by green synthesis interact with the cell membrane of bacteria. Ag NPs incorporated with surfactants and polymers had advantages in antibacterial activities against Gram-positive and Gram-negative bacteria.

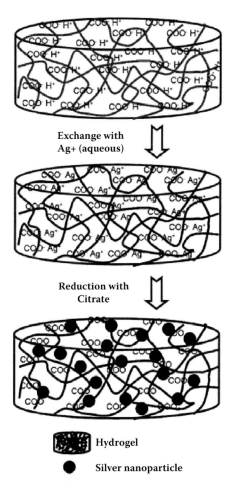

FIGURE 18.3
Formation of silver nanoparticles within the swollen copolymeric network. (Reproduced with permission from Thomas, V., et al., *J. Colloid Interface Sci.*, 315, 389, 2007. Copyright Elsevier.)

18.2.2 Titanium Dioxide or Titania (TiO$_2$) Nanoparticles

TiO$_2$ exhibits a strong antibacterial activity against *E. coli* when exposed to near-UV light (with wavelengths of less than 385 nm).[17] It has also been shown that the TiO$_2$ photocatalytic reaction results in continued antibacterial activity after the UV illumination terminates. In a report by Li and co-workers[18] on the antimicrobial effect of surgical masks coated with nanoparticles consisting of a mixture of silver nitrate and titanium dioxide, antimicrobial activity of the masks was quantified according to the procedure of AATCC 100-1999. The resulting 100% reduction in viable *E. coli* and *S. aureus* was observed in the coated mask after 48 h of incubation. Rana and co-workers[19] reported on the antimicrobial function of Nd^{3+}-doped anatase titania-coated nickel ferrite composite material, using *E. coli* as a microbial. It turns out that doping the titania shell with neodymium significantly enhanced the photocatalytic and antimicrobial function of the core-shell composite nanoparticles without influencing the magnetic characteristics of the nickel ferrite core. The increased performance was related to the inhibition of electron-hole recombination and a decrease in the band gap energy of titania.

In terms of temperature, when Amin and co-workers[20] synthesized a TiO_2-Ag nanocomposite by the sol-gel method and investigated its antibacterial activity against *E. coli*, they showed that a nanocomposite sample calcined at 300°C was ideal to reach the maximum *E. coli* inhibition growth. The calcined TiO_2-Ag powder at 300°C was able to inhibit bacterial growth even in the absence of UV irradiation; however, this antibacterial activity was not observed for the amorphous sample and the sample calcined at 500°C.

Kangwansupamonkon and co-workers[21] investigated the nontoxicity to humans of apatite-coated TiO_2 against four types of bacteria (*S. aureus*, *E. coli*, methicillin-resistant *S. aureus* (MRSA), and *M. luteus*). The cotton fabrics coated with apatite-coated TiO_2 by the dip coating technique were shown to be nontoxic to normal human dermal fibroblasts and had an antimicrobial effect, which inhibited the growth of contaminating microorganisms.

18.2.3 Other Metallic Oxides: ZnO, MgO, CaO, and CeO$_2$

Other metallic oxides, such as ZnO, MgO, CaO, and CeO_2, also exhibited antibacterial activity.[22–26] Antibacterial activities of ZnO, MgO, and CaO powders with mean particle sizes of 2.6, 3.7, and 2.7 μm, respectively, against *S. aureus* and *E. coli* were quantitatively evaluated by measuring the change in electrical conductivity of the growth medium caused by bacterial metabolism.[25] CaO was the most effective against *E. coli*, followed by MgO and ZnO. On the other hand, ZnO was the most effective against *S. aureus*. It has been shown that the antibacterial activity of MgO nanoparticles increases with decreasing particle size.[26]

When Zhang et al.[27,28] investigated the antibacterial behavior of suspensions of ZnO nanoparticles (ZnO nanofluids) against *E. coli*, they found that the antibacterial activity increased with increasing particle concentration, as well as increased with decreasing particle size. The use of dispersants (polyethylene glycol (PEG) and polyvinlypyrolidone (PVP)) did not affect the antibacterial activity of ZnO nanofluids, but instead enhanced the stability of the suspensions. SEM analyses of the bacteria before and after treatment with ZnO nanofluids showed that the presence of ZnO nanoparticles damaged the membrane wall of the bacteria. In another, similar investigation by Tam and co-workers[29] on the antibacterial activity of ZnO nanorods prepared by a hydrothermal method, it was also found that ZnO nanorods showed good activity against a Gram-negative bacteria *E. coli* and a Gram-positive bacteria *B. atrophaeus*.

Sustainability in antimicrobial activity was evaluated using a colony count method with *E. coli* bacteria on a nutrient agar medium (36°C/24 h) in a Na-P-buffer solution.[30] The positively charged CeO_2 nanoparticles at neutral pH were tested on *E. coli*.[31] The nanoparticles displayed a strong electrostatic attraction toward bacterial outer membranes. Nano-MgO crystallites with different particle sizes prepared using four different methods under controlled reaction conditions[32] were active even in the absence of irradiation. They particularly showed excellent bactericidal activity against *S. aureus* ATCC 6538, and their efficiency against *B. subtilis* ATCC 9372 increased with decreasing size of MgO nanoparticles.

Antimicrobial activities have also been tested on CaO nanoparticles in Gram-positive and Gram-negative bacteria by the paper disc method. These calcium oxide nanoparticles were synthesized by the thermal decomposition process using calcium nitrate and sodium hydroxide as starting materials and ethylene glycol as a medium.[33] Synthesized CaO nanoparticles in toluene can inhibit the growth of three kinds of bacteria: *E. coli*, *S. aureus*, and *S. epidermidis*. Table 18.1 shows the effect of calcium oxide nanoparticles on the bacteria. The inhibition zone was defined as the distance from the rim of the filter paper to the edge

TABLE 18.1

The Effect of Calcium Oxide Nanoparticles on the Bacteria *E. coli*, *S. epidermidis*, and *S. aureus*

CaO Nanoparticle	Inhibition Zone Diameter (cm)		
	E. coli	*S. epidermidis*	*S. aureus*
0 ppm	0	0	0
1,000 ppm	0	0	0
3,000 ppm	0.60	0.53	0.35
5,000 ppm	0.83	0.68	0.45

Source: Channei, D., et al., Antimicrobial Activity of Nano Calcium Oxide Synthesized by the Thermal-Decomposition Method, paper presented at Proceedings of the 26th Microscopy Society of Thailand (MST) Annual Conference, Chiang Mai, Thailand, January 28–30, 2009.

of the clear zone, as shown in Figure 18.4. The CaO concentrations at 3,000 and 5,000 ppm were found to effectively inhibit bacterial growth.

In our studies,[34–42] zinc oxide (ZnO), titanium dioxide (TiO$_2$), and silver (Ag) nanoparticles have been synthesized by wet chemical methods such as precipitation, thermal decomposition, sol-gel, and evaporation to dryness processes. Nanoparticles have been characterized by x-ray diffraction (XRD), specific surface area analysis, scanning electron microscopy (SEM), and transmission electron microscopy (TEM). Antimicrobial activities have been tested on the ZnO nanoparticles in Gram-positive (*S. aureus*) and Gram-negative (*E. coli*) bacteria. In our investigation on the effect of nanoparticle sizes on the antimicrobial efficacy, we found that smaller-sized (~20 nm) ZnO nanoparticles had antimicrobial activities toward both Gram-positive and Gram-negative bacteria, while larger-sized ZnO nanoparticles only affect the Gram-negative bacteria. The antimicrobial activities toward bacteria *E. coli* ATCC 25922 and *S. aureus* ATCC 25923 of zinc oxide and titanium dioxide nanoparticles are shown in Table 18.2. Interestingly, TiO$_2$ nanoparticles showed no antimicrobial activity in the absence of UV light.

In practice, zinc oxide is advantageous when compared to titanium dioxide in terms of cost of production. We also attempted to further reduce the cost in our experiment by using the paper disc method with cotton fabric instead of filter paper. It was found that zinc oxide nanoparticles with an average size of 20 nm inhibited the growth of four kinds of fungi, *Rhizopus oligosporus*, *Pennicillium oxalicum*, *Aspergillus awamori*, and *Aspergillus fumigatus*, and three kinds of bacteria, *Escherichia coli*, *Bacillus subtilis*, and *Staphylococcus aureus*. The antimicrobial activity of zinc oxide nanoparticles toward both fungi and bacteria is shown in Table 18.3. The growth inhibition of zinc oxide nanoparticles toward *A. awamori* is shown in Figure 18.5. Putting the experiment into actual use, ZnO nanoparticles were padded on white and green fabric used for a medical protective garment with the appropriate concentrations of 0.3 w/v% with polyethyleneimine as mordant and 0.2 w/v% for white and green fabric, respectively.

Also proposed is the mechanism of the interaction between cotton and zinc oxide nanoparticles:[34] the hydroxylic groups on cotton lose hydrogen atoms, leaving the negatively charged O$^-$, which repels the electron on zinc oxide nanoparticles. In this model, polyethyleneimine (PEI), used as a binder with amino groups, would react with O$^-$ on cotton to form strong electrostatic forces, as shown in Figure 18.6.

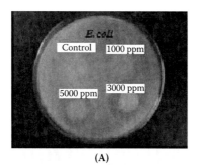

(A)

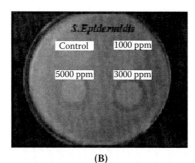

(B)

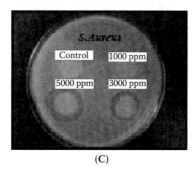

(C)

FIGURE 18.4
Antibacterial activity of CaO nanoparticles in toluene against *E. coli* (A), *S. Epidermidis* (B), and *S. aureus* (C) with concentrations of 1,000, 3,000, and 5,000 ppm. (From Channei, D., et al., Antimicrobial Activity of Nano Calcium Oxide Synthesized by the Thermal-Decomposition Method, paper presented at Proceedings of the 26th Microscopy Society of Thailand (MST) Annual Conference, Chiang Mai, Thailand, January 28–30, 2009.)

Based on our observation and shown in Table 18.4, there was little difference in the diameter across the inhibition zone between 100 µl of 50 and 100 ppm suspension of ZnO nanoparticles. The minimum inhibitory concentration of ZnO nanoparticles toward four kinds of fungi and three kinds of bacteria was found to be 50 ppm by the paper disc method. Growth inhibitions toward *A. fumigatus* and *B. subtilis* by 50 and 100 ppm ZnO with a volume of 100 µl are also shown in Figures 18.7 and 18.8, respectively.

After a washing treatment of the green fabric according to AATCC 135 for 5, 10, and 30 times, the green fabric showed various antimicrobial activities. The fabric that was washed five times showed the widest clear zone. The lowest concentration for ZnO nanoparticles to maintain the antimicrobial activity using the paper disc method was found to be 50 µg/

TABLE 18.2

Antimicrobial Activity toward Bacteria *E. coli* ATCC25922 and *S. aureus* ATCC25923 of Zinc Oxide and Titanium Dioxide Nanoparticles

	Diameter across the Inhibition Zone (mm)[a]	
	E. coli ATCC25922	*S. aureus* ATCC25923
TiO$_2$ (N1) size 40–100 nm	0	0
TiO$_2$ (N2) size 40–100 nm	0	0
TiO$_2$ (N3) size 15–20 nm	0	0
ZnO (J1) size 40–60 nm	16	0
ZnO (J2) size 20–50 nm	11	22
ZnO (T1) size 10–20 nm	8	17
ZnO (T1) size 5–20 nm	11	18

Source: Phanichphant, S., et al., Final Report on Development of Medical Textile Using Nanotechnology: Production of Medical Protective Garment, submitted to Thai Textile Institute, Ministry of Commerce, Thailand, 2006 (in Thai, unpublished).

[a] Average value from two trials.

TABLE 18.3

Antimicrobial Activity of Zinc Oxide Nanoparticles

Microbial	Diameter Across the Inhibition Zone (mm)
Rhizopus oligosporus (mold)	13
Penicillium oxalicum (mold)	13
Aspergillus fumigatus (mold)	15
Aspergillus awamori (mold)	16
Escherichia coli (Gram –)	14
Bacillus subtilis (Gram +)	28
Staphylococcus aureus (Gram +)	13

Source: Phanichphant, S., et al., Final Report on Development of Medical Textile Using Nanotechnology: Production of Medical Protective Garment, submitted to Thai Textile Institute, Ministry of Commerce, Thailand, 2006 (in Thai, unpublished).

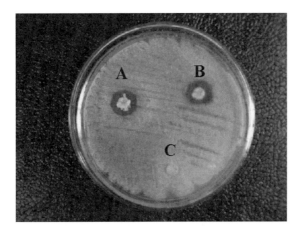

FIGURE 18.5
Growth inhibition of zinc oxide nanoparticles toward *A. awamori*: A and B; C as control. (From Phanichphant, S., et al., Final Report on Development of Medical Textile Using Nanotechnology: Production of Medical Protective Garment, submitted to Thai Textile Institute, Ministry of Commerce, Thailand, 2006 in Thai, unpublished.)

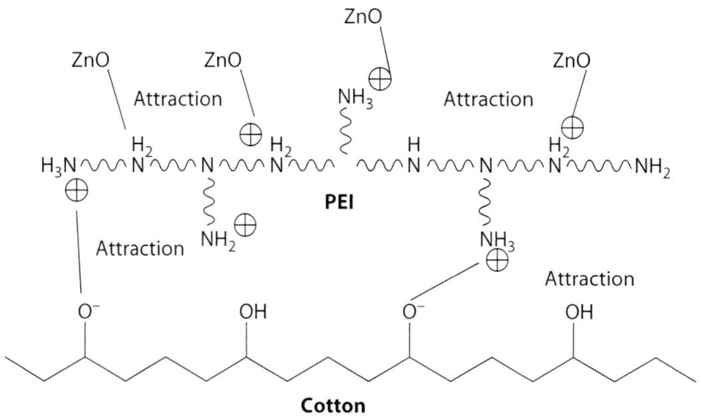

FIGURE 18.6
Schematic diagram showing the interaction between white cotton-PEI-zinc oxide nanoparticles. (From Phanichphant, S., et al., Final Report on Development of Medical Textile Using Nanotechnology: Production of Medical Protective Garment, submitted to Thai Textile Institute, Ministry of Commerce, Thailand, 2006 (in Thai, unpublished).)

ml, or 50 ppm. Then the amount of zinc oxide nanoparticles adsorbed onto the green fabric before washing was determined using atomic absorption spectrometry. The minimum effective dose of ZnO nanoparticles was found to be 0.025 g/m^2. This concentration of ZnO nanoparticles was used to pad the green fabric to produce a medical protective garment for industrial uses.[34]

Copper and silver nanoparticles were also tested against *E. coli* (four strains), *B. subtilis*, and *S. aureus* (three strains).[43] The study showed that the antimicrobial action of silver nanoparticles was superior to that of copper nanoparticles, while a combination of silver and copper oxide (CuO) nanoparticles gave a more complete bactericidal effect against a mixed bacterial population.[44]

In separate research, CuO nanoparticles generated by thermal plasma technology were characterized and investigated with respect to potential antimicrobial applications. The

TABLE 18.4

Growth Inhibition of Zinc Oxide Nanoparticles with 50 and 100 ppm Concentration toward Several Microbes

	Diameter Across the Inhibition Zone (mm)	
	100 μl of 50 ppm ZnO	100 μl of 100 ppm ZnO
Rhizopus oligosporus (mold)	13	17
Penicillium oxalicum (mold)	13	17
Aspergillus fumigatus (mold)	15	18
Aspergillus awamori (mold)	16	20
Escherichia coli (Gram −)	14	16
Bacillus subtilis (Gram +)	28	32
Staphylococcus aureus (Gram +)	13	14

Source: Phanichphant, S., et al., Final Report on Development of Medical Textile Using Nanotechnology: Production of Medical Protective Garment, submitted to Thai Textile Institute, Ministry of Commerce, Thailand, 2006 (in Thai, unpublished).

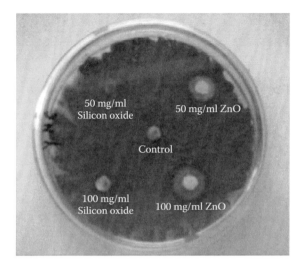

FIGURE 18.7
Growth inhibition toward *A. fumigatus* by 50 and 100 ppm ZnO with a volume of 100 µl. (From Phanichphant, S., et al., Final Report on Development of Medical Textile Using Nanotechnology: Production of Medical Protective Garment, submitted to Thai Textile Institute, Ministry of Commerce, Thailand, 2006 (in Thai, unpublished).

suspension of CuO nanoparticles showed activity against *S. aureus* and *E. coli*, with minimum bactericidal concentrations (MBCs) ranging from 100 to 5,000 µg/ml.[45] The ability of CuO nanoparticles to reduce a bacterial population to zero was enhanced in the presence of silver nanoparticles below the MBCs value. Copper oxide nanoparticles were synthesized and subsequently deposited on the surface of cotton fabric using ultrasound irradiation.

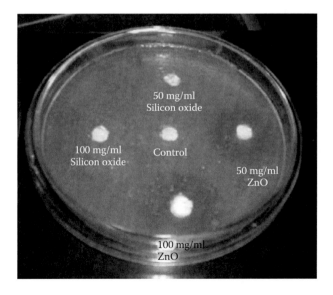

FIGURE 18.8
Growth inhibition toward *B. subtilis* by 50 and 100 ppm ZnO with a volume of 100 µl. (From Phanichphant, S., et al., Final Report on Development of Medical Textile Using Nanotechnology: Production of Medical Protective Garment, submitted to Thai Textile Institute, Ministry of Commerce, Thailand, 2006 (in Thai, unpublished).)

The antibacterial activities of the CuO-fabric composite were tested against *E.coli* (Gram-negative) and *S. aureus* (Gram-positive) cultures. These coated fabrics can have potential applications in wound dressing, bed linens, and as active bandages.

18.3 Natural Antimicrobial Agents

18.3.1 Chitosan Coated on Cotton Fabric

Textile goods, especially those made from natural fibers, are often considered to be more vulnerable to microbial attack than man-made fibers because of their large surface area and ability to retain moisture. Therefore, the use of antibacterial agents to prevent or retard the growth of bacteria is becoming a standard finishing for textile goods; unfortunately, many antibacterial agents are toxic chemicals.[46] Thus, an ideal textile antimicrobial finishing should be safe and environmentally benign, in addition to killing undesirable microorganisms.[47] Chitosan, a natural biopolymer, has a combination of many unique properties, such as biodegradablility, nontoxicity, cationic nature, and antibacterial activity.[48] Because of its polycationic nature, chitosan possesses good antimicrobial properties against various bacteria and fungi through ionic interaction at the cell surface, which eventually kills the bacterial cells.[48] Direct coating of chitosan on cotton fabric was invented to enhance the water solubility of chitosan over the entire pH range using the quaternary ammonium salt, *O*-acrylamidomethyl-*N*-[(2-hydroxy-3 trimetylammonium)propyl] chitosan chloride (NMA-HTCC).[49] The cotton treated with NMA-HTCC by the cold-pad batch method in the presence of an alkaline catalyst at a concentration of 1% on weight of fabric showed 100% *S. aureus* reduction. The effectiveness was maintained over 99% even after being exposed to 50 consecutive home laundering washes. Another improvement to chitosan coating on cotton fabric employed chitosan-based core-shell particles, with chitosan as the shell and a soft polymer as the core.[50] The core-shell particles were synthesized via a graft copolymerization of *n*-butyl acrylate (PBA) from chitosan in aqueous solution. The cotton fabric coated with these PBA-chitosan particles using the pad-dry-cure method showed excellent antibacterial activity, with more than 99% reduction of *S. aureus*.

When Holappa and co-workers[51] studied the effect of the degree of substitution on the antimicrobial activity of water-soluble quaternary chitosan *N*-betainates against *E. coli* and *S. aureus*, they found that the antimicrobial activity increased with decreasing degrees of substitution in acidic conditions (pH 5.5). In a similar investigation by Alonso and co-workers,[52] it was found that the use of citric acid as the cross-linking agent and NaH_2PO_4 as the catalyst over a previously UV-irradiated raw cotton sample gave the highest chitosan incorporation of 27 mg/g of functionalized textile. The control against growth of *P. chrysogenum* and *E. coli* directly related to the amount of chitosan incorporated into the fiber.

Nanochitosan is a natural material with excellent physicochemical properties. It has been used to improve the strength and washability of textiles and to confer antibacterial effects.[53] Nanochitosan (NCH) was prepared from low molecular weight chitosan (LMCS) by Yang and co-workers.[54] The wool fabric was treated with various types of chitosan. The antibacterial and shrink-proofing properties of the treated fabric were ranked as follows: NCH > LMCS > chitosan. The properties also increased as the concentration of NCH increased. In addition, the nanochitosan-treated wool fabric possessed better antibacterial and shrink-proofing properties after washing for 20 times.

18.3.2 Other Natural Antimicrobial Agents

A novel bactericidal fabric coating with potent in vitro activity against meticillin-resistant *S. aureus* (MRSA) was reported by O'Hanlon and Enright.[55] A new N-halamine precursor, 3-(2,3-dihydroxypropyl)-7,7,9,9-tetramethyl-1,3,8-triazaspirol[4,5]decane-2,4-dione (TTDD diol), was synthesized and bonded onto cotton fabrics using a 1,2,3,4-butanetetracarboxylic acid (BTCA) cross-linker at different concentrations.[56] The coated fabric was loaded with various amounts of chlorine upon exposure to dilute household bleach. The biocidal tests against *S. aureus* and *E. coli* showed that the chlorine loadings and surface hydrophobicities influenced the antimicrobial efficacies. Hashem and co-workers[57] reported on enhancing antimicrobial properties of dyed and finished cotton fabrics where 1,2,3-benzothiazole-7-thiocarboxylic acid-S-methylester (commercially known as Actigard® AM-87) was utilized to impart cotton fabric-durable antimicrobial properties. The cotton fabric was padded with 6% Actigard in aqueous solution at pH 5. Treatment of cotton fabric with Actigard turned out to improve its antimicrobial properties toward *S. aureus* and *E. coli*, and also decreased fabric tensile strength, elongation at break, roughness, and whiteness index, while both wettability and the crease recovery angle remained practically intact. Chitosan was also used to increase the henna dye uptake of wool fabric and significantly improved the antimicrobial activity of the dyes.[58] These fabrics were good against washing, light, and perspiration, yet they are nontoxic and eco-friendly.

18.4 Method for Stabilizing Nanoparticles on Textile Surfaces

A pilot-scale sonochemical coating of nanoparticles onto textiles to produce biocidal fabrics was designed by Abramov and co-workers.[59] This pilot system could coat up to 50 m of continuous cotton fabric per run with CuO or ZnO nanoparticles. The coating was homogeneous, stable, and retained its biocidal properties through at least 20 washing cycles. Microwave curing is a possible alternative to conventional curing for improving the mechanical properties of cross-linked textile materials.[60] This method can generate heat uniformly throughout the textile substrate. Microwave radiation is absorbed by molecules having resonant frequencies in the microwave spectral region. When an electric field is applied at microwave frequencies, polar molecules rotate to align their dipole moment with the charging electric field. Energy is absorbed and heat is generated by the internal friction between the rotating molecules. A new microwave curing system was also developed to assist cross-linking of cotton fabric, resulting in easy care and antibacterial properties.[61] Water-soluble chitosan can also be incorporated in the finishing bath in order to produce the antibacterial activity of the fabric. Compared to the conventional curing system, the microwave curing system was found to be advantageous in the production of cotton fabrics with longer-lasting antibacterial properties and without high losses in strength properties.

Dastjerdi and co-workers[62] also presented a novel idea to achieve permanent antibacterial activity with no negative effect on other properties, such as comfort and strength. PET fabric samples were treated with cross-linkable polysiloxane in various ways: simultaneously or after treatment with various concentrations of nanosized colloidal silver. Antibacterial activity was then evaluated and compared with that of *K. pneumoniae* and *S. aureus* according to AATCC 100.

18.5 Mechanism of Action of Nanomaterials

Nanoparticles have an extremely large relative surface area, thus increasing their contact with bacteria or fungi, and vastly improving their bactericidal and fungicidal effectiveness. Silver nanoparticles show excellent antimicrobial activity by binding to (1) microbial DNA, preventing bacterial replication, and (2) the sulfhydryl groups of the metabolic enzymes of the bacterial electron transport chain, causing their inactivity. When silver nanoparticles enter the bacterial cell, they form a low molecular weight region in the center of the bacteria to which the bacteria conglomerate, thus protecting the DNA from the silver ions. The nanoparticles preferably attack the respiratory chain, undergoing cell division and finally leading to cell death. The nanoparticles release silver ions in the bacterial cells, which enhance their bactericidal activity.[8,14,15,27,31] ESR and chemical photoluminescence analyses showed that the radical oxygen of superoxide (O_2^-) originated from the surface of ZnO might exhibit an antibacterial activity even under dark conditions.[30] Several mechanisms have been proposed to explain the antimicrobial activity of chitosan, but its exact mechanism of activity is still unknown. Chitosan is a positively charged molecule, and the target of its antimicrobial activity is the negatively charged cell wall of bacteria, where it binds and disrupts the normal functions of the membrane, e.g., by promoting the leakage of intermolecular components and also by inhibiting the transport of nutrients into the cells.[51] The antimicrobial activity of chitosan increases with decreasing pH.[63] This is due to the fact that the amino groups of chitosan become ionized at pH values below 6 and carry a positive charge. Unmodified chitosan is not antimicrobially active at pH 7 since it does not dissolve, and it does not contain any positive charge on the amino groups.[63] However, the antimicrobial activity of chitosan increases with increasing degree of deacetylation, due to the increasing number of ionizable amino groups.[64] Direct coating of chitosan on cotton fabric was developed by enhancing the water solubility of chitosan over the entire pH range using the quaternary ammonium salt, *O*-acrylamidomethyl-*N*-[(2-hydroxy-3 trimetylammonium)propyl] chitosan chloride (NMA-HTCC).[49] The mechanism of antimicrobial activity of 1,2,3-benzothiazole-7-thiocarboxylic acid-S-methylester (commercially known as Actigard AM-87) was proposed by Hashem and co-workers[57] in that Actigard inhibits or preferably kills microorganisms by a number of different mechanisms that act around the cell wall of the microorganisms. In summary, the following chemical approaches can be utilized by antimicrobial finishes to inhibit or kill microorganisms: cell wall damage, alteration of cytoplasmic membrane permeability, alteration of the physical or chemical state of proteins and nucleic acids, inhibition of enzyme action, and inhibition of protein or nucleic acid synthesis.

18.6 Nanomaterial Toxicity

The toxicities of nanosized and bulk ZnO, CuO, and TiO_2 to bacteria Gram-negative rod-shape *Vibrio fischeri* and crustaceans *Daphinia magna* and *Thamnocephalus platyurus* were studied by Heinlaan and co-workers.[65] The innovative approach based on the combination of traditional ecotoxicology methods and metal-specific recombinant biosensors allowed the researchers to clearly differentiate the toxic effects of metal oxide nanoparticles and solubilized metal ions. It was shown that metal oxide nanoparticles do not necessarily have to enter the cells to cause toxicity. Rather, it seems to be the intimate contact between the bacterial cell wall and the particles. This may cause changes in the microenvironment in the vicinity of the organism-particle contact area, and either increase the solubilization of metals or generate extracellular reactive oxygen species (ROS) that may damage cell membranes. Their results on nano- and bulk metal oxides are in agreement with the SCENIHR Report (2007),[65] which states: "Not all nanoparticle formulations have been found to induce a more pronounced toxicity than the bulk formulations of the same substance." This suggests that the evaluation of nanoparticle formulations should be carried out on a case-by-case basis.

The toxicities of nanoscaled aluminium, silicon, titanium, and zinc oxides to bacteria (*B. subtilis, E. coli*, and *P. fluorescens*) were examined and compared to those of their respective bulk counterparts.[66] All nanoparticles except TiO_2 exhibited signs of toxicity to the tested bacteria, compared to the control. The toxicity of nanoparticles was not only from the dissolved metal ions, but also from their greater tendency to attach to the cell walls than to aggregate together. The sensitivity of bacteria to each type of nanoparticles varied with the species. The ZnO nanoparticles were the most toxic of the nanoparticles studied. The effect of natural environmental parameters on nanoparticle toxicity and the exact toxicity mechanism need to be studied thoroughly.

A thorough review on in vitro assessments' nanomaterial toxicity was reported by Jones and Grainger.[67] The review discussed current methodologies used to assess nanomaterial physicochemical properties and their in vivo effects. Therefore, physicochemical nanomaterial assays need to be improved to provide accurate exposure risk assessments and genuine predictions of in vitro behavior and therapeutic value.

18.7 Conclusions

This chapter elaborates on the use of nanomaterials, particularly metal oxide nanoparticles, polymer composites, and dyes in the textile industry, especially as antimicrobial agents. Metal oxide nanoparticles consist of silver, titanium dioxide, zinc oxide, magnesium oxide, calcium oxide, copper oxide, and cerium oxide nanoparticles. Silver nanoparticles are considered an effective agent against microorganisms. Titanium dioxide nanoparticles require activation by UV light and are less attractive than zinc oxide nanoparticles in the sense that the latter still have antimicrobial activity even in the dark. Considering the production cost, nano-zinc oxide is the most effective. Therefore, the future trend of using nano-zinc oxide in the textile industry is imminent.

Chitosan, chitosan derivatives, and henna dyes are utilized as natural antimicrobial agents in the textile industry. Since most antibacterial agents are toxic chemicals, an ideal textile antimicrobial should be safe and environmentally friendly or benign, in addition to the killing of unintentional microorganisms. Thus, natural antimicrobial agents would be better suited for these purposes.

Methods for stabilizing nanoparticles on textile surfaces, mechanisms of action of nanomaterials, and their toxicities were also mentioned.

Acknowledgments

The National Rearch University Project under the Office of the Higher Education Commission, Ministry of Education, Thailand and Thai Textile Institute, Ministry of Commerce, Thailand are gratefully acknowledged.

References

1. Nanomaterials: a big market potential, *Chemical Week*, October 16, 2002, p. 17.
2. D.R. Forrest, The future impact of molecular nanotechnology on textile technology and on the textile industry; presentation at Discover Expo '95, Industrial Fabric & Equipment Exposition Charlotte, NC, 1995.
3. D.R. Forrest and the Industrial Fabrics Association International, http://www.salsgiver.com/people/forrest/IFAI text.html.
4. http://www.nanophase.com/applications/textile_fibers.asp (accessed May 6, 2006).
5. J. Chen, N. Tsubokawa, Electric properties of conducting composite from poly(ethylene oxide) and poly(ethylene oxide)-grafted carbon black in solvent vapor, *Polymer*, 2000, 32(9), 729.
6. S.M. Hartley, H. Axtell, *The next generation of chemical and biological protective material utilizing reactive nanoparticles*, Gentex Corporation, Carbondale, PA 18407, presented at the Advanced Technology of Materials Synthesis and Processing Information Gatways (ATMSP 2003) conference, May 9, 2003.
7. T. Yuranova, A.G. Rincon, A. Bozzi, S. Parra, C. Pulgarin, P. Albers, J. Kiwi, Antibacterial textiles prepared by RF-plasma and vacuum-UV mediated deposition of silver, *J. Photochem. Photobiol. A Chem.*, 2003, 161, 27–34.
8. I. Sondi, B. Salopek-Sondi, Silver nanoparticles as antimicrobial agent: a case study on *E. coli* as a model for Gram-negative bacteria, *J. Colloid Interface Sci.*, 2004, 275, 177–182.
9. K.H. Cho, J.E. Park, T. Osaka, S.G. Park, The study of antimicrobial activity and preservative effects of nanosilver ingredient, *Electrochem. Acta*, 2005, 51, 956–960.
10. W.K. Son, J.H. Youk, W.H. Park, Antimicrobial cellulose acetate nanofibers containing silver nanoparticles, *Carbohyd. Polym.*, 2006, 65, 430–434.
11. S.T. Dubas, P. Kumlangdudsana, P. Potiyaraj, Layer-by-layer deposition of antimicrobial silver nanoparticles on textile fibers, *Colloid Surf. A*, 2006, 289, 105–109.
12. V. Thomas, M.M. Yallapu, B. Sreedhar, S.K. Bajpai, A versatile stategy to fabricate hydrogel-silver nanocomposites and investigation of their antimicrobial activity, *J. Colloid Interface Sci.*, 2007, 315, 389–395.
13. A. Petica, S. Gavriliu, M. Lungu, N. Buruntea, C. Panzaru, Colloidal silver solutions with antimicrobial properties, *Mater. Sci. Eng. B*, 2008, 152, 22–27.
14. V. Llić, Z. Šaponjić, V. Vodnik, B. Potkonjak, P. Jovančić, J. Nedeljković, M. Radetić, The influence of silver content on antimicrobial activity and color of cotton fabrics funtionalized with Ag nanoparticles, *Carbohyd. Polym.*, 2009, 78, 564–569.
15. M. Rai, A. Yadav, A. Gade, Silver nanoparticles as a new generation of antimicrobials, *Biotechnol. Adv.*, 2009, 27, 76–83.
16. V.K. Sharma, R.A. Yngard, Y. Lin, Silver nanoparticles: green synthesis and their antimicrobial activities, *Adv. Colloid Interface. Sci.*, 2009, 145, 83–96.
17. Z. Huang, P.C. Maness, D.M. Blanke, E.J. Wolfrum, S.L. Smolinski, W.A. Jacoby, Bactericidal mode of titanium dioxide photocatalysis, *J. Photochem. Photobiol. A Chem.*, 2000, 130,163–170.
18. Y. Li, P. Leung, L. Yao, Q.M. Song, E. Newton, Antimicrobial effect of surgical masks coated with nanoparticles, *J. Host. Infect.*, 2006, 62, 58–63.

19. S. Rana, J. Rawat, M.M. Sorensson, R.D.K. Misra, Antimicrobial function of Nd^{3+}-doped anatase titania-coated nickel ferrite composite nanoparticles: a biomaterial system, *Acta Biomater.*, 2006, 2, 421–432.

20. S.A. Amin, AM. Pazouki, A. Hosseinnia, Synthesis of TiO$_2$–Ag nanocomposite with sol-gel method and investigation of its antibacterial activity against *E. coli*, *Powder Technol.*, 2009, 196, 241–245.

21. W. Kangwansupamonkon, V. Lauruengtana, S. Surassmo, U. Ruktanonchai, Antibacterial effect of apatite-coated titanium dioxide for textiles applications, *Nanomedicine*, 2009, 5, 240–249.

22. J. Sawai, S. Shoji, H. Igarashi, A. Hashimoto, T. Kohugan, M. Shimizu, H. Kojima, Hydrogen peroxide as an antibacterial factor in zinc oxide powder slurry, *J. Ferment. Bioeng.*, 1998, 86, 521–522.

23. J. Sawai, H. Kojima, H. Igarashi, A. Hashimoto, S. Shoji, M. Shimizu, Bactericidal action of calcium oxide powder, *Trans. Mater. Res. Soc. Jpn.*, 1999, 24, 667–670.

24. J. Sawai, H. Kojima, H. Igarashi, A. Hashimoto, S. Shoji, T. Sawaki, A. Hakoda, E. Kawada, T. Kokugan, M. Shimizu, Antibacterial characteristics of magnesium oxide powder, *World. J. Microbiol. Biotechnol.*, 2000, 16, 187–194.

25. J. Sawai, Quantitative evaluation of antibacterial characteristics of metallic oxide powders (ZnO, MgO and CaO) by conductometric assay, *J. Microbiol. Meth.*, 2003, 54, 177–182.

26. L. Huang, D.Q. Li, Y.J. Lin, M. Wei, D.G. Evans, X. Duan, Controllable preparation of nano-MgO and investigation of its bactericidal properties, *J. Inorg. Biochem.*, 2005, 99, 586–593.

27. L. Zhang, Y. Jiang, Y. Ding, M. Povey, D. York, Investigation into the antibacterial behaviour of suspensions of ZnO nanoparticles (ZnO nanofluids), *J. Nanopart. Res.*, 2007, 9, 497–489.

28. L. Zhang, Y. Ding, M. Povey, D. York, ZnO nanofluids—A potential antimicrobial agent, *Prog. Natl. Sci.*, 2008, 18, 939–944.

29. K.H. Tam, A.B. Djurišić, C.M.N. Chan, Y.Y. Xi, C.W. Tse, Y.H. Leung, W.K. Chan, F.C.C. Leung, D.W.T. Au, Antibacterial activity of ZnO nanorods prepared by a hydrothermal method, *Thin Solid Films*, 2008, 516, 6167–6174.

30. K. Hirota, M. Sugimoto, M. Kato, K. Tsukagoshi, T. Tanigawa, H. Sugimoto, Preparation of zinc oxide ceramics with a sustainable antimicrobial activity under dark condition, *Ceram. Int.*, 2010, 36, 497–506.

31. A. Thill, O. Zeyons, O. Spalla, F. Chauvat, J. Rose, M. Auffan, A.N. Flank, Cytotoxicity of CeO$_2$ nanoparticles for *Escherichia coli* physico-chemical insight of the cytotoxicity mechanism, *Environ. Sci. Technol.*, 2006, 40, 6151–6156.

32. L. Huang, D.Q. Li, Y.J. Lin, M. Wei, D.G. Evans, X. Duan, Controllable preparation of nano-MgO and investigation of its bactericidal properties, *J. Inorg. Biochem.*, 2005, 99, 986–993.

33. D. Channei, S. Phanichphant, N. Pimpra, Antimicrobial activity of nano calcium oxide synthesized by the thermal-decomposition method, paper presented at Proceedings of the 26th Microscopy Society of Thailand (MST) Annual Conference, Chiang Mai, Thailand, January 28–30, 2009, collected by the Microscopy Society of Thailand.

34. S. Phanichphant et al., Final report on development of medical textile using nanotechnology: production of medical protective garment, submitted to Thai Textile Institute, Ministry of Commerce, Thailand, 2006, in Thai, unpublished.

35. N. Wetchakul and S. Phanichphant, Synthesis and characterization of titanium dioxide nanoparticles coated on fly ash, *Int. J. Nanosci.*, 2006, 5(4–5), 657–662.

36. C. Liewhiran, S. Seraphin, S. Phanichphant, Synthesis of nano-sized ZnO powders by thermal decomposition of zinc acetate using *Broussonetia papyrifera* (L.) Vent pulp as a dispersant, *Curr. Appl. Phys.*, 2006, 6, 499–502.

37. C. Dechakiatkrai, J. Chen, C. Lynam, S. Phanichphant, G.G. Wallace, Photocatalytic degradation of methanol using titanium dioxide/single-walled carbon nanotube composite, *J. Electrochem. Soc.* 2007, 154(5), A407–A411.

38. N. Wetchakun, S. Phanichphant, Effect of temperature on the degree of anatase-rutile transformation in titanium dioxide nanoparticles synthesized by the modified sol-gel method, *Curr. Appl. Phys.*, 2008, 8, 343–346.

39. K. Wetchakun, N. Wetchakun, S. Phanichphant, Enhancement of the photocatalytic performance of Ru-doped TiO_2 nanoparticles, *Adv. Mater. Res.*, 2008, 55–57, 65–68.

40. C. Dechakiatkrai, J. Chen, C. Lynam, N. Wetchakul, S. Phanichphant, G.G. Wallace, Direct growth of carbon nanotubes onto titanium dioxide nanoparticles, *J. Nanosci. Nanotechnol.*, 2009, 9, 955–959.

41. C. Dechakiatkrai, C. Lynam, K.J. Gilmore, J. Chen, S. Phanichphant, D.V. Bavykin, F.C. Walsh, G.G. Wallace, Single-walled carbon nanotube/trititanate nanotube composite fibers, *Adv. Eng. Mater.*, 2009, 11(7), B55–B60.

42. P. Pookmanee, S. Phanichphant, Titanium dioxide powder prepared by a sol-gel method, *J. Ceram. Proc. Res.*, 2009, 10(2), 167–170.

43. G. Ren, D. Hu, E.W.C. Chen, M.A. Vargas-Reus, P. Reip, R.P. Allaker, Characterisation of copper oxide nanoparticles for antimicrobial applications, *Int. J. Antimicrob. Agent*, 2009, 33, 587–590.

44. J.P. Ruparelia, A.K. Chatterjee, S.P. Duttagupta, S. Mukherji, Strain specificity in antimicrobial activity of silver and copper nanoparticles, *Acta Biomater.*, 2008, 4, 707–716.

45. I. Perelshtein, G. Applerot, N. Perkas, E. Wehrschuetz-Sigl, A. Hasmann, G. Guebitz, A. Gedanken, CuO-cotton nanocomposite: formation, morphology, and antibacterial activity, *Surf. Coat. Technol.*, 2009, 204, 54–57.

46. T.L. Vigo, Protection of textiles from biological attack, in M. Lewin, S.B. Sello, eds., *Handbook of fiber science and technology: chemical processing of fibers and fabrics, functional finishes*, Vol. II, Part A, Marcel Dekker, New York, 1983, pp. 367–426.

47. L.M. Corcoran, *Determining the processing parameters and conditions to apply antimicrobial finishes on 100% cotton and 100% polyester dyed knit fabrics*, Institute of Textile Technology, Charlottesville, VA, 1998.

48. S.H. Lim, S.M. Hudson, Review of chitosan and its derivatives as antimicrobial agents and their uses as textile chemicals, *J. Macro. Sci. Polym. Rev.*, 2003, C43(2), 223–269.

49. S.H. Lim, S.M. Hudson, Application of a fiber-reactive chitosan derivative to cotton fabric as an antimicrobial textile finish, *Carbohyd. Polym.*, 2004, 56, 227–234.

50. K.F. El-tahlawy, M.A. El-bendary, A.G. Elhendawy, S.M. Hudson, The antimicrobial activity of cotton fabrics treated with different crosslinking agents and chitosan, *Carbohyd. Polym.*, 2005, 60, 421–430.

51. J. Holappa, M. Hjálmarsodóttir, M. Másson, O. Rúnarsson, T. Asplund, P. Soininen, T. Nevalainen, T. Järvinen, Antimicrobial activity of chitosan N-betainates, *Carbohyd. Polym.*, 2006, 65, 114–118.

52. D. Alonso, M. Gimeno, R. Olayo, H.Vázquez-Torres, J.D. Sepúlveda-Sánchez, K. Shirai, Cross-linking chitosan into UV-irradiated cellulose fibers for the preparation of antimicrobial-finished textiles, *Carbohyd. Polym.*, 2009, 77, 536–543.

53. D.R. Ting, Y. Shen, Antifinishing with chitosan derivatives and their nani-particles, *Dyeing Finishing*, 2005, 14, 12–14.

54. H.C. Yang, W.H. Wang, K.S. Huang, M.H. Hon, Preparation and application of nanochitosan to finishing treatment with anti-microbial and anti-shrinking properties, *Carbohyd. Polym.*, 2010, 79, 176–179.

55. S.J. O'Hanlon, M.C. Enright, A novel bactericidal fabric coating with potent *in vitro* activity against meticillin-resistant *Staphylococcus aureus* (MRSA), *Int. J. Antimicrob. Agent*, 2009, 33, 427–431.

56. X. Ren, A. Akdag, H.B. Kocer, S.D. Worley, R.M. Broughton, T.S. Huang, N-Halamine-coated cotton for antimicrobial and detoxification applications, *Carbohyd. Polym.*, 2009, 78, 220–226.

57. M. Hashem, N.A. Ibrahim, W.A. El-Sayed, S. El-Husseiny, E. El-Enany, Enhancing antimicrobial properties of dyed and finished cotton fabrics, *Carbohyd. Polym.*, 2009, 78, 502–510.

58. V.R. Giri Dev, J. Venugopal, S. Sudha, G. Deepika, S. Ramakrishna, Dyeing and antimicrobial characteristics of chitosan treated wool fabrics with henna dye, *Carbohyd. Polym.*, 2009, 75, 646–650.

59. O.V. Abramov, A. Gedanken, Y. Koltypin, N. Perkas, I. Perelshtein, E. Joyce, T.J. Mason, Pilot scale sonochemical coating of nanoparticles onto textiles to produce biocidal fibrics, *Surf. Coat. Technol.*, 2009, 204, 718–722.

60. S. Vukusic, D. Katovic, Influence of microwave on nonformaldehyde DP finishing dyed cotton fabric, *Textile Res. J.*, 2003, 73, 733–738.
61. M.M.G. Fouda, A. El Shafei, S. Sharaf, A. Hebeish, Microwave curing for producing cotton fabrics with easy care and antibacterial properties, *Carbohyd. Polym.*, 2009, 77, 651–655.
62. R. Dastjerdi, M. Montazer, S. Shahsavan, A new method to stabilize nanoparticles on textile surfaces, *Colloid. Surf. A*, 2009, 345, 202–210.
63. T.C. Yang, C.C. Chou, C.F. Li, Antimicrobial activity of *N*-akylated disaccharide chitosan derivatives, *Int. J. Food Microb.*, 2005, 97, 237–245.
64. X.F. Liu, Y.L. Guan, D.Z. Yang, Z. Li, K.D. Yao, Antimicrobial action of chitosan and carboxymethylated chitosan, *J. Appl. Polym. Sci.*, 2001, 79, 1324–1335.
65. M. Heinlaan, A. Ivask, I. Blinova, H.C. Dubourguier, A. Kahru, Toxicity of nanosized and bulk ZnO, CuO and TiO_2 to bacteria *Vibrio fischeri* and crustaceans *Daphinia* and *Thamnocephalus platyurus*, *Chemosphere*, 2008, 71, 1308–1316 and references therein. Because in this reference cited "SCENIHR (Scientific Committee on Emerging and Newly Identified Health Risks) report. The appropriateness of the risk assessment methodology in accordance with the technical guidance documents for new and existing substances for assessing the risks of nanomaterials. Approved for public consultation on 29 March 2007. http://ec.europa.eu/health/ph_risk/committees/04_scenihr/docs/scenihr_o_004c.pdf
66. W. Jiang, H. Mashayekhi, B. Xing, Bacterial toxicology comparison between nano- and microscaled oxide particles, *Environ. Pollut.*, 2009, 157, 1619–1625.
67. C.F. Jones, D.W. Grainger, *In vitro* assessments of nanomaterial toxicity, *Adv. Drug Deliv. Rev.*, 2009, 61, 438–456.

19

Recent Advances in Water Desalination through Biotechnology and Nanotechnology

Shaurya Prakash, Karen Bellman, and Mark A. Shannon

Department of Mechanical Engineering, Ohio State University, Columbus, Ohio, USA

CONTENTS

19.1 Introduction

Recent reports and various publications have identified the importance of clean water for not only public health, but also for energy and security needs.[1–4] Consider the following statistics that present a sobering picture for the current state of water supplies and the importance for advancing the science and technology to increase freshwater supplies. Based on World Health Organization (WHO) reports, nearly 2.4 million people die every year due to contaminated water, and a child under the age of 5 dies every 20 s. The most common waterborne diseases, affecting millions of people worldwide, are malaria, cholera,

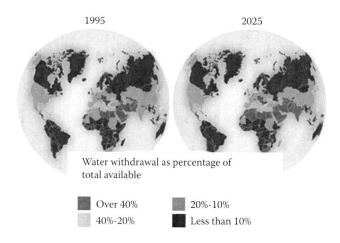

1995   2025

Water withdrawal as percentage of
total available

■ Over 40%   ■ 20%-10%
  40%-20%   ■ Less than 10%

FIGURE 19.1
(See color insert.) Water withdrawal rate as a percentage of total available water, projected until 2025. (From Service, R.F., *Science*, 313, 1088–1090, 2006. With permission.)

and diarrhea. In addition, it has been estimated that over 20 million people are affected by Arsenic poisoning in the Bengal region of the Indian subcontinent (including Bangladesh). Given these brief statistics, it is not surprising that there is a renewed focus on developing new technologies for water purification.[2,5]

Among the many challenges that exist toward generating high-quality potable water, desalination of saltwater has been identified as a key step toward making progress for developing sustainable sources for freshwater.[6] However, to create effective desalination and subsequent distribution strategies, it is important to review how the water is distributed around the planet. The total available water on earth is approximately 1.4 $\times 10^{21}$ L (or about 3.3 $\times 10^8$ mi^3). Of this seemingly enormous water supply, over 99% of the water is currently inaccessible to human use, and more than 97% of the earth's water exists in oceans, bays, seas, and saline aquifers as large reservoirs of saltwater. In fact, several estimates place the approximate supply of usable freshwater at ~0.7%, or about 9.8 $\times 10^{18}$ L. While this may appear to be a large number, accounting for population increases, demands on freshwater for agriculture and industry, potentially changing climate, declining freshwater quality from worldwide contamination via industrial, municipal, and agricultural discharge, and increasing energy needs leads to a rather bleak picture for future availability of clean water for human use. Consider a recent editorial review[6] that reports data from the World Meteorological Organization (WMO): the combination of uneven population and water distribution is causing rapidly increasing water withdrawal rates as a fraction of total available water (Figure 19.1), and it is projected that by 2025 (only 15 years from now) most of the world's population will be facing serious water stresses and shortages. Therefore, the ability to affordably and sustainably desalinate water can resolve many of the impending and projected water crises. The purpose of this chapter is to provide a succinct literature review of the current and emerging state of science and technology for water desalination, with a particular focus on achieving the highest possible energy efficiency in water desalination.

19.2 Review of Existing Water Desalination Methods

A huge variety of water desalination methods exist. In this section, a brief review of the most common methods is presented; therefore, not all techniques or references are included in this review. However, the reader is pointed to several review articles and books, along with a broad variety of literature, to allow further reading given the reader's specific interests.

19.2.1 Theoretical Minimum Energy Requirement for Water Desalination

Consider an equilibrium analysis for the process of water desalination. From the second law of thermodynamics, for a reversible process, the amount of energy used for such a process is independent of the method used.[7] As a consequence, given the starting salinity of water and the target salinity of desalted water, it is possible to calculate a theoretical minimum amount of energy required for water desalination. This exercise may appear academic, but is of value since the energy-water nexus[8,9] is a major consideration toward evaluating the current state of water desalination technologies and identifying new technologies (see Section 19.3). Consider an ideal compressor for moving water vapor from a tank of seawater (typically assumed at ~35,000 ppm or mg/L salinity) to a tank of freshwater (typically assumed at ~500 ppm salinity).[7] Using these conditions, the minimum energy requirement for water desalination at 25°C with a recovery rate of zero (i.e., a negligibly small amount of water produced from a near-infinite amount of seawater) is 0.70 kWh/m³.[7] The recovery rate is defined as the ratio of freshwater produced to the inlet saltwater. For a viable system, recovery ratios must be maximized in contrast to the waste or brine streams. An increase of the recovery rates to 25, 50, and 75% requires theoretical energy minimums of 0.81, 0.97, and 1.29 kWh/m³ of freshwater product, respectively.[2] Therefore, in principle it should be possible to remove salt from water in an efficient manner with low energy consumption. However, several challenges exist to achieving these theoretical limits.

19.2.2 Challenges to Desalination Processes

Inherent irreversibilties present in real systems typically drive the energy requirements higher than the theoretical minima. Some of these irreversibilities relate to the presence of organics and particulates, and varying pH and salinity of source waters. Others relate to the operation of mechanical and electrical equipment at varying energy efficiencies. The varying source water content can lead to scale formation and deposition or membrane fouling in desalination plants.[10] In order to mitigate these problems, desalination plants often employ extensive pretreatment steps involving intensive chemical treatment processes, including precipitation, flocculation, lime softening, ion exchange columns, or mechanical processes such as aeration and sedimentation. For example, in order to minimize scaling, pretreatment of feedwater by introducing an acid, followed by CO_2 degassing, has been shown to be an effective method for preventing alkaline scale formation.[10] Antiscalants are particularly popular due to their effectiveness at low concentrations; the chief chemical families from which antiscalants have been developed are condensed polyphosphates, organophosphonates, and polyelectrolytes. Of these three classes of compounds, polyphosphates are the most economical, while effectively retarding scale formation and offering corrosion protection.[10] Organophosphonates are suitable for a wider range of operating pH and temperature conditions than polyphosphates.[10] The main consequence of all pretreatment processes is that they all involve increased energy consumption and material costs, regardless of the specific desalination methodology employed. There is also growing concern that the rejection of these pretreatment chemicals poses environmental threats due to their toxicity. The subsequent treatment, if required, to remove the antiscalants also consumes energy.

19.2.3 Common Separation Methods

About 88% of the available desalination production capacity employs either reverse osmosis (RO) or multistage flash (MSF) distillation processes for freshwater production from either brackish water or seawater.[11] Other main technologies include multieffect distillation (MED), vapor compression (VC), and electrodialysis (ED).[11] The primary energy requirement for multistage flash, multieffect distillation, and vapor compression is in the form of thermal energy, while reverse osmosis requires primarily mechanical energy, and electrodialysis requires primarily electrical energy. Other methods, such as solar distillation, freezing, gas hydrate processes, membrane distillation, capacitive deionization, and ion exchange, are used for desalination, but current technology levels for these processes find limited use and are not commercially viable on a worldwide scale for widespread implementation.

19.2.3.1 *Multistage Flash (MSF) Distillation*

The MSF process is well suited for highly saline or contaminated waters[12] because flashing of water vapor from the top of brine pools allows for minimal scale formation, as the precipitates resulting from evaporation form in liquid rather than the critical surfaces of heat transfer.[7] The desalination procedure begins by heating incoming seawater by condensing steam contained in a set of tubes running through the brine heater. The MSF process elements are depicted schematically in Figure 19.2. To reduce energy costs, MSF plants are

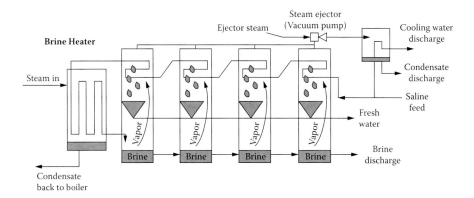

FIGURE 19.2
A block diagram schematic showing the basic elements of a multistage flash (MSF) distillation process. (Based on ref. 12; Trieb, F., *Concentrating Solar Power for Water Desalination*, ed. G.A.C. (DLR) and I.o.T. Thermodynamics, Federal Ministry for the Environment, Nature Conservation and Nuclear Energy, Stuttgart, Germany, 2007; Spiegler, K.S., *Salt water purification*, New York: John Wiley & Sons, 1962.)

often combined with steam cycle power plants, allowing for the utilization of the cooled steam from such production facilities[12] in combined heat and power (CHP) cycles. The incoming water is often pretreated with antiscalants and heated to 90–110°C (194–230°F), with higher temperatures being avoided due to concerns of scale development,[13] particularly of calcium sulfate and calcium carbonate.[7] The heated saltwater is processed through a series of chambers at decreasing pressures, causing the water to "flash," or immediately boil upon entry into the chamber[13]; i.e., for a given temperature, the chamber pressure is matched to the vapor pressure to induce boiling. The number of successive chambers used can be as high as 40, although many systems employ around 20 such distillate collection areas.[12] MSF plants are rated with performance factors such as the gained output ratio (GRO), which is the mass of desalinated water produced to the mass of steam.[14] For a 20-stage MSF plant, a conventional GRO is 8, with a typical heating requirement of about 290 kJ/kg product water.[14] Although the highest in terms of energy needed per unit product water produced, the MSF desalination process provides proven reliability as well as the ability to deal with highly contaminated feedwaters. The first commercial multistage flash plants were developed as early as the 1950s, and the method is most popular in Middle Eastern countries, particularly in Saudi Arabia, Kuwait, and the United Arab Emirates, due to MSF's ability to operate with the highly saline and particulate-laden waters of the Persian Gulf.[6,14]

19.2.3.2 Multieffect Distillation (MED)

MED processes for commercial water distillation were also introduced in the 1950s.[14] This technology borrows from the plants that were first developed to produce sugar from sugar cane juice or salt through evaporation.[13] MED systems were employed for solutions at lower temperatures than MSF, and thus have found use for processes that have left-over steam heat below 100°C, although today many dedicated MED systems are used for water desalination alone. It has been estimated in the literature that the MED process offers superior thermal performance to the MSF processes, but the scaling problems within the plants were noted to be higher.[12,14] Current systems resolve some of the scaling and corrosion issues by operating at a maximum brine temperature of approximately 70°C, and in some cases systems employ a maximum brine temperature of 55°C, allowing for utilization of low-grade waste heat.[12] The MED process begins by distributing (usually by spraying) the preheated saline feedwater onto a heat exchange surface as a thin film to encourage boiling and evaporation of water[13] through a surface area enhancement for improved heat transfer. The vapor phase of the water is then condensed in a lower-pressure chamber, much like the MSF design,[12] while simultaneously creating vapor to be fed into the next chamber, as shown in Figure 19.3. The MED process continues for 8 to 16 effects or cycles, depending on plant design.[13] It is not uncommon for MED plants to be integrated with additional heat inputs between stages, usually by thermal vapor compression (TVC) or mechanical vapor compression (MVC).[12] Energy consumption in MED plants incorporated with the cold end of a steam cycle requires about 145–390 kJ/kg in process steam withdrawn from a steam turbine and 1.5–2.5 kWh/m³ electricity for control and pumping processes.[12] Due to the relatively low operating temperatures of the MED process, a large amount of surface area is required by the system, thus requiring large areas for MED production facilities.[13] The integration of MED plants with TVC reduces surface area and the number of effects needed per plant capacity.[13]

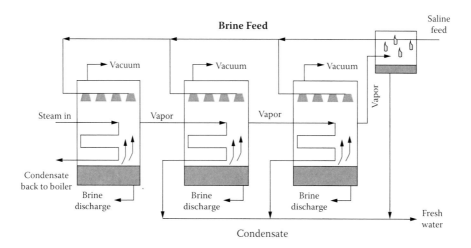

FIGURE 19.3

A block diagram schematic showing the basic elements of a multeffect distillation (MED) process. (Based on ref. 12; Trieb, F., *Concentrating Solar Power for Water Desalination*, ed. G.A.C. (DLR) and I.o.T. Thermodynamics, Federal Ministry for the Environment, Nature Conservation and Nuclear Energy, Stuttgart, Germany, 2007.)

19.2.3.3 Vapor Compression (VC) Processes

By themselves, vapor compression processes (both thermal and mechanical) offer simple, consistent operation and are usually used for small- to medium-scale desalination units. They are often utilized for applications such as resorts, industries, or drilling sites with a lack of direct access to freshwater supply.[12] Figure 19.4 shows a schematic representation of the unit processes involved in vapor compression cycles for water desalination. Mechanical vapor compression (MVC) units are generally single-stage plants with a production capacity up to 3,000 m^3/day. MVC systems typically operate as single-stage units, as only the capacity, and not the efficiency, is increased with additional stages.[14]

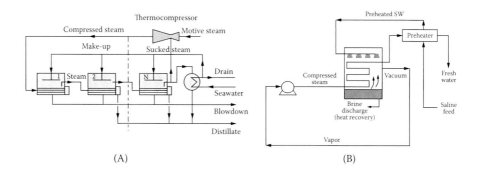

FIGURE 19.4

Schematic representations of basic unit processes involved in vapor compression cycles for water desalination with (A) the thermal vapor compression system and (B) a single-stage mechanical vapor compression system. (Based on ref. 12; Trieb, F., *Concentrating Solar Power for Water Desalination*, ed. G.A.C. (DLR) and I.o.T. Thermodynamics, Federal Ministry for the Environment, Nature Conservation and Nuclear Energy, Stuttgart, Germany, 2007.)

In the case of thermal vapor compression (TVC) plants, multiple stages are used due to increased efficiency, with designed capacities approaching 36,000 m³/day.[12,14] Vapor compression systems generate the heat for evaporation by compressing feed vapor as the heat source for the heat exchanger (i.e., changes in density), as opposed to mechanically produced heat steam, as in MSF and MED.[13] In thermal vapor compression units, steam ejectors are used for vapor compression, whereas in mechanical vapor compression units, a mechanical compressor is used.[12] Vapor compression is often coupled with MED systems, which increases the efficiency of the system while raising the steam pressure requirement.[12]

19.2.3.4 Reverse Osmosis (RO)

In contrast to the thermal processes discussed so far, another class of water desalination systems arises from filtration technologies relying on the use of polymeric membranes. Membrane systems[15] are becoming increasingly popular due to the low cost of polymeric membranes and relatively high-energy efficiencies. The water industry extensively uses membrane processes for pretreatment filtration, depending on particle size (e.g., micro-, nano- or ultrafiltration). Conceptual treatment of ions as hard spheres allows for filtration to work with ionic salts, similar to particles, if small enough pores are used in membranes. This model led to the development of reverse osmosis (RO) desalination. In RO, a semipermeable membrane separates the feedwater from the effluent stream. RO works by pressurizing the salinated feedwater to a pressure higher than the osmotic pressure of the solution, causing the semipermeable membranes of the system to reject most of the solute (in this case, mostly salts), while allowing solvent (freshwater) to pass through.[12] A schematic of the RO process is shown in Figure 19.5. Since osmotic pressure is determined by the salinity of the feedwater, the pressure requirements for brackish water desalination are much lower (15–25 bar) than those for seawater desalination (54–80 bar).[14] It should be noted that although the osmotic pressure of seawater is approximately 25 bar,[16] higher operating pressures are required to achieve practical flows, as well as balance the increasing salinity of the feedwater and overcome concentration polarization across the membranes as the

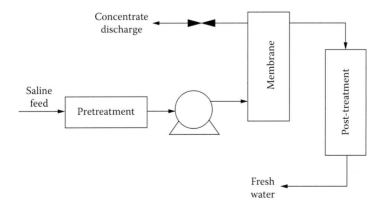

FIGURE 19.5
A simple schematic representation of the RO process showing the essential unit processes. (Based on refs. 12, 20; Trieb, F., *Concentrating Solar Power for Water Desalination*, ed. G.A.C. (DLR) and I.o.T. Thermodynamics, Federal Ministry for the Environment, Nature Conservation and Nuclear Energy, Stuttgart, Germany, 2007.)

desalination process progresses.[14,17] Concentration polarization on the external surfaces of the membrane results from the concentration imbalance created by the solution dilution on one side of the membrane and the solution concentration on the other during separation processes. It should be noted that while progress has been made in developing high-flux membranes,[18,19] increasing flux often leads to increasing polarization.[2] Conceptually, it is useful to imagine the concentration polarization region as a pseudo-membrane in series with the physical membrane, impeding flow and adding to the overall energy losses in the system.

Due to the increasing pressure requirements with feedwater salinity increase, 40% is a common product water recovery rate for these systems,[14] although recent systems often run at 50%–55% recovery at the expense of slightly higher (~0.5 Wh/L) energy consumption. Higher recovery ratios reduce the size of the system needed, as well as the amount of pre-treatment of the feedwater. Feedwater quality is of high concern in RO systems, as a given membrane material operates within a fixed set of operating parameters relating to pH, organics, algae, bacteria, particulates, and other foulants found in salinated water.[12] As a consequence, RO systems are typically characterized by several pre-treatment stages. From an energy standpoint, RO systems come closest to approaching the thermodynamic limits, requiring approximately 2.5 Wh/L of water product for the RO process alone; however, it should be noted that the total process, including pretreatment, recirculation, and distribution, requires additional energy.[5] Of concern in many areas with RO is the crossover of boron (usually in the form of boric acid), which in concentrations above 0.6 µg/L (ppb) can be toxic to plants. Thus, for source waters high in boron, and for desired uses of the desalinated waters to be below 0.6 ppb, a second RO stage is added to take approximately 20–30% of the product water from the first stage. The pH of the first-stage product water is raised above 10 to transform the boric acid to borate, which is more successfully rejected by the RO membranes. The two remaining product water streams are blended to bring the boron concentrations down to acceptable levels. The overall salt concentration is also reduced, but the usage of a partial second stage results in higher overall energy usage. Recent improvements in the RO process have been brought about by the use of more selective membranes and the implementation of energy recovery devices that concentrate the stream exiting the pressure vessels.[13] The minimum energy for a given amount of product water is a complicated function of minimizing the cost of pre- and post-treatment, the energy required, and capital costs. Therefore, each location and system typically operates slightly differently.

19.2.3.5 Electrodialysis (ED)

Electrodialysis was introduced in industrial desalination applications in the 1960s as a membrane desalination technique suited to the removal of ions from brackish water.[20] The conceptual idea is based on the filtration mechanism used in medical systems and the principles for artificial kidney operation, with these systems presenting the first evidence of the influence of bionanotechnology on commercial water desalination processes. The ED process shown in Figure 19.6 removes electrically charged salt molecules of saline water by employing alternating stacks of cationic or anionic selective membranes, with ionic flux driven by an applied electric field.[14] As the cations and anions are pulled through the membranes, the salinity of the water drops at each stage and freshwater is produced (Figure 19.6). Energy requirements of ED systems are based on the concentration of ions within the water; thus, electrodialysis can be competitive with RO processes at concentrations up to 3,000 ppm.[20] A full membrane stack for this process can be composed of a bank of membranes comprising hundreds of layers of alternating anionic and cationic

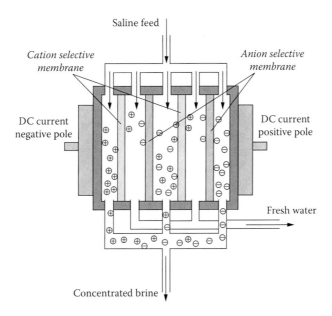

FIGURE 19.6
Schematic representation of a bank of membranes being used for electrodialysis. (Based on ref. 12; Trieb, F., *Concentrating Solar Power for Water Desalination*, ed. G.A.C. (DLR) and I.o.T. Thermodynamics, Federal Ministry for the Environment, Nature Conservation and Nuclear Energy, Stuttgart, Germany, 2007.)

selective membranes.[14] One of the main challenges to ED is the high energy consumption rates needed to obtain high flux rates of desalted water. High flux demands higher flow rates, which are determined by the applied DC potential at the electrodes. The power consumption, P, for a given salt solution of resistance R varies with the square of the current, I, flowing through the system. To move the ions across the solution and membranes, a DC potential at the electrodes has to be added to move ions through the solution, $\eta_{solution}$, across the membrane, $\eta_{membranes}$, and over the potentials needed to overcome the Faradaic losses at the electrodes (i.e., the transformation of electrons in the metal to ions in solution, usually via the electrolysis of water), $\eta_{overpotential}$. The power consumed can be expressed as

$$P = I\left(\eta_{solution} + \eta_{membranes} + \eta_{overpotential}\right) \qquad (19.1)$$

The potentials across the solution and membranes (to the first order) are simply proportional to the current, or $\eta_{solution} = C_{J}I$. However, the overpotentials at the electrodes can grow exponentially with the current, such that $\eta_{overpotential} = C_{F}I^{(n+1)}$, where $n \geq 1$ and $C_{F} \gg C_{J}$ for concentrated salts. Therefore, the power consumption for desalination by ED can increase rapidly, such that

$$P = C_{J}I^{2} + C_{F}I^{(n+2)} \qquad (19.2)$$

so that when either high fluxes are desired or concentrated solutions are being desalted, or both, the power consumption can greatly exceed both RO or thermal methods. However, at low fluxes and relatively low salt concentrations (<3,000 ppm), ED can be among the most efficient methods.

Other challenges in ED are similar to those with all membrane processes and relate to membrane fouling, membrane response to changing pH conditions when H^{+} and OH^{-} ions are transported across membrane surfaces, and concentration polarization.[14,21] To mitigate fouling issues, the electrodialysis reversal (EDR) process was developed, which changes the polarity of the applied biases, allowing the system to be flushed periodically.[13] One key advantage of the ED process is its ability to deal with higher levels of uncharged particles in comparison with RO processes, as these particles do not travel though the membranes.[14] ED recovery rates for brackish water range from 80% to 90%, while RO recovery rates for this type of water range from 65% to 75%.[22] Energy consumption of ED systems ranges from 1.2 to 2.5 kWh/m^3 for brackish water.[22]

Variants to ED are also being developed. For example, it has been projected by Siemens that in 2030 and beyond, Siemens will provide nearly 75% of the urban water demand in Australia through continuous electrodeionization (CEDI) technologies, reaching an estimated population of 25 million people at energy consumption levels of 1 MWh/ml.[23] ED and electrodialysis reversal (EDR)-based technologies are also being investigated and employed by GE in many of its water desalination plants.[24] The promise of low energy usage has also encouraged use of other electrical methods, such as capacitive deionization, which is being used by companies (e.g., PROINGESA) in Europe.[25]

19.3 Emerging Water Desalination Technologies

The idea of generating freshwater from saltwater is not new, and as discussed above, a variety of methods and processes already exist to desalinate water.[26,27–29] However, with water, there has been a renewed interest in developing new technological platforms that can provide solutions for the next generation of water desalination needs.[2,30]

As discussed above, membrane processes are becoming increasingly popular for water desalination. The minimum thermodynamic energy needed for desalination assumes complete energy recovery, minimal polarization impedance, no membrane viscous and fouling losses, and no energy losses for membrane cleaning. Furthermore, these thermodynamic estimates assume that the source water stream is comprised only of salt and water; i.e., the energy required for all pre-filtration steps for removal of pathogens, organic waste, colloids, etc., is not included.

There are two main approaches being targeted for developing new water desalination methods. First is to improve existing technologies by integration with well-established separation or distillation technologies not previously used with water, or revisiting older methods with advanced technologies. Examples include using forward osmosis,[31] humidification-dehumidification cycles,[32] and trapping solar energy for thermal distillation processes.[33] The second approach relies on exploiting transport phenomena at the nanoscale, and therefore employs use of nanofluidics, nanotechnology, bio-inspired methods, and nanomaterials.[5,18,34–43] Both approaches present their own sets of pros and cons, including challenges in energy consumption, management of waste streams, fabrication and cost considerations, and eventual output flux for given water quality. Of the various methods being considered, the bio- or nanotechnology-driven approaches are considered to be particularly promising, as these approaches work at length scales where fundamental physical processes occur. Consequently, with recent advances in fabrication techniques, it is possible to develop systems that can manipulate these physical processes, and thereby develop the next generation of water desalination systems. Next, a discussion of some of the emerging bio- and nanotechnology approaches is presented.

19.3.1 Microfluidic and Nanofluidic Concentration Polarization for Desalination

Concentration polarization in most water desalination systems is considered to be a loss term, and immense resources and effort have been employed in better understanding and mitigating polarization effects.[15,44,45] Recent research has revealed that the imbalances created by concentration polarization (enrichment and depletion regions) can be sustained indefinitely when triggered across a nanochannel connecting two microchannels.[46] Utilizing this fact, researchers demonstrated the ability to desalinate water using a setup as shown in Figure 19.7, thereby exploiting polarization for separations. However, for efficient separation, pre-treatment was needed for the removal of Ca^{2+} ions, and physical filtration for elimination of precipitation and large debris. The membrane-less system that was tested was primarily driven by electrokinetic flows and achieved a 99% salt rejection ratio.[5] In addition to achieving seawater desalination, this system was shown to remove most solid particles, microorganisms, and biomolecules due to their charged nature.[5] The energy consumption of this system for the actual separation process is modest due to the deflection of ions, rather than physical displacement across a membrane, and the low flow resistance of microchannels, with an energy consumption of approximately 5 Wh/L for flow rates of 0.25 µl/min.[5] Although considerable challenges lie ahead for the scaling up of a system using the concentration polarization technique, massive parallelization of the proof-of-concept device is estimated to deliver approximately 180–288 ml/min for a small-scale system.[5]

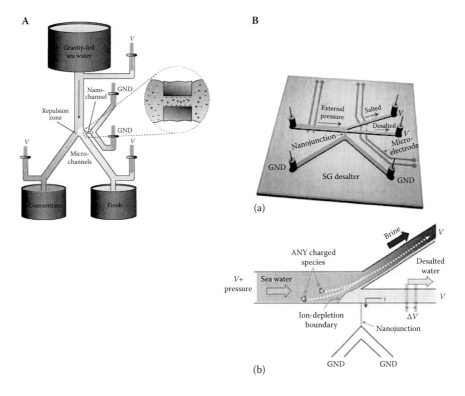

FIGURE 19.7
(A) Small-scale setup for concentration polarization desalination. (From Shannon, M.A., *Nature Nanotechnology*, 5, 248–250, 2010.) (B) Schematic of microchannel-nanochannel prototype tested for water desalination using concentration polarization. (From Kim, S.J., et al., *Nature Nanotechnology*, 5, 297–301, 2010.)

19.3.2 Advanced Membranes

In approximately the past 5 years many new classes of membranes with functional nanoscale components, or those inspired by biological systems, have been developed as discussed in a recent review article.[47] One kind of advanced membranes is the fouling-resistant membrane[48,49] with integrated polymeric brushes and nanostructures, which allow it to have both hydrophilic and hydrophobic regions that mitigate fouling. A second kind of advanced membrane, composed of aligned carbon nanotubes (CNTs), exhibits super-high flux for water transport.[18,19] The measured fluxes were found to be up to three orders of magnitude higher than those predicted by conventional Hagen-Poiseuille flow theory. These aligned CNT membranes were shown to have average tube diameters of 1.6 nm[18] and ultra-smooth, hydrophobic walls that do not follow the basic no-slip condition, thereby letting fluids slide along the wall in an essentially frictionless configuration, permitting high flux through the CNTs. While water flows nearly effortlessly through the aligned CNTs, the energy cost is expended in the water molecules entering the CNTs, and in the modification of the mouths of the CNTs to reject small salt ions, which increases the energy needed. The final energy usage for these membranes is still to be determined for concentrated salt solutions with a high rejection potential.

19.3.3 Forward Osmosis

Desalination by the process of forward or direct osmosis (FO) employs a highly concentrated solution (also referred to as the draw solution) to create an osmotic pressure that extracts freshwater from salinated water across a semipermeable membrane.[50] Unlike the RO processes, because forward osmosis uses osmotic pressure as the driving force, this method operates at small or no hydraulic pressure, which reduces the amount of observed fouling.[17,51] The reliance on osmotic pressure necessitates that a major factor in the choice of the draw solution is the osmotic pressure of the draw solution.[17] The lack of hydraulic pressure allows forward osmosis setups to be uncomplicated, with relatively little membrane support, as the only force the flow has to overcome is the few bars provided by the membrane itself[17] as the physical barrier. Possibilities for the draw solution are numerous; for water desalination solutions of sulfur dioxide, aluminum sulfate, glucose, potassium nitrate, mixtures of glucose and fructose, and mixtures of ammonia (NH_3) and carbon dioxide (CO_2) gases have all been suggested as draw solution candidates.[17] However, many draw solutions add an after taste or an odor requiring post-treatment. Sugar (glucose and fructose primarily) has been a popular draw solution due to the minimal need for extensive post-treatment of desalinated water. The major energy used in FO is to remove the draw solution. For draw solutions of NH_3 and CO_2, which both are soluble in water, low-quality thermal energy can be used to decrease the solubility and decrease the energy needed to volatilize the NH_3 and CO_2, to remove as a gas and reuse upstream. A major hindrance to this technology has been the development of an appropriate membrane.[52] Ideally, FO membranes should be able to allow a high flux of water while retaining a high rejection rate of dissolved solids, demonstrating compatibility with both the feedwater and the draw solution, and handling the mechanical stresses imposed by the osmotic pressures experienced during the desalination process.[52] Currently available commercial membranes, such as cellulose triacetate (CTA) membranes, are not compatible with many of the preferred draw solutions. However, due to the low pressure across the membrane, fragile membranes made from aquaporins, which are nature's perfect nanochannel and only permit water to flow through, can potentially be used for FO.

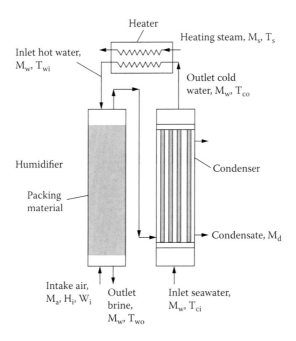

FIGURE 19.8
Schematic of conventional humidification-dehumidification process. (From Ettouney, H., *Desalination*, 183, 341–352, 2005. With permission.)

19.3.4 Humidification-Dehumidification Desalination

Humidification-dehumidification (HDH) desalination processes rely on distillation to achieve desalination.[53] Simple technology is required due to their ability to function at atmospheric pressure; generally, HDH units require two exchangers, an evaporator to humidify air and a condenser to collect the freshwater generated, as shown in the basic system schematic in Figure 19.8. The process begins by allowing dry air to absorb vapor from heated feedwater in one chamber, transferring the humid air to another chamber, where it comes in contact with a cool surface, thus allowing freshwater collection.[27] Energy consumption in such a system is required by the heat source as well as the pumps and blowers to move the water and vapor. Recovery rates for these systems range from 5 to 20%, and heat sources such as solar or geothermal supplies are common choices, especially since the rates of freshwater production are suited for demands of a few m^3/day. Temperature differences between 2 and 5°C are required, necessitating heat consumption of approximately 20.9 kJ/m^3.[53]

19.3.5 Microbial Desalination Cells

Microbial fuel cells convert biowaste to electricity through microbial activity[54] by generating electrons available for harvesting in an external load circuit. In an innovative approach, a three-cell system was converted into a water desalination system (Figure 19.9). In a proof-of-concept study it was shown that the three-cell design with an integrated anion exchange membrane (AEM) and cation exchange membrane (CEM) could generate a small amount of potential (typically less than 1 V) and desalinate water. The bacteria grow on the anode side and discharge protons into the water; however, the protons cannot pass

A

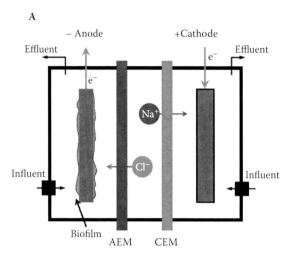

FIGURE 19.9
(A) Basic schematic representing a three-chamber cell design for coupling a microbial fuel cell with a water desalination system. (B) Microbial desalination cell prototype. (From Cao, X., et al., *Environmental Science Technology*, 43(18), 7148–7152, 2009. With permission.)

through the anion exchange membrane, so negatively charged ions from the salinated water flow through the AEM to balance the positive charges produced. A similar process takes place on the cathode side, except protons are consumed, requiring positively charged ions from the salinated water to cross the CEM to correct the charge imbalance. Current testing of these systems has been conducted with NaCl solutions at concentrations of 5, 20, and 35 g/L, which is consistent with concentrations seen from brackish to seawater. The amount of salt removed from each concentration was at least 88% ± 2% for the 5 g/L case, and up to 94% ± 3% for the 20 g/L case.[54] Comparison of the electrons harvested and NaCl removed revealed a charge transfer efficiency of almost 100% due to negligible effects of electrolysis (because of the low current generated) and insignificant back diffusion of ions from the electrode chambers to the desalination chamber.[54]

19.3.6 Asymmetry-Driven Ion Pumps

As with nanotechnology, biology has also provided clues to developing advanced water desalination systems by following the ideas of ion transport in biological ion channels and ion pumps.[55] While existential proof for the working of these systems has been around in living organisms for a long time, implementing technologies for practical systems has continued to be a challenge due to several gaps in engineering these systems, and due to a lack of mechanistic understanding of how to manipulate ion transport in artificial systems. However, some recent studies have shown promise in advancing the state of knowledge toward both the science and technology of these bioinspired systems.[36,55–58] In particular, mimicking ion channel gating behavior for transport and control of ions has generated significant interest, leading to the development of fluidic transistors[37] and conical nanopores for ion pumps for the potassium ion[59] (Figure 19.10). One advantage of bioinspired

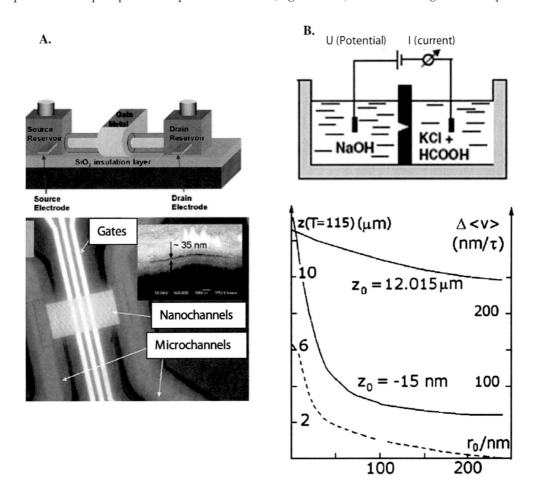

FIGURE 19.10
(A) Schematic and SEM image showing the development of a device being used as a fluidic transistor allowing electrostatic control of ions. (From Karnik, R., et al., *Nano Letters*, 5(5), 943–948, 2005. (with permission.)) (B) Schematic depicting formation of a conical nanopore by electrochemical etching at potential U with current I, leading to data that show pumping of the potassium ions against a concentration gradient as a function of pore diameter on the narrow side of the conical nanopore. (From Siwy, Z., and Fulinski, A., *Physical Review Letters*, 89(19), 19803-1–19803-4, 2002. With permission.)

systems is that the fundamental processes occur at the nanoscale, and engineering these systems could allow for rapid development of bionanotechnology as a core area for water desalination research. Further, with most biological systems operating near thermodynamic energy minima, such systems hold the promise of delivering the most energy-efficient water desalination systems ever conceived. However, significant questions still remain open and would likely need extensive investigations.

19.4 Summary and Conclusions

In this broad-brushed review of the current state of knowledge of water desalination methods, a succinct comparison between commonly used methods and opportunities for further research and growth have been discussed. Advantages and disadvantages of each approach were discussed with an eye on the need for further understanding in either basic science or engineering systems to implement these technologies, with the primary focus being on energy efficiency for water desalination. While biotechnology and nanotechnology provide interesting clues on developing novel systems for ion transport with applications in water desalination, many scientific and technological questions remain open. It is likely that further investigation is needed for most all the emerging areas listed above, especially the ones that rely on engineering a nanoscale system based on inspiration from biology.

References

1. UNICEF and WHO. *Water for life: making it happen*. United Nations, Geneva, 2006.
2. Shannon, M.A., et al. Science and technology for water purification in the coming decades. *Nature*, 452: 301–310, 2008.
3. Torcellini, P., N. Long, and R. Judkoff. *Consumptive water use for U.S. power production*. National Renewable Energy Laboratory, Golden, CO, 2003.
4. Yeston, J., et al. A thirsty world. *Science*, 313: 1067, 2006.
5. Kim, S.J., et al. Direct seawater desalination by ion concentration polarization. *Nature Nanotechnology*, 5: 297–301, 2010.
6. Service, R.F. Desalination freshens up. *Science*, 313: 1088–1090, 2006.
7. Spiegler, K.S., *Salt water purification*, New York: John Wiley & Sons, 1962.
8. Veerapaneni, S., et al. Reducing energy consumption for seawater desalinations. *Journal of the American Water Works Association*, 99: 95–106, 2007.
9. U.S. Department of Energy. *Energy demands on water resources*. Sandia National Laboratory, Livermore, CA, pp. 1–80, 2006.
10. Hasson, D., and R. Semiat. Scale control in saline and wastewater desalination. *Israel Journal of Chemistry*, 46: 97–104, 2006.
11. Zhou, Y., and R.S.J. Tol. Evaluating the costs of desalination and water transport. *Water Resources Research*, 41(W03003): 1–10, 2005.
12. Trieb, F. *Concentrating solar power for water desalination*, G.A.C. (DLR) and I.o.T. Thermodynamics, eds. Federal Ministry for the Environment, Nature Conservation and Nuclear Energy, Stuttgart, Germany, 2007.

13. Buros, O.K. *The ABCs of desalting*. International Desalination Association, 2000, http://www.idadesal.org/pdf/ABCs.pdf
14. Miller, J.E. *Review of water resources and desalination technologies*. Sandia National Laboratories, Livermore, CA, 2003.
15. Laksminarayanaiah, N. *Transport phenomena in membranes*. Academic Press, New York, 1969.
16. Reid, C.E. Principles of reverse osmosis. In *Desalination by reverse osmosis*, U. Merten, ed. MIT Press, Cambridge, MA, pp. 1–14, 1966.
17. Cath, T.Y., A.E. Childress, and M. Elimelech. Forward osmosis: principles, applications, and recent developments. *Journal of Membrane Science*, 281: 70–87, 2006.
18. Holt, J.K., et al., Fast mass transport through sub-2nm carbon nanotubes. *Science*, 213: 1034–1037, 2006.
19. Majumder, M., et al. Experimental observation of enhanced liquid flow through aligned carbon nanotube membranes. *Nature*, 438: 44–45, 2005.
20. Fritzmann, C., et al. State-of-the-art reverse osmosis desalination. *Desalination*, 216: 1–76, 2007.
21. Forgacs, C., et al. Polarization at ion-exchange membranes in electrodialysis. *Desalination*, 10: 181–214, 1972.
22. Pilat, B. Practice of water desalination by electrodialysis. *Desalination*, 139: 385–392, 2001.
23. Siemens. *What will the future of salt water desalination look like?* Case study. 2010. (http://aunz.siemens.com/PicFuture/Documents/WaterCasestudy_desalination.pdf).
24. Allison, R.P. *Electrodialysis treatment of surface and waste waters*. Technical paper, GE Water and Process Technologies Paper TP1032EN. (http://www.gewater.com/pdf/Technical%20Papers_Cust/Americas/English/TP1032EN.pdf).
25. Innovations. *Electrochemical capacitors for water desalination*, January 22, 2009. (http://www.innovations-report.com/html/reports/energy_engineering/electrochemical_capacitors_water_desalination_125885.html).
26. Bhattacharya, A., and Ghosh, P. Nanofiltration and reverse osmosis membranes: theory and application in separation of electrolytes. *Reviews in Chemical Engineering*, 20: 111–173, 2004.
27. Bourouni, K., M.T. Chaibi, and L. Tadrist. Water desalination by humidification and dehumidification of air: state of art. *Desalination*, 137: 167–176, 2001.
28. Darwish, M.A., and H. El-Dessouky. The heat recovery thermal vapour-compression desalting system: a comparison with other thermal desalination processes. *Applied Thermal Engineering*, 16: 523–527, 1996.
29. Glueckauf, E. Seawater desalination—in perspective. *Nature*, 211: 1227–1230, 1966.
30. Darwish, M.A. Desalting: fuel energy cost in Kuwait in view of $75/barrel oil price. *Desalination*, 208: 306–320, 2007.
31. McCutcheon, J.R., and M. Elimelech. Modeling water flux on forward osmosis: implications for improved membrane design. *AIChE Journal*, 53: 1736–1744, 2007.
32. Enezi, G.L., E. Hisham, and N. Fawzy. Low temperature humidification dehumidification desalination process. *Energy Conversion and Management*, 47: 470–484, 2006.
33. Jackson, R.D., and C.H.M. van Bavel. Solar distillation of water from soil and plant materials: a simple desert survival technique. *Science*, 149: 1377–1378, 1965.
34. Daiguji, H., Y. Oka, and K. Shirono. Nanofluidic diode and bipolar transistor. *Nano Letters*, 5(11): 2274–2280, 2005.
35. Daiguji, H., Yang, P. and A. Majumdar. Ion transport in nanofluidic channels. *Nano Letters*, 4(1): 137–142, 2004.
36. Gong, X., et al. A charge-driven molecular water pump. *Nature Nanotechnology*, 2: 709–712, 2007.
37. Karnik, R., et al. Electrostatic control of ions and molecules in nanofluidic transistors. *Nano Letters*, 5(5): 943–948, 2005.
38. Kemery, P.J., J.K. Steehler, and P.W. Bohn. Electric field mediated transport in nanometer diameter channels. *Langmuir*, 14(10): 2884–2889, 1998.
39. Qiao, R., and N.R. Aluru. Charge inversion and flow reversal in a nanochannel electroosmotic flow. *Physical Review Letters*, 92(19): 198301-1-198301-4, 2004.

40. Qiao, R., and N.R. Aluru. Atomistic simulation of KCl transport in charged silicon nanochannels: interfacial effects. *Colloids and Surfaces A*, 267: 103–109, 2005.
41. Qiao, R., and N.R. Aluru. Surface charge induced asymetric electrokinetic transport in confined silicon nanochannels. *Applied Physics Letters*, 86: 143105–143107, 2005.
42. Qiao, R., J.G. Georgiadis, and N.R. Aluru. Differential ion transport induced electroosmosis and internal recirculation in heterogeneous osmosis membranes. *Nano Letters*, 6(5): 995–999, 2006.
43. Shannon, M.A. Fresh for less. *Nature Nanotechnology*, 5: 248–250, 2010.
44. Gray, G.T., R.L. McCutcheon, and M. Elimelech. Internal concentration polarization in forward osmosis: role of membrane orientation. *Desalination*, 197: 1–8, 2006.
45. Pedley, T.J. Calculation of unstirred layer thickness in membrane transport experiments: a survey. *Quarterly Review of Biophysics*, 16: 115–150, 1983.
46. Pu, Q., et al. Ion-enrichment and ion-depletion effect of nanochannel structures. *Nanoletters*, 4(6): 1099–1103, 2004.
47. Prakash, S., M.B. Karacor, and S. Banerjee. Surface modification in microsystems and nanosystems. *Surface Science Reports*, 64: 233–254, 2009.
48. Asatekin, A., et al. Antifouling nanofiltration membranes for membrane bioreactors from self-assembling graft copolymers. *Journal of Membrane Science*, 285: 81–89, 2006.
49. Asatekin, A., et al. Anti-fouling ultrafiltration membranes containing polyacrylonitrile-graft-poly(ethlene oxide) comb copolymer additives. *Journal of Membrane Science*, 298: 136–146, 2007.
50. Lange, K.E. Get the salt out. *National Geographic*, 217(4): 32–35, 2010.
51. Holloway, R.W., et al. Forward osmosis for concentration of anaerobic digester centrate. *Water Research*, 41: 4005–4014, 2007.
52. Yip, N.Y., et al. High performance thin-film composite forward osmosis membrane. *Environmental Science Technology*, 44: 3812–3818, 2010.
53. Ettouney, H. Design and analysis of humidification dehumidification desalination processes. *Desalination*, 183: 341–352, 2005.
54. Cao, X., et al. A new method for water desalination using microbial desalination cells. *Environmental Science Technology*, 43(18): 7148–7152, 2009.
55. Hille, B. *Ionic channels of excitable membranes*. Sinauer Associates, Sunderland, MA, 1992.
56. Prakash, S., et al. Development of a hydrogel-bridged nanofluidic system for water desalination. Paper presented at 233rd National Meeting and Exposition. American Chemical Society, Chicago, 2007.
57. Prakash, S., et al. Characterization of ionic transport at the nanoscale. *Proceedings of the Institution of Mechanical Engineers, Part N: Journal of Nanosystems and Nanoengineering*, 220(2): 45–52, 2007.
58. Hinds, B.J. A blueprint for a nanoscale pump. *Nature Nanotechnology*, 2: 673–674, 2007.
59. Siwy, Z., and A. Fulinski. Fabrication of a synthetic nanopore ion-pump. *Physical Review Letters*, 89(19): 19803-1–19803-4, 2002.

20

Liposomal pH-Sensitive Nanomedicines in Preclinical Development

Maria Jose Morilla and Eder Lilia Romero

Programa de Nanomedicinas, Universidad Nacional de Quilmes, Buenos Aires, Argentina

CONTENTS

20.1 Introduction: Nanomedicines Are the Key to Intracellular Targeting

In order to increase the therapeutic efficacy of a drug or to reduce its toxicity, biodistribution and intracellular traffic have to be changed, since the sole modification of its pharmacokinetics is insufficient. Nanomedicines (nano-objects loaded with small drugs or macromolecules) are powerful tools developed in the framework of the application of nanotechnology to medicine[1] that, functioning as nano-drug delivery systems, are capable of modifying the pathway followed by molecules.[2–5] Pharmacokinetics, biodistribution, and intracellular traffic of loaded molecules no longer depend on their chemical structures but on the size, shape, and chemical structure of the nano-object.[6] Liposomes, for instance, are the best known example of nano-objects, recently clasified as nanoparticules (with their three dimensions in the nanoscale (<200–300 nm)).[7] Different from conventional drug delivery systems, and depending on the biodegradability of the nano-object, nanomedicines can cross anatomical and phenomenological barriers.[8,9] They can be administered by parenteral, transcutaneous, or mucosal vias,[10] but changes in biodistribution can only be achieved upon parenteral administration[11–22] (Figure 20.1). An exclusive feature of nanomedicines is their uptake by phagocytic or pynocitic mechanisms upon cell recognition.[23,24] The structure of the nanomedicine is responsible for its own recognition by a given mechanism of cellular uptake.[25] Each uptake mechanism leads to well-defined intracellular traffic mediated by vesicles, which ends up in different cellular compartments[26] (Figure 20.2).

The control of biodistribution and kinetics of selective delivery into cell compartments is the key to generate more efficient and less toxic therapeutic effects. Tissue and cell targeting, as well as controlled intracellular traffic, depend on the size and structure of the nano-object, which must be administered parenterally.[27]

Many therapeutic small molecules or macromolecules, such as enzymes or peptides, antisense oligonucleotides, or corrective genes, must be delivered to the cytosol for therapeutic effect. However, upon capture by clathrin-mediated endocytosis, molecules loaded into nano-objects must exit the endolysosomal route before being degraded in lysosomes. To that aim, nano-objects have to escape as a response to the growing intravesicular acidity[28] by experiencing structural changes leading to the delivery of their content to the cytoplasm.[29] Alternatively, nano-objects can respond to stimuli found in ischemic tissues or infection sites, primary or metastasized tumors that exhibit lower extracellular pH than normal tissue. For instance, the extracellular pH 60 h after the onset of inflammatory reactions drops from 7.4 to 6.5.[30] The high level of proteinases,[31] increased glycolysis, and plasma membrane proton pump activity of tumor cells lowers the extracellular pH to between 6.5 and 7.2.[32,33] Hence, nano-objects capable of experiencing site-specific structural changes triggered by extracellular acidity could contribute to design more selective therapies.

Since the first article published in 1980,[34] the idea of tailoring liposomes capable of experiencing structural changes triggered by acidity (pH sensitivity) has evolved from simple lipid mixtures to complex lipid matrices combined with intelligent polymers. In this chapter we will survey the different strategies developed to improve the performance of pH-sensitive liposomes and their current preclinical applications.

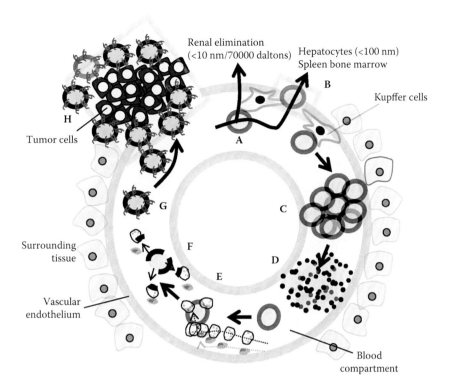

FIGURE 20.1

Schematic representation of the events occurring in the blood circulation. A: pH-sensitive liposomes in blood circulation upon intravenous (i.v.) (or intramuscular (i.m.)/intraperitoneal (i.p.) administration if <100 nm diameter) administration. Liposomes remain confined in continuous endothelium vasculature. B: Normal (nonpathological) fenestrate endothelium in liver allows the extravasation of liposomes to reach hepatocytes. Kupffer cells in light of liver vasculature phagocyte opsonized/aggregated pH-sensitive liposomes. C: Nonpegylated pH-sensitive liposomes (e.g., DOPE/OA) aggregate in blood circulation and block pulmonary capillaries. D: Intravascular aggregation of liposomes (e.g., Dioleoylphosphatidylethanolamine/oleic acid, (DOPE/OA)) leads to phase transition and inner content leakage. E: Nonpegylated pH-sensitive liposomes are covered by opsonins (opsonization) and albumin. Opsonization is a previous step to phagocytosis.[11] Interaction with plasma proteins leads to loss of pH sensitivity, with the exception of pH-sensitive liposomes bearing poly(NIPA-co-MAA-co-DODA). F: Blood albumin eliminates OA from DOPE/OA pH-sensitive liposomes, and intravascular leakage is produced. G: Pegylated pH-sensitive liposomes have reduced liposome-liposome interactions and are refractory to opsonization.[15–18] Pegylated liposomes avoid uptake by reticuloendothelial system (RES) macrophages. H: Pegylated pH-sensitive liposomes are long circulating and can extravasate at pathological sites presenting enhanced permeation retention (EPR; resultant from increased vascular permeability plus impaired lymphatic drainage).[12,13] Most gaps in the leaky vasculature fall below 400 nm.[14] The site-specific extravasation is known as passive targeting. A previous occurrence of passive targeting is required for active targeting of liposomes with surfaces bearing ligands or antibodies. The passive targeting can be limited due to the presence of poly(ethylene glycol) (Peg). Pegylated surfaces can prevent interaction between liposomes and the cell surface.[19,20] The reduced interactions inhibit effective uptake of the payload by the tumor cells.[21] Alternatively, the impaired uptake should occur above a threshold value; for instance, clathrin-dependent and clathrin-independent endocytosis are not appreciably affected by the presence of <10% surface coverage of Peg, although internalization is affected at higher (>20%) concentrations in the plasma membrane.[22]

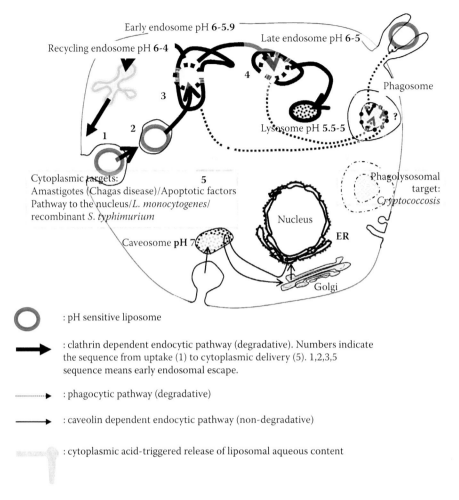

Cytoplasmic targets:
Amastigotes (Chagas disease)/Apoptotic factors
Pathway to the nucleus/*L. monocytogenes*/
recombinant *S. typhimurium*

FIGURE 20.2
Cell uptake mechanisms.

20.2 DOPE-Containing Liposomes: Acid Triggering the Lα → H$_{II}$ Phase Transition

The phase behavior of lipids depends on their molecular shape.[35,36] Cylindric lipids self-assemble into bilayers (lamellar phase Lα), and those with a cone shape (transversal area of the headgroup higher than the transversal area of acyl chains) self-assemble into micelles. In particular, the unsaturated dioleoylphosphatidylethanolamine (DOPE) is an inverted cone (small and poorly hydrated headgroup, with a minor transversal area compared to that of acyl chains) (Figure 20.3). The inverted cone shape favors the formation of hydrogen bonds between headgroups of DOPE (amino group and phosphate group of adjacent DOPE) that diminish interfacial hydration and favor its assemblage into an inverted hexagonal phase (H$_{II}$). At physiological pH DOPE molecules do not form bilayers, and above the phase transition temperature (T$_H$ = 10°C), they adopt the H$_{II}$ phase.[35,37]

The first pH-sensitive liposomes were prepared combining DOPE with lipids that possess titrable acid groups (negatively charged at physiological pH), such as palmitoylhomocysteine (PHC),[38] n-succinyldioleoylphosphatidylethanolamine,[39] oleic acid (OA),[40–43] dipalmitoylsuccinylglycerol (DPSG), dioleoylsuccinylglycerol (DOSG),[44] or cholesteryl hemisuccinate (CHEMS).[45] These liposomes possess electrostatically charged bilayers with high interfacial hydration that repel each other, impairing interaction between DOPE molecules. Such repulsion prevents the formation of the H_{II} phase and stabilizes the $L\alpha$ phase at physiological pH and temperature.[40,46] When the pH is decreased, the protonation of carboxilic groups neutralizes the negative charge and dehydrates the interfacial region. The adjacent bilayers get close, allowing interaction between DOPE molecules, with subsequent destabilization of the $L\alpha$ phase and conversion to the H_{II} phase.[45,47]

Once captured by cells, liposomes made of DOPE/OA, DOPE/PHC, DOPE/DPSG,[44] and DOPE/CHEMS[48] release their aqueous content to the cell cytoplasm within 5 and 15 min. This speed is compatible with release occurring between early and late endosomes, since the transport up to lysosomes takes nearly 30 min.[49,50]

The liposomal escape is thought to be mediated by fusion between the liposomal and endosomal bilayers.[51,52] The observation by freeze fracture of lipid particles in the point of union between membranes in fusion has been interpreted as highly curved fusion intermediaries (inverted micelles,[53] or "stalk"[54]). These intermediaries, in the facing monolayers of two contacting membranes, favor the incorporation of lipids with a negative membrane curvature,[55] such as DOPE. In the case of liposomes, the two contacting membranes in question will be the outer monolayer of the vesicles and the inner monolayer of the endosomal membrane.

The pH sensitivity of these liposomes depends on the pK_a of the anionic lipid that stabilizes the lamellar phase; varying their molar ratio slightly modifies the pH sensitivity. For instance, DOPE/OA liposomes are the most labile, becaming unstable at pH values between 6.9 (8:2 mol:mol) and 6.5 (7:3),[56] whereas the other liposomes destabilize at lower pH: DOPE/PHC (8:2) destabilizes at pH 6.25, DOPE/CHEMS (6:4) at 5.5, and DOPE/DPSG (8:2) at 5.[44,45] In theory, appropiate formulations with different pH sensitivities should be chosen according to the different minimal intravesicular pH values exhibited by different cells: 4.6 for macrophages,[57,58] 5.5 for fibroblasts, and 5.6 for epithelioid CV-1 cells.[59,60]

20.2.1 Plasma Stability

Since pH-sensitive liposomes are mainly administered i.v., structural stability must be maintained during blood circulation, until liposomes reach the target site, where they will be captured. Leakage of inner content, as well as aggregation that prematurely leads to phase transition, must be minimized. In addition, plasma proteins should not reduce the pH sensitivity. However, in blood, DOPE/titrable anionic lipids become pH insensitive and experience leakage that varies according to the nature of the anionic lipid. Formulations of liposomes containing OA are less stable. The addition of cholesterol and Peg does not impair the loss of pH sensitivity, but Peg reduces leakage.

DOPE/OA liposomes extensively destabilize (content release and aggregation) in mouse[61] and human plasma[62] because of the albumin extraction of OA. DOPE/DPSG and DOPE/DOSG liposomes are more stable, releasing less than 20% calcein upon overnight incubation in human plasma.[63,64] However, after only 3 h incubation in plasma, the pH sensitivity of these liposomes is diminished (instability is triggered at pH 4.2), due to the extraction of DPSG.[65]

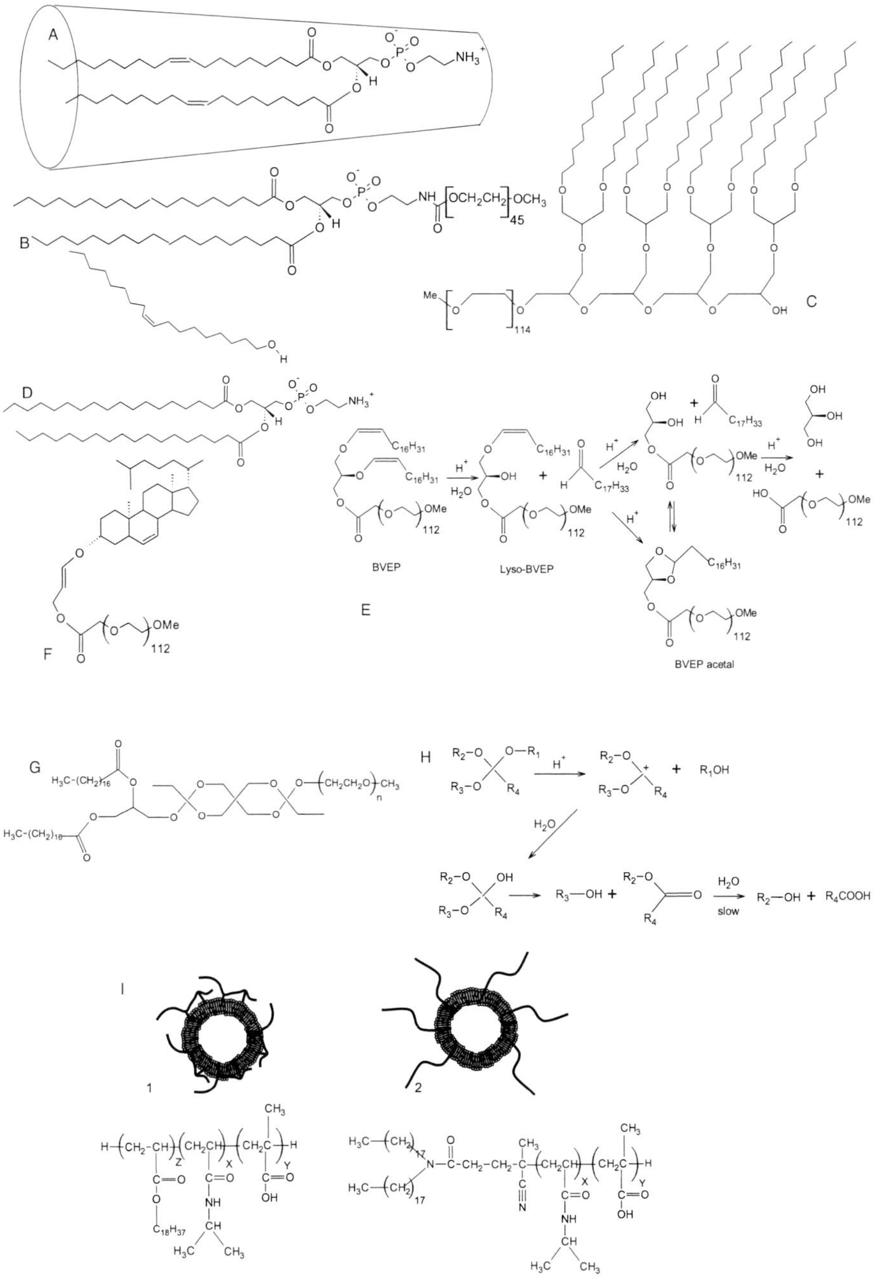

FIGURE 20.3 (Continued)

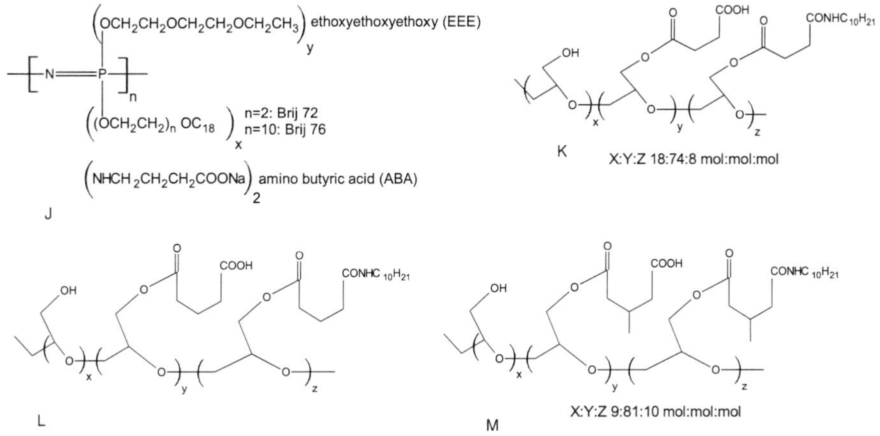

FIGURE 20.3 (Continued)
Chemistry structures of (A) dioleoylphosphatidylethanolamine (DOPE), (B) DSPE-Peg$_{2000}$, (C) 1,3-didodecyloxy-2-glycidyl glycerol with 114 ethylene oxide monomer units of the Peg chain (DDGG)$_4$EO$_{114}$ (Mw = 6,800), (D) oleyl alcohol (OAlc) forming a hydrogen bond through its hydroxyl to an oxygen atom on the phosphate group on the phosphatidylcholine molecule, (E) proposed hydrolysis pathways for 1,2-di-O-(1′Z,9′Z-octadecadienyl)-glyceryl-3-(ω-methoxy-poly(ethylene glycolate)) (BVEP; Mw = 5,000), indicating the sequential cleavage of the two alkenyl chains, (F) 1′-(4′-cholesteryloxy-3′-butenyl)-ω-methoxy-poly(ethylene glycolate) (CVEP), (G) Peg-diorthoester-distearoyl glycerol conjugate (POD), and (H) mechanism of POD hydrolysis. The quick degradation of POD at a pH of 5–6 is claimed to occur with a higher sensitivity than that of acetals, vinyl ethers, and polyorthoesters. Such higher sensitivity of POD is attributed to the following two factors. First, the dialkoxy carbocation intermediate of orthoester hydrolysis has four lone pairs of electrons over which to distribute the positive charge from the carbon, and hence is much more stable than the monoalkoxy carbocation intermediate, stabilized by two lone pairs of electrons, in the case of acetals and vinyl ethers. Second, the Peg headgroup of POD is more hydrophilic than the functional groups of the reported polyorthoesters, allowing better hydration and faster proton transfer to the diorthoester linkage. (I) Schematic illustration of the randomly (1) and terminally alkylated (2) poly(NIPA-co-MAA-co-DODA) copolymers anchored to liposomes. (J) Poliphosphazene grafted with up to three different substituents: ethoxyethoxyethoxy (EEE) groups, stearyl chains, and amino butyric acid (ABA). (K) SucPG. (L) GluPG. (M) MGluPG.

The addition of cholesterol (Chol) does not diminish the leakage and also decreases the pH sensitivity of DOPE liposomes. In vitro, DOPE/OA/Chol liposomes show a biphasic release of calcein (aqueous content marker): 95% of calcein is released within 1 h, with more than 60% OA being transferred to albumin; the remaining 5% is released in the following 8 h.[66] Seventy-five percent of the i.v. injected dose of DOPE/OA/Chol liposomes in mice is degraded in the first 5 min, wheres the remaining 25% is mainly distributed between the liver and spleen.[67]

The addition of 40 mol% cholesterol to DOPE/PHC liposomes also diminishes the pH sensitivity and fusogenic activity.[38]

On the other hand, the addition of poly(ethylene glycol) (Peg) to DOPE/CHEMS liposomes reduces the leakage, but the pH sensitivity is diminished.[65,68,69] DOPE/CHEMS liposomes lose 36% of calcein after 1 h in pH 7.4 buffer, whereas in 95% rat plasma 47% of the aqueous content marker ^{111}In is released after 1 h, increasing to 69% after 19 h. In buffer at pH 5.5, the calcein release is 84%, but in culture media supplemented with 10% fetal bovine serum (FBS) it diminishes to 29%. DOPE/CHEMS/PE-Peg (5 mol%) liposomes release 5% of calcein in pH 7.4 buffer, and 20% of ^{111}In after 1 h, increasing to 32% after 19 h, in rat plasma. However, liposomes having very small amounts of PE-Peg (e.g., 1 mol%) hamper the L$\alpha \rightarrow$ H$_{II}$ transition at low pH. For instance, pegylated liposomes release 5.8 and 6.8% in buffer and plasma at pH 5.5, respectively.[68]

It is noteworthy that though the uptake of pegylated liposomes by differentiated THP-1 cells was lower that that of nonpegylated liposomes, the capacity of intracellular release of calcein by DOPE/CHEMS/Peg-PE liposomes (measured as the calcein/rhodamine-PE ratio) was the same as that of DOPE/CHEMS liposomes. Inside the endosome, despite the absence of pH sensitivity, Peg behaves as a dehydrating agent and promotes fusion by a synergistic effect with DOPE, while in the presence of distearoylphospatydilcholine (DSPC), instead of DOPE, the effect of Peg alone is insufficient to promote destabilization and cytoplasmic release.[70]

An alternative to the use of PE-Peg that causes loss of pH sensitivity is steric stabilization by novel nonionic copolymers (DDGG)$_4$(EO)$_{114}$, bearing short blocks of lipid-mimetic units and different Peg molecular weights (Figure 20.3).[71] The incorporation of up to 10 mol% (DDGG)$_4$(EO)$_{114}$ into DOPE/CHEM liposomes does not decrease the pH sensitivity. In 25% human plasma, the calcein release of 5 mol% (DDGG)$_4$(EO)$_{114}$-containing liposomes is less than 20% after 5 h (vs. 35% for DOPE/CHEM liposomes). The intracellular delivery of calcein to EJ cells is similar to that achieved by DOPE/CHEMS liposomes.

The leakage from pH-sensitive liposomes depends on the chemical structure of the molecules loaded into their inner aqueous space. For instance, weak bases such as quinine, chloroquine, mitoxantrone, epirubicin, daunorrubicin, vincristin, and doxorubicin (DRX) are incorporated into liposomes by active loading. DXR (pKa = 8.1) is uncharged at neutral pH and is membrane permeable. To protonate and accumulate DRX inside liposomes, the inner aqueous space must be at an acidic pH (below 5.5). However, DOPE/CHEMS/PE-Peg 5 mol% liposomes are not stable at that pH, and between 80 and 100% of the DRX that is loaded in the uncharged form rapidly escapes in 90% human plasma.[72] The inclusion of cholesterol and hydrogenated soy PC increases the liposomal retention of DXR (10 and 60% release along 12 and 24 h, respectively).[73]

20.2.2 Pharmacokinetics and Biodistribution

The main feature of DOPE/OA (7:3) liposomes when administered i.v. is the aggregation and accumulation in pulmonary capillaries (as well as important leakage in circulation resulting from the extraction of OA by albumin) that occur within the first minutes[61,65,74]; in 15 min liposomes devoid of their aqueous content accumulate in the liver and spleen.

On the other hand, DOPE/CHEMS (6:4) liposomes show little leakage, do not aggregate or accumulate in lungs, and are almost completely captured by the liver and spleen after 30 min. The addition of 5 mol% PE-Peg increases the half-life to 11 h in rats, with a sustantial proportion of liposomes (8.5%) remaining in circulation at 24 h. Remarkably, the aqueous content of these liposomes, independently of pegylation, accumulates more in the liver and spleen and is excreted less in urine than the content of pegylated pH-insensitive liposomes. This was attributed to the fact that in pH-sensitive liposomes the content exits the endolysosomal route and is less excreted from the cell.[68]

The replacement of PE-Peg by 5 mol% (DDGG)$_4$(EO)$_{114}$ in DOPE/CHEMS liposomes increases its half-life in circulation, in comparison to that of nonpegylated liposomes (that are cleared from circulation within 1 h postinjection). Also, (DDGG)$_4$(EO)$_{114}$ liposomes exhibit even slower elimination than pegylated liposomes, as ~15% of the initial dose is still in circulation 48 h after i.v. administration in Wistar rats.[71] Compared to pegylated, (DDGG)$_4$(EO)$_{114}$ liposomes increase the area under the plasma concentration vs. time curve (AUC) 3.31-fold (30.6-fold compared to DOPE/CHEMS liposomes) and reduce the volume of distribution and plasma clearance 1.4- and 4.5-fold, respectively. Compared to DOPE/CHEMS liposomes, the liver and spleen accumulation is reduced by a factor of 1.3 and 2.2 for the pegylated and $(DDGG)_4(EO)_{114}$ liposomes, respectively.

20.3 Tunable pH-Sensitive Liposomes: Plasma-Stable Liposomes

A drawback associated with DOPE/titrable anionic lipid liposomes is that the pH at which the $L\alpha \rightarrow H_{II}$ transition occurs is determined by the pKa of the ionizable component and is, therefore, not easily tuned; only discrete tuning can be obtained by using different ionizable components.[44] On the contrary, liposomes made of a molar excess of ionizable anionic lipid, CHEMS, and a permanently charged cationic lipid *N,N*-dioleoyl-*N,N*-dimethylammonium chloride (DODAC) exhibit $L\alpha \rightarrow H_{II}$ transition at a pH value that can be tuned continuously.[75] At neutral pH, CHEMS is negatively charged and associates with great amounts of contra-ions and hydration water. These interactions make CHEMS's shape cilindric and stabilize the lamellar phase. As the pH is decreased, the negative charges are neutralized and CHEMS's shape turns from cylindric to an inverted cone, therefore resulting in the $L\alpha \rightarrow H_{II}$ transition. The addition of varying amounts of DODAC competes with the contra-ions, and the acidity required to induce the phase transition is a monotonically increasing function of the DODAC concentration. The phase transition occurs when the molar fraction of the negatively charged CHEMS equals the molar fraction of DODAC, at a pH that can be calculated employing a modified Henderson-Hasselbach equation.[76] These liposomes can also be stabilized by a strong bilayer-forming lipid, such as phosphatidylcholine.[77]

The possibility of tuning the pH of the phase transition could be used to trigger site-specific extracellular release from liposomes, but currently remains unexplored. Only a slight increase of plasma stability has been reported, in comparison to plasma stability of nonpegylated DOPE/CHEMS liposomes. Nonetheless, liposomes made of anionic/cationic lipids (eggPC/CHEMS/DDAB (dimethyldioctadecylammonium bromide), 25:49:25) require the addition of 1 mol% Tween 80 to decrease the calcein release from 38% to 4% in buffer containing 10% FBS. Tween 80 is thought to increase liposomal stability and avoid aggregation, providing steric stabilization by its polyoxyethylene headgroup. The pH sensitivity of these liposomes in buffer is similar to that of DOPE/CHEMS liposomes (50% release at pH 5.5), and remains at the same value in 90% plasma, while the pH sensitivity of DOPE/CHEMS liposomes decreases to 25%.[78]

Similarly, pH-sensitive liposomes made of phosphatidylcholine from egg yolk (eggPC)/CHEMS/OAlc (oleyl alcohol)/Tween 80 (50:50:80:2) exhibit excellent stability at pH 7.4 and undergo rapid destabilization upon acidification, as shown by calcein release and particle size increase.[79] Higher Tween 80 content improves stability but reduces liposomal pH sensitivity. The degree of pH sensitivity could be conveniently tuned by altering the amount of OAlc in the formulation, since higher OAlc content exhibits greater content release in response to low pH. The unsaturated fatty alcohol OAlc is capable of forming a hydrogen bond through its hydroxyl to an oxygen atom on the phosphate group on the phosphatydilcholine (PC) molecule, resulting in the formation of a complex with geometry similar to that of DOPE (Figure 20.3). This could result in a lowering of the energy barrier for the $L\alpha \rightarrow H_{II}$ transition. The observed pH-dependent leakage among these liposomes entrapping calcein might primarily be due to membrane destabilization, without inducing bilayer fusion.

These two types of liposomes have been used to increase the intracellular delivery of 1-β-arabinofuranosylcytosine (ara-C) on KB human oral cancer cells in vitro. Ara-C acts specifically on the process of DNA synthesis, and most of it is deaminated to an inactive form in the first 20 min after i.v. administration to animals or humans.[80] Also, N-glycosidic linkage of ara-C is degraded by lysosomal hydrolases being inactivated within the lysosomes. The folate (f) receptor (FR), on the other hand, which is amplified in many types of human tumors, has been shown to mediate the internalization of f-derivatized liposomes into an acidic intracellular compartment. The incorporation of f-Peg derivatives into eggPC/DDAB/CHEMS/Tween 80/PE-Peg$_{350}$-f (25:25:49:1:0.1) and eggPC/CHEMS/Tween 80/OAlc/Chol-Peg-f (50:50:2:80:0.5) liposomes enhances the cytotoxicity of entrapped ara-C, as shown by an 11-fold reduction in the 50% inhibitory concentration (IC$_{50}$) (for cationic formulation),[78] and 17 times greater cytotoxicity (for OAlc formulation)[79] on KB cells than on FR-targeted pH-insensitive liposomes.

In a recent application, pH-sensitive liposomes made of DMPC/DDAB/sodium oleate/ Tween 80 (180:10:14:2) were used to release the water-soluble cationic Fe-porphyrin on the gastric cancer cell line MKN28. Fe-porphyrin acts as a superoxide dismutase (SOD) mimic, converting O$_2$*$^-$ in H$_2$O$_2$, to further generate OH* radical, and inducing stronger cytotoxicity for cancer cell lines than for normal cells without any phototoxic effect. A sixfold decrease of 50% lethal dose (LD$_{50}$) with liposomal Fe-porphyrin was achieved, in comparison to the free porphyrin. The cytotxicity was reduced when the cells were pretreated with chloroquine, indicating the action of a pH-dependent mechanism.[81]

20.4 DOPE pH-Sensitive Liposomes + Cleavable Peg

These strategies are based on DOPE transition to a nonlamellar phase that is triggered by acid hydrolisis of labile linkages, leading to the detachment of the Peg layer that impairs interbilayer interaction.

The first labile linkages were designed to profit from the reductive environments on the cell surface, endosomes, and tumor cells.[82] For instance, pH-sensitive DOPE/CHEMS liposomes bearing Peg 3–6 mol% possessing the disulfide linkage dithiobis(propanoic dihydrazide)[69] or the benzyl carbamate linkage substituted with a disulfide in the *para* or *ortho* position[83] showed promising in vitro results. However, in humans, despite the oxidizing environment in circulation, reduction of disulfide bonds does occur due to the presence of low concentrations of cysteine (~8 µM) and glutathione (~2 µM). In vivo, pegylated pH-sensitive liposomes with redox-sensitive linkage are leaky and are cleaved in circulation.[72,84]

The pH-sensitive linkages were shown to be more stable in circulation. In theory, the pH-sensitive linkage would allow the detachment of Peg both in the periphery of acid sites (therefore increasing the site-selective uptake) and intracellularly (increasing the pH sensitivity of the phase transition). However, up to now the design of extracellular pH-sensitive linkages responding to a slight pH decrease in the periphery of tumors has failed, and only polymeric micelles have successfully made pH sensitive to extracellular tumors.[85,86] Indeed, the hydrolysis at pH 5–6 of the hydrazone (Hz) linkage in eggPC/Chol (6:3) liposomes containing up to 18 mol% PE-Peg$_{5000}$-Hz and 0.5–1 mol% PE-Peg-TAT (TAT is a cell-penetrating peptide that enters cell cytoplasm directly)[87] detaches the shielding Peg, revealing TAT peptide, and liposomes enter the cells more efficiently than pH-insensitive liposomes.[88] However, the pH responsiveness is shown at pH 5, which is much lower than the pH of the tumoral extracellular matrix.

The stronger intracellular acidity is sufficient to hydrolize pH-sensitive linkages, but former assays in the absence of cells led to the observation that Peg chains remain associated with the bilayers, impairing the $L\alpha \rightarrow H_{II}$ transition.

For instance, though stable for 24 h in plasma, the acid-catalyzed hydrolysis of the vinyl ether linkage from the Peg-conjugated lipid (1,2-di-*O*-(1′Z,9′Z-octadecadienyl)-glyceryl-3-(ω-methoxy-poly(ethylene glycolate)), Mw = 5,000; BVEP; Figure 20.3) incorporated at 3 mol% in DOPE liposomes generates a very slow pH-sensitive response ($t_{50\%release}$ at pH 4.5 ≈ 4 h). Time-resolved cryoscopic-TEM shows that significant liposome collapse into dense particles (attributed to H_{II} phase) occurs within 3 h after exposure at pH 4.5. However, lipid-mixing assays (FRET/N-nitrobenzoxadiazole phosphatidylethanolamine (NBD-PE)/N-rhodamine phosphatidylethanolamine (Rh-PE)) indicate that membrane fusion at endosomal pH does not occur.[89]

On the other hand, the acid-catalyzed hydrolysis of the vinyl ether linkage of the Peg-conjugated lipid 1′-(4′-cholesteryloxy-3′-butenyl)-ω-methoxy-poly(ethylene glycolate) (CVEP)[112] (Figure 20.3) requires only a single vinyl ether cleavage event.[90] The slow pH-sensitive response of the BVEP linkage was adjudged to the requirement of two cleavage events, and it was expected that CVEP should lead to a faster response. However, the observed release kinetics of DOPE/CVEP/PE-Peg-f (5:95:0.5) liposomes is significantly slower than that of BVEP. Similar to BVEP liposomes, the occurrence of fusion between CVEP liposomes and endosomal bilayers at endosomal pH (as judged by the inconclusive FRET/NBD-PE/Rh-PE lipid mixing assay) is doubtful. Further monolayer experiments ruled out the influence of the number of cleavable linkages on pH sensitivity and confirm that the detached Peg remains adsorbed on the liposomal surface, thus preventing the $L\alpha \rightarrow H_{II}$ transition (fusion between bilayers). Curiously, f-targeted KB cell DOPE/CVEP/PE-Peg-f liposomes exhibit significant cytoplasmic delivery of calcein. This signifies that once endocytosed, the liposomes and endosomal bilayers act as a sink for the detached Peg chains. Clearly, the detachment experiments performed in the absence of cells are not predictive of the complex interactions occurring inside the cells.

Another example is the orthoester linkage in Peg-orthoester-distearoylglycerol lipid (POD) (Figure 20.3), which is stable at neutral pH for more than 3 h and completely degradates in 1 h at pH 5.[91] When acid- (pH 5–6) catalyzed hydrolysis of the orthoesters lowers the mol percentage of POD on the liposome surface below 2.3, liposomes aggregate and intervesicular lipid mixing is initiated, resulting in a burst of content release. The leakage is biphasic, occurring in a lag and a burst phase. During the lag phase, less than 20% of liposomal content is released. During the burst phase, the leakage rate is dependent on interbilayer contact. As the liposomes are subjected to more acidic environments, the lag phase shortens (12 h at pH 7.4, 60 min at pH 6, less than 5 min at pH 5). At pH 7.5, DOPE/POD (10 mol%) liposomes are relatively stable without aggregation for more than 10 h. At pH 6.2, the liposome size starts to increase after 1 h, and at pH 5, extensive aggregation is observed within 10 min. Thus, the kinetics of the liposome aggregation correlates with that of pH-dependent POD degradation. These data are consistent with the stalk theory to describe the $L\alpha \rightarrow H_{II}$ transition and set a lower bound of ~16 DOPE lipids on the external monolayer as the contact site required for lipid mixing between two bilayers.[92] DOPE/POD liposomes experience POD hydrolysis at pH 5–6, followed by extensive aggregation and leakage within 10–100 min. POD's usefulness for intracellular delivery remains to be determined, and it will depend on the detachment kinetics, considering that the transit through the endosomes in cells occurs about 10–30 min before the endosomal contents are delivered into the lysosome.[49,50]

The same as DOPE/PE-Peg liposomes, pH-sensitive DOPE/POD liposomes are stable in 75% FBS in the first 12 h, but POD-containing liposomes rapidly release 85% of the inner content during the following 4 h, indicating a liposome destabilization process specifically mediated by POD hydrolysis. DOPE/POD liposomes administrated i.v. to female ICR mice showed a one-compartment clearance kinetics, similar to pH-insensitive DOPE/ PE-Peg liposomes (nonpegylated pH-insensitive liposomes usually exhibit a two-compartment elimination kinetics), with a half-life of nearly 3 h, mainly accumulating in liver and intestine.[91,93]

20.5 Liposomes Made of Hydrolyzable Lipids

In these strategies the liposomal destabilization is not based on a phase transition but on the acid hydrolysis of special lipids composing the bilayer.

The acid hydrolysis of the vinyl ether lipid diplasmenylcholine (1,2-di-*O*-(*Z*-1'-hexadecenyl)-sn-glycero-3-phosphocholine, DPPlsC)[94] produces glycerophosphocholine and fatty aldehydes. The permeability of DPPlsC liposomes is enhanced when >20% of the vinyl ether lipid is hydrolyzed, with $t_{50\% \text{ release}} \sim 1$–4 h between pH 4.5 and 5.5. The stability of DPPlsC/dihydrocholesterol (9:1) liposomes in 50% serum is higher than that of DOPE-based liposomes. DPPlsC/PE-Peg$_{3350}$-f (99.5:0.5) loaded with ara-C exhibit ~6,000-fold enhancement in cytotoxicity on KB cells compared to free ara-C[95], and chloroaluminum phthalocyanine tetrasulfonate (AlPcS$_4^{4-}$; a water-soluble sensitizer used in photodynamic therapy) enhances both tumor selectivity and phototoxicity.[96–98] However, the difficulty and prohibitive cost of synthesizing DPPlsC made necessary the development of an alternative approach that would effect cytoplasmic delivery of liposomal contents but require only small quantities of synthetic vinyl ether lipids.

On the other hand, the acid hydrolysis of the orthoester phosphocholine (OEPC) converts OEPC from a double-chain amphiphile to a single-chain lyso-phosphocholine (lyso-PC). OEPC liposomes submitted to acid hydrolysis transform into lyso-phosphocholine micelles. The acid hydrolysis consists of a lag and a burst phase. The lag phase is pH dependent: the lower the pH, the shorter the time. The pH-dependent leakage of OEPC liposomes is similar to that of POD liposomes, but with a relatively shorter lag time. For the POD liposomes, when the amount of surface POD is lowered to a critical level, the Lα → H$_{II}$ transition is initiated, and the liposomes aggregate and collapse rapidly. However, the OEPC liposomes do not spontaneously collapse under acidic conditions. Unlike DOPE/POD vesicles, they do not show a concentration-dependent release rate at low pH, as evidence for a non-contact-mediated content release.[99] Upon acid hydrolysis, OEPC liposomes are transformed into leaky metastable vesicles that rapidly collapse in the presence of albumin. When the amount of lyso-phosphocholine reaches a critical level, pores form on the surface of the vesicle, resulting in rapid content release. Eventually, the metastable liposomes can be completely destroyed when the lyso-phosphocholine and dodecanol are removed from the bilayer by interaction with plasma. OEPC significantly enhances the in vitro transfection efficiency compared with pH-insensitive phosphocholine when formulated with cationic lipids (8.24 × 10^5 vs. 3.95 × 10^4 RLU (relative light units)/mg protein).[100] To date, the cytotoxicity, pharmacokinetics, and biodistribution of OEPC liposomes are unknown.

20.6 Liposomes Bearing pH-Sensitive Polymers

20.6.1 Liposomes Bearing NIPA Polymers

Water-soluble nonbiodegradable amphiphiles N-isopropyl acrylamide polymers (poly(NIPA)) experience a first-order volume phase transition in response to a change of temperature.[101,102] Poly(NIPA) presents a typical lower critical solution temperature (LCST) or cloud point, a temperature under which the polymer is hydrated and extended as a random coil, where polymer-aqueous solvent interactions are predominant. Above the LCST, polymer-polymer hydrophobic interactions predominate, the hydrogen bonds are disrupted, and water is expelled from the polymer coils, which start to collapse to a globular form due to increased interactions between hydrophobic moieties. This process results in aggregation, formation of compact globules, and precipitation.[103] In aqueous solution the LCST of poly(NIPA) is approximately 32°C and is independent of its molecular weight and concentration.[104,105] LCST can be shifted by modifying the polymer hydrophilic-hydrophobic balance. Copolymerization with hydrophobic comonomers decreases the LCST, whereas hydrophilic comonomers have the opposite effect.[106–108] Copolymerization with methacrylic acid (MAA) (Figure 20.3) increases the LCST of the copolymer above that of poly(NIPA) as the number of acidic units is increased, at all pHs. This is because both the acid and ionic states of the MAA units are more hydrophilic than NIPA units. The degree of ionization of MAA (and therefore the hydrophilic-hydrophobic balance of the copolymer) changes with the pH: the COOH form that predominates at acid pH is less hydrophilic than the COO– form, which predominates at neutral pH. Hence, the LCST of the copolymer is also changed with the pH.[109–111] As the pH decreases, the degree of protonation (hydrophobicity) of the copolymer increases; the transition peak is narrowed and the cooperativity of the transition increases. As an example, poly(NIPA-co-MAA)'s (≥20.4 mol%) LCST decreases from 53.7°C at pH 7.4 to 39.2°C at pH 4.0.

Above the LCST and mixed with liposomes, poly(NIPA-co-MAA) destabilizes the bilayer.[112] The collapse of the polymer in acid media introduces a curvature in the bilayer plane, which induces membrane defects[113–116] and triggers the release of the entrapped molecules.[113,117] The poly(NIPA-co-MAA) effect on lipid bilayers would not be at the molecular level, but would instead perturb the liposome at the supramolecular level.[118] When liposomes bearing poly(NIPA-co-MAA) are captured by cells by clathrin-dependent endocytosis, the collapse is triggered by the low pH inside the endosomes, and the entrapped molecules inside acidic organelles are released.[113,119]

Poly(NIPA-co-MAA) can be anchored to the liposomal bilayer by terminal or random alkylation with hydrophobic anchors,[120] with a binding efficiency that is proportional to the number of hydrophobic anchors.[121] However, randomly alkylated poly(NIPA-co-MAA) (5 mol%; Mw = 29,000) containing 2–4 mol% octadecylacrylate (ODA)[120] is desorbed from the lipid bilayer upon interaction with plasma, leading to a partial loss (15–25%) of pH responsiveness.[117–119] In addition, randomly alkylated poly(NIPA-co-MAA) slightly increases the circulation time of liposomes. The absence of steric stabilization is due to its partial dehydration at neutral pH, the random coil conformation being incomplete.[68,119,122] However, the addition of Peg to liposomes bearing randomly alkylated poly(NIPA-co-MAA) makes the liposomes long circulating but diminishes their pH sensitivity.[114,118,123]

The i.v. administration to male Sprague-Dawley rats of liposomes made of eggPC/Chol (3:2) with 3 and 6 mol% Peg-PE (180–230 nm) containing 0.3 w/w poly(NIPA-co-MAA-co-N-vinyl-2-pyrrolidone (VP)-co-ODA) copolymer (90:4:2:4 molar ratio; Mw = 22,600) shows that the presence of Peg increases plasma $AUC_{0-\infty}$ above that of nonpegylated liposomes.[118] However, in spite of the addition of VP to maintain the polymer's solubility in water, the pH sensitivity decreases between 45 and 90% in the presence of 3–6 mol% Peg. The decreased pH sensitivity could be a consequence of the impaired aggregation and dehydration of membrane surfaces caused by Peg. Moreover, Peg may also prevent interchain aggregation of the poly(NIPA)copolymer, reducing the extent of polymer collapse within the liposomal membrane.

The terminal alkylation of poly(NIPA-co-MAA) employing dioctadecylamine (DODA) (Figure 20.3) allows its anchorage to the lipid bilayer, proportioning an optimal mobility to its polymeric chain,[115] but neither enhances the liposome circulation time in rats.[117] However, the pH sensitivity of poly(NIPA-co-MAA-co-DODA) pegylated liposomes can be preserved in the presence of human plasma.[123] The combination of both the terminally alkylated NIPA copolymer and a pegylated lipid in the vesicle structure was found to provide liposomes with both strong pH-responsive properties and a long half-life.[123,124]

Since the nonbiodegradability of the polymer may be a serious limitation to its widespread use in the biomedical field, the molecular cutoff for renal filtration of poly(NIPA-co-MAA-co-DODA) was recently estimated at 32,000,[124] a value comparable to that of different polymers like Peg (Mw = 30,000)[125] and poly(2-hydroxypropyl)methacrylamide (PHPMA; Mw = 40,000–45,000).[126] It was observed that NIPA polymers with LCST below physiological temperature are highly captured by the organs of RES, and the presence of terminal alkyl moieties at the polymer chain induces the formation of micelles.[127] In the context of designing a clinically viable pH-responsive liposome, high Mw NIPA copolymers (Mw = 30,000–50,000) with a long circulation time and low apparent volume of distribution, but poorly excreted in the urine and feces, should be avoided to minimize the risk of accumulation following multiple dosing. Unlike randomly alkylated poly(NIPA-co-MAA) that rapidly detaches from pegylated liposomes in circulation, it is sufficient to use a terminally alkylated poly(NIPA-co-MAA) with a Mw of 10,000 to produce a drastic change in liposomal pharmacokinetics parameters, indicating that the polymer remains anchored in the phospholipid bilayer in the bloodstream.[124]

Recently, pH-sensitive immunoliposomes (eggPC/Chol/PE-Peg/PE-Peg-mAb) (3:2:0.17:0.09) containing 0.1-0.3 w/w poly(NIPA-*co*-MAA-co-DODA) have been employed to deliver ara-C to HL60 leukemic cells.[128] Ara-C is used alone or in association with anthracycline agents in the treatment of acute myeloid leukemia (AML). The incorporation of ara-C into pegylated pH-insensitive liposomes has been found to substantially increase its therapeutic effect on L1210/C2 leukemia.[80] However, when incorporated into pH-insensitive liposomes, ara-C is delivered within lysosomal organelles, where it degrades into its inactive form.[41,129,130] The murine anti-CD33 p67.6 monoclonal antibody (mAb) as targeting ligand binds the CD33 receptor with great avidity and is currently used in the clinic to treat AML. These poly(NIPA-co-MAA-co-DODA) immunoliposomes exhibited the highest cytotoxicity selectively against HL60 cells. However, they were less efficient than DOPE-based liposomes in augmenting the intracellular bioavailability of ara-C, probably because the copolymers are devoid of membrane fusion activity.[114]

The poly(NIPA-co-MAA-co-DODA) liposomes can be pegylated without losing pH sensitivity, and this is undoubtedly an advantage over DOPE-based liposomes.[123] However, studies with the Langmuir balance technique on lipid monolayers show that poly(NIPA-co-MAA-co-DODA) does not seem strongly anchored in the lipid membrane at neutral pH, as it can be expelled from the phospholipid monolayer at surface pressures above 30 mN/m.[115] Even if these results are cannot be extrapolated to in vivo, the relatively weak association of poly(NIPA-co-MAA-co-DODA) to bilayers raises doubts about its permanence on long-circulating liposomes. This fact, together with the excellent intracellular pH-sensitive response of the DOPE/CHEMS liposomes, indicates the need for comparative in vivo studies.

20.6.2 Liposomes Bearing Polyphosphazenes: Speed of Release

Polyphosphazenes (PPZs) are biodegradable inorganic polymers with backbones consisting of alternating nitrogen and phosphorus atoms linked by alternating single and double bonds (Figure 20.3).[131]

Amphiphilic temperature- and pH-sensitive poly(organophosphazenes) with varying ratios of ethylene oxide, alkyl chains, and free acid units synthesized by living cationic polymerization exhibit an LCST that is dependent on their composition.[132] The same as poly(NIPA) copolymers, the hydrophilic comonomers (carboxilic acids) increase the LCST from 32°C to 44°C and make it pH sensitive. Stimuli-responsive liposomes that release their content upon a change in temperature or pH can be obtained when alkylated copolymers (for instance, (ethoxyethoxyethoxy)$_{1.72}$(amino butyric acid)$_{0.18}$ (C$_{18}$(EO)$_{10}$ = polyoxyethylen stearyl ether (Brij) 76)$_{0.1}$ polymer; Mw = 38,000; 9.2 mol% acid) are anchored in 0.05–0.1% w/w into eggPC/Chol (3:2) liposomes. Lowering the pH to 5–6 causes polymer dehydration, and within 1 min a virtually complete release of entrapped HPTS (hydrophilic dye pyranine) is triggered. Such a rate is thought to be related to the high cooperativity of the phase transition, which favors polymer collapse. The short time span over which release occurs is an important feature of the system since, upon endocytosis, the transit time from uptake to degradation in lysosomes is generally about 30 min.[133] However, the formulations still require optimization to minimize leakage at pH 7.4.

20.6.3 Liposomes Bearing Polyglycidols

Polyglycidol (PG) polymers are biocompatible[134] flexible hydrophilic aliphatic polyether polyols that can be prepared in branched or linear forms, and share structural similarities with Peg (Figure 20.3). PG polymers become pH sensitive by incorporating varying amounts of carboxylic acids as side chains to their backbones. The hydrophobic-hydrophilic balance of the carboxylated PG polymer changes as the pH shifts from neutral (where the hydrophilic COO$^-$ form predominates) to low (where the more hydrophobic COOH form predominates). Accordingly, the carboxylated PG polymer conformation changes from an extended form at neutral pH to a collapsed form at low pH.

As the hydrophobicity of the carboxylic residue increases, polymer collapse and precipitation occur at higher pH: SucPG (succinic anhydride-modified PG) collapses at pH 4, GluPG (glutaric anhydride-modified PG) at pH 4.7, MGluPG (3-methylglutaric anhydride-modified PG) at pH 5, and CHexPG (1,2- cyclohexanedicarboxylic anhydride-modified PG) at pH 5.4, which is consistent with the number of carbon atoms of these polymers' side chains.[135]

Free polymers collapse and precipitate at a pH that is lower than the pH that triggers liposomal destabilization upon mixing with polymers. For instance, after being mixed with liposomes, only CHexPG polymer induces a marked release of the aqueous content at pH 6.5. However, polymers inserted in the bilayer through hydrophobic anchors increase the liposomal pH sensitivity. Liposomes with anchored SucPG-C$_{10}$, GluPG-C$_{10}$, and MGluPG-C$_{10}$ exhibit a similar pH response, retaining HPTS at pH 7.4, with an almost complete release at pH 5.5.

Polymers SucPG having 56% Suc residues combined with long alkyl chains as anchors to eggPC liposomes remain stable at neutral pH but induce strong fusion at pH 4.0. Liposomes bearing the polymer fuse more intensively with decreasing pH and with an increasing amount of bound polymer.[136] A high percentage (early 70–80%) of Suc COO⁻ must lose their negative charges by protonation to induce the content release of liposomes.[133]

EggPC liposomes can be associated with a maximum proportion of 30% w/w of SucPG (PG mean Mw = 4,600 and calculated Mw of SucPG = ca. 10,000). The uptake of SucPG liposomes by CV-1 cells is inversely proportional to the SucPG content, but the calcein release into cytoplasm is directly proportional to the SucPG content (30 > 20 > 10%).[137] Fusion assay (FRET NBD to Rho) suggests the occurrence of fusion between the polymer-modified liposomes and endosomal or lysosomal membranes. Liposomes with a higher SucPG content show a higher percent fusion after internalization. Hence, the high cytoplasmic delivery at a higher content of pH-sensitive polymer overcomes its low rate of uptake.

PH sensitivity of SucPG liposomes (eggPC/SucPG, 20% w/w, 130 nm) coupled to ~20 anti-BCG monoclonal antibody molecules per liposome showed ≥90% calcein release triggered between pH 4.0 and 4.5; it decreases below 30% above pH 5. Fluorescence microscopic observation indicates that BCG-sucPG immunoliposomes bind to colon 26 tumor cells and induce receptor-mediated endocytosis at 37°C. Fusion assay (FRET NBT-PE/ Rh-PE) suggests that fusion between BCG-SucPG immunoliposomes and endosomal or lysosomal membrane does occur, without impairment caused by the presence of anti-BCG monoclonal antibodies.[138]

pH-sensitive SucPG imunoliposomes (eggPC/SucPG, 30% w/w in 5% glucose solution) were recently employed as a nonviral vector of transfection in vitro.[139] SucPG liposomes are negatively charged and form electrostatic complexes with positively charged lipoplexes (cationic liposomes made of 3,5-dipentadecyloxybenzamidine hydrochloride (TRX-20)/dilauroylphosphatidylcholine (DLPC)/DOPE, 0.26:0.26:0.52) plus plasmid DNA encoding the firefly luciferase sequence, imparting fusion ability to them. Complexes prepared in an aqueous glucose solution have small diameters and much higher transfection activity toward HeLa cells than do complexes prepared in phosphate buffer saline (PBS).[140] Transferrin (t), a glycoprotein transporting iron into cells, is incorporated into the surface of SucPG liposomes to overcome the low affinity by cell membranes caused by the shielding of positive charges upon complexation. Transferrin receptors (TRs) are expressed on the cell surface and are constitutively taken up through clathrin-dependent endocytosis.[141,142] After internalization, t and its TR are recycled to the cellular surface via the recycling pathway.[143] The combination of cytoplasmic delivery of DNA after complex fusion with the endosomes, and the addition of t results in a higher efficiency of transfection. However, the extent of the t-induced enhancement is dependent on the cell line. Conjugation of t increases the transfection activity on HeLa (human cervical cancer cell line) and KB (human epidermoid adenocarcinoma line) cells, although it only slightly enhances transfection on HT1080 (human fibrosarcoma line), HepG2 (human hepatoma line), and K562 (human leukemia line) cells.[144,145] Remarkably, TR in HeLa and KB cells are internalized quickly and actively, whereas those in HT1080, HepG2, and K562 cells are internalized slowly. Hence, the transfection mediated by the ligand-attached hybrid complex does not correlate with the amount of TR in the cell surface, but correlates with the activity of internalization of TR into the cells. Toward further improvement of their transfection activity, selection of the ligand might be important to facilitate cellular internalization and acid-triggered activation of SucPG complexes.

20.7 Preclinal Applications of Pegylated pH-Sensitive Liposomes

20.7.1 Tumor Therapy/Cisplatin

Cis-diamminedichloroplatinum(II) (cisplatin) is a platinum complex employed in the treatment of ovary, lung, testicle, head, and neck carcinoma that is mainly limited by its severe nephrotoxicity. Additionally, it generates intrinsic and acquired resistance. A sterically stabilized liposomal formulation of cisplatin was developed by ALZA Pharmaceuticals (Mountain View, California), but phase I–II clinical studies between 2002 and 2006 were unsuccessful. The lack of efficacy was attributed to the low bioavailability and slow release kinetics of cisplatin, in such a way that the cisplatin concentration fails to exceed the threshold for therapeutic effect.[146–148]

Aiming to increase the cisplatin bioavailability in tumors, cisplatin-pegylated pH-sensitive liposomes (DOPE:CHEMS:PE-Peg, 5.7:3.8:0.5, 110 nm) were administered as a single i.v. bolus in solid Ehrlich tumor-bearing mice.[149] Cisplatin-pegylated pH-sensitive liposomal treatments resulted in a 2.1-fold higher blood AUC and 2.6-fold higher tumor AUC than free cisplatin. The tumor AUC/blood AUC ratio is higher for cisplatin-pegylated pH-sensitive liposomes than for free cisplatin (1.51 vs. 1.23). However, the kidney (the major site of toxicity for cisplatin) AUC produced by pH-sensitive liposomes is 1.5-fold higher than that corresponding to free cisplatin.

Recently the same researchers evaluated the acute toxicity in mice upon a single i.p. administration of cisplatin pegylated pH-sensitive liposomes (7, 12, 30, 45, and 80 mg/kg) and free cisplatin (5, 10, and 20 mg/kg).[150] Mice treated with the higher dose of free cisplatin (20 mg/kg) showed strong weight loss and a longer time of recovery than those receiving the liposomal drug. The LD_{50} values in male and female mice for liposomal cisplatin were 2.7- and 3.2-fold higher, respectively, than those for free cisplatin. No hematological toxicity is registered for liposomal cisplatin, and neither were histopathological alterations observed after either treatment. Free cisplatin, however, leads to an appearance of mild anemia and a reduction in total white blood cell counts, as well as pronounced alterations in the blood urea and creatinine levels of mice. In contrast, these parameters are slightly altered only after liposomal cisplatin treatment at a dose of 30 mg/kg. Microscopic analysis of kidneys from mice treated with the liposomal drug showed no morphological alteration.

20.7.2 Tumor Therapy/Gemcitabine

Recently, pegylated antibody-conjugated pH-sensitive liposomes (DOPE/CHEMS/PE-Peg-PDP, 6:4:0.1, 150 nm) were developed to overcome the short in vivo half-life, lack of targetability, and cytoplasmic delivery of gemcitabine (GEM).[151] GEM (2′,2′-difluoro-2′-deoxycytidine) is a deoxycytidine nucleoside analogue, with cytotoxic activity in non-small-cell lung cancer cells. It induces an S-phase arrest and inhibits DNA synthesis, but has shown hematotoxicity with other toxic effects. pH-sensitive liposomes containing PDP-Peg-PE (a derivative where α-amino-ω-hydroxy Peg is coupled with SPDP (N-succinimidyl-3-(2-pyridyldithio) propionate), and then with DOPE) were conjugated to a maleimidophenyl-butyrate-derivatized monoclonal antibody against epidermal growth factor receptor (EGFR) that is overexpressed in solid tumors. The antiproliferative effect of conjugated and nonconjugated pH-sensitive liposomes in A549 cells is twofold higher than for pH-insensitive liposomes (DPPC/Chol) or the free drug, and ~10% apoptosis for pH-sensitive liposomes with or without EGFR vs. ~1% for free drug or pH-insensitive liposomes. The authors claim that pegylated targeted pH-sensitive liposomes retain the pH sensitivity.

Administered on non-small-cell lung cancer in a nude mice model[152] at 160 mg GEM/kg (8 i.v. doses, days 0, 3, 6, 10, 13, 16, 20, and 28), these liposomes diminish the growth of tumor (20–2.26 folds volume reduction after 1 month), improving the performance of the nontargeted pH-sensitive liposomes and the free drug. The cell proliferation was reduced in direct relation with the increased apoptotic index.

20.7.3 Scintigraphic Diagnostic

A particular application as diagnostic radiopharmaceuticals has been recently published, where the biodistribution and ability of pegylated pH-sensitive liposomes (DOPE/CHEMS/PE-Peg, 6.5:3.0:0.5) were radiolabeled with ^{99m}Tc-hexamethylpropylene amine oxime (^{99m}Tc-HMPAO) to identify inflammatory regions in a rat focal inflammation model upon i.v. administration.[153] For identification of an inflammation region, the target-to-no target ratio must be at least 2, allowing for the acquisition of improved scintigraphic imaging quality. ^{99m}Tc-stealth pH-sensitive liposomes offer a high target-to-no target ratio at earlier times than other pH-insensitive liposomes. It was possible to detect 0.43% of the injected dose/g at only 30 min postinjection of ^{99m}Tc-pegylated pH-sensitive liposomes. In contrast, the same uptake level was obtained using nonpegylated and pH-insensitive liposomes, composed of DSPC, DMPG, and Chol, 24 h after injection,[154] while using pegylated pH-insensitive liposomes resulted in the same absolute uptake at only 2 h after administration.[155] The value of the inflammatory and noninflammatory site radioactivity counting ratio was greater than 5 at 2 h postadministration.

20.8 Preclinical Applications of pH-Sensitive Liposomes

Long circulating liposomes extravasate at the sites of existing enhanced permeability and retention (EPR) effect and justify the need for absence of early drug leakage and aggregation. However, there are many therapeutic targets other than EPR sites. Examples are diseases affecting cells that are accesible from blood circulation, such as those from the RES, like Kupffer cells and hepatocytes in the liver, or spleen and bone marrow macrophages. To target these cells, nonpegylated pH-sensitive liposomes are suitable vehicles, and long circulation is not essential. In particular, nonpegylated pH-sensitive liposomes administered to the skin have shown promising results in therapeutics of melanoma, and also in targeting skin antigen-presenting cells (Langerhans and dendritic cells) to elicit immune local responses. The complexation of nonpegylated pH-sensitive liposomes with cationic lipoplexes was successfully applied to in vivo gene therapy. In the following section, the latest preclinical application of nonpegylated pH-sensitive liposomes will be discussed.

20.8.1 Chagas Disease

Chagas disease is caused by the parasite *Trypanosoma cruzi*. In humans, *T. cruzi* colonizes cells of the RES, and also cardiac and nerve cells. Parasites can be found as trypomastigotes, an infective extracellular circulating form that predominates in the acute clinical form, or as intracellular cytoplasmic amastigotes, which predominate in the indeterminate and chronic clinical forms of the disease (Figure 20.2). Current trypanocidal therapy based on the first-line benznidazole (BNZ) fails to eliminate the amastigote forms, contributing to the perpetuation of the infection.[156] In 2005, our research group in Argentina employed pH-sensitive liposomes (DOPE/CHEMS, 6:4, ~400 nm) to deliver the hypoxic cell radio-sensitizer 2-nitroimidazole etanidazole (ETZ) to an acute murine model of Chagas disease. The pH-sensitive liposomes should increase the efficiency of treatment by allowing the delivery of trypanocidal drugs to the cytoplasm. BNZ, a 2-nitroimidazol that dissolves in liposomal bilayers, was replaced by the hydrophilic ETZ, which allows its dissolution in the inner aqueous space of liposomes. It was previously determined that ETZ is not a suitable trypanocidal agent, as judged by its LD_{50} on trypomastigotes and IC_{50} on intracellular amastigotes upon 48 h incubation: ETZ was 8.2-fold less active on RA trypomastigotes, and 23.5-fold (amastigote-infected Vero cells) and 15-fold (amastigote-infected J774 cells) less active than BNZ.[157] However, when incorporated in DOPE/CHEMS liposomes, both in vitro and in vivo trypanocidal activities of ETZ dramatically increased above those of free ETZ and BNZ. While free ETZ (200 µg/ml) lacks antiamastigote activity (aa) when loaded in pH-sensitive liposomes, the aa is increased to 72% on amastigote-infected J774 cells after only 2 h incubation. ETZ pH-sensitive liposomal treatment (i.v. administration of 0.56 mg of ETZ/kg, starting 5 days postinfection (p.i.), 3 days a week over 3 weeks) significantly decreased the parasitemia on days 12, 19, 21, and 23 of Balb/c mice infected with 50 trypomastigotes of the RA strain. A one hundred eighty-fold higher dose of free ETZ does not decrease parasitemia.[158] Moreover, ETZ pH-sensitive liposomal treatment (3.2 mg of ETZ (293 µg of liposomal lipid)/mouse) leads to the complete elimination of parasitemia of mice infected with 10^2 trypomastigotes of the Tulahuen strain. The same dose of empty liposomes or free ETZ had no effect on parasitemia (unpublished results). ETZ pH-sensitive liposomal therapy remains to be tested against higher-infecting inoculums from different *T. cruzi* strains, in both the acute and the indeterminate and chronic stages of the disease.

20.8.2 Systemic Cryptococcosis

Cryptococcosis is a serious and potentially fatal fungal disease caused by the opportunistic fungal pathogen *Cryptococcus neoformans*. *C. neoformans* is found almost exclusively in macrophages of hosts with chronic or latent infection. Upon *C. neoformans* being taken up by macrophages, phagolysosomal fusion occurs. The phagolysosomal membrane becomes leaky and its inner pH is alkalinized. Cells containing *C. neoformans*-filled phagosomes retain phagocytic capacity and motility. Phagosomes of infected cells became distended, and filled with yeast cells and polysaccharides that finally kill the macrophage (Figure 20.2).[159]

The polyene antifungal nystatin (Nys) is a hydrophobic macrolide that shares a chemical structure with amphotericin B (AMB), employed in the treatment of systemic cryptococcosis. Similar to AMB, the slight hydrosolubility of Nys at neutral pH is increased as the pH is reduced.

The group of Owais in India empoyed pH-sensitive liposomes (DOPE/CHEMS, 6:4) to deliver Nys against systemic murine cryptococcosis.[160] Balb/C mice were challenged systemically with 7.10^5 spores of *C. neoformans* and received intraperitoneal (i.p.) treatment on days 1, 2, and 3 postinfection. Nys pH-sensitive liposomal treatments (5 mg of Nys/kg) presented 80% survival at day 40 postinfection, while treatments with Nys pH-insensitive (eggPC/Chol, 49:21) liposomes or free Nys resulted in 50 and 10% survival, respectively. At lower doses (3 mg of Nys/kg), the survival of mice treated with Nys pH-sensitive liposomes was 50%, whereas treatments with Nys pH-insensitive liposomes or free Nys resulted in 20 and 0% survival, respectively. The pH-sensitive liposomes were significantly more effective than pH-insensitive liposomes and free Nys in eliminating the fungi from the liver and brain at day 9 postinfection, as well as in eliminating the residual fungi in surviving animals (insignificant number of colony-forming units in liver vs. 50–350 and 25–160 in pH-insensitive liposomes and free Nys, respectively). Since the phagocytic activity is maintained upon infection, probably the released acid-solubilized Nys diffuses to the fungi across the leaky membrane of phagolysosomes.

20.8.3 Skin Papilloma

In 2007, the same group in India studied the potential of diallyl sulfide (DAS), a water-insoluble active component of garlic, topically applied as DAS pH-sensitive liposomes against skin cancer in a murine model of dimethyl benzanthracene (DMBA)-induced skin papilloma. Swiss albino mice were exposed to DMBA dissolved in acetone and applied topically three times a week for 12 weeks.[161] The tumor suppressor gene p53 can induce cell cycle arrest, DNA repair processes, and apoptosis; in 80% of human malignnant tumors this gene is mutated (p53mut). Exposure to DMBA induces downregulation of p53wt protein and upregulation of p53mut protein. One hour after DMBA exposure, pH-sensitive liposomes containing DAS (250 µg) were applied on the shaven dorsal skin. DAS pH-sensitive (DOPE/CHEMS, 6:4) liposomes and DAS pH-insensitive (eggPC/Chol, 49:21) liposomes are more efficacious in delaying the induction of tumors (for more than 1 week) and in reducing the total number of tumors (3.92, 5.84, 10.6, and 20 for pH-sensitive, pH-insensitive, free DAS, and nontreated animals, respectively). The mean tumor volume was significantly lower for pH-sensitive than pH-insensitive liposomes, free DAS, and nontreated animals (62, 169, 309, and 1,041 mm^3, respectively). The survival (animals completely free of tumor incidences upon treatment) of treated animals was 34 and 23% for pH-sensitive and pH-insensitive liposomes, respectively. Inhibition of tumor growthe (calculated from the average size of the tumor induced) was significantly higher for pH-sensitive liposomes (94%) than for pH-insensitive (84%) and free DAS (70%). Treatment with the liposomal formulations ensured an upregulation of p53wt (147, 94, and 23% for pH-sensitive and pH-insensitive liposomes and free DAS, respectively), while levels of p53mut expression were reduced (normal levels and 64% and 40% reduction for pH-sensitive, pH-insensitive, and free DAS, respectively).[162] The tumors' induction by DMBA/acetone generates ulcerations and wounding. The observed chemoprotectant activity of DAS is exerted by modulation of apoptotic factors at the cytoplasmic level. Probably, the absence of the stratum corneum enables the liposomal formulations to access the tumor cells in the deep epidermal layer for cytoplasmic delivery of DAS.

20.8.4 Immune Therapy/Vaccination

20.8.4.1 *Induction of Cytotoxic T Lymphocyte (CTL) Responses*

Tumor therapy can be mediated by CTLs that recognize tumor-associated antigens generally derived from mutants of self-proteins or of viral origin. It is generally accepted that CTL responses are induced when antigen is delivered into the cytosol (Figure 20.2). It has been reported that humoral and cellular immunity were induced simultaneously in Balb/c mice by the HIV *gag* protein or the 15-mer peptide from the HIV V3 loop encapsulated in pH-sensitive liposomes.[163,164] pH-sensitive liposomes (POPE/CHEMS/MPL (monophosphoryl lipid A), 7:3:0.01; 1,100 nm) containing a CTL epitope peptide from Hantaan virus nucleocapsid protein (M6) were i.m. administered to C57BL/6 and boosted after a week. It was found that pH-sensitive liposomes more efficiently generate peptide-specific CTL responses than the free peptide. One week after the immunization, the mice were challenged with 1×10^5 M6-transfected B16 melanoma cells. Seventy percent of the immunized mice rejected the tumor development, whereas the mice immunized with irrelevant peptide in pH-sensitive liposomes suffered the growing tumors and died at day 45. These results indicate that the CTL response could block tumor formation in an antigen-specific manner. On the other hand, mice were first inoculated with M6-transfected melanoma cells subcutaneously, and 3 days later immunized with pH-sensitive liposomes and boosted after a week. All the mice immunized with irrelevant peptide in liposomes died on day 32 after tumor cell inoculation, and 40% of the mice immunized with M6 in liposomes remained alive for 40 days after tumor inoculation. It was also observed that the pH-sensitive liposomes induced Th1-type cytokine-mediated CTL responses.[165]

20.8.4.2 Induction of Cellular Immunity

Recently, SucPG and MGluPG liposomes (eggPC/DOPE, 1:1, 30% w/w SucPG or MGluPG) containing ovalbumin (OVA) were employed as carriers for induction of cellular immunity.[166] Confocal laser scanning microscopy shows that PG liposomes are taken up efficiently by DC2.4 cells (a murine dendritic cell line), thereafter delivering their contents into the cytosol, probably through fusion with endosomal membranes. Murine bone marrow-derived dendritic cells treated with OVA PG liposomes stimulate CD8-OVA1.3 cells more strongly than OT4H.1D5 cells, indicating that the liposomes induce major histocompatibility complex (MHC) class I-restricted presentation.

Nasal immunization with PG liposomes (100 mg OVA, female C57BL/6 mice on days 0 and 14) results in stronger cellular immune responses than the OVA pH-insensitive liposomes (eggPC/DOPE). Splenocytes of the immunized mice exhibit high toxicity toward OVA-presenting E.G7-OVA cells, but almost no toxicity toward their parent EL4 cells, indicating that both liposomes are able to induce OVA-specific CTLs. Remarkably, the polymer-modified liposomes activate cellular immunity in the absence of additional immunostimulating molecules such as monophosphoryl lipid A (MPL), in a fashion comparable to that of complete Freund adjuvant (which is a toxic adjuvant only used in veterinary). Hence, their potential use in the production of efficient vaccines for immunotherapy requires further studies to guarantee safety for human use.

20.8.5 Gene Therapy

Nonviral gene therapy still suffers from limited efficacy in order to be involved in more clinical trials. The major problems are degradation in biological fluids in vitro and in vivo by nucleases, the weak intracellular penetration, and poor cytoplasmic delivery of genes or antisense oligonucleotides (AOs). pH-sensitive liposomes could be used to increase nucleic acids' stability or to improve their cell penetration.

The efficacy of AOs loaded in DOPE/CHEMS/Chol (7:4:2, 100–200 nm) to suppress lipopolysaccharide (LPS)-induced production of tumor necrosis factor (TNF)-α in Kupffer cells of Sprague-Dawley rats has been determined.[167] The in vivo effects of the liposomal AO are time dependent, exhibiting a maximal inhibition of 30% of TNF-α production at 48 h post-i.v. injection. Two daily doses result in a cumulative increase in tissue levels of the AO, which is also associated with a significant inhibition of LPS-induced TNF-α production, in both the liver (50%) and plasma (68%). On the other hand, no inhibitory effect on LPS-induced TNF-α production is observed after AO administration in pH-insensitive liposomes (DPPC/DPPG/Chol, 4:1:5). In liver tissue, the decreased production of TNF-α is also associated with a decrease (35%) in the levels of TNF-α mRNA. This suggests that the reduced production of TNF-α is due in part to the oligonucleotide-induced degradation of mRNA by antisense mechanisms.

Previously, it was also demonstrated that the i.v. administration of DOPE/OA/Chol delivers a high amount of intact AOs in the liver and spleen.[67] Nonetheless, the development of pH-sensitive formulations that are able to avoid recognition by the RES is a very important challenge for the successful i.v. administration of AOs or genes.

For instance, anionic pegylated complexes were prepared by adding anionic pegylated Ph-sensitive liposomes (DPPC/(CCTC or CHEMS)/Chol-Peg$_{110}$), where CCTC is a cholesterol derivative bearing four carboxylate moieties (titrable anionic lipids) at <30 mol% lipids, to cationic lipoplexes (DMAPAP/DOPE liposomes + DNA), where DMAPAP is the cationic lipid dimyristylamidocarbamoylmethylaminopropyl-diaminopropylamine, which possesses two primary, one secondary, and one tertiary amine of low toxicity and high reproducibility as in vitro transfection agents.[168] Anionic pegylated complexes do aggregate, and their Z potential becomes positive at low pH. The i.v. administration of a plasmid encoding for chloramphenicolacetyltransferase in anionic pegylated complexes to mice-bearing 3LL tumors results in low gene expression in the tumor, but CCTC shows a slightly higher level than the CHEMS-containing formulation. There is no increase in gene expression in the liver and spleen, or in the lung, showing that anionic pegylated complexes are not retained in the lung due to nonspecific interactions, as often observed with cationic complexes.

20.8.6 Salmonellosis

Gentamicin (GEN) is a hydrosoluble polycation aminoglycoside antibiotic with broad-spectrum antibacterial activity. After i.v. or i.m. administration, the majority of GEN remains in extracellular locations. Intracellular infections, caused by pathogens such as *Salmonella* (typhoid fever and salmonelosis), *Listeria* (meningitis and septicemia), and *Mycobacterium* species, require that antibiotics reach therapeutic levels at the intracellular site of infection. The in vitro efficacy of GEN pH-sensitive liposomes (DOPE/N-succinyl-DOPE, 70:30; DOPE/N-glutaryl-DOPE, 70:30) containing small amounts of Peg-ceramide C$_{20}$ (0.5 mol; added to reduce the aggregation generated upon incorporating the polycation GEN to anionic liposomes) against intracellular infections is superior to the efficacies of free and GEN pH-insensitive liposomes.[169] The activity of GEN pH-sensitive liposomes is similar against bacteria residing in phagolysosomes (*S. typhimurium)* or in cytoplasm (recombinant hemolysin-expressing *S. typhimurium* strain and *L. monocytogenes*). However, correcting the values of bactericidal activity by substituing the activity of free GEN and dividing by the total nanomols of lipids taken up by 10^6 cells renders a higher activity for pH-sensitive liposomes (that is, higher on cytoplasmic than on phagolysosomal infections) than for pH-insensitive liposomes. The addition of cholesterol to DOPE/N-succinyl-DOPE/Chol/ceramide-Peg (35:30:30:5) liposomes reduces the rate of drug release in plasma upon administration to healthy mice. In the infection model of *Salmonella enterica* serovar *Typhimurium* infection, the pH-sensitive liposomes were more rapidly eliminated from blood than pH-insensitive liposomes, and both formulations increased the plasma AUC of GEN, compared to that of free drug. Liposomal GEN was redirected to the liver and spleen, and its accumulation in the kidney was reduced. Finally, an i.v.-administered monodose of GEN pH-sensitive or GEN pH-insensitive liposomes reduces 10^3- to 10^4-fold the number of resident bacteria from liver and spleen macrophages.[170]

20.8.7 Complement Activation

Recently, it was determined that pH-insensitive liposomes (DOPC/Chol) activate a complement system with lower amounts of phospholipid than pH-sensitive (DOPE/CHEMS) liposomes. The presence of phosphorylcholine groups (that bind to C-reactive protein) in the DOPC molecules, the higher fluidity of DOPC/Chol liposomal membranes (fluid membranes are more susceptible to binding with C3 fragments (C3b)), and the use of CHEMS instead of Chol (the hydroxyl group of Chol constitutes an activated C3b) could be the reasons. Whereas pegylated pH-insensitive liposomes activated complement, pegylated pH-sensitive liposomes proved to be poor activators.[171]

20.9 Conclusions

After three decades of basic research, in the last 5 years the first preclinical studies employing pH-sensitive liposomes have been published. This is probably a consequence of their major drawback—lability in the bloodstream—caused by the extraction of lipids from the pH-sensitive bilayers or the loss of the attached pH-sensitive polymers.

The concept of tunable pH-sensitive liposomes was born in the search for fine-tuning the pH response that is poorly achieved with DOPE/CHEMS liposomes. However, up to the moment, only improved in vitro plasma stability has been achieved, and there is no evidence for a better capacity of tuning in vivo.

Pegylation of DOPE/CHEMS liposomes results in lack of pH sensitivity in plasma; hence, their capacity to execute simultaneously multiple functions (long circulation and pH sensitivity) is arguable. Recently synthesized nonionic polymers were capable of replacing the steric stabilization provided by Peg in DOPE/CHEMS liposomes. However, nothing is known about their biocompatibility. On the other hand, the lower cell uptake of pegylated pH-sensitive liposomes occurring above a threshold of Peg content should be compensated by a higher cytoplasmic delivery. In this controversial scenario, a recent series of preclinical studies has undoubtedly shown the feasibility of tumor therapeutics and diagnosis by employing pegylated pH-sensitive DOPE/CHEMS liposomes.

As an alternative to the mechanism of pH sensitivity mediated by the transition to a nonlamellar phase triggered by CHEMS protonation, the transition of DOPE liposomes is triggered by the hydrolysis of pH-sensitive linkages, leading to the detachment of the Peg layer that avoids liposomal aggregation. The hydrolysis of the vinyl ether linkages BVEP and CVEP is not accompanied by membrane fusion. The performance of pH sensitivity of liposomes bearing Peg attached by these labile linkages remains to be compared to that of DOPE/CHEMS liposomes in vivo. This would decide if their higher plasma stability overcomes the lower efficiency of cytoplasmic delivery with DOPE/CHEMS liposomes. The higher rate of acid hydrolysis of the orthoester linkage showed by POD was determined in the absence of cells, and only pharmacokinetics and biodistribution were determined in mice. A common disadvantage of labile linkages is the need for an ad hoc synthesis.

Liposomes that destabilize as a consequence of an acid trigger phase transition induced on an attached polymer such as poly(NIPA) and PG possess a complex architecture. In theory, the destabilization pH can be tuned by the percentage and degree of hydrophilicity of carboxilic acids added as comonomers or linked to the polymer backbone. Remarkably, the pegylation does not reduce the pH sensitivity in plasma of liposomes bearing poly(NIPA-co-MAA-co-DODA). Their performance in vivo remains unknown. Probably the shorter the poly(NIPA) chains, the more biodegradable and biocompatible the polymer will be. However, multiple dosages of liposomes bearing poly(NIPA) and PG polymers deserve special attention, in view of the previous experience with pegylated liposomal doxorubicin that caused severe toxic reactions due to the accelerated blood clearance effect.

Clearly, the major field of preclinical application of pH-sensitive liposomes is represented by those targets that are not achieved upon long circulation and accumulation at EPR effect sites. DOPE/CHEMS pH-sensitive liposomes can be easily prepared, employing commercially available components. Though the scale-up feasibility, sterilization, and shelf life of these formulations remain to be determined, their biocompatibility at multiple dosages is well known and positions them as potential candidates to enter clinical studies. In sum, the equation of drawbacks of simplicity vs. advantages of complexity will be solved when in vivo performance, biocompatibility, and scale-up feasibility are determined.

References

1. Wagner, V., et al., The emerging nanomedicine landscape, *Nat. Biotechnol.*, 24, 1211, 2006.
2. Farokhzad, O.C., and Langer, R., Impact of nanotechnology on drug delivery, *ACS Nano*, 3, 16, 2009.
3. Bawarski, W.E., et al., Emerging nanopharmaceuticals, *Nanomed. Nanotechnol. Biol. Med.*, 4, 273, 2008.
4. Duncan, R., Nanomedicine gets clinical, *Nano Today*, 16, 2005.
5. High-Level Group and European Technology Platform Nanomedicine, *Strategic research agenda for nanomedicine: nanomedicine nanotechnology for health*, 2006.
6. Caruthers, S.D., Wickline, S.A., and Lanza, G.M., Nanotechnological applications in medicine, *Curr. Opin. Biotechnol.*, 18, 26, 2007.
7. International Organization for Standardization, *Nanotechnologies—terminology and definitions for nano-objects—nanoparticle, nanofibre and nanoplate*, ISO/TS 27687: 2008. (http://www.iso.org/iso/catalogue_detail.htm?csnumber=44278&utm_source=ISO&utm_medium=RSS&utm campaign=Catalogue).
8. Oberdörster, G., Oberdörster, E., and Oberdörster, J., Nanotoxicology: an emerging discipline evolving from studies of ultrafine particles, *Environ. Health Perspect.*, 113, 823, 2005.
9. Prokop, A., and Davidson, J.M., Nanovehicular intracellular delivery systems, *J. Pharm. Sci.*, 97, 3518, 2008.
10. Alonso, M.J., Nanomedicines for overcoming biological barriers, *Biomed. Pharmacother.*, 58, 168, 2004.
11. Alexis, F., et al., Factors affecting the clearance and biodistribution of polymeric nanoparticles, *Mol. Pharm.*, 5, 505, 2008.
12. Greish, K., Enhanced permeability and retention of macromolecular drugs in solid tumors: a royal gate for targeted anticancer nanomedicines, *J. Drug Targeting*, 15, 457, 2007.
13. Maeda, H., Bharate, G.Y., and Daruwalla, J., Polymeric drugs for efficient tumor-targeted drug delivery based on EPR-effect, *Eur. J. Pharm. Biopharm.*, 71, 409, 2009.

14. Gabizon, A.A., Shmeeda, H., and Zalipsky, S., Pros and cons of the liposome platform in cancer drug targeting, *J. Liposome Res.*, 16, 175, 2006.
15. Owens, D.E., and Peppas, N.A., Opsonization, biodistribution, and pharmacokinetics of polymeric nanoparticles, *Int. J. Pharm.*, 307, 93, 2006.
16. Howard, M.D., et al., PEGylation of nanocarrier drug delivery systems: state of the art, *J. Biomed. Nanotechnol.*, 4, 133, 2008.
17. Oku, N., et al., Evaluation of drug targeting strategies and liposomal trafficking, *Curr. Pharm. Des.*, 6, 1669, 2000.
18. Dos Santos, N., et al., Influence of poly(ethylene glycol) grafting density and polymer length on liposomes: relating plasma circulation lifetimes to protein binding, *Biochim. Biophys. Acta*, 1768, 1367, 2007.
19. Hong, R.L., et al., Direct comparison of liposomal doxorubicin with or without polyethylene glycol coating in C-26 tumor-bearing mice: is surface coating with polyethylene glycol beneficial? *Clin. Cancer Res.*, 5, 3645, 1999.
20. Mishra, S., Webster, P., and Davis, M.E., PEGylation significantly affects cellular uptake and intracellular trafficking of non-viral gene delivery particles., *Eur. J. Cell Biol.*, 83, 97, 2004.
21. Hatakeyama, H., et al., Development of a novel systemic gene delivery system for cancer therapy with a tumor-specific cleavable PEG-lipid, *Gene Ther.*, 14, 68, 2007.
22. Baba, T., et al., Clathrin-dependent and clathrin-independent endocytosis are differentially sensitive to insertion of poly(ethylene glycol)-derivatived cholesterol in the plasma membrane, *Traffic*, 2, 501, 2001.
23. Gaoa, J., and Xu, B., Applications of nanomaterials inside cells, *Nano Today*, 4, 37, 2009.
24. Sanvicens, N., and Marco, M.P., Multifunctional nanoparticles—properties and prospects for their use in human medicine, *Trends Biotechnol.*, 26, 425, 2008.
25. Alberola, A.P., and Rädler, J.O., The defined presentation of nanoparticles to cells and their surface controlled uptake, *Biomaterials*, 30, 3766, 2009.
26. Di Marzio, L., et al., pH-sensitive non-phospholipid vesicle and macrophage-like cells: binding, uptake and endocytotic pathway, *Biochim. Biophys. Acta*, 1778, 2749, 2008.
27. Faraji, A.H., and Wipf, P., Nanoparticles in cellular drug delivery, *Bioorganic Med. Chem.*, 17, 2950, 2009.
28. Ganta, S., et al., A review of stimuli-responsive nanocarriers for drug and gene delivery, *J. Control. Release*, 126, 187, 2008.
29. Perret, E., et al., Evolving endosomes: how many varieties and why? *Curr. Opin. Cell Biol.*, 17, 423, 2005.
30. Hunt, C.A., MacGregor, R.D., and Siegal, R.A., Engineering targeted *in vivo* drug delivery. I. The physiological and physicochemical principles governing opportunities and limitations, *Pharm. Res.*, 3, 333, 1986.
31. Egeblad, M., and Werb, Z., New functions for the matrix metalloproteinases in cancer progression, *Nat. Rev. Cancer*, 2, 161, 2002.
32. Gerweck, L.E., and Seetharaman, K., Cellular pH gradient in tumor versus normal tissue: potential exploitation for the treatment of cancer, *Cancer Res.*, 56, 1194, 1996.
33. Montcourrier, P., et al., Breast cancer cells have a high capacity to acidify extracellular milieu by a dual mechanism, *Clin. Exp. Metastasis*, 15, 382, 1997.
34. Yatvin, M.B., et al., pH-sensitive liposomes: possible clinical implications, *Science*, 210, 1253, 1980.
35. Cullis, P.R., and De Kruijff, B., Lipid polymorphism and the functional roles of lipids in biological membranes, *Biochim. Biophys. Acta*, 559, 399, 1979.
36. Israelachvili, J.N., Marcelja, S., and Horn, R.G., Physical principles of membrane organization, *Q. Rev. Biophys.*, 13, 121, 1980.
37. Gruner, S.M., et al., Lipid polymorphism—the molecular-basis of nonbilayer phases, *Annu. Rev. Biophys. Biophys. Chem.*, 14, 211, 1985.
38. Connor, J., Yatvin, M.B., and Huang, L., pH-sensitive liposomes: acid-induced liposome fusion, *Proc. Natl. Acad. Sci. USA*, 81, 1715, 1984.

39. Schroit, A.J., Madsen, J., and Nayar, R., Liposome-cell interactions: *in vitro* discrimination of uptake mechanism and *in vivo* targeting strategies to mononuclear phagocytes, *Chem. Phys. Lipids*, 40, 373, 1986.
40. Düzgünes, N., et al., Proton-induced fusion of oleic acid-phosphatidylethanolamine liposomes, *Biochemistry*, 24, 3091, 1985.
41. Connor, J., and Huang, L., pH-sensitive immunoliposomes as an efficient and target-specific carrier for antitumor drugs, *Cancer Res.*, 46, 3431, 1986.
42. Huang, L., and Liu, S.S., Acid induced fusion of liposomes with inner membranes of mitochondria, *Biophys. J.*, 45, 72a, 1984.
43. Straubinger, R.M., Duzgunes, N., and Papahadjopoulos, D., pH-sensitive liposomes mediate cytoplasmic delivery of encapsulated macromolecules, *FEBS Lett.*, 179, 148, 1985.
44. Collins, D., Maxfield, F., and Huang, L., Immunoliposomes with different acid sensitivities as probes for the cellular endocytic pathway, *Biochim. Biophys. Acta*, 987, 47, 1989.
45. Ellens, H., Bentz, J., and Szoka, F.C., pH-induced destabilization of phosphatidylethanolamine-containing liposomes: role of bilayer contact, *Biochemistry*, 23, 1532, 1984.
46. Lai, M.Z., Düzgünes, N., and Szoka, F.C., Effects of replacement of the hydroxyl group of cholesterol and tocopherol on the thermotropic behavior of phospholipid membranes, *Biochemistry*, 24, 1646, 1985.
47. Düzgünes, N., Membrane fusion, *Subcell. Biochem.*, 11, 195, 1985.
48. Collins, D.S., Findlay, K., and Harding, C.V., Processing of exogenous liposome-encapsulated antigens *in vivo* generates class I MHC-restricted T cell responses, *J. Immunol.*, 148, 3336, 1992.
49. Schmid, S.L., Toward a biochemical definition of the endosomal compartment. Studies using free flow electrophoresis, *Sub-Cell. Biochem.*, 19, 1, 1993.
50. Murphy, R.F., Powers, S., and Cantor, C.R., Endosome pH measured in single cells by dual fluorescence flow cytometry: rapid acidification of insulin to pH 6, *J. Cell. Biol.*, 98, 1757, 1984.
51. Collins, D., *pH-sensitive liposomes as tools for cytoplasmic delivery*, ed. Philippot, J.R., and Schuber, F., CRC Press, Boca Raton, FL, 1995, p. 1.
52. Ropert, C., Malvy, C., and Couvreur, P., *pH sensitive liposomes as efficient carriers for intracellular delivery of oligonucleotides*, ed. Gregoriadis, G., Plenum Press, New York, 1995, p. 290.
53. Cullis, P.R., and Hope, M.J., Effects of fusogenic agent on membrane structure of erythrocyte ghosts and the mechanism of membrane fusion, *Nature*, 271, 672, 1978.
54. Siegel, D.P., The modified stalk mechanism of lamellar/inverted phase transitions and its implications for membrane fusion, *Biophys. J.*, 76, 291, 1999.
55. Hafez, I.M., and Cullis, P.R., Roles of lipid polymorphism in intracellular delivery, *Adv. Drug Del. Rev.*, 47, 139, 2001.
56. Drummond, D.C., Zignani, M., and Leroux, J.C., Current status of pH-sensitive liposomes in drug delivery, *Prog. Lipid Res.*, 39, 409, 2000.
57. de Duve, C.T., et al., Commentary. Lysosomotropic agents, *Biochem. Pharmacol.*, 23, 2495, 1974.
58. Ohkuma, S., and Poole, B., Fluorescence probe measurement of the intralysosomal pH in living cells and the perturbation of pH by various agents, *Proc. Natl. Acad. Sci. USA*, 75, 3327, 1978.
59. Tyko, B., and Maxfield, F.R., Rapid acidification of endocytic vesicles containing alpha 2-macroglobulin, *Cell*, 28, 643, 1982.
60. Straubinger, R.M., Papahadjopoulos, D., and Hong, K., Endocytosis and intracellular fate of liposomes using pyranine as a probe, *Biochemistry*, 29, 4929, 1990.
61. Connor, J., Norley, N., and Huang, L., Biodistribution of pH-sensitive immunoliposomes, *Biochim. Biophys. Acta*, 884, 474, 1986.
62. Liu, D., and Huang, L., Interaction of pH-sensitive liposomes with blood components, *J. Liposome Res.*, 987, 47, 1994.
63. Chu, C.J., and Szoka, F.C., pH-sensitive liposomes, *J. Liposome Res.*, 4, 361, 1994.
64. Collins, D., Litzinger, D.C., and Huang, L., Structural and functional comparisons of pH-sensitive liposomes composed of phosphatidylethanolamine and three different diacylsuccinylglycerols, *Biochim. Biophys. Acta*, 1025, 234, 1990.

65. Liu, D., and Huang, L., pH-sensitive, plasma-stable liposomes with relatively prolonged residence in circulation, *Biochim. Biophys. Acta*, 1022, 348, 1990.

66. Liu, D., and Huang, L., Role of cholesterol in the stability of pH-sensitive large unilalellar liposomes prepared by the detergent dialysis method, *Biochim. Biophys. Acta*, 981, 254, 1989.

67. De Oliveira, M.C., et al., Improvement of *in vivo* stability of phosphodiester oligonucleotide using anionic liposomes in mice, *Life Sci.*, 67, 1625, 2000.

68. Slepushkin, V.A., et al., Sterically stabilized pH-sensitive liposomes. Intracellular delivery of aqueous contents and prolonged circulation *in vivo*, *J. Biol. Chem.*, 272, 2382, 1997.

69. Kirpotin, D., et al., Liposomes with detachable polymer coating: destabilization and fusion of dioleoylphosphatidylethanolamine vesicles triggered by cleavage of surface-grafted poly(ethylene glycol), *FEBS Lett.*, 388, 115, 1996.

70. Simões, S., et al., On the mechanisms of internalization and intracellular delivery mediated by pH-sensitive liposomes, *Biochim. Biophys. Acta*, 1515, 23, 2001.

71. Momekova, D., et al., Long-circulating, pH-sensitive liposomes sterically stabilized by copolymers bearing short blocks of lipid-mimetic units, *Eur. J. Pharm. Sci.*, 32, 308, 2007.

72. Ishida, T., et al., Targeted delivery and triggered release of liposomal doxorubicin enhances cytotoxicity against human B lymphoma cells, *Biochim. Biophys. Acta*, 1515, 144, 2001.

73. Ishida, T., et al., Development of pH-sensitive liposomes that efficiently retain encapsulated doxorubicin (DXR) in blood, *Int. J. Pharm.*, 309, 94, 2006.

74. Liu, D., et al., Interactions of serum proteins with small unilamellar liposomes composed of dioleoylphosphatidylethanolamine and oleic acid: high-density lipoprotein, apolipoprotein A1, and amphipathic peptides stabilize liposomes, *Biochemistry*, 29, 3637, 1990.

75. Hafez, I.M., and Cullis, P.R., Cholesteryl hemisuccinate exhibits pH sensitive polymorphic phase behavior, *Biochim. Biophys. Acta*, 1463, 107, 2000.

76. Hafez, I.M., Maurer, N., and Cullis, P.R., On the mechanism whereby cationic lipids promote intracellular delivery of polynucleic acids, *Gene Ther.*, 8, 1188, 2001.

77. Hafez, I.M., Ansell, S., and Cullis, P.R., Tunable pH-sensitive liposomes composed of mixtures of cationic and anionic lipids, *Biophys. J.*, 79, 1438, 2000.

78. Shi, G., et al., Efficient intracellular drug and gene delivery using folate receptor-targeted pH-sensitive liposomes composed of cationic/anionic lipid combinations, *J. Control. Release*, 80, 309, 2002.

79. Sudimack, J.J., et al., A novel pH-sensitive liposome formulation containing oleyl alcohol, *Biochim. Biophys. Acta*, 1564, 31, 2002.

80. Allen, T.M., et al., Stealth liposomes: an improved sustained release system for 1-beta-d-arabinofuranosylcytosine, *Cancer Res.*, 52, 2431, 1992.

81. Hiraka, K., et al., Preparation of pH-sensitive liposomes retaining SOD mimic and their anti-cancer effect, *Colloids Surf. B Biointerfaces*, 67, 54, 2008.

82. Saito, G., Swanson, J.A., and Lee, K.D., Drug delivery strategy utilizing conjugation via reversible disulfide linkages: role and site of cellular reducing activities, *Adv. Drug Del. Rev.*, 55, 199, 2003.

83. Zalipsky, S., et al., New detachable poly(ethylene glycol) conjugates: cysteine-cleavable lipopolymers regenerating natural phospholipid, diacyl phosphatidylethanolamine, *Bioconjugate Chem.*, 10, 703, 1999.

84. Thorpe, P.E., et al., New coupling agents for the synthesis of immuno toxins containing a hindered disulfide bond with improved stability *in vivo*, *Cancer Res.*, 47, 5924, 1987.

85. Lee, E.S., Gao, Z., and Bae, Y.H., Recent progress in tumor pH targeting nanotechnology, *J. Control. Release*, 132, 164, 2008.

86. Kale, A.A., and Torchilin, V.P., Design, synthesis, and characterization of pH-sensitive PEG-PE conjugates for stimuli-sensitive pharmaceutical nanocarriers: the effect of substitutes at the hydrazone linkage on the pH stability of PEG-PE conjugates, *Bioconjugate Chem.*, 18, 363, 2007.

87. Caron, N.J., et al., Intracellular delivery of a Tat-eGFP fusion protein into muscle cells, *Mol. Ther.*, 3, 310, 2001.

88. Sawant, R.M., et al., "SMART" drug delivery systems: double-targeted pH-responsive pharmaceutical nanocarriers, *Bioconjugate Chem.*, 17, 943, 2006.

89. Boomer, J.A., et al., Acid triggered release from sterically-stabilized fusogenic vesicles via a hydrolytic dePEGylation strategy, *Langmuir*, 19, 6408, 2003.
90. Boomer, J.A., et al., Cytoplasmic delivery of liposomal contents mediated by an acid-labile cholesterol-vinyl ether-PEG conjugate, *Bioconjugate Chem.*, 20, 47, 2009.
91. Guo, X., and Szoka Jr., F.C., Steric stabilization of fusogenic liposomes by a low-pH sensitive PEG-diortho ester-lipid conjugate, *Bioconjug. Chem.*, 12, 291, 2001.
92. Guo, X., MacKay, J.A., and Szoka, F.C., Mechanism of pH-triggered collapse of phosphatidyle-thanolamine liposomes stabilized by an ortho ester polyethyleneglycol lipid, *Biophys. J.*, 84, 1784, 2003.
93. Palatini, P., Disposition kinetics of phospholipid liposomes, *Adv. Exp. Med. Biol.*, 318, 375, 1992.
94. Rui, Y., et al., Diplasmenylcholine-folate liposomes: an efficient vehicle for intracellular drug delivery, *J. Am. Chem. Soc.*, 120, 11213, 1998.
95. Qualls, M.M., and Thompson, D.H., Chloroaluminum phthalocyanine tetrasulfonate delivered via acid-labile diplasmenylcholine-folate liposomes: intracellular localization and synergistic phototoxicity, *Int. J. Cancer*, 93, 384, 2001.
96. Gerasimov, O.V., Rui, Y., Thompson, D.H., and Rosoff, M., Vesicles, ed. Rosoff, M. Marcel Dekker, New York, 1996.
97. Kim, J.M., Thompson, D.H., and Texter, J., *Surfactanl Science Series: Reactions & Synthesis in Surfactant Systems* ed. J. Texter, Marcel Dekker, New York, 2001.
98. Gerasimov, O.V., Schwan, A., and Thompson, D.H., Acid-catalyzed plasmenylcholine hydroly-sis and its effect on bilayer permeability: a quantitative study, *Biochim. Biophys. Acta*, 1324, 200, 1997.
99. Ellens, H., Bentz, J., and Szoka, F.C., H+- and Ca+2-induced fusion and destabilization of lipo-somes, *Biochemistry*, 24, 3099, 1985.
100. Huang, Z., et al., Acid-triggered transformation of diortho ester phosphocholine liposome, *J. Am. Chem. Soc.*, 128, 60, 2006.
101. Heskins, M., and Guillet, J., Solution properties of poly(N-isopropylacrylamide), *J. Macromol. Sci. A Pure Appl. Chem.*, 2, 1441, 1968.
102. Rzaev, Z.M.O., Dincer, S., and Piskin, E., Functional copolymers of N-isopropylacrylamide for bioengineering applications, *Prog. Polym. Sci.*, 32, 534, 2007.
103. Zhu P.W., and Napper, D.H., Entanglement knotting in globule to-coil transitions of poly(N-iso-propyacrylamide) at interfaces, *J. Colloid Interface Sci.*, 168, 380, 1994.
104. Heskins, M., and Guillet, J.E., Solution properties of poly(N-isopropylacrylamide), *J Macromol Sci Chem A2*, 1441, 1968.
105. Fujishige, S., Kubota, K., and Ando, I., Phase transition of aqueous solutions of poly(N-isopro-pylacrylamide) and poly(N-isopropylmethacrylamide), *J. Phys. Chem.*, 93, 3311, 1989.
106. Chen, G., and Hoffman, A.S., A new temperature- and pH-responsive copolymer for possible use in protein conjugation, *Macromol. Chem. Phys.*, 195, 1251, 1995.
107. Kuckling, D., et al., Temperature and pH dependent solubility of novel poly(N-isopropylacrylamide)-copolymers, *Macromol. Chem. Phys.*, 201, 273, 2000.
108. Principi, T., et al., Solution properties of hydrophobically modified copolymers of N-isopropylacrylamide and N-glycine acrylamide: a study by microcalorimetry and fluores-cence spectroscopy, *Macromolecules*, 33, 2958, 2000.
109. Shimizu, T., and Minakata, A., Effect of divalent cations on the volume of a maleic acid copoly-mer gel examined by incorporating lysozyme, *Eur. Polym. J.*, 38, 1113, 2002.
110. Feil, H., et al., Effect of comonomer hydrophilicity and ionization on the lower critical solution temperature of N-isopropylacrylamide copolymers, *Macromolecules*, 26, 2496, 1993.
111. Kesim, H., et al., Bioengineering functional copolymers. II. Synthesis and characterization of amphiphilic poly(N-isopropylacrylamide-co-maleic anhydride) and its macrobranched deriva-tives, *Polymer*, 44, 2897, 2003.
112. Yessine, M.A., et al., Characterization of the membrane-destabilizing properties of different pH-sensitive methacrylic acid copolymers, *Biochim. Biophys. Acta*, 1613, 28, 2003.

113. Francis, M.F., et al., *In vitro* evaluation of pH sensitive polymer/niosome complexes, *Biomacromolecules*, 2, 741, 2001.

114. Zignani, M., et al., *In vitro* characterization of a novel polymeric-based pH-sensitive liposome system, *Biochim. Biophys. Acta*, 1463, 383, 2000.

115. Pétriat, F., et al., Study of molecular interactions between a phospholipidic layer and a pH-sensitive polymer using the Langmuir balance technique, *Langmuir*, 20, 1393, 2004.

116. Meyer, O., Papahadjopoulos, D., and Leroux, J.C., Copolymers of N-isopropylacrylamide can trigger pH sensitivity to stable liposomes, *FEBS Lett.*, 42, 61, 1998.

117. Roux, E., et al., Steric stabilization of liposomes by pH-responsive N-isopropylacrylamide copolymer, *J. Pharm. Sci.*, 91, 1795, 2002.

118. Roux, E., et al., On the characterization of pH-sensitive liposome/polymer complexes, *Biomacromolecules*, 4, 240, 2003.

119. Roux, E., et al., Polymer-based pH-sensitive carriers as a means to improve the cytoplasmic delivery of drugs., *Int. J. Pharm.*, 242, 25, 2002.

120. Leroux, J.C., et al., N-Isopropylacrylamide copolymers for the preparation of pH-sensitive liposomes and polymeric micelles, *J. Control. Release*, 72, 71, 2001.

121. Kono, K., et al., Thermosensitive polymer-modified liposomes that release contents around physiological temperature, *Biochim. Biophys. Acta*, 1416, 239, 1999.

122. Holland, J.W., et al., Poly(ethylene glycol)-lipid conjugates regulate the calcium-induced fusion of liposomes composed of phosphatidylethanolamine and phosphatidylserine, *Biochemistry*, 35, 2618, 1996.

123. Roux, E., et al., Serum-stable and long-circulating, PEGylated, pH-sensitive liposomes, *J. Control. Release*, 94, 447, 2004.

124. Bertrand, N., et al., Pharmacokinetics and biodistribution of N isopropylacrylamide copolymers for the design of pH-sensitive liposomes, *Biomaterials*, 30, 2598, 2009.

125. Yamaoka, T., Tabata, Y., and Ikada, Y., Distribution and tissue uptake of poly(ethylene glycol) with different molecular weights after intravenous administration to mice, *J. Pharm. Sci.*, 83, 601, 1994.

126. Seymour, L.W., et al., Effect of molecular weight (Mw) on N-(2-hydroxypropyl)methacrylamide copolymers on body distribution and rate of excretion after subcutaneous, intraperitoneal, and intravenous administration to rats, *J. Biomed. Mater. Res.*, 21, 1341, 1987.

127. Chung, J., et al., Effect of molecular architecture of hydrophobically modified poly(N-isopropylacrylamide) on the formation of thermoresponsive core-shell micellar drug carriers, *J. Control. Release*, 53, 119, 1998.

128. Simard, P., and Leroux, J.C., pH-sensitive immunoliposomes specific to the CD33 cell surface antigen of leukemic cells, *Int. J. Pharm.*, 381, 86, 2009.

129. Huang, A., Kennel, S.J., and Huang, L., Interactions of immunoliposomes with target cells, *J. Biol. Chem.*, 258, 14034, 1983.

130. Rustum, Y.M., et al., Inability of liposome encapsulated 1-beta-d-arabinofuranosylcytosine nucleotides to overcome drug resistance in L1210 cells, *Eur. J. Cancer Clin. Oncol.*, 17, 809, 1981.

131. Allcock, H.R., Pucher, S.R., and Scopelianos, A.G., Poly[(amino acid ester)phosphazenes] as substrates for the controlled release of small molecules, *Biomaterials*, 15, 563, 1994.

132. Couffin-Hoarau, A.C., and Leroux, J.C., Report on the use of poly(organophosphazenes) for the design of stimuli-responsive vesicles, *Biomacromolecules*, 5, 2082, 2004.

133. Mukherjee, S., Ghosh, R.N., and Maxfield, F.R., Endocytosis, *Physiol. Rev.*, 77, 759, 1997.

134. Kainthan, R.K., et al., Biocompatibility testing of branched and linear polyglycidol, *Biomacromolecules*, 7, 703, 2006.

135. Sakaguchi, N., et al., Preparation of pH-sensitive poly(glycidol) derivatives with varying hydrophobicities: their ability to sensitize stable liposomes to pH, *Bioconjug. Chem.*, 19, 1040, 2008.

136. Kono, K., Zenitani, K., and Takagishi, T., Novel pH-sensitive liposomes: liposomes bearing a poly(ethylene glycol) derivative with carboxyl groups, *Biochim. Biophys. Acta*, 1193, 1, 1994.

137. Kono, K., Igawa, T., and Takagishi, T., Cytoplasmic delivery of calcein mediated by liposomes modified with a pH-sensitive poly(ethylene glycol) derivative, *Biochim. Biophys. Acta*, 1325, 143, 1997.

138. Mizoue, T., et al., Targetability and intracellular delivery of anti-BCG antibody-modified, pH-sensitive fusogenic immunoliposomes to tumor cells, *Int. J. Pharm.*, 237, 129, 2002.

139. Sakaguchi, N., et al., Effect of transferrin as a ligand of pH-sensitive fusogenic liposome-lipoplex hybrid complexes, *Bioconjug. Chem.*, 19, 1588, 2008.

140. Sakaguchi, N., et al., Generation of highly potent nonviral gene vectors by complexation of lipoplexes and transferrin-bearing fusogenic polymer-modified liposomes in aqueous glucose solution, *Biomaterials*, 29, 1262, 2008.

141. Conner, S.D., and Schmid, S.L., Regulated portals of entry into the cell, *Nature*, 422, 37e, 2003.

142. Wagner, E., Curiel, D., and Cotton, M., Delivery of drugs, proteins and genes into cells using transferrin as a ligand for receptor-mediated endocytosis, *Adv. Drug Del. Rev.*, 14, 113, 1994.

143. Sheff, D.R., et al., The receptor recycling pathway contains two distinct populations of early endosomes with different sorting functions, *J. Cell Biol.*, 145, 123, 1999.

144. Kono, K., et al., Novel gene delivery systems: complexes of fusogenic polymer-modified liposomes and lipoplexes, *Gene Ther.*, 8, 5, 2001.

145. Sakaguchi, N., et al., Enhancement of transfection activity of lipoplexes by complexation with transferrin-bearing fusogenic polymer-modified liposomes, *Int. J. Pharm.*, 325, 186, 2006.

146. Bandak, S., et al., Pharmacological studies of cisplatin encapsulated in long-circulating liposomes in mouse tumor models, *Anticancer Drugs*, 10, 911, 1999.

147. White, S.C., et al., Phase II study of SPI-77 (sterically stabilised liposomal cisplatin) in advanced non-small-cell lung cancer, *Br. J. Cancer*, 95, 822, 2006.

148. Harrington, K.J., et al., Phase I–II study of pegylated liposomal cisplatin (SPI-077™) in patients with inoperable head and neck cancer, *Ann. Oncol.*, 12, 493, 2001.

149. Júnior, A.D., et al., Tissue distribution evaluation of stealth pH-sensitive liposomal cisplatin versus free cisplatin in Ehrlich tumor-bearing mice, *Life Sci.*, 80, 659, 2007.

150. Leite, E.A., et al., Acute toxicity of long-circulating and pH-sensitive liposomes containing cisplatin in mice after intraperitoneal administration, *Life Sci.*, 84, 641, 2009.

151. Kim, M.J., et al., Preparation of pH-sensitive, long-circulating and EGFR-targeted immunoliposomes, *Arch. Pharm. Res.*, 4, 539, 2008.

152. Kim, I.Y., et al., Antitumor activity of EGFR targeted pH-sensitive immunoliposomes encapsulating gemcitabine in A549 xenograft nude mice, *J. Control. Release*, 140, 55, 2009.

153. Carmo, V.A., et al., Biodistribution study and identification of inflammation sites using 99mTc-labelled stealth pH-sensitive liposomes, *Nucl. Med. Commun.*, 29, 33, 2008.

154. Goins, B., et al., Biodistribution and imaging studies of technetium-99m-labeled liposomes in rats with focal infection, *J. Nucl. Med.*, 34, 2160, 1993.

155. Oyen, W.J.G., et al., Detecting infection and inflammation with technetium-99m-labeled stealth liposomes, *J. Nucl. Med.*, 37, 1392, 1996.

156. Maya, J.D., et al., Mode of action of natural and synthetic drugs against *Trypanosoma cruzi* and their interaction with the mammalian host, *Comp. Biochem. Physiol. A*, 146, 601, 2007.

157. Petray, P.B., et al., *In vitro* activity of etanidazole against the protozoan parasite *Trypanosoma cruzi*, *Mem. Inst. Oswaldo Cruz*, 99, 233, 2004.

158. Morilla, M.J., et al., Etanidazole in pH-sensitive liposomes: design, characterization and *in vitro/in vivo* anti-*Trypanosoma cruzi* activity, *J. Control. Release*, 103, 599-607, 2005.

159. Tucker, S.C., and Casadevall, A., Replication of *Cryptococcus neoformans* in macrophages is accompanied by phagosomal permeabilization and accumulation of vesicles containing polysaccharide in the cytoplasm, *Proc. Natl. Acad. Sci. USA*, 99, 3165, 2002.

160. Nasti, T.H., Khan, M.A., and Owais, M., Enhanced efficacy of pH-sensitive nystatin liposomes against *Cryptococcus neoformans* in murine model, *J. Antimicrob. Chemother.*, 57, 349, 2006.

161. Verma, A.K., Conrad, E.A., and Boutwell, R.K., Differential effects of retinole acid and 7,8-benzoflavone on the induction of mouse skin tumors by the complete carcinogenesis process and by the initiation-promotion regimen, *Cancer Res.*, 42, 3519, 1982.

162. Khan, A., et al., Potential of diallyl sulfide bearing pH-sensitive liposomes in chemoprevention against DMBA-induced skin papilloma, *Mol. Med.*, 13, 443, 2007.

163. Luo, L., et al., Induction of V3-specific cytotoxic T lymphocyte responses by HIV gag particles carrying multiple immunodominant V3 epitopes of gp120, *Virology*, 240, 316, 1998.

164. Chang, J.S., et al., Immunogenicity of synthetic HIV-1 V3 loop peptides by MPL adjuvanted pH-sensitive liposomes, *Vaccine*, 17, 1540, 1999.

165. Chang, F.H., et al., Development of Th1-mediated CD8+ effector T cells by vaccination with epitope peptides encapsulated in pH-sensitive liposomes, *Vaccine*, 19, 3608, 2001.

166. Yuba, E., et al., pH-sensitive fusogenic polymer-modified liposomes as a carrier of antigenic proteins for activation of cellular immunity, *Biomaterials*, 31, 943, 2010.

167. Ponnappa, B.C., et al., *In vivo* delivery of antisense oligonucleotides in pH-sensitive liposomes inhibits lipopolysaccharide-induced production of tumor necrosis factor-α in rats, *J. Pharmacol. Exp. Ther.*, 297, 1129, 2001.

168. Mignet, N., et al., Anionic pH-sensitive pegylated lipoplexes to deliver DNA to tumors, *Int. J. Pharm.*, 361, 194, 2008.

169. Lutwyche, P., et al., Intracellular delivery and antibacterial activity of gentamicin encapsulated in pH-sensitive liposomes, *Antimicrob. Agents Chemother.*, 42, 2511, 1998.

170. Cordeiro, C., et al., Antibacterial efficacy of gentamicin encapsulated in pH-sensitive liposomes against an *in vivo Salmonella enterica* Serovar Typhimurium intracellular infection model, *Ant. Agents Chemother.*, 44, 533, 2000.

171. Carmo, V.A.S., et al., Physicochemical characterization and study of *in vitro* interactions of pH-sensitive liposomes with the complement system, *J. Liposome Res.*, 18, 59, 2008.

21

Bio- and Nanotechnology Offer High Performance for Water Cleanup

J. Richard Schorr, Suvankar Sengupta, and Richard Helferich

MetaMateria Technologies LLC, Columbus, Ohio, USA

CONTENTS

21.1 The Problem

Clean water is, of course, an essential element for human existence, and increasingly, in many parts of the world, demand is fast exceeding supply.[1] While water is plentiful, clean, fresh water for human consumption is often limited. It is estimated that only two drops of every gallon of water on the planet are available for human consumption. *Access to potable water* is recognized as an important and growing priority, and a major challenge in many parts of the world. Eighty percent of all the diseases you could name would be wiped out if you just gave people clean water, and it has been suggested that in the 21st century, clean water will be delivered by entrepreneurs.

Drinking water is being used faster than it is replenished. Population increases of 80 million people per year make water treatment a critical component of the overall water supply. Population growth is estimated at double the production of new water resources. The United Nations describes serious water shortages in more than 20 countries, and this will increase in the coming decades since 85% of the world population lives in the drier countries. Even in the United States critical water shortages are projected to occur in some municipalities by 2030. According to the United Nations, more than 1 billion people on earth already lack access to clean drinking water.[1] Limited availability, with conventional cleanup technologies often constrained by the cost of equipment and the energy required to operate it, combined with a growing population, is a recipe for conflict.[2]

Cleanup of water is essential. In addition to natural contaminants, anthropogenic contaminants ranging from disease to modern chemicals are found in supplies we all draw from for drinking water. The virtual elimination of typhoid in first world countries—cited as a medical miracle—is actually due to the engineering treatment of public water

415

supplies. At the same time, groundwater supplies in the United States, and especially in some third world countries, have serious contaminants. Arsenic is found to be a natural groundwater contaminant in many areas of the world (especially in Bangladesh, where 25% of the population is at high risk). Arsenic levels in the United States exceed the new standard in thousands of communities.[2] Lack of access to clean water and sanitation in developing countries remains a high development priority,[3] and increased emphasis is being given to point-of-use water treatment methods.

Traditional approaches for water treatment include filtration to remove trash, sinkers, and floaters; chemical flocculation to remove suspended solids (and to change dissolved solids into suspended solids for removal); ion exchange/sorption to remove low-concentration materials that flocculants cannot address cost-effectively; aeration to remove dissolved gases and high vapor pressure dissolved liquids; and disinfectation (and dechlorination). Specialty treatments may include desalination (brine and salt water for drinking) or demineralization (lab and industrial processes). While these conventional approaches work, they are often expensive, especially for smaller communities. Demand for water reuse is rapidly increasing, and this trend will continue, especially for nonpotable applications.

New approaches are in demand and are continually being examined to supplement traditional water purification methods, where the cost for using conventional water treatment facilities is not sustainable in the future. Infrastructure in some municipalities is aging and contributes to up to 50% of water being lost during transport. Better solutions need to be lower in overall cost, durable, and more effective for the removal of contaminants from water, either in situ or in water purification systems. It is in this context that both higher-performance biofiltration and nano-enabled technologies are being examined. These consist of filtration materials that have enhanced performance arising from high concentrations of beneficial facultative bacteria, higher specific surface area, and an abundance of surface state electrons found in films and parts containing nanomaterials. These bio and nano-enabled technologies will play a significant role in water treatment in the future.

21.2 Bioremediation

Bioremediation (using beneficial bacteria) has historically offered an environmentally elegant method for dealing with waste streams where longer contact times can be used, especially for management of nitrogen compounds. For waste gas cleanup, bioremediation is favored over more expensive and environmentally damaging technologies, such as thermal incineration and chemical treatment.

For improved bioremediation systems, bacteria carriers are needed that provide higher surface areas to allow significantly more room for growth of active biocolonies of beneficial bacteria than traditional media available today. More surface area available generally increases the rate of biofiltration action. Conventional media, such as urethane foam, plastic materials, or many ceramics, have only moderate surface areas available for bacteria growth, which limits their effectiveness, often leading to undesireable clogging due to excessive bacteria growth. Alternatively, a highly porous ceramic containing a hierarchy of interconnected pores, ranging from millimeters to nanometers in size, is available for both bioremediation and nano-enabled filtration. This ceramic typically has over 5 times the surface area available, compared with open urethane foams, and frequently 20 times

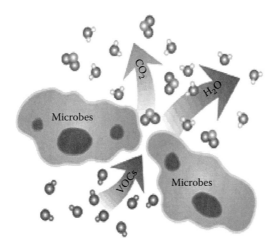

FIGURE 21.1
Schematic illustrating how VOCs are converted by bacteria into CO_2 and H_2O.

more area than plastic media and 70 times more than a traditional sand filter. Water flow through the porous ceramic media, either in aggregrate or monolith form, is more than five times higher than that of a sand filter for a given pressure drop.

Bioremediation is commonly used to remove/convert:

- Ammonia in aquaculture, industrial, and on-site wastewater
- Nitrites and nitrates commonly found in industrial, aquaculture, agriculture, and municipal wastes
- Hydrogen sulfide and volatile organic compounds (VOCs) in waste gases (Figure 21.1) from industrial and municipal sources

High surface area porous ceramics have proven more effective than other media for these applications, requiring fewer media for equivalent performance.

21.3 Nanotechnology

Nanotechnology is the manipulation of individual atoms and molecules in ways that create materials and devices having vastly different properties. *Nano* means 1 billionth, in this case a nanometer (nm). A nanometer particle contains only a few atoms. Typically nanomaterials are less than 100 nm. For comparison, a virus is typically 100 nm in size, and a human hair is 70,000 nm thick.

Nanomaterials are available today in the form of activated materials like carbon or alumina.[4] These are materials with high surface areas; however, because of the fine pores, not all of the water can easily reach the active surfaces and fine pores are easily plugged. This highlights a major challenge in using nanomaterials—their small size. These fine particles or fibers cannot just be added to drinking water, but rather, they must be incorporated into filtration media in ways that allow for the contaminants in water (or waste gases) to readily come in contact with the active media. Water flow through fine materials is extremely slow and requires considerable pressure to obtain acceptable flow rates. Attention is being given by many organizations to the development of filters and media that take advantage of the properties of nanomaterials for removal of contaminants from water, particularly carbon nanotubes, which are being examined for cleanup of brackish water or the selective absorption of contaminants.[4] While such nanomaterials have been shown to be extremely effective, cost and application on larger-scale systems are still major issues.

Other approaches[5] described are ligand nano-coatings capable of binding multiple layers of heavy metal contaminants. One end of the ligand attaches to a high surface area substrate (such as surface-modified zeolites), and the other binds to the contaminant. Once contaminants have saturated the surface area, they often can be removed by additional processing and the media regenerated for re-use. Laboratory testing has demonstrated that a wide range of organics, metals, and inorganics can be removed, as well as some bacteria and viruses.

MetaMateria utilizes nanomaterials and proprietary process technology to prepare particles, whiskers, and platelets for nano-enabled products that can take advantage of the unique properties afforded by these nanomaterials for the removal of contaminants from drinking water. Figure 21.2 shows nanoscale whiskers. These nanomaterials, when properly prepared, can have 1,000 times the surface area of a typical sand filter bed.

In order for one to take advantage of nanomaterials, they need to be part of a system through which contaminated water will easily flow and allow contaminants to come into contact with the specific nano-filtration media selected for water cleanup, such as for removal of metal ions (e.g., As, Pb, Mn, Se, Fe) or hydrocarbons, or nutrients such as phosphates. Porous materials are found to be highly effective carriers for nanomaterials selected for removal of specific contaminants from water.

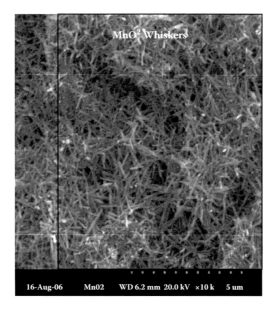

FIGURE 21.2
Nanoscale MnO_2 whiskers grown within a porous structure.

21.4 Filtration Media

Particularly suitable as a base for the application of nanomaterials are ceramics having interconnected porosity, where nanomaterials can be grown directly in the material. Ceramics are ideal because they can withstand the chemical treatment for growing nano-materials and are durable. Many different shapes of nanocrystals can be grown directly on to the pore wall. These are stable and not removed by water passing into or through the porous material.

MetaMateria uses a novel process for making porous ceramics, which allows formation of simple or complex shapes using inexpensive processing methods. These can be made as aggregates or monoliths, depending upon the requirements of the cleanup system. Inorganic hydrogels are used to strongly bond together additives such as zeolite or iron-containing compounds. This approach allows for control of material composition, pore structure, and pore size distribution. Shown in Figure 21.3 are examples of different product shapes (21.3a), microscopic close up of product surface (21.3b) internal interconnected pores of different sizes (21.3c), and nanomaterials grown on the pore surfaces (21.3d).

This porous ceramic is sold for aquaculture and waste gas cleanup applications where facultative bacteria are used for the removal of ammonia, nitrites, nitrate, and related con-taminants, or the removal of odorous gases (H_2S and VOCs). These products are used in systems where water is circulated and reused, such as in aquaculture and where waste-water or waste gases require intensive and specialized treatment. These same materials, when prepared from iron-containing materials, have been shown to be effective filters for water remediation,[5] such as removal of phosphates.

These porous materials can be made from nearly any fine-grained (even waste) materi-als, and flexibility exists for controlling porosity without firing the ceramic, which lowers cost. Pore sizes can be controlled from submicron to 250 microns or larger. An intercon-necting pore network is formed that provides a large surface area for the growth of bac-teria or deposition of nanomaterials. The typical surface area of these porous materials is over 650,000 m^2/m^3, compared with 12,000 m^2/m^3 found in a typical sand filter bed. This provides for a significant increase in active surface area and is also a good platform for making filter materials tailored for a wide number of applications, including remediation and biofiltration.

The typical porosity of these ceramics is around 85%, and because of the hierarchy of pores, water easily flows through the material under a low hydrostatic pressure, as seen in Figure 21.4. In fact, 140 gallons/minute can pass through these porous ceramics

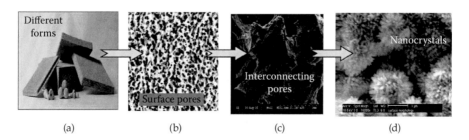

(a) (b) (c) (d)

FIGURE 21.3
Examples showing products(a), product surface(b), internal pore structure(c) and nanocrystals on pore surfaces (d).

FIGURE 21.4
Water being poured through porous plate.

under essentially no pressure, whereas flow through a sand filter is less than 10 gallons/minute.

Media can be provided as granules for replacement of other granular filtration media, or designed as monoliths into smaller, high-performance systems (Figure 21.5). These high surface area, porous ceramics have also been used effectively for in situ remediation. Blocks or granules of the porous ceramic provide for the same effectiveness. Surface area, density, and strength can be adjusted to withstand geostatic loads.

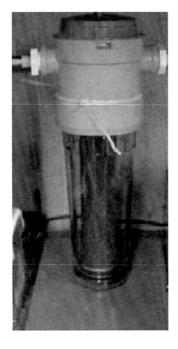

FIGURE 21.5
Example of a small filter containing a nano-enhanced porous ceramic cylinder.

21.5 Nano-Enabled Approach

The typical surface area of the porous ceramics used can be significantly increased through the growth of nanomaterials in the pores of this filter material. This can increase the surface area by 50 to 200 times and even higher, thus providing a very effective medium that only requires short contact times between the contaminant and the active filter agent. This in turn greatly enhances the filtration media performance and reduces the amount of media needed.

How effective the nano-enabled media will be frequently determines both the cost and footprint of the system. For a given set of conditions and contaminant levels, key factors are

- *How long must the water remain in contact with the filter media?* This is typically referred to as empty bed contact time (EBCT), and *shorter is better* since it reduces the size of the system and its capital cost. Effective filtration media need shorter contact times.

    Metals (e.g., Se, As, Pb) are removed in 1–2 minutes vs. a normal 10- to 20-minute time.

    Removing contaminants more quickly lowers system capital cost by 40% or more.

- *How long will the filter media last?* This is typically referred to as the number of bed volumes (BV) of contaminated water passed through a system before the contaminant in the water being discharged (effluent) exceeds the allowable limit. *More BV is better* since it determines how frequently the media must be replaced or regenerated.

    Nanomedia for removing Se and P lasted 10 times longer than conventional media.

- *What are the operating costs of a system using a particular media?* Operating costs can include energy in the form of electricity or pressure, as well as replacement cost of the media. High flow rates without pressure increases are also better and desirable. If pressure to flow water through the media bed is higher, the operating costs will increase.

The type of nanomaterial, along with the methods used to prepare and grow these materials onto the porous substrate, and also the composition of the porous material, will largely determine performance for removal of specific contaminants. Active agents like manganese oxide can be used to change the valence state of metal ions in water (e.g., using manganese oxide to oxidize arsenic from the +3 to the +5 state), which can be more easily removed by reaction with nano-iron oxide hydoxide compounds. Nano-iron oxide compounds are widely used to break down organics, pesticides, and such materials into nonhazardous compounds. Often reactions are enhanced through the addition of other nanomaterials, such as copper or copper oxide.

Coupled with application-specific nano-engineered agents, the resulting media can effectively remove phosphates, heavy metals, lead, arsenic, and other pollutants as water flows through or into the porous filter media. Porous ceramics also can be made to support thin, nanostructured membranes with thicknesses under 100 nm. Figure 21.6 shows a cross section of a membrane supported on a porous alumina substrate. These ceramic membrane filters can provide efficiency benefits and flow rates shown in the laboratory to be 100 times higher than those of conventional reverse-osmosis filtration membranes, which have the potential to reduce the waste of significant quantities of water now disposed of during water cleanup with the membrane technology.

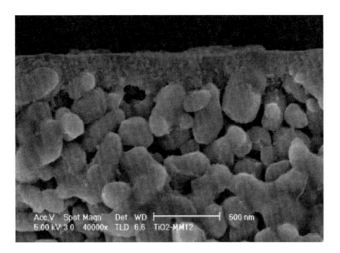

FIGURE 21.6
Cross section of ceramic a 100 nm membrane supported on a porous alumina substrate.

21.6 Application Examples

The following brief examples demonstrate the great flexibility that bio- and nano-enabled media have in addressing a wide range of contaminant removal problems.

- *Conversion of nitrates* into nitrogen gas can be done at rates three times faster than those of other media. With a contact time of 3 minutes, testing shows that over 30 g of nitrate-nitrogen/L/day is achieved, compared with 10 g/L/day reported for other systems. This is done by controlling the surface area and the composition to allow large colonies of heterotrophic and autotrophic bacteria to develop. Nitrates come mainly from nitrites, ammonia, and organic nitrogen compounds and are a common form of inorganic nitrogen found in all wastewater treatment systems, industrial wastewater, aquacultures, and runoff from farms where fertilizer or manure handling exists. Nitrate-nitrogen compounds are found in brackish, salt, and fresh water systems. While nitrates are considered a plant nutrient, concentrations greater than 5 mg/L cause excessive algae growth, leading to accelerated eutrophication that lowers oxygen and kills aquatic life. Levels above 10 mg/L in drinking water can harm infants.

- *Phosphorous* in water is recognized as a major impairment to streams and lakes. Efficient, low-maintenance technologies are needed to lower discharge from sources such as animal farms, septic systems, field runoff, and industrial wastewater. Few phosphorus removal approaches exist, and most are based upon use of iron compounds. The most efficient commercial product is iron activated alumina (AA). Nano-enabled iron, carried on a porous ceramic, has been found to remove significantly more phosphorous and last longer when used to treat water collected from an on-site waste treatment system. The media had a capacity of 55 mg of phosphorus/g media vs. the AA of 18 mg/g. Phosphorous was still being removed after 500 days, while the AA no longer removed phosphorous after 50 days. Phosphorous nutrients come from septic systems, agricultural products, and industrial and municipal wastes, and cause algae growth in water.

- *Selenium* (Se) represents a primary inorganic pollutant that can occur in U.S. West Coast groundwater and is found in oil refinery-produced water, irrigation, and mining wastewater, and in power plant ash ponds. Se concentration in oil refinery-produced water varies from ~50 ppb to over 2,000 ppb with different species (selenide (Se^{-2}), selenite (Se^{+4}), selenate (Se^{+6})). Selenate (Se^{6+}, SeO_4^{2-}) and selenite (Se^{4+}, SeO_3^{2-}) are found in produced water. While selenium is an essential nutrient for humans and animals, it can be toxic in higher doses.

  Surface-functionalized (modified) porous ceramic with nano-iron was found to remove Se from synthetic water samples and actual challenge water samples. The synthetic water samples contained selenite and selenate (50/50); the challenge water samples from oil refinery water had 50 ppb selenium (selenite and selenate) and a high concentration (~400 ppm) of sulfate (SO_4^{2-}), which is considered a primary competition ion for selenate due to the similar molecular structures. Hydrocarbons and high concentrations of solids in the challenge water samples also provide obstacles to selenium removal. The media proved to be effective in lowering Se from 50 ppb to 5 ppb (detectable limit) with bed volumes (BV) of over 40,000 (a long life) at a contact time of 2 minutes. No other media are commercially available with a similar performance. Media were also shown to lower Se from 220 ppb to 50 ppb with a 1-minute contact time and a useful life of over 100,000 BV. In comparison, testing reported by the U.S. Environmental Protection Agency (EPA) on activated alumina reduced selenium in groundwater from 200 ppb to 50 ppb at a 15-minute contact time, but it only had a life of 1,200 BV. Selenium is considered dangerous to wildlife and people, and the EPA limit for drinking water is 5 ppb.

- *Arsenic* is a serious contaminant that needs to be lowered in many sources for drinking water. While many methods exist, they are often cost-prohibitive. Nanomaterials have great potential for use in preparing effective filtration media. Nano-enabled media were shown to lower arsenic from over 50 ppb to 5 ppb with a 2-minute contact time. The Safe Drinking Water Act requires arsenic levels to be 10 ppb.

  It has been shown that a combination of nano-manganese oxide fibers combined with nano-iron oxide (shown in Figure 21.7) is exceptionally effective at removing arsenic (and lead) from water.[6] This technology was prepared on the porous ceramic media and provides a cost-effective approach with a very fast removal rate of arsenic and lead from water. Small column tests were done at a contact time of 0.5 minute using National Sanitation Foundation (NSF) "challenge water" containing 50 ppb arsenic. This medium was almost equivalent in efficiency for the removal of As(III) and As(V). As(III) is much more toxic and harder to remove than As(V); however, the manganese oxide fibers cause preoxidation of As(III) to As(V) prior to sorption of the As(V) by nano-iron, thus removing the need for a preoxidation step during pretreatment. The life of the media was twice that of a commercially available iron-containing medium.

  More than 4,000 municipal communities obtaining water from wells need to reduce arsenic too, as well as many commercial businesses relying on well water.

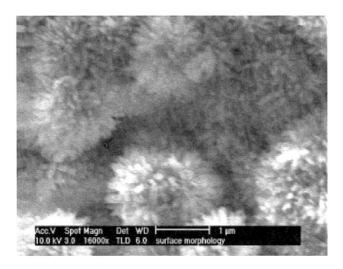

FIGURE 21.7
Clusters of nano-iron oxide crystals used for arsenic removal.

- *Lead* that leaches from plumbing must be below an EPA level of 15 ppb. Using the same nano-enabled media for arsenic, lead was lowered to 0 from 50 ppb, and long life was demonstrated in laboratory testing.
- *Dissolved hydrocarbons* were removed from oil well "produced water" using a new, high surface area nanomedia that was treated to make it hydrophobic. Contact times were 4 minutes. Dissolved oil/grease hydrocarbons went from 1,460 to 190 mg/L, as seen in Figure 21.8. This same approach may be useful for cleanup of dissolved organics from industrial wastewater.

Influent   Effluent

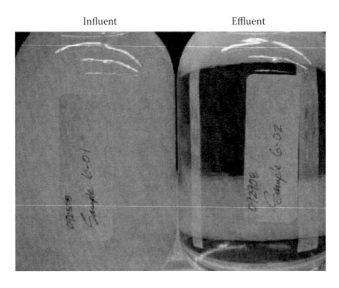

FIGURE 21.8
Example of a cloudy influent containing dissolved organics and the clean effluent.

- *Perchlorates* are found in groundwater, primarily coming from military and industrial waste. A surfactant modified and nano-enhanced porous ceramic substrate was found to remove up to 88% of perchlorates (from 400 ppb to 47 ppb) from contaminated groundwater.

- *Iron, manganese, and hydrogen sulfide* removal from groundwater has also been demonstrated in field tests utilizing nano-manganese oxide and nano-iron.

- *Adsorption of heavy metals* can be addressed using ferrous and some other metal oxides that sorb heavy metals and radionuclides.[8] Sometimes, combinations of metal and oxides are needed. Deposition of zeolite type structures in the pores of the ceramic can provide a "caging process" that allows a significantly high surface area without blocking the pores.[5,7]

21.7 Conclusion

Both bio- and nanotechnology offer many great opportunities to address the critical, global issue of supplying clean water. The increased use of nanomaterials will be seen in the coming decade because of the tremendous performance advantages these materials offer and the affordability of simple, yet effective filters and components that will incorporate these new materials. One of the most exciting approaches is the use of porous ceramic media to carry the active nanomaterials and facultative bacteria for bioremediation. Proper preparation of porous carriers enhances bacteria populations and their ability to break down contaminants. Similarly, selective use of proper nanomaterials has been shown effective in the removal of contaminants, such as metal ions, like selenium, arsenic, lead, and phosphate; removal of noxious gases, such as hydrogen sulfide; sorption of hydrocarbons; and breakdown or removal of some organic contaminants. While field trials are needed to observe the long-term advantages of these nanomedia products, the initial results are highly encouraging.

References

1. UN World Water Development Report 3. *Water in a changing world.* 2009.
2. "The wars of the next century will be about water." Ismail Serageldin, vice president of the World Bank.
3. *"Analysis of Arsenic" Water Conditioning & Purification.* September 2006.
4. Meridian Institute Workshop paper. *"Global dialogue on nanotechnology and the poor: opportunities and risks."* Background paper on water treatment technologies, October 2006.
5. Lisa M. Farmen. Removal of biological and chemical contaminants from water with surface modified minerals. *Ultrapure Water,* January/February 2007.
6. U.S. Department of Energy. *Improvements in permeable reactive barriers to remediate groundwater contaminates by metals and radionuclides.* Report, MSE Technology Applications, 1998.
7. U.S. Patents 6,162,530 and 6,517,802.
8. U.S. Department of Energy. *Surface altered zeolites as permeable barriers for in situ treatment of contaminated groundwater.* Contract DE-AR21-95MC32108.

22

Nanoscience and Engineering
for Robust Biosolar Cells

Ponisseril Somasundaran,[1] **Michael Chin,**[1] **Urszula Tylus Latosiewicz,**[2]
Harry L. Tuller,[3] **Bernardo Barbiellini,**[4] **and Venkatesan Renugopalakrishnan**[2,5]

[1]*Department of Earth and Environmental Engineering, Columbia University, New York, New York, USA*

[2]*Department of Chemistry and Chemical Biology, Northeastern University, Boston, Massachusetts, USA*

[3]*Department of Materials Science and Engineering, MIT, Cambridge, Massachusetts, USA*

[4]*Department of Physics, Northeastern University, Boston, Massachusetts, USA*

[5]*Children's Hospital, Harvard Medical School, Boston, Massachusetts, USA*

CONTENTS

22.1 Introduction

In this chapter we will explore science and engineering involved in utilizing and integrating photoactive proteins with man-made solid-state materials for electronic applications, with a focus on developing a photovoltaic device based on integrating photoactive proteins onto solid-state electrodes.

Increasing energy demands and growing environmental concerns are driving the search for new, sustainable sources of energy. Global energy consumption, now running at 13.5 TW, is estimated to reach 27 TW by 2050. Nearly 85% of today's energy usage originates from fossil fuels. Imminent shortages of fossil fuels[1] and CO_2 induced global warming are driving the demand for new sustainable energy sources. While no single source is likely to satisfy all energy needs, solar energy, with 120,000 TW of solar energy striking the surface of the earth at all times, is by far the most abundant clean energy source available. However, only 0.04% of energy needs are presently generated by photovoltaics. An increase to 20 TW peak power would require covering 0.16% of the surface of the earth with 10% efficient solar cells,[2] an area nearly equivalent to the entire states of Arizona, New Mexico, and Utah combined.[3] Clearly, large-scale solar cell deployment will become feasible only if it becomes possible to supply electricity, cheaply and efficiently, while keeping in mind the environmental impact of making and recycling the solar cell materials.

In a short while, man-made solar technology has made significant progress in harnessing the power of the sun with efforts in solar-thermal, photohydrogen, and photovoltaics. However, as is often the case, there are analogs in nature that already perform these feats of physics and electrochemistry, and have been doing so efficiently and sustainably for hundreds of millennia. Photosynthesis is one such example. Through millions of years of evolution, nature has crafted elaborate photosynthetic machinery for harnessing solar energy that is responsible for essentially all life functions on the earth's surface, from simple single-celled organisms to all plant matter, which in turn fuel the world's foods chains. It does this with a number of photoactive proteins and a series of complex biochemical reactions, harnessing solar energy and storing it as chemical energy in the form of biomass.

There is a distinct similarity in the physics behind solid-state solar cells and particular photoactive proteins responsible for light-activated metabolic processes. The parallel that is of the most fundamental and practical significance is that both man-made solar cells and natural photosynthesis rely on photonic energy to create excited electronic states. Likewise, natural and man-made systems also face similar challenges and limitations, such as limited photon adsorption, maintaining separation of charges, and transport of charge species to reaction sites. Ultimately, it is in the logistics of environmental sustainability and longevity where both man-made and synthetic solar cells differ, suffering from different shortcomings. Man-made solar panels are designed to produce energy only, but also result in many environmental problems with toxic by-products. Natural systems have evolved primarily for survival, so they convert the electronic potential derived from the sun into a biomass that they can store, and unlike man-made solar devices, they are not robust, static systems, but constantly regenerate components to maintain photosynthetic function—again, focusing on survival vs. total energy output.

However, a new technological paradigm in the form of biohybrid systems is emerging, and it holds the promise of merging the longevity of man-made devises, crafted by intentional design, with the functionality of biological components perfected through millions of years of natural selection that are still too intricate for us to duplicate. Biosolar is a prime example of one such technology emerging from this recent philosophy, and with it we ask ourselves a fundamental question: Can we integrate the biological components responsible for turning sunlight into energy into a man-made device? And in doing so, can we harness their functionality to produce an energy-producing device that does not have the environmental shortcomings of conventional solar energy?

In this chapter, we will discuss the engineering aspects of producing such a hybrid device. We will start by briefly comparing how different classes of solar-voltaic devices operate, comparing the similarities and differences between silicon solid-state, dye-sensitized solar, and biosensitized solar technologies. Next, we will describe in greater detail the components and physical considerations associated with the biohybrid solar cell: the structure and functions of photoactive proteins, the conformation of solid-state substrates, and a consideration of physical interactions that occur at the bio-nanostructure interfaces. Finally, we will report on engineering strategies aimed at building and maintaining protein adsorption and stability on a solid electrode.

22.2 Solar Cell Review: From Silicon to Dye to Protein

22.2.1 Silicon Solar Cell

Regardless of whether they are man-made or naturally occurring, deriving and then the process of capitalizing excited electrons via photonic means can be broken down into four fundamental elements:

1. Absorption of photons
2. Excitation of electronic states
3. Charge separation and transport through a load
4. Returning of electrons to the original donor materials

Solid-state silicon-based photovoltaics operate on manipulating the electroenergy states of semiconductors by doping the material with impurities, making it more prone to accept or donate electrons. Silicons doped with electron donors (i.e., phosphorus) are known as n-type materials, while those infused with electron acceptors (boron) are known as p-type materials.

By itself, an n-type doped silicon material can absorb sunlight to promote a stationary electron into a delocalized state (steps 1 and 2). However, the pathway of this electron is random and will quickly refill the positively charged valence that was coproduced upon

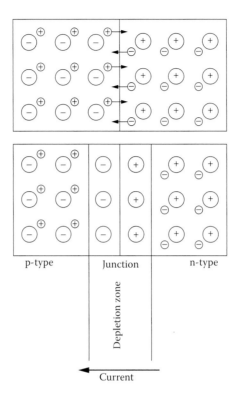

FIGURE 22.1
A p-n junction and creation of a depletion zone. The charge imbalance creates an electric field that drives electrons only after they have obtained sufficient energy to exceed the energy barrier.

excitation. To generate an electric field, an n-type material is introduced to the p-type one. Then, the excess electrons and holes, in their respective n- and p-type doped silicons (Figure 22.1), migrate toward the oppositely charged material (i.e., electrons from n-type will occupy adjacent regions of the p-type material, and vice versa), creating a charge gradient at this junction.

This depleted zone at the p-n junction creates an electrical bias that moves liberated electrons away from their positively generated hole counterparts, fulfilling the third photovoltaic requirement of preventing charge recombination. The fourth and final part of the cycle occurs after the electron flow passes through the circuit and returns the electrons to the p-type material to be excited and start the cycle anew (Figure 22.2).

Because this singular doped semiconductor material is responsible for all four of the listed fundamental functions, a huge challenge in solid-state solar cell technology is finding the right materials that will best fulfill these conditions. These constraints limit the kinds of materials that can be used and the impurity tolerances. Furthermore, the manufacturing processes necessary for photovoltaic applications are energetically intense and environmentally harmful (Table 22.1).

While there are systems in place to mitigate the release of harmful materials into the environment and populace, the management of these waste materials will always be a concern, especially considering that solar cells are posed to be a significant source of alternative energy. Currently the expected lifetime for a commercial solid-state solar cell is approximately 25 years. But not all applications require such long lifetimes, and high

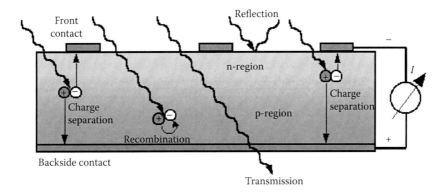

FIGURE 22.2
Current generation upon photonic absorption at the p-n junction. (From http://fourier.eng.hmc.edu/e84/lectures/ch4/node1.html.)

energy and environmental requirements warrant cheap, technically flexible, and benign solar-voltaic technologies.

22.2.2 Dye-Sensitized Solar Cells

An important milestone toward this goal was the invention of excitonic solar cells, commonly refered to as dye-sensitized solar cells (DSSCs).[4] DSSCs are currently the most efficient third-generation solar cell technology (as high as 11%), while promising low cost and ease of manufacture. Instead of the potentially costly and toxic semiconductors used in other thin-film technologies, DSSCs utilize nanocrystalline semiconducting metal oxides (SMOs) such as TiO_2, the low-cost, inert pigment used in paint. In DSSCs, light active synthetic dyes (organic or organometallic) (Figure 22.3) are bound to nanostructured wide-band-gap metal oxide semiconductors and used as photosensitizers to harvest the solar energy and generate excitons, thus fulfilling the first and second requirements of the photovoltaic scheme (see Figure 22.4). Operationally, photon excitation of the dye results in transfer of electrons from the dye to the semiconductor, with the holes remaining in the dye. Although it is energetically possible for electrons to recombine with holes in the dye, the rate at which this occurs is very slow compared to the rate that the dye regains an electron from the surrounding electrolyte. This fulfills the charge separation component, the physical mechanism by which electrons are driven to differ from the p-n junction, and will be further described in Section 22.4. The oxidized species in the electrolyte are then reduced at the counterelectrode, with electrons supplied, via the outer circuit,

TABLE 22.1

Common By-Products of PV-Grade Silicon Treatment

Material Used in Si Solar Cell Manufacturing	Environmental and Human Health Effects
Silicon tetrachloride	An extremely toxic substance that reacts with water, causes skin burns, and is a respiratory, skin, and eye irritant
Sulfur hexafluoride	Used to clean reactors in silicon production; known to be the most potent greenhouse gas molecule
Waste silicon dust	Created by sawing Si wafers; causes inhalation problems

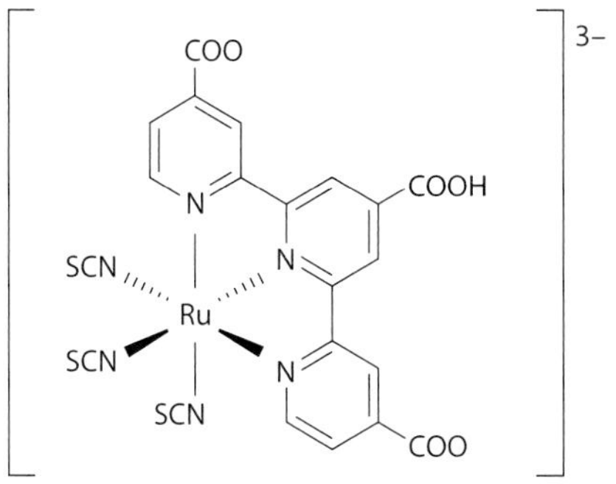

$$\left[\begin{array}{c} \text{COO} \\ \text{N} \\ \text{COOH} \\ \text{SCN} \\ \text{SCN} \\ \text{Ru} \\ \text{SCN} \\ \text{N} \\ \text{COO} \end{array} \right]^{3-}$$

FIGURE 22.3

"Black dye": Common rubidium-based dye used for DSSC applications. Note conjugated π-bonding system and central coordinated metal.

from the photoanode. Due to the very low electron-hole recombination probability in the semiconductor, in contrast to, e.g., Si-based solar cells, DSSCs work even under low-light conditions, e.g., under cloudy skies.

22.2.3 Biosensitized Solar Cells

Biosensitized solar cells (BSSCs) derive their inspiration from the relatively less understood and underappreciated phenomenon of photon-triggered electron ejection by light-activated proteins. DSSCs and BSSCs differ in the electron donor–synthetic (ruthenium II complex) dye in the former and light-activated biomolecules in the latter. Similar to DSSCs, in BSSCs the electron acceptor remains an SMO, such as TiO_2 or ZnO in various

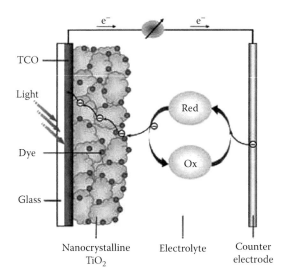

FIGURE 22.4

Schematic of dye-sensitized solar cell. TCO is transparent conductive oxide. (Modified from http://www.uni-ulm.de/fileadmin/website_uni_ulm/nawi.inst.190/fotos_Farbstoffe/Farbstoffe_Figure_1_klein.JPG.)

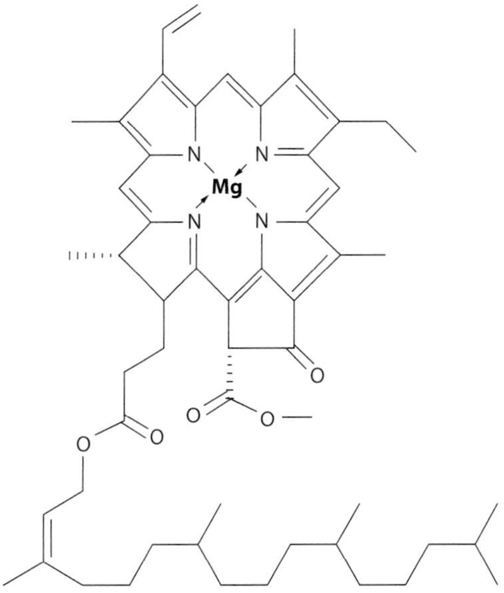

FIGURE 22.5
Chlorophyll chromophore structure. Variations of this light-harvesting dye are at the core of many photoactive proteins. Note the similarity of conjugated π-bonding with coordinated metal, similar to Ru-based dye.

morphologies. Investigations on the possibility of employing naturally occurring organic molecular materials in energy conversion devices for charge generation and separation have been ongoing for some time, especially with a focus on chlorophyll derivatives as the sensitizer in TiO_2-based solar cells.[5] An alternative to Ru-based dyes, which are highly effective but potentially in short supply, is needed. Importantly, the sensitizer should exhibit thermal robustness and maintain its functional activity long term while absorbing solar radiation capable of producing charge generation/separation efficiently.

In considering alternative dyes, replacement of the Ru dye in a DSSC by environmentally friendly and lower-cost light-harvesting proteins is most attractive. A rational approach is to use a light-activated protein as the source of photogenerated carriers. They are biologically synthesized and require neither high temperature nor any rare elements to produce. Biodegradability and renewability are implicit in their design, and because the pigmentation is responsible for plant coloration, this fits the very definition of "green technology."

Aside from concerns of sustainability and cost, millions of years of selective evolution have yielded nanoscale protein structures with intricate functionalities that are still irreproducible by man-made technology. If fact, much evidence suggests that photosynthetic proteins can only operate at the nanoscale by taking advantage of quantum mechanical properties such as quantum coherence and entanglement[6] to maximize photon adsorption and energy transfer. As ubiquitous as plants and photosynthesis are, human understanding of the quantum world has only begun to develop in the past several decades. While practical application of quantum mechanics promises to bring about a new era of computing and information technology, plants and bacteria have been relying on these exotic properties for millions of years. With this in mind, it is not unreasonable to use biological systems as nanoscale components that perform functions that we are currently unable to simulate.

22.3 Photoactive Protein Systems

Considering the wide variety of living systems in existence, different proteins responsible for photoactivity are surprisingly uniform from species to species. In this section we will give a brief overview of common antenna proteins photosystems I and II, major motivators in photosynthesis of almost all plant life. We will also describe the structure and use of bacteriorhodopsin, a photoactive protein found in bacteria that exhibits remarkable structural integrity.

22.3.1 Antenna Proteins

Many photoactive protein systems consist of several proteins bound together, each performing specific tasks associated with light absorption, charge transport, stabilization, and chemical reactions. Antenna proteins are typically associated with the initial stages of photosynthesis and are assigned the task of collecting solar radiation; as such, they are of particular interest for BSSC applications for which they exhibit several desirable traits. First, they can harvest photons at low intensities of illumination by efficient funneling of the delocalized states. Second, during high light intensities and excited state formation, the antenna proteins may dissipate the superfluous energy, protecting the system from thermal deterioration.[7] Light-harvesting antenna systems are based on chromophore-protein complexes, where the protein acts as a scaffold responsible for the orientation of the chromophore molecule. The chromophore orientation is responsible for both light adsorption and excitonic transfer, providing an efficient transport (~100% of illuminating photon energy is trapped) of the antenna's excitation energy toward the lower-energy reaction center (RC),[8] where the electron potential is used to motivate biochemical reactions.

The sophisticated antenna's structure provides absorption among broad-spectrum regions and increases the effective molecular absorptivity. This is one advantage that these complex systems have over conventional silicon-based solar cells, which are most active under a narrow range of photon energies. The antenna's chromophores are carotenoids and chlorophylls. The main function of the former is photoprotection against harmful excessive energy, and to provide structural stabilization of the antenna. Second, carotenoids also act as light harvesters.

A classical example of naturally occurring light-harvesting antenna systems is purple bacteria, whose pigments (bacteriochlorophylls and carotenoids) can harvest the light energy in the near-infrared (NIR) and green regions of the solar spectrum.[9] In most cases, it contains two types of light-harvesting complexes, LH1 and LH2. The first is a part of the RC-LH1 core complex situated in the inner side of the photosynthetic units. In comparison, the LH2 complexes are situated around the LH1-RC core complex (see Figure 22.6).

Each type is associated with specific bacteriochlorophyll molecules absorbing different light intensities. The bacterial chlorophyll associated with LH2 absorbs light with a shorter wavelength vs. the chromophores associated with the LH1. This structure allows the absorbed energy by LH2 to be funneled to the LH1, and finally to the RC. LH2 consists of two rings of bacterial chlorophyll molecules: B800 with an absorption maximum at 800 nm, and B850 absorbing light at 850 nm.[10] These complex energy transfer schemes are commonly seen in photoactive structures and are responsible for the efficient energy transfers required to move excitons from antenna structures to reaction centers. The main

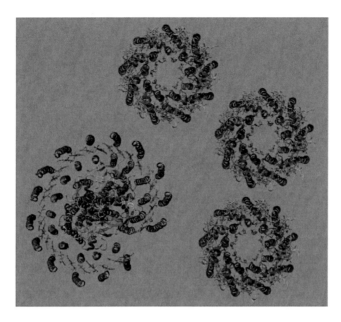

FIGURE 22.6
(See color insert.) Structure of antenna protein system LH1-RC complex surrounded by LH2.

electron-accepting role is played by one of the two quinines that are present in one of the subunits.

22.3.2 Photosynthetic Reaction Center

Like their antenna counterparts, photosynthetic reaction centers (RCs) are pigment-protein complexes that are another subunit present in photosynthetic units of all green plants, as well as in many bacteria and algae, where the key step of solar energy transformation takes place. The molecular excitations resulting from the sun's irradiation are followed by a number of electron transfer reactions finalized in a stable charge-separated radical pair. Conceptually, the reaction centers can be seen as the point of transition where the photon-generated charge carriers are utilized to spur chemical reactions required for biological metabolism.

RC proteins bind a number of photoactive pigments, bacteriochlorophyll (BChl) molecules in the anoxygenic bacteria and chlorophylls (Chls) in oxygen-evolving organisms, and pheophytins (chlorophyll-like molecules without the Mg cation), which are responsible for light absorption and electron excitation catalyzing transfer of the electron within the chromophore molecules.[7] Reaction centers of bacteria are better understood for their less complicated structure (e.g., purple bacteria). The more sophisticated RCs of green plants, photosystems I and II (PS I and PS II), with a very complex arrangement of multisubunit protein, are still under study. Giardi and Pace[8] described two main groups of the photosynthetic reaction centers: (1) PS II, containing pheophytin as the primary electron acceptor and quinone as the secondary acceptor, and (2) PS I, with sulfur clusters serving as early acceptors. Both RC types occur in eukarya and bacteria.[11]

Among bacterial photosynthetic reaction centers, the RC of the purple bacterium *Rhodobacter sphaeroides* has been known to be the simplest of structures. It exists in anaerobic environments in deeper layers of lakes, and it consists of three protein subunits:

- L, through which electron transport takes place.
- M and H coordinating six pigment molecules, a bacteriochlorophyll dimer known as the special pair.
- P, two monomer bacteriochlorophylls (BL and BM), two bacteriopheophytins (HL and HM), and two quinines (Qa and Qb). The quinone Qb acts as the main electron acceptor in the complex.[12]

Due to its much simpler structural arrangement, compared to plant photosystems, the bacterial RC can be easily isolated from its natural culture, sustaining its stability against denaturation. This feature makes bacterial reaction centers an attractive component for BSSC applications. However, manipulation of orientation and conformation of the photosynthetic complexes at the surface of electrodes has been a significant challenge for obtaining an optimum speed of electron transport across the device.[13] It has been observed that immobilization of the light-harvesting complex of purple bacteria on a solid semiconducting surface (i.e., TiO_2) results in an impaired effectiveness of the protein as a photosensitizer. One proposed explanation is that the biomolecules undergo interfacial interaction-induced structural deformation of their light-harvesting complex. The possible forces responsible for this deformation are discussed in Section 22.5. The photosynthetic reaction center of purple bacteria has been studied as a light harvester by Lu's research group.[13] The RC immobilization effectiveness was measured for various morphologies of the TiO_2-WO_3 surface. It was found that the most efficient RC immobilization was achieved when a three-dimensional (3D) worm-like pore structure of the semiconductor was used.

22.3.3 Photosystem I (PS I)

Photosystem I is an efficient natural light-harvesting device. It consists of various protein subunits, including antenna and reaction proteins to absorb photons of a wide-spectrum range and transfer that energy to reaction centers, ultimately to reduce NADP+ (nicotinamide adenine dinucleotide phosphate) into NADPH, which is used in the "dark reactions" of phothosynthesis. PS I absorbs light in the spectral range between 700 and 645 nm.[14,15] It consists of an absorption band at the red spectrum (~715 nm), which is the unique feature of PS I. Its reaction center itself is complex, comprising up to 15 subunits, among which 3 subunits are found only in higher plants (PS I-G, PS I-H, and PS I-N).[11] This complexity yields some unique challenges to utilizing multiunit structures as biosensitizers in BSSC applications, namely, how to isolate only the subunits necessary for photon absorption and charge carrier generation, as well as those subunits responsible for overall stabilization.

PS I consists of light-harvesting antennae whose diverse structure ensures very efficient use of illuminating light under various environmental conditions with sustained light-harvesting functionality over a long period of time.[16,17] The PS I core antenna complex consists of 2 protein subunits binding a number of cofactors, including 96 chlorophyll a molecules, 22 B-carotenes, 4 lipids, 3 iron-sulfur clusters, and 2 phylloquinones.[17,18] Unlike other photosynthetic systems, whose structure involves distinct complexes of antennae

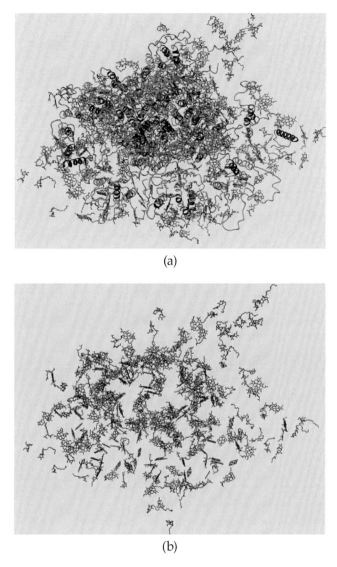

(a)

(b)

FIGURE 22.7
(a) Photosystem I with the protein. (b) Photosystem I without the protein. The complex array of chlorophyll chromophores can be visualized.

and reaction centers, the PS I light-harvesting chlorophyll molecules are associated with the same protein that binds the redox cofactors present in its reaction center. All pigments are packed with high density and specific orientation, allowing maximum adsorption of a decoherent solar source and extremely fast hopping of the excitations throughout the PS I network. This excitation energy transfer between the Chl chromophores occurs in a very short time (200–600 fs).[19] The optimal design of site energies and orientations of chlorophyll molecules in the reaction center ensures very efficient excitation dynamics of PS I. Figure 22.7 illustrates the complex arrangement of chlorophyll chromospheres with and without protein scaffolding. Without the protein overlay, one can see the complex array of chlorophyll chromophores.

The complexes and pigments of the PS I peripheral antenna have developed a specific structure that allows it to play two significant functions, light harvesting, prompt and efficient delivery of the excited energy to the RC, and photoprotection of the photosynthetic unit.[19]

PS I sensitization of a semiconducting solid surface has been recently studied by Frolov's and Sepunaru's research groups. Frolov et al.[47] built a junction between specially engineered unique cysteine mutants of PS I and GaAs, a semiconductor of current interest. This PS I-GaAs system sustained its stability for a few months. Moreover, an electron transfer from PS I to the semiconductor was observed when 0.3 and −0.47 V (vs. normal hydrogen electrode (NHE)) of photovoltage was obtained with PS I/p-GaAs and PS I/n-GaAs, respectively. In a later study of the PS I/GaAs system, Sepunaru et al.[48] observed that the negative photovoltage in the PS I/p-GaAs junction was a result of the electron transfer, similar to the one in the original PS I—from the primary electron donor, chlorophyll chromophore (Chl), to the final acceptor, FeS. This was found to be opposite with the n-GaAs structure, where positive photovoltage was observed. This was a result of fast electron transfer between the photoexcited reaction center of the PS I and its primary electron acceptor (Chl) to the n-GaAs structure. Sepunaru et al. reported that this occurrence was a consequence of a 0.9 V energy gap generation, resulting from the light-induced charge separation. These researchers reported that the generated energy gap forced the electrons from the exited PS I reaction center to the n-GaAs conducting band, causing a decrease in band bending.

22.3.4 Photosystem II (PS II)

Contrary to its numeric designation, photosystem II is actually where the first steps of photosynthetic pathways in modern cyanobacteria, algae, and higher plants occur.[20] Working in conjunction with photosystem I, several light-activated reactions take place to produce factors necessary to motivate the "dark" portions of photosynthesis. Such factors include excitation of electrons, establishment of a proton gradient, and the reduction of NADP+. Being the first step in photosynthesis, photosystem II is responsible for the light-driven splitting of water as a source of electrons and protons; as a result, much work has gone toward elucidating the natural mechanisms responsible for efficient use of light to deconstruct water's stable bonds. Similar to PS I, PS II is a transmembrane protein complex that consists of multiple polypeptide subunits, each responsible for a particular function. Some structures of interest include the antenna assemblies designated CP47 and CP43 and reactions centers D1 and D2 (see Figure 22.5).[21] CP47 and CP43 are thought to be involved in the resonance energy adsorption and transfer; as such, it is these subunits that are the subject for possible application in BSSC devices. X-ray diffraction and electron microscopic characterization illustrate that the interior of CP47 houses 14 chlorophyll molecules held in very specific conformations within 17 Å of each other, theoretically optimizing them for photon adsorption and transfer. From these antennae subunits, the absorbed energy is transferred into the reaction center through a noncovalently bonded Chl a.[22] Subunits D1 and D2 house these reaction centers, where the primary electron donor, a special pair of chlorophylls called P680, and water-cleaving components reside. Here, charge separation occurs when the P680 is excited by either direct absorption of a photon or, more likely, a transfer from antenna chlorophyll molecules. The strongly oxidizing P680+ has its electrons replaced by a secondary electron donor identified as a specific Tyr amino acid, located on D1 and D2 structures,[23] which in turn replenishes its electrons from a 4 Mn atom cluster, the ultimate electron source and site of water oxidation.

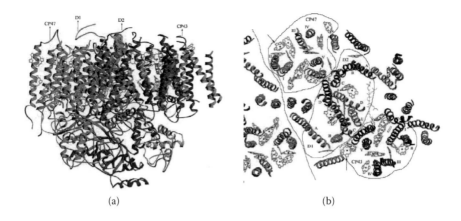

FIGURE 22.8
(a) Structure of PS II as viewed from along the membrane plane. Antenna units CP43 and CP47 are situated on either side of the D1 and D2 reaction centers. (b) PS II as seen perpendicular to the membrane plane. P680 reaction center chromophores can be viewed situated within the D1 and D2 structures. (From Shinkarev, V. P., *Biophys. J.*, 85, 435, 2003.)

This water-splitting mechanism, more commonly referred to as the oxygen evolution cycle, has been the subject of intense study. While the precise structure of the Mn cluster is still not known, several theories describing the overall pathways have been proposed to explain how the structure creates enough potential to overcome the chemical bonds of water. In 1970, Kok and colleagues[25,26] hypothesized that with every photonic excitation, the Mn clusters donate electrons toward the P680+ cluster, and there is a succession of oxidizing equivalents that accumulate on the metal atoms. Only when the Mn ions reach a +4 oxidative state do they extract four electrons from two water molecules, producing O_2, $4H^+$, and returning to their original state to start the cycle again. This makes PS II of interest in biohybrid devices for various reasons. Like the PS I system, it houses the intricate array of antennae choromophores capable of capturing a wide range of the visible spectrum. Moreover, its ability to directly cleave water is an attractive feature: hydrogen production directly from solar energy.

22.3.5 Bacteriorhodopsin (bR)

Chlorophyll and its derivatives, such as photosystems I and II, are sensitive to extreme thermal conditions once extracted from their source, an in vitro environment. Alternatively, bacteriorhodopsin (bR), a natural light-activated protein, found in *Halobacterium salinarum*, is more robust, holding much promise for solar energy conversion[27] with high quantum efficiency.

Upon absorption of visible light, the bR protein undergoes a photocycle during which protons are pumped from the cytoplasmic side of the membrane to the extracellular side, producing a proton gradient. This proton gradient leads to the formation of adenosine triphosphate (ATP) molecules. Due to its natural harsh environment in salt marshes, bR has became a robust protein with a photosensitivity to illumination that is higher than that of synthetic photochromic materials. In the dry stage, bR occurs in a 2D crystal structure, providing exceptionally high stability toward chemical and thermal degradation.[28] bR is very stable at high salinities (up to 5 M NaCl) and high temperature, even up to 140°C. Furthermore, bR sustains its functionality in a broad range of pH values (5–11).[29] These highly stable properties of bR are developed by acquiring its specific structural features by forming a hydrogen-bonding network that involves several water molecules and amino acid functional groups in the extracellular region of the bR membrane.[30] Nevertheless, when the crystal structure is disrupted by solubilization into a monomeric or trimeric bR form or some of the amino acid residues from the bR structure, the protein loses its robustness.

Because of its long-term stability granted by its ability to pack into 2D crystalline phases on surfaces, bR has been the predominant focus of biohybrid electronic applications. The range of potential applications of bR includes the construction of nanoscale devices that are capable of continuous, renewable conversion of light to chemical or electrical energy. Interestingly, Cahen and collaborators[31] have recently shown that bR presents a relatively simple and stable biological system to explore electronic transport with a biological material that can be manipulated with all of the tools known to modern chemistry and biology. Futhermore, recent simulations[32] suggest that bR could have a photoinduced metal-insulator transition (due to conformational changes) that is attractive for biosolar applications.

Recently, bR films deposited on metals or metal oxides were targeted for biooptoelectronic applications.[33] The construction of an excitonic solar cell based on bR triple mutant E9Q/E194Q/E204Q (3Glu) as a photosensitizer has recently been demonstrated.[34] Preliminary studies have examined the photoelectrical response of the constructed biocell (bR/TiO$_2$ film) in response to light illumination. Molecular dynamic calculations have been carried out to describe the possible binding site(s) of a genetically engineered bR mutant onto the surface of TiO$_2$.

22.3.5.1 bR Proton Pumping Action and Electron Transfer Mechanism on TiO$_2$ Surfaces

The process of photoreactive ATP synthesis in bR occurs in several steps, where the retinal chromophore plays the main role in solar energy harvesting and conversion. It isomerizes upon light absorption, causing several further changes in the protein structure. This feature has been used in solar energy conversion device development, where the retinal molecule is the photoactive part of the bR-TiO$_2$-based (or other semiconducting structure) system. The chromophore acts as a photosensitizer that, upon photoexcitation, injects electrons to the semiconductor structure. In such systems, bR undergoes its natural photocycle steps, but only the first step of the photocycle is believed to be needed for TiO$_2$ sensitization to occur. Further bR transformations following the first step are believed to be competing with the electron transfer from the retinal to the electrode surface, slowing down this process and at the same time impairing the efficiency of the electron injection into the TiO$_2$ surface. The bR protein, however, plays the role of a protecting scaffold, just like chlorophyll molecules; retinal loses its robustness once separated from its natural environment. Gai et al.[35] studied the role of the protein and retinal on the efficiency of the photocycle, verifying the time of retinal photoisomerization in its natural environment of the bR protein vs. the retinal extracted from the protein in a methanol solution. The results suggested that the protein plays a catalytic role in the photoisomerization of retinal. It was found that the efficiency and selectivity of photoisomerization of retinal in methanol solution are very low. The protein surrounding the chromophore, on the contrary, has a positive impact on the photoreaction, ensuring both high efficiency and high selectivity of this process. When the Schiff base is protonated, the retinal present in bR selectively photoisomerizes to the 13-*cis* form with an efficiency of 64%, and it is highly decreased when the chromophore is extracted from the protein. It has also been reported[36,37] that the overall isomerization rate of retinal need not be fast for the efficient proton pumping of bR. Any structural modifications or analogs that elevate the height of the isomerization barrier slow down the isomerization and, as a consequence, further photocycle steps. For the same reasons, the slowed rate of retinal isomerization may have a positive effect on the electron injection to TiO_2 in the bR-sensitized solar cell. In summary, the faster retinal isomerization provides the higher efficiency of the whole photocycle. In order for TiO_2 sensitization to occur, the retinal chromophore needs to absorb light and inject an excited electron in the semiconductor surface; bR photocycle steps, including chromophore isomerization, seem to be competitive side steps in regard to the bR-to-TiO_2 electron injection. Hence, it can be stated that to improve the dynamics of the electron injection to the electrode, structural modifications of the retinal molecule or its surroundings, to allow slowing down of retinal photoisomerization, could be considered.

22.4 Design, Characterization, and Optimization of the Semiconducting Metal Oxide Network

Referring back to the four essential steps of photovoltage generation, a functioning solar cell must be capable of performing the following actions:

1. Absorption of photons
2. Excitation of electronic states
3. Charge separation and transport through a load
4. Returning of electron to the original donor

We have already discussed how photoactive proteins are capable of absorbing photons to generate excited electronic states, thus fulfilling the first two required procedures. In this section we will describe the solid-state substrate onto which these proteins are adsorbed—these materials are largely responsible for step 3: accepting the newly excited electron, keeping it separated from the electron's hole counterpart, and transporting the charge through the load circuit. In another major departure from the silicon-based solar paradigm, the electrons are returned to their donors via an electrolyte that shuttles negatively charged species from the cathode to anode, thus fulfilling the fourth requirement of photovoltaic devices. A novel method of using gold as a solid-state replacement for the aqueous electrolyte process will be discussed here as well.

22.4.1 Current Generation via Conduction Band Energy Gradient

Electronic properties are closely related to the relative energy levels of the highest occupied molecular orbitals (HOMOs) and lowest unoccupied molecular orbitals (LUMOs), or the work functions required to liberate an electron from a valence shell. In the context of repeating atomic configurations—as is the case of solid crystalline materials—these energy states combine to form continuous bands, known as the valence and conduction bands respective to the HOMO and LUMO energy states. The energy differences between the valence and conduction bands are what classify most materials as conductors, semiconductors, or insulators, as they dictate electron concentration and mobility. These electrical

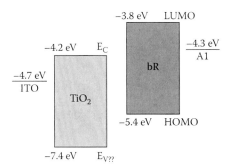

FIGURE 22.9
Fermi level diagram (chemical potentials) of the sensitizer, substrate, and conductive electrode contact. An energy gradient creates a voltage that drives the movement of delocalized charge carriers.

properties and energy band gaps become especially pertinent to solar cell technology because they are directly responsible for their optical properties, thereby dictating how light affects the materials' electronic states.

This is why semiconducting materials are a fundamental element in solid-state solar cell technologies today. As was discussed earlier, doped, semiconducting silicon is the material of choice of commercial applications of high-output photovoltaics because its band gap energies are suitable for operation under the sun's visible spectrum.

In BSSC and DSSC applications, photoactive proteins or dye take the responsibility of photon absorption and electron generation, and the substrate they are absorbed onto assumes the task of maintaining charge separation and charge carrier transport to the circuit electrodes. Because it is no longer responsible for light adsorption, this solid-state substrate can be chosen from a much wider variety of materials that do not exhibit electron excitation under the visible light spectrum. They still, however, need to be semiconducting; again, this is directly attributed to the band gap energies inherent in conducting vs. semiconducting materials. In excitonic solar cells, these band gap differences between the charge-producing proteins and dyes and the charge-accepting substrates are what dictate electron flow, and therefore current.

Figure 22.9 is representative of the energy differences between bacteriorhodopsin, its TiO_2 substrate, and an indium tin oxide (ITO) conductive electrode.

The HOMO energy state of bR is representative of the preexcited electron at –5.4 electron volts and –3.8 eV volts after photon absorption. When paired with TiO_2 with a conduction band energy at –4.2 eV, the excited electron will naturally seek out the state of lower energy in the metal oxide conduction band. Furthermore, the valence band of the metal oxide semiconductor is filled, disallowing the free electrons in the conduction band to fall into the valence states. Finally, when placed in contact with the ITO electrode, electron movement is generated by this stepwise gradient of energy states. This is in contrast to silicon-based solid-state solar cell, which relies on a p-n junction to create an electric field to move electrons.

Understanding these intricate energy level relationships reveals the importance of the substrate design. The foremost advantage that this excitonic solar cell design demonstrates is that it is no longer reliant on a flat p-n junction, as is the case in conventional solar cells. Because of this, the substrate can be created into unique 3D architectures that maximize surface area, and thus transferred electrons per unit volume of material. This section will review elements of the solid-state substrate design, discussing the charge transfer properties of nanoscale metal oxide structures to gold surfaces for electrolyteless biosensitized solar cells.

22.4.1.1 Dimension Solid-State Matrices

The TiO$_2$ nanoparticle network in the Grätzel cell is the recipient of injected electrons from optically excited dye molecules adsorbed onto its surface. Indeed, as a result of the SMO's small particle size (~20 nm), the surface area is more than a thousand times that of a flat electrode of the same size. The SMO network then provides the conductive pathway from the site of electron injection to the transparent collecting electrode. Lastly, it provides the interconnected network of pores through which the electrolyte percolates, allowing the redox carrier (e.g., I$_3^-$) to be reduced by the collected electrons at the counterelectrode. The combination of low cost, chemical stability, and reasonably high solar-to-electrical energy conversion efficiency (~11%) has made the anatase form of TiO$_2$ the SMO material of choice for most DSSC designs.

Here, consideration is directed toward replacing costly ruthenium-based dyes with the environmentally friendly and lower-cost light-harvesting bR protein. This requires that several requirements be satisfied. The first is to ensure that the SMO surface is receptive to bR dye adhesion and offers efficient charge transfer from dye to SMO, but not from SMO to dye or electrolyte. A second objective is to improve the conductive SMO pathway, limited by electron diffusivities several orders of magnitude lower than those in single crystal anatase. A third objective is to create a hierarchical porous 3D SMO network offering a high surface area for dye attachment, reduced electron pathways, and the opportunity for tandem cell operation. These are discussed below.

22.4.1.2 Charge Transfer between Dye and SMO

The success of the DSSC is the ability of the dye to absorb in the visible portion of the solar spectrum and transfer the excited electron to the SMO conduction band while trapping the hole until it can be used to oxidize an ionic species in the electrolyte. Given the small number of suitable SMOs so far identified for DSSC operation, the options for optimizing cell performance by replacing the dye remain limited. As a consequence, means for tuning the properties of suitable SMOs to achieve higher DSSC performance need to be investigated.

A high charge-injection efficiency requires that the distribution of dye excited states lie above the SMO conduction band edge. Raising the energies of the dye's excited states is likely to reduce its light-harvesting efficiency. A better option is to shift the SMO conduction band edge energy to lower values. The core-shell approach, as discussed by Diamant et al., offers some promise.[38]

In the core-shell approach, the TiO$_2$ network is dipped in precursor solutions of the candidate oxides (e.g., SrTiO$_3$, Nb$_2$O$_5$, SnO, ZnO, etc.), followed by sintering at temperatures of ~400–500°C. For example, an SrTiO$_3$ shell can be applied onto the TiO$_2$ core for shifting the TiO$_2$ conduction band in the negative (upward) direction to achieve an increased open-circuit photovoltage. On the other hand, a shell of SnO$_2$ can be applied to induce a positive shift in the conduction band to enable the use of dyes with excited states too low for efficient coupling to bare TiO$_2$. If recombination of electrons with the oxidized dye or redox mediator should become a problem, then a shell material with electron affinity more negative than that of the core (e.g., Nb$_2$O$_5$ on TiO$_2$) could be applied to generate an energy barrier for the reaction.

22.4.1.3 Conductive SMO Pathway

Under operating conditions, the electrons are required to diffuse several microns in the SMO surrounded by electron acceptors (dye, electrolyte) only several nanometers away. Fortuitously, recombination is slow, but overall efficiency remains limited by the slow effective diffusivity of electrons within, e.g., the TiO_2 conductive pathways. These diffusivities of ~10^{-5} to 10^{-4} cm^2/s are at least several orders of magnitude lower than those reported for single crystal anatase. The low D values are attributed to multiple trapping of carriers, with the trap states exhibiting an exponential energy distribution. While the traps are suspected to be due to charged defects, grain boundaries, and surface states, only limited evidence supports these choices. Because the SMO networks are composed of arrays of nano-sized particles connected either three dimensionally or in the form of pseudo 1D wires, each electron must traverse large numbers of grain boundaries before reaching the transparent electrode. Small changes in grain boundary chemistry or structure can have a large impact on the density and distribution of traps at the boundaries. Correlations between boundary chemistry/structure and trap therefore need further investigation.

22.4.1.4 Hierarchical Porous 3D SMO Network

Electron transport dynamics modeled using simulated mesoporous random nanoparticle TiO_2 films by random-walk diffusion find that (1) the average number of particles visited by electrons, and as a result the pathway, increases significantly with higher porosity, and (2) films composed of ordered particles with similar porosity show a reduced path length.

Indeed, the use of TiO_2 nanofibers promises a directed electron transport medium as well as enhanced light scattering.[39] Electrospinning is a cost-effective and scalable technique for the fabrication of 1D nanostructures, with reported efficiencies of electrospun DSSCs as high as 5.8%.[40] Tuller and colleagues have reported the fabrication of electrospun TiO_2 fibers with a hierarchical porous structure achieved by controlling the ratio of the polyvinyl acetate (PVAc)/TiO_2 precursor solution used in spinning (see Figure 22.5).[41] Here one obtains fiber-like structures controllable at several length scales, e.g., a bundle structure composed of sheaths of 200 to 500 nm and their cores filled with aligned 10 nm fibrils.

22.4.2 Gold Surfaces

The present scheme of BSSCs relies on the SMO to provide a conducting medium that will keep the excited electrons from recombining with charged holes in the dye. The electrons will travel to the cathode, where they reduce the electrolyte, which in turn diffuses back through the separator membrane and resupplies electrons to the photoactive adsorbent. Recent experiments with metal-semiconductor Schottky barrier thin films suggest that the liquid electrolyte can be replaced by thin films of gold on the metal oxide.[42] It has been shown for photoactive dye adsorbed onto 10–50 nm of Au coated onto 200 nm of TiO_2 that the photoexcited electrons have sufficient ballistic momentum to transverse the gold layer, into the TiO_2 portion, before being backscattered to the dye or stopped at the Schottky barrier formed at the metal-semiconductor junction. Once within the SMO material, the electrons conduct as they would be in the traditional DSSC scheme, where they are

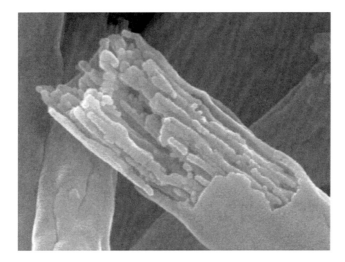

FIGURE 22.10
Bundle structure composed of sheaths of 200–500 nm; cores filled with 10 nm fibrils.

transported to the Ti anode, through the load, and reintroduced to the original gold thin film. In this conformation the gold thin film essentially acts as the solid-state electrolyte, shuttling electrons back to the photoactive adsorbents so that the process can start over.

This introduces several engineering and operational benefits. For one, chemical reduction, diffusion, and oxidation of an electrolyte species are no longer involved, thus eliminating one potentially rate-limiting step. Additionally, liquids sealed in enclosed spaces such as those found in solar cells risk rupturing their containment as a result of thermal expansion or evaporation when exposed to high temperatures. Finally, gold has been shown to be an ideal surface for protein adsorption, promoting higher protein crystallization at the electrode surface, and thus improving overall cell performance.

22.5 Protein Dynamics at Nanoscale Surfaces

Because proteins are essentially polymers self-assembled into very specific conformations, minute changes in the conformation of these biopolymer aggregates can lead to a loss of protein function. This is particularly true in antennae proteins where Forster resonant energy transfer (FRET)[43] between the chromophore molecules is highly dependent on spatial relations. So, it is not surprising that the foremost concern of BSSCs is the stability of biological components in the presence of the solar cell's high-energy environment. Combining significant temperature fluxuations, high surface energy materials, and an abundance of surface area, the solid-state interface of an exitonic junction solar cell can be a hostile environment for proteins. A keen understanding of the forces and the resulting protein folding dynamics will be a key aspect in the future designs of such biohybrid devices, as well as addressing unknown interactions in biological systems for efforts to accurately map the effects of nanotoxicity.

22.5.1 Factors That Influence Protein Structure at Surfaces

22.5.1.1 Boundaries Shaping the Interface

The nano-bio interface is comprised of three dynamic interaction components:

1. Nanoparticle surface: Characteristics are determined by its physicochemical composition.
2. Solid-liquid/solid-gas interface: Changes occur when the particle interacts with components in the surrounding medium.
3. Solid-liquid/gas interface: Contact zone with biological substrates.

In a given medium, the most important nanoparticle characteristics that determine surface properties are the material's chemical composition, surface fictionalization, shape, angle of curvature, porosity, and surface crystallinity. Other quantifiable properties, such as effective surface charge (zeta potential), particle aggregation, state of dispersion, stability/biodegradability, dissolution characteristics, hydration, and valence of the surface layer, influence its interaction with biological components. Conversely, protein properties that influence interactions include hydrophobicity/hydrophilicity, nonspecific attractive forces (electrostatics, Van der Waals (VDW)), specific binding (covalent), protein size, hydration, side group charges, protein crystallization, and intraprotein forces. Many of these properties can be influenced by the solution environment, such as pH, ionic strength, temperature, and the presence of large organic molecules like detergents (Table 22.2).

22.5.2 Forces at the Nano-Bio Interface

At first glance, interactions between nanoparticles and cells seem to embody some of the same principles as those between colloidal particles. VDW, electrostatic, solvation, solvophobic, and depletion forces still apply, but they require special consideration for events occurring at the nanoscale. For example, because nanoparticles contain relatively few atoms, their VDW forces are dependent on the positioning of their surface atoms and their standard bulk permittivity functions.

Typical interactions between SiO_2 particles in water involve VDW, electrostatic, and solvation forces. VDW forces result from the quantum mechanical flux of the electrons; at any moment, their fluctuations produce a small but important dipole in the particle, thereby inducing a dipole moment in the atoms of the adjacent SiO_2 particle and causing an attractive force. The electrostatic force in the system results from surface charges that inevitably arise on the SiO_2 particles. In contact with water, silanol (Si-OH) groups dissociate to yield negative surface charges, which will generate, at least transiently, repulsive electrostatic forces between particles; under constant surface potential conditions, such repulsion might be mitigated by surface regulation. For SiO_2 particles, the sum of the attractive VDW and repulsive electrostatic forces yields the well-established Derjaguin-Landau-Verwey-Overbeek (DLVO) theory of colloid science[44,45] (Figure 22.6).

22.5.2.1 Ionic Strengths

In an I_3^- electrolyte solution, the ionic strength is quite high relative to biological environments, meaning that the electrostatic forces are most likely to be screened within a few nanometers of the surface. The high ionic strength also obscures the zero-frequency

TABLE 22.2

Main Forces Governing the Interfacial Interactions between Nanomaterials and Biological Systems

Force	Origin and Nature	Range (nm)	Possible Impact on the Interface
Hydrodynamic interactions	Convective drag, shear, lift, and Brownian diffusion are often hindered or enhanced at nanoscale separations between interacting interfaces	10^2–10^6	Increase the frequency of collisions between nanoparticles and other surfaces responsible for transport
Electrodynamic interactions	VDW interactions arising from each of the interacting materials and the intervening media	1–100	Universally attractive in aqueous media; substantially smaller for biological media and cells owing to high water content
Electrostatic interactions	Charged interfaces attract counter-ions and repel co-ions through Coulombic forces, giving rise to the formation of an electrostatic double layer	1–100	Overlapping double layers are generally repulsive, as most materials acquire negative charge in aqueous media, but can be attractive for oppositely charged materials
Solvent interactions	Lyophilic materials interact favorably with solvent molecules	1–10	Lyophilic materials are thermodynamically stable in the solvent and do not aggregate Lyophobic materials are spontaneously expelled from the bulk of the solvent and forced to aggregate or accumulate at an interface
Steric interactions	Polymeric species adsorbed to inorganic particles or biopolymers expressed at the surfaces of cells give rise to spring-like repulsive interactions with other interfaces	1–100[a]	Generally increase stability of individual particles but can interfere in cellular uptake, especially when surface polymers are highly water soluble
Polymer bridging interactions	Polymeric species adsorbed to inorganic particles or biopolymers expressed at the surfaces of cells containing charged functional groups can be attracted by oppositely charged moieties on a substrate surface	1–100	Generally promote aggregation or deposition, particularly when charge functionality is carboxylic acid and dispersed in aqueous media containing calcium ions

[a] Depending on the length of adsorbed or expressed polymeric species.

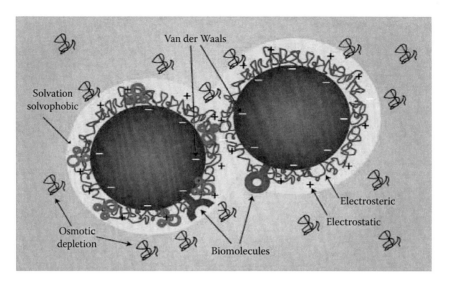

FIGURE 22.11
Interactions between nanoparticles. Traditional forces for colloidal fabrication (for example, electrostatic, VDW, covalent) and other important interactions (for example, solvation, solvophobic, biomolecular, depletion) that occur when particles are suspended in biological media and come into contact with proteins.

contribution of the VDW forces, (Figure 22.11), whereas higher-frequency dispersion interactions remain operative. Moreover, solvation is another phenomenon that is important for inorganic and other hydrophilic nanoparticles. The water molecules adhere to the particles with sufficient energy to form steric bumper layers on their surfaces, making it difficult for pairs of particles to touch or adhere. Thus, solvation forces increase particle stability through hydration pressure or hydrophilic repulsion. Alternatively, rapid dehydration and aggregation will occur if the relative affinity of two interacting surfaces for water molecules is much lower than that between the water molecules themselves—a hydrophobic attraction or hydrophobic effect. All of these forces, which change little or slowly over time, can be predicted or estimated using known theories.[46] These experimental measurements also tend to be fairly consistent.

22.5.2.2 Particle Dissolution

So far it has been assumed that proteins are the only component that would be affected in such a bioinorganic interface. However, it has been demonstrated that proteins and organic substances increase the dissolution rates of particles of ZnO, CdSe, iron oxides, aluminum oxides, and oxyhydroxides. The negative effects of this dissolution on BSSC efficacy can be inferred; increased dissolution of the nanoscale metal oxide structures will eventually lead to an Ostwald ripening process resulting in a loss of surface area. Furthermore, such a dynamically changing landscape has significant effects on the sensitizer adsorption kinetics.

This bioenhanced dissolution has been shown to occur through at least two mechanisms: aqueous complexation (that is, aqueous species complexing free ions released from the material's surface) and ligand-enhanced dissolution (adsorbed natural organic material and organic acids extracting surface metal atoms from nanoparticle surfaces). The latter mechanism has been demonstrated for iron and aluminum oxides and oxyhydroxides, and is also likely to occur for ZnO.

22.5.2.3 Induced Protein Aggregation

Although the exact trigger of protein unfolding at the particle surface is unknown, it might involve contact forces, such as the release of free surface energy through surface reconstruction. An example might be relaxation of the particle crystal structure through protein binding.[47] Similarly, it is possible that electron confinement or the formation of electron-hole pairs at the material surface could lead to cleavage of structural bonds or covalent cross-linking of protein domains.

In one studied experimental model, nanoparticles act as catalysts exposing protein interaction domains that induce aggregation through hydrogen bonding and the formation of fibrils. For example, human β2 microglobulin fibrillation occurs on the surfaces of cerium oxide and copolymer nanoparticles and Carbon Nano Tabes (CNTs).[48] Such aggregation would have serious implications for maintaining protein functionality on the surface of a BSSC electrode. It remains to be determined, however, whether these experimental conditions can be duplicated in vivo, where competitive binding in complex biological fluids may screen the nucleation surface. Extrapolating this in the context of BSSCs, the use of large molecule coadorbants is one strategy toward stabilizing photoactive proteins at the solid-state electrode surface. Such approaches are further explained in Section 22.6.

22.5.2.4 Particle Curvature

Particle curvature and total surface area can also play a significant role in how solid surfaces affect protein folding. It is commonly assumed that the greatest available surface area would induce more biopolymer unfolding; however, recent studies have shown that larger nanoparticles with lower radii of curvature exhibited greater interactions with protein, resulting in more pronounced denaturization.

In studies done by Lundqvst et al.,[49] the conformational states of the enzyme human carbonic anhydrase I (HCAI) were measured by circular dichromisim, nuclear magnetic resonance (NMR), AFM, and gel permeation as they were introduced to food-grade silica nanoparticles of 6, 9, and 15 nm independently. It was shown that the curvature of the solid surface influenced the degree of conformational change induced in the secondary structure of HCAI. The larger, 15 nm diameter, induced conformational changes approximately sixfold higher than the enzymes introduced to the 6 nm suspensions. In contrast, the tertiary structures remained very similar for all particles and not as affected by the silica nanoparticle curvatures. This effect was also noted by the Somasundaran group when studying adsorption characteristics of bovine serum albumin on hematite nanoparticles.

It is currently hypothesized that the relatively smaller radii of curvature afforded by the larger particles allows more interactions between the particle surface and portions of the protein, while a smaller particle has a smaller particle/protein contact area. This seems to suggest that, in BSSC applications, protein degradation may be retarded simply by using smaller nanostuctures as electrode supports. However, these studies have been done at protein concentrations between 0.5 µm and 0.3 mM, and at protein-to-nanoparticle ratios ranging from 0.25:1 to 1:1 in suspension. While these studies yield informative results on protein-nanoparticle interactions, they are still a distant model of a dye- or biosensitized electrode surface. The exact mechanisms of interaction between nanoparticles and proteins remain to be determined. This is even more so in the case of long-range solid-state surfaces with unique nanoscale curvatures. These conditions yield environments on which the conformational dynamics of photoactive proteins have yet to be explored.

22.6 Engineering Strategies in BSSCs

Clearly, one of the greatest challenges in BSSCs, or in any future biohybrid technology endeavor, will be to preserve the function of the integrated proteins. Strategies toward reaching this goal now involve an eclectic range of disciplines, including genetic engineering, colloid science, and surface chemistry.

Integrating photosynthetic proteins poses a particularly difficult challenge, as they are intramembraneous biomolecules. With large hydrophobic, alpha helix structures, these proteins are found embedded in the similarly solvophobic environment within the cells' lipid bilayers. The complex photosystems extracted from higher plant forms are especially delicate.

22.6.1 Protein Engineering: Genetic Modification

However, photoactive proteins from bacteria such as bacterial reaction centers (RCs) or bacteriorhodopsin (bR) extracted from *Rhodobacter sphaeroides* and *Halobacterium salinarium*, respectively, are robust and remain functional out of their native environments. bR is particularly resilient at high temperatures and extreme pH levels. This may be attributed to the harsh environmental conditions the *Halobacterium salinarium* have evolved to thrive in, resulting in a relatively simple proton-pumping structure. Because of its long-term resistance to photo, thermal, and chemical degradation, bR is commonly of interest as a component in technical applications.

Enhancement of thermostability of bR has been an area of intense research in our laboratory. One method of choice has been rational site-directed protein engineering of bR. Advancements in thermostable vectors, antibiotic resistance genes, and the genetic characterization of extreme thermophiles have prompted the development of in vivo thermoselection systems to optimize mesophilic protein bR for device applications. *Thermus thermophilus* has been a useful in vivo screening platform for bR mutants; a versatile, heat-stable expression vector is required. Others have constructed a bifunctional vector system (pMKE1) capable of expression in both extremely thermophilic (*T. thermophilus*) and mesophilic (*E. coli*) microorganisms. Mutants that retain structural stability at elevated temperatures have been used as starting points for additional rounds of mutagenesis and thermoselection. Several iterations of thermoselection may be required before a bR variant with adequate thermostability suitable for device application can be identified.

Thavasi et al.[50] also studied wild bR and its modification (3Glu) as an assembly on the TiO_2 surface. The modifications were done by eliminating three negatively charged Glu residues. It was found that this modification of the protein improved performance of the solar cells. Thavasi's[50] group also observed that both the wild bR and the 3Glu bR responded to light illumination, but the triple mutant (3Glu) gave better photoelectric performance. The cells assembled with the triple mutant showed 0.038 mA/cm² of short-circuit photocurrent density (J_{SC}), while only 0.0269 mA/cm² was achieved with the wild bR. This improved cell performance was attributed to the more efficient assembly and binding nature of the mutated protein (3Glu) to TiO_2. It was found that the elimination of the three negative charges from the extracellular (EC) site of the bR structure resulted in a dramatic change of the surface potential map of the protein that could facilitate its better binding to TiO_2.

22.6.2 Lipid/Surfactant Coabsorbants

While genetic modification can be used to modify the protein itself to improve surface binding or charge transfer characteristics, surfactant molecules can be used to pacify the excess energies presented at the protein–metal oxide interface. Fortunately, finding ways to maintain the protein's integrity has long been a topic of interest for those performing structural studies on them, and so a number of techniques have been and are being developed to stabilize biomacromolecules for spectroscopy methods such as X-Ray Diffraction (XRD), Fourier Transform Infrared (FTIR), and NMR. Many of these techniques are being adapted as interest in biomolecular hybrid technology grows.

Crystallizing transmembrane proteins for x-ray characterization is one area where surfactants have been used to preserve protein integrity outside of the biological membrane[51]— ideally through a selection of surfactants and liquid crystal phases that closely mimic the proteins' native surroundings. It is not unreasonable to apply these same techniques to stabilizing proteins onto solid surfaces. One novel application of this scheme is forming a lipid-protein self-assembled monolayer onto an Attenuated Total Reflectance (ATR) sensor surface for FTIR, making it possible to study protein orientation, conformation, and dynamics in the presence of water.[53] Being able to study functioning proteins adsorbed onto a solid-state crystal has far-reaching implications on the viability of BSSC applications.

In particular, the photosynthetic systems of higher plants, such as PSs I and II, consist of a very complicated structure; their isolation from the natural environment has faced significant challenges due to the tendency to aggregate, denaturation, and alteration of their functional conformation. All these changes impair the electron transfer ability of the light-harvesting system when utilized on photosensitization studies of DSSC devices. These challenges have been addressed using surfactants (i.e., chemicals, lipids, peptides, and peptide-chemical hybrid detergents). Among all the surfactants, lipids and peptides have been found to have the highest efficacy, although their interactions with membrane proteins are still under investigation.[51,53] Matsumoto et al.'s[19] work on PS I stabilization by peptide surfactant molecules with specific amino acid sequences proposed an interaction of those molecules with the PS I membrane. They also proposed that while the peptide surfactants act like the standard chemical surfactants, stabilizing the transmembrane protein by aligning themselves around the membrane protein, they also encapsulate the PS I complex into their own macrostructure by self-assembling into a larger, 3D structural network, protecting PS I for long periods of time. In accordance with their observations, Matsumoto et al. proposed the crucial structural features that are needed for the peptides to be effective in stabilization of the transmembrane system. In addition to their amphiphilic properties, peptide surfactants should contain an acetylated N-terminus, a short hydrophobic tail with six consecutive hydrophobic amino acids, one or two positively charged amino acids (i.e., lysine or arginine) in the C-terminus (not the N-terminus), and lastly, an amidated C-terminus.

Similarly, advances in PS II structure determination in the past several decades are partially due to the evolution of techniques for protein isolation and crystallization. However, these techniques become increasingly challenging to implement when working with transmembrane structures that have portions meant to be stabilized in both the hydrophobic bilayer interstitial and aqueous intercellular environments. To preserve PS II structural integrity and functionality, design and implementation of protein isolation, purification, and crystallization will have to take these heterogeneities into account.[54] Since the 1970s, detergent molecules have been a major tool used to dissolve the cellular lipid bilayers and stabilize the proteins for analysis. Since then, understanding surfactant-protein interactions has become a field in itself, since the dynamics between micelle formations, surfactant structure, and protein stability are codependent and can be strongly affected by minor chemical changes. Extrapolating from this line of thought, it is not inconceivable that maintaining protein stability and functionality will also be an important aspect in BSSC design. Protein interactions with nanoscale surfaces of varying composition, surface potential, curvature, and size are currently an area of study, typically in the context of nanotoxicity. It must be noted that conformation, including coiled vs. stretched vs. semi-coiled, might play a critical role.

22.6.3 Quantum Dot Enhancement

Aside from stabilizing protein structures through surface chemistry, recent advances in understanding and manipulating the unique quantum mechanics of nanodots have been applied to enhancing the capabilities of photoactive proteins. Thus far, thin films of CdSe/ZnS quantum dots and bR protein have exhibited enhanced performance as a photoactive sensor device.[55] Because the enhanced activity is observed only at distances below 10 nm, it is hypothesized that photonic energy is also being absorbed by the quantum dot and transferred not through fluorescent emission, but through radiationless FRET—not unlike that seen in complex antenna protein arrays.

Currently, efforts are being made to combine this quantum dot enhancement with the electrolyteless gold substrate presented in Section 22.4 to create the next generation of quantum-enhanced electrolyteless BSSC devices. Moreover, the nonradiative FRET transfer mechanisms revealed may be one step toward reproducing the complex mechanism found in natural photoactive proteins.

22.7 Conclusions

During a long process of evolution, naturally occurring photosynthetic machineries like antenna proteins, photosystems I and II, bacteriorhidopsin, and biochromophores acquired special structural features that allow them to effectively capture the available solar energy and utilize it with an incredibly high light-to-electron conversion efficiency. Moreover, they are able to sustain their functionality for long periods of time. Biosensitized solar cell technology integrates these naturally efficient biochemical/mechanical systems into a man-made device with the intention of generating energy at a low cost with minimal environmental impact.

Nevertheless, these biomolecules' extraordinary performance often suffers when the structures are extracted from their natural environment and immobilized on an electrode. This decline in performance is proposed to occur due to the undesirable structural and conformational changes impairing electron injection onto the electron acceptor. These antagonistic alterations during the extraction process are more severe with more complicated systems, like photosystems I and II. Their highly diverse structures make them sensitive to chemical and conformational changes. This requires a new field of nanoscale engineering to successfully meld the artificial and natural in a way that would prevent these changes. This is a several-fold challenge requiring an acute knowledge of the intricate protein structures, precise manipulation of nanoscale substrates, and creative use of coabsorbants, such as surfactants, polymers, and lipids. In this chapter we have discussed a few of the considerations and complications inherent in producing biohybrid systems.

Acknowledgments

P.S. acknowledges support from the National Science Foundation Industry/University cooperative Research Center and the Environmental Protection Agency (Cooperative Agreement Number EF 0830117). V.R. expresses his thanks to NSF, the Wallace H. Coulter Foundation, USAFOSR, ONR, NIH, and Harvard Medical School. H.L.T. thanks the Chesonis Family Foundation for its support. B. B. supported by U.S. Dept. of Energy, Office of Science, Basic Energy Sciences contracts DE-FG02-07ER46352 and DE-FG02-08ER46540 (CMSN).

References

1. (a) D. Kuciauskas, J. E. Monat, R. Villahermosa, H. B. Gray, N. S. Lewis, J. K. McCusker, *J. Phys. Chem. B* 2002, 106, 9347; (b) L.-Y. Luo, C.-F. Lo, C.-Y. Lin, I.-J. Chang, E. W.-G. Diau, *J. Phys. Chem. B* 2006, 110, 410.
2. (a) Y. Tachibana, J. E. Moser, M. Grätzel, D. R. Klug, J. R. Durrant, *J. Phys. Chem.* 1996, 100, 20056; (b) G. Benkö, J. Kallioninen, J. E. I. Korppi-Tommola, A. P. Yartsev, V. Sundström, *J. Am. Chem. Soc.* 2002, 124, 489; (c) N. A. Anderson, T. Lian, *Annu. Rev. Phys. Chem.* 2005, 56, 491; (d) C.-W. Chang, C. K. Chou, I.-J. Chang, Y.-P. Lee, E. W.-G. Diau, *J. Phys. Chem. C* 2007, 111, 13288.
3. (a) C.-Y. Lin, C.-F. Lo, L. Luo, H.-P. Lu, C.-S. Hung, E. W.-G. Diau, *J. Phys. Chem. C* 2009, 113, 755; (b) C.-W. Chang, L. Luo, C.-K. Chou, C.-F. Lo, C.-Y. Lin, C.-S. Hung, Y.-P. Lee, E. W.-G. Diau, *J. Phys. Chem. C* 2009, in press.
4. M. Graetzel, *Acc. Chem. Res.* 2009, 42, 1788.
5. A. Kay, R. Humphry-Baker, M. Graetzel, *J. Phys. Chem.*, 1994, 98, 952.
6. M. Sarovar, A. Ishizaki, G. R. Fleming, K. B. Whaley, *Nature Phys.* 2010, 6, 462.
7. R. Cogdell, A. T. Gardiner, *Microbiol. Today* 2001, 28, 120.
8. M. T. Giardi, E. Pace, *Trends Biotechnol.* 2005, 23, 257.
9. T. Pullerits, V. Sundström, *Annu. Rev. Phys. Chem.* 2008, 59, 53.
10. X. H. Chen, L. Zhang, Y. X. Weng, L. C. Du, M. P. Ye, G. Z. Yang, R. Fujii, F. S. Rondonuwu, Y. Koyama, Y. S. Wu, J. P. Zhang, *Biophys. J.* 2005, 88, 4262.
11. C. Theiss, I. Trostmann, S. Andree, F. J. Schmitt, T. Renger, H. J. Eichler, H. Paulsen, G. Renger, *J. Phys. Chem. B* 2007, 111, 13325.
12. T. A. Moore, A. L. Moore, D. Gust, *Phil. Trans. R. Soc. Lond. B* 2002, 357, 1481.
13. Y. Lu, M. Yuan, Y. Liu, B. Tu, C. Xu, B. Liu, D. Zhao, J. Kong, *Langmuir* 2005, 21, 4071.
14. P. E. Jensen, R. Bassi, E. J. Boekema, J. P. Dekker, S. Jansson, D. Leister, C. Robinson, H. V. Scheller, *Biochim. Biophys. Acta* 2007, 1767, 335.
15. H. Naver, A. Haldrup, H. V. Scheller, *J. Biol. Chem.* 1999, 274, 10784.
16. A. N. Melkozernov, J. Barber, R. E. Blankenship, *A. Chem. Soc.* 2006, 45.
17. R. J. Cogdell, T. D. Howard, R. Bittl, E. Schlodder, I. Geisenheimer, W. Lubitz, L. M. Utschig, M. C. Thurnauer, *Acc. Chem. Res.* 2004, 37, 439.
18. G. G. Prive, *Curr. Opin. Struct. Biol.* 2009, 19, 379.

19. K. Matsumoto, M. Vaughn, B. D. Bruce, S. Koutsopoulos, S. Zhang, *J. Phys. Chem. B* 2009, 113, 75.

20. D. L. Nelson, M. M. Cox, *Lehninger principles of biochemistry*, 2002.

21. B. Kê, Photosynthesis, in *Photobiochemistry and photobiophysics*, Vol. 10, New York: Springer, 2000.

22. A. F. Collings, C. Critchley, *Artificial photosynthesis: From basic biology to industrial application*, New York, Wiley-VCH, 2005.

23. K. H. Rhee, *Ann. Rev. Biomol. Struct.* 2001, 30, 307.

24. V. P. Shinkarev, *Biophys. J.* 2003, 85, 435.

25. B. Forbush, B. Kok, M. McGloin, *Photochem. Photobiol.* 1971, 14, 307.

26. N. Kamiya, J.-R. Shen, *Proc. Natl. Acad. Sci. USA* 2003, 100, 98.

27. U. Haupts, J. Tittor, D. Oesterhelt, *Annu. Rev. Biophys. Biomol. Struct.* 1999, 28, 367.

28. T. Polívka, V. Sundström, *Chem. Rev.* 2004, 104, 2021.

29. Padrós, E., C. Sanz, T. Lazarova, M. Márquez, F. Sepulcre, X. Trapote, F.-X. Muñoz, R. González-Moreno, J. L. Bourdelande, and E. Querol. Extracellular mutants of bacteriorhodopsin as possible materials for bioelectronic applications. In Bioelectronic Applications of Photochromic Pigments. A. Dér, and L. Keszthelyi, ed. (2001), pp. 120–136.

30. K. T. Sapra, J. Doehner, V. Renugopalakrishnan, E. Padrós, D. J. Muller, *Biophys. J.* 2008, 95, 3407.

31. Y. Jin, N. Friedman, M. Sheves, T. He, D. Cahen, *Proc. Natl. Acad. Sci. USA* 2006, 103, 8601.

32. E. Alfinito, L. Reggiani, *Europhys. Lett.* 2009, 85, 68002.

33. Y. Jin, N. Friedman, M. Sheves, Tao He, D. Cahen, *Proc. Natl. Acad. Sci. USA* 2006, 103, 8601.

34. K. T. Sapra, J. Doehner, V. Renugopalakrishnan, E. Padrós, D. J. Muller, *Biophys. J.* 2008, 95, 3407.

35. F. Gai, K. C. Hasson, J. C. McDonald, P. A. Anfinrud, *Science* 1998, 279, 1886.

36. K. J. Kaufmann, V. Sundstrom, T. Yamane, P. M. Rentzepis, *Biophys. J.* 1978, 121.

37. V. I. Prokhorenko, A. M. Nagy, S. A. Waschuk, L. S. Brown, R. R. Birge, R. J. D. Miller, *Science* 2006, 313.

38. Y. Diamant, S. Chappel, S. G. Chen, O. Melamed, A. Labon. Coordination Chemistry Reviews 245(2004) 1271–1276.

39. Frank, A. J., Kopiclakis, N., and Van de Lagemant, J. 2004. "Coord. Chem. Rev." 248 1165–1179

40. Fujihara, K., Kumar A., Jose R., Ramakrishna, S., and Uchida S., 2007. "Nanotechnology" vol. 18, 365704–365804.

41. Kim, I. D., Rothschild, A., Lee B. H., Kim D. Y., Jo, S. M., Tuller H. L., 2006. *Nano Letters* 6, 2009–2013

42. E. W. McFarland, J. Tang, *Nature* 2003, 421.

43. Forster T.; ANNALEN Der PHYSIK, 2, 55-75, 1998.

44. Y. Min, M. Akbulut, K. Kristiansen, Y. Golan, J. Israelachvili. *Nature Mater.* 2008, 7, 527.

45. D. Velegol, *J. Nanophoton.* 2007, 1, 012502.

46. H. Y. Kim, J. O. Sofo, D. Velegol, M. W. Cole, A. A. Lucas, *Langmuir* 2007, 23, 1735.

47. B. Gilbert, F. Huang, H. Zhang, G. A. Waychunas, J. F. Banfield, *Science* 2004, 305, 651.

48. S. Linse et al., *Proc. Natl. Acad. Sci. USA* 2007, 104, 8691.

49. M. Lundqvist, I. Sethson, B.-H. Jonsson, *Langmuir* 2004, 20, 10639.

50. V. Thavasi, T. Lazarova, S. Filipek, M. Kolinski, E. Querol, A. Kumar, S. Ramakrishna, E. Padrós, V. Renugopalakrishnan, *J. Nanosci. Nanotechnol.* 2008, 8, 1.

51. T. J. Knowles, R. Finka, C. Smith, Y. P. Lin, T. Dafforn, M. Overduin *J. Am. Chem. Soc.* 2009, 131, 7484.

52. L. Frolov, Y. Rosenwaks, S. Richter, C. Carmeli, I. Carmeli, *J. Phys. Chem. C* 2008, 112, 13426.

53. Vanclerbassche G., Clercx A., Cursteclt T., Jöhansson J., Jörnvall H, Raysschaert J–M., *Euro, J. Biochem.* 203, 201-209, 1992.

54. L. Sepunaru, I. Tsimberov, L. Forolov, C. Carmeli, I. Carmeli, Y. Rosenwaks, *Nano Lett.* 2009, 9, 2751. B) V. Cherezov, J. Clogston, Y. Misquitta, W. Abdel-Gawad, M. Caffrey, *Biophys. J.* 2002, 83, 3393.

55. M. H. Griep, K. A. Walczak, E. M. Winder, D. R. Lueking, C. R. Friedrich, *Biosensors Bioelectronics* 2010, 25, 1493.

23

Electrospinning in Drug Delivery

Pitt Supaphol[1], Orawan Suwantong[2], and Pakakrong Sansanoh[1]

[1]*The Petroleum and Petrochemical College, Chulalongkorn University, Pathumwan, Bangkok, Thailand*

[2]*School of Science, Mae Fah Luang University, Tasud, Muang, Chang Rai, Thailand*

CONTENTS

23.1 Introduction to Electrospinning

Over the past decade electrospinning, or the electrostatic spinning technique, has been recognized as the most promising method used for fabricating continuous fibers on a large scale, with diameters adjustable from microns down to nanometers, to form a nonwoven or alignment structure. The first publication of this process was originally pioneered by Formhals in 1934.[1] Despite these early discoveries, the procedure was not utilized commercially until 2003, as a part of the nonwoven industry for filter applications.[2] In contrast to conventional fiber production, like dry spinning and melt spinning, electrospinning is a relatively simple and versatile method to produce nanofibers, driven by the coloumbic interactions between charged elements on the surface of the fiber jet. First, polymers will be surveyed as fiber-forming materials. Later, other materials, such as metals, ceramics, and glass, will be considered as fiber precursors.

Electrospinning gained substantial attention due mainly to its potential for fabricating highly porous fibrous membranes with the fiber diameters in the range of micro- to nanoscale. Interestingly, the ranges of fiber diameters that can be achieved are two to three orders of magnitude smaller than those formed by conventional fiber production. Moreover, the electrospun fibers can have many different morphologies, such as porous-surface,[3–7] core-sheath,[1,8] and side-by-side[9] structure. The extremely unique characteristics of electrospun nanofibers make them very useful in a wide range of advanced applications, such as non-wovens for filtration,[10] membranes for aerosol purification,[1] thin coatings for defense and protection,[12] structures incorporated in composites,[13] and tissue scaffold, wound healing, and drug delivery devices in biomedical applications.[14,15]

In the past several years, the usage of electrospun fiber mats for biomedical applications has attracted a great deal of attention because the fiber mats have been considered suitable substrates for tissue engineering,[16–19] wound dressing,[20,21] and immobilized enzymes.[22–25] In addition, they have also been used as barriers for the prevention of post-operative-induced adhesion[26,27] and vehicles for controlled drug or gene delivery.[28–32] In tissue engineering, the electrospun fibers can mimic the nanofibrous features of an extracellular matrix (ECM) with suitable mechanical properties. They should also be able to promote cell adhesion, spreading, and proliferation.[33] For wound dressing, the nanofibrous scaffold should not only serve as a substrate for tissue regeneration, but also may deliver suitable bioactive agents, including drugs (e.g., antibiotic agent), in a controlled manner during healing.[33]

23.2 Electrospinning Process and Variables

In the electrospinning process, a high voltage is used to create a strong electrostatic field across a conductive capillary attached to a reservoir containing a polymer solution or melt and a screen collector. Upon increasing the electrostatic field strength up to a critical value, charges on the surface of a pendant drop destabilize its shape, from partially spherical to conical, known as the Taylor cone. Beyond a critical value at which the electrostatic field strength overcomes the surface tension of polymer solution or melt, a charged polymer jet is eventually ejected from the apex of the cone. The fiber jet undergoes an instability and elongation process, which allows the jet to become very long and thin. When the jet travels, the solvent evaporates or solidifies to finally leave ultra-fine fibers on the collector. For a typical electrospinning setup, the following are required: (1) a polymer reservoir attached to a capillary or needle, (2) a high-voltage power supply, and (3) a screen collector, as shown in Figure 23.1.

Although conceptually a simple process, electrospinning has significant challenges. A number of parameters can greatly influence the formation and structure of the obtained fibers. In principle, these parameters can be divided into two major ones: system parameters and process parameters. By appropriately varying all or some of these parameters, fibers are successfully electrospun. To understand the electrospinning process, the different parameters that affect the process have to be considered.

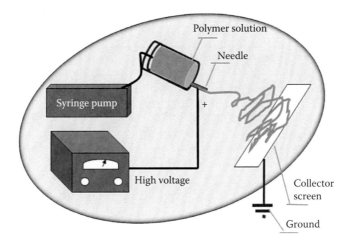

FIGURE 23.1
Schematic drawing of electrospinning apparatus.

23.2.1 Process Parameters

Despite the electrospinning technique's relative ease of use, there are a number of process parameters that can greatly affect fiber formation and structure. Grouped in order of relative impact to the electrospinning process, they are the applied voltage, polymer flow rate, and capillary-collector distance. All three parameters can influence the formation of bead defect nanofibers.

23.2.1.1 Applied Voltage

The fiber diameters can be controlled by the applied voltage, of which the achieved results vary strongly with the polymer system. The strength of the applied electric field controls formation of fibers from several microns in diameter to tens of nanometers. Suboptimal field strength could lead to bead defects in the electrospun fibers, or even failure in jet formation. Based on previous work,[34,35] it is evident that there is an optimal range of electric field strengths for a certain polymer-solvent system, as either too weak or too strong a field will lead to the formation of beaded fibers.

23.2.1.2 Polymer Flow Rate

Polymer flow rate also has an impact on fiber size, and additionally can influence fiber porosity as well as fiber shape.[36–38] Also, at high flow rates, significant amounts of bead defects are noticeable, due to the inability of fibers to dry completely before reaching the collector. Incomplete fiber drying also leads to the formation of ribbon-like (or flattened) fibers, compared to fibers with a circular cross section.[36,39]

23.2.1.3 Capillary-Collector Distance

While playing a much smaller role, the distance between capillary tip and collector can also influence fiber size. The fiber diameter decreased with increasing distances from the Taylor cone to the collector.[10,40] Additionally, morphological changes can occur upon decreasing the distance between the syringe needle and the substrate by forming beaded as-spun fibers, which can be attributed to inadequate drying of the polymer fiber prior to reaching the collector.[38]

23.2.2 System Parameters

In addition to the process parameters, a number of system parameters play an important role in fiber formation and structure. System parameters involve molecular weight, molecular weight distribution, the architecture of the polymer, and solution properties. Solution properties play especially an important role in the electrospinning process. In relative order of their impact on the electrospinning process, these include polymer concentration, solvent volatility, and solution conductivity.

23.2.2.1 Polymer Concentration

Polymer concentration influences both the viscosity and the surface tension of the solution. The formation of nanofibers through electrospinning is based on the uniaxial stretching of a viscoelastic solution.[5] For instance, the polymer solution must have a concentration high enough to cause polymer entanglements. However, the viscosity should not be so high as to prevent polymer motion induced by the electric field, and not so low as to prevent the charged jet from collapsing into droplets before the solvent has evaporated.[41] Further investigations on polymer concentration and viscosity have observed that fibers become more uniform and assume a cylindrical shape with increasing polymer concentration in solution, and fiber diameters also increase significantly with increasing polymer concentration. At lower concentrations, increasingly thinner fibers are formed, with additional beads along the fiber axis. However, at very high dilution, fiber formation no longer takes place.[42] It is not possible to make a general recommendation for particular concentrations and the resulting viscosities, because the ideal values of these parameters vary considerably with the polymer-solvent system.

23.2.2.2 Solvent Volatility

The choice of solvent is also critical as to whether fibers are capable of forming, as well as determining, surface topography.[36,39] If the fluid jet is collected prior to complete solvent evaporation, the deformable fibers may either flatten upon impact with the surface of a collector or adhere to other fibers. If the arriving jet or fiber lands on previously collected fibers, the still fluid material can merge and coalesce at a crossing point to create a conglutinated network, as shown in Figure 23.2. Conglutinated networks appear to be useful in some situations where a well-established network is desirable.

23.2.2.3 Solution Conductivity

Solution conductivity can influence fiber size within one to two orders of magnitude.[36] Solutions with high conductivity will have a greater charge-carrying capacity than solutions with low conductivity. Therefore, the fiber jet of highly conductive solutions will be subjected to a greater tensile force in the presence of an electric field than will a fiber jet from a solution with low conductivity. Highly conductive solutions were extremely unstable in the presence of strong electric fields, which led to a dramatic bending instability as well

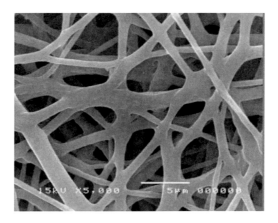

FIGURE 23.2
Conglutinated networks of electrospun polycaprolactone (PCL).

as a broad diameter distribution.[43] However, semiconducting and insulating liquids such as paraffinic oil produced relatively stable fibers. By adding a small amount of salt or ion, the electric force exerted on the jet also increased, and attributed to a decrease in the mean fiber diameter. Generally, the radius of the fiber jet is inversely related to the cube root of the solution conductivity.[44]

23.3 Electrospinning in Drug Delivery Applications

Since electrospinning has great flexibility in material selection, several biocompatible polymeric materials, either biodegradable or nonbiodegradable, have been electrospun to produce fiber matrices for drug or bioactive agent delivery. Release of the therapeutic molecules can be controlled, whether it occurs via diffusion alone or both diffusion and material degradation. A number of therapeutic agents, including biological active agents (i.e., anticancer drugs, anti-inflammatory drugs, antibiotics, and proteins), herbal extracts, and genes (i.e., DNA), can be easily loaded and delivered to the targets. Heat-sensitive drugs are also successfully delivered with retained properties. This is one of the advantages of the electrospinning process over the conventional melt process. There are various techniques used for loading and delivering a variety of therapeutic agents from the electropun fiber mat, as reported in several studies.

23.3.1 Biological Active Agents

The application of electrospun matrices as carriers for the controlled release of various types of biological active molecules, including anticancer drugs,[45–51] anti-inflammatory drugs,[52–59] antibiotics,[30,60–63] vitamins,[64,65] and protein,[66–68,70,72–79] has been examined.

23.3.1.1 Anticancer Drugs

A drug-loaded electrospun fiber mat is a promising controlled-release formulation in biomedical applications, especially in postoperative local chemotherapy. This method provides various advantages, such as improved therapeutic effect, reduced toxicity, and handling ease. Zeng et al.[45,46] investigated the potential to use electrospun poly(L-lactic acid) (PLLA) fiber mats as drug carriers. Various types of drugs, including rifampin (a drug for tuberculosis), paclitaxel (an anticancer drug), and doxorubicin hydrochloride (an anticancer drug), were loaded in the fiber mat. The release characteristics of drug-loaded PLLA fiber mats were studied by immersion of these mats in 0.05 M Tris-HCl buffer solution containing proteinase K at 37°C. Burst release of the drugs was not observed, implying perfect incorporation of drugs inside the fibers. In the presence of proteinase K, PLLA could be degraded and the drugs were subsequently released. In this case, the release was merely attributed to the PLLA degradation, not to the diffusion or permeation of drugs through the PLLA carrier.[45] In Zeng et al.'s recent work,[46] the influence of solubility and the compatibility of drugs in the drug-polymer system were studied. The results showed that there was good compatibility of paclitaxel and doxorubicin with PLLA, whereas doxorubicin hydrochloride was found on the surface of PLLA fibers, which resulted in the burst release of doxorubicin hydrochloride. The compatibility of the drugs in the drug-polymer system was found to be an important factor for controlled release in the drug delivery system.

The molecular weight of the polymer used to fabricate electrospun matrices is another important factor affecting the release mechanism of the loaded drugs. Xu et al.[47] developed PEG-PLLA diblock copolymer fiber mats for the controlled release of 1,3-bis(2-chloroethyl)-1-nitrosourea (BCNU) at loading concentrations of 5–20 wt%. The release of BCNU from the fiber mats was investigated in phosphate buffered saline (PBS; pH 7.4) at 37°C. The burst release was found within 30 min, and the release rate of BCNU increased with the amount of BCNU loaded. The release profile of BCNU from the electrospun PEG-PLLA fiber mats was mainly controlled by the diffusion mechanism, and not by degradation of the fiber mat. A combination of two releasing mechanisms was found in the case of electrospun amphiphilic PEG-PLLA containing doxorubicin hydrochloride (Dox) prepared via water-in-oil emulsion-electrospinning. The diffusion mechanism was predominant in the early stage of release, and enzymatic degradation became predominant subsequently. This phenomenon could be explained by the encapsulation of Dox inside and was well distributed throughout the electrospun fibers in the case of emulsion-electrospinning. Moreover, well-sustained release of Dox was observed for emulsion-electrospinning, compared with suspension-electrospinning. To investigate the release kinetics of drugs incorporated into fiber mats, different amounts of Dox were loaded into the PEG-PLLA fiber mat. The higher content of Dox in the emulsion-electrospinning resulted in a thinner core and a thicker sheath of fibers. Thus, Dox diffused more slowly through the relatively thicker PEG-PLLA sheath before being released from the fibers. The release behaviors of Dox in this system showed a three-stage diffusion-controlled mechanism in which the release rate of the first stage was slower than that of the second stage due to the reservoir-type structure of the fibers formed.[49] The release behaviors of both hydrophobic and hydrophilic drugs, paclitaxel (PTX) and doxorubicin hydrochloride (Dox), were affected by the solubility and distribution of the drugs in the drug-polymer system. The more hydrophilic Dox was compared to PTX, the easier it was to diffuse into buffer solution, leading to a higher release rate. Interestingly, the release rate of PTX was accelerated by released Dox from the coloaded fiber mats, but the release rate of Dox was not affected by the released PTX.[50]

In 2008, Ranganath and Wang observed the effect of structure and geometry on the release behavior of electrospun-loaded materials.[51] Electrospun poly(D,L-lactide-co-glycolide) (PLGA)-loaded paclitaxel (PTX) was prepared, and the release behaviors of paclitaxel for postsurgical chemotherapy against glioma were investigated. PLGA copolymer with a ratio of 85:15 was used to prepare a microfiber disc (MFD) and a microfiber sheet (MFS), and a ratio of 50:50 was used to prepare a submicrofiber disc (SFD) and a submicrofiber sheet (SFS). The release characteristics of these materials were studied in phosphate buffer saline (PBS; pH 7.4) solution containing 0.05% w/v Tween 80. The paclitaxel release was sustained for more than 80 days for all material forms. Submicrofiber discs/sheets showed faster release than microfiber discs/sheets due to higher glycolic acid content in PLGA 50:50 compared to in PLGA 85:15, resulting in a faster bulk degradation rate. In comparison with the geometry of materials, MFD and SFD exhibited a low initial burst release due to their multilayered structure, whereas MFS and SFS exhibited a relatively higher initial burst release due to their less compact structure.[51]

23.3.1.2 Anti-Inflammatory Drugs

Nonsteroidal anti-inflammatory drugs (NSAIDs) are used for controlling pain and inflammation in rheumatic disease. Kenawy et al.[52] examined the release characteristics of partially and fully hydrolyzed electrospun poly(vinyl alcohol) (PVA) containing 5 wt% of ketoprofen. In order to study the effect of stabilization of fiber mats, these as-spun fibers were treated with methanol for 1 and 24 h. The release behaviors were carried out in PBS (pH 7.4) at 37°C and 20°C against corresponding cast films. Interestingly, the burst release of ketoprofen was eliminated by treatment with methanol, and the releasing rate increased as the temperature of the medium increased. The degree of hydrolysis of the PVA was also affected by the release rate of ketoprofen.[52] In recent work, Kenawy et al.[53] also studied the release characteristics of ketoprofen from electrospun polycaprolactone (PCL), polyurethane (PU), and their blends at various ratios (i.e., 75/25, 50/50, and 25/75). They also found an influence of medium temperature on releasing characteristics. However, release rates from PCL, PU, and their blend had similar trends.

The incorporation of NSAIDs into electrospun cellulose acetate (CA) fiber mats was investigated by Tungprapa et al.[54] Four different types of NSAIDs (with varying water solubility), including naproxen (NAP), indomethacin (IND), ibuprofen IIBU), and sulindac (SUL), at a fixed concentration of 20 wt%, based on the weight of CA, were used to investigate release characteristics in acetate buffer solution (pH 5.5) at 37°C. The rapid release of NAP from the NAP-loaded CA fiber mats was observed due to the lack of interaction between the drug and the CA matrix and the high degree of swelling and high solubility of the drug-loaded fiber mat in the medium. The maximum release of the drugs from drug-loaded materials could be ranked as follows: NAP > IBU > IND > SUL, whereas the molar mass of the drugs could be ranked as follows: IBU < NAP < SUL ≈ IND. The ability of the drug to release from the polymer matrix depends on many factors, such as the solubility of the drug in the polymer matrix, the solubility of the drug in the medium, the swelling and solubility of the polymer matrix in the medium, the diffusion of the drug from the polymer matrix to the medium, etc.[54]

The synthesis of biodegradable pH-sensitive polymers containing ortho ester groups was successfully done by Qi et al.[55] The acid-labile segments were synthesized through 3,9-dimethylene-2,4,8,10-tetraoxaspiro undecane (DMTU) and 1,10-decanediol or poly(ethylene glycol) (PEG) to obtain POE_D and POE_P, respectively. These polymers were further copolymerized with D,L-lactide (D,L-LA), resulting in the triblock polymers PLA-POE_D-PLA (PLAOE_D) and PLA-POE_P-PLA (PLAOE_P), respectively. Qi et al. investigated the release characteristics of paracetamol from these pH-sensitive polymers in buffer solutions of pH 7.4, 5.6, and 4.0, respectively, in order to observe the acid responsiveness. The initial burst release of paracetamol was observed, followed by a gradual release during the immersion time.

Additionally, similar results were also observed for PLA fiber mats after immersion in buffer solutions with different pH values. In the case of $PLAOE_P$ fiber mats, a higher release rate was exhibited than with $PLAOE_D$ fiber mats under acid buffer solutions. This could be due to the hydrophilic surface of $PLAOE_P$ fiber mats helping in the diffusion of water and drug in and out of the fiber mats. Hence, Qi et al. concluded that the amount of drug release and the release rate of drug from the fiber mats were controlled by the inclusion of acid-labile segments with different hydrophilicities and the acid environment.[55]

To investigate the effect on the degradation and drug release behaviors of the pH environment, the synthesis of biodegradable pH-sensitive polymer through introducing acetal groups into the backbone of poly(D,L-lactide)-poly(ethylene glycol) (PELA) was carried out by Cui et al.[56] The acetal groups were introduced by reacting PEG with benzaldehyde to obtain the prepolymer PBE, which was further copolymerized with lactic acid monomer (LA) to obtain PBELA with a biodegradable backbone. The content of acetal groups of PBELA was adjusted with the inclusion of PBE into the copolymerization process. The release behaviors of paracetamol from these polymers were also investigated in different buffer solutions (i.e., pH 7.4, 5.5, and 4.0) in order to study the effect of the pH environment on the pH-sensitive polymers. The amounts of burst release were higher for PBELA with higher contents of acid-labile segments and for polymers incubated in medium at lower pH values. Hence, the amount of initial burst release and the time period of sustained release from fiber mats were also controlled by the matrix polymer with different contents of acid-labile segments and the local acid environment.[56] In 2006, electrospun poly(D,L-lactide) (PDLLA) fiber mats incorporated with paracetamol at different loadings (i.e., 2, 5, and 8 wt%) were prepared, and the drug release characteristics and degradation of fiber mats as a function of fiber characteristics were investigated.[57] The effect of different diameters and drug contents on the drug release characteristics and degradation of fiber mats was also studied. The results showed that the release characteristics of paracetamol from these fiber mats showed an initial burst release or steady release phase followed by a plateau or gradual release during the rest of the time. From all results, Cui et al. concluded that the drug release behaviors from fiber mats could be controlled by fiber size and amount of drug content. Peng et al.[58] also used paracetanol, which is a widely used analgesic and antipyretic drug, as a model drug. They prepared electrospun poly(ethylene glycol)-co-poly(D,L-lactide) (PELA) fiber mats with different amounts of PEG (i.e., 0–20 wt%) containing paracetanol at a fixed concentration of 5 wt% by electrospinning. They reported that the drug release rate increased as the PEG content decreased. Moreover, the drug burst release behavior was mainly related to drug-polymer compatibility, and the following sustained release phase depended on polymer degradation.

Meloxicam (MX) is one of the NSAIDs used for controlling pain and inflammation in rheumatic diseases. MX is insoluble in water, but soluble in organic solvents; thus, it has low dissolution and skin permeability. Hence, to develop a transdermal or dermal formulation of MX, meloxicam-loaded poly(vinyl alcohol) (PVA) fiber mats were prepared by electrospinning.[59] The amounts of meloxicam loaded in PVA solution were 2.5, 5, 10, and 20 wt% based on the weight of PVA. Comparisons were made against films. The obtained diameters of fibers ranged between 121 and 185 nm. The skin permeation characteristics of the meloxicam from the electrospun PVA fiber mats and films were carried out by transdermal diffusion through shed snake skin. The results showed that all of the meloxicam-loaded PVA fiber mats exhibited higher amounts of permeated meloxicam than the corresponding films. This could be due to the highly porous structure of the fiber mat contributing to higher swelling in an aqueous medium. Moreover, the permeation flux of meloxicam in both materials increased with an increase in the meloxicam content. This could be due to an increase in the meloxicam contents of the polymer providing a reduction of the relative amount of polymer as a diffusion barrier, resulting in an increase of meloxicam release.[59]

23.3.1.3 Antibiotic Drugs

Kenawy et al.[60] prepared electrospun fiber mats from either poly(lactic acid) (PLA) or poly(ethylene-co-vinyl-acetate) (PEVA), with a blend ratio of 50:50, as carriers for delivery of tetracycline hydrochloride. The release characteristics from the electrospun fiber mats were compared to a commercially available drug delivery system (Actisite®), as well as to the corresponding cast films. Drug release from the PEVA fiber mats showed a higher release rate than that from the 50/50 PLA/PEVA and pure PLA fiber mats. Moreover, the PEVA and 50/50 PLA/PEVA fiber mats exhibited relatively smooth release of the drug over about 5 days, and also had a release rate similar to that of Actisite. Hence, the melt processing of Actisite could be avoided by using electrospining, which is important for heat-sensitive drugs. In addition, the PLA fiber mats exhibited some instantaneous release of drug due to the partial crystallinity of PLA, which limits the diffusion of water into the polymer matrix and diffusion of the drug from the polymer matrix.[60]

Zong et al.[61] fabricated electrospun biodegradable amorphous poly(D,L-lactide) (PDLA) and semicrystalline poly(L-lactic acid) (PLLA) fiber mats for biomedical application. They investigated the effect of various parameters in electrospinning on the structure of these fiber mats. They also used an antibiotic drug (Mefoxin®) incorporated in these fiber mats to study release behaviors. The results showed that the effect of Mefoxin on the obtained fiber was a uniform structure with an average diameter of 160 nm. The burst release of Mefoxin was observed within 3 h, which is an ideal drug release profile for the prevention of post-operation-induced adhesion, since most infections occur in the first few hours after surgery.[61] In 2008, Thakur et al.[62] fabricated dual drug release PLLA fiber mats containing lidocaine hydrochloride (an anesthetic drug) and mupirocin (an antibiotic drug) by using dual spinneret electrospinning. Lidocaine hydrochloride-loaded PLLA electrospun fiber mats showed an initial burst release followed by a plateau after the first few hours, whereas mupirocin exhibited only 5% release in the first hour followed by a sustained release for over 72 h. The dual spinneret electrospinning process was able to achieve the required dual release profiles used in wound dressing applications by allowing lidocaine chloride to crystallize in other PLLA fibers and maintaining mupirocin in the noncrystallized form within the PLLA fibers.

Highly water-insoluble drugs such as itraconazole and ketanserin were used to incorporate polyurethane (PU) fiber mats into a segment for use in topical drug administration and wound healing.[63] The release behaviors of PU fiber mats containing either itraconazole (i.e., 10 or 40 wt%) or ketanserin (i.e., 10 wt/%) were investigated. The results showed that both poorly water-soluble itraconazole and ketanserin were released from the water-insoluble PU fiber mats into aqueous solutions. Moreover, ketanserin was released more rapidly from the PU fiber mats within the first 4 h than itraconazole. The burst release of ketanserin could be due to several factors, including higher drug solubility in the polymer and increased drug diffusivity in the polymer. In addition, the imperfection on the fiber of ketanserin resulted in the formation of aggregate on the fiber surface, which might dissolve more rapidly than the drug, which is embedded in the fiber.[63] The antibiotic drug cefoxitin sodium (Mefoxin), a hydrophilic drug used for the prevention of infection after surgery, was used in electrospun poly(lactide-co-glycolide) (PLGA)-based fiber mats.[30]

There were two base materials used in this work: PLGA, which controls the degradation rate of the material, and poly(lactide) (PLA), which provides mechanical strength to the material. Moreover, to control the drug release profile, an amphiphilic poly(ethylene glycol)-b-poly(lactide) (PEG-b-PLA) diblock copolymer was added to the polymer solutions to encapsulate the hydrophilic drug (cefoxitin sodium). The addition of cefoxitin sodium to the polymer solutions caused the morphology of the fiber to change from bead-and-string-like to fibrous-like. Interestingly, the addition of an amphiphilic PEG-b-PLA block copolymer into the polymer reduced the amount of released cefoxitin at earlier time points and extended the drug release profile to longer times.[30]

23.3.1.4 Vitamins

In 2007, Taepaiboon et al.[64] prepared electrospun cellulose acetate (CA) fiber mats and investigated their potential for use as drug carriers to deliver the model vitamins all-trans retinoic acid or vitamin A acid (Retin-A) and α-tocopherol or vitamin E (Vit-E). The release characteristics of these vitamin-loaded CA fiber mats and films were studied by immersion in an acetate buffer solution containing either 0.5 vol% Tween 80 or a mixture of 0.5 vol% Tween 80 and 10 vol% methanol. The results showed that the vitamin-loaded CA fiber mats exhibited a gradual release over the time periods, whereas the vitamin-loaded CA films exhibited a burst release of the vitamins. Poly(lactic-co-glycolic acid) (PLGA) fiber mats were also used as carriers to incorporate all-trans retinoic acid (RA).[65] Comparisons were made against the PLGA films. The release characteristics were carried out by immersion of these materials in phosphate buffer solution (pH 7.4) at 37°C. The results showed that the release rate of RA from the PLGA fiber mats was sustained due to the preserved fibrous structure of the PLGA fiber mats after 4 months. The release rate of RA from the PLGA films, however, showed a triphasic release profile. This was due to the different structures and degradation behaviors of the two drug delivery systems.[65]

23.3.1.5 Proteins

Zeng et al.[66] prepared protein-loaded (bovine serum albumin (BSA) or luciferase) poly(vinyl alcohol) (PVA) fiber mats by electrospinning. To control the release characteristics of protein from PVA fiber mats, poly(p-xylylene) (PPX) was coated to protein-loaded fiber mats with different thicknesses of PPX by chemical vapor deposition (CVD). They found that the uncoated PVA fiber mats showed a burst release of protein. On the other hand, the PPX-coated PVA fiber mats exhibited a significantly retarded release of protein, depending on the coating thickness of PPX. Moreover, the release of luciferase activity retained its enzyme activity after electrospinning. Hence, this was the prerequisite for the application of enzymes or other sensitive agents released from the electrospun fiber mats.[66] The encapsulation of human β-nerve growth factor (NGF), which was stabilized in carrier protein, bovine serum albumin (BSA) in a copolymer of ε-carpolactone and ethyl ethylene phosphate (PCLEEP), was prepared by conventional electrospinning.[67] A sustained release of NGF from PCLEEP fiber mats was obtained for up to 3 months. A PC12 neurite outgrowth assay also confirmed that the bioactivity of NGF-loaded PCLEEP fiber mats was retained throughout the period of sustained release. Hence, these fiber mats had efficiency for use in peripheral nerve regeneration applications.[67] Heparin-loaded poly(ε-caprolactone) (PCL) fiber mats were also prepared by conventional electrospinning.[68] Heparin is used to prevent vascular smooth muscle cell (VSMC) proliferation, which can lead to graft occlusion and failure. Luong-Van et al.[68] studied the effect of heparin incorporation on fiber morphology and release characteristics. They found that increasing heparin concentration caused the fiber diameter to decrease. Normally, the proliferation of VSMC cells is higher in the first week after injury and can continue for 14–28 days, depending on the severity of injury. Thus, the sustained release of heparin from these fiber mats over at least 14 days was suitable for vascular repair. Moreover, these fiber mats did not induce an inflammatory response in macrophage cells in vitro, and the release of heparin was effective in preventing the proliferation of VSMC cells.[68] In 2007, Kim et al.[69] developed a blend mixture of poly(ethylene oxide) (PEO) and poly(L-lactic acid) (PLLA), poly(ε-caprolactone) (PCL), or poly(D,L-lactic-co-glycolic acid) (PLGA) at a ratio of 7/3 wt% and loaded it with lysozyme in order to prepare lysozyme-loaded fiber mats. PEO was used to improve compatibility with the encapsulated lysozyme because it is a hydrophilic and biocompatible polymer. Among the three blend fibers, the PEO/PCL fiber mat showed the lowest amount of burst release, with a more sustained release profile during a 1-week period, while the PEO/PLLA and PEO/PLGA fiber mats showed a higher amount of burst release with less sustained release profiles. Additionally, the PEO/PLGA fiber mat exhibited a very large dimensional change in aqueous medium, whereas the PEO/PCL fiber mat maintained its original dimension. Hence, the PEO/PCL fiber mat was used for further investigations. The PEO/PCL blends with various ratios (i.e., 9/1, 7/3, and 5/5) were further prepared, and the release behaviors of lysozyme were investigated from these blends. Kim et al. found that the amount of lysozyme release was higher with an increased amount of PEO in blends. Moreover, the amount of initial burst release was higher when the blended fiber mats contained a higher amount of PEO. This implied that the lysozyme release rate was controlled by the dissolution rate of PEO domains that were separated in the fiber phase.[69]

However, the major disadvantage of conventional electrospinning shows the initial burst release, which leads to reducing the effectiveness of the formulation.[70] In addition, the loss of bioactivity during the preparation of the spinning solution and electrospinning process due to the long time exposure of biological agents to harsh organic solvents may be a concern. To overcome these problems, coaxial electrospinning is employed to encapsulate drug into polymeric materials. This technique provides some advantages, such as easily incorporating the drug into the polymeric matrix with high loaded efficiency and capacity, mild preparation conditions, and relatively steady release characteristics. Jiang et al.[71] fabricated an electrospun poly(ε-caprolactone) (PCL) core-shell structure and incorporated bovine serum albumin (BSA) and lysozyme containing poly(ethylene glycol) (PEG) into a core part by using the coaxial electrospinning technique. The thickness of the core and the shell could be controlled by varying the feed rate of the inner dope. The obtained diameters of both the shell and core were increased as the feed rate increased. The release behaviors of protein from the fiber mats were carried out in 0.05 M phosphate buffer solution (pH 7.4) at 37°C. A slight burst release was observed on the first day, followed by a relatively steady release. The result showed an increase in protein release when the feed rate increased. This implies that the release rate can be controlled by varying the feed rate of the PEG/protein solution. In 2009, Yan et al.[72] also investigated core-shell structure fiber mats prepared by coaxial electrospinning from PLLACL solution for the shell and phosphate buffered saline (pH 7.4) solution containing BSA or nerve growth factor (NGF) for the core for use in the nerve tissue engineering application. They compared the release characteristic of BSA from the mix PLLACL fiber mats with that from the coaxial PLLACL fiber mats. The results showed that the release characteristics of BSA from the mix PLLACL fiber mats exhibited two stages: an initial fast release followed by a constant release. On the other hand, the release characteristics of BSA from the coaxial PLLACL fiber mats showed a relatively stable release. However, there was some

BSA on the surface of PLLACL fiber mats because the inner and outer BSA solutions could not mix together completely. Thus, there were also initial stages in the release profiles of the coaxial PLLACL fiber mats, but they were quite different from those of the mixed PLLACL fiber mats.[72]

Zhang et al.[73] also studied the encapsulation of a model protein, fluorescein isothiocyanate-conjugated bovine serum albumin (FITC-BSA), along with poly(ethylene glycol) (PEG), within biodegradable poly(ε-caprolactone) (PCL) fiber mats, by using coaxial electrospinning. They varied the inner flow rates from 0.2 to 0.6 ml/h, with a constant outer flow rate of 1.8 ml/h, and the amount of FITC-BSA loading from 0.85 to 2.17 mg/g of fiber mats, and investigated the release characteristics of FITC-BSA from fiber mats. For comparing the effectiveness of the coaxial electrospinning technique, the same materials were prepared into the fiber mats by using the conventional single-fluid electrospinning method. The release characteristics of FITC-BSA from fiber mats were studied by immersion in phosphate buffered saline (PBS; pH 7.4) at 37°C. They found that the release of BSA from the core-shell structure fibers was continuous over a period of more than 5 months. Moreover, the release rate of smaller-sized fibers was faster than that of larger-sized fibers. Compared with the fiber mats from conventional single-fluid electrospinning, the core-shell fiber mats could retard the initial burst release.[73] Lu et al.[74] prepared composite fiber mats composed of cationized gelatin-coated polycaprolactone (PCL) by coaxial electrospinning. Cationized gelatin (CG) was used as the shell materials for fabricating core-shell fiber mats. They investigated the adsorption behaviors of FITC-labeled bovine serum albumin (FITC-BSA) or FITC-heparin onto the fiber mats. Moreover, vascular endothelial growth factor (VEGF) was impregnated into the fiber mats through specific interactions with the adsorbed heparin in the outer CG layer. The CG fiber mats were cross-linked by exposing the fiber mats in glutaraldehyde (GA) vapor. Lu et al. found that the release characteristics of BSA or heparin from core-shell fiber mats exhibited a slight burst release within the first 2 h, and there was negligible desorption from the fiber mats for both BSA and heparin. Moreover, the release rate of heparin could be controlled by varying the cross-linking density of the outer CG fiber mats. Sustained release of VEGF could be achieved from the heparin-adsorbed fiber mats for more than 15 days.[74]

Core-shell structure devices are generally used for incorporating proteins because they can preserve an unstable biological agent from harsh environments and functionalize the surface of materials without affecting the core materials. Emulsion-electrospinning was the one novel process to prepare core-shell structures. It is remarkable that unlike coaxial electrospinning, which needs a special apparatus, this technique uses basic equipment, such as a single needle. The composite fibers with beads-in-string structures were successfully prepared by emulsion-electrospinning.[75] Ca-alginate microspheres were loaded with bovine serum albumin (BSA) by W/O emulsion, and then were embedded into poly(L-lactic acid) (PLLA) fibers by electrospinning. For in vitro release study, the release of BSA from composite fibers exhibited prolonged release profiles and lower initial burst release rates than those from naked Ca-alginated microspheres.[75] Maretschek et al.[76] also prepared cytochrome C (hydrophilic model protein)-loaded fiber mats based on PLLA by emulsion-electrospinning. The release behaviors of protein-loaded PLLA fiber mats derived from 1, 2, or 3 wt% PLLA solution were studied in PBS (pH 7.4). PLLA fiber mats from 2 and 3 wt% solutions exhibited no burst release, which implied perfect inclusion of the protein within the fiber mats. The slight increase in the protein release from 1 wt% PLLA solutions might be due to the decrease in thickness of the fiber mats, which leads to better wettability. To study the effect of wettability on cytochrome C release, releasing media having different surface tensions, achieved by adding small amounts of Tween 20 (i.e., 0.001, 0.005, and 0.01%), were used. Higher amount of Tween 20 caused a faster release of protein. To improve the release behaviors of the hydrophobic PLLA fiber mats, hydrophilic polymers (i.e., poly(ethylene imine) and poly(L-lysine)) were added. With the increased amount of hydrophilic polymers, the hydrophobicity of PLLA fiber mats decreased and the protein release rate increased. Hence, the release behaviors of the hydrophobic PLLA fiber mats were controlled by blending with different amounts of hydrophilic polymers.[76] Yang et al.[77] prepared lysozyme encapsulated within the core-shell structure fibers by emulsion-electrospinning. The lysozyme was loaded in poly(D-lactic acid) (PLLA) and methyl cellulose (MC) solutions at concentrations of 1, 3, and 5 wt%. The release behaviors of the lysozyme loaded in core-shell PLLA/MC fiber mats were studied in phosphate buffer solution (pH 7.4) containing 0.02% sodium azide as a bacteriostatic agent. The release kinetics for all the fiber mats could be shown in two stages: an initial fast release followed by a constant linear release. An increase of lysozyme loading caused a higher amount of lysozyme release. Thus, the core-shell-structured fibers could reduce the initial burst release, sustain the release period dependent on the protein loading amount, and protect the structural integrity and bioactivity of encapsulated lysozyme during incubation in medium.[77] Rhodamine B and BSA were successfully loaded into the poly(L-lactide-co-carpolactone) (PLLACL) fiber mats by emulsion-electrospinning.[8] The release characteristics of the BSA and rhodamine B from these fiber mats were studied by immersion in phosphate buffered saline (pH 7.4) for various time points. The results showed that the release characteristics of rhodamine B and BSA were influenced and varied by their incorporated types. When these drugs were incorporated in the core of fibers, the release profile was sustained. On the other hand, when these drugs were located in the surface of fibers, there was a burst release in the beginning.[78]

23.3.2 Herbal Substances

There has been much interest in the electrospinning process for use in drug delivery systems for over 10 years. In recent years, the use of electrospun fiber mats as a carrier for herbal substances has attracted a great deal of attention due to their ability to produce ultra-fine fibers with diameters in the range of nanometers to submicrometers that exhibit a high surface area-to-volume or mass ratio.[9] Moreover, the use of synthetic drugs in clinical treatments of some diseases has limitations in developing countries due to their accessibility and relatively high cost. Hence, the use of herbal substances has been given attention as alternative treatments in these countries.

Raspberry ketone (RK) is a major aromatic compound of red raspberry (*Rubus idaeus*). It is composed of an abundance of sugars, vitamins, minerals, and polyphenols, and is widely used in cosmetics and foodstuffs. In addition, it has whitening and anti-inflammation properties.[80] Yang et al.[79] investigated the use of gelatin/PVA nanofibers as carriers for controlled release of RK. They studied the effect of the amount of RK loading (i.e., 2 and 5 wt%), the pH of releasing media (i.e., 1, 4.8, and 7), the cross-linking time (i.e., 0, 2, and 5 h), and the gelatin/PVA ratio (i.e., 10/0, 7/3, 5/5, 3/7, 1/9, and 0/10) on the releasing characteristic. RK-loaded gelatin/PVA fiber mats could be easily dissolved in aqueous solutions due to all of the water-soluble molecules. However, this property made them exhibit a high initial burst release at short times, and it reached a plateau after 2 h. The percentage of release of RK in a buffer solution of pH 1 was higher than those in buffer solutions of pH 4.8 and 7. This could be from the protonation of amino groups of gelatin causing positive charges on the polymer. Thus, the positive charges made molecular chains repulse each other, and the networks became looser, which provided the diffusion of drug molecules into the buffer solutions. With regard to the effect of cross-linking time, increasing the cross-linking time resulted in decreasing the RK release rate and the percentage of release of RK. In addition, a gelatin/PVA ratio of 7/3 had a slightly higher release rate than a gelatin/PVA ratio of 9/1. This could be due to the higher solubility of PVA in buffer solution. From these results, Yang et al. concluded that the RK release rate could be tailored by changing the content of RK in gelatin/PVA nanofibers, the ratio of gelatin and PVA, and the cross-linking time by glutaraldehyde vapor.[79]

Curcumin from the plant *Curcuma longa* is widely used for antitumor, antioxidant, and anti-inflammatory applications. Electrospun cellulose acetate fiber mats loaded with curcumin (i.e., 5–20 wt%) were fabricated.[81] The average diameters of the obtained curcumin-loaded CA fibers were between ~314 and ~340 nm. The release characteristics of curcumin from the curcumin-loaded CA fiber mats were investigated by total immersion and transdermal diffusion through a pig skin method, in comparison with solvent-cast films. In both methods, the amount of curcumin released from the electrospun fiber mats was higher than that from the as-cast films. Moreover, the free radical scavenging ability of the curcumin-loaded CA fiber mats was retained, even though they had been subjected to a high electrical potential during electrospinning.[81] Sikareepaisan et al.[82] fabricated electrospun gelatin fiber mats containing a methanolic crude extract (mCA) of *Centella asiatica* (L.) Urban. Different amounts of the herbal extracts (i.e., 5–30 wt%) were incorporated in the gelatin fiber mats. The obtained as-spun fibers had an average diameter in the range of 226–232 nm. Due to the instability of gelatin fibers in releasing medium, cross-linked mCA-loaded electrospun gelatin fibers were needed. The release characteristics of asiaticoside from the mCA-loaded gelatin fiber mats were investigated by using the total immersion method in an acetate buffer and a buffer that contained 10 vol% of methanol. The results were compared with those from corresponding solvent-cast films. Sikareepaisan et al. reported the cumulative release profile of asiaticoside from mCA-loaded materials in two different manners: based on the unit weight of the actual amount of asiaticoside present in the specimens and based on the unit weight of the specimens. They found that based on the unit weight of the actual amount of asiaticoside present in the specimens, the amount of asiaticoside released from the fiber mat specimens was lower than that from the films, while based on the unit weight of the specimens, the opposite trend was observed.[82] Moreover, Suwantong et al. prepared electrospon cellulose acetate (CA) fiber mats loaded asiaticoside (AC) from the plant cetella asiatica L. in the form of either pure substance (PAC) or crude extract (CACE).[84] The amount of asiaticoside in the form of either PAC or CACE was fixed at 40 wt% based on the weight of CA. They studied the potential to use these fiber mats as carriers for topical/transdermal delivery of AC from the fiber mats that contain either PAC or CACE. Various properties (i.e., morphological, water retention, weight loss, and cytotoxicity) of these fiber mats and release characteristics of AC from the herb-loaded CA fiber mats in two types of media (i.e., acetate and phosphate buffer solutions containing methanol) were investigated, and comparisons were made against the corresponding solvent-cast CA films. They found that the fibers obtained were smooth without any aggregates on surface, and the average diameter ranged between 485 and 545 nm. Due to the highly porous structure of the fiber mats, the amount of AC released from the fiber mats was also higher than that from the films in both types of release assays (i.e., the total immersion and the transdermal diffusion through pig skin). Moreover, the PAC- and CACE-loaded CA fiber mats and the corresponding as-cast films were assessed by investigating the cytotoxicity of these materials against normal human dermal fibroblasts for use as topical/transdermal or wound dressing patches. The results showed that only the extraction media from the CACE-loaded fiber mats and films at the extraction ratios of 5 and 10 mg/ml were toxic to the cells. This might be due to the toxicity of other triterpenoid compounds that are present in the CACE extract.[83]

Due to hydrophilicity, good chemical stability, and good thermal stability of poly(vinyl alcohol) (PVA),[25] it was selected to be used as the drug carrier. A PVA nanofiber incorporated with extracts from the fruit hull of mangosteen was prepared by electrospinnig.[84] Mangosteen (*Garcinia mangostana* Linn) (GM) has been known as "the queen of fruits" because of its excellent taste compared to that of other fruits. GM has been used in traditional medicine for the treatment of diarrhea, skin infection, wounds, and chronic ulcers. It also has antimicrobial, antioxidant, and anti-inflammatory properties.[85–87] The morphology of the GM extract-loaded PVA fiber mats, the actual amount of GM extract within the fiber mats, the swelling and weight loss behavior of these fiber mats in an aqueous medium, and the release characteristics of the extract from the GM extract-loaded PVA fiber mats were investigated by antioxidant activity. Opanasopit et al. incorporated GM extract at various contents (i.e., 2.5, 5, and 10 wt%) into the PVA fiber mats. The obtained morphologies of these GM extract-loaded PVA fiber mats ranged from 140 to 180 nm. The release characteristics of these GM-loaded PVA fiber mats were carried out by the total immersion method. The results showed that the amount of GM extract released from the PVA fiber mats exhibited burst release due to the high swelling ability of PVA. In addition, the amount of GM extract released from the PVA fiber mats was greater than that from the film counterparts because the as-spun fiber mats had a very porous structure that contributed to the high swelling ability in the medium.[84]

Various types of herbal substances that have antioxidant activity have been reported. Among them, gallic acid, which is a naturally occurring polyphenol in a variety of fruits and vegetables, such as grapes, cherries, tea leaves, and longan seeds, has been given more attention recently.[88] Gallic acid has been shown to exhibit antioxidant, anti-inflammatory, anticarcinogenic, and antifungal properties.[89] Chuysinuan et al.[90] have incorporated this herbal substance in poly(L-lactic acid) fiber mats at a concentration of 40 wt% based on the weight of PLLA by using electrospinning. The gallic acid-loaded PLLA fibers obtained were smooth, with average diameters of 843 nm. For use as carriers for topical or transdermal patches, the gallic acid-loaded PLLA fiber mats were investigated by the total immersion method in three different types of releasing medium (an acetate buffer, a citrate-phosphate buffer, and a normal saline). The amount of gallic acid released from the gallic acid-loaded PLLA fiber mats into each medium increased rapidly during the initial time, followed by a gradual increase, and reached a plateau value at the longest submersion time. At any given submersion time point, the amount of gallic acid released into the normal saline was the greatest, followed by those released into the citrate-phosphate and the acetate buffer solutions, respectively. This could be due to the greatest value of water retention of gallic acid-loaded fiber mats in the normal saline, followed by those in the citrate-phosphate and the acetate buffer solutions, respectively. In addition, the difference between the pKa value of gallic acid and the pH of the normal saline was greater than those of the other two media, which caused the dissociation of the carboxylic acid chain ends of PLLA into carboxyl groups and helped to promote the diffusion of gallic acid out of the fibers.[90]

Han et al.[91] fabricated poly(ε-caprolactone) (PCL)/poly(trimethylene carbonate) (PTMC) fiber mats as drug carriers to incorporate the herbal medicine shikonin isolated from the plant *Lithospermum erythrorhizon*. This provided some interesting properties involving antitumor, antioxidant, antibacterial, and anti-inflammatory properties.[92–94] The various ratios (i.e., 9:1, 7:3, and 5:5 w/w) of the PCL/PTMC blends containing 1 and 5 wt% shikonin were prepared by electrospinning, and the release characteristics were investigated. They found that shikonin-loaded as-spun fiber mats showed an initial burst release followed by a gradual release. In addition, the amount of shikonin released from the fiber mats increased with increasing the amount of PTMC in the blends. This could be due to the lower diameter of PCL/PTMC blend fibers at a ratio of 5:5, giving them a higher surface area to release a larger amount of drug. Moreover, PTMC being more hydrophilic than the semicrystallized PCL resulted in an increase of water absorption, and thus the fibers easily dissolved in the medium. The antioxidant and antibacterial activities of these fiber mats were also studied for use in biomedical applications. It was found that after being subjected to a high electrical potential, the biological activity of these fiber mats was retained.[91]

23.3.3 Genes

The major goal of gene therapy is to reconstitute a missing or defective gene with a correct copy in the host genome.[95] This can be achieved by introducing the missing genes, promoting expression of the existing genes, or stopping diseased gene expression. Since naked DNA has shown low transfection toward target cells, effective delivery of the desired genes into target cells is possible when biocompatible gene carriers are utilized.[96,97] Electrospinning techniques have been used to fabricate fibrous scaffolds for genetic materials delivery. In 2003, Luu et al. developed electrospun scaffolds of PLGA/PLA-PEG block copolymers loaded with pCMVβ plasmid encoding β-galactosidase DNA.[98] The amount of the plasmid released was as high as 80% of the initial load of DNA. The PLGA/PLA-PEG ratio affected the amount, rate, and efficiency of DNA release. Their results also showed that the released DNA remained intact, even when passing through high voltage during electrospinning. Moreover, the released DNA could transfect the preosteoblastic cell line, MC3T3-E1, and successfully encoded β-galactosidase protein in vitro. However, the transfection efficiency was still low. Since the gene carriers must deliver the genetic materials through several extra- and intracellular barriers to reach the nucleus of the target cells, including crossing the plasma membrane, surviving the acidic and degradative endosomal-lysosomal system, traveling to the nucleus, and releasing the genes into the nucleus, the direct incorporation of the plasmid into the electrospun fiber might not be enough to achieve high efficiency. Apart from electrospun fibers, the gene carriers commonly used to date can be divided into two categories: viral and nonviral.[99] Viral vectors have been used in current clinical trials mostly because of their innate ability to overcome the mentioned obstacles, thus resulting in high efficiency of gene delivery. Retroviruses and adeno/adeno-associated viruses are regarded as the most effective carriers of therapeutic genes for a prolonged gene expression period.[100] Long-term gene expression involves integration of viral genes into the host genome. Gene delivery using the viral vectors associated with electrospinning was recently studied. Coaxial electrospinning, a setup where an aqueous solution containing drugs, proteins, or genetic materials, forming the inner jet, is coelectrospun with a polymer solution forming the outer jet,[101] was used to fabricate a sustained viral gene delivery scaffold. Liao et al.[102] were among the first group to produce core-shell fibrous scaffolds via coaxial electrospinning to efficiently and locally deliver viral vectors into mammalian cells. They found that close to 90% of HEK 293 cells were transfected in the first 2 weeks, and then dropped to 0% in the following weeks, when the cells were exposed to the scaffold supernatant. On the other hand, the cells seeded directly onto the virus-encapsulated scaffolds expressed a high level of transgene expression for over 1 month. This indicated the prolonging and localizing cell transfection ability of the scaffolds in vitro. In addition, the coaxial electrospun scaffolds could reduce virus dissemination and its corresponding inflammatory cytokine secretion.

Although viral carriers primarily efficiently infect cells and prolong gene expression, they are associated with important drawbacks. They can potentially cause undesirable consequences, such as toxicity, uncontrolled proliferation of modified cells, inflammation, immunological reactions, and oncogenicity.[103,104] These concerns have limited the viral carrier application in clinical use. Nonviral vectors have been of great interest because they have been determined as a clinically safer alternative due to the lack of those viral limitations. Many biomaterials have been developed to create synthetic gene delivery vectors. The majority of the biomaterials are cationic polymers, cationic lipids, liposomes, chitosans, inorganic nanoparticles, and dendrimers.[99,105–108] The use of eletrospinning associated with these gene carriers to deliver a genetic portion to the target cell has been studied. Nie and Wang[99] studied the delivery of bone morphogenetic protein-2 plasmid DNA into human marrow stem cells (hMSCs) in vitro using electrospun poly(lactide-co-glycolide) (PLGA)/hydroxyapatite (HAp) composite scaffolds. They loaded the plasmids into the fibrous scaffolds in three different ways: coated naked DNA onto the fibers, coated DNA/chitosan nanoparticles onto the fibers, and encapsulated DNA/chitosan nanoparticles into scaffold by mixing with PLGA/HAP solution prior to electrospinning. In vitro studies show that the electrospun scaffolds could release the plasmids or plasmid/chitosan nanoparticles, and the sustained release could be up to 45–55 days. The amount of HAp loaded in the scaffold also affected the release rate. Moreover, the released DNA retained its structural and biological properties, even after 60 days of releasing. High transfection efficiency could be sustained with the aid of a chitosan carrier, especially in the latter way of loading the plasmids into the scaffold. In 2009, Nie et al.[109] also investigated the release of BMP-2 plasmid from the previously studied scaffolds in vivo using nude mice as an animal model. Animal experiments showed that the bioactivity of the BMP-2 plasmid released from the scaffolds was maintained and could improve new bone formation and healing of bone defects of mice tebia in vivo. The scaffolds containing DNA/chitosan nanoparticles transfected cells more efficiently than naked DNA. The in vivo test results in bone healing were well correlated with the in vitro release profiles of plasmid in their previous work.

23.4 Conclusions

The application of electrospinning to therapeutic molecule delivery holds great promise. Selection of appropriate polymers for specific therapeutics allows for controlled drug or bioactive molecule release rate and behavior.

References

1. A. Formhals, Process and apparatus for preparing artificial threads, U.S. Patent (1934) 1975504.
2. M. Jacobsen, *Chemiefasern/Textilind* (1991) 36–41.
3. P.X. Ma, R. Zhang, Synthetic nano-scale fibrous extracellular matrix, *J. Biomed. Mater. Res. A* 46 (1999) 60–72.

4. C.J. Ellison, A. Phatak, D.W. Giles, Melt blown nanofibers: Fiber diameter distributions and onset of fiber breakup, *Polymer* 48 (2007) 3306–3316.

5. W.E. Teo, S. Ramakrishna, A review on electrospinning design and nanofibre assemblies, *Nanotechnology* 17 (2006) R89–R106.

6. B.K. Gu, M.K. Shin, K.W. Sohn, Direct fabrication of twisted nanofibers by electrospinning, *Appl. Phys. Lett.* 90 (2007) 263902.

7. P.K. Panda, S. Ramakrishna, Electrospinning of alumina nanofibers using different precursors, *J. Mater. Sci.* 42 (2007) 2189–2193.

8. G. Kim, W. Kim, Formation of oriented nanofibers using electrospun electrospinning, *Appl. Phys. Lett.* 88 (2006) 233101.

9. J. Doshi, D.H. Reneker, Electrospinning process and applications of electrospun fibers, *J. Electrostat.* 35 (1995) 151–160.

10. J.M. Deitzel, J. Kleinmeyer, J.K. Hirvonen, T.N.C. Beck, Controlled deposition of electrospun poly(ethylene oxide) fibers, *Polymer* 42 (2001) 8163–8170.

11. P. Gibson, H. Schroeder-Gibson, D. Rivin, Transport properties of porous membranes on electrospun nanofibers, *Colloids Surf. A* 187–188 (2001) 469–481.

12. P. Gibson, Production and characterization of patterned electrospun fibrous membranes, *Fiber Soc. Meeting Proceedings* (2003) 42–49.

13. Z.M. Huang, Y.Z. Zhang, M. Kotaki, S. Ramakrishna, A review on polymer nanofibers by electrospinning and their applications in nanocomposites, *Composites Sci. Technol.* 63 (2003) 2223–2253.

14. M. Jaffe, S. Shanmukasundaram, A. Patlolla, K. Griswold, S. Wang, J. Mantilla, R. Walsh, T. Arinzeh, C.J. Prestigiacomo, L.H. Catalani, The benefits of nanofibers in biomedical applications, *Fiber Soc. Fall Meeting Proceedings* (2004) 8–9.

15. J.B. Park, J.D. Bronzino, *Biomaterials: Principles and applications*, CRC Press, Boca Raton, FL, 2003.

16. M.-S. Khil, S.R. Bhattarai, H.-Y. Kim, S.-Z. Kim, K.-H. Lee, Novel fabricated matrix via electrospinning for tissue engineering, *J. Biomed. Mater. Res. B Appl. Biomater.* 72B (2005) 117–124.

17. Z. Ma, M. Kotaki, R. Inai, S. Ramakrishna, Potential of nanofiber matrix as tissue-engineering scaffolds, *Tissue Eng.* 11 (2005) 101–109.

18. S.A. Riboldi, M. Sampaolesi, P. Neuenschwander, G. Cossu, S. Mantero, Electrospun degradable polyesterurethane membranes: Potential scaffolds for skeletal muscle tissue engineering, *Biomaterials* 26 (2005) 4606–4615.

19. F. Yang, R. Murugan, S. Wang, S. Ramakrishna, Electrospinning of nano/micro scale poly(L-lactic acid) aligned fibers and their potential in neural tissue engineering, *Biomaterials* 26 (2005) 2603–2610.

20. M.-S. Khil, D.-I. Cha, H.-Y. Kim, I.-S. Kim, N. Bhattarai, Electrospun nanofibrous polyurethane membrane as wound dressing, *J. Biomed. Mater. Res. B Appl. Biomater.* 67B (2003) 675–679.

21. H.Y. Kim, B.M. Lee, I.S. Kim, T.H. Jin, K.H. Ko, Y.J. Ryu, Fabrication of triblock copolymer of poly(r-dioxanone-co-L-lactide)-block-poly(ethylene glycol) nonwoven mats by electrospinning and applications for wound dressing, *PMSE* preprints 91 (2004) 712–713.

22. H. Jia, G. Zhu, B. Vugrinovich, W. Kataphinan, D.H. Reneker, P. Wang, Enzyme-carrying polymeric nanofibers prepared via electrospinning for use as unique biocatalysts, *Biotechnol. Progr.* 18 (2002) 1027–1032.

23. Y. Wang, Y.-L. Hsieh, Enzyme immobilization via electrospinning of polymer/enzyme blends, *Polymer* preprints 44 (2003) 1212–1213.

24. Y. Wang, Y.-L. Hsieh, Enzyme immobilization to ultra-fine cellulose fibers via amphiphilic polyethylene glycol spacers, *J. Polym. Sci. A Polym. Chem.* 42 (2004) 4289–4299.

25. L. Wu, X. Yuan, J. Sheng, Immobilization of cellulase in nanofibrous PVA membranes by electrospinning, *J. Membrane Sci.* 250 (2005) 167–173.

26. X. Zong, D. Fang, K. Kim, S. Ran, B.S. Hsiao, B. Chu, C. Brathwaite, S. Li, E. Chen, Nonwoven nanofiber membranes of poly(lactide) and poly(glycolide-co-lactide) via electrospinning and application for antiadhesions, *Polymer* preprints 43 (2002) 659–660.

27. X. Zong, S. Li, E. Chen, B. Garlick, K.-S. Kim, D. Fang, J. Chiu, T. Zimmerman, C. Brathwaite, S. Hsiao Benjamin, B. Chu, Prevention of post-surgery-induced abdominal adhesions by electrospun bioabsorbable nanofibrous poly(lactide-co-glycolide)-based membranes, *Ann. Surg.* 240 (2004) 910–915.

28. H.L. Jiang, D.F. Fang, B.J. Hsiao, B.J. Chu, W.L. Chen, Preparation and characterization of ibuprofen-loaded poly(lactide-co- glycolide)/poly(ethylene glycol)-g-chitosan electrospun membranes, *J. Biomater. Sci. Polym. Ed.* 15 (2004) 279–296.

29. D.S. Katti, K.W. Robinson, F.K. Ko, C.T. Laurencin, Bio-resorbable nanofiber-based systems for wound healing and drug delivery: Optimization of fabrication parameters, *J. Biomed. Mater. Res. B Appl. Biomater.* 70B (2004) 286–296.

30. K. Kim, Y.K. Luu, C. Chang, D.F. Fang, B.S. Hsiao, B. Chu, M. Hadjiargyrou, Incorporation and controlled release of a hydrophilic antibiotic using poly(lactide-co-glycolide)-based electrospun nanofibrous scaffolds, *J. Control. Release* 98 (2004) 47–56.

31. K.S. Kim, C. Chang, X. H. Zong, D.F. Fang, B.S. Hsiao, B. Chu, M. Hadjiargyrou, Incorporation of an antibiotic drug in electrospun poly(lactide-co-glycolide) non-woven nanofiber scaffolds, *Abstr. Pap. Am. Chem. Soc.* 226 (2003) U437.

32. Y.K. Luu, K. Kim, B.S. Hsiao, B. Chu, M. Hadjiargyrou, Development of a nanostructured DNA delivery scaffold via electrospinning of PLGA and PLA-PEG block copolymers, *J. Control. Release* 89 (2003) 341–353.

33. D. Liang, B.S. Hsiao, B. Chu, Functional electrospun nanofibrous scaffolds for biomedical applications, *Adv. Drug Deliver. Rev.* 59 (2007) 1392–1412.

34. J.M. Deitzel, J. Kleinmeyer, D. Harris, N.C.B. Tan, The effect of processing variables on the morphology of electrospun nanofibers and textiles, *Polymer* 42 (2001) 261–272.

35. C. Meechaisue, R. Dubin, P. Supaphol, V.P. Hoven, J. Kohn, Electrospun mat of tyrosine-derived polycarbonate fibers for potential use as tissue scaffolding material, *J. Biomater. Sci. Polym. Ed.* 17 (2006) 1039–1056.

36. T.J. Sill, H.A. von Recum, Electrospinning: Applications in drug delivery and tissue engineering, *Biomaterials* 29 (2008) 1989–2006.

37. G.I. Taylor, Electrically driven jets, *Proc. Roy. Soc. London Ser. London* A, 313 (1969) 453–475.

38. S. Megelski, J.S. Stephens, D.B. Chase, J.F. Rabolt, Micro- and nanostructured surface morphology on electrospun polymer fibers, *Macromolecules* 22 (2002) 8456–8466.

39. J. Lannutti, D. Reneker, T. Ma, D. Tomasko, D. Farson, Electrospinning for tissue engineering scaffolds, *Mater. Sci. Eng.* C27 (2007) 504–509.

40. R. Jaeger, M.M. Bergshoef, C.M.I. Batlle, H. Schonherr, G.J. Vancso, Electrospinning of ultrathin polymer fibers, *Macromol. Symp.* 127 (1998) 141–150.

41. J. Venugopal, Y.Z. Zhang, S. Ramakrishna, Electrospun nanofibres: Biomedical applications, *Proc. IMechE N J. Nanoeng. Nanosyst.* 218 (2005) 35–45.

42. A. Greiner, J.H. Wendorff, Electrospinning: A fascinating method for the preparation of ultrathin fibers, *Angew. Chem. Int. Ed.* 46 (2007) 5670–5703.

43. I. Hayati, A.I. Bailey, T.F. Tadros, Investigations into the mechanisms of electrohydrodynamic spraying of liquids: Effect of electric-field and the environment on pendant drops and factors affecting the formation of stable jets and atomization, *J. Colloid Interface Sci.* 117 (1987) 205–221.

44. P. Baumgarten, Electrostatic spinning of acrylic microfibers, *J. Colloid Interface Sci.* 36 (1971) 71–79.

45. J. Zeng, X. Xu, X. Chen, Q. Liang, X. Bian, L. Yang, X. Jing, Biodegradable electrospun fibers for drug delivery, *J. Control. Release* 92 (2003) 227–231.

46. J. Zeng, L.Yang, Q. Liang, X. Zhang, H. Guan, X. Xu, X. Chen, X. Jing, Influence of the drug compatibility with polymer solution on the release kinetics of electrospun fiber formulation, *J. Control. Release* 105 (2005) 43–51.

47. X. Xu, X. Chen, X. Xu, T. Lu, X. Wang, L. Yang, X. Jing, BCNU-loaded PEG-PLLA ultrafine fibers and their *in vitro* antitumor activity against glioma C6 cells, *J. Control. Release* 114 (2006) 307–316.

48. X. Xu, L. Yang, X. Xu, X. Wang, X. Chen, Q. Liang, J. Zeng, X. Jing, Ultrafine medicated fibers electrospun from W/O emulsions, *J. Control. Release* 108 (2005) 33–42.

49. X. Xu, X. Chen, P. Ma, X. Wang, X. Jing, The release behavior of doxorubicin hydrochloride from medicated fibers prepared by emulsion-electrospinning, *Eur. J. Pharm. Biopharm.* 70 (2008) 165–170.

50. X. Xu, X. Chen, Z. Wang, X. Jing, Ultrafine PEG-PLA fibers loaded with both paclitaxel and doxorubicin hydrochloride and their *in vitro* cytotoxicity, *Eur. J. Pharm. Biopharm.* 72 (2009) 18–25.

51. S.H. Ranganath, C.-H. Wang, Biodegradable microfiber implants delivering paclitaxel for post-surgical chemotherapy against malignant glioma, *Biomaterials* 29 (2008) 1996–3003.

52. E.-R. Kenawy, F.I. Abdel-Hay, M.H. El-Newehy, G.E. Wnek, Controlled release of ketoprofen from electrospun poly(vinyl alcohol) nanofibers, *Mater. Sci. Eng. A* 459 (2007) 390–396.

53. E.-R. Kenawy, F.I. Abdel-Hay, M.H. El-Newehy, G.E. Wnek, Processing of polymer nanofibers through electrospinning as drug delivery systems, *Mater. Chem. Phys.* 113 (2009) 296–302.

54. S. Tungprapa, I. Jangchud, P. Supaphol, Release characteristics of four model drugs from drug-loaded electrospun cellulose acetate fiber mats, *Polymer* 48 (2007) 5030–5041.

55. M. Qi, X. Li, Y. Yang, S. Zhou, Electrospun fibers of acid-labile biodegradable polymers containing ortho ester groups for controlled release of paracetamol, *Eur. J. Pharm. Biopharm.* 70 (2008) 445–452.

56. W. Cui, M. Qi, X. Li, S. Huang, S. Zhou, J. Weng, Electrospun fiber of acid-labile biodegradable polymers with acetal groups as potential drug carriers, *Int. J. Pharm.* 361 (2008) 47–55.

57. W. Cui, X. Li, X. Zhu, G. Yu, S. Zhou, J. Weng, Investigation of drug release and matrix degradation of electrospun poly(DL-lactide) fibers with paracetamol inoculation, *Biomacromolecules* 7 (2006) 1623–1629.

58. H. Peng, S. Zhou, T. Guo, Y. Li, X. Li, J. Wang, J. Weng, *In vitro* degradation and release profiles for electrospun polymeric fibers containing paracetamol, *Colloids Surf. B* 66 (2008) 206–212.

59. T. Ngawhirunpat, P. Opanasopit, T. Rojanarata, P. Akkaramongkolporn, U. Ruktanonchai, P. Supaphol, Development of meloxicam-loaded electrospun polyvinyl alcohol mats as a transdermal therapeutic agent, *Pharm. Dev. Technol.* 14 (2009) 70–79.

60. E.-R. Kenawy, G.L. Bowlin, K. Mansfield, H. Layman, D.G. Simpson, E.H. Sanders, G.E. Wnek, Release of tetracycline hydrochloride from electrospun poly(ethylene-co-vinylacetate), poly(lactic acid), and blend, *J. Control. Release* 81 (2002) 57–64.

61. X. Zong, K. Kim, D. Fang, S. Ran, B.S. Hsiao, B. Chu, Structure and process relationship of electrospun bioabsorbable nanofiber membranes, *Polymer* 43 (2002) 4403–4412.

62. R.A. Thakur, C.A. Florek, J. Kohn, B.B. Michniak, Electrospun nanofibrous polymeric scaffold with targeted drug release profiles for potential application as wound dressing, *Int. J. Pharm.* 364 (2008) 87–93.

63. G. Verreck, I. Chun, J. Rosemblatt, J. Peeters, A.V. Dijck, J. Mensch, M. Noppe, M.E. Brewster, Incorporation of drugs in an amorphous state into electrospun nanofibers composed of a water-insoluble, nonbiodegradable polymer, *J. Control. Release* 92 (2003) 349–360.

64. P. Taepaiboon, U. Rungsardthong, P. Supaphol, Vitamin-loaded electrospun cellulose acetate nanofiber mats as transdermal and dermal therapeutic agents of vitamin A acid and vitamin E, *Eur. J. Pharm. Biopharm.* 67 (2007) 387–397.

65. D. Puppi, A.M. Piras, N. Detta, D. Dinucci, F. Chiellini, Poly(lactic-co-glycolic acid) electrospun fibrous meshes for the controlled release of retinoic acid, *Acta Biomater.* 6 (2010) 1258–1268.

66. J. Zeng, A. Aigner, F. Czubayko, T. Kissel, J.H. Wendoff, A. Greiner, Poly(vinyl alcohol) nanofibers by electrospinning as a protein delivery system and the retardation of enzyme release by additional polymer coating, *Biomacromolecules* 6 (2005) 1484–1488.

67. S.Y. Chew, J. Wen, E.K.F. Yim, K.W. Leong, Sustained release of proteins from electrospun biodegradable fibers, *Biomacromolecules* 6 (2005) 2017–2024.

68. E. Luong-Van, L. GrØndal, K.N. Chua, K.W. Leong, V. Nurcombe, S.M. Cool, Controlled release of heparin from poly(ε-caprolactone) electrospun fibers, *Biomaterials* 27 (2006) 2042–2050.

69. T.G. Kim, D.S. Lee, T.G. Park, Controlled protein release from electrospun biodegradable fiber mesh composed of poly(ε-caprolactone) and poly(ethylene oxide), *Int. J. Pharm.* 338 (2007) 276–283.

70. X. Huang, C.S. Brazel, On the importance and mechanisms of burst release in matrix-controlled drug delivery systems, *J. Control. Release* 73 (2001) 121–136.

71. H. Jiang, Y. Hu, Y. Li, P. Zhao, K. Zhu, W. Chen, A facile technique to prepare biodegradable coaxial electrospun nanofibers for controlled release of bioactive agents, *J. Control. Release* 108 (2005) 237–243.

72. S. Yan, L. Xiaoqiang, T. Lianjiang, H. Chen, M. Xiumei, Poly(L-lactide-co-ε-caprolactone) electrospun nanofibers for encapsulating and sustained release proteins, *Polymer* 50 (2009) 4212–4219.

73. Y.Z. Zhang, X. Wang, Y. Feng, J. Li, C.T. Lim, S. Ramakrishna, Coaxial electrospinning of (fluorescein isothiocyanate-conjugated bovine serum albumin)-encapsulated poly(ε-caprolactone) nanofibers for sustained release, *Biomacromolecules* 7 (2006) 1049–1057.

74. Y. Lu, H. Jiang, K. Tu, L. Wang, Mild immobilization of diverse macromolecular bioactive agents onto multifunctional fibrous membranes prepared by coaxial electrospinning, *Acta Biomater.* 5 (2009) 1562–1574.

75. H. Qi, P. Hu, J. Xu, A. Wang, Encapsulation of drug reservoirs in fibers by emulsion electrospinning: Morphology characterization and preliminary release assessment, *Biomacromolecules* 7 (2006) 2327–2330.

76. S. Maretschek, A. Greiner, T. Kissel, Electrospun biodegradable nanofiber nonwovens for controlled release of proteins, *J. Control. Release* 127 (2008) 180–187.

77. Y. Yang, X. Li, M. Qi, S. Zhou, J. Weng, Release pattern and structural integrity of lysozyme encapsulated in core-sheath structured poly(DL-lactide) ultrafine fibers prepared by emulsion electrospinning, *Eur. J. Pharm. Biopharm.* 69 (2008) 106–116.

78. S. Yan, L. Xiaoqiang, L. Shuiping, M. Xiumei, S. Ramakrishna, Controlled release of dual drugs from emulsion electrospun nanofibrous mats, *Colloids Surf. B* 73 (2009) 376–381.

79. D. Yang, Y. Li, J. Nie, Preparation of gelatin/PVA nanofibers and their potential application in controlled release of drugs, *Carbohydr. Polym.* 69 (2007) 538–543.

80. C. Morimoto, Y. Satoh, M. Hara, S. Inoue, T. Tsujita, H. Okuda, Anti-obese action of raspberry ketone, *Life Sci.* 77 (2005) 194–204.

81. O. Suwantong, P. Opanasopit, U. Ruktanonchai, P. Supaphol, Electrospun cellulose acetate fiber mats containing curcumin and release characteristic of the herbal substance, *Polymer* 48 (2007) 7546–7557.

82. P. Sikareepaisan, A. Suksanrarn, P. Supaphol, Electrospun gelatin fiber mats containing a herbal—*Centella asiatica*—extract and release characteristic of asiaticoside, *Nanotechnology* 19 (2008) 015102.

83. O. Suwantong, U. Ruktanonchai, P. Supaphol, Electrospun cellulose acetate fiber mats containing asiaticoside or *Centella asiatica* crude extract and the release characteristics of asiaticoside, *Polymer* 49 (2008) 4239–4247.

84. P. Opanasopit, U. Ruktanonchai, O. Suwantong, S. Panomsuk, T. Ngawhirunpatm C. Sittisombut, T. Suksamran, P. Supaphol, Electrospun poly(vinyl alcohol) fiber mats as carriers for extracts from the fruit hull of mangosteen, *J. Cosmet. Sci*, 59 (2008) 233–242.

85. W. Weecharangsan, P. Opanasopit, M. Sukma, T. Ngawhirunpat, U. Sotanaphun, P. Siripong, Antioxidative and neuroprotective activities of extracts from the fruit hull of mangosteen (*Garcinia mangostana* Linn.), *Med. Prine. Pract.* 15 (2006) 281–287.

86. H.A. Jung, B.N. Su, W.J. Keller, R.G. Mehta, A.D. Kinghorn, Antioxidant xanthones from pericap of *Garcinia mangostana* (mangosteen), *J. Agric. Food. Chem.* 54 (2006) 2077–2082.

87. K. Nakatani, M. Atsumi, T. Arakawa, K. Oosawa, S. Shimura, N. Nakahata, Y. Ohizumi, Inhibitions of histamine release and prostaglandin E_2 synthesis by mangosteen, a Thai medicinal plant, *Biol. Pharm. Bull.* 25 (2002) 1137–1141.

88. N. Rangkadilok, S. Sitthimonchai, L. Worasuttayangkurn, C. Mahidol, M. Ruchirawat, J. Satayavivad, Evaluation of free radical scavenging and antityrosinase activities of standardized longan fruit extract, *Food Chem. Toxicol.* 45 (2007) 328–336.

89. J.H. Kim, N.J. Kang, B.K. Lee, K.W. Lee, H.J. Lee, Gallic acid, a metabolite of the antioxidant propyl gallate, inhibits gap junctional intercellular communication via phosphorylation of connexin 43 and extracellular-signal-regulated kinase 1/2 in rat liver epithelial cells, *Mutat. Res.* 638 (2008) 175–183.

90. P. Chuysinuan, N. Chimnoi, S. Techasakul, P. Supaphol, Gallic acid-loaded electrospun poly(L-lactic acid) fiber mats and their release characteristic, *Macromol. Chem. Phys.* 210 (2009) 814–822.

91. J. Han, T.-X. Chen, C.J. Branford-White, L.-M. Zhu, Electrospun shikonin-loaded PCL/PTMC composite fiber mats with potential biomedical applications, *Int. J. Pharm.* 382 (2009) 215–221.

92. B. Singh, M.K. Sharma, P.R. Meghwa, P.M. Sahu, S. Singh, Anti-inflammatory activity of shikonin derivatives from *Arnebia hispidissima*, *Phytomedicine* 10 (2003) 375–380.

93. J. Han, X. Weng, K. Bi, Antioxidants from a Chinese medicinal herb—*Lithospermum erythrorhizon*, *Food Chem.* 106 (2008) 2–10.

94. C.C. Shen, W.J. Syu, S.Y. Li, C.H. Lin, G.H. Lee, C.M. Sun, Antimicrobial activities of naphthazarins from *Arnebia euchroma*, *J. Nat. Prod.* 65 (2002) 1857–1862.

95. a. W.F. Anderson, Gene therapy—The best of times, the worst of times, *Science* 288 (2000) 627.
    b. W.F. Anderson, Human gene therapy, *Science* 256 (1992) 808.

96. V. Incani, E. Tunis, B.A. Clements, C. Olson, C. Kucharski, A. Lavasanifar, H. Uludağ, Palmitic acid substitution on cationic polymers for effective delivery of plasmid DNA to bone marrow stromal cells, *J. Biomed. Mater. Res.* 81A (2007) 493–504.

97. G.Y. Wu, P.L. Zhan, L.L. Sze, A.R. Rosenberg, C.H. Wu, Incorporation of adenovirus into a ligand-based DNA carrier system results in retention of original receptor specificity and enhances targeted gene expression, *J. Biol. Chem.* 269 (1994) 11542–11546.

98. Y.K Luu, K. Kim, B.S. Hsiao, B. Chu, M. Hadjiargyrou, Development of a nanostructured DNA delivery scaffold via electrospinning of PLGA and PLA-PEG block copolymers, *J. Control. Release* 89 (2003) 341–353.

99. K.A. Partridge, R.O.C. Oreffo, Gene delivery in bone tissue engineering: Progress and prospects using viral and nonviral strategies, *Tissue Eng.* 10 (2004) 295–307.

100. G.R. Whittaker, A. Helenius, Nuclear import and export of viruses and viral genomes, *Virology* 246 (1998) 1–23.

102. H. Jiang, Y. Hu, Y. Li, P. Zhao, K. Zhu, W. Chen, A facile technique to prepare biodegradable coaxial electrospun nanofibers for controlled release of bioactive agents, *J. Biomat. Sci. Polymer Ed.* 108 (2005) 237–243.

102. I.C. Liao, S. Chen, J.B. Liu, K.W. Leong, Sustained viral gene delivery through core-shell fibers, *J. Control. Release* 139 (2009) 48–55.

103. K.B. Kohn, M. Sadelaim, J.C. Glorioso, Occurrence of leukaemia following gene therapy of X-linked scid, *Nat. Rev. Cancer* 3 (2003) 883.

104. M.A. Kay, J.C. Glorioso, L. Naldini, Viral vectors for gene therapy: The art of turning infectious agents into vehicles of therapeutics, *Nat. Med.* 7 (2003) 33–40.

105. A. Neamnark, O. Suwantong, R. Bahadur, K.C., C.Y.M. Hsu, P. Supaphol, and H. Uludağ, Aliphatic lipid substitution on 2 kDa-polyethylenimine improves plasmid delivery and transgene expression, *Pharmaceutics* 6 (2009) 1798–1815.

106. C. Kneuer, M. Sameti, U. Bakowsky, T. Schiestel, H. Schirra, H. Schmidt, C.M. Lehr, A nonviral DNA delivery system based on surface modified silica-nanoparticles can efficiently transfect cells *in vitro*, *Bioconj. Chem.* 11 (2000) 926–932.

107. H.L. Fu, S.X. Cheng, X.Z. Zhang, R.X. Zhou, Dendrimer/DNA complexes encapsulated functional biodegradable polymer for substrate-mediated gene delivery, *J. Gene Med.* 10 (2008) 1334–1342.

109. H. Nie, C.H. Wang, Fabrication and characterization of PLGA/Hap composite scaffolds for delivery of BMP-2 plasmid DNA, *J. Control. Release* 120 (2007) 111–121.

109. H. Nie, M.L. Ho, C.K. Wang, C.H. Wang, Y.C. Fu, BMP-2 plasmid loaded PLGA/Hap composite scaffolds for treatment of bone defects in nude mice, *Biomaterials* 30 (2009) 892–901.

24

Noble Metal Nanoparticles in Drug Delivery

Robert T. Tshikhudo and Ndabenhle M. Sosibo

Nanotechnology Innovation Centre, Mintek, Randberg, South Africa

CONTENTS

24.1 Metal Nanoparticles in Drug Discovery

There is a general need for improvements in current drug discovery methodologies motivated by less than ideal screening and validation efficiencies of targets.[1] Potential useful targets have been unearthed via proteomic and genomic techniques with microarray technology, allowing for the extraction of further information from smaller sample volumes. However, problems associated with the biological and chemical aspects of targets selected have been shown to be a constant hurdle during this process. These challenges include poorly validated targets failing at various stages or failures in lead assessments. High-throughput screening (HTS) technologies have thus far been used in candidate selection of leads and address the quantitative aspect of this process. Further improvements are required in order for quality to be ensured in minimal time and high efficiency.[2] This would ensure the decrease in time to market of possible drug candidates, as secondary screening tends to require more time in optimizing via synthetic chemistry techniques and preclinical evaluation in animal models. The application of nanotechnology in this area could introduce a component of miniaturization, and thus improve the ability to fabricate massive microarrays in small spaces using microfludic technology.

24.2 The Nano-Bio Interface

The size comparability of the nanoparticles (1–100 nm) with those of most biomolecules and subcellular components makes it logical for the interfacing of nanotechnology and molecular biology. Nanoparticles have received a lot of attention recently in the development of imaging, targeting, diagnostic, and drug delivery and discovery in both in vivo and in vitro systems.[3,4] Numerous formulations of these entities have raced into the forefront, including dendrimers, nanocages, micelles, molecular conjugates, liposomes, quantum dots, and nanoshells.[5] The design of stealth, multiplexed drug delivery and imaging systems has started to benefit immensely from the unique versatility of nanoparticles in terms of functionalities and impressive optical properties.[6] In vitro and in vivo labeling using quantum dots has also been reported.[7] Intracellular delivery of cargoes into the nucleus of the cells can be manipulated by conforming the nanoparticle systems to a few guidelines: (1) the nanoparticles must enter the cell via receptor-mediated endocytocytosis (RME); (2) they must evade the endosomal-lysosomal pathways; (3) they must possess a nuclear localization signal (NLS) for successful internalization; and (4) they must be small enough to cross the nuclear membrane.[8] Attention has also been given to the toxicity effects of these entities inside a living organism. For slightly larger particles (≥200 nm), clearance has been reported through opsonization, with the large size assisting in recognition of the nanoparticles by macrophages.[5] The incorporation of surface functionalities such as polyethylene glycol (PEG) polymer renders these nanoparticles highly biocompatible and lowers their cytotoxicity.

24.2.1 Targeted Drug Delivery Systems

Concerted efforts have of late been focused on developing novel drug carriers that can selectively be used to target, label, and destroy clinical conditions in a reproducible manner. For instance, cancer targeting would be such that delivery of therapeutics targets the tumors, killing only the infected cells and sparing the normal cells. These strategies can therefore be employed in new and existing drugs, thereby improving their therapeutic index. Three common requirements have been identified for a successful carrier-based delivery strategy: (1) prolonged blood circulation, (2) sufficient tumor accumulation, and (3) controlled drug release and uptake by the tumor cells. The release profile of these hybrid delivery systems must also be designed such that it matches the pharmacodynamics of the drug.[9] Nanoparticulate carriers, such as those of noble metals, have been demonstrated as a viable alternative for the delivery of targeted therapeutics. The surface modifications of nanocarriers allow for controlled biological properties of these entities for the timely execution of therapeutic and diagnostic functions. Furthermore, temporally and spatially dependent properties, such as the half-life of the carriers, biodistribution, local physiological stimuli responsiveness, and the ability to serve as imaging and contrast agents for various imaging modalities (magnetic resonance imaging (MRI), computed tomography, etc.), can all be engineered through surface functionality carriage of the nanocarriers.[10] Despite milestones in the therapeutic medical sector, there are still outstanding issues. For instance, issues related to the delivery drugs not being stable in the environment of applications have been shown as a major drawback. As an example, the oral, noninvasive delivery of drugs fails to deliver enough quantities due to acids in the stomach and also the resistance of the intestines, which alter and reduce the absorption of drugs. Also, the circulation of the drugs in the system for enough time is critical, requiring further strategies in drug formulations. Another issue has been the development of drugs that target only specific areas in the body. This would avoid the toxicity of drugs that occurs as a result of their accumulation in healthy tissue. In general, there is still not enough information regarding the interactions of drugs with their targets, and more research is required in this area. Studies of the fundamental interactions with cells, tissue, and in vivo models are still required.

24.2.2 Nanoparticle-Based Targeting Strategies

Poor pharmacokinetics and biopharmaceutical properties of the majority of potential therapeutics have shown the need for the development of delivery, biolabeling, and diagnostic systems with improved efficacy and efficiency.[11] The noninvasive manner through which preliminary findings have indicated nanoparticles interact with biomolecules on the surface of cells and inside, leaving all biochemical behavior intact, has strengthened the case for a nanoparticle-based therapeutic "toolbox."[12] A careful understanding of the processes involved in the interactions of nanoparticles with cells, leading to their uptake (or rejection) and their subsequent release, is of crucial importance. Nanoparticles offer a rare opportunity to develop multifunctional therapeutics, integrating diagnostics and therapeutic functions in one toolbox, simultaneously labeling, monitoring, and delivering drug materials, controlling both spatial and temporal release and monitoring.[13]

24.2.2.1 Mechanism of Uptake

Normally the uptake of particulate matter into cells is by either phagocytosis or endocytosis.[6] Phagocytosis is usually involved in the uptake of particles larger than 500 nm, a pathway observed in macrophages, dendritic cells, monosites, and basophils. Most inorganic nanoparticles are synthesized to within the 1–100 nm range, and there they are sought to be endocytosed into the cell, since the size remains at less than 200 nm regardless of surface modifications. Endocytosis can be either receptor mediated or nonreceptor mediated, which either enhances or retards the uptake.[14,15] The uptake can be viewed as a result of the contest between the thermodynamic driving force for wrapping, which refers to the amount of free energy needed to drive the nanoparticles into the cell, and the receptor diffusion kinetics, which refers to the kinetics of recruitment of receptors to the binding cell.[16] Targeting of cellular diseases such as cancer in cells by nanoparticle-based systems is subdivided into two pathways: (1) passive and (2) active targeting. Passive targeting employs pharmacokinetic manipulation and nanoparticle size reduction, whereas active targeting involves specifically targeting areas on a disease cell by surface modifications of the nanoparticles. This targeting seeks to address the current shortfalls, such as those observed with conventional drugs that show less than desired accumulations on the target or disease area.

24.2.2.1.1 Passive Targeting

This approach occurs via the extravasation of the nanoparticles at the disease site, where the fast-growing tumor microvasculature is leaky with defective architecture, which is highly permeable to macromolecules relative to normal tissue.[17] The dysfunctional lymphatic drainage system of the disease cell allows for the enhanced permeation and retention (EPR) effect,[18] as a result of which nanoparticles accumulate at the tumor site. For this mechanism to take place, the size and surface of the nanoparticles should be engineered such that they circumvent uptake by the reticuloendothelial system (RES).[19] One way of achieving passive targeting is the use of direct local delivery of the agents into the affected cell. This method is advantageous since it preserves the drug from systemic circulation, but is also very invasive since it involves injections and surgical procedures. The size of open interendothelial gap junctions and transendothelial channels determines the extent of the nanoparticle extravasation, with a cutoff of between 400 and 600 nm reported elsewhere.[20] One of the biggest challenges for passive targeting is the inability to accumulate large enough amounts of the nanoparticles on the tumor cells, resulting in low therapeutic efficacy and drawing unwanted systemic adverse effects.[21] Extended circulation time is also required for passive targeting to allow the drug-loaded nanoparticles to pass the disease site repeatedly to deposit effectively.

24.2.2.1.2 Active Targeting

Localized diseases such as cancer and inflammation overexpress some epitopes or receptors that can be used as specific targets.[17] The use of nanoparticles involves the conjugation of the targeting molecules on the surface that recognizes and attaches to specific receptors that are unique to a particular disease cell. This targeting approach is based on interactions such as lectin-carbohydrate, ligand-receptor, and antibody-antigen. Active targeting, which is ligand dependent, is particularly invaluable to therapeutics that are not taken up easily by cells. It has also been demonstrated that active targeting improves the distribution of nanomedicines within the tumor interstitium.[22] A factor to be considered when selecting the targeting ligand is its immunogenicity. If, for example, antibodies expose their constant regions on the liposomal surface, they become more susceptible to Fc-receptor-mediated phagocytosis by the mononuclear phagocytic system (MPS).[23] This method is particularly useful for primary tumors that have not yet metastasized. One of the major issues in chemotherapy is multiple drug resistance (MDR), which often starts from the overexpression of the plasma membrane P-glycoprotein (Pgp) in cancer cells.[24] Pgp acts as an efflux pump to extrude positively charged xenobiotics, which include many anticancer drugs, out of the cells. Glycoproteins are known to be unable to remove polymer-drug conjugates that enter the cell via endocytosis, and this renders the active targeting mechanism an alternative route to overcoming MDR.[25]

24.3 The Importance of Noble Metal Nanoparticles or Monolayer-Protected Clusters (MPCs)

24.3.1 Essential Properties

It appears that gold and silver nanoparticles could probably be the most widely studied materials in nanotechnology to date, due to their stability, ease of preparation, and unique optical properties.[26–33] In fact, the unusual optical properties of gold and silver nanoparticles that are dependent on size, shape, etc., have long been exploited, for instance, in decorative pigments during the Roman period. They exhibit a strong UV-vis absorption band not present in the spectrum of the bulk material, near IR. The origin of the light absorption by these noble metal nanoparticles is the collective oscillation of the conduction band electrons induced by the interacting electromagnetic field. The absorption band arises when the incident photon frequency is resonant with the collective oscillation of the conduction electrons and is commonly known as surface plasmon resonance (SPR). In addition to the transverse plasmon band for Au nanoparticles observed at regions around 520 nm, an additional, stronger band at longer wavelengths arising from longitudinal plasmon oscillation is observed for gold nanorods.[34] The longitudinal plasmon resonance can be shifted into the NIR region by increasing the nanorod aspect ratio.[35]

These optical properties, in particular those of gold colloids, have long been used as biomolecular labels in electron microscopy[36–38] because they are much more electron dense than organic molecules, and hence yield excellent contrast in comparison to cells and tissues. Exploitation of gold nanoparticles as optical markers in dark field microscopy of biological samples can be regarded as a more recent development, as this area has always been dominated by fluorescence dyes. These, however, have limitations, such as photobleaching and often a relatively high detection threshold, while light scattering by gold nanoparticles even has the potential for single-particle detection.[39] This opens the ability to track single molecules within a cell, and hence to monitor single molecular binding events in biological systems.[39] Owing to the limitations accompanying the use of fluorescent dyes, gold and silver nanoparticles, and in particular semiconductor quantum dots, have now gained popularity as alternative biolabeling probes.[40–44] Apart from their unique optical properties, gold and silver nanoparticles are also known to act as strong Raman enhancers,[45,46] and may also amplify fluorescence under certain circumstances.[47] An important requirement for their biological applications, apart from making them stable in aqueous environments, is the ability to attach biomolecules of interest.

24.3.2 Nanomaterials Preparation

The preparation of gold and silver nanoparticles has indeed become an area of considerable interest in both applied and fundamental research, largely due to their fascinating properties and attractive potential applications in electronics,[48] optics,[49–53] biomedical and bioanalytical areas, etc.[54–57] It is documented that ever since Faraday[58] published a report on the synthesis and properties of colloidal gold, a reasonable insight into the nature of gold sols (nucleation, growth, and kinetics of coagulation) has developed.[59] Considerable attention has been drawn during the last few decades to develop and optimize methods for the preparation of gold[60–63] and silver[64,65] nanoparticles. Ongoing research in this field has primarily been focused on developing stable monodisperse particles of controllable size and shape, as it is generally believed that current and novel potential applications of nanostructured materials are dependent on these properties.

The terms *colloids* and *clusters* are often used interchangeably in the literature. The criteria to distinguish clusters (particles described herein as ligand-stabilized or monolayer-protected clusters (MPCs)) from colloids (electrostatically stabilized; note that this can be confusing since others prefer "charge and ligand stabilized") are based on the notion that clusters are isolable, reproducible molecules with a precise composition and structure, while colloids are much less defined and typically have a distribution of size (Figure 24.1).[66] Basically, the preparation of ligand-stabilized nanoclusters dates

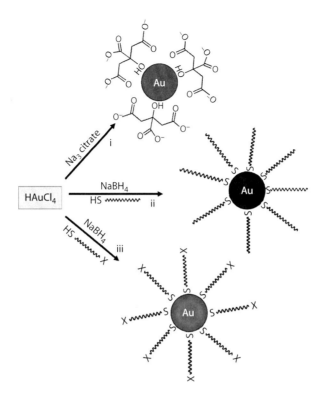

FIGURE 24.1
Synthesis of gold nanoparticles by the reduction of the gold(III) salt by (i) citrate reduction/capping or colloids, (ii) the Brust biphasic method, and (iii) the direct reduction route in the presence of a bi-/multifunctional
capping ligand. Both i and ii are regarded as ligand-stabilized or monolayer-protected clusters in this case.
However, note that others would prefer i as ligand-stabilized particles. Silver nanoparticles are prepared in the
same way.

back to 1981 to the work that was carried out at the University of Essen in Germany by
Schmid et al. for the preparation and characterization of $Au_{55}(PPh_3)_{12}Cl_6$ nanoclusters[67]
(Figure 24.1). These findings have led to the development and exploitation of ligand-
stabilized metal nanoclusters in diverse areas, including catalysis,[68] electronics, etc. This
approach had some drawbacks and limitations, including lack of stability. However, it
was later realized that organic sulfur compounds form stable self-assembled monolayers on gold surfaces,[69] and interestingly, it was also discovered that thiols can be used to
stabilize gold colloids.[70]

In 1994, the first facile synthesis of very stable, isolable gold clusters was developed in
Liverpool (Figure 24.1).[71] The evolution of this work has led to the preparation of metal
nanoparticles of defined size and shape, and has been exploited in diverse applications.
This approach takes advantage of using tetraoctyl ammonium bromide as a phase transfer catalyst, to transfer aqueous chloroauric acid to the organic phase (toluene) to form
an intensely orange Au(III) complex (tetraoctylammonium tetrabromoaurate). Reducing
this complex in the presence of thiols ligands, mostly alkanethiols, by $NaBH_4$ allows the
formation of small clusters (1–5 nm), the diameter of which is influenced by a number
of factors, such as the thiol:gold ratio. It was found that the use of a larger thiol:gold
molar ratio produces very small clusters,[72] which are characterized by their unique dark

brown color and damped plasmon absorption band in the UV-vis spectrum. This plasmon absorption band even disappears completely when very small clusters are formed. Fast addition of the reducing agent and cooling the reaction mixture again lead to the formation of small clusters, with improved monodispersity.[72,73] In order to isolate a high yield of very small clusters (~2 nm), the reaction must be carefully terminated immediately after reduction.[74,75]

Templeton et al. demonstrated the usefulness of this method by carrying out extensive investigations, particularly to show that this method tolerates various modifications, to include straight-chain alkanethiols, glutathione, tiopronin, thiolated PEG, *p*-mercaptophenol, aromatic alkane thiol, phenyl alkane thiol, and mercaptopropyl trimethoxysilane as stabilizing agents.[76] Murray's group also characterized these materials to gain insight into their physical and chemical properties.

Upon understanding the three-dimensional (3D) self-assembly process of these materials, in contrast to the two-dimensional (2D) self assembled monolayers (SAMs) of ligand molecules adsorbed on the metal surface, Murray coined the name *monolayer-protected clusters* (MPCs) to distinguish them from 2D SAMs. Many laboratories today employ this original protocol for the size-selective formation of MPCs by varying the ratio between Au and HS-R, including the preparation of very small Au MPCs.[77] Of particular importance again has been the ability to control the size distribution of these MPCs, since it is known that their optical properties depend on size and shape. This has been achieved by using a number of techniques, such as solvent fractionation,[73] continuous free-flow electrophoresis,[73,78] size exclusion chromatography,[79] and high-performance liquid chromatography (HPLC).[80] Successive fractionation appears to be a reliable means of obtaining narrow size distributions by taking advantage of allowing bigger particles to aggregate first due to the stronger van der Waals interaction between the larger spheres.[81] The use of chromatographic techniques for the separation and determination of the composition of MPCs appears more attractive, since it is practically not possible, using TEM, to distinguish the size and composition of the ligand shell, but size exclusion chromatography does.[79] As shown in Figure 24.2, the assessment of ligand attachment in different proportions of mixed neutral and carboxylated (charged) ligands on gold nanoparticles was investigated using horizontal agarose gel electrophoresis. Increasing the carboxylated ligand from 1% (lane 1) to almost 100% (lane 4) carboxy demonstrates the different degrees of mobility on gel in line with the

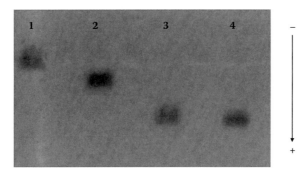

FIGURE 24.2
A typical example of agarose gel electrophoresis of 14 nm gold nanoparticles stabilized by a mixture of neutral and carboxylated ligands at different ratios (lanes: 1 = 1%, 2 = 10%, 3 = 50%, 4 = 100% carboxylated ligand). The migration of gold MMPCs on the gel increases relative to the amount of charge on their surface.

composition of mixed MPCs. This shows that gel electrophoresis is the most powerful tool to determine the degree of functionalization on monolayer-protected clusters or mixed monolayer-protected clusters.

The scope of MPC synthesis is not limited to the 1–5 nm region, since it was realized that the basic two-phase protocol in the absence of thiols leads to the formation of somewhat larger, stable nanoparticles. The stability of these particles is provided by the bromides and tetraoctyl ammonium ions, which adsorb to the surface of the particles. These tetraoctyl ammonium bromide (TOABr)-stabilized particles are in the 5–8 nm size region. Increasing the dielectric constant of the medium, e.g., by adding alcohols to the particles, leads to the formation of irreversible aggregates; however, they have been used as readily available starting materials for the construction of self-assembled mono- and multilayers on different substrates since the ionic capping can be easily removed by thiols and amine functionalities.[82]

Apart from the two-phase solution synthesis, single-phase methods have also been developed, to obtain MPCs of comparable sizes, including the use of $NaBH_4$ to reduce gold in an alcoholic solution-containing thiols.[83] However, this is only applicable when the thiols to be used are soluble in alcohols. In essence, particles with both hydrophobic and hydrophilic characteristics are obtainable, as was demonstrated using mercaptoundecanoic acid (MUA) as a stabilizing/capping agent.[84]

Again, the 5–8 nm particles are not limited to the two-phase approach; MPCs of about 6 nm can also be prepared by first preparing borohydride-stabilized particles, which can be directly converted to the desired MPCs. Even larger particles (10–40 nm) can be prepared following the classical Frens method[60] of using citrate as the reducing and stabilizing agent. By this widely used approach, gold and silver MPCs of different sizes and shapes can be prepared. The adsorption of citrate ions to the cluster surface is essential to electrostatically overcome the van der Waals attraction, and hence to preserve the stability of the metal nanoparticles. These particles can be easily converted to MPCs by a self-assembly process, as demonstrated by Weisbecker et al.[85]

Since the ultimate goal in preparing Au and Ag MPCs is to explore their unique properties[86–91] in biological applications, these materials require more than water solubility. Key issues, such as functionality, stability, and biocompatibility, have to be addressed in particular silver known for its propensity to oxidatively corrode[92,93] and aggregate in electrolytic solutions.

24.3.3 Attachment of Biomolecular Functionality on Nanoparticles

The rationale to attach biomolecules on nanoparticle surfaces is mainly to use nanoparticles as probes to investigate biological and intracellular processes. Many applications of nanoparticles in biology and medicine have been reviewed by Salata,[94] including (1) biodetection of pathogens, (2) detection of proteins, (3) drug and gene delivery, (4) tumor destruction via heating (hyperthermia), (5) probing DNA structure, and (6) fluorescent biological labels. All these applications require the attachment of a biomolecule of interest to the nanoparticle. This bioconjugation step must be performed under mild conditions to retain the biological activity of the resulting nanohybrid material. Many strategies have been attempted and utilized to date to attach biomolecules to nanoparticles. Reviews by Kumar[95] and Katz and Willner[96] outlined relevant bioconjugation strategies that can be categorized as electrostatic binding,[97] covalent coupling,[98–101] and specific recognition.[102–104] Some of these strategies are shown in Figure 24.3.

Adsorption of biomolecules to the nanoparticles by electrostatic interaction is illustrated in Figure 24.3A and B. Gold and silver hydrosols prepared by the citrate reduction method,

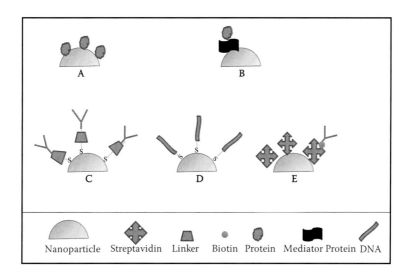

FIGURE 24.3
Strategies commonly used for biomolecular functionalization: (A) direct physisorption of biomolecules to nanoparticles, (B) assisted physisorption using prebound biomolecules, (C) bioconjugation via cross-linkers on the chemisorbed functional ligand on nanoparticles, (D) direct conjugation of thiolated biomolecules to nanoparticles, and (E) attaching biotinylated molecules to avidin physisorbed on nanoparticles. (From Kumar C.S.S.R., *Biofunctionalisation of nanomaterials* (Nanotechnologies for the Life Sciences 1), Wiley-VCH, Weinheim, 2005).

for instance, are negatively charged, allowing simple adsorption of positively charged proteins[97] (Figure 24.3A). This has been commonly used to prepare bioconjugates for immunolabeling since the 1960s.[105] Multilayer assemblies of proteins, as shown in Figure 24.3B, can be constructed by attaching a polyelectrolyte polymer to the protein-coated nanoparticles, which then allows the deposition of a second protein layer.[98,99] The major drawbacks of protein adsorption by electrostatic interaction are that (1) it is very difficult to control the number of biomolecules attached, (2) the particles lack stability, and (3) the activity of biomolecules generally decreases gradually with time.

Chemisorption of biomolecules to nanoparticles is shown in Figure 24.3C and D. In the simplest case, proteins can be chemisorbed to the nanoparticles through thiol or amine groups that can be engineered[106,107] or are already present in the protein.[108,109] Either way, the biological activity of the resulting bioconjugates usually decreases dramatically due to conformational change upon adsorption.[96,109] In addition, it has been shown that proteins are often denatured, leading to the loss of their biological activity. To avoid some of these problems, bifunctional linkers containing a thiol group to link to the particle and functional groups such as carboxylic acid to link to the protein have been used[101,110,111] (Figure 24.3C). Particles engineered in this way are stable; however, site-specific attachment of biomolecules by covalent means is generally a substantial synthetic challenge.[113] Nucleic acids have been widely used following the approach illustrated in Figure 24.3D. Since they can be synthesized terminated with alkane thiol groups, they have been used to directly stabilize gold nanoparticles.[113–115]

Figure 24.3E shows physisorption of biomolecules to nanoparticles and the immobilization of other biomolecules by the biotin-avidin system. Avidin and antibodies can be adsorbed to nanoparticles, which are then used to bind to biotinylated functionalities and antigens, respectively. This is an attractive and frequently used strategy; however, the particles often lack stability.[116,117]

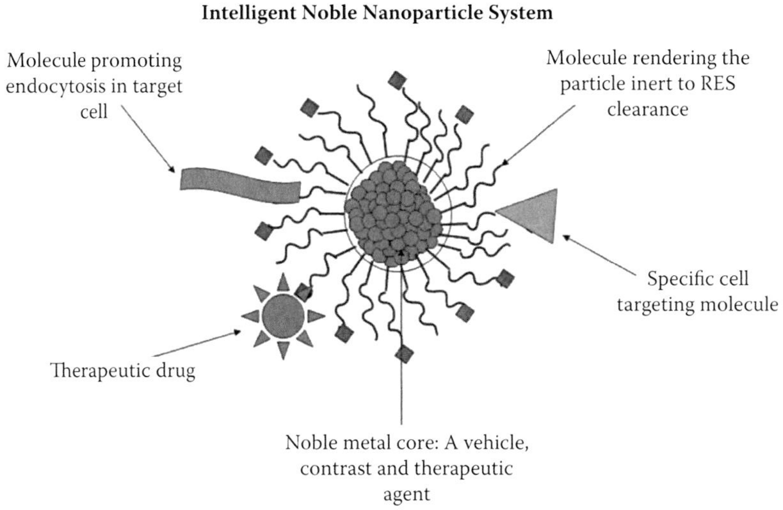

Intelligent Noble Nanoparticle System

FIGURE 24.4
A generic design of highly functional mixed monolayer protected clusters or tool box showing the incorporation of multiple functionalities on the surface of the metal nanoparticle core. Each of the functionalities in the
multiplexed system serves a specific function in biological applications and they could be attached or replaced
at various stages of synthesis.

24.3.4 Key Properties of the Ideal MPC Bioconjugate System

The primary objective in attaching a biomolecule to a nanoparticle is to develop a functional MPC bioconjugate system possessing the properties shown in Figure 24.4. Such an
"intelligent" nanoparticle system should have properties that can, in principle, be exploited
for biolabeling purposes, in either optical or electron microscopy. The idea is not to make
particles with only one specific biological recognition function, but instead to provide a
generic surface chemistry that will easily allow the incorporation of a practical variety
of desired recognition motifs. The desired biogold/silver system, as schematically shown
in Figure 24.4, is designed and developed to address many biological and pharmaceutical applications. The final system should also accommodate the following criteria: water
soluble, extremely stable in biological environments, functionalized, and hence versatile to
address a wide range of biological and related applications.

24.4 The Use of Noble MPCs in Therapeutics

24.4.1 Cancer Targeting Strategies Using Metal Nanoparticles

Strategies for the targeted therapy of malignancies can be achieved by the bioconjugation of biomolecules such as peptides, nucleic acid aptamers, carbohydrates, and small
molecules. These, in addition to the widely accepted antibody-based techniques, present
a multifunctional toolbox when coupled to noble metal nanoparticles in addressing active

delivery challenges. Nanoparticulate materials, including biodegradable polymeric nano-particles, dendrimers, nanoshells, spheres, and others, can all be actively conjugated to specific biomolecules for delivery. Grafting of amphiphilic polymers such as PEG improves the systemic circulation times and bioresistance properties. Factors such as the biological environment, target cell surface receptors, molecular mechanisms, and pathobiology of the disease should all be considered in the delivery hybrid materials. These factors translate to material dose and retention time, stability in the target environment (cell, tissue, or organ), disease progression (alterations in the drug target composition), and biocompatibility and efficient release kinetics. A cross-sectional preamble into the different targeting molecules (in addition to antibodies discussed above) is briefly discussed. Aptamers, which rival the commonly abundant use of antibodies in diagnostics and therapeutics, have also been explored in the diagnostic/detection strategies for cancer.[118] These molecules are DNA or RNA oligonucleotides that fold by intramolecular interactions into unique conformations with ligand-binding characteristics.[119] An A10 aptamer has been reported conjugated with nanoparticles for the specific targeting of prostate-specific membrane antigen (PSMA),[120] a transmembrane protein upregulated in many cancer types. In vivo evaluation for LNCap prostate cancer, expressing PMSA antigens, showed tumor size regression in a 109-day intratumor single-injection experiment. Other aptamer-nanoparticle conjugates have been reported for colometric assays for the detection of platelet-derived growth factor (PDGF) and its receptor using Au-NP-aptamer conjugates.[121] A strategy based on oligopeptides has been identified as a powerful route for targeting strategies. As an alternative to antibod-ies, the small-sized, low-immunogenic, and highly stable short peptides have garnered exceptional attention, with combinatorial libraries leading to the identification of short peptides with varied capabilities for binding targeted proteins, cells, or tissue with high specificity.[122] A cyclic peptide arginine-glycine-aspartic acid (RGD) that binds specific inte-grins on cancer cell surfaces is currently in phase II trials for the treatment of non-small-cell lung cancer and pancreatic cancer.[123] Another strategy, based on the overexpression of specific epitopes on cancerous cells, is the small vitamin folic acid (folate), which binds to surface folate receptors (FRs) with high affinity (KD = ~10^{-9} M). This allows for derivatiza-tion of different delivery nanoparticulate materials with folic acid for delivery into cancer-ous cells without harming healthy populations, via FR-mediated endocytosis.[124] Due to the expression of the folate receptors in other, healthy epithelial cell types, such as in the choroids plexus, placenta, lung, intestine, and kidney, FR-mediated nanoparticle delivery techniques are still subject to refinement for increased selectivity. Hybrid combinations of these different targeting platforms with noble metal particles of nominal sizes present a viable and versatile toolbox for the targeted delivery of cargoes into cancerous cells, tis-sues, or organs.

24.4.2 Photothermal Cancer Therapy

The mainstream approaches in the treatment of cancer, such as surgery, chemotherapy, and radiotherapy, present some persistent drawbacks in their line of operation. Surgery can only be effective for a physically accessible tumor, chemotherapy has numerous side effects, and radiotherapy is highly invasive to healthy cells in the line of radiation. Therefore, there is a need for the development of alternative methods in the optical heating arena by depos-iting photoabsorbers within cancer cells and applying optical heating for tumor ablation.[125] Gold nanoparticles are known as fast-rate converters of light energy into heat.[126] With the gold nanoparticles possessing much higher absorption cross sections than conventional dyes (indocyanine green), they can be employed for selective photothermal treatment of cancer. Specific cell targeting functions can be embedded by the incorporation of targeting molecules, and the SPR absorption and scattering of the nanoparticles can be exploited for a dual-function imaging/therapy approach. El-Sayed et al. demonstrated this effect by employing high-scattering gold nanoparticles (40 nm) for the photothermal destruction of oral squamous carcinoma cell lines, by incorporating anti-EGFR antibody on the nanoparticles and laser-irradiating (continuous-wave (CW)) labeled cells.[127] Trypan blue cell assays showed that the cancerous cells were compromised, whereas the healthy cells were intact. Laser energy thresholds for total cell annihilation are also reached at much lower strengths for diseased cells than for healthy normal cells.[128] The need for deeper-tissue in vivo photothermal strategies requires light energy at the near-infrared (NIR) region, compared with visible light for in vitro and skin and surface cancer applications. This region presents the highest transmittivity for hemoglobin and water (prominent tissue components), with NIR light penetration possible for up to a few centimeters depth.[129] The use of gold nanoparticles for NIR imaging and photothermal therapy of cancer can be achieved by changing the optical properties of the Au nanoparticles by changing their shape.[130]

24.4.3 Antiangiogenic Gold Nanoparticles

Angiogenesis as a regulatory process for structural and functional integrity of organisms plays an important role in the maintenance of function and structure. There are two types of angiogenesis. The first is the normal angiogensis that is responsible for positive developments, such as the growth of a fetus to an adult, and thus the maintenance of the organism. The second type, abnormal angiogenesis, is known to be the driving force behind the formation and proliferation of many melanomas and cancers, which employ the angiogenesis process to advance their supply of blood, leading to tumor growth and proliferation. Gold nanoparticles have been shown to inhibit angiogenesis by inhibiting the VEGF165-induced proliferation in HUVEC cells.[131] On top of this, gold nanoparticles were shown to be nontoxic compared with untreated cells. Various cell lines also showed similar profiles, with the presence of the heparin-binding domain on the VEGF165 shown as a requisite factor in the inhibition by gold nanoparticles. This observation allows for the application of gold nanoparticles as possible therapies for many cancer types and carcinomas.

24.5 Fate of Nanoparticles in Live Systems

The use of modified nanoparticles for drug delivery systems maximizes the localization of the delivery agents and their subsequent accumulation only at the target site.[132] The EPR effect (for passive targeting) and surface derivatization (for active targeting) are important in the specificity of the delivery systems.[21] Nanoparticle-based delivery systems are a good method for the circumvention of adverse effects due to the migration of the nanoparticles to other sensitive parts of the body.[133] The exact mechanism and pathways of how the body rids itself of nanoparticles are still the subject of intense investigation, with the liver most likely the destination of the spent nanoparticles, with their design such that they pass freely in other organs but can be retained in the liver and spleen.[5]

24.6 Toxicity Issues

Definite metabolic and immunogenic responses to induction abilities by nanomaterials are currently poorly understood. This is in contrast to common chemical compounds (drugs and therapeutics) that are routinely subjected to well-established toxicological testing prior to public release, a scenario missing for nanomaterials.[134] In light of these shortfalls and the sporadic conflicting reports of similar nanomaterials reported as both toxic[135] and nontoxic,[136] concerted efforts have therefore been dedicated to the elucidation of nanomaterials toxicity. The toxicity of nanomaterials has been approached in a three-pronged strategy:[137]

1. Composition of the nanoparticles: The nanomaterials can be composed of toxic materials such as CdSe, which upon corrosion inside an organism can release toxic ions that will eventually poison the cell.[138] Compared to the corresponding bulk materials, partial decomposition and release of ions is highly likely for nanosized CdSe due to the large surface-to-volume ratio.

2. Chemical properties of the nanoparticles: Nanoparticles have been shown to adhere to cell membranes[139] and also be ingested by cells.[140] The breaching of the cell membrane and intracellular storage may have a negative effect on cells, regardless of the toxicity of the particles and their subsequent functionality.

3. Nanoparticle morphology: Distinctly shaped materials such as carbon nanotubes can rip cells like needles.[141] This in turn suggests that nanomaterials of the same composition would have different biologic responses for different morphologies.

The use of conventional in vivo techniques in the study of biologic effects and cytotoxicity caused by nanomaterials is highly informative but also inherently costly and labor-intensive.[142] This has rendered in vivo techniques ill-suited for systematic and routine biologic investigations in light of the sheer number of diverse nanomaterials currently in development. Various mammalian cell culture models have long been considered as a simpler, more cost-effective, and convenient alternative in the assessment of nanomaterials toxicity, compared to the use of live animal models.[143] Advantages of the in vitro approach include simplicity, consistency of the experimental setup, and reproducibility of the experimental results.[144] The in vitro approach encompasses the testing of viability or fluctuations in a designated inherent biological pathway selected against the nanomaterials of interest. Such testing is also motivated by the ever-increasing scope of applications of engineered nanomaterials requiring the design and development of effective and reliable in vitro assays.[145,146] Common assays for toxicity studies use measurements of triozonium salt cellular metabolic activities such as 3-(4,4-dimethylthiozol-2-yl)-2,5 diphenyl tetrazonium bromide test (MTT), cellular plasma membranes such as lactate-dehydrogenase (LDH) release assay, or the activities of other housekeeping enzymes.[147] In vitro assays are robust in nature and have been used extensively in the elucidation of mechanisms of toxicity, as well as for demonstrating the relevant biological processes that are involved in the response to specific test materials. In vitro measurements of apoptosis as a form of toxicity elucidation have been reported in the literature.[148] The approach was based on the measurements of DNA fragmentation occurring in apoptotic cells. Alternatively, the distinct morphological changes accompanying the process of apoptosis can be used to measure apoptosis as a function of the light scattering properties of the cells (flow cytometry).

Information such as changes in cell size and granularity can be elucidated in this way. Due to the overall loss of size for apoptotic cells through water loss and nuclear fragmentation, apoptotic cells give lower forward scatter and higher side scatter values than viable and necrotic cells, indicating the smaller size as well as different nucleus/cytoplasm consistency.[149] The neutral red assay procedure is another method for the identification of surviving/viable cells after treatment with the materials of choice. The assay is based on the ability of viable cells to incorporate and bind neutral red (3-amino-*m*-dimethylamino-2-methyl-phenazine hydrochloride), a weakly cationic supravital dye that penetrates the cell membrane through a nonionic diffusion process and then subsequently accumulates in the lysosomes.[150] Changes in the cell surface and lysosomal membrane lead to decreased neutral red uptake and binding, thus making it possible to distinguish between damaged and viable cells through spectrophotometric measurements.[151] The quantification of dye extracted from the treated cells has been shown to be linear with cell numbers, with cytotoxicity expressed as a concentration-dependent reduction of the neutral red uptake, giving a robust, sensitive signal for both cell integrity and growth inhibition.[152] Staining techniques have also been shown to effectively quantify the DNA loss as a function of apoptosis, for instance, the use of intercalating DNA dyes such as propidium iodide (PI) and 7-aminoactinomycin D (7-AAD).[153] Hypotonic solutions of PI dye have been shown to be among the easiest and most rapid methods of DNA staining for apoptosis measurements, strengthening their use in large-scale in vitro investigations.[154]

24.7 Summary and Outlook

Noble metal nanoparticles or mixed monolayer-protected clusters (MMPCs) are attractive candidates for use in the development of nanoparticle-based systems for diagnostic and therapeutic purposes due to their ease of fabrication and native biocompatibility. However, a lot still needs to be done in order to address the health, safety, and environmental issues of nanomaterials in general. If the development of future advanced MMPCs could focus on ensuring that these systems are more stable, robust, biocompatible, and versatile enough to be used in vivo, it is clear that the future therapeutic platform would rely on noble metal nanoparticles. It must be emphasized that the benefits of nanotechnology have been overstated in some respects, and thus could lead to overreliance and appearing as a quick fix to some of the long-standing scientific challenges. Again, if this is true, and lives could be improved (and thus extended considerably) through nanotechnology, then other socioeconomic factors need to be factored in.

References

1. Jain K.K., *Drug Discov. Rev.*, 2005, 10 (21), 1435.
2. Brown D. and Supertia-Furga G., *Drug Discov. Today*, 2003, 8, 1067.
3. Dobrovolskaa M.A. and McNeil S.E., *Nature Nanotechnol.*, 2007, 2, 469.
4. Azzazy H.M.E., Mansour M.M.H., and Kazmierczak S.C., *Clin. Chem.*, 2006, 52(7), 1238.
5. Morghimi S.M., Hunter A.C., and Murray C.J., *FASEB J.*, 2005, 19, 311.
6. Pei R., Yang X., and Wang E., *Analyst*, 2001, 126, 4.

7. Parak W.J., Pellegrino T., and Plank C., *Nanotechnology*, 2005, 16, R9.

8. Tkachenko A.G., Xie H., Coleman D., Glomm W., Ryan J., Anderson M.F., Franzen S., and Feldheim D.L., *J. Am. Chem. Soc.*, 2003, 125, 4700.

9. Andresen T.L., Jensen S.S., and Jorgenses K., *Progr. Lip. Res.*, 2005, 44, 68.

10. Torchllin V.P., *Adv. Drug Deliver. Rev.*, 2006, 58, 1532.

11. Braydon D.J., *Drug Discov. Today*, 2003, 8, 876.

12. Yang H. and Xia Y., *Adv. Mater.*, 2007, 19, 3085.

13. Kawasaki E.S. and Player A., *Nanomed. Nanotechnol. Biol. Med.*, 2005, 1, 101.

14. Derfus A.M., Chan W.C., and Bhatia S.N., *Nano Lett.*, 2004, 4, 11.

15. Kirchner C., Liedl T., Kudera S., Pellegrino T., Javier A.M., Gaub H.E., Stolzle S., Fertig N., and Parak W.J., *Nano Lett.*, 2005, 5, 331.

16 Bao G. and Bao X.R., *Proc. Natl. Acad. Sci. USA*, 2005, 9, 997.

17. Koo O.M., Rubinstein I., and Onyuksel H., *Nanomed. Nanotechnol. Biol. Med.*, 2005, 1, 193.

18. Matsumura Y. and Maeda H., *Cancer Res.*, 1986, 46, 6387.

19. Gref R., Minamitake Y., Peracchia M.T., Trubetskoy V., Torchilin V., and Langer R., *Science*, 1994, 263, 1600.

20. Yuah F., *Cancer Res.*, 1995, 55, 3752.

21. Ferrari M., *Nature Rev. Cancer*, 2005, 5, 161.

22. Drummond D.C., Meyer O., Hong K., Kirpotin D.B., and Papahadjopoulos D., *Pharm. Rev.*, 1991, 51, 691.

23. Metselaar J.M., Mastrobattista E., and Storm G., *Mini Rev. Med. Chem.*, 2002, 2, 319.

24. Links M. and Brown R., *Expert Rev. Mol. Med.*, 1999, 1, 1.

25. Bennis S., Chapey C., Robert J., and Couvreur P., *Eur. J. Cancer*, 1994, 30(1), 89.

26. Haes A.M., Stuart D.A., Nie S., and Duyne P., *J. Fluorescence*, 2004, 355.

27. Mulvaney P., *Langmuir*, 1996, 12, 788.

28. Pinchuk A., von Plessen G., and Kreibig U., *J. Phys. D Appl. Phys.*, 2004, 37, 3133.

29. Pinchuk A., Kreibig U., and Hilger A., *Surf. Sci.*, 2004, 557, 269.

30. Pinchuk A. and Kreibig U., *New J. Phys.*, 2005, 5, 151.

31. Liz-Marzan L.M. and Mulvaney P., *J. Phys. Chem. B*, 2003, 107, 7312.

32. Hilger A., Cuppers N., Tenfelde M., and Kreibig U., *Eur. Phys. J. D*, 2000, 115, 10.

33. Kreibig U., Bour D., Hilger A., and Gartz M., *Phys. Stat. Sol.*, 1999, 175, 351.

34. Weisseleder R., *Nat. Biotechnol.*, 2001, 19, 316.

35. Link S. and El-Sayed M.A., *J. Phys. Chem. B*, 1999, 103, 8410.

36. Horisberger M. and Rosset J., *J. Histochem. Cytochem.*, 1997, 25, 295.

37. Takizawa T., *J. Histochem. Cytochem.*, 1999, 47, 569.

38. Thoolen B., *J. Histochem. Cytochem.*, 1990, 38, 267.

39. Raschke G., Kowarik S., Franzl T., Sonnichsen C., Klar T.A., and Fieldmann J., *Nano Lett.*, 2003, 3, 935.

40. Hayat M.A., *Colloidal gold: Principles, methods and applications*, Academic Press, New York, 1989.

41. Pellengrino T., Kudera S., Liedl T., Javier A.M., Manna L., and Parak W.J., *Small*, 2005, 1, 48.

42. Rosi N.L. and Mirkin C.A., *Chem. Rev.*, 2005, 105, 1547.

43. Alivisatos P., *Nat. Biotechnol.*, 2004, 22(1), 47.

44. Tkachenko A.G., Xie H., Coleman D., Glomm W., Ryan J., Anderson M.F., Franzen S., and Fieldheim D.L., *J. Am. Chem. Soc.*, 2003, 125, 4700.

45. Kneipp K., Kneipp H., and Kneipp J., *Acc. Chem. Res.*, 2006, ACS ASAP article.

46. Cao Y.C., Jin R., Nam J., Thaxton S., and Mirkin A., *J. Am. Chem. Soc.*, 2003, 125, 14676.

47. Schultz S., Smith D.R., Mock J.J., and Schultz D.A., *Proc. Natl. Acad. Sci USA*, 2000, 97, 996.

48. Feldheim D.L. and Keating C.D., *Chem. Soc. Rev.*, 1998, 27, 1.

49. Henglein A., *Chem. Rev.*, 1989, 89, 1861.

50. Schmid G., *Chem. Rev.*, 1992, 92, 1709.

51. Rechberger W., Hohenau A., Leitner A., Krenn J.R., Lamprecht B., and Aussenegg F.R., *Optics Commun.*, 2003, 220, 137.

52. Qu S., Du C., Song Y., Wang Y., Gao Y., Liu S., Li Y., and Zhu D., *Chem. Phys. Lett.*, 2002, 356, 403.

53. Prodan E., Nordlander P., and Halas J., *Nano Lett.*, 2003, 3, 1411.

54. Niemeyer C.M., *Angew. Chem. Int. Ed.*, 2001, 40, 4128.
55. Raschke G., Kowarik S., Franzi T., Sonnichsen C., Klar T.A., and Fieldmann J., *Nano Lett.*, 2003, 3, 935.
56. Katz E. and Willner I., *Angew. Chem. Int. Ed.*, 2004, 43, 6042.
57. Parak W.J., Gerion D., Pellegrino T., Zanchet D., Michael C., Larabell C.A., and Alivisatos A.P., *Nanotechnology*, 2003, 14, 15.
58. Faraday M., *Phil. Trans. Roy. Soc.*, 1857, 147, 145.
59. Turkevich J. and Enustun B.V., *J. Am. Chem. Soc.*, 1963, 85, 3317.
60. Frens G., *Nature (Lond.) Phys. Sci.*, 1973, 241, 20.
61. Morris R.H. and Milligan W.O., *Morphol. Colloidal Gold*, 1964, 3461.
62. Duff D.D. and Baiker A., *Langmuir*, 1993, 9, 2301.
63. Brust M., Bethell D., Schriffrin D.J., and Kiely C.J., *Adv. Mater.*, 1995, 7(9), 795.
64. Heath J.R., Knobler C.M., and Leff D.V., *J. Phys. Chem. B,* 1997, 101, 189.
65. Korgel B.A., Fullam S., Connolly S., and Fitzmaurice D., *J. Phys. Chem. B,* 1998, 102, 8379.
66. Schmid G. and Peschel S., *New J. Chem.*, 1998, 22, 669.
67. Schmid G., Pfeil R., Boese R., Bandeman F., Meyer S., Calis G.H.M., and van der Felden J.W.A., *Chem. Ber.*, 1981, 114, 3634.
68. Haruta M., Yamada N., Kobayashi T., and Lijima S., *Catalysis*, 1989, 115, 301.
69. Nuzzo R.G. and Allara D.L., *J. Am. Chem. Soc.,* 1983, 105, 4481.
70. Giersig M. and Mulvaney P., *Langmuir*, 1993, 9, 3408.
71. Brust M., Walker M., Bethell D., Schiffrin D.J., and Whyman R., *J. Chem. Soc. Chem. Commun.,* 1994, 801.
72. Hostetler M.J., Wingate J.E., Zhong C.J., Harries J.E., Vachet R.W., Clark M.R., London J.D., Green S.J., Stokes S.J., Wignal G.D., Glish G.L., Porter M.D., Evans N.D., and Murray R.W., *Langmuir*, 1998, 14, 17.
73. Wetten R.L., Khoury J.L., Alvarez M.M., Murthy S., Vezmar I., Wang Z.L., Stephan P.W., Cleverland C.L., Luedtke W.D., and Landman U., *Adv. Mater.,* 1996, 8, 428.
74. Schaaff T.G., Shafigullin M.N., Khoury J.T., Vezmar I., Whetten R.L., Cullen W., First P.N., Gutierrez-Wing C., Ascensio J., and Jose-Yacaman M.J., *J. Phys. Chem. B*, 1997, 101, 7885.
75. Alvarez M.M., Khoury J.T., Schaaff T.G., Shafigullin M.N., Vezmar I., and Whetten R.L., *Chem. Phys. Lett.*, 1997, 266, 91.
76. Templeton A.C., Wuelfing W.P., and Murray R.W., *Acc. Chem. Res.*, 2000, 33, 27.
77. Jimenez V.L., Georganopoulou D.G., White R.J., Harper A.S., Mills A.J., Lee D., and Murray R.W., *Langmuir,* 2004, 20, 6864.
78. Hicks J.F., Templeton A.C., Chen S., Sheran K.M., Jasti R., Murray R.W., Debord J., Scaaff T.G., and Whetten R.L., *Anal. Chem.*, 1999, 71, 3703.
79. Wilcoxon J.P., Martin J.E., and Provencio P., *Langmuir*, 2000, 16, 9912.
80. Song Y., Jiminez V., McKinney C., Donkers R., and Murray R.W., *Anal. Chem.*, 2003, 75, 5088.
81. Ohara P.C., Leff D.V., Heath J.R., and Gelbart W.M., *Phys. Rev. Lett.*, 1995, 75, 3466.
82. Fink J., Kiely C.J., Bethell D., and Schiffrin D.J., *Chem, Mater.*, 1998, 10, 922.
83. Brust M., Fink J., Bethel D., Schiffrin D.J., and Kiely C.J., *J. Chem. Soc. Chem. Commun.,* 1995, 1655.
84. Kim Y., Johnson R.C., and Hupp J.T., *Nano Lett.*, 2001, 1, 166.
85. Weisbecker C.S., Merritt M.V., and Whitesides G.W., *Langmuir*, 1996, 12, 3763.
86. Yguerabide J. and Yguerabide E.E., *Anal. Biochem.*, 1998, 262, 137.
87. Vlckova B., Tsai D., Gu X., and Moskovits M., *J. Phys. Chem.*, 1996, 100, 3169.
88. Cao Y.C., Jin R., and Mirkin C.A., *Science*, 2002, 297, 1536.
89. Nie S. and Emory S.R., *Science,* 1997, 275, 1102.
90. Lee P.C. and Mesisel D., *J. Phys. Chem.*, 1982, 86, 3391.
91. Zhang J. and Lakowicz, J.R., *J. Phys. Chem. B.*, 2005, 109, 8701.
92. Mulvaney P., Linnert T., and Henglein A., *J. Chem. Phys.*, 1991, 95, 7843.
93. Henglein A., *J. Phys. Chem.*, 1993, 97, 5457.
94. Salata O.V., *J. Nanobiotechnol.*, 2004, 2, 3.

95. Kumar C.S.S.R., *Biofunctionalisation of nanomaterials* (Nanotechnologies for the Life Sciences 1), Wiley-VCH, Weinheim, 2005.
96. Katz E. and Willner I., *Angew. Chem. Int. Ed.*, 2004, 43, 6042.
97. Shenton W., Davis A.S., and Mann S., *Adv. Mater.*, 1999, 11, 449.
98. Caruso F., *Adv. Mater.*, 2001, 13, 11.
99. Caruso F. and Mohwald H., *J. Am. Chem. Soc.*, 1999, 121, 6039.
100. Taton T.A., Mirkin C.A., and Letsinger R.L., *Science*, 2000, 289, 1757.
101. Niemeyer C.M., *Angew. Chem. Int. Ed.*, 2001, 113, 4254.
102. Gestwicki J.E., Strong L.E., and Kisseling L.L., *Angew. Chem. Int. Ed.*, 2000, 112, 4742.
103. Okano K., Takahashi S., Yasuda K., Imai K., and Koga M., *Anal. Biochem.*, 1992, 202, 120.
104. Soukka T., Harma H., Paukunen J., and Lovgren T., *Anal. Chem.*, 2001, 73, 2254.
105. Hermanson G.T., *Bioconjugate techniques*, Academic Press, London, 1996, p. 593.
106. Hong H.G., Bohn P.W., and Sligar S.G., *Anal. Chem.*, 1993, 65, 1635.
107. Hong H.G., Jiang M., Sligar S.G., and Bohn P.W., *Langmuir*, 1994, 10, 153.
108. Hainfield J.F. and Powell R.G., *J. Histochem. Cytochem.*, 2000, 48, 471.
109. Aubin-Tam M. and Hamad-Schifferli K., *Langmuir*, 2005, 21, 12080.
110. Dreshler U., Erdogan B., and Rotello V.M., *Chem. Eur. J.*, 2004, 10, 5570.
111. Shenhar R. and Rotello V.M., *Acc. Chem. Res.*, 2003, 36, 549.
112. Manjula B.N., Tsai A., Upadhya R., Perumalsamy, Smith P.K., Malavalli A., Vanderiff K., Winslow R.M., Intaglieta M., Prabhakaran M., Friedman J.M., and Archarya A.S., *Bioconjugate Chem.*, 2003, 14, 46.
113. Elghanian R., Storhoff J.J., Mucic R.C., Letsinger R.L., and Mirkin C.A., *Science*, 1997, 277, 1078.
114. Rosi N.L. and Mirkin C.A., *Chem. Rev.*, 2005, 105, 1547.
115. Mirkin C.A., *Inorganic Chem.*, 2000, 39(11), 2258.
116. Goldman E.R., Balighaian E.D., Mattoussi H., Kuno M.K., Mauro J.M., Tran P.T., and Anderson G.P., *J. Am. Chem. Soc.*, 2002, 124, 6378.
117. Baeumie M., Stamou D., Segura J.M., Hovius R., and Vogel H., *Langmuir*, 2004, 314, 529.
118. Gold L., *J. Biol. Chem.*, 1995, 270, 13581.
119. Wilson D.S. and Szostak J.W., *Annu. Rev. Biochem.*, 1999, 68, 611.
120. Farokhzad O.C., et al., *Cancer Res.*, 2004, 64, 7668.
121. Huang C.C., et al., *Anal. Chem.*, 2005, 77, 5735.
122. Shukla G.S. and Krag D.N., *Oncol. Rep.*, 2005, 13, 757.
123. Benns J.M., et al., *J. Controlled Release*, 2002, 79, 255.
124. Hilgenbrink A.R. and Low P.S., *J. Pharm. Sci.*, 2005, 94, 2135.
125. Nosloe C.P., et al., *Radiology*, 1993, 187, 333.
126. El-Sayed M.A., *Acc. Chem. Res.*, 2001, 34, 257.
127. El-Sayed, et al., *Nano Lett.*, 2005, 5, 829.
128. Huang X., et al., *Photochem. Photobiol.*, 2006, 82, 412.
129. Weisseleder R., *Nat. Biotechnol.*, 2001, 19, 316.
130. Link S. and El-Sayed M.A., *J. Phys. Chem. B*, 1999, 103, 8410.
131. Bhattacharya R. and Mukherjee P., *Adv. Drug Deliv. Rev.*, 2008, 60, 1289.
132. Pissuwan D., Valenzuela S.M., and Cortie M.B., *Trends Biotechnol.*, 2006, 24, 62.
133. Salata O.V., *J. Nanobiotechnol.*, 2004, 1, 1.
134. Brunner T.J., Wick P., Manser P., Spohn P., Grass R.N., Limbach L., Bruinink A., and Stark W.J., *Environ. Sci. Technol.*, 2006, 40, 4374.
135. Connor E.E., Mwamuka J., Gole A., Murphy C.J., and Wyatt M.D., *Small*, 2005, 1, 325.
136. Goodman C.M., McCusker C.D., Yilmaz T., and Rotello V.M., *Bioconjugate Chem.*, 2004, 15, 897.
137. Kirchner C., Liedl T., Kudera S., Pellegrino A.M.J., Gaub H.E., Stolzle S., Fertig N., and Parak W.J., *Nano Lett.*, 2005, 5(2) 331.
138. Goyer R.A., *Am. J. Clin. Nutr.*, 1995, 61, 646.
139. Ghuitescu L. and Fixman A., *J. Cell Biol.*, 1984, 99, 639.
140. Parak W.J., Boudreau R., Gros M.L., Gerion D., Zanchet D., Micheel C.M., Williams S.C., Alivisatos A.P., and Larabell C.A., *Adv. Mater.*, 2002, 14, 882.

141. Warheit D.B., Laurence B.R., Reed K.L., Roach D.H., Reynolds G.A.M., and Webb T.R., *Toxicol. Sci.*, 2004, 77, 117.

142. Shaw S.Y., Westly E.C., Pitter M.J., Subramanian A., Schreiber S.L., and Weissleder R., *Proc. Natl. Acad. Sci. USA*, 2008, 105 (21), 7387.

143. Judson R., Richard A., Dix D.J., Houck K., Martin M., Kaylock R., Dellarco V., Henry T., Holderman T., Sayre P., Tan S., Carpenter T., and Smith E., *Environ. Health Perspect.*, 2008, 1, 1.

144. Ying Z. and WenXin L., *Sci. China Ser. B Chem.*, 2008, 51(11), 1021.

145. Daniel C.-M. and Astruc D., *Chem. Rev.*, 2004, 104, 293.

146. Long T.C., Saleh N., Tilton R.D., Lowry G.V., and Veronesi B., *Environ. Sci. Technol. Technol.*, 2006, 40, 4346.

147. Tarantola M., Schneider D., Sunnick E., Adam H., Pierrat S., Rosman C., Breus V., Sonnichsen C., Basche T., Wagener J., and Janshoff A., *ACSNANO*, 2009, 3(1), 213.

148. Matzinger P., *J. Immunol. Methods*, 1991, 145, 185.

149. Carbonari M., Cibati M., Cherich M., Sbarigia D., Pesce A.M. Dell'Anna L., Modica A., and Fiorilli M., *Blood*, 1994, 83, 1268.

150. DeRenzis F.A. and Schechtman A., *Technol. J. Microtechnol. Histochem.*, 1973, 48, 135.

151. Borenfreund E. and Puerner J.A., *J. Tissue Culture Methods*, 1984, 9, 7.

152. a. Borenfreund E. and Puemer J.A., *Toxicol. Lett.*, 1985, 24, 119.
    b. Borenfreund E. and Puemer J.A., *Toxicology*, 1986, 39, 121.

153. Elford W.G., King L.E., and Franker P.J., *Cytometry*, 1992, 13, 137.

154. Nicoletti I., Migliorati G., Pagliacci M.C., Grignani F., and Riccardi C., *J. Immunol. Methods*, 1991, 139, 271.

25

Photonic Structures in Nature

Peter Vukusic and Joseph Noyes

School of Physics, University of Exeter, Exeter, United Kingdom

CONTENTS

For hundreds of years, humans have been drawing inspiration from the natural world, from the hook-and-loop design in Velcro to the examination of the wing tips of eagles for a method to improve wing efficiencies. As we often consider photonic structures to require careful manufacture within sterile environments, it is easy to ignore examples within nature as a source for inspiration when creating new designs. The diversity evident within even a small number of organisms demonstrates methods applicable to enhancing the quality of diffuse white reflection, emission from OLEDs, and the creation of three-dimensional (3D) photonic crystals. The earliest example of their inspiration is the observations of reflection from a series of iridescent natural samples in the late 1800s and early 1900s that led to developments in the understanding of how electromagnetic waves interact with thin films.[1,2]

    Despite the constant examination of a variety of natural samples since this time, and the application of transmission electron microscopy to natural samples in the 1940s, it is still possible to observe new and unique structures in studies conducted today, or to further our understanding by applying new techniques. In this chapter we aim to provide a small insight into the field of natural photonics, and thereby demonstrate its relevance to future developments within the field of photonics.

25.1 Simple Structures, One-Dimensional Periodicity

The simplest and most widely observable photonic structures are those based on a variation in refractive index in one dimension, typically produced by multiple alternating
layers of material. The most basic form is that observed in the epicuticle of a number of
beetle families, including the scarabs and buprestid jewel beetles, such as *Chrysochroa raja*.
Variations are also largely present in iridescent butterflies and moths, from a simple multilayer within the wing scales of moths in the *Urania* family,[3,4] to discontinuous layers in
the superficial structures on the scales of *Morpho* butterflies.[5]

25.1.1 *Chrysochroa raja*

Chrysochroa raja is an excellent example of the multilayers observed in iridescent beetles.
At normal incidence it appears a metallic green with a red stripe along the center of the
elytral wings (Figure 25.1a). The inner surface of the elytra also displays a metallic green
color, although observation of the cross section by optical microscopy demonstrates no
discernible iridescence (Figure 25.1b). Examination by transmission electron microscopy
(TEM) reveals that there is a uniform multilayer structure within the top 2 μm of the epicuticle. It has long been known that for a semi-ideal multilayer stack, where the structure
is a repeating stack of identical layer pairs, the peak wavelength can be calculated using a
relatively simple formula:

$$\lambda_{max} = 2(n_a d_a \cos(\theta_a) + n_b d_b \cos(\theta_b)) \tag{25.1}$$

where n is the refractive index, d is the thickness, and θ is the angle of incidence for the
different layers, a and b, with θ related to θ by Snell's law.

While the variations between the alternating layer pairs in *C. raja* are approximately
consistent, they make it impossible to treat the structure as an ideal multilayer structure,
and thereby prevent the reduction of the complexity of the optical model.[6,7]

Furthermore, the surface of the cuticle is highly irregular and incapable of specular reflection. This makes it impossible to measure the absolute reflection spectra of the system at a

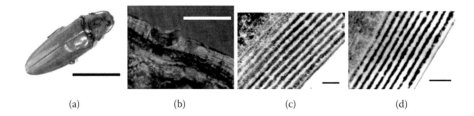

(a)   (b)   (c)   (d)

FIGURE 25.1
(a) Picture of *Chrysochroa raja*. The red regions on *C. raja* are visible on both elytral wings, with the majority of
the wing being green. It is also possible to see the variation in hue toward the edge of the wings, as they tilt
away from the camera. (b–d) Cross-sectional images from the elytra from *C. raja*. (b) Optical image of cross
section showing the surface contour. High-resolution TEM image showing the structure in the green region (c)
and red region (d). Typical layer thicknesses are around 100 nm, although there is significant variation between
50 and 110 nm in the layer thicknesses observed throughout the structure. Scale bars: 20 mm (a), 50 μm (b),
500 nm (c, d).

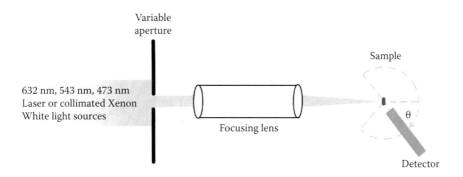

FIGURE 25.2
Schematic of the apparatus used to determine accurately the reflectivity of a surface. The sources used were a series of lasers and an Ocean Optics HPX-2000 white light. By combining white light spectra with absolute values obtained for single wavelengths, it is possible to remove features caused by short- and medium-range irregularities in the sample geometry.

specified angle using the apparatus shown in Figure 25.2. In order to obtain absolute measurements, the white light source was replaced with a series of laser sources. This made it possible to measure the proportion of the incident light that was reflected by the sample, and this information could be used to calibrate and then validate the data obtained using the white light source.

Regions of the sample were found to typically contain 17 layers, although areas were found where the number ranged from as few as 8 to as many as 23 layers. TEM measurements showed that the layers varied in thickness from approximately 80 nm to 130 nm, with the red region demonstrating dimensions proportionally larger than those in the green region. The use of least squares algorithms to match experimentally measured absolute reflectance profiles to theoretical values generated using recursive calculations of Fresnel formulae for the individual layers allowed the real and imaginary parts of the refractive indices to be established.

Previous studies have regularly concluded refractive indices in the region of 1.56, with other work proposing values for the individual layers varying from 1.3 to 2. This sample also demonstrated refractive indices in this region, with values of $n = 1.68$ and $k = 0.03$ for the light contrasted layers and $n = 1.55$ and $k = 0.14$ for the dark contrasted layers. While these values are reliable, they are unable to be used to define the material that comprises the layers within these structures.

Despite many years of study, the composition of the layers observed in the epicuticle of iridescent beetles is still largely unknown. The material is often reported to contain a high proportion of chitin,[7,8] although it has also been reported that chitin is restricted to the bulk, with the epicuticle remaining chitin-free.[9] Chitin is a polymer composed of N-acetyl-D-glucosamine and D-glucosamine monomer subunits linked by β-1,4-glucosidic bonds,[9,10] and it forms three types of highly ordered crystalline structures, termed α, β, and γ, dependent on the orientation of the chitin chains.[11] However, with very few exceptions,[12] chitin is not found in the animal kingdom in a pure crystalline form, but rather as a complex matrix involving significant contributions of proteins, minerals, and other organic compounds,[9,10] with the insect cuticle shown to contain 40%–76% protein.[12,13] This binding of proteins, minerals, and pigments can result in significant changes to the physical and optical properties of resulting chitin-based materials. It has been shown that the presence of uric acid within these materials can significantly increase the measured refractive index.[14]

25.1.2 *Papilio palinurus* and *Papilio ulysses*

Papilio palinurus and *Papilio ulysses* are two brightly colored swallowtail butterflies. Like most butterflies, their bright colors are based in the scales on their wings. In these examples, the lamellae found in the scales of these butterflies are located in the bulk of the scale. However, unlike the layers in *C. raja*, where the multilayer is broadly continuous, these scales have additional regular orthogonal modulations imposed on them, and the lamellae follow the curvature of the resulting surface, often maintaining the gap between subsequent layers (Figure 25.3).

In both *P. palinurus* and *P. ulysses*, this gives rise to the appearance of an array of shallow flat-bottomed concavities across the surface of the scale, although the overall structure is

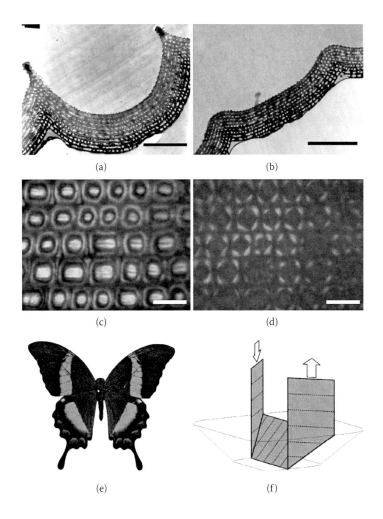

(a)   (b)

(c)   (d)

(e)   (f)

FIGURE 25.3

(a) TEM of the concavity observed in the scales of *P. palinurus*. (b) TEM of the concavity observed in the scales of *P. ulysses*. (c) Microscope image of the concavities on the surface of a scale from *P. palinurus*. (d) The same section as shown in (c), observed through cross-polarizers, demonstrating the polarization rotation in the reflections from the blue region. (e) Image of the butterfly, *P. palinurus*, demonstrating its overall emerald green color. (f) Schematic showing the method of rotation responsible for the blue reflection. Scale bars: 1 μm (a, b), 6 μm (c, d). (From Vukusic, P., et al., *Applied Optics*, 40(7), 1116–1125, 2001. With permission.)

slightly different in each case. In *P. ulysses*, the concavities are broad and shallow with a multilayer containing 10 layers of air pockets, similar to those seen in the feathers of birds. They generally measure in the range of 90–100 nm deep and follow the contours of the concavity. Furthermore, a form of ridges is present between the concavities running along the length of the scales, and the concavities undulate along the length of the scale rather than having any ribs. This structure has a primary reflection peak of 460 nm. Similarly to *Chrysiridia ripheus*, the concavities in the scales of *P. palinurus* contain a two-dimensional (2D) retro-reflector that allows for limited polarization conversion around an axis parallel to the ridges.

In contrast, the scales on *P. palinurus*, while very similar, have concavities that are approximately hemispherical. These are separated along the length of the scale by ribs of construction similar to that of the ridges running down the scale, although these ridges do not contain lamella, as seen in scales in the butterflies of the *Morpho* family. Unlike the scales of *P. ulysses*, it is possible for the concavities in the scales of *P. palinurus* to rotate polarized light of any initial orientation through 90° due to multiple reflections from the high-angled sides of the concavities that have an almost circular symmetry.[15]

The yellow reflection from the base of the concavity has a spectrum centered around 540 nm. The blue reflection from the sides is caused by the opposing sides acting as retro-reflecting surfaces, with both reflections occurring at a higher angle of incidence, around 45°, such that the peak is significantly blue shifted. The result is a combination of yellow spots with a blue annulus that appears macroscopically green due to a pointillistic mixing effect.[16]

25.1.3 *Morpho Rhetnor* and *Morpho didius*

Not all butterflies have color producing structures within the bulk of their scales. An excellent example of scales with surface structures is from the members of the *Morpho* family, which are found in South and Central America. Due to their distinctive blue color, which is so bright that it can be seen a quarter mile off,[17] they are one of the most popular butterflies to be used in fashion jewellery and butterfly presentation cases.

In the case of the butterfly *Morpho rhetenor*, the scales display the classic Christmas tree structure, with a series of lamellae that project perpendicularly to the spine and approximately parallel to the base of the ridge, with the size of the lamellae decreasing as the ridge tapers toward the top. The surface of the scales has parallel ridges that are in the region of 500 nm apart and 1,500 nm high. Each ridge is about 70 nm deep and has, on average, 10 lamellae on each side, which are separated by a similarly sized air gap. While this system behaves in a manner similar to the simple multilayer observed in *C. raja*, the discontinuity of the multilayers limits the angles over which the brightest reflections can be observed, and thereby limits the range of color visible in the iridescence. Furthermore, the flexibility of the ridge design allows the peak reflectance to be distributed over a wider angle. Macroscopically, the scaled nature of the wing results in the reflected light being non-specularly reflected, with the peak reflection distributed over a range of angles. However, this is also true of the reflection from a single scale that can be shown to be reflected across a broader angle than would be expected if the scales were not curved and the ridges were not subject to some side-to-side bending or tilting.

Another example of a butterfly with a similar scale is *M. didius*. This butterfly differs from *M. rhetenor* in that it has two sets of different overlapping scales on its inner wing surface, the upper being a highly transparent scale and the lower being an iridescent scale (Figure 25.4). The lower of the two is structurally and dimensionally very similar to those of *M. rhetenor*, with a periodicity of approximately 500 nm, although the height of these

(a)   (b)

(c)   (d)

FIGURE 25.4
(See color insert.) Optical and TEM images of the butterflies *Morpho rhetenor* (a, b) and *Morpho didius* (c, d). Both optical images (a, c) show the brilliant blue that is synonymous with this family of butterflies. The two TEM images (b. and d.) show the substructure of the butterfly scales. TEM image of a section from a *M. rhetenor* scale (b.) shows the multilayer within the ridges responsible for the brilliant colour, while the image from a section of *M. didius* (d.) shows the two discretely configured multilayers. Scale bars: 20 mm (a), 1.8 μm (b), 20 mm (c), 1.3 μm (d). (From Vukusic, P., and Sambles, J.R., *Nature*, 424, 852–855, 2003. With permission.)

ridges is in the region of 950 nm, allowing room for only six or seven lamellae. The upper layer has fewer ridges that are 1,700 nm apart and 1,200 nm high with four or five lamellae present toward the top of the ridge. Macroscopically the effect is very similar to that of *M. rhetenor*. However, the semitransparent upper scale strongly diffracts the transmitted light, and this enhances the observed scattering. The result is a classically high-reflecting scale that is viewable over a wide angle.

25.1.4 *Ancyluris meliboeus* and *Hypolimnas bolina*

In the *Morpho* butterflies, the lamellae run virtually parallel to the surface of the scale, and therefore the wing. In a number of other butterflies, the scales contain lamellae that are oriented at an angle that is between perpendicular and parallel to the base of the ridges. This reduces the number of layers that act in the multilayer stack, as well as the total reflectivity that can be achieved from the structure, but it also has the effect of creating a "dark zone" within the observation hemisphere, in which there is an angle range where no iridescence is observed.

Both the butterflies *Ancyluris meliboeus* and *Hypolimnas bolina* have scales with this form of structure (Figure 25.5.) In the case of *A. meliboeus*, the scales have what can be loosely described as ridges that are split along their length into plates that are inclined to around 30° from parallel to their base. These lamellae act as a multilayer, and its incline results in

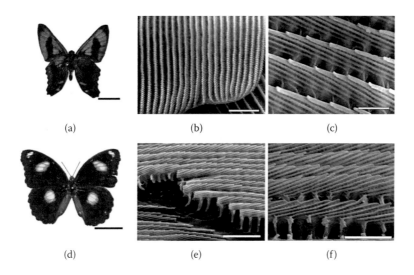

(a) (b) (c)

(d) (e) (f)

FIGURE 25.5

(a) *Ancyluris meliboeus* has highly iridescent blue scales that "flash" when observed at a shallow angle behind the butterfly. (b) SEM image of a scale from *A. meliboeus* demonstrating the rows of plates along the scale. (c) Side-view SEM showing the separation between the plates. (d) The butterfly *Hypolimnas bolina* has a region around the white spots where there are iridescent scales that display a low-angle violet reflection. (e) SEM images again show that there are rows of plates on the scales. (f) The separations between these create a shallow-angle multi-layer effect similar to that of *A. meliboeus*. Scale bars: 20 mm (a, d), 3 μm (b, e), 700 nm (c), 2 μm (f).

an angle at which the iridescence is not observed. Single scale measurements and optical imagery show that for 120° in the axis parallel to the ridges, there is a strong iridescence showing a wide range of hues, from blue to orange.[19,20]

This form of scales is also found on the male *Hypolimnas bolina*. The butterfly is predominantly black with four white spots, one on each wing, with the iridescent scales located around the edges of the white spots. The scales are not visible from normal incidence and display a purple hue at lower angles in one direction. The scales responsible for this color have ridges similar to those observed in *Ancyluris meliboeus*, with a periodicity along the scales of around 600 nm.[21] The periodicity perpendicular to the lamellae is around 120 nm, with the lamellae themselves averaging around 50 nm thick, with the separation being maintained by interlamella spacers that do not extend across the entire width of the lamellae.[21]

The most extreme example of these types of scales is found on the yellow and black butterfly *Troides magellanus*, the yellow regions of which flash a blue-green color when viewed from the outermost edge at near-grazing incidence. The iridescent scales have ridges around 2.2 μm high, with lamellae running from the base to the top at an angle of 36° from normal to the scale base layer. These lamellae form a multilayer with a periodicity of approximately 260 nm when viewed at near-grazing incidence.[22]

25.1.5 *Papilio zalmoxis* and *Papilio nireus*—Nature's OLEDs

While many of these structures have intricate designs that can demonstrate simple principles, perhaps the most amazing is when the structures are combined to great effect. The butterfly *Papilio zalmoxis* was often held to be an example of Tyndall scattering, a system by which shorter wavelengths are more preferentially coherently scattered than lower wavelengths by small particles, thereby resulting in the separation of different wavelengths of

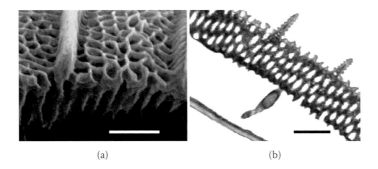

(a)   (b)

FIGURE 25.6
Electron microscopy images showing the substructure of the scales in *P. nireus*. The upper surface has an array of hollow cylinders that act as a 2D photonic crystal to enhance emission from the structure in the upward direction. Any emission that occurs in the downward direction is reflected back upward by a three-layer reflector in the substructure. (a) SEM image of the upper 2D photonic crystal. (b) TEM image of a cross section from *P. sesostris* bisecting the cylinders and showing the lower structure that separates the 2D photonic crystal from the emitter. Scale bars: 1 μm. (From Vukusic, P., and Hooper, I., *Science*, 310, 1151, 2005. With permission.)

light and the structural color of the material. However, recent studies have demonstrated that the blue color is the result of a fluorescent pigment, and that both *P. zalmoxis* and *P. nireus* contain a structure that acts in a manner similar to that of recent LED structures.[23,24]

The structure in *P. nireus* (examined by electron microscopy in Figure 25.6) consists of two sections. At the top is a series of approximately vertical tubes around 240 nm in diameter that extend 2 μm from the upper to lower surface of the scales. The cylinders are approximately 340 nm long and were shown by Fourier analysis to be not spatially independent of one another, which is a requirement for incoherent effects such as Tyndall scattering. Fluorescent emission occurs in the 2D photonic crystal formed by the cylinders.

At the bottom of the structure is a distributed Bragg reflector separated from the tube array by a 1.5 μm gap. Any emission that is initially directed downward from the tube array is reflected upward by the reflecting structure. This produces a highly efficient directional fluorescent source.[23]

25.2 Three-Dimensional Crystals

Electronic band engineering has achieved great success in controlling and designing the electronic properties of materials. One of the ultimate goals in the field of photonic materials engineering is to define completely the optical properties of a crystal by controlling and designing its photonic bands, thereby defining its optical properties. The process of defining electronic band gaps is now well understood, and advances within this field have come by reducing the dimensionality of the structures from operating in three dimensions to sizes sufficiently small to remove the effects of the extra dimensions. In contrast, advances in optical systems have come from increasing the dimensionality, starting with simple solids and multilayers, such as those used in thin-film filters and fiber optics, and moving toward more elaborate systems.

The same is true in nature; in the majority of iridescent insects, the structural color is produced by 1D photonic crystals. However, there is a minority that has a common trait in that they frequently produce color through a 3D photonic crystal. Since these 3D crystals potentially manipulate the flow of light in all directions by the production of partial photonic band gaps, these structures represent one of the most advanced photonic systems observed in structural color production.

25.2.1 *Eupholus magnificus* and *Eupholus schoenherri pettiti*

The weevils of the *Eupholus* genus are herbivorous beetles from the Curculionoidea family and have all the hallmark characteristics of other insects of the order Coleoptera, although the morphology is slightly different from that of other beetles in that the head is elongated into a snout and the legs are proportionally longer when compared to the abdomen. As with all beetles, weevils have four wings, two of which are flight wings and the other two forming a hard protective shell, the elytra. The iridescent colors displayed by these beetles are typically generated in extracuticular scales that cover the majority of their bodies, with the underlying cuticle appearing virtually black, greatly increasing the apparent vibrancy of the scales.

When removed from the beetle and imaged under an optical microscope in reflection and transmission (Figure 25.7a and b, respectively), it was observed that colors of individual scales in reflection and transmission were complementary, which is consistent with the presence of only minor absorption. Manipulation of these images to evaluate the color of absorption demonstrated a total hue consistent with the presence of a small amount of melanin within the cuticle of the scale.

On illuminating the scale with linearly polarized light and observing with a crossed linear analyzer, domains that demonstrated a reflection of shorter-wavelength light were still brightly colored, which would only be observed through the rotation of the polarization vector

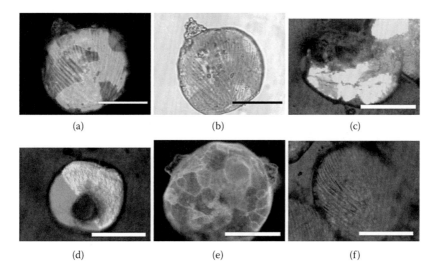

(a) (b) (c)

(d) (e) (f)

FIGURE 25.7
(See color insert.) White illuminated microscope images to show the (a) reflection and (b) transmission colors of a single scale from the blue region of *E. magnificus*. This scale was selected to show the large size of the domains within these scales, and the consistency of the hue that they produce. By comparison, the scales from other regions demonstrate multiple domains and colors (see Figure 25.9). (c–f) Four images of domained scales from other regions: (c) yellow region, *E. schoenherri pettiti*; (d) green region, *E. schoenherri pettiti*; (e) blue region, *E. magnificus* (dark field image); (f) green region, *E. magnificus*. Scale bars: 30 µm.

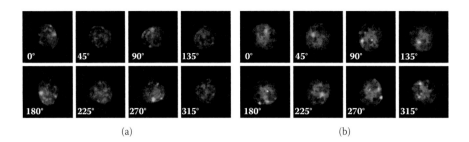

FIGURE 25.8

White illuminated microscope images of an individual scale from *E. magnificus* while it is rotated under static crossed polarizers. Images shown for (a) a blue scale from approximately 0° to 315° at approximately 45° increments, and (b) a green scale at similar orientations. In both sets it is possible to see domains that fade in and out. While the majority of the blue domains in both scales fade in this manner, there is also an orange region clearly visible in the upper region of the (a) 270° that also changes in intensity.

of incident light by 90°. When the sample was rotated (Figure 25.8), the reflected intensity was observed to fade in and out for the majority of the domains at 90° intervals. This demonstrates that the reflection was caused by a double reflection within the crystal. This requires that two optical planes within the crystal reflect at right angles to the incident light and are oriented at right angles to one another. This way, the reflected light undergoes a second reflection from the other plane and is reflected with the polarization rotated by 90°. While this effect is comparable to those observed in the *Papilio* butterfly,[16] the continuous nature of the photonic crystal allows for a more uniform reflection that is only subject to the domaining of the scales.

There appears to be no change in hue as the sample is rotated, with only the intensity appearing to change. Other domains within the scale can be observed through crossed polarizers to reflect over the full rotation, thereby indicating that some domains consistently scatter light with a converted polarization state. As there is only one structure type that has been observed to display a color, it is unlikely that there are domains with a significantly different structure that would be responsible for this feature. It is therefore more credible that these observations are caused by the scattering of light from the structure with a random polarization, or that the rotational symmetry is of a higher order. By reflecting light with a randomized polarization state, the domain will consistently reflect light of all polarizations, regardless of the illumination. Alternatively, by having a higher degree of rotational symmetry, any right-angle-oriented planes capable of producing the double-bounce effect will be evident at most angles, such that the variation in intensity appears constant.

By taking optical images of the samples immediately before the deposition of the gold layer, it was possible to match the structure to the colors in the regions under observation. In a scale from *E. schoenherri pettiti*, which was imaged with both optical microscopy and SEM, shown in Figure 25.9, it is demonstrated that the observed colors are not directly linked to a change in the lattice dimensions. In this example, there are three different colors present in a single scale. The periodicities in each section were measured using the image processing package ImageJ over a number of periods to reduce the random error natural variation. Multiple measurements were taken from the image, and the maximum variations in these measurements are quoted.[25] The periodicity seen in the yellow region (Figure 25.9b) was measured to be 283 ± 24 nm, the periodicity within the green area (Figure 25.9c) was measured to be 288 ± 28 nm, and within the blue region (Figure 25.9d) the periodicity was 330 ± 31 nm, with the air gap diameter in each region being 150 ± 30 nm.

(a) (b)

(c) (d)

FIGURE 25.9
Picture of a fractured scale from *E. schoenherri pettiti*, with closeup images of the differently colored regions within the scale showing that the differences between the colors observed are not the periodicity of the crystal. (a) A single predominantly yellow scale from the light green region of *E. schoenherri pettiti*. Subsequent images are higher-magnification pictures of the colored regions seen in the picture. (a, b) Yellow region with a measured periodicity of 283 nm. (c) Green area with a periodicity measured to be 288 nm. (d) Blue region with a periodicity of 330 nm. Scale bars: 20 μm (a), 3 μm (b–d).

While this qualitative observation does lead to the conclusion that similarly periodic domains can result in different colors, it does not explain how this arises or what other factors lead to changes in the observed color. In order to examine this, singles scales were mounted onto the head of a fine needle and examined using a dark field microscope (Nikon TE2000-U) attached to a spectrometer (Acton SpectraPro 2500i), which allowed the reflection from a single domain to be observed and measured concurrently (Figure 25.10).

A single scale from *E. magnificus*, mounted on a pin, was placed at the focal point of the microscope. Although this illuminates the entirety of the scale as well as the pin upon which the scale is mounted, the spectrometer can be configured to calculate the spectrum from a limited area of the internal imaging plate, thereby measuring the reflected spectra from individual sections of individual domains while the scale is rotated azimuthally around the axis of the pin in small increments.

The most noticeable feature of these results is that while the intensity of the reflected light changes, the color of the observed reflection does not change significantly with small changes in angle. This appears to demonstrate that in this optical geometry the photonic crystal within the weevil scale interacts with the incident light with the same optical characteristics over a range of angles. As a feature, this allows for the reflection from the surface of the scale to be enhanced over the reflection by a pigment by the presence of the photonic crystal while allowing the striking color to be evident over a wider range of angles.

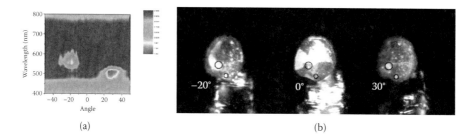

(a)   (b)

FIGURE 25.10

(See color insert.) (a) Graph showing the reflection measured when a scale from *E. schoenherri pettiti* is rotated around the azimuthal θ axis, with the reflection spectra recorded by a microspectrometer. (b) Image of the scale examined in 1.10 at three angles of incidence: –20°, 0°, and 30°. The larger black circle encompasses the region that the spectra in Figure 25.10a was measured on. The smaller black circle shows another domain that displays a similar color progression. Scale bars: (b) 80 μm. (With permission.)

25.2.2 *Parides sesostris*

The butterfly *P. sesostris* has scales of type IIIa that are colored by an inverse-opal structure. They comprise a 3D lattice of cuticle consisting of an array of spherical holes in a matrix of cuticle. The order of this system leads to strong scattering of favored wavelengths. As with most structures of this type, the lattices are arranged into small, irregular, but distinct domains that neighbor each other and bear slightly different characteristic wavelengths, the constant color effect across the wing being the result of spatial averaging.[26]

In *P. sesostris*, the scales of the bright green patches on the wings of the butterfly display a series of domains when examined under a microscope. These domains display a series of similar colors, which creates a constant color effect. Examination by electron microscope shows a highly ordered inverse-opal structure within the bulk of the scale, and that the domains are caused by their being ordered into different orientations (Figure 25.11.)

Illuminating the scales with polarized light and observing with a crossed polarizer again reveals polarization conversion of lower wavelengths from different domains. As the scale is rotated while observed with a crossed polarizer setup, the intensity of reflection from different domains changes. This is caused by a system similar to that observed

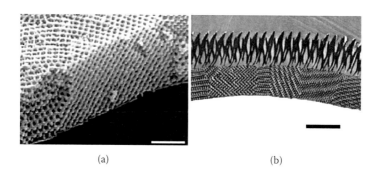

(a)   (b)

FIGURE 25.11

(a) SEM images from the wing of *P. sesostris*. (b) TEM image from *P. sesostris* showing the domain within the scales. Scale bars: 1.2 μm (a), 750 μm (b). (From Vukusic, P., and Sambles, J.R., *Nature*, 424, 852–855, 2003.) (With permission)

in *P. palinurus* and the weevils discussed earlier, where the multiple reflections are from differently oriented planes within the scales.[27]

25.2.3 Whiteness

It has long been known that there is no naturally occurring pigment that causes a white color; white is more simply the lack of color selection. As such, whiteness in nature is caused by irregularly scattering wavelength-independent structures in otherwise transparent materials.[28]

Good examples of this are the *Pierid* butterflies, such as the small cabbage white, *Pieris rapae*, or the white patches on the black jezebel, *Delias nigrina*, which have small beads around 150 nm wide and 200 nm high that scatter the light. It can be seen by removing these beads that they are responsible for the white of the scales.[29,30]

However, the brightest whites have been observed in the scales of the beetles *Lepidiota stigma* and *Cyphochilus* spp. (Figure 25.12a). As with the weevil, the scales contain a structure that is responsible for the observed color, although in these scales the substructure has been shown to consist of a highly irregular fibrous structure that has no periodicity (Figure 25.12b and c).[31]

The sections, as seen in Figure 25.12, show a disordered system inside the scale, with an average volume occupancy of 61.3%, varying from 58.5% to 64.5%. A volume occupancy of this range has been shown to maximize scattering ability while avoiding overcrowding that would otherwise decrease the scattering efficiency. The fiber size is typically 300 nm, varying from 250 nm to 315 nm.[31]

Reflection spectra were collected using a Datacolor ELREPHO® photospectrometer. This spectrophotometer is based off a 152 mm diameter integrating sphere that has two measurement devices. The first is a reference device focused on the barium sulfate-coated inner wall of the sphere, and the other is focused on the sample. Both detectors are oriented such that they only detect light from the targeted areas, and the signals from each are fed into a SP2000 dual 256 diode array.

The light is from a CIE D65 daylight source based off a xenon bulb that is passed through a pulsed shutter, an optical filter, and then into the sphere. This configuration has a spectral working range of 360–700 nm. The device is specifically designed to measure the reflectivity of white substances.

Measurements taken from the Datacolor photospectrometer allow calculations of the whiteness and brightness of the sample under examination. These measurements showed

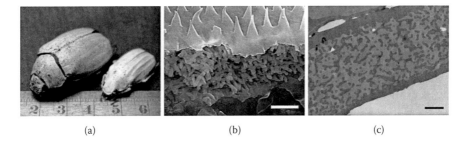

(a)   (b)   (c)

FIGURE 25.12
(a) Photograph of two white beetles, *Lepidiota stigma* (left) and *Cyphochilus* sp. (right). (b) Scanning electron microscopy image of a scale from the surface of *Cyphochilus* sp. (c) TEM section from a scale from *Lepidiora stigma*. Scale bars: 3 μm, 1 μm. (From Vukusic, P., et al., *Science*, 315(5810), 348, 2007.) (With permission.)

that the scales from the *Lepidiota stigma* have a whiteness value of 60 and a brightness of 65. For scales that are 5 nm thick, this equates to a brilliant whiteness exceeding that of most other natural samples, such as milk teeth and the cabbage white butterfly, *Pieris rapae*, and rivaling a number of man-made samples, such as tissue paper.

The link between the whiteness and the structure was firmly established by comparing the Fourier transform of the structure with the observed diffraction pattern. Despite the finite size of the laser spot and the limited area of the TEM image, the two data sets correlated strongly. It was also observed that both Fast Fourier Transform (FFT) and optical images were devoid of any periodic structures, which reaffirms the absence of any real-space periodicity in the structure.[31]

Further modeling of the internal structure of the scale has been undertaken by Hallam et al.,[32] who are modeling the fibrous system as a series of small particles with a lognormal distribution of sizes around a given sphere diameter. With the refractive index assumed to be 1.57 and the surrounding medium assumed to be air, it was demonstrated that the maximum scattering efficiency for red light occurs at a mean particle size of 275 nm; for blue light, 200 nm; and for green light, 250 nm. This implies that the structure size is optimized for scattering visible light, while the packing fraction is optimized for high reflectivity, thereby demonstrating that the overall structure is optimized to produce a very bright white.[32]

25.3 Two-Dimensional Structures

Two-dimensional structures are less commonly seen in natural samples than their 1D and 3D counterparts. The best examples are the keratin barbs present in the feathers of a large number of brightly colored parrots, such as those of the peach-faced lovebird, *Agapornis rosicollis*. As with a number of other species of bird, the greens are produced by a combination of pigmentary yellow or brown barbules with an additional structural blue color component to give the perception that their feathers are a vibrant green or olive color (depending on the pigment present).[33,34] This blue has been commonly attributed to Rayleigh or Tyndall scattering, the process that makes clear skies blue,[35–37] although other work has consistently demonstrated that this was not the case and that the blue coloration is an interference effect.[2,38]

Further examples include the nanotubes in the spine of the sea mouse *Aphrodita* sp. The nanotubes run along the length of the spine in a close-packed hexagonal array. Each nanotube has an internal diameter that increases along the length of the stack, although the external diameter is 510 nm. These then demonstrate a diffraction effect longitudinally and act as a waveguide along the length.[39] Similar designs are exploited in photonic crystal fibers as an improvement over photonic crystal designs.[40]

However, the most intriguing examples are the highly intricate patterns observed in the marine diatom.

25.3.1 Marine Diatoms

As biological entities, diatoms are ubiquitously present in nearly all aquatic environments, and despite making up a small percentage of the global green biomass, they are responsible for over 20% of the primary CO_2 conversion through photosynthesis.[41,42] They are eukaryotic algal organisms that comprise only one cell, are photosynthetic, and have a long scientific history of study,[43,44] which is partly due to the astonishingly intricate design and architecture of their silica-based cell walls.

The dimensions and periodicity of the structures within these walls are such that they would be expected to produce strong interaction with light of visible wavelengths.[45] However, this alone was not responsible for making diatoms the subject of scientific study. There are three features that make them of significant interest, the first being the species-related range of cell wall design that is controlled by a relatively limited genome-making DNA manipulation to produce tuned structures.

The second reason for scientific interest is that the growth of diatoms is sufficiently well understood to allow them to be manipulated into biomineralizing a range of materials beyond silica,[46] or to be used as templates in the manufacture of nanopatterned structures made of other materials.[47–49]

Additionally, the surface of the diatoms may be chemically altered by the addition of biologically active components that, when combined with the other techniques, would allow for the production of finely tuned potential microdevices and technological components from collections of ordinary diatoms.[50] All these factors add together to make an examination of the optical effects within diatoms of great scientific interest.

It was observed by optical microscopy that the surface of the valve appears structurally colored by a mechanism that is distinct from thin-film interference and more consistent with a diffracting structure (Figure 25.13b).[18,51] Optical transmission data through the same single valves were recorded with a single white light source and the apparatus described in Figure 25.2.

Strong diffraction can be clearly discerned in these data (Figure 25.14): the zero-diffracted order running vertically up the graph at 0° transmission (detector) angle, and two first diffracted orders running diagonally (Figure 25.14.) These experimental diffraction data have been overlaid with theory generated using the equation for a standard diffraction grating.[52,53] The angle and wavelength features in the data obtained in this section very closely match the theoretically predicted diffraction pattern (Figure 25.14).

To discern the spatial distribution of the light scattered from and through each valve, the white light source was changed to a series of lasers, and hemispherical screens were separately placed in front and behind the sample. The screens allowed the intensity of this scattered light to be spatially sampled. The resulting screen-imaged spatial intensity

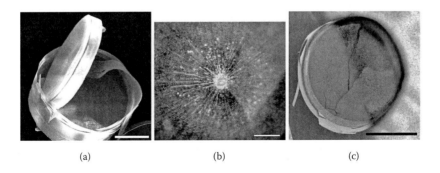

(a) (b) (c)

FIGURE 25.13
(a) SEM images of a complete diatom showing the two valves and the internal structures. (b) Optical microscope images of the inside of a valve face of the diatom, *C. wailesii*. The image shows the periodic structure that is responsible for the diffraction colors that are also visible. (c) A composite image of optical and electron microscope images of the same valve. The same features are clearly visible on both SEM and optical images. Scale bars: 100 μm (a), 20 μm (b), 150 μm (c).

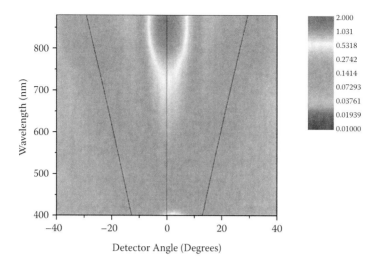

FIGURE 25.14
(See color insert.) Angle-dependent transmission data for a single valve of a *C. wailesii* diatom using the equipment shown in Fgure 25.2. The black lines superimposed on this image correspond to the theoretical zero- and first-order diffraction lines for the 2 μm periodic structure within the valve. Intensity in arbitrary units.

distribution profiles were digitized and used, in conjunction with single-wavelength angle-scan data from the same samples, to calculate the absolute proportion of power in any portion of reflected or transmitted solid angle.

Intense diffracted orders with sixfold symmetry were observed when laser light was transmitted through the valve shell plates (Figure 25.15d). Significantly less intense, but spatially analogous, diffraction was observed in reflection. The spatial distribution of this diffraction pattern is entirely predictable from the sixfold symmetry of the variation in refractive index of the valve ultrastructure. This is confirmed by 2D fast Fourier transforms (FFTs) of the periodic ultrastructure of the internal surface geometry of the diatom, which yield the same sixfold momentum-space geometries as the diffraction patterns (see Figure 25.16) [54]. Measurements of angular separation of observed diffracted orders for each laser (e.g. 14.1° (±0.5) for the 1st diffracted orders using 472 nm) further confirmed that the 2 μm periodicity was the source of the observed diffraction.

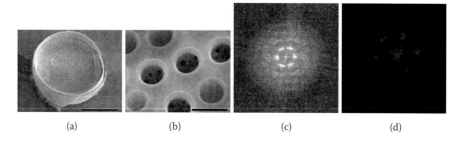

(a)   (b)   (c)   (d)

FIGURE 25.15
SEM images of a diatom, *C. wailesii*. (a) A complete frustule. (b) High-magnification image of the inner wall, through the apertures of which it is possible to see the outer plate. (c) Fast Fourier transform of the structure observed within a *C. wailesii* frustule. (d) Diffraction pattern of 472 nm of light diffracted by a single *C. wailesii* valve. Scale bars: 100 μm (a), 2 μm (b).

These data also show not only that the structure of the diatom has a strong diffracting effect, but also that the efficiency of transmission of light depends strongly on the wavelength. Calculations were performed to evaluate the absolute diffracted intensity in each peak in each diffracted order by integrating over the area under the curve. These values were then compared with the initial incident light intensity in order to obtain the absolute values. For wavelengths in the blue and green, the transmission is relatively low, with mean absolute totals (measured using valves from five separate diatoms) of 22 (±1) % for 472 nm incident light and 29 (±1) % for 543 nm incident light. However, consistently and significantly higher absolute transmission, 78 (±2) %, was measured for red light (632 nm).

The strong diffraction exhibited by light passing through the diatom valve, identified in these experiments, is a mechanism by which the available light intensity is more evenly spatially distributed over the internal volume of *C. wailesii* (Figure 25.16). It has a significant effect: it facilitates diffraction-assisted diffusion of the radiant energy throughout a larger proportion of the diatom's internal volume. Not only would this minimize the likelihood of photobleaching-related damage due to high incident intensity in any one unit volume of the diatom, but in environments where the available light intensity is much lower, but still nonetheless exceeds the photosynthetic activity threshold, such diffraction would assist with the distribution of light throughout the internal diatom volume, thereby increasing the number of chloroplasts within the diatom that have sufficient incident energy to photosynthesize. Furthermore, if the diatom itself is not optimally oriented, the diffraction of the incident light redirects a proportion of the light energy such that the first- and second-order diffraction peaks will pass through the central volume of the diatom.

These combined effects would therefore likely cause the chloroplasts within the valve to be subjected to net elevated light intensities that would trigger or enhance photosynthetic productivity despite their inferior intradiatom location, thereby increasing the efficiency of the diatom and offering a distinct advantage in the wide ranges of lighting conditions that are associated both with weather-related variable sunlight and with the variation in their water depth due to turbulent mixing and bloom cycles.[55]

Furthermore, the measurement presented here of preferential transmission of red and far-red wavelengths through the valve ultrastructure of *C. wailesii* (Figures 25.14.b and 25.16) appears to confirm experimentally several findings from a previous theoretical study of a very closely related centric diatom, *C. granii*.

This diatom has an ultrastructure entirely analogous to that observed in *C. wailesii*. In their paper, Fuhrmann et al.[56] calculated the band structure and thereby showed that the optical properties of the *C. granii* shell strongly influenced incoming light by coupling it into distinct waveguide modes.

For a variety of these modes, the waveguide-coupled light is distinctly blue-green and is absent of red, implying that red light is preferentially transmitted. The experimental measurement in this investigation into *C. wailesii*—that there is as much as a fourfold increase in the transmission of red compared to blue light into the diatom—highlights the extent of the interaction of incident light with the centric diatom form of ultrastructure.[56]

This phenomenon also appears congruous with separate and detailed biological evidence that suggests the importance of red and far-red wavelengths in some marine environments has been grossly underestimated. By studying diatom responses to informational light signals, Falciatore and Bowler identified extraordinary sensitivity to red and far-red wavelengths,[57] a phenomenon that infers the existence of a phytochrome-type photoreceptor in marine diatoms. The evolved ultrastructure of *C. wailesii* therefore appears particularly well designed to enhance the intravalve delivery of specifically appropriate radiant

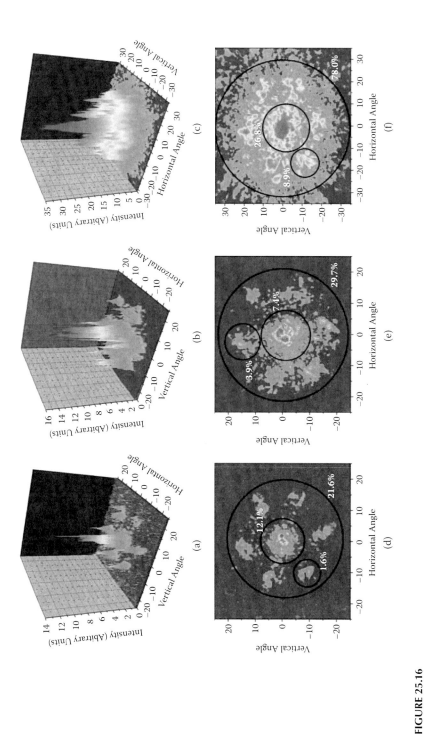

FIGURE 25.16
Images showing the transmitted power in the diffraction pattern for different wavelengths of light: (a, d) blue, 473 nm; (b, e) green, 543 nm; (c, f) red, 632 nm. Comparing the transmission ratios, it is evident that there is enhanced transmission in the red region, with approximately 78 % transmitted compared to 21 % for blue laser light diffraction patterns. Measurements of angular separation of observed diffracted orders for each laser (e.g., 14.1δ(δ0.5) for the first diffracted orders using 472 nm) further confirmed that the 2 μm periodicity was the source of the observed diffraction. (From Lipson, H.S., *Optical Transforms*, Academic Press, New York, 1972.)

energy to potentially indiscriminately situated chloroplast-housed photoreceptors comprising phytochrome.

However, even if biological implications of this structure are ignored, its importance in the future should not be underestimated. Recent work has examined the method of production of these structures and has established a very likely mechanism of the way in which the structures are formed,[58,59] although the formal understanding of the entire process is not quite complete. Studies are still examining how a simple life form, such as a diatom, is able to absorb dissolved silicate from its marine environment and transform it into elaborate geometric nanopatterned structures. Other studies are attempting to map the genome of the diatom with the goal of ascertaining how diatoms can be manipulated in order to suit scientific needs. With the understanding of how to control the crystal production, and possibly subsequent cell-culture-related replication,[60,61] insight into the production of new ceramics and composite materials will emerge.[48,62]

References

1. Lord Rayleigh. On the reflection of light from a regularly stratified medium. *R. Soc. Proc. A*, 93:565, 1917.
2. C.W. Mason. Structural colors in bird feathers. *J. Phys. Chem.*, 27:201–251, 401–447, 1923.
3. P. Vukusic. *Structural colors in biological systems: principles and applications*. Osaka University Press, Osaka, Japan, 2005, chap. 3.2, pp. 95–111.
4. S. Yoshioka and S. Kinoshita. Polarization-sensitive color mixing in the wing of the madagascan sunset moth. *Optics Express*, 15:2691–2701, 2007.
5. P. Vukusic, J.R. Sambles, C.R. Lawrence, and R.J. Wootton. Quantified interference and diffraction in single *Morpho* butterfly scales. *Proc. R. Soc. London Ser. B*, 266:1403–1411, 1999.
6. A.F. Huxley. A theoretical treatment of the reflextion of light by multilayer structures. *J. Exp. Biol.*, 48:227–245, 1968.
7. M.F. Land. The physics and biology of animal reflectors. *Progr. Biophys. Mol. Biol.*, 24:75–106, 1972.
8. A.R. Parker, D.R. McKenzie, and M.C.J. Large. Multilayer reflectors in animals using green and gold beetles as contrasting examples. *J. Exp. Biol.*, 201:1307–1313, 1998.
9. R.F. Chapman. *The insects: structure and function*, 4th ed. Cambridge University Press, Cambridge, 1998.
10. R.A.A. Muzzarelli. Biochemical modifications of chitin. In *The insect integument*, ed. H.R. Hepburn. Elsevier, New York, 1976.
11. K.M. Rudall. Molecular structure of arthropod cuticles. In *The insect integument*, ed. H.R. Hepburn. Elsevier, New York, 1976.
12. G.A.F. Roberts. *Chitin chemistry*. MacMillan Press, London, 1992.
13. R.H. Hackman. The interactions of cuticular proteins and some comments on their adaptation to function. In *The insect integument*, ed. H.R. Hepburn. Elsevier, New York, 1976.
14. S. Caveney. Cuticle reflectivity and optical activity in scarab beetles; the role of uric acid. *Proc. R. Soc. Lond. B*, 178:205–225, 1971.
15. P. Vukusic, J.R. Sambles, and H. Ghiradella. Optical classification of microstructure in butterfly wing-scales. *Photonic Sci. News*, 6:61–66, 2000.
16. P. Vukusic, J.R. Sambles, C.R. Lawrence, and G. Wakely. Sculpted-multilayer optical effects in two species of *Papilio* butterfly. *Applied Optics*, 40(7):1116–1125, 2001.
17. H.W. Bates. New species of butterflies from Guatemala and Panama (continued). *Entomologist's Monthly Mag.*, 1(4):81–85, September 1864.
18. P. Vukusic and J.R. Sambles. Photonic structures in biology. *Nature*, 424:852–855, 2003.

19. P. Vukusic, J.R. Sambles, C.R. Lawrence, and R.J Wootton. Structural colour: now you see it, now, you don't. *Nature*, 410:36, 2001.

20. P. Vukusic, J.R. Sambles, and C.R. Lawrence. Limited-view iridescence in the butterfly *Ancyluris meliboeus*. *Proc. R. Soc. Lond. B*, 269:7–14, 2002.

21. S. Luke and J. Noyes. Unpublished SEM and optical data taken from the butterfly *Ancyluris meliboeus* and subsequent private communication. Work in progress, 2008.

22. C.R. Lawrence, P. Vukusic, and J.R. Sambles. Grazing-incidence iridescence from a butterfly. *Appl. Opt.*, 41:437–441, 2002.

23. P. Vukusic and I. Hooper. Directionally controlled fluorescence emission in butterflies. *Science*, 310:1151, 2005.

24. R.O. Prum, T. Quinn, and R.H. Torres. Anatomically diverse butterfly scales all produce structural colours by coherent scattering. *J. Exp. Biol.*, 209:748–765, 2006.

25. V. Girish and A. Vijayalakshmi. Affordable image analysis using NIH image/imageJ. *Ind. J. Cancer*, 41:47, 2004.

26. K. Michielsen and D.G. Stavenga. Gyroid cuticular structures in butterfly wing scales: biological photonic crystals. *J.R. Soc. Interface*, 5(18):85–94, 2008.

27. P. Vukusic. Natural coatings. In *Optical interference coatings*. Berlin, Springer-Verlag, 2003, chap. 1.

28. C.W. Mason. Structural colors in insects I. *J. Phys. Chem.*, 30:383–395, 1926.

29. D.G. Stavenga, S. Stowe, K. Siebke, J. Zeil, and K. Arikawa. Butterfly wing colours: scale beads make white pierid wings brighter. *Proc. R. Soc. Lond. B*, 271:1577–1584, 2004.

30. N.I. Morehouse, P. Vukusic, and R. Rutowski. Pterin pigment granules are responsible for both broadband light scattering and wavelength selective absorption in the wing scales of pierid butterflies. *Proc. R. Soc. Lond. B*, 274:359–366, 2007.

31. P. Vukusic, B. Hallam, and J. Noyes. Brilliant whiteness in ultrathin beetle scales. *Science*, 315(5810):348, 2007.

32. B.T. Hallam, A.G. Hiorns, and P. Vukusic. Private communication on the modelling of fibres in the scales of *Cyphochilus* sp. as similarly sized spheres. Unpublished, 2008.

33. F. Frank. Die farbung der vogelfeder durch pigment und struktur. *J. Ornithol.*, 87:426–523, 1939.

34. J. Dyck. Structure and spectral reflectance of green and blue feathers of the rose-faced lovebird (*Agapornis roseicollis*). *Biol. Skrifter*, 18(2):5–65, 1971.

35. J. Tyndall. On the blue colour of the sky, the polarization of skylight, and on the polarization of light by cloudy matter generally. *Lond. Edinb. Dubl. Phil. Mag.*, 37:384–394, 1869.

36. C.H. Greenewalt, W. Brandt, and D.D. Friel. Iridescent colors of hummingbird feathers. *J. Opt. Soc. of Am.*, 50(10):293–301, 1960.

37. C.L. Ralph. The control of color in birds. *Am. Zool.*, 9:521–530, 1969.

38. C.V. Ramen. The origin of colours in the plumage of birds. *Proc. Ind. Acad. Sci. A*, 1–7, 1935.

39. A.R. Parker. A vision for natural photonics. *Philos. Trans. Math. Phys. Eng. Sci.*, 362(1825):2709–2720, 2004.

40. J.C. Knight. Photonic crystal fibres. *Nature*, 424:847–851, 2003.

41. P.G. Falkowski, Z. Dubinsky, and K. Wyman. Light harvesting and utilization by phytoplankton. *Plant Cell Physiol.*, 27:1335–1349, 1986.

42. C. Lopez, L. Vazquez, F. Meseguer, R. Mayoral, M. Ocana, and H. Miguez. Photonic crystal made by close packing SiO_2 submicron spheres. *Superlattices Microstructures*, 22(3):399–404, 1997.

43. J. Rheinberg. On an addition to the methods of microscopical research, by a new way of optically producing colour-contrast between an object and its background, or between definite parts of the object itself. *J. R. Microscop. Soc.*, 19:373, 1896.

44. S.D. Wilson. A reflection-diffraction microscope for observing diatoms in colour. *Applied Optics*, 5(10):1683–1684, 1966.

45. M.C. Hutley. *Diffraction gratings*. Academic Press, New York, 1982.

46. C. Sanchez, H. Arribart, and M.M.G. Guille. Biomimetism and bioinspiration as tools for the design of innovative materials and systems. *Nat. Mater.*, 4:277–288, 2005.

47. D. Losic, G. Triani, P.J. Evans, A. Atanacio, J.G. Mitchell, and N.H. Voelcker. Controlled pore structure modification of diatoms by atomic layer deposition of TiO_2. *J. Mater. Chem.*, 16:4029–4034, 2006.

48. U. Kusari, Z. Bao, Y. Cai, G. Ahmad, K.H. Sandhage, and L.G. Sneddon. Formation of nanostructured, nanocrystalline boron nitride microparticles with diatom-derived 3-d shapes. *Chem. Commun.*, 11:1177–1179, 2007.

49. Z. Bao, M.R. Weatherspoon, S. Shian, Y. Cai, P.D. Graham, S.M. Allan, G. Ahmad, M.B. Dickerson, B.C. Church, Z. Kang, H.W. Abernathy III, C.J. Summers, M. Liu, and K.H. Sandhage. Chemical reduction of three-dimensional silica micro-assemblies into microporous silicon replicas. *Nature*, 446:172–175, 2007.

50. L. De Stefano, A. Lamberti, L. Rotiroti, and M. De Stefano. Interfacing the nanostructured biosilica microshells of the marine diatom *Coscinodiscus wailesii* with biological matter. *Acta Biomater.*, 4:126–130, 2008.

51. M. Srinivasarao. Nano-optics in the biological world: beetles, butterflies, birds and moths. *Chem. Rev.*, 99:1935–1961, 1999.

52. C. Kittel. *Introduction to solid state physics*. John Wiley & Sons, New York, 1995.

53. M. Born and E. Wolf. *Principles of optics*, 7th (expanded) ed. Cambridge University Press, Cambridge, 2001.

54. H.S. Lipson. *Optical transforms*. Academic Press, New York, 1972.

55. M.J. Furnas. *In situ* growth rates of marine phytoplankton: approaches to measurement, community and species growth rates. *J. Plankton Res.*, 12:1117–1151, 1990.

56. T. Fuhrmann, S. Landwehr, M. El Rharbi-Kucki, and M. Sumper. Diatoms as living photonic crystals. *Appl. Phys. B Lasers Optics*, 78:257–260, 2004.

57. A. Falciatore and C. Bowler. The evolution and function of blue and red light photoreceptors. *Curr. Topics Dev. Biol.*, 68:13–63, 2005.

58. M. Sumper. A phrase separation model for the nanopatterning of diatom biosilica. *Science*, 295:2430–2433, 2002.

59. M. Sumper and E. Brunner. Learning from diatoms—nature's tools for the production of nanostructured silica. *Adv. Funct. Mater.*, 16:17–26, 2006.

60. S.H. Liu, C. Jeffryes, G.L. Rorrer, C.H. Chang, J. Jiao, and T. Gutu. *Biological and bio-inspired materials and devices*, Vol. 873E. Materials Research Society Symposia Proceedings, Warrendale, PA, 2005.

61. T. Gutu, L.F. Dong, J. Jiao, G.L. Rorrer, C.H. Chang, C. Jeffryes, and Q. Tian. Characterization of silicon-germanium oxide nanocomposites fabricated by the marine diatom *Nitzschia frustulum*. *Proc. Microsc. Microanal.*, 11:1958–1959, 2005.

62. C.E. Hamm, R. Merkel, O. Springer, P. Jurkojc, C. Maier, K. Prechtel, and V. Smetacek. Architecture and material properties of diatom shells provide effective mechanical protection. *Nature*, 421:841–843, 2003.

26

Bionanotechnology Applications in Plants and Agriculture

Stephen R. Wilson and Barry S. Flinn

Transformation Nanotechnologies LLC, and Institute for Advanced Learning and Research, Danville, Virginia, USA

CONTENTS

26.1 Introduction

For thousands of years humans have tried to modify plants to improve their performance and useful properties. Long before any scientific knowledge of inheritance, the techniques of breeding, crossing, and selecting plants were widely practiced with the goal to discover better crops,[1] more beautiful flowers,[2] and beneficial medical treatments.[3] During the last century, scientific efforts were directed toward improving plants, and in recent decades, the tools of molecular biology have been brought to bear. Now our understanding of the genomics of plants and their complex gene expression choreography is impressive. A recent review[4a] surveys the breadth and strength of our understandings of plant genomes and lays out a pathway for creating plants of the future. In the past 25 years, biotechnology has driven a major revolution in agricultural practice and crop production. Genetically engineered crops with improved agronomic traits have made the transition from laboratory benches and greenhouses to fields all over the world, where they are being grown commercially.[4b]

The merger of nanotechnology and biotechnology over the last decade has led to productive synergy and creation of the new field of *bionanotechnology*, but only very recently has it been recognized that this technology could be applied to plants.[5] Plants and agriculture are extremely important to commerce, particularly food crops. However, renewed efforts to create drugs from plants and the emerging global biofuels industry are poised for growth.

This review will not cover plant application of basic nanotechnology tools such as atomic force microscopy (AFM)[6] and gene-chip assays[7]—topics that have already been so widely used that they are beyond the scope of this review. Instead, we focus this survey on direct studies of nanomaterial interactions with plant systems and the beginnings of commercial exploitation. The variety of core nanomaterials available, coupled with tunable surface properties, make nanoparticles an excellent platform for a broad range of biological and biomedical applications.[8,9]

The global activity in bionanotechnology has finally come to agroresearch. Published work is expanding in the area of study and use of synthetic nanoparticles for gene delivery, to deliver plant protection substances, and to enhance other useful plant properties. We will begin with a short description of the plant as a biological system, and then review the published literature on interactions of nanomaterials with plants, with particular emphasis on the potential for beneficial application in agriculture and related commercial areas.

26.2 How Plant Cells Differ from Animal Cells

26.2.1 The Chloroplast

A plant cell differs from an animal cell in two important ways. The plant cell contains photosynthetic machinery and has a cell wall (Figure 26.1).

The plant's photosynthetic process takes place in the chloroplast, within the stacks of thylakoids, the membranous suborganelles, which are arranged in stacks called grana, and involves two complex protein units, photosystem I and photosystem II. Photosystem I (PSI) contains a reaction center of up to 14 protein subunits, as well as four membrane-associated antenna complex proteins (Lhca1–4) that capture light and guide its energy to the reaction center. This complex architecture of protein scaffolding and chlorophyll molecules provides the light-harvesting antenna and electron transport pathway for conversion

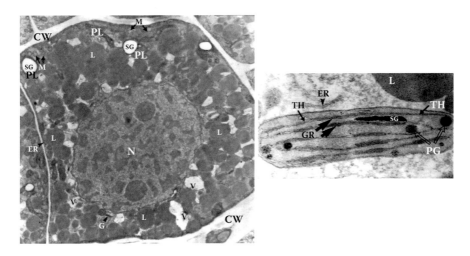

FIGURE 26.1
The architecture of a plant cell. Transmission electron micrographs showing a plant cell (left panel) and a developing chloroplast (right panel). The plant cell is surrounded by a thick cell wall (CW) and is highly cytoplasmic, with a large nucleus (N), several small vacuoles (V), lipid droplets (L), golgi bodies (G), mitochondria (M), endoplasmic reticulum (ER), and plastids (PL) containing large starch grains (SG). The developing chloroplast is stained to detect carbohydrates, such that the starch grain appears black. Thylakoid membranes (TH) are well developed, with some arranged into granal stacks (GR). Small lipid deposits, or plastoglobuli (PG), are also present with the chloroplast.

of photons into electric potential. In 2003, the complete structure of the PSI protein complex from a pea was determined by x-ray crystallography.[10] The structure of PSI is shown in Figure 26.2.

The PSI protein (in red in Figure 26.2) complexes and organizes approximately 200 chlorophyll pigment molecules (shown in blue in Figure 26.2). Despite its complexity, PSI is highly efficient, and almost every photon absorbed results in excitation of the special chlorophyll pair that converts the photon energy into +/− charge separation. In the plant, this potential is used to create ATP. Progress has been made toward harnessing photosynthesis to create hybrid nanodevices and electric potential. Some of this work is discussed in Section 26.7.

26.2.2 The Cell Wall

The cell wall consists primarily of cell wall polysaccharides, with cellulose microfibrils representing the primary structural units, embedded in a matrix of hemicellulose and pectin, as well as secondary wall lignin, with a smaller, but functionally relevant, complement of proteins. This tough barrier often presents an obstacle for transport of materials into the plant cell. Some unique features of nanomaterials that can penetrate this barrier will be highlighted in later sections. An alternative biochemical technology for studying plant biochemistry involves removing or digesting this cell wall with cellulase, hemicellulase, and pectinase enzymes, leaving a cell membrane more reminiscent of an animal cell, which contains the cytoplasm, nucleus, mitochondria, chloroplasts, and other organelles. The resulting package is called a protoplast, and in this state, the plant cell biochemical machinery can still function. Since there is only a cell membrane, most of the usual animal cell methods for material transfer across the membrane can be employed. After experimental drug delivery, gene insertion, or other protocols, regrowth of the cell wall to an

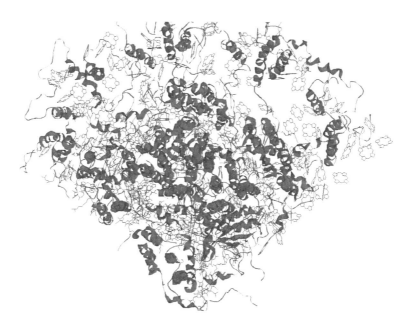

FIGURE 26.2
(See color insert.) The x-ray structure of plant photosystem I (PSI) at 4.4 A° from the protein data bank (code 1qzv). The protein backbone is in red, and the chlorophyll pigments are shown in blue. The four light-harvesting proteins (Lhca1–4) are at the top, the region where light is absorbed.

intact plant cell can be triggered. The balance of this chapter discusses work on intact plants, seeds, whole plant cells, and protoplasts.

26.3 Plants as Test Systems for Nanomaterial Environmental Toxicology

Many studies have appeared on the environmental toxicity of nanomaterials, particularly in the last decade, as engineered nanoparticles entered the market. Exotic new materials like fullerenes and carbon nanotubes, as well as other unique nanomaterials, such as quantum dots (typically prepared from cadmium selenide (CdSe)), captured the public interest. Nano-versions of well-known inorganic materials like non-nano-titanium dioxide (TiO_2) and other nano-metals eventually emerged as the largest production materials of concern in the environment. Several reviews on general aspects of environmental nanotoxicology are available.[11–13]

While mechanisms allowing nanoparticles to pass through cell walls and membranes are as yet poorly understood, once inside cells, nanoparticles can directly affect alterations in membranes and other cell structures.[14] The surface properties of engineered nanoparticles are of essential importance for their aggregation behavior, and resultant aqueous mobility and transport. While ecotoxicological literature on engineered nanoparticles shows toxic effects on fish and invertebrates, as of a 2008 review by Handy et al., little data on terrestrial plant species were available.[15] While the last decade saw great advances in nanotechnology and a corresponding increase in the use of nanomaterials in products, another 2008 review by Klaine et al. also criticized the lack of knowledge base for behavior, disposition, and toxicity of manufactured nanomaterials in plants.[16]

TABLE 26.1

Phytoxicity of Various Nanoparticles on Plants

| | MWNT | | Aluminum | | Alumina | | Nano-Zinc | | Nano-ZnO | |
| | 10–20 nm diameter | | 60 nm | | 20 nm | | 18 nm | | 35 nm | |
	Seed	Root	Seed	Root	Seed	Root	Seed	Root	Seed	Root
Radish	+	+	+	+	+	+	+	50 mg/L	+	50 mg/L
Rape	+	+	+	+	+	+	+	20 mg/L	+	20 mg/L
Ryegrass	+	+	+	+	+	+	2 g/L	20 mg/L	+	20 mg/L
Lettuce	+	+	+	+	+	+	+	2 g/L	+	2 g/L
Corn	+	+	+	+	+	+	+	2 g/L	2 g/L	2 g/L
Cucumber	+	+	+	+	+	+	+	2 g/L	+	2 g/L

Note: MWNT = multiwalled carbon nanotubes, + = no observed toxicity, numerical value = IC_{50}.

In a relatively early publication appearing in 2007, Lin and Xing reported a study of five types of nanoparticles in six different plant systems.[17] Phytotoxicity studies of multiwalled carbon nanotubes (MWNTs), as well as nanoparticle aluminum, alumina, zinc, and zinc oxide, on seed germination and root growth of six plant species (radish, rape, ryegrass, lettuce, corn, and cucumber) were examined (Table 26.1).

Most nanoparticles showed little effect in these systems, except for nano-Zn and nano-ZnO.[17] Seed germination was not affected, except for the inhibition by nano-Zn on ryegrass and nano-ZnO on corn at 2,000 mg/L. On the other hand, inhibition of root growth varied greatly among nanoparticles and plants. The 50% inhibitory concentrations (IC_{50}) of nano-Zn and nano-ZnO were estimated to be near 50 mg/L for radish and 20 mg/L for rape and ryegrass. These results suggested significant environmental effects could result from inappropriate use and disposal of such engineered nanoparticles.

A more recent study reported a comprehensive survey of a number of different nanoparticle materials in comparison with its non-nano bulk material.[18] This study compared some data from five typical nanoparticles using three different test systems to evaluate toxicity. The first test examined the effects on zucchini seed germination of the presence or absence of test compounds, using both the nanoparticles and the corresponding bulk materials (Table 26.2). In data from the paper, a wide variation was observed when the nanomaterials were dispersed using a surfactant. There seemed to be more of an effect on the surfactant sodium dodecyl sulfate (SDS) than on the compounds tested, and difficulty in testing water-insoluble materials.

A second test involved effects of test materials on root growth, with few effects observed for any material. The final test used overall biomass to assay effects on plant growth. These data were somewhat difficult to assess quantitatively because of the difficulty in determining the concentrations of nanoparticle metals, compared with metal ions. One interesting result, however, is shown in Figure 26.3, wherein the effect of multiwall carbon

TABLE 26.2

Effect of Nanomaterials on Plants: Percent Zucchini Seed Germination of Bulk vs. Nanomaterials

Water Control	MWNT	Activated Carbon	ZnO Powder	ZnO 5 nm	ZnO 10 nm	Si Powder	Si 100 nm	Cu Powder	Cu 50 nm	Ag Powder	Ag 100 nm
90	60	40	67	86	67	40	0	40	53	67	33

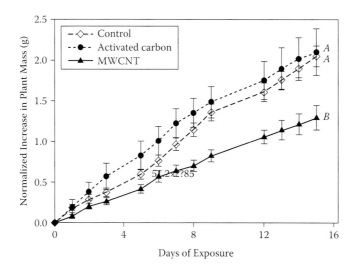

FIGURE 26.3

Effect of bulk carbon (activated carbon) vs. multiwall carbon nanotubes (MWNTs) on growth of zucchini seedlings in batch hydroponic assays. MWNTs (1 mg/ml) were added to 4-day-old seedlings in 25% Hoagland solution under controlled conditions and weight was monitored daily for 15 days. (Reprinted from Stampoulis, D., et al., *Environmental Science and Technology*, 43(24), 9473, 2009. With permission of the American Chemical Society.)

nanotubes vs. activated carbon on total biomass of zucchini seedlings was evaluated. Generally, the data seem to indicate that carbon nanotubes retard zucchini seedling growth. Interestingly, this result contrasts with carbon nanotube studies in tomato, which is discussed in Section 26.4.

26.3.1 Nanoparticle Titanium Dioxide

The effects of nano- and non-nano-TiO_2 in various forms were studied with aged spinach seed, and significant effects on plant growth were observed. It was shown that the physiological effects are related to the nanometer-size particles at a 0.25%–4.0% concentration. The mechanism by which nano-TiO_2 improves the growth of spinach seeds still needs further study.[19]

Another report explored the effects of nano-TiO_2 on willow trees, testing the possible toxicity of TiO_2 nanoparticles using the so-called willow tree transpiration test. The findings suggested that willow trees were not sensitive to exposure to TiO_2 nanoparticles. Effects were, however, observed for zinc and zinc oxide particles, but these effects were determined to be due to heavy metal toxicity and were not nano-size specific.[20]

26.3.2 Nanoparticle Iron

A phytotoxicity study was carried out with *Cucurbita maxima* (pumpkin). The authors showed that plants grown in an aqueous medium containing magnetite nanoparticles (nano-Fe_3O_4) can absorb, translocate, and accumulate nanoparticles in the plant tissues. No particular toxic effects were observed, but the effective transport of the nanoparticle iron led investigators to extend their studies (see Section 26.4) and use magnetite nanoparticles as a potential vehicle to deliver compounds to plants.[21]

26.3.3 Nanoparticle Zinc Oxide

A study by Lin and Xing[22] examined the internalization and upward translocation of ZnO nanoparticles by *Lolium perenne* (ryegrass) and compared the results to Zn^{+2} salts. Uptake and phytotoxicity of nanoparticles were visualized by light and electron microscopy. In the presence of ZnO nanoparticles, ryegrass biomass was significantly reduced, root tips shrank, and root epidermal and cortical cells collapsed. The effects of Zn^{+2} salts at similar concentrations were less. Therefore, the phytotoxicity of ZnO nanoparticles was due to the nanoparticles and not their dissolution in the bulk nutrient solution. The authors observed that ZnO nanoparticles adhered to the root surface, and individual ZnO nanoparticles were observed in apoplast and protoplast tissues of the root endodermis and stele. Finally, their conclusions also showed that little (if any) ZnO nanoparticles would translocate up the ryegrass stems.

26.3.4 Nanoparticle Copper

Another study explored bioaccumulation of non-water-soluble nanoparticle copper on the growth of plant seedlings.[23] The plant species examined were *Phaseolus radiatus* (mung bean) and *Triticum aestivum* (wheat). Growth inhibition of seedlings exposed to different concentrations of Cu nanoparticles was examined in plant agar media, and data suggested that all concentrations tested were toxic to both plants. Experiments showed that cupric ion released from Cu nanoparticles had negligible effects in the concentration ranges of the present study, and that the apparent toxicity clearly resulted from nanoparticles. The authors proposed that the plant agar test is a good protocol for testing the phytotoxicity of non-water-soluble nanoparticles.

26.3.5 Nanoparticle Silver

A study of the uptake and sequestration of silver nanoparticles was reported.[24] The authors examined the uptake of silver nanoparticles using two common metallophytes, Indian mustard (*Brassica juncea*) and alfalfa (*Medicago sativa*). Indian mustard accumulated up to 12 wt% silver when exposed to an aqueous substrate containing 1,000 ppm $AgNO_3$ for 72 h, while alfalfa accumulated up to 14 wt% silver when exposed to an aqueous substrate containing 10,000 ppm $AgNO_3$ for 24 h. In both cases the silver was stored as discrete nanoparticles, with a mean size of 50 nm.

26.3.6 Carbon Nanotubes

Possible toxic effects of multiwalled carbon nanotubes (MWNTs) on rice cells (*Oryza sativa* L.) were investigated.[25] When a rice cell suspension was cultured with MWNTs, an increase in reactive oxygen species (ROS) and a decrease in cell viability were observed. When the antioxidant ascorbic acid was introduced into the medium, the ROS content decreased and cell viability increased. Transmission electron microscopy revealed individual tubes in contact with cell walls. The rice suspension cells with individual MWNTs associated with their cell walls seemed to undergo a hypersensitive response known as the ROS defense response cascade, which led to decreased viability.

26.3.7 Mechanistic Studies of Nanoparticle Toxicity

The phytotoxicity of alumina nanoparticles loaded with and without the chemical phenanthrene was investigated by means of root elongation experiments.[26] Five plant species, *Zea mays* (corn), *Cucumis sativus* (cucumber), *Glycine max* (soybean), *Brassica oleracea* (cabbage), and *Daucus carota* (carrot), were used in the study of phytotoxicity. The surface characteristics of phenanthrene-loaded and nonloaded nanoparticles were investigated using Fourier transform infrared spectroscopy. It was found that when the phenanthrene monolayer concentration increased, the degree of the root elongation inhibition caused by the particles decreased. When mixed with a known free hydroxyl radical scavenger, dimethyl sulfoxide at concentrations from 0.5 to 1.0%, the nonloaded particles also showed decreased inhibition of root elongation. The authors propose that the surface characteristics of the particles play an important role in the phytotoxicity of alumina nanoparticles.

26.4 Plant-Nanomaterial Interactions—Uses of Nanoparticles for Compound Delivery to Plants

A general review of nanoparticles in higher plants appeared in Monica and Cremonini.[27] Another article focuses on the importance of bionanotechnology and its potential impact on crops and agriculture.[28]

26.4.1 Silica Breaks through in Plants—The First Direct Plant Application

Despite all the plant work on environmental toxicity of nanomaterials, recognition of the real impact of nanotechnology on plant science was not recognized until 2007, when the delivery of DNA to plants by nanoparticle carriers was reported. The paper described the use of engineered mesoporous silica nanoparticles (MSNs) to deliver DNA into tobacco plant protoplasts.[29] MSNs were coated with plasmid DNA coding for green fluorescent protein (GFP). MSNs and DNA formed a stable complex, as deduced by the detection of no free DNA in solution. Treatment of tobacco plant protoplasts with the above MSN/DNA nanoparticles gave up to 7% transient expression of the GFP. In addition, the MSN/DNA nanoparticles were coated on gold particles and introduced into maize using biolistics technology (see Section 26.4.5.1). A paper by Torney and co-workers was widely cited and could be said to have launched the interest in engineered nanomaterials and application of bionanotechology in the agriculture realm.[29,30] Shortly after the initial study, several other papers began to appear that used nanoparticles as delivery systems to plants.

26.4.2 Magnetic Particles as Delivery Systems

First, magnetic nanoparticles were used to carry materials into plants, and indeed to use their magnetic properties to guide transport and localization. *Cucurbita pepo* plants (marrow) were cultivated in vitro and treated with carbon-coated Fe nanoparticles. Localization of the magnetic nanoparticles was determined using light, confocal, and electron microscopy. Penetration and translocation of magnetic nanoparticles in whole living plants could be observed.[31]

A study by Corredor et al.[32] studied penetration and movement of iron-carbon nanoparticles in plant cells and in living plants of *Cucurbita pepo*. Carbon-iron nanoparticles were produced by the Huffman-Kratchmer arc discharge method. The soot contained a mixture of carbon-coated iron particles as well as metal particles. The mixture was treated with HCl, which removed the metal and partly functionalized the carbon-coated iron nanoparticles with –COOH groups. This material was suspended in sodium dodecyl sulfate (SDS) and applied using two different methods: injection and spraying. When the material was injected, after 48 h, nanoparticles were observed to migrate toward the interior of the stem parenchyma, evidently assisted by cellular structures, but now directly associated with the vascular system. A second method involving spraying led to no nanoparticles in the plant other that in the outer epidermal layer.

Treated plants were examined using light and electron microscopy. Evidence suggested localization of nanoparticles after injection, as well as migration from the application point. Placement of a magnet on the stem of the plant indicated long-range movement of the particles and concentration near the magnet.

26.4.3 Carbon Nanotubes as Delivery Systems

In 2009, Liu and co-workers used carbon nanotubes to directly deliver chemicals to tobacco BY-2 cells.[33] The investigators showed the capability of single-walled carbon nanotubes (SWNTs) to penetrate the cell wall and cell membrane. Confocal fluorescent microscopy revealed cellular uptake of SWNTs conjugated to the dye FITC, as well as SWNT-dye-DNA conjugates. Confocal microscopy showed SWNT-fluorescein conjugates internalized in BY-2 cells (see Figure 26.4), and demonstrated that SWNTs "hold great promise as nano-transporters for walled plant cells."

The effects of multiwalled carbon nanotubes (MWNTs) were studied in tomato by Khodakovskaya et al.[34] MWNTs were produced using $Fe/Co/CaCO_3$ catalysts and purified to >98% purity as assessed by TEM and Raman spectroscopy. Seeds grown in media containing MWNTs exhibited faster seed germination and faster seedling growth, leading to greater biomass. MWNTs were found to penetrate tomato seeds and affect their germination. The germination was found to be dramatically higher for seeds that germinated on medium containing MWNTs (10–40 µg/ml) than for the control. Analytical methods indicated that the MWNTs are able to penetrate the thick seed coat and support water uptake

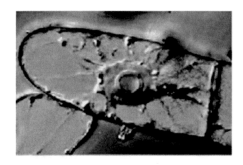

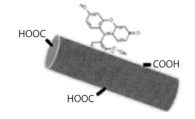

FIGURE 26.4

Top: Confocal microscope image of whole tobacco plant cells BY-2 incubated with SWNT-FITC compound. Bottom: Structure representation of SWNT-FITC compound used in Liu et al.[33] (Reprinted from Liu, Q., et al., *Nano Letters*, 9(3), 1007, 2009. With permission of the American Chemical Society.)

Control   Carbon nanotubes

FIGURE 26.5

(See color insert.) Tomato seedlings grown with and without ~40 mg/L MWNT additive. (Reprinted from Khodakovskaya et al. *ACS Nano* 3(10) 3221 (2009). With permission of the American Chemical Society.)

inside seeds, a process that can affect seed germination and enhance the growth of tomato seedlings, as shown in Figure 26.5.[34]

The effects of functionalized and nonfunctionalized single-walled carbon nanotubes on root elongation were studied in several select crop species that are generally used in phyto-toxicity testing (cabbage, carrot, cucumber, lettuce, onion, and tomato).[35] Nanotubes were functionalized with poly-3-aminobenzenesulfonic acid and compared to nonfunctional-ized controls. Root growth was measured at 0, 24, and 48 h following exposure by scanning electron microscopy. In general, nonfunctionalized carbon nanotubes affected root length more than functionalized nanotubes. Nonfunctionalized nanotubes inhibited root elon-gation in tomato and enhanced root elongation in onion and cucumber. Functionalized nanotubes inhibited root elongation in lettuce. Cabbage and carrots were not affected by either form of nanotubes. The effects observed following exposure to carbon nanotubes tended to be more pronounced at 24 than at 48 h. Microscopy images showed the presence of nanotube sheets on the root surfaces, but no visible uptake by the cells. Canas reported the effect of MWNTS on root elongation.[35]

26.4.4 Nanoparticles as Protective Systems

Hong and co-workers examined protective effects of nanoparticles on the photosynthetic apparatus which could lead to a useful application.[36] Nano-TiO_2-treated chloroplasts of spinach were irradiated, and results showed that spinach treated with 0.25% nano-TiO_2 appeared to be protected from light-induced redox damage. The nano-TiO_2 treatment sig-nificantly increased the activities of superoxide dismutase, catalase, and peroxidase. The authors measured decreased accumulation of reactive oxygen radicals and lower levels of malondialdehyde, a marker for membrane redox damage.

26.4.5 Nanoparticles as Tools for Plant Biology

26.4.5.1 The Gene Gun

A common application of nano- and micron-sized gold or tungsten particles is widely used in plant genetic modification: the gene gun.[37,38] This process, called biolistics, involves shooting nanoparticles at a plant cell. The nanoparticles are coated with DNA of interest, and a certain percentage of the DNA can get through the cell wall and into the nucleus, and be expressed.

26.4.5.2 Nanoparticles for Gene Delivery

Another approach that employed starch nanoparticles was reported in 2008 by Liu et al.[39] The material was synthesized as a water-in-oil microemulsion and coated with poly-L-lysine. The surface of the starch nanoparticles was combined with fluorescence dye and then conjugated with pEGAD plasmid DNA coding for green fluorescence protein (GFP). Mediated by ultrasound, plant suspension cells were treated with the DNA-nanoparticle complexes, which could pierce the cell wall and the cell membrane and enter the nucleus. The expression of the GFP was observed at a high frequency of >5%.

26.4.5.3 Fullerene Nanoparticles for Gene Delivery

The Wilson/Flinn team is focused on the application of nanoscale materials and assemblies to aid in the development of novel bioenergy crops such as switchgrass and *Miscanthus*. The collaboration between Transformation Nanotechnologies and the Institute for Sustainable and Renewable Resources uses proprietary fullerene nanoparticle plant transformation technology, with the ultimate goal being the creation of bioenergy feedstocks with superior qualities.[40]

26.4.5.4 Fullerenes as Reagents

Degradation of pollen grains of *Cycas rumphii* (palm) was achieved using a novel reagent C60 fullerene in benzene solution. Transmission electron microscopic observations were presented showing unique characteristics of this new nanoparticle reagent compared to the usual fixative agents, such as OsO_4.[41]

26.5 Plants as Nanobiosensors

Several other interesting approaches for chemical detection systems for pollutants and pathogens have been reported.[42–44] The strategy involves using plants as biosensors by inserting signaling genes that can respond to external molecular events.

26.6 Plants and Microbes as Nano-Factories

Another area of future interest is the potential use of plants to create nanoparticles of nanomaterials via biosynthesis.[45] Ankamwar et al.[45] report the extracellular synthesis of gold and silver nanoparticles using *Emblica officinalis* (amla, Indian gooseberry) fruit extract as the reducing agent to synthesize Ag and Au nanoparticles, their subsequent phase transfer to an organic solution, and the transmetallation reaction of hydrophobized silver nanoparticles with hydrophobized chloroaurate ions. On treating aqueous silver sulfate and chloroauric acid solutions with *Emblica officinalis* fruit extract, they observed a rapid reduction of the silver and chloroaurate ions, leading to the formation of highly stable silver and gold nanoparticles in solution. Transmission electron microscopy analysis of the silver and gold nanoparticles indicated that they ranged in size from 10 to 20 nm and 15 to 25 nm, respectively. Ag and Au nanoparticles thus synthesized were then phase transferred into an organic solution using a cationic surfactant, octadecylamine. A transmetallation reaction between hydrophobized silver nanoparticles and hydrophobized chloroaurate ions in chloroform resulted in the formation of gold nanoparticles.

Despite progress in the biological synthesis of nanomaterials, the molecular mechanism of synthesis of bionanomaterials remains largely unknown. Parikh et al. reported the extracellular synthesis of crystalline silver nanoparticles (AgNPs) by bacterial *Morganella* sp., and showed evidence of the synthesis mechanism.[46] The AgNPs created were 20 ± 5 nm in diameter and were highly stable at room temperature. The kinetics of their formation was investigated. The formation of particles was linked to three gene homologues, *silE*, *silP*, and *silS*, identified in silver-resistant *Morganella* sp. The genes showed 99% sequence similarity with the previously reported gene, *silE*, which encodes a periplasmic silver-binding protein.

In 2003 Sastry et al.[47] reviewed the progress and concepts for biological synthesis of nanomaterials. While they proposed the possibility of using plants and microorganisms to create new nanomaterials, the study they discussed employed fungi and two species of bacterial actinomycetes (*Verticillium* sp. and *Thermomono sporia*). They could demonstrate direct nanoparticle formation from metallic ions by these microorganisms. The authors discussed and highlighted some of the challenges in this emerging field.

The extension of nanoparticle biosynthesis to plant systems has also been reported recently by Sathishkumar and co-workers.[48] Silver (Ag) nanoparticles can be prepared from silver precursors using the bark extract of *Cinnamon zeylanicum*, the small evergreen tree that is the source of the common spice. Water-soluble organics present in the plant materials were mainly responsible for the reduction of silver ions to nano-silver particles. TEM and x-ray diffraction were used to prove the presence of nanocrystalline Ag particles. Bark extract produced more Ag nanoparticles than the powder, which is attributed to the large availability of the reducing agents in the extract.

26.7 Quantum Dots and Solar Cells

26.7.1 Quantum Dots

In 2009, Lin et al. reported the effects of quantum dots on algal photosynthesis.[49] Using *Chlamydomonas* sp. (algae that have chloroplasts and a cell wall), the authors examined the effects of quantum dots on the microorganisms' growth and photosynthesis behavior. Microscopy examination showed strong interaction of CdSe/Zn/Se quantum dots with the cell wall. There was no evidence of uptake, but they did observe a decrease in the fluorescence intensity of bound quantum dots. This decrease was attributed to inefficient light absorption in the multilayer-bound quantum dots, and not due to fluorescence resonance energy transfer (FRET) quenching from the algal chlorophylls. Decreased photosynthesis was observed and attributed to a physiochemical phenomenon causing disruption of gas flow and nutrient uptake by bound nanoparticles. This inhibited photosynthetic activity of the algae exposed to the quantum dots suggested a potential environmental impact of nanoparticles on plant species.

26.7.2 Plant-Derived Solar Cell Devices

The photosynthesis machinery of plants responsible for nature's 90-terawatt solar energy conversion process has long been an admired target for mimicry. With today's expanding interest in solar energy, it has been the dream of bionanotechnology designers to replicate the photosynthesis system to create more efficient solar cells. Despite huge challenges, there has been progress. Several reports have appeared that integrate the PSI protein discussed in Section 26.2 with solid-state electronics to produce biological solar cells. We will mention only two examples, but the work is just beginning and bears close watching.

First, the group of Das and co-workers[50] created a hybrid solid-state device containing photosystem I (PSI) isolated from spinach chloroplasts as the solar collector. In this device, the PSI protein was coated as a thin layer on a gold electrode. Electronic integration of the devices was achieved by self-assembling the protein complex in an oriented monolayer, stabilizing and hydrating the complex with surfactant peptides, and then coating this layer with a protective organic semiconductor. The resulting photovoltaic cell resulted in a respectable internal quantum efficiency of almost 12%.

A more recent study demonstrated that the PSI protein's remarkable functionality can be accessed in wet or dry solid-state nanoelectronic devices. Ciesielski and co-workers[51] showed that the PSI protein complex retains its photonic energy conversion functionality after covalent attachment to the nanoporous gold electrode surface (Figure 26.6a). Modification of the gold surface with an aldehyde thiol linker provides a versatile functional

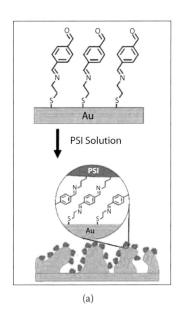

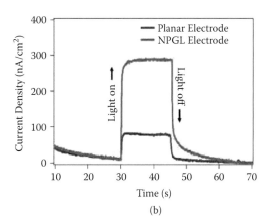

(a)   (b)

FIGURE 26.6
a: Conjugation of PSI protein to porous gold surface can be accomplished by Schiff-based formation. b: Photocurrent of immobilized PSI on 2D planer and 3D nanoporous gold photodevice. (Reprinted from Ciesielski, P.N., et al. *ACS Nano* 2(12) 2465 (2008). With permission of the American Chemical Society.)

surface, so that PSI can be conjugated to the surface, providing a high density of attached PSI protein, as shown in Figure 26.6. Furthermore, the additional PSI-electrode interfacial area resulting from the three-dimensional (3D) nanoporous gold layer–NPGL electrode arrangement allows for an increase in PSI-mediated electron transfer with respect to an analogous thin-film (2D) planer electrode device. Measurement of photocurrent vs. light intensity showed considerable improvement over previous devices.

Photoamperometric measurements were taken in phosphate buffer solution containing dichloroindophenol as an electrochemical mediator and sodium ascorbate as a sacrificial reagent. Upon irradiation, photon energy is collected by PSI protein and transferred to the "special pair" reaction center, subsequently reducing the iron-sulfur complex, which can be transferred to an electrochemical mediator in solution. A current response is measured by the potentiostat and is shown in Figure 26.6b. The difference between the photocurrents measured in 2D (planer) and 3D (NPGL) cell arrangements shows increased efficiency with the design.

26.8 Summary and Conclusions

Bionanotechnology has recently evolved as a hybrid discipline, and with this review we hope to have shown that continued integration of the materials and tools of nanotechnology in plant science is a field with an exciting future. Table 26.3 summarizes the current range of plant studies already completed and discussed in this brief chapter. While relatively few key papers have appeared so far, hopes for great benefits to agriculture from bionanotechnology are creating excitement.[52]

TABLE 26.3

Summary of Plant/Nanomaterial Studies

Test Plant	Nanomaterial	Reference
Alfalfa	Au	24
Cabbage	Alumina-phenanthrene	26
	Unfunctionalized SWNT	35
	Functionalized SWNT	35
Carrot	Alumina-phenanthrene	26
	Unfunctionalized SWNT	35
	Functionalized SWNT	35
Cinnamon zeylarnicun	Au	48
Corn	MWNT, Al, Al_2O_3, Zn, ZnO	17
	Alumina-phenanthrene	26
Cucumber	MWNT, Al, Al_2O_3, Zn, ZnO	17
	Alumina-phenanthrene	26
	Unfunctionalized SWNT	35
	Functionalized SWNT	35
Indian gooseberry	Au	45
Indian mustard	Au	24

TABLE 26.3 (CONTINUED)

Summary of Plant/Nanomaterial Studies

Test Plant	Nanomaterial	Reference
Lettuce	MWNT, Al, Al_2O_3, Zn, ZnO	17
	Unfunctionalized SWNT	35
	Functionalized SWNT	35
Maize	Mesoporous silica	29
Morganella sp.[a]	Au	45
Mung bean	Cu	23
Miscanthus	Fullerene nanoparticles	40
Onion	Unfunctionalized SWNT	35
	Functionalized SWNT	35
Palm	C60	41
Pumpkin	Nano-Fe_3O_4	31
	Carbon-coated iron	32
Radish	MWNT, Al, Al_2O_3, Zn, ZnO	17
Rape	MWNT, Al, Al_2O_3, Zn, ZnO	17
Rice	MWNT	25
Ryegrass	MWNT, Al, Al_2O_3, Zn, ZnO	17
	ZnO	22
Soybean	Alumina-phenanthrene	26
Spinach	TiO_2	19, 36
Switchgrass	Fullerene nanoparticles	40
Tobacco	Mesoporous silica	29
	SWNT-dye	33
	SWNT-DNA	33
Tomato	MWNT	34
	Unfunctionalized SWNT	35
	Functionalized SWNT	35
Wheat	Cu	23
Willow tree	TiO_2, Zn, ZnO	20
Zucchini	MWNT, ZnO, Si, Cu, Ag	18

[a] Bacteria, not plant.

References

1. Rindos, D., *The origins of agriculture: An evolutionary perspective*, Academic Press, New York, 1987.
2. Kreig, M.B., *Green medicine: The search for plants that heal*, Rand McNally & Co, Chicago, 1964.
3. Stewart, A., *Flower confidential*, Algoquin Books, Chapel Hill, NC, 2008.
4. a. Yuan, J.S., et al., Plant systems biology comes of age, *Trends in Plant Science*, 13(4), 165 (2008).
   b. Nap, J.P., et al., The release of genetically modified crops into the environment. Part I. Overview of current status and regulations. *Plant Journal*, 33(1), 1 (2003).

5. Moeller, L., and Wang, K., Engineering with precision: Tools for the new generation of transgenic crops, *Bioscience*, 58(5), 391 (2008).
6. Kirby, A.R., et al., Visualization of plant cell walls by atomic force microscopy, *Biophysical Journal*, 70, 1138 (1996).
7. Grotewold, E., Plant functional genomics, in *Methods in molecular biology*, Humana Books, Totowa, NJ, 2003, p. 236.
8. De, M., Ghosh, P.S., and Rotello, V.M., Applications of nanoparticles in biology, *Advanced Materials*, 20, 4225 (2008).
9. Wilson, S.R., The biology of fullerenes p 437, in *Fullerenes: Chemistry, physics, and technology*, Kadish, K., and Ruoff, R., eds., John Wiley & Sons, New York, 2000.
10. Ben-Shen, A., Frolow, F., and Nelson, N., Crystal structure of plant photosystem I, *Nature*, 426, 630 (2003).
11. Linkov, I., Satterstrom, F.K., and Corey L.M., Nanotoxicology and nanomedicine: Making hard decisions, *Nanomedicine: Nanotechnology, Biology, and Medicine*, 4(2), 167 (2008).
12. Oberdorster, G., Stone, V., and Donaldson, K., Toxicology of nanoparticles: A historical perspective, *Nanotoxicology*, 1(1), 2 (2007).
13. Ray, P.C., Yu, H.T., and Fu, P.P., Toxicity and environmental risks of nanomaterials: Challenges and future needs, *Journal of Environmental Science and Health Part C—Environmental Carcinogenesis and Ecotoxicology Reviews*, 27(1), 1 (2009).
14. Navarro, E., et al., Environmental behavior and ecotoxicity of engineered nanoparticles to algae, plants, and fungi, *Ecotoxicology*, 17(5), 372 (2008).
15. Handy, R.D., et al., The ecotoxicology and chemistry of manufactured nanoparticles, *Ecotoxicology*, 17(4), 287 (2008).
16. Klaine, S.J., et al., Nanomaterials in the environment: Behavior, fate, bioavailability, and effects, *Environmetal Toxicology and Chemistry*, 27(9), 1825 (2008).
17. Lin, D., and Xing, B., Phytotoxicity of nanopartibles: Inhibition of seed germination and root growth, *Environmental Pollution*, 150(2), 250 (2007).
18. Stampoulis, D., Singh, S. S., and White, J.C., Assay-dependent phytotoxicity of nanoparticles to plants, *Environmental Science and Technology*, 43(24), 9473 (2009).
19. Zheng, L., et al., Effect of nano-TiO_2 on strength of naturally and growth aged seeds of spinach, *Biological Trace Element Research*, 104(1), 83 (2005).
20. Seeger, E., et al., Insignificant acute toxicity of TiO_2 nanoparticles to willow trees, *Journal of Soils and Sediments*, 9(1), 46 (2009).
21. Zhu, H., et al., Uptake, translocation, and accumulation of manufactured iron oxide nanoparticles by pumpkin plants, *Journal of Environmental Monitoring*, 10(6), 713 (2008).
22. Lin, D.H., and Xing, B.S., Root uptake and phytotoxicity of ZnO nanoparticles, *Environmental Science and Technology*, 42(15), 558 (2008).
23. Lee, W.M., et al., Toxicity and bioavailability of copper nanoparticles to the terrestrial plants mung bean (*Phaseolus radiatus*) and wheat (*Triticum aestivum*): Plant agar test for water-insoluble nanoparticles, *Environmental Toxicology and Chemistry*, 27(9) 1915 (2008).
24. Harris, A.T., and Bali, R., On the formation and extent of uptake of silver nanoparticles by live plants, *Journal of Nanoparticle Research*, 10(4), 691 (2008).
25. Tan, X., Lin, C., and Fugetsu, B., Studies on toxicity of multi-walled carbon nanotubes on suspension rice cells, *Carbon*, 47(15), 3479 (2009).
26. Yang, L., and Watts, D.J., Particle surface characteristics may play an important role in phytotoxicity of alumina nanoparticles, *Toxicology Letters*, 158(2), 123 (2005).
27. Monica, R.C., and Cremonini, R., Nanoparticles and higher plants, *Caryologia*, 62(2), 161 (2009).
28. Lin, S.J., et al., Potential impacts of engineered nanoparticles on crop productivity and quality, *Hortscience*, 44(4), 1143 (2009).
29. Torney, F., et al., Mesoporous silica nanoparticles deliver DNA and chemicals into plants, *Nature Nanotechnology*, 2, 295 (2007).
30. Galbarith, D.W., Silica breaks through in plants, *Nature Nanotechnology*, 2, 272 (2007).

31. Gonzalez-Melendi, P., et al., Nanoparticles as smart treatment-delivery systems in plants: Assessment of different techniques of microscopy for their visualization in plant tissues, *Annals of Botany*, 101(1), 187 (2008).

32. Corredor, E., et al. Nanoparticle penetration and transport in living pumpkin plants: *In situ* subcellular identification, *BMC Plant Biology*, 9(45), (2009) (http://www.biomedcentral.com/1471-2229/9/45).

33. Liu, Q., et al., Carbon nanotubes as molecular transporters for walled plant cells, *Nano Letters*, 9(3), 1007 (2009).

34. Khodakovskaya, M., et al., Carbon nanotubes are able to penetrate plant seed coat and dramatically affect seed germination and plant growth, *ACS Nano*, 3(10), 3221 (2009).

35. Canas, J.E., et al., Effects of functionalized and nonfunctionalized single-walled carbon nanotubes on root elongation of select crop species, *Environmental Toxicology and Chemistry*, 27(9), 1922 (2008).

36. Hong, F.H., et al., Influences of nano-TiO_2 on the chloroplast aging of spinach under light, *Biological Trace Element Research*, 104(3), 249 (2005).

37. Lundin, K.E., et al., Nanotechnology approaches to gene transfer, *Genetica*, 137, 47 (2009).

38. Sanford, J.C., The development of the biolistic process, *In Vitro Cellular and Developmetal Biology. Plant*, 36, 303 (2000).

39. Liu, J., et al., Preparation of fluorescence starch-nanoparticle and its application as plant transgenic vehicle, *Journal of Central South University of Technology*, 15, 768 (2008).

40. Wilson, S.R., Transformation Nanotechnologies LLC (www.transformationnanotechnologies.com). Flinn, B., IALR's Institute for Sustainable and Renewable Resources (ISRR) (http://www.ialr.org/research/isrr).

41. Tripathi, S.K.M., Kumar, M., and Kedves, M., Advantages of the use of C60 fullerene/benzol solution in ultrastructure investigations: A case study of *Cycas rumphii* Miq. Pollen, *Current Science*, 87(6), 769 (2004).

42. Kovalchuk, I., and Kovalchuk, O., Transgenic plants as sensors of environmental pollution genotoxicity, *Sensors*, 8, 1539, (2008).

43. Mazarei, M., et al., Pathogen phytosensing: Plants to report plant pathogens, *Sensors*, 8, 2628 (2008).

44. Paul, A.L., et al., Deployment of a prototype plant GFP imager at the Arthur Clarke Mars Greenhouse of the Haughton Mars Project, *Sensors*, 8, 2762 (2008).

45. Ankamwar, B., et al., Biosynthesis of gold and silver nanoparticles using *Emblica officinalis* fruit extract, their phase transfer and transmetallation in an organic solution, *Journal of Nanoscience and Nanotechnology*, 5(10), 1665 (2005).

46. Parikh, R.Y., et al., Extracellular synthesis of crystalline silver nanoparticles and molecular evidence of silver resistance from *Morganella* sp.: Towards understanding biochemical synthesis mechanism, *Chembiochem*, 9(9), 1415 (2008).

47. Sastry, M., et al., Biosynthesis of metal nanoparticles using fungi and actinomycete, *Current Science*, 85(2), 162 (2003).

48. Sathishkumar, M., et al., Cinnamon zeylanicum bark extract and powder mediated green synthesis of nano-crystalline silver particles and its bactericidal activity, *Colliods and Surfaces B: Interfaces*, 73(2), 332 (2009).

49. Lin, S., et al., Effects of quantum dots adsorption on algal photosynthesis, *Journal of Physical Chemistry C*, 113(25), 10962 (2009).

50. Das, R., et al., Integration of photosynthetic protein molecular complexes in solid-state electronic devices, *Nano Letters*, 4(6), 1079 (2004).

51. Ciesielski, P.N., et al., Functionalized nanoporous gold leaf electrode films for the immobilization of photosystem I, *ACS Nano*, 2(12), 2465 (2008).

52. Knauer, K., and Bucheli, T., Nano-materials—The need for research in agriculture, *Agrarforschung*, 16(10), 390 (2009).

Index

A

AA, *see* Activated alumina
Absorption spectrum, 15
Actigard AM-87, 358
Activated alumina (AA), 422
Active targeting, 161
AEM, *see* Anion exchange membrane
AFM, *see* Atomic force microscopy
Agriculture, *see* Plants and agriculture
Ancyluris meliboeus, 502–503
Anion exchange membrane (AEM), 377
Anisotropic nanoparticles, biosynthesis of,
181–193
    biosynthesis of nanoparticles, 182–186
        bacterial production of nanomaterials,
        182–183
        nanoparticle formation by
        actinomycetes, 183–184
        use of fungi in nanoparticle synthesis,
        184–186
        use of yeast in nanoparticle synthesis,
        183
    control of growth parameters to
        manipulate nanoparticle
        characteristics, 186–187
    mechanism of biological nanoparticle
        synthesis, 187–189
    use of templates for nanoparticle synthesis,
        189–190
Antenna proteins, 434–435
Antibody-conjugated nanoparticles, *see*
        Biodegradable polymers, antibody-
        conjugated nanoparticles of
Anticancer drugs, 459–461
Antigen-presenting cells (APCs), 286
Antiretroviral treatment (ART), 127
Antiviral therapeutics, *see* Nanoviricides
APCs, *see* Antigen-presenting cells
ART, *see* Antiretroviral treatment
Aspergillus
    awamori, 351
    flavus, 185
    fumigatis, 185, 351

Atomic force microscopy (AFM), 3, 305, see also
        Malaria-infected erythrocytes, atomic
        force microscopy to investigate
AuNPs, see Gold nanoparticles (AuNPs) in
        biomedicine

B

Bacillus
    licheniformis, 188
    subtilis, 182, 350
Bacteriopheophytins, 436
Bacteriorhodopsin (bR), 439–441, 450
Bed volumes (BV), 421, 423
Benznidazole (BNZ), 401
BER, see Bit error rate
Biodegradable polymers, antibody-conjugated
        nanoparticles of (targeted drug
        delivery), 155–180
    antibody-conjugated nanoparticles for
        targeting, 161–166
        assessment of HER2 status, 163
        clinical efficacy of trastuzumab, 165
        functionalization of nanoparticles for
        targeting, 161–162
        HER2-targeted therapy, 162–163
        mechanism of action of trastuzumab, 164
        monoclonal antibody, 162
        trastuzumab (herceptin), 163–164
        trastuzumab-conjugated nanoparticles,
        162–165, 165–166
        tyrosine kinase inhibitors, 161–162
    characterization of trastuzumab-conjugated
        nanoparticles, 168–171
        drug encapsulation efficiency, 170
        in vitro docetaxel-release kinetics, 171
        size and size distribution, 168–169
        surface charge, 169–170
        surface chemistry, 170–171
        surface morphology, 169
    in vitro evaluation of trastuzumab-
        conjugated nanoparticles, 172–174
        in vitro cellular uptake, 172–173
        in vitro cytotoxicity, 173–174

Printed and bound by CPI Group (UK) Ltd, Croydon, CR0 4YY

18/10/2024

01776270-0020